World Scientific Lecture Notes in Physics

I0046939

The QCD Vacuum, Hadrons and Superdense Matter

Second Edition

Edward V Shuryak

State University of New York, USA

World Scientific

NEW JERSEY · LONDON · SINGAPORE · BEIJING · SHANGHAI · HONG KONG · TAIPEI · CHENNAI

Published by

World Scientific Publishing Co. Pte. Ltd.

5 Toh Tuck Link, Singapore 596224

USA office: 27 Warren Street, Suite 401-402, Hackensack, NJ 07601

UK office: 57 Shelton Street, Covent Garden, London WC2H 9HE

British Library Cataloguing-in-Publication Data
A catalogue record for this book is available from the British Library.

World Scientific Lecture Notes in Physics — Vol. 71
THE QCD VACUUM, HADRONS AND SUPERDENSE MATTER
Second Edition

Copyright © 2004 by World Scientific Publishing Co. Pte. Ltd.

ISBN-13 978-981-238-573-4
ISBN-10 981-238-573-8

ISBN-13 978-981-238-574-1 (pbk)
ISBN-10 981-238-574-6 (pbk)

Typeset by Stallion Press
Email: enquiries@stallionpress.com

Contents

Preface   *xv*

Chapter 1   Theoretical Introduction   **1**

1.1. Gauge Fields   2

    1.1.1. Gauge "symmetry"   2

    1.1.2. Gauge fields on the lattice   4

    1.1.3. Fixing the gauge   7

    1.1.4. Hamiltonian quantization   8

    1.1.5. Rotator — a toy model for periodic coordinates   9

1.2. Road to QCD 12

    1.2.1. Quarks and the "colors" 12

    1.2.2. Renormalizability and asymptotic freedom 12

1.3. Path Integrals and Euclidean Time 16

    1.3.1. Green functions and Feynman path integrals 16

    1.3.2. Perturbation theory and Euclidean path integrals 20

    1.3.3. Numerical evaluation of Euclidean path integrals 23

1.4. Gauge Fields on the Lattice 24

    1.4.1. Renormalization group and asymptotic freedom 26

    1.4.2. Continuum limit of lattice gauge theory 28

    1.4.3. Path integrals for fermions 30

1.5. Light Quarks and Symmetries of QCD 32

    1.5.1. Exact and approximate symmetries of QCD 32

    1.5.2. Chiral anomalies, the UV approach 34

    1.5.3. Chiral anomalies, the IR approach 38

    1.5.4. Other applications of chiral anomalies 39

    1.5.5. Scale anomaly 40

1.6. Heavy Quarks, New Symmetry and Effective Theory 42

1.7. Changing the Number of Colors N_c 45
    1.7.1. Large number of colors 45
    1.7.2. QCD with the smallest ($N_c = 2$) number of colors 49

Chapter 2 Phenomenology of the QCD Vacuum   51

2.1. Phenomenology of the Hadronic World 52
    2.1.1. Brief history 52
    2.1.2. The "usual" hadrons 55
    2.1.3. The "unusual" mesons 56
    2.1.4. The exotic hadrons 59
    2.1.5. Remarks about highly excited states 61
2.2. Models of Hadronic Structure 64
    2.2.1. Generalities 64
    2.2.2. MIT bag 66
    2.2.3. Skyrmions 68
    2.2.4. Chiral bags 70
    2.2.5. Evolving views on the nature of the spin forces 71
2.3. Models of the QCD Vacuum: An Overview 77
    2.3.1. Condensates and scales 77
    2.3.2. Condensate factorization and stochastic vacuum model 78
    2.3.3. An example of a highly inhomogeneous model:
        the instanton vacuum 80
2.4. Chiral Symmetry Breaking and Effective Low Energy Theory 83
    2.4.1. Spontaneous breaking of the chiral symmetry 83
    2.4.2. The Goldstone modes: oscillations of the quark condensate . . . 84
    2.4.3. Quark condensate and Dirac eigenvalue spectrum 86
    2.4.4. Elements of chiral perturbation theory 88
    2.4.5. Effective chiral Lagrangian 90
    2.4.6. Nambu–Jona–Lasinio model 93
2.5. Color Confinement 95
    2.5.1. Static potential 95
    2.5.2. Dual superconductivity 97
    2.5.3. Structure of flux tubes 98
    2.5.4. Interaction of flux tubes 101

Chapter 3 Euclidean Theory of Tunneling:
From Quantum Mechanics to Gauge Theories   105

3.1. Tunneling in Quantum Mechanics 105
    3.1.1. Brief history of tunneling 105
    3.1.2. Double-well problem and instantons 106
    3.1.3. Pre-exponent and zero modes 109
    3.1.4. Instanton gas 111
    3.1.5. Two-loop quantum corrections 113

3.2. A Digression: Tunneling Versus Perturbative Series 115
    3.2.1. Convergence of perturbative series 115
        3.2.1.1. Dyson instability 115
        3.2.1.2. Perturbative series in high orders 116
        3.2.1.3. Semiclassical evaluation of Dyson's instability 117
        3.2.1.4. High orders of perturbative series in field theories . . . 118
    3.2.2. Instanton–anti-instanton interaction and one more
        correction to the ground state energy 118
3.3. Fermions Coupled to the Double-Well Potential 121
3.4. Instantons in Gauge Theories 124
    3.4.1. Topologically nontrivial objects 124
    3.4.2. Topologically distinct pure gauge configurations 125
    3.4.3. Digression: spherically symmetric Yang–Mills fields 127
    3.4.4. Static magnetic configurations and their minimal energy 128
3.5. Tunneling and BPST Instanton 132
    3.5.1. Instanton solution 132
    3.5.2. Theta vacua 136
    3.5.3. Tunneling amplitude 138

Chapter 4   Instanton Ensemble in QCD   **141**
4.1. Brief History of Instantons 141
    4.1.1. Discovery and early applications 141
    4.1.2. Phenomenology leads to a qualitative picture 142
    4.1.3. Technical development during 1980s 143
    4.1.4. Recent progress 144
    4.1.5. Instantons at finite temperatures and chiral restoration 145
    4.1.6. Instantons and color superconductivity at high densities 145
4.2. Tunneling and Light Quarks 146
    4.2.1. Relating gauge field topology to the axial charge 146
    4.2.2. Fermionic zero modes 147
    4.2.3. The 't Hooft effective interaction 149
    4.2.4. Baryon number violation in the standard model 151
4.3. Instanton Ensemble 152
    4.3.1. Qualitative discussion of the instanton ensembles 152
    4.3.2. Mean field approximation: pure glue 156
    4.3.3. Quark condensate in the mean field approximation 159
    4.3.4. The single instanton approximation 165
4.4. The Interacting Instanton Liquid Model 170
    4.4.1. Screening of the topological charge 178
4.5. Instantons for Larger Number of Colors 180
    4.5.1. Naive counting and expectations 181
    4.5.2. Mean field arguments and the chiral condensate 182

4.5.3.   Fluctuations in the interacting instanton liquid 184

4.5.4.   Do instantons cluster at large N_c? 186

Chapter 5   Lattice QCD   193

5.1. Generalities 193

    5.1.1.   Brief history 193

    5.1.2.   Lattice limitations 195

    5.1.3.   Mesoscopic regime and the random matrix theory 198

    5.1.4.   Art of numerical simulation of multi-dimensional integrals 201

5.2. Fermions on the lattice 204

    5.2.1.   Fermionic doublers 204

    5.2.2.   Wilson fermions 205

    5.2.3.   Ginsparg–Wilson relation and lattice chiral symmetry 206

    5.2.4.   Known solutions to GW relation 206

    5.2.5.   Domain wall fermions 207

5.3.   Hadronic spectroscopy on the lattice 208

    5.3.1.   Glueballs in gluodynamics 209

    5.3.2.   Light quark spectroscopy in quenched approximation 210

    5.3.3.   Spectroscopy with dynamical quarks 212

5.4. Topology on the lattice 214

    5.4.1.   Quantum-mechanical topology and perfect actions 214

    5.4.2.   Naive and geometric methods for gauge fields 218

    5.4.3.   Are the lowest Dirac eigenstates locally chiral? 221

    5.4.4.   Testing the large N_c limit on the lattice 225

Chapter 6   QCD Correlation Functions   227

6.1. Generalities 227

    6.1.1.   Why the correlation functions? 227

    6.1.2.   Different representations of the correlation functions 230

    6.1.3.   Quantum numbers and inequalities 232

    6.1.4.   Correlators with chirality flips 234

6.2. Phenomenology of Mesonic Correlation Functions 236

    6.2.1.   Vector and axial correlators 236

    6.2.2.   Comparing axial and vector channels 244

    6.2.3.   Pseudoscalar $\mathbf{SU}(3)$ octet (π, K, η) channels 246

    6.2.4. $\mathbf{SU}(3)$ singlet pseudoscalars 248

    6.2.5.   Hadron–parton duality 250

6.3. Operator Product Expansion and QCD Sum Rules 252

    6.3.1.   Brief history and overview 252

    6.3.2.   Separation of scales 254

    6.3.3.   OPE in a background field 255

    6.3.4.   Sum rules for heavy–light mesons 260

6.3.5. OPE for light quark baryons 262
6.3.6. OPE for mesons made of light quarks 265
6.4. Instantons and the Correlators: Analytic Results 270
6.4.1. Propagator in the field of a single instanton 270
6.4.2. First order in the 't Hooft effective vertex 271
6.4.3. Propagator in the instanton ensemble 273
6.4.4. Propagator in the mean field approximation 275
6.4.5. Correlators in the random phase approximation 276
6.5. Correlators in the Instanton Liquid 279
6.5.1. Quark propagator in the instanton liquid 279
6.5.2. Mesonic correlators 280
6.5.3. Baryonic correlation functions 284
6.5.4. Comparison to correlators on the lattice 287
6.5.5. Gluonic correlation functions 288
6.6. Hadronic Structure and n-Point Correlators 293
6.6.1. Wave functions 294
6.6.2. Form factors 295

Chapter 7   High Energy Hadronic Collisions   301
7.1. Introduction 301
7.1.1. Reggions and the Pomeron 301
7.1.2. High energy collisions in pQCD and its "phases" 303
7.1.3. Evolving descriptions of soft Pomeron dynamics 307
7.2. Instanton-Induced Processes at High Energies 310
7.2.1. Toward the "holy grail" 310
7.2.2. Exciting a quantum system from under the barrier 311
7.2.3. Semiclassical production of sphaleron-like clusters 312
7.2.4. Explosion of the turning states 314
7.2.5. Semiclassical evaluation of the cross section 316
7.2.6. Semiclassical Wilson lines 317
7.2.7. Pomeron from instantons 323
7.3. Pomeron Structure and Interactions 330
7.3.1. Clustering in inclusive pp collisions 330
7.3.2. Inclusive production of clusters in double-Pomeron
       processes 331
7.3.3. Exclusive production of hadrons in double-Pomeron
       processes 334

Chapter 8   QCD at Finite Temperatures   337
8.1. Introduction 337
8.1.1. Brief history and the basic scales 337
8.1.2. From field theory to thermodynamics 339

8.1.3.   A quantum particle at finite T 340
8.1.4.   Gauge and fermion fields at finite T 346
8.2. QCD at High Temperatures 350
8.2.1.   Screening versus anti-screening 350
8.2.2.   Thermodynamical potential in the lowest order 353
8.2.3.   Ring diagram re-summation 354
8.2.4.   IR divergences in general 356
8.2.5.   Are perturbative series useful in practice? 357
8.2.6.   HTL re-summations and the quasiparticle gas 359
8.2.7.   Viscosity of the QGP 362
8.3. Hadronic Matter 363
8.3.1.   Pion gas at low T 363
8.3.2.   Resonance gas 364
8.3.3.   Pion liquid 366
8.4. QCD Phase Transitions at Finite T 368
8.4.1.   Deconfinement 368
8.4.2.   Chiral symmetry restoration 371
8.4.3.   Static quark potential at high T 374
8.4.4.   Equation of state in the transition region 376
8.5. Instantons at Finite T 381
8.5.1.   Finite temperature field theory and the caloron solution . . . 381
8.5.2.   Instanton density at high temperature 382
8.5.3.   Instantons at low temperature 386
8.5.4.   Chiral symmetry restoration and instantons 387
8.5.5.   Instanton ensemble in the phase transition region 389
8.5.6.   Critical behavior in the instanton liquid 392
8.6. Hadronic Correlation Functions at Finite Temperature 394
8.6.1.   Screening masses 395
8.6.2.   Temporal correlation functions 397
8.6.3. $U(1)_A$ breaking at high T 399

Chapter 9   Excited Hadronic Matter in Heavy Ion Collisions   403
9.1. Introduction 403
9.1.1.   Toward the macroscopic limit 403
9.1.2.   "Little Bang" versus Big Bang 406
9.1.3.   Experimental centers, present and future 408
9.1.4.   Mapping the phase diagram 409
9.2. Relativistic Hydrodynamics 410
9.2.1.   Equations of the ideal hydrodynamics 410
9.2.2.   Dissipative terms 413
9.2.3.   The Bjorken solution 414

9.2.4. Further simplifications and solutions 416
9.2.5. Singularities: shocks, rarefaction waves, and
the "explosive edge" 417
9.3. Chemical and Thermal Freezeouts 421
9.3.1. Why two freezeouts? 421
9.3.2. Chemical freezeout 423
9.3.3. Between chemical and kinetic freezeouts 423
9.3.4. Thermal freezeout 428
9.3.5. After freezeout 431
9.3.6. Freezeout of resonances and nuclear fragments 432
9.3.7. Resonance modification 433
9.4. Hydrodynamic Description of Heavy Ion Data 436
9.4.1. A long road to unveiling the transverse flow 436
9.4.2. Qualitative effects of the QCD phase transition 440
9.4.3. Solutions to hydrodynamic equations 442
9.4.4. Radial flow at SPS and RHIC 445
9.4.5. Elliptic flow 449
9.4.6. Limits to ideal hydrodynamics 455
9.5. Interferometry of Identical Secondaries or HBT Method 459
9.5.1. Main idea of the method 459
9.5.2. Correlator and random source model 461
9.5.3. Issue of "coherency" and long-lived resonances 463
9.5.4. Expanding sources and regions of homogeneity 464
9.6. Correlation of Non-Identical Hadrons 466
9.6.1. Ordering the production time for all hadronic species 467
9.6.2. Balance functions 467
9.7. Event-by-Event Fluctuations 469
9.7.1. Fluctuations of hadronic cross sections are
surprisingly large 469
9.7.2. All heavy ion collisions are (about) the same! 470
9.7.3. Critical opalescence near the tricritical point 472
9.7.4. Can QGP charge fluctuations survive the hadronic phase? 475

Chapter 10   Early Diagnostics of Hadronic Matter   **477**
10.1. Penetrating Probes: Dileptons and Photons 477
10.1.1. Basic rates and space–time profile 477
10.1.2. Dilepton data versus expectations 480
10.1.3. Direct photon production 486
10.2. Quarkonia in Heavy Ion Collisions 488
10.2.1. Charmonium suppression, the mechanisms 488
10.2.2. Charmonium suppression, the data 489
10.2.3. ϕ-related puzzles 491

10.3. Evolving Views on the Initial Stage 492

     10.3.1. Perturbative processes and minijets 492

     10.3.2. Classical fields in heavy ion collisions 493

     10.3.3. Non-perturbative equilibration ánd topological clusters 495

     10.3.4. Dilemma of weakly versus strongly coupled QGP 496

10.4. Jet Quenching 497

     10.4.1. Jet quenching in experiment 498

     10.4.2. Azimuthal asymmetry at large p_t 503

     10.4.3. Radiation in matter 505

     10.4.4. Synchrotron-like QCD radiation 507

Chapter 11   QCD at High Density   **519**

11.1. From Nuclear to Quark Matter 519

     11.1.1. Nuclear matter 519

     11.1.2. Other phases of nuclear matter? 523

     11.1.3. From nuclear to quark matter 523

     11.1.4. Chiral waves and chiral crystals 526

11.2. Compact Stars 531

     11.2.1. Brief introduction 531

     11.2.2. Phases of matter in compact stars 534

11.3. Color Superconductivity in Very Dense Quark Matter 536

     11.3.1. Brief introduction to superconductivity 536

     11.3.2. BCS pairing and Gorkov abnormal Green functions 539

     11.3.3. Three mechanisms of quark pairing 540

     11.3.4. Magnetic pairing in asymptotically dense matter 542

     11.3.5. Instanton-induced color superconductivity 545

     11.3.6. Two-flavor QCD: 2SC phase 546

     11.3.7. Three-flavor QCD: CFL phase 546

     11.3.8. Excitations of color superconductor 548

     11.3.9. Quark matter with charge neutrality and realistic m_s 550

     11.3.10.Open questions 550

Chapter 12   A Wider Picture   **553**

12.1. Hadronic World in Alternative or Changing Universe 553

12.2. Increasing the Number of Quark Flavors: The First Window to Confor-
     mal World 555

12.3. $\mathcal{N} = 1$ Supersymmetric Theories 557

     12.3.1. Instantons and exact beta function 559

     12.3.2. N_c–N_f phase diagram: $\mathcal{N} = 1$ SUSY versus QCD 562

12.4. $\mathcal{N} = 2$ Supersymmetric Theories 565

12.5. $\mathcal{N} = 4$ Supersymmetric Theories and AdS/CFT Correspondence 570

    12.5.1. Conformal field theory 570

    12.5.2. A window to the string world: strong coupling 571

    12.5.3. AdS/CFT duality at weak coupling and instantons 574

Appendix A   Notations   **575**

A.1: Some Abbreviations Used 575

A.2: Units 575

A.3: Space–Time and Other Indices, Standard Matrices 576

A.4: Properties of η Symbols 576

A.5: Gauge Fields 577

A.6: Quark Fields 578

A.7: QCD Feynman Rules 579

Appendix B   Basic Instanton Formulae   **581**

B.1: Instanton Gauge Potential 581

B.2: Fermion Zero Modes and Overlap Integrals 582

B.3: Group Integration 583

Appendix C   A Sample Program for Numerical Simulation
    of the Euclidean Quantum Paths   **585**

Bibliography   **587**

Index   **617**

Preface

This book is an extensive set of lecture notes on various aspects of Non-perturbative Quantum Chromodynamics, the fundamental theory of strong interaction on which nuclear and hadronic physics is based.

Its first edition, written in the mid-1980's, was more of a review style. Only a general outline remains the same, while the text is completely rewritten and extended. Apart from new developments during these years, this edition benefits from several graduate courses I taught at Stony Brook over the last decade. It is now complemented by exercises and contains in total about 1000 references to major works, arranged according to subject. I hope it can be used as a basis for several different courses.

More generally, this book is intended to put the reader from the first encounter with the subject at the front lines of research, as quickly as possible (but not quicker than that, as Einstein once remarked). Therefore, the pedagogical considerations take priority above the historic ones: e.g. many quantum field theory issues are first illustrated using quantum mechanics. (We discuss in this way Feynman perturbation theory in Chapter 1, the semiclassical instantons in Chapter 3 and the finite temperature formalism in Chapter 6.) A large fraction of this material appears in print for the first time. I will use two marginal signs in the texts, the exclamation mark means that special attention is needed, while the latter E indicates that there is an exercise related to this point.

$\boxed{!}$

$\boxed{E}$

Quantum Chromodynamics (QCD), the theory of interacting quarks and gluons, was discovered 30 years ago. As a theoretical construction QCD is even more logical and much more self-contained that its older relative, Quantum Electrodynamics (QED), which provided an amazingly accurate description of all atomic phenomena. So, in the 1970's the theorists were very optimistic about the rapid progress in this field: they finally had the fundamental theory at hand, supplemented by a huge amount of data on hadronic spectroscopy. What more could be needed?

However it soon became clear that such optimism was rather naive. The Lagrangian written on the blackboard did not by itself explain the data. For "hard processes" one could use familiar perturbative language, the Feynman diagrams and the like, but in order to understand the non-perturbative phenomena one had a lot of physics to do. The QCD vacuum is a very complicated matter, made of strongly interacting quark and gluon fields, and one can hardly understand the "elementary particles" — the hadrons — without understanding the vacuum first. Furthermore, QCD has other phases as well, which are revealed at extreme conditions: the Quark–Gluon Plasma phase at high temperatures and Color Superconducting phases at high density. A lot of technical development had to be made before we even got to know the right degrees of freedom in terms of which these phases could be explained. In this book we will, in particular, emphasize the development of the semiclassical methods, based on the fundamental topological solutions of Yang–Mills equations known as *instantons*. Another methodological tool which produced a lot of progress is numerical simulations by supercomputers, in the framework of the so called *lattice gauge theories*.

The book also covers the phenomenological aspect of the related physics involved. The voluminous "Particle Data Tables" do not necessarily contain the information we really need to test our theoretical ideas. The "elementary particles" like the proton or the pion are in fact closer to the phonons in condense matter. By measuring the phonon spectra and properties it is possible, although not easy, to understand that phonons are just displacements of atoms arranged in a certain way. Similarly, in order to go from the spectra of the lowest excitations to understanding of the QCD vacuum we need special tools, such as *QCD correlation functions*, a bridge between the theory and the experiment.

Two chapters of the book cover in significant detail the *Little Bang* of exploding hadronic matter, studied in heavy ion collisions. As we will discuss, those experiments recreate some moments of the cosmological Big Bang, with many amusing similarities between the two. Recent data from the Relativistic Heavy Ion Collider are discussed, with a detailed explanation as to why we think those experiments have produced a new phase of matter called *Quark–Gluon Plasma*.

Like atomic and nuclear physics before, the hadronic physics is no longer at the edge of the highest available energies. Nevertheless, there is no doubt that the understanding of the non-perturbative QCD continues to provide a significant intellectual challenge, to theorists, phenomenologists and experimentalists. And when we understand how it works, we may very possibly find our current extrapolations, to the Grand Unification and even the Plank scale, to be extremely naive.

Another change which took place between the two editions was a "deconfinement transition" in Russia. I have been working at Stony Brook ever since 1990, which is a university with a short history but strong traditions on first-rate research. I am much indebted to Gerry Brown for many things, but especially for maintaining the invaluable habit of daily common lunches with on-going physics discussions. Many points in the book stem from repeated, often heated debates at the large blackboard in our common room. Many things appearing in this new edition for the first time I learned from my collaborators, especially Jac Verbaarschot, Thomas Schafer and Ismail Zahed.

Edward V. Shuryak
Stony Brook, May 2003

CHAPTER 1

Theoretical Introduction

Although the main goal of this book is to bring the reader to the front lines of research related to the nonperturbative phenomena in Quantum Field Theories (QFT), and in Quantum Chromodynamics (QCD) in particular, we have to start with a brief summary of well established facts. Most of the material of Chapter 1 can be found in other textbooks, and in more extended form. However probably even expert readers may have a look at it and see if they find something new to them: this section includes the absolute minimum of facts to be used later.

Section 1.1 starts with the discussion of the general meaning of the local gauge symmetry and its discretized lattice formulation. The issue of a gauge fixing is discussed, in connection with Hamiltonian quantization and Gribov copies. We also discuss general issues of angle-like and/or periodic coordinates in quantum theory, using the simplest toy models.

Then in Section 1.2 we turn to major discoveries which lead to QCD, such as the color quantum number of quarks and the "asymptotic freedom". Section 1.3 deals with quantum mechanics: we use it as an example to introduce Feynman path integrals and other methods to be used below for quantization of gauge theories. In particular, we discuss Feynman perturbation theory in quantum mechanical contexts (which, to our knowledge, are missing from pedagogical literature for no reason), as well as numerical lattice-based methods for the quantum mechanical problems. In Section 1.4 we generalize it to a "constructive definition" of non-perturbative gauge theories, as the appropriate limit of the lattice gauge theories.

Another ground stone of QFT is symmetries: those associated with light fermions in QCD-related physics are considered in Section 1.5.1. The main attention is paid to the so called "chiral symmetries", existing in gauge theories with massless fermions. Although in the real world there are no massless quarks, the masses of u, d quarks are only about a few MeV, and the QCD Lagrangian is chirally symmetric to high accuracy. The U(1) part of the chiral symmetry is however violated by "chiral anomalies" and we discuss their origin and meaning. Another anomaly is related with scale invariance, present in classical gauge theory but absent from the quantized version of it. Using the path integral formulation, one can in fact get a direct and elegant derivation of the anomalies, due to Fijikawa [49], which can be understood as non-invariance of the integration measure with respect to these symmetries, and the r.h.s. as stemming from the nonzero Jacobians of it following the symmetry transformations. However I decided on a more traditional discussion, which follows the historical route and is closer to our ultimate applications.

Section 1.7 is devoted to a kind of a "theoretical model": it is QCD in the limit of either large number of colors, $N_c \to \infty$, or its smallest value, $N_c = 2$. It was suggested by a number of authors that those cases are simpler than the real world, and therefore they may be solved first.

1.1.   Gauge fields

1.1.1.   *Gauge "symmetry"*

Local gauge invariance is the most important principle of modern quantum field theory: all fundamental interactions we know today — electromagnetic, weak, strong and gravitational — obey it. One would not be able to discover most of those, without it.

On the other hand, local gauge symmetry is not a "symmetry" in the usual sense, because it is indeed quite different from, say, translational, rotational or flavor symmetries we are used to. All physical observables are singlets under gauge rotations, and therefore are not really transformed. Thus there cannot be any relations between them to be deduced from this symmetry.

Rather gauge symmetry tells us which degrees of freedom of the fields are unphysical, or *redundant*, variables. Those are used simply due to the particular description of the system. However the principle that physical quantities should be independent of those variables is quite powerful and restrictive. Furthermore, in contrast to other symmetries, the gauge symmetry cannot be considered as approximate or slightly broken: it is either present or absent.

In order to explain its general meaning it is useful to trace an analogy with Einstein's general relativity. As special relativity postulates that all inertial reference frames can equally be used, general relativity goes further and suggests also making arbitrary coordinate transformations. This means one can also use different reference frames at different space–time points. Similarly, in gauge theories from "global symmetries" (rotations in the space of internal quantum numbers) the physicists came to local gauge invariance, based on *arbitrary* rotations at different space–time points.

The first "principle of local gauge invariance" was formulated by H. Weyl in 1919 [2]. His idea was that the length (or "gauge") of all physical quantities should also be allowed to be arbitrary in different places. For a small step dx_μ he introduced a vector field $A_\mu(x)$ correcting the scale or "gauge" by the amount $A_\mu dx_\mu$. The original intention was to identify this field with the electromagnetic potential, but such a theory was clearly wrong.

Nevertheless, Weyl has attracted attention to the invariance of the electromagnetic field strength,

$$F_{\mu\nu} = \partial_\mu A_\nu - \partial_\nu A_\mu, \tag{1.1}$$

under the gauge transformation of the potentials,

$$A_\mu \to A_\mu + (i/e)\partial_\mu \phi. \tag{1.2}$$

This implied that the gradient-like part of the vector potential is redundant and can be chosen at will.

Soon after the discovery of quantum mechanics, in 1927, it was noted by Fock [3] and London [4] that *the phase of the wave function can be chosen arbitrarily at all space–time points,*

$$\psi \to \psi e^{i\phi(x)}. \tag{1.3}$$

Of course, in quantum mechanics the momentum operator $p_\mu = i\partial/\partial_\mu$ would have to be modified, since the derivative of $\phi(x)$ is non-zero. However this derivative can be *compensated* by a gauge transformation of the electromagnetic field, with a change of the redundant part of it above. Thus, when the momentum is re-defined in terms of the *covariant* (or "long") derivatives,

$$i\partial_\mu \to iD_\mu = i\partial_\mu + gA_\mu, \tag{1.4}$$

one gets gauge invariant quantum mechanics, or QED (if the action of the gauge field is added and it is also quantized).

The principle of local gauge symmetry was extended by C.N. Yang and R. Mills in 1954 [6]. Let the matter field $\psi_i(x)$ have some internal quantum number[1] $i = 1 \dots N$. Let the axis in this "internal" space be chosen arbitrarily in any space–time point. The phase rotation (1.3) is generalized to the unitary matrix rotation

$$\psi_i(x) \to \Omega_{ij}(x)\psi_j(x), \tag{1.5}$$

with a local symmetry, in which the matrix $\Omega(x)$ is now different at different points. By analogy to electromagnetism, in order to maintain the invariance, Yang and Mills introduced vector *compensating fields*, which are matrix-valued $A_\mu = A_\mu^a T^a$, with the T being the generators[2] of the corresponding group. We would like it to

---

[1] As a particular example of "internal quantum number" Yang and Mills have considered the isotopic spin, so their gauge particle was supposed to be an ρ meson. It became the so called weak isospin with $N = 2$ in electroweak theory, and color with $N = 3$ in QCD.

[2] Their specific matrix form depends on the particular representation of the group which is appropriate. In QCD with quarks in the fundamental representation of SU(3), $T = t^a/2$, where t represents the, Gell-Mann SU(3) matrices. The 1/2 is historical, like a relation between spin and Pauli matrices, for the fundamental (spin 1/2) representation of SU(2).

complement the ordinary derivatives in a "covariant" combination

$$iD_\mu = i\partial_\mu + gA_\mu^a T^a. \tag{1.6}$$

The expected property is that this combination can be transformed with just a rotation,

$$D_\mu \psi \to \Omega(D_\mu \psi). \tag{1.7}$$

This requirement is met by the following non-Abelian gauge transformation of the potential

$$A_\mu \to \Omega A_\mu \Omega^{-1} - (2i/g)(\partial\Omega)\Omega^{-1}, \quad A_\mu = A_\mu^a T^a. \tag{1.8}$$

1.1.2. *Gauge fields on the lattice*

In order to explain the geometric meaning of these "compensating fields", to prepare the reader for lattice gauge theory, and finally to discuss the issue of complete gauge fixing, we will immediately introduce a space–time lattice. For simplicity, let it be a cubic $4d$ lattice, and we start with a tiny world consisting of only one small square (or plaquette), shown in Fig. 1.1.

There are 4 (vector-valued) matter fields ψ_i defined at its corners, numbered $1\ldots4$. The gauge field is described by 4 (unitary matrix-valued) "relative orientation matrices"[3] $U_{ij}(l, m)$, associated with the *links* between the two sites, e.g. the matrix for the $(2, 3)$ link is shown in the figure. Since the purpose of U is the same as of the vector field, to undo (or compensate) for arbitrary rotations of matter fields at all corners, the obvious transformation law for the $U(l, m)$ is

$$U(l, m) \to \Omega(l)U(l, m)\Omega^{-1}(m). \tag{1.9}$$

For example a combination with 2 links,

$$\psi^+(2)U(2, 3)U(3, 4)\psi(4)$$

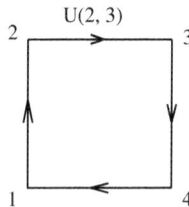

Fig. 1.1. A small plaquette with matter field at its corners. The link variables U have arrows, needed to indicate their directions.

---

[3]For clarity, here l, m are not the indices of the matrix, which we imply without showing below, but just a label of the link.

would then be independent of any possible rotations at point 3. It is clear, one can make an "open string" of any length from a product of such link variables, and its gauge transformation would be rotations at its ends. Finally, one may consider "closed strings", the trace of the product of link matrices over any closed path — those are obviously *gauge invariant* quantities.

How are the link variables related to the vector potentials? The formal relation is

$$U(1,2) = P \exp \left(ig \int_1^2 A_\mu^a(x) T^a dx_\mu \right) \tag{1.10}$$

where the definition of the path-ordered exponent of a matrix-valued variable is as follows. If the length of the path 1–2 Δx_μ is small, one can use an expansion of the exponent as $1 + igA_\mu^a(x)T^a \Delta x_\mu + \cdots$. If it is not small, it can be divided into a series of small steps, with multiplication of the rotation matrices at each step. The $P \exp$ is obviously defined as the limit of such a procedure, as the step size goes to zero.

It is instructive to recall at this point what the link variable is in electrodynamics. Since there are no matrices, U is the usual exponent of the integrated vector potential. In this case A_μ obtains its gradient contribution $A_\mu \to A_\mu + \partial_\mu \phi$. The integral in the exponent is then the difference $\phi(2) - \phi(1)$, which compensates for the phase difference at the two end points.

Now we are ready to discuss the following question. Suppose we have a 4d lattice and at each link there is some particular matrix $U(l, m)$; we will call such a set a gauge field configuration. One would like to use the gauge freedom "gauging away" as many U links as possible, setting the corresponding U to 1, the unit matrix. Starting at point 1 and going to point 2 we meet $U(1, 2)$, and by rotating the matter field at point 2 by $U^{-1}(1, 2)$ we can set $U(1, 2) = 1$ (or equivalently eliminate the vector potential) at the 1–2 link. Then we may come from 2 to 3 and do it again, with the rotation at point 3 eliminating $U(2, 3)$ at the link 2–3 etc. The third time, it also works on link 3–4, but not on the last link 4–1. We have used all the gauge freedom, so at the link 4–1 the vector field cannot be eliminated any more! So we learned that the product $W = U(1, 2)U(2, 3)U(3, 4)U(4, 1)$ cannot be put equal to 1. Furthermore, the reader may easily check that its trace is gauge invariant, as it has not changed at all, with all those manipulations.

The analogy with the Abelian case (electrodynamics) is now obvious, the closed loop integral does not depend at all on the redundant gradient-like part of the gauge field. Naturally, it cannot be eliminated by a gauge transformation. Furthermore, in this case one can use the Stokes theorem and relate the circulation of A_μ over the closed contour 1–2–3–4–1 with the *field flux* through the plaquette

$$\int_C A_\mu dx_\mu = \int F_{\mu\nu} d\sigma_{\mu\nu}, \tag{1.11}$$

where $d\sigma_{\mu\nu}$ is the element of the area. As the field flux cannot be eliminated, it is a physical quantity.

This analogy naturally leads to the definition of the non-Abelian field strength (or the "curvature" in our "internal space"). For a small plaquette $d\sigma_{\mu\nu}$

$$(g^2/8)(G_{\mu\nu}d\sigma_{\mu\nu})^2 + O(d\sigma^3) = \operatorname{Re}\operatorname{Tr}(1 - UUUU), \qquad (1.12)$$

where we suppressed links and indices of the four U, taken around the plaquette. Translating this to vector field, one gets the expression

$$G_{\mu\nu} = \partial_\mu A_\nu - \partial_\nu A_\mu + ig[A_\mu, A_\nu]. \qquad (1.13)$$

It is useful to explain by a simple example why its nonlinear part is the commutator term. Consider as an example $A_\mu(x)$ to be independent of x. Then the U are also constant matrices over all space, so that all the derivatives vanish. Constructing the product of 4 links over the plaquette one can see that in fact there are two different matrices instead of four because $U(1,2) = U^{-1}(3,4)$ and $U(2,3) = U^{-1}(4,1)$. They appear as $e^\lambda e^\mu e^{-\lambda} e^{-\mu}$, if we call these two matrices λ, μ. Note that they stand in such an order that their cancellation (and the product being the unit matrix) is possible if their *commutator* is also zero. One can use the Campbell–Baker– Hausdorff formula, in the form $\boxed{E}$

$$e^\lambda e^\mu = \exp\left(\lambda + \mu + \tfrac{1}{2}[\lambda, \mu] + \tfrac{1}{12}[(\lambda + \mu), [\lambda, \mu]] + \cdots\right), \qquad (1.14)$$

from which a non-Abelian version of Stokes' theorem for small contours follows,

$$e^\lambda e^\mu e^{-\lambda} e^{-\mu} = \exp([\lambda, \mu] + \cdots). \qquad (1.15)$$

Note also, that the field strength has the homogeneous transformation law

$$G_{\mu\nu} \to \Omega G_{\mu\nu} \Omega^{-1}. \qquad (1.16)$$

Now we have all the necessary ingredients for the formulation of a gauge invariant theory, generalizing electrodynamics to the case of an N-dimensional "internal space" with any symmetry group.

The action of the non-Abelian gauge theory looks in brief notation the same as in QED, if the matter field (quarks) is the Dirac fermion

$$S = \int d^4x \left[-\tfrac{1}{4}(G_{\mu\nu}^a)^2 + \bar\psi(iD_\mu\gamma_\mu - m)\psi \right]. \qquad (1.17)$$

Note that the first term can be written in lattice variables as a sum over traces as was done above for all plaquettes.

Exercise 1.1. Derive (1.12) for A_μ^a constant in space, by expanding the trace of the product of four exponents up to a^4 term and verify that it is indeed the square of the field commutator.

Exercise 1.2. Same as in Exercise 1.1 but in a general case, with a space-dependent field. Consider the size of the plaquette small and expand in it, first the field in Taylor series starting from the plaquette center, then each exponent, and then collecting all terms $O(a^4)$ which produce the non-zero color trace.

Exercise 1.3. For constant A_μ derive the next terms in the expansion for small plaquette to $O(a^6)$ order, see that the result is zero, and go to the next order. *Hint*: Use the Campbell–Baker–Hausdorff formula 1.14. *Comment*: Using OPE methods and certain simplifications, Shifman [139] has derived expressions for many orders in powers of the size of any small-size contour C.

1.1.3. *Fixing the gauge*

Let us now return to the issue of the gauge choice. Suppose we insist on doing calculations using only "physical" degrees of freedom, eliminating the redundant ones. The reader is obviously familiar with the practice to use some additional conditions imposed on A_μ to "fix the gauge". Popular examples of gauge conditions are

$$A_0 = 0 \quad \text{the temporal or the Weyl gauge}$$
$$\partial_m A_m = 0 \quad \text{the Coulomb gauge, } m = 1, 2, 3$$
$$\partial_\mu A_\mu = 0 \quad \text{the Landau gauge}$$
$$n_\mu A_\mu = 0 \quad \text{the axial gauge, where } n \text{ is some unit 4-vector}$$
$$x_\mu A_\mu = 0 \quad \text{the Fock–Schwinger gauge}$$

Examples of the Feynman rules for the covariant and Coulomb gauges are given in Appendix A.7.

Indeed, one can use gauge freedom, e.g. by setting $A_0 = 0$, as for all temporal plaquettes, it is possible to set 2 links going in the time direction to zero. However the preceding discussion should have prepared the reader for the conclusion that the gauge freedom is *far from being completely used by any such conditions*. Indeed, even for a single plaquette one was able to set 3 (rather than 2) links to zero.

In general, it is possible to define a "gauge tree" on the lattice, starting from a particular point and fixing as many links as possible. So, for lattice configurations there is no problem to fix the gauge completely, and in fact it can be done in very many ways. A simple example is shown in Fig. 1.2: in this case $A_1(x) = 0$ as in some axial gauge with $n = (1, 0)$, but some residual gauge transformation makes it also possible to put $A_2(x = 0) = 0$.

Unfortunately, there is no way to complete the gauge fixing in a continuum description in any practical way, so in fact only *partial* gauge fixing, under conditions

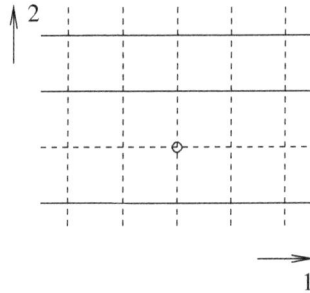

Fig. 1.2. Example of a gauge fixing tree in the $2d$ lattice. The solid lines have the nontrivial link variables U, while the dashed links indicate a "tree" starting from a root point (a small circle) where one can set $U = 1$ by using the available gauge freedom at the appropriate sites.

like those given above, is actually done in practice. A standard way out, used on the lattice, is to abandon gauge fixing altogether and define the partition function for gauge fields integrating over *all* configurations

$$Z = \int \prod_{\mu, x} DA_\mu(x) \exp(iS_g[A]), \qquad (1.18)$$

where S_g is the gauge field action. It is the only existing meaningful non-perturbative definition of QCD.

1.1.4. *Hamiltonian quantization*

Although we would be using it marginally, let me comment on the Hamiltonian approach. Can one eliminate the gauge freedom and then perform the usual quantization, as is done in quantum mechanics?

The most natural starting point for such a program is to take the temporal gauge $A_0 = 0$, in which three spatial components of $A_m(x)$, $m = 1, 2, 3$ play the role of "coordinates", while the components of the electric field strength,

$$E_m = G_{0m} = \dot{A}_m, \qquad (1.19)$$

are the conjugate "momenta". So the electric part of the energy is just quadratic in time derivatives and can be viewed as the kinetic term. The non-linear magnetic energy is a potential term. In discretized $3d$ lattice with $\vec{A}$ variables one may consider gauge theory as a system of coupled non-linear oscillators. One may think that it can be quantized in a standard way, and a finite number of them can even be solved numerically.

The remaining problem is that such momenta are not independent: there exists a constraint,

$$D_m E_m = j_0, \qquad (1.20)$$

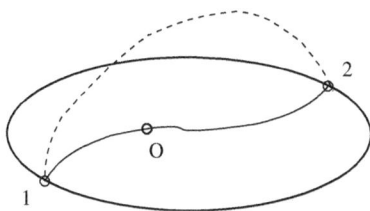

Fig. 1.3. The interior of the ellipsoid is a schematic representation of the fundamental domain. Two points at its boundary, marked 1 and 2, are gauge equivalent, and should be identified (connected by the dashed line).

known as the "Gauss law", which in this formalism is an additional condition on states, consistent but unrelated to the Hamiltonian, to be enforced "by hand".

More generally, any gauge fixing scheme has a problem, as pointed out by Gribov [16]. The gauge copies of the same configuration may be rather nontrivial to recognize as such, in a general non-perturbative context.

Some progress in the Hamiltonian approach had been made, especially in a finite volume. The reader is invited to read a very nice pedagogical review on the subject by van Baal [257]. Without going too deeply into this complicated issue, let me still try to explain here what the problem is.

Suppose that some discretized set of $A_m(x)$ in a finite volume can indeed be solved, and that the Hamiltonian eigenvectors and eigenvalues can be determined. The problem is that only solutions possessing proper symmetries are allowed, and their identification is quite difficult.

The manifold of all field configurations on which a quantization can be made should ideally include only one gauge copy of each configuration only, this is the so called *fundamental domain*. Due to the "Gribov copy phenomenon", this means that this domain may possess quite complicated boundaries and even a nontrivial topology.

For example, imagine that we start with a zero A_μ configuration, and move to some different directions, see Fig. 1.3. At the boundaries of the domain, marked by the thick ellipse, it is no longer true that configurations have a single gauge copy. Let those two copies be two points marked 1 and 2. If so, the coordinate along the path should be treated as a kind of an angle, with the ends identified (this is what the dashed line means). The wave function of the system should satisfy certain periodicity conditions relating $\Psi(A_m^1(x))$ to $\Psi(A_m^2(x))$. What those conditions may possibly be, is discussed in the next subsection, using a toy model.

1.1.5. *Rotator — a toy model for periodic coordinates*

The main issue we address in this subsection is a distinction between the angle-like and the periodic coordinates. This will be important in Chapter 3 dealing with the topological phenomena in gauge theories, especially with respect to the so called theta-vacua.

Another point we will make is that classical equations of motion do not contain sufficient information for quantization, of which the so called Aharonov–Bohm effect [5] is a good example.

We use a very simple toy model, which consists of a particle of mass m rotating on a circle of radius R, or "rotator" for short. Its classical action is

$$S = \frac{mR^2}{2} \int dt \, (\dot{\alpha})^2, \tag{1.21}$$

Classically the problem is trivial since the solution of the equation of motion $\ddot{\alpha} = 0$ is the free motion $\alpha(t) = \omega t + \alpha(0)$.

Its standard quantum mechanical treatment is also simple. Introducing the operator of angular momentum, conjugated to angle α, $\hat{L} = i\hbar \partial/\partial\alpha$, and the moment of inertia $I = mR^2$ one can write a simple Hamiltonian

$$H = (1/2I)\hat{L}^2. \tag{1.22}$$

Obviously, the eigenfunctions or $\hat{L}$ are $\psi_l(\alpha) = \exp(-il\alpha)$. Because the variable α is an *angle*, so that 0 and 2π are physically the same location, the wave function should be periodic $\psi(2\pi) = \psi(0)$. This leads to the quantization of the angular momentum $l =$ integer (in units of $\hbar$, of course).

The nontrivial question is whether we really should have treated the coordinate α (i) as an *angle*, as we have done thus far, or (ii) as a *periodic coordinate*. The standard example of the latter case is a position x of an electron in a crystal. The point shifted by a period is not physically the same location, just the physical conditions at this point are the same as in the original one.

In the latter situation quantization is done quite differently, as all texts on quantum mechanics explain. Now only the *physical observables* (such as the wave function modulus) are required to be periodic, but not the "unphysical" phase ϕ of the wave function. The condition

$$\psi(x + \text{period}) = \psi(x) \exp(i\phi) \tag{1.23}$$

leaves ϕ arbitrary, there is *no* quantization condition and the spectrum is continuous. The eigenvectors are denumerated by a continuous (quasi) momentum $p = \phi/\text{period}$.

Finally, there is the third option (iii), an angle-like coordinate but with an external field acting on it in a specific way. The model would now be an *electrically charged* rotator with a narrow Aharonov–Bohm solenoid [5], see Fig. 1.4. The solenoid is assumed to be infinite, so all the magnetic flux is confined inside it, and there is no magnetic field at the circle. So, classically there is no Lorentz force on the rotator, and one may expect that the field inside the solenoid cannot affect its motion.

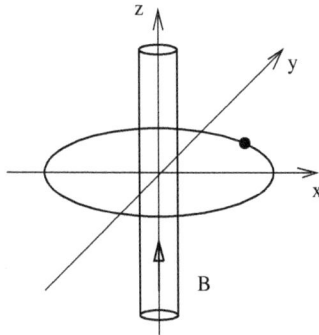

Fig. 1.4. Rotator, a small charged sphere rotating on a circle in the x–y plane. Along the vertical z axis there is an Aharonov–Bohm solenoid with a certain flux of magnetic field B.

Quantum mechanics suggests otherwise. Formally, in the presence of electro-magnetic field, in the angular momentum operator $\hat{L} = iD_\alpha$ the ordinary derivative should be changed to a covariant one. The derivative with respect to the angle α is appended by a term containing the corresponding component of the vector potential. It is easy to see from Stokes' theorem that outside of the solenoid[4]

$$A_\alpha = \Phi/2\pi R, \qquad (1.24)$$

where Φ is the magnetic flux, its other components are zero. After a circle, an additional phase $-e\Phi$ is added to the wave function, correcting (former integer) values of the angular momentum. The spectrum is still discrete, but now it is

$$E_l = \frac{1}{2I}(l - e\Phi)^2. \qquad (1.25)$$

In the path integral quantization the role of the solenoid can be included by adding a new term to the action, proportional to the "winding number", the number of rotations made by the rotator,

$$S_{\text{topological}} = \theta \left(\int dt \, \frac{d\alpha}{dt} \right). \qquad (1.26)$$

Here θ is some constant, proportional to the magnetic flux in the example above. Note that adding a topological term like this "winding number" would not be seen by classical equations of motion. Indeed, the topological integral is not changed if the path is slightly modified, and so it has no variation. Quantum-mechanically, this term would still add a non-trivial phase to the topologically non-trivial "winding" paths, and thus affect the path integral. We will later see that the Yang–Mills field possesses similar properties.

---

[4]Note that naively one may think this field has zero curl and produces no magnetic field, but an accurate treatment reveals a delta-function-like field at the origin.

1.2.   Road to QCD

1.2.1.   *Quarks and the "colors"*

The "internal space" of the fields we speak about in QCD is not the isospin discussed by Yang and Mills, it is in this case substituted by another quantum number named "color" by M. Gell-Mann. In the early days of the quark model of hadrons it was realized that there were difficulties connected with the Pauli principle for quarks. There exist baryons, Δ^{++}, Δ^-, and the famous Ω^-, consisting of three quarks of the same kind (or "flavor"): uuu, ddd and sss, respectively. Moreover, all quarks have the same spin orientation ($S = 3/2$), and identical coordinate wave functions. In order to avoid a contradiction with the Pauli principle, it was assumed in Refs. [7, 8] that there exists some triplet quantum number (the "color") which distinguishes these three quarks.

QCD was born when "colors" were combined with the Yang–Mills (YM) theory, see Refs. [9–11]. The non-Abelian gauge field A_μ^a, mediating the strong interaction, is called the "gluon field" (because it "glues" quarks together into hadrons). Their index a counts the generators of the SU(3) "color" group, therefore it runs over 1–8.

The fermionic part of the QCD Lagrangian includes the sum over quark "flavors", $f = u, d, s, c, b, t$. In fact for nearly all phenomena to be discussed in this book the last three heavy quarks are quite irrelevant since their mass is large compared to the characteristic energy scale.

1.2.2.   *Renormalizability and asymptotic freedom*

We have formulated above the classical YM theory. Now we will ask whether it can be promoted to a good quantum field theory. In particular, is it renormalizable? Can all infinities be grouped into renormalization of its single parameter, the gauge coupling constant g? Those questions were answered positively by 't Hooft and Veltman [17] after the Feynman rules were finally[5] established by Faddeev and Popov [15]. We will not discuss these highly technical points.

Note that the gauge field has no mass term since this would simply kill the gauge invariance. As a first physical application, the Yang–Mills fields were connected with the intermediate bosons mediating weak interactions. We are not going to discuss in this book how the Weinberg–Salam model had become a "Standard Model". We will only discuss electroweak instantons and sphalerons when appropriate, and now just remind the reader that the internal quantum number in this case is the "left-handed isospin", and the group is SU(2). Weinberg and Salam used the Higgs mechanism or "soft mass" generation to avoid confrontation with gauge symmetry.

---

[5]The reader interested in the history of the question and the difficulties on the way toward such rules should read the Feynman paper on the subject [14].

The next breakthrough was the discovery of *asymptotic freedom*, a feature of non-Abelian theory which is quite different from Abelian QED. It has a rather dramatic history. Three approaches, which I will call *magnetic*, *electric* and *covariant*, have evolved independently. The final result is of course the same, the universal charge renormalization is driven by the same beta function in all situations. The fact that the charge is "anti-screened" rather than screened (which is the case in QED and any scalar theory studies previously, and believed to be unavoidable) was truly shocking.[6] The reader can find a detailed discussion of its history in Ref. [1].

The first approach was *magnetic* derivation in the early work by Vanyashin and Terentiev [18] (also later Savvidi *et al.* [137] and many others have followed these ideas), and it is probably the easiest to understand.[7] So we will discuss it below in more detail, after making some brief remarks about the other two.

The *electric* path to asymptotic freedom was worked out by I.B. Khriplovich in Ref. [21], using a (non-covariant) Coulomb gauge.[8]

The dates of both early works — 1965 and 1968, respectively — are very important, they took place *before* the famous SLAC deep inelastic experiments of 1969 which confirmed the Björken scaling and soon lead to Feynman's parton model. Therefore at that time the surprising negative sign of the beta function could not possibly be related with strong interactions.

The *electric* approach by Khriplovich considers a field of a static charge. In zeroth order the scalar potential A_0 has a propagator $1/\vec{q}^{\,2}$, which is simply a Fourier transform of the Coulomb potential. Corrections to a static charge come entirely from the polarization operator $\Pi_{00}(\vec{q}, q_0 = 0)$ which comes from magnetic gluons (those can be treated best in the Coulomb gauge $\partial_m A_m = 0$ so as not to interfere with the initial Coulomb field) and quarks. The logarithmic contributions from the three diagrams shown in Fig. 1.5 are sufficient to obtain the invariant charge renormalization, as in this gauge there is no renormalization of the Coulomb vertex function. Including the usual contribution from quarks (analogous to that in QED) plus these three gluonic diagrams one gets their respective contributions to the gluon polarization operator,

$$\Pi_{00}(\vec{q}) = \frac{g^2 \vec{q}^{\,2}}{\pi^2} \log(\vec{q}^{\,2}/\Lambda^2) \left[\tfrac{1}{3} N_f + \tfrac{1}{2} - 6 + 0 \right]. \qquad (1.27)$$

Note that the only negative term comes from the unusual diagram without the imaginary part, with one magnetic gluon produced out of the Coulomb ones. Explicit calculation shows that it has a logarithmic behavior $\log(\vec{q}^{\,2})$ which contains not the usual 4-dimensional momentum squared but its spatial components only. Clearly it

---

[6]In retrospect however, the early works of the 1960s have attracted surprisingly little attention, and were not even emphasized much by the authors themselves.

[7]Among many summaries we would recommend e.g. a discussion in a historical recollection by Wilczek [19].

[8]For quantization in this gauge see Refs. [12, 13].

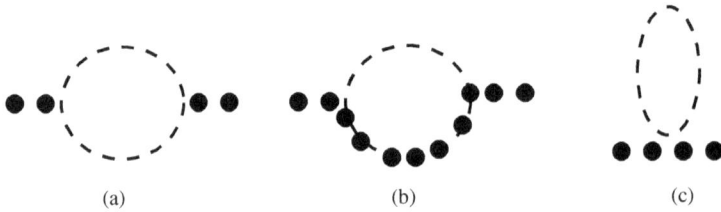

Fig. 1.5. Three diagrams for the gluon polarization operator in the Coulomb gauge, in the lowest g^2 order. The dashed lines indicate the transverse (magnetic) gluon propagators, while the dotted lines are for the Coulomb field ones.

has nothing to do with unitarity and physical intermediate states, so its sign is not restricted by it, unlike e.g. the signs of terms 1 and 2.

Thus the corrected field is

$$(1/\vec{q}^2) + (1/\vec{q}^2)\Pi_{00}(1/\vec{q}^2) + \cdots \tag{1.28}$$

$$= \frac{1}{\vec{q}^2}\left(1 + \frac{g^2}{\pi^2}\log(\vec{q}^2/\Lambda^2)\left(\tfrac{1}{3}N_f + \tfrac{1}{2} - 6\right)\right)^{-1}, \tag{1.29}$$

which is the asymptotic freedom formula. We will return to a more detailed discussion of this work in the part devoted to pQCD at finite temperature, since the Coulomb gauge used in it is also convenient for these particular applications.

Two *covariant* derivations were given in 1973 simultaneously, by H.D. Politzer, as well as by D. Gross and F. Wilczek [23, 24]. By that time the SLAC data on deep inelastic electron scattering and the parton model were widely known, so the connection had been made in these famous works, and the asymptotic freedom was forever connected the Yang–Mills theory of colored quarks with strong interactions. The covariant derivation of asymptotic freedom is discussed in all QFT textbooks, so we will not discuss it here. This derivation of the renormalized charge is in fact more complicated since one should calculate not only loop corrections to propagators but also the vertex corrections; only then their particular combination gives the renormalization of the invariant charge g.

Now we return to the *magnetic* approach to the beta function. I think this is the one W. Pauli and L.D. Landau would have liked, more than anyone else; it is very close to the other effect in the magnetic field which they have studied back in the early days of quantum mechanics. Let me remind the reader that in a metal, approximated by an ideal Fermi gas of free electrons, there are two effects: (i) Pauli paramagnetism due to different occupation of spin $\pm 1/2$ states; and (ii) Landau diamagnetism due to Landau levels and electron rotation.

The non-relativistic Hamiltonian of this textbook problem is

$$H = \frac{(i\vec{\partial} - e\vec{A})^2}{2m} - \frac{\mu}{s}(\vec{s}\vec{H}). \tag{1.30}$$

Selecting e.g. $A_x = -Hy, A_z = A_y = 0$ we get plain waves in the x–y plane and oscillator motion in y. The Landau levels are

$$E_{n,s} = (eH/m)\left(n + \tfrac{1}{2} + \sigma\right), \qquad (1.31)$$

where $n + \tfrac{1}{2}$ comes from the orbital harmonic oscillator and $\sigma = \pm 1/2$ from the spin (magnetic moment) term. In combination, electron levels are twice degenerate integers, except for the *zero* energy one which is only *once* degenerate. Note that the spin term compensates the positive zero-point energy associated with electron localization at the Landau level, this is our first example of a *fermionic zero mode* which will repeatedly appear in this book.

Now we move toward QCD by including a massive vector charged particle. Its magnetic moment is twice as large as that of the electron because the spin is 1, not $1/2$. This makes σ in the last formula $\sigma = \pm 1$ and the lowest level becomes negative! What it implies is that a classical gluo-magnetic field is intrinsically unstable, creating favorable negative energy levels for quantum gluons with the appropriate polarization.

In order to see how all of this can be related to the beta function one has only to rewrite the magnetic energy in zeroth order as $E_{\text{magnetic}} = (1/2g^2)\tilde{B}^2$ by renormalizing the field strength. Now

$$\tilde{G}^a_{\mu\nu} = \partial_\mu A^a_\nu - \partial_\nu A^a_\mu - f^{abc} A^b_\mu A^c_\nu$$

and the charge moves in front.

In the next order (to be explained below) when the vacuum gets polarized and responds to the field, the answer is

$$E_{\text{magnetic}} = \left[\frac{1}{2g^2(\Lambda)} - \frac{\eta}{2}\ln\left(\frac{\Lambda^2}{\tilde{B}}\right)\right]\tilde{B}^2 \qquad (1.32)$$

and it can be rewritten simply as $E_{\text{magnetic}} = (1/2g^2(\tilde{B}))\tilde{B}^2$, with the correction finally attributed to the running or renormalized charge $g(B)$.

The calculations resemble the Landau treatment of the problem very closely, there is a summation over Landau levels, which is partly integrals over 2-momenta and partly sums over n. For any spin one can write an oscillator-type equation. But first one can consider momenta large compared to B and write just kinetic energy plus the spin term

$$E = \int^\Lambda \frac{d^2k}{2(2\pi)^2}\left[\sqrt{k^2 + g_m sB} + \sqrt{k^2 - g_m sB} - 2k\right]$$
$$\approx -B^2(g_m sB)^2 \log(\Lambda^2/B). \qquad (1.33)$$

The orbital part gives for a scalar particle a positive energy $(B^2/2)(\log(\Lambda^2)/96\pi^2)$, while the fields with different spins get the same, simply multiplied by the number of components.

The central thing to understand is the magnitude of the crucial g-factor, $g_m = 2$. Recall that for spin-1/2 particles the g-factor follows from the squared Dirac equation

$$[(D^\mu \gamma_\mu)^2]\psi = [D^2 + \tfrac{1}{2}\sigma_{\mu\nu}F_{\mu\nu}]\psi = 0. \tag{1.34}$$

For quantum gluons this statement follows from the Yang–Mills equation of motion $D_\mu G_{\mu\nu} = 0$. We have to decompose the gauge field into the background classical field $A_\mu^{\text{classical}}$, generating B, plus small quantum one a_μ^{quantum}. For simplicity, one can take the SU(2) group (so $a, b, c = 1, 3$ and $f^{abc} = \epsilon^{abc}$) and assume that the background field has color 3 only, and the quantum field is $a_\mu = A^1$ only and it is in the "background gauge" $D_\mu a_\mu = 0$. After some algebra,

$$D_\mu D_\mu a_\nu^1 + 2i(\partial_\mu A_\nu^3 - \partial_\nu A_\mu^3)a_\mu^1 = 0. \tag{1.35}$$

The first term is the same as for the spinless particle. Taking the magnetic field in the z direction and taking $a_\mu^1 = (1/\sqrt{2})(1, \pm i, 0)$ we see that indeed $g_m = 2$.

Finally, since the gluon has color charge twice as large as the quark, and its spin is twice as large as well, the paramagnetic spin-related part has a chance to win over the orbital one. As a result, the magnetic field gets destabilized due to the production of negative energy quantum modes. Savvidi *et al.* even argued [137] that when the coupling constant gets big and the second term may win over the initial energy, the field will be created *spontaneously*. The problem however is that one should not trust the perturbative calculation if the correction is as big as the first term.

1.3.   Path integrals and Euclidean time

1.3.1.   *Green functions and Feynman path integrals*

Standard texts on quantum mechanics discuss the wave-function (Schrödinger) and, occasionally, the matrix (Heisenberg) formulations. In the simplest case of motion in a time independent potential $V(x)$ the objects discussed are the stationary states $\psi_n(x)$, their energies E_n and sometimes the matrix elements of the relevant operators between those states. Modern approaches to Quantum Field Theories use instead another approach due to Feynman, who described quantum theory in terms of the so called *path integrals*.[9] Although there is an extensive discussion of that approach in the literature, let us recall its main points.

Another point we would like to emphasize in this section is the similarity between quantum and statistical mechanics. Qualitatively, both quantum and statistical mechanics deal with variables that are subject to random fluctuations (quantum or thermal), so that only ensemble averaged quantities make sense. Formally the

---

[9]The classical presentation of the path integral method can be found in the book by Feynman and Hibbs [153].

connection comes from the similarity of the partition function $\text{tr}[\exp(-\beta H)]$ and the generating functional for quantum dynamics which we will introduce below.

The central object of the consideration is not the wave function but rather the so called *Green function* $G(x, y, t)$, the amplitude for the particle to propagate from point x to point y during the time t. As we will show below, many features of quantum and even statistical mechanics can be obtained if the Green function is known.

There are two ways to define it, which we would like to compare. The "old fashioned" one uses the Schrödinger stationary states. Indeed, these states have the simplest time evolution. Sandwiching the full set of stationary states $|n\rangle$ on both sides of the exponent, and noticing that $\psi_n(x) = \langle n|x \rangle$ one gets the expression[10]

$$G(x, y, t) = \langle y| \exp(-iHt)|x \rangle = \sum_n \psi_n(x)\psi_n^*(y) \exp(-iE_n t). \qquad (1.36)$$

As an example [153], one may calculate it for a free motion, using the usual plane waves $\psi_p(x) \sim \exp(ipx)$, $E_p = p^2/2m$. The (1-dimensional) answer is

$$G(x, y, t) = \sqrt{\frac{m}{2\pi it}} \exp\left(\frac{im(x-y)^2}{2t}\right). \qquad (1.37)$$

Thus a free quantum particle spreads "diffusionally", $(x-y)^2 \sim t$. In principle, one can use it in other cases, although even for the harmonic oscillator (for which the wave functions are Hermite polynomials) the summation over all states is not so easy to make.

Another definition (in which we are most interested) is the celebrated Feynman *path integral* expression

$$G(x, y, t) = \int_{x(0)=x}^{x(t)=y} Dx(t) \exp(-iS[x(t)]). \qquad (1.38)$$

This implies that a quantum system takes all possible paths, and to calculate the probability amplitude of its behavior one has to add up all the path amplitudes with the proper phase factors. By definition the integral should be taken over all paths $x(t)$ with fixed ends, starting at point x at time zero and ending at point y at time t. Here S is the usual classical *action* of the problem, e.g.

$$S = \int dt \left[\frac{m}{2}\left(\frac{dx}{dt}\right)^2 - V(x)\right] \qquad (1.39)$$

for a particle of mass m in a static potential $V(x)$ provides the weight of the paths in (1.38).

One way to calculate the path integral is just to use its definition. If one cuts the t interval into N steps, $a = t/N$, the path integral is the N-dimensional integral

---

[10]We remind the reader that we use natural units $\hbar = h/2\pi = 1$ and $c = 1$. Mass, energy and momentum all have dimension of inverse length.

taken over all the intermediate positions of the particle x_i, $i = 1 - N$. After it is calculated, the limit $N \to \infty, a \to 0, t = a * N = fixed$ should be taken.

It is convenient to view the path as a $1d$ "lattice", with the x_i being variables defined at its sites. The simplest discretized action approximates the derivative by the difference[11]

$$S_E = \sum \left(im \frac{(x_n - x_{n-1})^2}{2a} - iaV(x_n) \right), \tag{1.40}$$

which can be interpreted as a $1d$ Ising lattice, with only the nearest neighbors interacting.[12]

As the standard example, consider the quantum harmonic oscillator with the quadratic potential $V(x) = m\Omega^2 x^2/2$ for which all the integrals are Gaussians, so that the finite-N and then the infinite-N limit can both be nicely calculated [153]. The result is

$$G_{\text{osc}}(x, y, t) = \left(\frac{m\Omega}{2\pi i\hbar \sin \Omega t} \right)^{1/2}$$

$$\times \exp \left[-\left(\frac{im\Omega}{2\hbar \sin \Omega t} \right) [(x^2 + y^2) \cos \Omega t - 2xy] \right]. \tag{1.41}$$

As we will see below, this nice expression of Feynman's can be used to demonstrate many correct limits.

One can use the path integral in the general case when the potential is not quadratic and integrals are not Gaussian.

(i) For the *weakly anharmonic* oscillator one can use a perturbative expansion in the anharmonic couplings, which leads to Feynman diagrams.[13]

(ii) One can also use the *semiclassical approximation* selecting the paths with the smallest or stationary action. We will discuss this approach later.

(iii) Finally, one may try to use brute force and do multidimensional integrals over paths numerically.

Unlike many methods developed for quantum mechanics, all three approaches to the path integral mentioned have direct generalization in QFT in general and QCD in particular, namely (i) pQCD, (ii) instantons and semiclassical reactions which we will discuss later, (iii) lattice QCD. In all of these approximate methods one finds that the strongly fluctuating phase of the weight in (1.38) is not good for

---

[11] This form is of course not unique, other discretization with the same continuum limit can be more accurate, those are "improved lattice actions".

[12] What plays the role of the temperature in the initial quantum mechanical problem, causing the fluctuations? It is easy to see, that it is the Planck constant $\hbar$, to which the action S is normalized in the exponent. It is suppressed in our units in which $\hbar = 1$.

[13] It remains a mystery to me why Feynman himself has not described or developed this approach in quantum mechanics.

practical purposes. The trick one can use is analytic continuation of $G(x, y, t)$ into imaginary time, $\tau = it$, also called the Euclidean time.

In this case, the paths enter the sum with the non-oscillating weight $\exp(-S_E[x(\tau)])$, where the new *Euclidean action* is

$$S_E = \int dt \left[\frac{m}{2} \left(\frac{dx}{d\tau} \right)^2 + V(x) \right].$$
(1.42)

Note that the *relative sign* between the kinetic and potential energy has now changed.

The highest weight has the path $x = x_{\min}$, corresponding to the minimum of $V(x)$. This is the simplest "classical path", obviously minimizing the action. But, as emphasized by Feynman, the quantum particle is not just sitting at the bottom of the potential simply because there are many other paths. Thus its behavior is a compromise between the *action* which tries to enforce classical behavior, and the number of paths, or *entropy* of the configurations.

Certainly, path integrals in Euclidean time are easier to calculate, especially numerically. For the oscillator it is of course a simple analytic transformation of the result known previously

$$G_{\text{osc}}(x, y, \tau) = \left(\frac{m\Omega}{2\pi\hbar \sinh \Omega\tau} \right)^{1/2}$$

$$\times \exp\left[-\left(\frac{m\Omega}{2\hbar \sinh \Omega\tau} \right) [(x^2 + y^2) \cosh \Omega\tau - 2xy] \right]$$
(1.43)

but for the numerically found ones it is a hopeless task to perform an analytic continuation back to Minkowski time.

The main point we would like to make here is that this is not necessary because the Euclidean Green function has a physical meaning by itself.

Let us recall the first definition of the Green function (1.36). It is easy to see what it is in Euclidean time, the sum over states becomes a set of decreasing exponentials, $\exp(-E_n\tau)$. It leads us to conclusion No. 1:

(i) *At large Euclidean time duration the path ensemble describes the properties of the ground state.*

Now, let us set $x = y$ and integrate over it. Now the restriction is only that the particle should come to the same point after the time period τ, making *periodic paths*. Using the same "old fashioned" definition and the normalization condition of the wave function $\int |\psi_n(x)|^2 = 1$, one gets

$$\int dx\, G(x, x, \tau) = \sum_n \exp(-E_n\tau).$$
(1.44)

If one substitutes $\tau = \hbar/T$ here, it becomes evident that it is nothing else but the *statistical sum* for our system, at the (so called Matsubara) temperature T. This leads us to conclusion No. 2:

(ii) *Periodic paths with finite τ correspond to (the so called Matsubara tempera-ture) $T = 1/\tau$.*

This observation provides the formal link between an n-dimensional statistical system and Euclidean quantum (field) theory in $n - 1$ dimensions.

In particular, the diagonal $G(x, x, 1/T)$ has the physical meaning of the *probability distribution* over coordinates, at the temperature T. For example let us use Feynman's harmonic oscillator result mentioned above. Taking $y = x$ one finds that at any temperature the spatial distribution of the particles is Gaussian, with the width given by

$$\langle x^2 \rangle = \frac{1}{2m\Omega} \coth(\Omega/2T). \tag{1.45}$$

At low T it leads to the width corresponding to the quantum mechanical ground state wave function, while in the opposite limit of high T it leads to the *classical* result for Boltzmann statistics $\langle x^2 \rangle = T/m\Omega^2$.

Since for harmonic oscillator the total energy is twice the potential energy, we also have the mean energy, which can be re-written as

$$\langle E \rangle = \frac{\Omega}{2} + \Omega \frac{1}{e^{\Omega/T} - 1}, \tag{1.46}$$

with the obvious meaning of the two terms as the energy of the zero-point quantum oscillations plus the energy of thermal excitation with thermal Bose distribution.

1.3.2. *Perturbation theory and Euclidean path integrals*

Non-Gaussian path integrals cannot be done exactly, but if the nonlinearities are small, one can use perturbation theory. In QFT it leads to the widely used Feynman diagrams, and we would like to demonstrate now how it can also be done in a quantum mechanical setting. We follow the paper of Aleinikov and myself [149] where a number of such diagrams were calculated for the anharmonic oscillator, first for (trivial) perturbative background and then in the background of an instanton. In $1d$ all diagrams are never divergent, computed easily and the results can be easily verified by other means. So it provides very good pedagogical material for those who want to get used to Feynman diagrams in a setting as simple as possible.

Consider an anharmonic oscillator with (Euclidean) action

$$S_E = \int d\tau \left(\tfrac{1}{2}\dot{x}^2 + \tfrac{1}{2}\Omega^2 x^2 + \alpha x^3 + \beta x^4 \right), \tag{1.47}$$

where α, β are considered to be small and are used as expansion parameters. The nth order terms simply follow from an expansion of the exponent, so it is just the perturbation action to power n divided by n factorial. What appears is then a set of Gaussian-like integrals with different powers of x_i in the numerator, which can be easily evaluated by standard rules.

The basic Feynman rules for the anharmonic oscillator are then:

(i) the vacuum energy is given by the sum of all closed diagrams, with proper combinatorial factors;

(ii) α and β provide triple and quartic vertices;

(iii) all vertices are to be connected by "propagators";

(iv) integrals are to be taken over all frequencies in the loops, with the Matsubara summation over $w_n = 2\pi T n$ with integer n substituting for it if T is non-zero,

$$\int \frac{d\omega}{2\pi} \cdots \rightarrow T \sum_n \cdots$$

Alternatively, one can handle the integrals in the time representation, in which integrals over moments of time for each vertex should be performed.

The propagator for the harmonic oscillator is given by the same basic expression as in QFT, only without the momenta,

$$G_0(\omega) = \frac{1}{\omega^2 + \Omega^2}. \tag{1.48}$$

Note that there is a plus sign in the denominator, unlike the familiar $p^2 - m^2$ in QFT. This is because of the Euclidean formulation, in which there are no poles and thus no need for the notorious Feynman $i\epsilon$ prescription. In fact one can get it correctly by rotating the contour back to Minkowski frequencies.[14] In the time representation the propagator also has a simple expression

$$G_0(\tau_1, \tau_2) = \langle x(\tau_1) x(\tau_2) \rangle = \frac{1}{2\Omega} \exp(-\Omega|\tau_1 - \tau_2|). \tag{1.49}$$

Note that there is a modulus of time difference in the exponent, so the correlation between coordinates decreases in both directions. The discontinuity this modulus provides is quite natural, as the propagator satisfies the equations of motion with the delta function in the r.h.s. Note however that the discontinuity is in the derivative, while the Green function at the origin, $G_0(\tau = 0) = \frac{1}{2\Omega}$, is by itself perfectly finite.

The vacuum energy is given by the sum of all closed diagrams. At the one-loop order, there is only one diagram, the free particle loop diagram. At the two-loop order, there are two $O(\alpha^2)$ and one $O(\beta)$ diagrams, see [Fig. 1.6(a)]. The calculation of the diagrams is remarkably simple. Since the propagator is exponentially suppressed for large times, everything is finite. Summing all the diagrams, we get

$$\langle 0| \exp(-H\tau)|0\rangle = \sqrt{\frac{\Omega}{\pi}} \exp\left(\frac{\Omega\tau}{2}\right) \left[1 - \left(\frac{3\beta}{4\Omega^2} - \frac{11\alpha^2}{\Omega^4}\right)\tau + \cdots\right] \tag{1.50}$$

[14]Of course, historically this is how Feynman got it right in the first place.

(a)

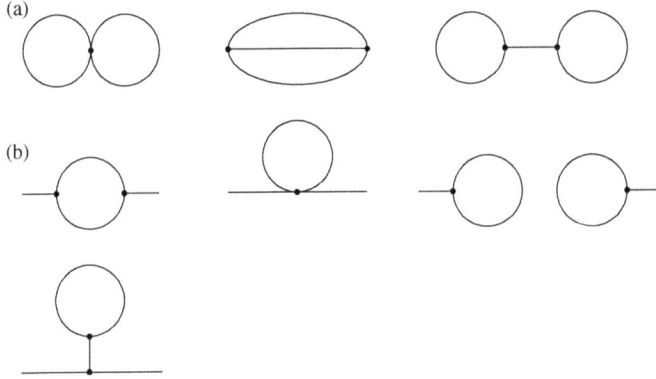

(b)

Fig. 1.6. Feynman diagrams for (a) the energy and (b) the Green function of the anharmonic oscillator.

For small α^2, β and τ not too large, we can exponentiate the result and read off the correction to the ground state energy

$$E_0 = \frac{\Omega}{2} + \frac{3\beta}{4\Omega^2} - \frac{11\alpha^2}{\Omega^4} + \cdots . \tag{1.51}$$

Of course, one can obtain the same results using ordinary Rayleigh–Schrödinger perturbation theory routinely used in quantum mechanics texts. The difference however is that ordinary perturbation theory cannot be formulated in terms of simple rules, valid for nth order directly. The Feynman version leads not only to simple "Feynman rules", but for Green functions the coefficients are even independent of n, allowing nice "graphical re-summations". This method is well proven to be much more powerful in the context of systems with many degrees of freedom and quantum field theory.

One more simple exercise is worth mentioning, the evaluation of the first perturbative correction to the Green function. The diagrams shown in [Fig. 1.6(b)] give

$$\Delta G_0(0, \tau) = \frac{9\alpha^2}{4\Omega^6} + \frac{\alpha^2}{2\Omega^6} e^{-2\Omega\tau} + \frac{15\alpha^2}{4\Omega^5} \tau e^{-\Omega\tau} - \frac{3\beta}{2\Omega^3} \tau e^{-\Omega\tau}. \tag{1.52}$$

Comparing the result with the decomposition in terms of stationary states,

$$G_0(0, t) = \sum_{n=0}^{\infty} e^{-(E_n - E_0)t} |\langle 0|x|n\rangle|^2, \tag{1.53}$$

we can identify the first (time independent) term with the square of the ground state expectation value $\langle 0|x|0\rangle$ (which is non-zero due to the tadpole diagram). The second term comes from the excitation of 2 quanta, and the last two (with extra factors of τ) are the lowest order "mass renormalization", or corrections to the zero order gap $E_1 - E_0 = \Omega$.

Exercise 1.4. (1) Derive (1.50) using the ordinary Rayleigh–Schrödinger perturbation theory. (2) Derive (1.50) using the Feynman rules and diagrams shown in [Fig. 1.6(a)].

1.3.3. *Numerical evaluation of Euclidean path integrals*

The straightforward numerical simulation of ensembles of paths in Euclidean time for quantum mechanical (and thermal) motion for systems with few degrees of freedom is these days quite easy to do on any computer. Since the weight function is positive, standard importance sampling algorithms like the Metropolis algorithm work well and a few hundred subdivision points are usually enough. Examples of extensive studies of this method of the double-well potential can be found in Refs. [155, 157], and I recommend as an exercises to write a program, or use the one provided in Appendix B.3 and to do so. Also simulations of the simplest atoms and nuclei have been demonstrated, e.g. by Zhirov and myself [156]. Although the Schrödinger formulation is better suited for the simplest problems with $1d$ or separable coordinates (e.g. the hydrogen atom), numerical studies of the paths are not restricted by the number of degrees of freedom. For example, it is possible to do He atoms with 2 electrons and get the correct binding in a very straightforward way.

After the ensemble of paths is generated, any operator can be averaged, e.g. the mean square of the coordinate reads

$$\langle x^2 \rangle = \frac{\int Dx(\tau) x^2(\tau_1) \exp(-S)}{\int Dx(\tau) \exp(-S)}. \qquad (1.54)$$

Longer Euclidean times β correspond to lower Matsubara temperatures $T = 1/\beta$, so if only the ground state and lowest excitations are of interest one should select a sufficiently small T.[15]

Another way to use the ensemble is make histograms of certain variables, e.g. "spins" of the Ising analog model themselves. If the total Euclidean time is long enough, the distribution $P(x)$ is nothing else but the probability distribution in the ground state $P(x) = |\psi_0(x)|^2$. Otherwise it corresponds to the ensemble at the non-zero temperature $T = 1/\tau$.

One more useful thing one may easily evaluate is the correlation function of the "spins"

$$K(\tau_1 - \tau_2) = \langle x(\tau_1) x(\tau_2) \rangle. \qquad (1.55)$$

---

[15]Naively, one should compare T to the gap between the ground state and the next excitation, but sometimes it is not so. For example, in simulation of atoms a very small T still can lead to their ionization. This happens because for the Coulomb potential there are very many levels close to zero.

At zero time difference it is the mean fluctuation considered above, while at larger values of τ it decays. In statistically analogous system it is natural to have finite "correlation length", so that $K(\tau) \sim \exp(-\tau/\tau_{\text{corr}})$. This correlation length is nothing else but the inverse gap, or the level spacing $\tau_{\text{corr}} = 1/(E_1 - E_0)$. The same method is used in lattice QCD, for measurements of hadronic masses, which are also "gaps" between the ground state and the first state which can be excited by an operator with the appropriate quantum numbers. Finally, one may hunt for topologically non-trivial objects, like solitons and instantons.

1.4.   Gauge fields on the lattice

In this section we are going to give the definition of what we mean by non-perturbative QCD, following the constructive lattice definition due to Wilson [251].

Let us consider a 4-dimensional lattice, with Euclidean time replacing the usual Minkowskian one. As we saw earlier for quantum mechanics, it eliminates phases and oscillation in the weight function.

Furthermore, the field variables are not defined on the lattice sites,[16] but on links between them. As we have discussed already, the "constructive definition" of the quantized Yang–Mills theory is given in terms of $N_c * N_c$ matrix-valued link variables. The traditional lattice notations are slightly different from what we used above, it is $U_\mu(x)$ where x is the position of the site from which the link originated and $\mu = 1$–4 is its direction. Recall that those are related to continuous vector potentials $A_\mu^a(x)$ by $U_\mu(x) = 1 + (iag)A_\mu(x) + O(a^2)$. Here $A_\mu(x) = T^a A_\mu^a(x)$ and matrices $T = t/2$ where t^a, $a = 1$–8 are the so called Gell-Mann matrices, the generators of the SU(3)$_c$ group.

The partition function of "gluodynamics" is formally defined by the integral over A_μ,

$$Z = \int DA_\mu(x) \exp(-S_E). \tag{1.56}$$

The field strength in the Euclidean action[17] is

$$S_E = -(1/4) \int d^4x \, (G_{\mu\nu}^a)^2. \tag{1.57}$$

Let me remind the reader, that this integral taken in infinite Euclidean space–time domain describes the ensemble of "field histories" which characterize the ground state of the theory, the vacuum of gluodynamics. If the time is finite and the

---

[16]This is the place reserved for the "matter fields", that is, for quarks in QCD, to be discussed later.

[17]Let me remind the reader that in Minkowski space–time the field strength should be $E^2 - B^2$, but in Euclidean one it becomes $E^2 + B^2$. It is like changing the sign of the kinetic energy for quantum mechanical action. Thus in Euclidean space–time there is no difference between time and any other coordinate.

field configurations are periodic (that is, time is compactified into a torus) with some period, the ensemble of fields would then correspond to a finite temperature $T = 1/\tau_{\text{period}}$.

Wilson's version of this integral, given in terms of link variables, is as follows:

$$Z = \int \prod_{\text{links}} DU_\mu(x) \exp\left[-\frac{2N_c}{g^2} \sum_\square \left(1 - \frac{1}{N_c} \operatorname{Tr} UUUU\right)\right], \qquad (1.58)$$

where N_c is the number of colors and the sum $\sum_\square$ is taken over all plaquettes, while $UUUU$ is our shorthand notation for the product of 4 link variables around a plaquette. Emphasizing the similarity between lattice gauge theory and $4d$ statistical mechanics, lattice practitioners routinely call the adjustable coefficient in the exponent $\beta = 2N_c/g^2$ as if it is the inverse temperature, $\beta = 1/T$. As the lattice becomes finer and finer, β grows and the fields on the plaquettes decrease in magnitude, becoming more and more perturbative.

The reader may wonder why the action contains only a single plaquette, since a product of U over any closed loop is gauge invariant. And indeed, there are so called *improved* lattice actions which contain e.g. products of 6 Us. The simplest of the new terms is a double plaquette in a plane, or a "chair" of two plaquettes with a common link, or a fancier "butterfly" with two mutually orthogonal plaquettes (possible in $4d$). Each term has its own coefficient, which can ideally be determined from a renormalization group flow; see discussion below.

For each link variable the integral runs over the whole gauge group volume.

With the ensemble generated by the weight $\exp(-S_E)$ one can also calculate the average values of some operators, make histograms or calculate more complicated things. We will be especially interested in *correlation functions* of some operators $O[U]$, e.g.

$$K(x) = \frac{\int DU \exp(-S) O(x) O(0)}{\int DU \exp(-S)}. \qquad (1.59)$$

They can provide us with masses and other parameters of the lowest excitations, with the quantum numbers of the operators used.

A remark about gauge freedom. Above we have integrated over all components of the A_μ field. However, as we know, some of them are not physical. Is it right to include them in the integral? Well, if only gauge invariant expressions are calculated, their presence causes no harm. For example, the correlation function $K(x)$ defined above is made of gauge-invariant operators and so all gauge copies will provide the same values. Thus, the contribution of gauge copies of each configuration will simply provide the same factor in the numerator and the denominator, to be canceled in the ratio.[18]

---

[18]Strictly speaking, this explanation is only convincing if we are sure that any configuration has the same number (or volume) of the gauge copies. We do not really know that. So it is possible that the Wilson definition and the hypothetical "fixed gauge" ones (which allows no gauge copies at all) would be different theories.

The last step in the definition of gluodynamics as QFT should be a precise definition of all parameters in the limit $a \to 0$, ideally with a good proof that this limit indeed exists and is non-trivial. But before we get to this, we have to remind the reader about the asymptotic freedom.

1.4.1. *Renormalization group and asymptotic freedom*

The beta function is discussed in all standard QFT courses, so this brief introduction can be omitted by people familiar with it. It was introduced by Gell-Mann and Low in 1954 and describes the renormalization of the charge. Its definition is the charge derivative over the scale at which it is defined,

$$\frac{\partial g}{\partial \log \mu} \equiv \beta(g). \tag{1.60}$$

If the beta function is known, this equation can be solved,

$$\log(\mu/\mu_0) = \int_{g_0}^{g} \frac{dg'}{\beta(g')}. \tag{1.61}$$

For small coupling values the beta function can be determined perturbatively, and its standard definition includes the so called first and second beta function coefficients

$$\beta(g) = -b\frac{g^3}{16\pi^2} - b'\frac{g^5}{(16\pi^2)^2} + \cdots , \tag{1.62}$$

which are computed from one and two-loop diagrams. We will discuss below what their values are in QCD-like theories and how they have been calculated. Before that we proceed with a number of simpler examples

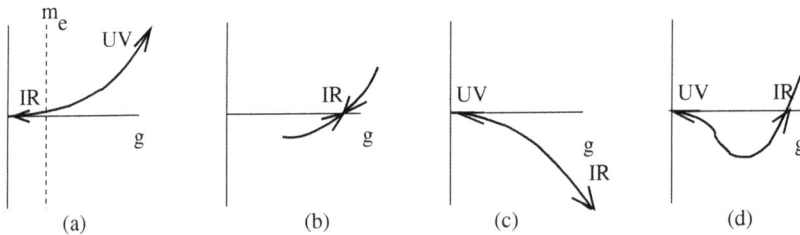

Fig. 1.7. Schematic behavior of the beta functions in the four examples considered: (a) QED, (b) second order phase transition, (c) QCD and (d) QCD with a sufficiently large number of light flavors. IR and UV with arrows indicate the direction of the charge motion at small and large momenta, respectively.

Example 0.   (A very special case which became famous recently due to the Maldacena duality, see Section 12.5.) The $\mathcal{N} = 4$ supersymmetric (SUSY) gluodynamics is a theory which has, on top of the gauge fields, four kinds of gluinos and six scalars, all in the same — adjoint — color representation. In this theory $\beta(g) = 0$ so the charge does not "run" with the scale at all. Therefore the $\mathcal{N} = 4$ SUSY glu-odynamics is also called the Conformal Field Theory (CFT).

A comment. The absence of a beta function and running charge does not mean that there is no finite renormalization of the charge, e.g. $g^2 = g_0^2 + f(g_0)$. An example of a non-trivial function f not seen in diagrams can be e.g. $\exp(-\text{const}/g_0^2)$. And indeed, there is a charge renormalization like that which appears due to instantons.

Example 1. QED in one loop is given by the so called vacuum polarization diagram:

$$\beta(e) = e^3/12\pi^2. \tag{1.63}$$

The solution for the charge is then

$$e^2(\mu) = \frac{e^2(\mu_0)}{1 - [e^2(\mu_0)/6\pi^2] \log(\mu/\mu_0)}, \tag{1.64}$$

and when μ grows (the charge is measured at smaller distances) it appears larger. Eventually one approaches the so called Landau pole, when the charge becomes of order 1 and we cannot trust this perturbative derivation any more. That is why QED is not a good theory: one cannot answer questions about strong coupling domain, or even in principle formulate how to get them.

Example 2. *Second order phase transitions.* In this case the beta function changes sign. Close to its zero it can be approximated by $\beta(g) = a(g - g^*)$. This leads to the infrared fixed point:

$$g(\mu) - g^* = [g(\mu_0) - g^*](\mu/\mu_0)^a. \tag{1.65}$$

Example 3. QCD: the ultraviolet fixed point (its history was covered above in detail) is

$$\beta(g) = -b\frac{g^3}{16\pi^2} - b'\frac{g^5}{(16\pi^2)^2} + \cdots, \qquad b = \frac{11}{3}N_c - \frac{2}{3}N_f. \tag{1.66}$$

Integrating, one has the famous asymptotic freedom formula

$$g^2(\mu) = \frac{g^2(\mu_0)}{1 + b(g^2(\mu_0)/8\pi^2) \log(\mu/\mu_0)}, \tag{1.67}$$

and at large μ (small distances) the charge goes to zero. It can be written as

$$g^2(L) = 8\pi^2 \left[\left(\frac{11}{3}N_c - \frac{2}{3}N_f \right) \log \left(\frac{1}{L\Lambda_{\text{QCD}}} \right) \right]^{-1}, \tag{1.68}$$

where the constant $g(a)$ is traded for a dimensional one, Λ_{QCD}, giving to the strong interaction theory its natural scale.

Example 4. QCD close to the $b = 0$ line has both infrared and ultraviolet fixed points (Banks and Zaks [870]):

$$b' = (34/3)N_c^2 - (13/3)N_c N_f + (N_f/N_c) \tag{1.69}$$

and so when $b = 0$ (for example, 3 colors and 16.5 flavors) b' is negative. Thus the beta function is zero at

$$\frac{g_*^2}{16\pi^2} = |b/b'|. \tag{1.70}$$

Note that it is small if b is small, in this case the charge is *always* small. Therefore hadrons cannot exist and the correlators are power-like at large distances.

1.4.2. *Continuum limit of lattice gauge theory*

The gauge theories are "asymptotically free". This means the charge value $g^2 = 0$ is the critical point. So, while increasing resolution of our microscope (that is, taking finer and finer lattices) we have to start each time with smaller and smaller $g(a)$, to get the same physics. Specifically, we need

$$g^2(L) = g(a)^2 \left[1 + \left(\frac{11}{3}N_c - \frac{2}{3}N_f \right) \frac{g(a)^2}{16\pi^2} \log\left(\frac{a}{L} \right) \right]^{-1} \tag{1.71}$$

which can be written as

$$g^2(L) = 8\pi^2 \left[\left(\frac{11}{3}N_c - \frac{2}{3}N_f \right) \log\left(\frac{1}{L\Lambda_{\text{QCD}}} \right) \right]^{-1} \tag{1.72}$$

where the constant $g(a)$ is traded for a dimensional one, Λ_{QCD}, giving to the strong interaction theory its natural scale.

In many applications one may consider only some dimensionless ratios (e.g. the ratios of hadronic masses, or hadronic masses in units given by the string tension).

All the dimensional quantities are expressed in terms of the lattice spacing a, but at the end of the day we would like to convert those into physical units, GeV and femtometers. The non-specialized reader often cannot understand precisely how it was done, especially because lattice practitioners routinely use "physical units" for nonphysical theories (e.g. the SU(2) gluodynamics) as well, for which physical units can be nothing more than pure convention.

The basic lattice mass unit was introduced as a lattice lambda parameter Λ_L. By definition, it is connected with the lattice spacing a and the bare coupling g in the Lagrangian by the following relation

$$a\Lambda_L = \left(\frac{bg^2}{16\pi^2} \right)^{-b'/16\pi^2 b} \exp\left(-\frac{8\pi^2}{bg^2} \right), \tag{1.73}$$

which follows from the 2-loop asymptotic freedom formula, and assumes that the standard Wilson action is used (for other lattice actions one introduces similar quantities). The relation between various lambda parameters in QCD can in principle be found only by the 2-loop calculations of the same physical quantity in both sets of notations. Other definitions are used in other physical applications, for example the (now standard in $pQCD$) $\Lambda_{\overline{MS}}$ is connected with it as follows

$$\Lambda_{\overline{MS}} \approx 39\Lambda_L \ (N_c = 3, N_f = 4). \tag{1.74}$$

The absolute value of the lambda parameter can only be fixed from experiment, see Review of Particle Properties for a detailed definition and current error bars. Experimental results are fitted to the solution of the renormalization group equation by,

$$\alpha_{\overline{MS}}(\mu) = \frac{1}{\beta_0} \left[t + d_1 \ln t + d_1^2 \frac{\ln t}{t} + \frac{d_2}{t} - \frac{d_1^3 \ln^2 t}{2 \ t^2} + \frac{\beta_1 \beta_2}{\beta_0^5} \frac{\ln t}{t^2} \right]^{-1}, \tag{1.75}$$

with $t = 2\ln(\mu/\Lambda_{\overline{MS}})$, $d_1 = \beta_1/\beta_0^2$ and $d_2 = d_1^2 - \beta_2/\beta_0^3$, which is consistent to order α_s^3. The results can be summarized now in terms of

$$\alpha_s(M_Z) = 0.117 \pm 0.002, \tag{1.76}$$

corresponding to

$$\Lambda_{\overline{MS}}^{(N_f=5)} = 216 \pm 25 \, \text{MeV}. \tag{1.77}$$

(The sixth t quark is heavier that Z and naturally it is not included, that is why $N_f = 5$.) The data range from historical data on scaling violation in deep-inelastic scattering to lattice studies of bottonium spectroscopy.

As soon as any of such lambdas are measured, all others are known and the question of putting the results of the lattice calculations into physical units is finished. In particular, the lattice Λ_L happens to be about 5 MeV.

The first serious lattice works in gauge theories were related to the confinement issue, and the string tension they gave can be related to the phenomenological value of the string tension $\sigma \approx (420 \, \text{MeV})^2$. As confinement seems to persist for all non-Abelian gauge theories, this value is traditionally used as the convention for the unit fixing in all of them. In particular, the pioneer works done by Creutz and Pietarinen for the pure SU(3) gluodynamics in the early 1980s have given

$$\Lambda_L/\sigma^{1/2} = (5 \pm 1.5, 7. \pm 2)10^{-3}, \tag{1.78}$$

respectively, while a few years later with larger lattices people got larger values for this ratio, ≈ 12 which, as is easy to check, is in perfect agreement with the latest results on QCD units mentioned above.

In principle, this agreement might not have happened, since one had to wait for realistic lattice QCD with dynamical light quarks. After those became feasible in

1990s, there appeared a possibility to fix lambda by means of other observables, e.g. the rho meson or nucleon masses.

The string tension is a quantity defined at large distances which is more difficult to get from lattice results than the tension at finite distance. The scale r_0 is defined by the so called Sommer parameter determined by

$$\frac{dV(r)}{dr}r^2\bigg|_{r=r_0} = 1.65. \tag{1.79}$$

The number in the r.h.s. is an arbitrary convention, I do not even know why 1 or 2 is not chosen instead. From quarkonium phenomenology one finds $r_0 \approx 0.5$ fm. Often lattice results are also given in units of Sommer's r_0.

Summarizing this discussion; the problem of matching the lattice and the physical units is basically a perturbative issue and is solved. However the current accuracy on the lambda is only at the 10 percent level, and so employing dimensionless ratios is still the standard way to present the lattice results.

1.4.3. *Path integrals for fermions*

The introduction of fermion fields into the path integral needs special definitions appropriate for Grassmannian (anti-commuting) variables. Those have been defined in a classic work by Berezin [280] and are usually discussed in any modern textbook on QFT. Additional subtleties appear with its transformation into the Euclidean space–time, in which $\bar{q}$ and q must be treated as independent variables.

I have to explain in what sense the integral over ψ should be defined. It is a fermionic variable, not just the ordinary field. They are called Grassman variables, or anti-commuting ones $\chi_1\chi_2 = -\chi_2\chi_1$. In particular, for such variables one has $\chi^2 = 0$, which represents the Pauli principle. These variables have funny rules for the integrals. They are given essentially by two basic integrals

$$\int d\chi = 0, \qquad \int \chi\, d\chi = 1. \tag{1.80}$$

One can then derive the following formula:

$$\int \exp(-\chi_k^* M_{k,l}\chi_l)\, d\chi_1^*\, d\chi_1 \cdots d\chi_N^*\, d\chi_N = \det M, \tag{1.81}$$

which can be proved in the eigenvector basis by decomposition of the exponent. Note, that the ordinary Gaussian integral with N variables of such type is equal to

$$\int \exp\left(-\sum_{k,l} \phi_k M_{k,l}\phi_l/2\right) d\phi_1 \cdots d\phi_N = \frac{(2\pi)^{N/2}}{\sqrt{\det M}} \tag{1.82}$$

and the determinant stands in the denominator. So, if one of the eigenvalues is zero, the fermionic integral vanishes while in the corresponding bosonic one diverges.

Note also in passing, that one fermionic integral can compensate two bosonic ones, if the eigenvalue spectra happen to be equal. This comment is important for supersymmetric theories, in which fermionic and bosonic integrals do indeed compensate each other, in the vacuum energy and many other observables.

Fortunately, fermionic integrals can be bypassed because a general integration over the fermions in QCD can be made in a simple form, considering those appear in the action only *linearly*. The master formula for it, we repeat, is

$$\int D\bar{q}\, Dq\, e^{\bar{q}Mq} = \det M, \tag{1.83}$$

and this factor (for each quark flavor) will appear in QCD partition function. Here M is in general a matrix in all fermionic indices and also possibly a differential operator.[19] In QCD $M = iD_\mu\gamma_\mu$ is the Dirac operator, with the color matrix in covariant derivative obviously in fundamental representation. As a final comment, since the operator $i\hat{D}$ is Hermitian, its eigenvalues λ are all real. However, one may still ask how it happens that at non-zero mass (entering with i in the Euclidean formulation)

$$\prod_f \det[i\hat{D}(A_\mu(x)) + im_f]$$

and non-positive λ, the ratio of the determinants happens to be positive (otherwise one cannot use probability language). We noticed that the former Dirac operator is hermitian, therefore its eigenvalues λ are real. Due to chiral symmetry, the eigenvalues go in pairs and therefore we always have

$$\prod_{\lambda>0}(\lambda + im)(-\lambda + im) = -\prod_{\lambda>0}(m^2 + \lambda^2),$$

so this factor is real too, even having definite sign, and can, with the appropriate definition, be made positive. This is important, as we want to ascribe to it the meaning of the probability of occurrence of the corresponding configuration in vacuum.

For completeness, the explicit definition of Euclidean and Minkowski fermionic notations we use is given in the Appendix. Note how the transition to Euclidean space is done for fermions. First, the Euclidean gamma matrices are defined by

$$\gamma_4^E = \gamma_4^M, \qquad \gamma_m^E = -i\gamma_m^M, \quad \text{here } m \text{ is index 1, 2, 3}$$

so their anti-commutator is simply $2\delta_{\mu\nu}$. We do not change the fermion field,

$$\psi_E = \psi_M,$$

but we do change its conjugate,

$$-i\bar{\psi}_E = \bar{\psi}_M. \qquad \text{here } M \text{ means Minkowski.}$$

The Euclidean action then reads (where we omit the subscript E)

$$S = \int d^4x\, \bar{\psi}(i\gamma_\mu D_\mu - im)\psi.$$

---

[19]To prove it imagine that the operator is diagonalized, the exponent is expanded and the "Pauli principle" in the form $q^2 = 0$ holds.

Let me explicitly mention the definitions used here:

$$\psi_E = \psi_M, \qquad \bar{\psi}_M = -i\bar{\psi}_E, \qquad \gamma_E^0 = \gamma_M^0, \qquad \gamma_E^m = -i\gamma_E^m.$$

We get

$$S_E^f = \int d^4x \, \bar{\psi}_E (i\hat{D} + im)\psi,$$

thus the complete partition function of QCD is

$$Z = \int DA_\mu(x) \, D\bar{\psi} \, D\psi \exp(-S_E - S_E^f).$$

We will now terminate our introductory discussion of lattice fermions, which will be continued in Section 5.2.

Exercise 1.5. How does one generate an invariant integration measure for link variables in practice? Outline a program generating random SU(3) matrices with the invariant distribution on the group. *Hint 1.* The SU(2) group is locally equivalent to the three-dimensional sphere, so it is easier to start from those. *Hint 2.* Write a SU(3) matrix as three mutually orthogonal complex vectors of unit norm. Selecting the first of them randomly corresponds to a $5d$ sphere; the second, to $3d$ sphere; and the last vector follows uniquely.

Exercise 1.6. Suppose we want to average out the product of link variables over some closed loop C, $W = \Pi_C(U \cdots U)$. Show that in the *strong coupling limit* $1/g^2 \to 0$ the only non-zero contribution appears if one includes all plaquettes in a surface with the boundary on C.

1.5. Light quarks and symmetries of QCD

1.5.1. *Exact and approximate symmetries of QCD*

The quark field $\psi_{\alpha,i,f}$ has three types of indices: $\alpha = 1\text{--}4$ is the spinor index, and gamma matrices act on them. $i = 1\text{--}3$ is the color index, and the corresponding Gell-Mann matrices we called t^a, $a = 1\text{--}8$, act on them. The flavor index $f = u, d, s, c, b, t$ but actually we will always ignore heavy c, b, t quarks and use the $SU(3)_f$ notations, usually calling the corresponding matrices λ^a, $a = 1\text{--}8$.

Let us first identify a set of possible transformations of the quark fields, related with the corresponding Noether currents:

- the colored vector current $\bar{\psi}\gamma_\mu t^a \psi$, corresponding to *color rotation* in $SU(3)_c$;
- $U(1)_V$ vector current $\bar{\psi}\gamma_\mu \psi$ (baryonic charge) corresponding to modification of the *common phase* of all quark fields;

- $SU(N_f)$ vector $\bar{\psi}\gamma_\mu\lambda^a\psi$ corresponding to an isospin (and its $SU(3)_f$ analog) rotation;
- The SU(3) singlet (or U(1)) axial current $\bar{\psi}\gamma_\mu\gamma_5\psi$, which corresponds to an *opposite phase rotation of left- and right-handed quarks*;
- $SU(N_f)$ axial current $\bar{\psi}\gamma_\mu\gamma_5\lambda^a\psi$, corresponding to the *opposite* $SU(N_f)$ *rotation of left- and right-handed quarks*.

The first color vector current is related to the *local gauge symmetry* of QCD, on which the whole theory is based. It is therefore exactly conserved and remains unbroken by any known interaction.

The second one, related with the global phase and the baryonic charge, is also exactly conserved in QCD. (It is however *not* conserved in the theory of weak interactions.)

There is significant difference between color $SU(3)_c$ and flavor $SU(3)_f$ symmetries — the former is exact, the latter approximate. In the early sixties people thought that the reason for $SU(3)_f$ is that quark masses are nearly equal: by that they thought of what we now call the constituent quark masses. However the real masses, originating from outside of QCD are far from being equal, their values are approximately[20]

$$m_u = 3 \pm 1\,\mathrm{MeV}, \qquad m_d = 7 \pm 1\,\mathrm{MeV}, \qquad m_s = 120 \pm 30\,\mathrm{MeV}.$$

The real reason for $SU(3)_f$ symmetry is therefore that all *these masses are too small to be important*. (This statement is not so obvious for the strange quark and actually not true for some observables.)

There can be two simple limits. The first simple case is when *all* quark masses are equal $m_u = m_d = m_s = m$ so that the $SU(3)_f$ symmetry becomes exact. Two comments are here in order:

(i) One should also switch off (or subtract the effect of) the electromagnetic interaction, e.g. p, n masses are not the same, even if quark masses are equal.

(ii) But even if all masses were the same, the $SU(3)_f$ symmetry is only "global". This means that one can arbitrarily change the axis in the flavor space, but once for all points.

The second simple limit we would study below is the so called *"chiral[21] limit"* in which all quark masses are set exactly to *zero*, $m_q = 0$. In this case QCD has additional symmetry, called the *chiral symmetry*.

---

[20]The quarks do not exist as free particles and thus this mass needs proper definition. The masses in the QCD Lagrangian have pQCD renormalization and the anomalous dimension, so they depend on the normalization scale. All of it, together with sources of information and current errors/allowed regions for all masses, can be found in the latest Review of Particle Properties [91].

[21]This term originates from the Greek word which means a "hand". Looking at a hand is the simplest way to tell/explain a difference between the left and the right.

Chirality itself is defined as two, left- and right-handed, components

$$\psi_{L,R} = P_{L,R}\psi = \left(\frac{1 \pm \gamma_5}{2}\right)\psi, \tag{1.84}$$

where we have introduced the projection operators P_L, P_R into two chiral components. As is easy to check from diagrams, any number of gluons emitted and absorbed leave quark chirality unchanged. In general the Lagrangian for massless fermions and *any* gauge field can be written as a sum of two independent parts, of right-handed and left-handed fermions.

Let me also clarify the definition of the chirality of the anti-quark: we use standard notations such that

$$\bar{\psi}_L = \psi\dagger P_L \gamma_0.$$

Let us now define two different *chiral* rotations in the SU(3)$_f$ space

$$\psi_{L,f} \to A_{f,k}\psi_{L,k}, \qquad \psi_{R,f} \to B_{f,k}\psi_{R,k}. \tag{1.85}$$

Alternatively, chiral rotations can be written in terms of full quark fields as the "axial" rotation

$$\psi \to \exp(i\beta\gamma_5 + i\beta^a \lambda^a \gamma_5)\psi, \tag{1.86}$$

is distinct from the ordinary vector rotations

$$\psi \to \exp(i\alpha + i\alpha^a \lambda^a)\psi. \tag{1.87}$$

Note, that we have here two different groups for each of transformation: two U(1) rotations α and β, and twice 8 parameters of the SU(3) rotations, in total 18. These symmetries have quite different fates, to be discussed in detail below.

We remind the reader that in spatial reflection γ_5 changes sign; thus the last transformation mixes states with opposite parity. This implies then, that *all physical states with the opposite parity should have equal masses.*

This statement seems to be in terrible disagreement with the real world. For example, we have hadrons like the ρ meson or the nucleon, which have quite definite parity, and are not at all degenerate with particles possessing the opposite one (e.g. A_1 meson for ρ). We return to the solution of this paradox later.

1.5.2. *Chiral anomalies, the UV approach*

The name of the phenomenon to be discussed in this section does not sound very attractive. However, it reflects well enough the historical attitude toward this phenomenon, which was found unexpectedly and, in spite of significant efforts by theorists to eliminate it by multiple formal tricks, has survived. Now, when its nature and consequences are well understood, the name "anomaly" became just a term

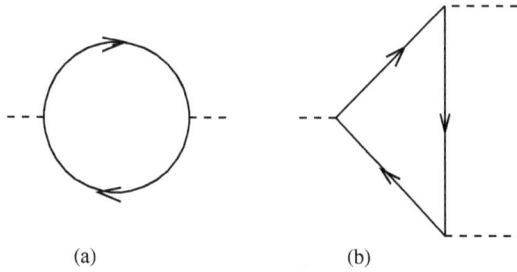

Fig. 1.8. One loop diagrams with (a) 2 and (b) 3 currents. The solid lines are quarks, while the definition of verticies and the dashed lines depend on the particular currents. Vector and axial currents correspond to γ_μ and $\gamma_\mu \gamma_5$, for gluons there are also color matrices $t^a/2$, for photons and W there appear quark charges (or flavor matrices).

which does not create any negative feelings. Moreover, their consequences are phenomenologically welcome.

Anomalies were discovered in the perturbative calculations of certain diagrams involving currents we identified in the preceding section. We argued there that since *classically* the Lagrangian is invariant under certain transformations, the corresponding neuter currents are *classically* conserved by the equations of motion. However *quantum* theory is an integral over all paths and it does not care about equations of motion, and thus the conservation of each current (and consequently, the symmetry it was derived from) should be reconsidered.

Naturally, one starts to do so in perturbation theory, diagram by diagram. It is a standard exercise in QFT courses to calculate the "vacuum polarization" one-loop diagram [Fig. 1.8(a)] for the correlator of two vector currents,

$$\Pi_{\mu\nu}(x, y) = \langle j_\mu(x) j_\nu(y)\rangle. \tag{1.88}$$

The result is in agreement with the current conservation, because

$$\frac{\partial}{\partial x_\mu}\Pi_{\mu\nu}(x, y) = 0. \tag{1.89}$$

In momentum space the same statement means that this tensor is purely "transverse", $\Pi(q)_{\mu\nu} = (q_\mu q_\nu - g_{\mu\nu} q^2)\Pi(q^2)$. One can show that going to more loops does not change this conclusion.

However the one-loop "triangular" diagrams with 3 currents [Fig. 1.8(b)] with one axial and two vector currents $\langle j_\mu(x) j_\nu(y) j_\sigma^5(z)\rangle$ have created troubles.[22] None of the attempts to regularize it, so that it becomes transverse for all 3 currents,

---

[22] The diagram first came out in 1949 in Steinberger paper [42], in which he considered the $\pi^0 \to 2\gamma$ decay to which we turn below. Therefore, instead of two gluons he has two photons. Also, he of course had no quarks — instead the fermions under consideration were "nucleons". This calculation was correct, but he probably was so disappointed by the lack of clarity, he switched to experiment, and later got the Nobel prize for two types of neutrino experiment.

work. On the other hand, in the chiral limit $m_q \to 0$ we expected the axial current to be conserved.

There are two ways of looking at this triangular diagram, using standard Feynman rules and momentum representation and Schwinger background field method. Let me briefly discuss both of them in succession.

It is easy to construct the corresponding expression for such diagrams using standard Feynman rules. The loop has one d^4k and three $\sim 1/k$ propagators, so it is linearly divergent. If one takes a divergence on the current we would like to focus on $\partial_\mu \Gamma_{\mu\nu\sigma}$, it has one more power of momentum, so in total it is a quadratic $O(k^2)$ divergence. Regularization of a divergent expression is always a tricky business. The indices of two vector currents are not yet convoluted. Two photons should be represented by two field strengths, not potentials, so one needs at least a product of two photon momenta, which just has the needed dimension.

One more observation is that the left- and the right-handed quarks give contributions of the *same* sign to the vector and of *opposite* signs to the axial current. As was shown by Bell–Jackiw [44] and especially Adler [44], these triangular graphs can not be regularized in such a way that these L or R contributions vanish. Therefore, vector and axial current cannot both be simultaneously conserved. They suggested to select "the lesser evil", and assume that it should be O.K. with vector currents, while the axial one is not conserved indeed.

In space–time language the same pion decay was first considered by Schwinger, in the famous paper [43] where he also calculated the anomalous magnetic moment of the electron. He has used the method of background field, or considered motion of the electron in some external electromagnetic field A_μ.

The current was given the following non-trivial definition in that work

$$j_\mu(x) = \lim_{\epsilon \to 0} \mathrm{Tr}[\gamma_\mu S_F(x, x + \epsilon)], \qquad (1.90)$$

where S_F is the quark propagator in the external field. One might think that the r.h.s. simply follows from the convolution of two fermion operators, but the necessity to define the limit at small ϵ makes this point non-trivial. In particular, the current in the l.h.s. is a physical local quantity, so we want it to be gauge invariant. In the r.h.s. there stands a bi-local quantity, the propagator, which is gauge-dependent.

Schwinger solved this problem by adding "by hand" into the r.h.s. of (1.90) a new factor $\exp\left(ie \int_x^{x+\epsilon} dx_\mu A_\mu\right)$, which makes it gauge invariant as well. Now, let us try to expand the propagator in powers of ϵ. The free propagator is very singular $\sim \hat{\epsilon}/\epsilon^4$, but this certainly should be subtracted. There are no photons, so it is purely a vacuum loop. There are two other dangerous terms: (i) the one with singularity $O(1/\epsilon^2)$, and (ii) terms like $O((\epsilon x)^2/\epsilon^2)$ which are formally of zeroth order but depend on the orientation of the epsilon vector. However, if one adopts the Schwinger "splitting prescription" both disappear and the following finite answer in the $\epsilon \to 0$ limit follows:

$$\partial_\mu j_\mu^{Axial} = \partial_\mu \mathrm{Tr}\left[\gamma_5 \gamma_\mu S_F(x, x)\right] = \frac{2\alpha}{\pi}(\vec{E}\vec{H}). \qquad (1.91)$$

This relation, discussed and better understood later, is now known as "Adler–Bardeen–Jackiw anomaly". The first application of this formula was related to π^0 meson decay. In the "effective nucleon" theory of the 1950s, with nucleon lines instead of quarks, one had the same answer for the triangular diagram except for the factor N_c, the number of colors. So the QCD calculation has an extra N_c^2 in the pion lifetime, and the experiments precisely confirmed this factor. So, ironically, the tricky anomaly and puzzling pion decay considered by Adler historically happened to become one of the most convincing arguments in favor of QCD with three colors.

The fact that the result for the triangular diagram is finite and independent of the fermion mass is actually very strange. In particular, we are used to the idea that heavy quarks are decoupled from light fields, while the explicit calculations suggested otherwise.

Of course, we do not know what happens at arbitrarily large masses and how many fermion fields can possibly exist at such scales. What we normally do in renormalized field theories is regularization at some large scale, which is anyway needed in order to make divergent diagrams meaningful. For example, in the method of Pauli and Villars heavy "regulator quarks" are introduced for each species, which should be included in any loop with the opposite sign. As a result, at momenta above the regulator mass both terms cancel out, but the results depend on the regulator mass.

Let us try to use the same idea for the triangular diagram, which is not divergent. Then the "anomalous" diagrams for, say, the top quark are exactly compensated by the top quark *regulator* contribution, so we can forget about both of them. The desired decoupling of QCD from heavy fermions is recovered.

Now consider a light quark contribution. Convergence took place at distances among the three points $\sim 1/m$ where m is the quark mass. For the heavy "regulator" quark the distances are indeed small, and pQCD is perfectly applicable. However, the inverse of the light quark mass is much larger than fm, the distance in the hadronic scale is large. Obviously any perturbative calculations are completely meaningless. Furthermore, in the calculation one finds the "current" mass in the numerator and may think that the massless quark contributes exactly zero. Thus, the only contribution of the light quark loop seems to be the contribution of its regulator. It gives the nonzero divergence of the singlet axial current in the presence of two gluons. Inserting proper color matrices for the gluons, one gets the following result for the axial anomaly of the SU(3) singlet current:

$$j_\mu^{\eta'} = (\bar{u}\gamma_\mu\gamma_5 u + \bar{d}\gamma_\mu\gamma_5 d + \bar{s}\gamma_\mu\gamma_5 s)/3^{1/2}, \tag{1.92}$$

$$\partial_\mu j_\mu^{\eta'} = \sqrt{3}(1/32\pi^2)\epsilon^{\alpha,\beta,\gamma,\delta}G_{\alpha,\beta}^a G_{\gamma,\delta}^a. \tag{1.93}$$

So, we found that this current is no longer conserved quantum-mechanically, so QCD does *not* have the U(1)$_A$ symmetry.

Of course, the non-singlet current is still conserved. In the corresponding triangular diagram, with it and two colored vector currents (2 gluons), there appears a single flavor SU(3) matrix λ^a, while gluons are flavor blind, and Tr $\lambda = 0$. For example, the π^0 current mentioned above is $\sim(\bar{u}\gamma_\mu\gamma_5 u - \bar{d}\gamma_\mu\gamma_5 d)$ and so the triangular diagrams with u and d quarks cancel out each other. (This should not be confused with the case of pion decay into two photons discussed above; in the present case u and d quark loops have squares of electric charges, 4/9 and 1/9 respectively, and therefore do not cancel out.)

This correlates with the phenomenological observation that the SU(3) singlet η' meson is not like all other pseudoscalars like pions and kaons — even in the chiral limit it remains massive. It is not a Goldstone mode because there is no U(1) axial symmetry to be broken! The problem is in principle solved (although one still has to calculate the η' mass).

The non-conservation of the axial current means that the "axial charge" defined by $Q_5 = \int d^3x\, j_0^5$, or the difference between the number of left- and right-quarks, can be changed. Furthermore the anomaly (1.93) gives a precise value for the variation of this quantity. The complete integral of it over space–time $\int d^4x\, \partial_\mu j_\mu^5$ is given by the r.h.s. which is nothing but the topological charge we will discuss in connection with instantons.

In fact the r.h.s. of the anomaly equation is a total derivative of some other current,

$$(1/32\pi^2)\epsilon^{\alpha\beta\gamma\delta}G^a_{\alpha\beta}G^a_{\gamma\delta} = 2\partial_\mu K_\mu, \tag{1.94}$$

where

$$K_\mu = (1/16\pi^2)\epsilon^{\mu\alpha\beta\gamma}(A^a_\alpha \partial_\beta A^a_\gamma + (1/3)f^{abc}A^a_\alpha A^b_\beta A^c_\gamma). \tag{1.95}$$

Therefore a proper combination of $\tilde{j}_{\mu,5}$ and K_μ is a *conserved* current, which however is *not* gauge invariant. The meaning of all of this will become clear when we discuss specific examples.

In summary, we first found that in an external gluonic or electromagnetic field of some appropriate structure, some amount of the chiral (L–R) charge somehow leaks below the UV regularization cutoff. We then concluded that the integral over K_0 is therefore a combination of the gauge field which happens to be locked in some rigid relation with the chiral charge.

1.5.3.   *Chiral anomalies, the IR approach*

Another view of the same phenomenon can be provided from the opposite end of the energy spectrum. We will use here constant Abelian E and B fields as the simplest example of the kind following Ambjorn *et al.* [48]. The final "demystification" of the phenomenon in its practically important form will be made later, in Section 4.2.2.

We will do this in two steps. First, let us switch on only the magnetic Abelian field B constant in space–time and directed along the z axis. The relativistic

Dirac equation for massless fermions is solved in the same way as is done non-relativistically in quantum mechanics textbooks. The relativistic Landau levels are in this case

$$\omega_n = (k_z^2 + eB(2n+1) \pm eB)^{1/2}, \tag{1.96}$$

where $\pm$ stands for spin projections. We see that there is a $k_z = 0, n = 0$ state with the negative spin term which has zero energy. Another zero energy state is formed for the antiparticle $e \to -e$ with the opposite spin direction.

As a second step we introduce an additional weak electric field E_z. This implies the potential $A_0 = E_z * z$, which means that the original levels are now slightly tilted along the z direction. The slightly negative energy states now start to move along the field direction z, accelerated by it to positive kinetic energy. Antiparticles or holes move in the opposite direction. This phenomenon does not produce total charge, but it does produce a non-zero *chiral charge*, because the particles and holes produced have opposite chiralities.

If one calculates correctly the density of the Landau levels, the total rate of the chirality production rate is actually

$$\frac{dQ_5}{d^4x} = \frac{e^2}{8\pi^2} \vec{E}\vec{B}, \tag{1.97}$$

which is the same as follows from the UV-based derivation of the anomaly.

We may further generalize the argument. The surface of the Dirac sea, a state with a zero energy, is not really special. In the external field with a non-zero $\vec{E}\vec{B}$ there is continuous *flow of states*, for all energies ranging from $\omega = -\infty$ to ∞. One chirality moves up, the other down.

The first consequence of this phenomenon would be in high density QCD; changing zero energy to finite Fermi energy at the surface of the Fermi sphere of occupied states does not change the phenomenon because the same amount of chiral states would cross it per unit volume per unit time.

The second consequence is an explanation of perturbative anomaly by quite similar level motion which happens near the UV regulator cutoff, namely states of one chirality sink below it and disappear, the states of the other chirality appear from "below the cutoff".

1.5.4. *Other applications of chiral anomalies*

Let me deviate for a while from our main topic, QCD, to weak interaction theory. The entire setting is the same with a vertex in which there are W^+W^- (instead of two photons), and the third current, vector or axial, in the third vertex. The weak interaction W boson interacts with left-handed fermions only. So there cannot be any difference now between the vector and axial current, and if the anomaly diagram gives a non-zero divergence (which is the case), then *neither* the vector *nor* the axial current is conserved! But the vector singlet current for quarks corresponds to the baryon number, and thus the baryon charge can be changed in weak interaction. The divergence is proportional to $(\vec{E}_W \cdot \vec{H}_W)$, where the index W means weak (W) field. Therefore, if one manages to create sufficiently strong W fields, to make the r.h.s. of the anomaly equation integrated over space–time equal to 1, one may indeed see a disappearance of the proton! We will discuss the probability for this to happen in Section 4.2.4.

One final comment: if we consider a triangle with ZW^+W^- quantum numbers and get non-conservation of the axial current, it is certainly a disaster, because the theory of weak interactions is based on the "left-handed SU(2) gauge symmetry" principle, similarly to the color gauge symmetry in QCD. If this symmetry is spoiled at the quantum level, it means the whole theory collapses. However, if we put in the triangle not a single fermion but a *doublet* of leptons (e and ν_e and another doublet — u, d quarks — the total answer can be made to vanish. The same thing happens if we add other generations of quarks and leptons. (Now we see why the last discovered top quark was absolutely necessary, otherwise the standard model of weak interactions would be inconsistent.)

Another deep consequence of the axial anomalies was pointed out by t'Hooft [46]. It relates in a nontrivial way confinement with the spontaneous breaking of chiral symmetry. Consider some gauge theory with the light quarks, and imagine that, unlike the QCD case, the chiral symmetry remains unbroken while confinement is there. If so, then the baryons are massless and the pseudoscalar mesons are heavy. One may then consider some effective Lagrangian for such light baryons at large length scale. The key idea is that the anomaly relations should be kept intact as they express non-conservation of certain charges. The same triangular diagrams can be written for the baryons, as for the quarks. If the coefficient is different, this regime is impossible on general grounds.

It can easily be seen that such a condition *cannot* be met for QCD because there are many more quarks than baryons in this theory. However, one may well imagine other theories where this argument may actually work. (For example, if quarks have the same color representation as the gluons, there are as many "qg baryons" as quarks.) In the next section we discuss how similar considerations may help to reconstruct the so called chiral effective Lagrangian for the pseudoscalar mesons, and now turn to a completely different subject.

1.5.5. *Scale anomaly*

The gluodynamics (or massless QCD) has also another symmetry, which is present at the classical level but is violated by the radiative corrections: it is the *scale invariance*. Indeed, this theory has a dimensionless coupling constant and, in the chiral limit, the classical theory is obviously scale invariant. For a general review on scale invariance see Ref. [50].

However, since the quantized QCD does describe the real world, the typical hadronic scale — 1 fm — should be fixed, and we have already discussed in Section 1.2 how the asymptotic freedom explains it. Indeed, the renormalized "running" charge, depends on the scale.

In this section we comment on more formal aspects of this phenomenon. Consider the dilatation transformation

$$x_\mu \to (1 + \delta\lambda)x_\mu, \qquad A_\mu \to (1 - \delta\lambda)A_\mu. \tag{1.98}$$

By general rules the action modification can be calculated and then written as

$$\frac{\delta S}{\delta \lambda} = \int dx \, \partial_\mu j_\mu^{\text{dil}} \tag{1.99}$$

where the "dilatational current" in the r.h.s. is connected with the energy–momentum tensor

$$j_\mu^{\text{dil}} = T_{\mu\nu} x_\nu. \tag{1.100}$$

As noted above, the classical gauge theory is scale invariant, so

$$\partial_\mu j_\mu^{\text{dil}} = T_{\mu\mu} = 0, \tag{1.101}$$

as one can easily check from the classical stress tensor of QED or QCD.

In order to see that in the quantum theory this current is *not* conserved, one can perform a straightforward calculation of the "triangular" diagrams [51, 52] as we did in the previous subsection for other currents. Another option (we will follow) is to look at more general arguments suggested by Collins *et al.* in Ref. [53].

Step number one is to change field units, from the "perturbative" normalization of the gauge field to the "non-perturbative" one, as

$$\tilde{A}_\mu = g A_\mu, \tag{1.102}$$

so that the charge disappears from the field strength

$$\tilde{G}_{\mu\nu} = \partial_\mu \tilde{A}_\nu - \partial_\nu \tilde{A}_\mu + i[\tilde{A}_\mu \tilde{A}_\nu], \tag{1.103}$$

and only appears in front of the action as a factor

$$S = -\frac{1}{4g^2} \int dx (\tilde{G})^2. \tag{1.104}$$

Now, suppose one makes a change of scale and now includes the running of the charge. By definition, its derivative over scale is the Gell–Mann–Low beta function, and therefore one gets the result

$$\partial_\mu j_\mu^{\text{dil}} = -(\tilde{G})^2 \frac{\beta(g)}{2g^3} \approx (\tilde{G})^2 \frac{b}{32\pi^2}, \tag{1.105}$$

where the last equality uses the beta-function up to the one-loop order. The same follows from the triangular diagram.

We return to this important relation in connection with the issue of "nonperturbative vacuum energy density" below.

1.6.   Heavy quarks, new symmetry and effective theory

The physics of heavy–light hadrons, especially of B meson decays, have become a separate vast field of research, with two recently completed "*B*-factories" in SLAC, Standford (CA) and KEK (near Tokyo, Japan) supplying a huge amount of data. We will not go deeply into this subject in these lectures, the interested reader is referred to pedagogical reviews, e.g. by Neubert [58] or Shifman [59].

The idea of *heavy quark symmetry* was first formulated in my two papers [54, 55], so let us simply start with the opening paragraphs of the latter paper.

"*The understanding of hadronic world began with the discovery of isotopic symmetry, and the (less accurate) $SU(3)_f$ symmetry of hadrons made of light u, d, s quarks. As now becomes clear, these symmetries are not due to a similarity of the quark masses, but to the fact that they are all too small to be important.*

A similar phenomenon takes place if some quark mass is too large to be important. Families of similar hadrons should exist, the only difference between their members being the substitution of one heavy quark by another. An attempt to put this idea on more quantitative grounds has been made in the previous work [54], where the limit $M_Q \to \infty$ (M_Q is the mass of a heavy quark) was discussed as well as $O(1/M_Q)$ corrections. In this limit hadrons with a heavy quark resemble the hydrogen atom with its fixed center, and many problems of current models of hadronic structure (e.g. that of c.m. motion) are made trivial. Mesons made of one very heavy and one very light quark are, in some sense, the simplest hadrons in which non-trivial QCD dynamics is essential, so study of them is of great importance . . ."

If heavy c, b quarks[23] are considered sufficiently heavy, one expects simple relations between properties of the corresponding hadrons containing them (e.g. $D = \bar{q}c$ and $B = \bar{q}b$ mesons) to exist.

Such statements, as important as they are for QCD, are not really new for physics in general, in atomic physics they have been known from the chemistry of the isotopes eversince the 19th century. Indeed, when we teach the "hydrogen atom" in a QM class, we imply *any* of the hydrogen isotopes. Whether the nucleus is p, d or t, it is just a point charge in the approximation used.

The realization of the existence of the heavy quark symmetry in the $m_Q \to \infty$ limit is only the *first* possible step. The *second* is to build a systematic expansion in $1/m_Q$. Again, one may expect that *coefficients* of this expansion, being themselves independent on m_Q would be the same for say charm and bottom quarks. This program was initiated in Ref. [55] as well and examplified for correlation functions for heavy–light hadrons (which we will consider in Section 6.3.4). The simplest issue addressed there was that of the heavy–light meson mass which can be expanded in

---

[23]The top t quark is *too* heavy, its weak decay lifetime is comparable to hadronic scale, and so it cannot form hadrons.

powers of the heavy quark mass[24]

$$M_H = m_Q + \Lambda_H + O(1/m_Q) + \cdots \tag{1.106}$$

with the second term being entirely due to the light quark. Strikingly, the value of the second term is empirically rather large, $\sim$500 MeV, significantly exceeding the naive "constituent quark mass".

Comparing this with the stress tensor of QCD in the massless limit one can formally express the second term via quantum correction due to the scale anomaly which we have derived in Section 1.5.2

$$\Lambda_H = \frac{1}{2M_H} \left\langle H \left| \frac{\beta(\alpha_s)}{4\alpha_s} G^2 \right| H \right\rangle. \tag{1.107}$$

Indeed, if momenta of the heavy (and light) quarks in heavy–light meson $p \sim \Lambda_{\rm QCD}$, the kinetic energy is only

$$T_Q \equiv E_Q - m_Q \sim \Lambda_{\rm QCD}^2/m_Q \ll \Lambda_{\rm QCD}, \tag{1.108}$$

and thus small compared to typical light quark energies and can, in first approximation, be neglected.

Since the magnetic moment of the heavy quark is very small, $O(1/m_Q)$, all spin splittings (such as the vector $S = 1$ and pseudoscalar $S = 0$ b-quark mesons B^*, B) can also in first approximation be ignored.

Formal transition to a non-relativistic description of the heavy quark propagator follows standard procedures, well known in QED and atomic physics. In particular, the heavy quark field should be redefined

$$Q(x) \rightarrow e^{-im_Q v_\mu x_\mu} \tilde{Q}(x) \tag{1.109}$$

where v_μ is velocity of heavy quark; it is (1,0,0,0) in its rest frame. The covariant momentum can be written as

$$iD_\mu Q = e^{-im_Q v_\mu x_\mu} (m_Q v_\mu + \pi_\mu) \tilde{Q}(x), \tag{1.110}$$

where the small π_μ is a covariant derivative acting on the non-relativistic spinor $\tilde{Q}$. The Euclidean propagator of a heavy quark with a non-relativistic energy E in external gluon fields is $\sim 1/(i\pi_0 + E)$, etc. From the squared Dirac equation for Q one obtains the standard non-relativistic equation

$$\pi_0 \tilde{Q} = -\frac{\vec{\pi}^2 + (i/2)\sigma_{\mu\nu} G_{\mu\nu}}{2m_Q} \tilde{Q}, \tag{1.111}$$

---

[24]Relations like this mass formula should be properly defined in QFT sense, since the quark mass is not a physical observable and depends on its definition and scale.

which may be complemented to higher order terms in $1/m_Q$, if needed. The non-relativistic Hamiltonian and Lagrangian generating all equations can also be copied from the corresponding QED textbooks.

Among the specific results of Ref. [55] was the expansion of the matrix element of the heavy–light currents[25]

$$f_Q = (2m_Q)^{1/2}(c_1 + c_2/m_Q + \cdots), \tag{1.112}$$

where

$$\langle 0|\bar{q}\gamma_\mu\gamma_5 Q|H\rangle = if_Q p_\mu. \tag{1.113}$$

The coefficients c_i were determined from the QCD sum rules.

An important further step has been made by Isgur and Wise [56] who have been discussing weak decays including D and B mesons. They have introduced the so called Isgur–Wise formfactor

$$\langle D|\bar{c}\gamma_\mu b|B\rangle = (2m_B 2m_D)^{1/2}[v + v']_\mu \xi_{\text{IW}}(vv'), \tag{1.114}$$

where v, v' are 4-velocities (not momenta!) of the two mesons. When there is no recoil, $v = v', vv' = 1$, the formfactor in the heavy quark symmetry limit is simply 1 because then the light quark wave functions of both mesons are the same.

Many results have been obtained in this formalism. Let me mention as an example a correction to total semileptonic width by Bigi *et al.* [57]

$$\Gamma = \frac{G_F^2 |V_{Qq}|^2 m_Q^5}{192\pi^3} \left(1 - \frac{3\mu_G^2 + \mu_\pi^2}{m_Q^2} + \cdots\right), \tag{1.115}$$

where the first term is just the lifetime of the quark and the second is the correction we are looking for. Two matrix elements which appear here are

$$\mu_\pi^2 = \frac{1}{2M_H}\langle H|\bar{Q}\vec{\pi}^2 Q|H\rangle \approx 0.5\,\text{GeV}^2, \tag{1.116}$$

$$\mu_G^2 = \frac{1}{2M_H}\langle H|\bar{Q}(i/2)\sigma G Q|H\rangle \approx 0.35\,\text{GeV}^2, \tag{1.117}$$

are universal objects, which appear in many other expressions. They describe the influence of the light quark on the heavy one: so the empirical numerical values given above tell us something about the light quark dynamics.

$\boxed{\text{E}}$

Exercise 1.7.   Show how the two matrix elements above enter the next term of the mass formula (1.106).

---

[25]The factors containing $\sqrt{2m}$ here and below are a simple consequence of standard relativistic spinor normalization.

1.7. Changing the number of colors N_c

1.7.1. *Large number of colors*

QCD does not have many parameters (and that is why why we like it so much). The number of colors, N_c, is one of them. Experience in theoretical physics shows that the limiting cases are simpler to understand. Simpler scalar theories or gauge theories in lower dimensions are a good example: the large N approximation for those has proven to be a powerful tool used with success.

It should also be useful for QCD: in the large N_c limit one clearly has fewer diagrams to deal with. On the other hand, we still have not solved/summed them. The reader who wants a more detailed pedagogical text can read Coleman's Erice lecture [78]. We will briefly discuss recent progress related with Maldacena duality and AdS/CFT correspondence in Section 12.5, in which the large N_c limit is also heavily used. In this section we introduce some ideas and general results which will be used later, in connection with lattice and instanton calculations.

The story started with the paper by 't Hooft [77] who looked at the N_c dependence of the Feynman diagrams. He introduced a very useful notation, depicting a gluon line as a double quark-antiquark line, and counting factors of N_c just by the total number of color sums. His remarkably simple conclusion is that the leading power is always given by *planar* diagrams, while the non-planar diagrams are suppressed by powers of $1/N_c$.

Furthermore, all planar diagrams are of the same order in the so called 't Hooft limit, in which $g \to 0, N_c \to \infty$ while the combination $\lambda = g^2 N_c$ is held to be fixed. For example, the diagram in Fig. 1.9 has 11 color loops and 20 vertices, so it is of the order of $\sim g^{20} N^{11} = N\lambda^{10}$. If $\lambda = O(1)$ it is comparable to the zeroth order empty loop without gluons, which is $\sim N$.

If one would like to study observables made of quarks, those can only generate the leading planar diagrams as one quark loop, with gluon lines attached to each other and the loop, see Fig. 1.9. In high orders one may think of it as a membrane

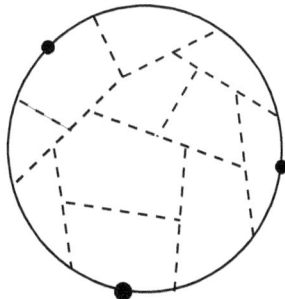

Fig. 1.9. An example of a high order diagram with a quark loop which is leading in the 't Hooft limit. Three small black circles on the quark lines indicate the external quark currents, so the diagram can be considered as contributing to the 3-meson vertex.

with a boundary at the quark loop: this configuration of course may be a hint to explain strings and confinement. It indeed works this way in the $2d$ QCD 't Hooft has solved. A search for an adequate representation of large N_c gauge theories had been going on for about 3 decades, with the hope that it would be a *string-based* representation of QCD naturally explaining confinement.[26]

Now let us think about correlation functions of several quark-based currents, or mesonic operators $M = (1/\sqrt{N_c})\bar{q}\Gamma q$ type, where we have included the square root of the number of colors for proper normalization of the state. In the leading planar diagrams all of the sources should be placed on the quark outer ring: the answer for k operators then is $\sim N_c^{1-k/2}f(\lambda)$. (The power 1 comes from the trace of the quark loop.) What this result means is that effective mesonic Lagrangian would have weak interaction in the large N_c limit. In particular, a decay of one meson into 2 (described by a $k = 3$ correlator) has a small amplitude, so their widths should be small compared to the mass. The same reasoning can be extended to glueballs.

One more set of consequences follows from these ideas with respect to flavor mixing. For example, each time an $\bar{s}s$ meson (like ϕ) decays into non-strange mesons (pions), a diagram with one quark loop is not enough, and at least two quark loops are needed. As those are sub-leading, we have derived in the large N_c limit, the so called Zweig rule.

Let us now move to an even more ambitious goal, following a program initiated by Witten. But before we come to its core, we have to prove a few things first. The first such general statement is: in the large N_c limit all "white" variables are nearly "homogeneous" in space–time, with small relative fluctuations. Indeed, the multi-color vacuum is very dense and the fluctuations of various components should be uncorrelated. The intuitive idea is based on statistical arguments. Consider some volume of a gas, containing N particles. The relative fluctuation of the number of particles in it is small

$$\langle (N - \langle N\rangle^2)^2\rangle^{1/2}/\langle N\rangle \sim 1/N^{1/2}, \tag{1.118}$$

if the particles move independently.

And indeed, consider the correlator of two white gluonic operators, such as $\phi \equiv (G_{\mu\nu})^2$. The statement is that

$$\langle \phi(x)\phi(y)\rangle \rightarrow \langle \phi\rangle^2 + O(1/N_c^2). \tag{1.119}$$

The proof is obvious from Fig. 1.10: the uncorrelated part (a) has 4 color traces and the correlated one (b) only 3.

Furthermore two coupling constants g^2 which enter diagram (b) add to the 't Hooft limit an extra $1/N_c$.

---

[26] Due to the remarkable Maldacena conjecture [893] to be discussed in Section 12.5 it looks as if such a representation seems to be finally found. However ironically the $\mathcal{N} = 4$ supersymmetric Yang–Mills theory is *not* a confining theory.

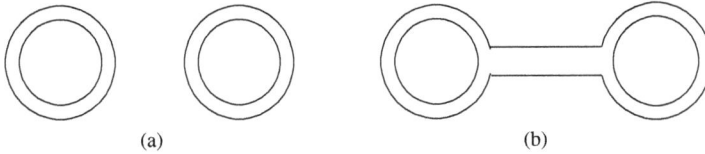

Fig. 1.10. (a) A disconnected and (b) a connected diagram in the large N_c limit for the correlator of the gluon condensate.

Looking at the diagram itself, one finds that actually the one-gluon exchange between colorless objects vanishes. However, one can always exchange 2 gluons (or more). That will add a bubble inside the bubble in [Fig. 1.10(b)], going from 3 to 4 color traces, but will also add another g^2, so that any number of connecting planar gluons will leave the total power of N_c unchanged. (Therefore the counting rule by itself does not solve the problem.)

Unfortunately, the statistical sum is defined not in terms of weakly interacting colorless objects like G^2 but strongly fluctuating A_μ^a. Multiple efforts to reformulate everything in terms of gauge invariant "colorless" objects (such as "loop dynamics") have yet to produce fruits.

Another related idea goes as follows: since fluctuations are small, it seems likely that there should be only one single field configuration, the so called *master field*, dominating the statistical sum. It was also considered likely that the space–time dependence of this field is not important in comparison to multicolor space; people therefore studied matrix models, with constant fields. We will see in Section 12.5 that a large cluster of many instantons may play such a role.

Although we do not have a formal proof, we think that further the masses of mesons and glueballs are $O(1)$ times some scale like Λ_{QCD}, whereas the masses of baryons are $O(N_c)$. Also, meson decay constants are $O(N_c^{1/2})$ and meson–meson scattering amplitudes are $O(1/N_c)$ and thus mesons are expected to be narrow in this limit. In general a large N_c QCD is expected to be a theory of weakly interacting mesons and glueballs, with very heavy (classical) baryons. In the same spirit people expected the fermionic determinant to play very little role as compared to gluonic dynamics: if so the quenched approximation in lattice QCD would be justified.

Witten [188] suggested that non-perturbative effects generate a θ-dependence in the pure gauge theory in the form of a function $E = N_c^2 F(\theta/N_c)$ and that consequently the topological susceptibility,

$$\chi_{\text{top}} = \left.\frac{d^2 E}{d\theta^2}\right|_{\theta=0},$$ (1.120)

is $O(1)$ in the large N_c limit [81,82]. If one follows perturbative rules,[27] one would expect that the contribution of fermions to χ_{top} is further sub-leading in $1/N_c$.

---

[27] It will be a simple exercise for the reader to find the suppression power.

Here however comes a puzzle: we argued above that for massive quarks one can introduce the theta term, and theta vacua have a real energy dependence on θ, while for a massless fermion it is not so. This implies that for a massless quark $\chi_{\text{top}} = 0$ and so quark masses do matter. How may sub-leading quark terms cancel a much larger gluonic contribution? Witten argued that this apparent contradiction can be resolved if the mass of the η' meson scales as $N_c^{-1/2}$ in the large N_c limit. Witten [188] and Veneziano [189] derived a relation between the mass of the η' and the topological susceptibility in pure gauge theory,

$$\frac{f_\pi^2}{2N_f} m_{\eta'}^2 = \chi_{\text{top}}. \tag{1.121}$$

Using $\chi_{\text{top}} = O(1)$ and $f_\pi^2 = O(N_c)$ we observe that indeed $m_{\eta'}^2 = O(1/N_c)$. This result implies that the $U(1)_A$ anomaly is effectively restored in the large N_c limit. We will see later that lattice/instantons[28] indeed support such a scaling.

The θ-dependence of the vacuum energy is related to the topological properties of QCD. In the semi-classical approximation these features can be described in terms of instantons. Later in this book we will show how this relation holds.

Finally we address a practical question. Is the multicolor QCD indeed close to the real world, with $N_c = 3$, or not? Many relations derived in the large N_c limit work very well, but others fail completely. It is instructive to mention a few of those.

(i) The masses of η', ρ, p scale as the powers of N_c equal to $-1/2, 0, 1$, but they all are in fact nearly equal to each other in the real world!

(ii) Most resonances are relatively narrow but some — like the σ peak in low energy $\pi\pi$ scattering — are not.

(iii) Most channels follow the Zweig rule and do not mix flavors, but all scalar and pseudoscalar channels do.

(iv) In weak $K \to \pi\pi$ decay the $\Delta I = 1/2$ amplitude is N_c-suppressed, but in reality it is about 20 times larger than the unsuppressed one.

(v) In the large N_c limit extra quark loops are suppressed. This implies that the so called "quenched" approximation on the lattice — ignoring the fermionic determinant or the sum of dynamical quark loops — should be a good one. It is not so.

Of course, it is not the N_c counting rules which are wrong: just some non-perturbative QCD phenomena managed to generate *other large parameters*, interfering with the powers of N_c and may in some cases lead to unexpected results. In the rest of the book we will show that most (if not all) of the cases in which N_c counting fails are directly related with instantons.

---

[28] Another famous argument of Witten's was that since small size instantons are suppressed in this limit, somehow all these relations should not be refer to instantons. We will discuss in detail in Section 4.5 how this problem is resolved.

Exercise 1.8. Check what happens in the large N_c limit with the asymptotic freedom relation for λ: is it indeed N_c-independent?

Exercise 1.9. Evaluate the N_c dependence of the correlator with k scalar glueball operator $G_{\mu\nu}^2$.

1.7.2. QCD with the smallest ($N_c = 2$) number of colors

The theory with the SU(2) color group has particular properties which differ from other QCD-like theories. However it is instructive to use it as a testing ground for ideas to be developed.

In the massless case this theory has an additional Pauli–Gursey symmetry [86], which incorporates mixing of quarks with anti-quarks. The reason for that is that the SU(2) gauge group is quasi-real, with no real distinction between quarks and antiquarks.[29] Therefore in this theory there are multiplets of *degenerate* hadrons, including both diquarks (= baryons of this theory) and mesons.

It may remind the reader what happens in supersymmetric QCD-like theories, in which baryons and mesons also form supermultiplets with equal masses. However in $N_c = 2$ QCD these degenerate baryons and mesons form different representations of flavor and spin groups, and therefore their numbers do *not* in general match. So this theory, $N_c = 2$ QCD, can be considered as some intermediate case between SUSY QCD and the ordinary QCD, in which no relation between baryons and mesons exists. Another obvious difference is that SU(2) baryons are diquarks, so they are not fermions but bosons.

Of specific interest are multiplets including massless Goldstone modes following from chiral symmetry breaking. The issue is rather technical, so we should not go into it and just mention the result. The correct group-theoretical analysis of possible patterns of (quark-antiquark) symmetry breaking was made by Peskin [87]. They are SU($2N_f$) → Sp($2N_f$), and the number of Goldstone modes for N_f quark flavors was given by Smilga and Verbaarschot [88]

$$N_{\text{goldstones}} = 2N_f^2 - N_f - 1. \tag{1.122}$$

Let us mention three cases specifically.

For $N_f = 1$ there are *no* Goldstone modes. As usual, due to the U(1) anomaly, the only pseudoscalar meson to be called η', is massive. The diquark, if it should be similar to what we need, cannot even exist due to Pauli principle.

For the most interesting case $N_f = 2$, the coset of the full initial symmetry group over the final (remaining) one is

$$K = \text{SU}(4)/\text{Sp}(4) = \text{SO}(6)/\text{SO}(5) = \text{S}^5, \tag{1.123}$$

---

[29]The simplest way to see that is the fact that in this theory there is an invariant tensor — epsilon — with two indices allowing us to raise or lower the tensor indices.

which means that Goldstone modes make a 5-dimensional sphere with 5 massless modes; three of those are pions, plus scalar diquark S and its anti-particle $\bar{S}$.

The next case $N_f = 3$ leads to 14 goldstones. It is easy to count them: mesons form the usual $SU(3)_F$ octet, plus 3 diquarks and and 3 anti-diquarks belonging to $\underline{3}$ and 3 representations because flavor indices are convoluted with ϵ_{ijk}.

CHAPTER 2

Phenomenology of the QCD Vacuum

In this chapter we discuss properties of the ground state of QCD and its lowest excitations as they were established empirically. We will start with a brief review of hadronic spectroscopy in Section 2.1; it is not meant to be complete in any sense. Instead of enumerating multiple regularities and successes of various models, its main aim is to introduce the puzzling observations to be addressed later in the book.[1]

We then proceed to a discussion of two simplest models of the QCD vacuum in Section 2.3. An example of a random model is that of "stochastic vacuum", to be contrasted with the much more structured "instanton liquid model".

Then we return to a more substantive discussion of the lowest excitations — the pions — and their interactions. A discussion of the spontaneous breaking of the $SU(N_f)_A$ chiral symmetry in Section 2.4 lead to the low energy effective chiral Lagrangians. We then proceed in Section 2.5 to discuss confinement, existence of flux tubes and their properties, as they appear from phenomenology and lattice studies. In general, at the moment we have a much better understanding of the former phenomenon than the latter. We will return to the microscopic theory of chiral symmetry breaking in Chapter 5; the microscopic theory of confinement is still lacking.

Some chapters can be omitted at a first reading; but this one (except maybe the introductory hadronic physics) is not one of them. Many issues we are going to discuss below in this book will make their appearance here, as a phenomenological material first.

As a small digression, let me make here a small tribute to phenomenology in general. Often people who have solved some complicated theoretical problem tend to look down on those who try to dig something new from the experimental data, or build primitive models to explain them. Let me give two reasons indicating why they are wrong. The first is a purely phenomenological fact; as the practice of

---

[1]Readers familiar with hadronic physics may well skip it. Readers who want to have a look at details can proceed to the official Review of Particle Properties [91] or the home page of Particle Data Group at http://www-pdg.lbl.gov.

the employment market shows, there are many more people who can solve well-posed problems than there are good phenomenologists. The second is based on the recognition people are given, such as Nobel prizes etc. For example, the greatest achievement of the 20th century physics — quantum mechanics — started by Planck and Bohr and continued by Born, Heisenberg and Schrödinger. None of them really "derived" their results, as there was no available basis for such derivation yet. They rather guessed parts of the future theory, and these successful guesses (based on data, at least for Planck and Bohr) were very highly praised. At the same time technical improvements (as an example let me mention Sommerfeld's generalization of the Bohr quantization idea) were treated as important steps but still very distinct from the "logical jumps" the founding fathers have made.

Why do we need phenomenologists now, the reader may ask, as the basic structure of QCD is firmly established? Why can't we derive all we need from its famous Lagrangian? Well, indeed in the 1970s, there were people who spoke about "solving QCD", but I have not heard this expression for a long time. I firmly believe that the old way of doing physics — thinking about the data trends, making a model and compare it with data, "real" or numerical, from lattice simulations, is still the way to go.

2.1. Phenomenology of the hadronic world

2.1.1. *Brief history*

Not counting the *proton*, known as a hydrogen nucleus for a long time, we start with the discovery of the *neutron* in 1932. Comparing their masses,

$$m_p = 938.256 \, \text{MeV}, \qquad m_n = 939.550 \, \text{MeV}, \qquad (2.1)$$

Heisenberg proposed to consider these two states as a SU(2) doublet, two different *isospin projections* of the same object — the nucleon. That was when new internal symmetries first came in. Let me also remind that the neutron is unstable under weak beta decay; $n \to p + e + \nu$ and its lifetime is about 15 min. It is however stabilized in nuclei since the binding energy can overcome 1 MeV of the mass difference.

The next important point was the discovery of the *pions*, in 1947. Their existence supported earlier ideas by Ukawa about light bosons carrying nuclear forces, as well as the reality of the *isospin* (the u–d or SU(2) flavor) symmetry; the pions π^+, π^-, π^0 indeed form a nice isomultiplet with masses for charged and neutral pions being 139.57 and 134.97 MeV, respectively. An attentive reader may notice that the $\pi^\pm - \pi^0$ mass difference is rather large compared to that for $p - n$, and other similar splittings. It is indeed true, and indicates that the pion size R_π is rather small compared to that of other hadrons, which enhances the $O(1/R_\pi)$ Coulomb electromagnetic contribution.

The neutral pion decay $\pi^0 \to \gamma\gamma$ showed that it cannot have spin 1 and the spin must be zero. The polarization measurements have shown that polarizations of two photons tend to be orthogonal to each other; the decay amplitude was well described by the $\vec{E}\vec{B}$ combination which implies that the pion is *pseudoscalar* and has negative *P*-parity, This in turn suggested that nuclear Ukawa forces should be modified from their original form for scalar mesons, with isospin-invariant interaction Lagrangian of the type

E

$$L = ig\bar{N}\gamma_5 \vec{\tau}\vec{\pi}N. \tag{2.2}$$

Exercise 2.1. Show that this Lagrangian leads to relations between processes, e.g. effective couplings for the processes indicated are related by $g_{pp\pi^0} = -g_{nn\pi^0} = (1/\sqrt{2})g_{pn\pi^+} = (1/\sqrt{2})g_{pn\pi^-}$.

The strange particles, *kaons and hyperons*, were found in early 1950s. Their main decay (e.g. $\Lambda^0 \to p + \pi^-$) is due to weak interactions, as its long lifetime $T_\Lambda = 2 \times 10^{-10}$ sec indicates. Thus lambdas can fly noticeable distance in the detectors and leave distinctive *V*-like decay pictures, with two charged tracks originating from a point outside the main interaction vertex. The same is true for the short-lived CP-even neutral kaon $K_S \to \pi^+\pi^-$.

The reason why kaons and hyperons have such long lifetime was not clear at first, till Gell–Mann and Nishijima argued in 1953 that they contain a new quantum number called "strangeness" conserved by strong interactions. That did not solve all puzzles related with kaons. The so called τ–θ paradox was that neutral kaons decay both to 2 and 3 pions, or states of opposite parity; that was clarified by the observation that there are two combinations of $K^0 \pm \bar{K}^0$, which have opposite CP parity. Further studies led to the discovery, in 1964, of the CP violating transitions between them.

Multiple hadronic resonances, such as the famous excitation of the nucleon called Δ, $S = 3/2$, $I = 3/2$ were also discovered during the 1950s. That has shown that baryons are composite, not elementary particles. Rapid proliferation of states were put to relative order only in 1964, when Gell–Mann and Neeman promoted the flavor symmetry from the SU(2) to SU(3), in the so called "eightfold way". (Named after the most popular — adjoint — representation of the SU(3) group, the octet.) In Gell–Mann's papers the word "*quark*" had appeared for the first time, as a purely bookkeeping concept devised to explain to physicists how to construct the representations of the unfamiliar SU(3) group.

Searches for quarks resulted in a surprise; in a way, they were indeed found in SLAC in 1969, but only as "*partons*", seen inside a nucleon in hard processes such as deep inelastic lepton–nucleon scattering.

Many people who had strong reservations against the quark model surrendered when the revolutionary discoveries of the first heavy flavor — *charm* — was initiated by a dramatic discovery of the J/ψ in October of 1974, in Brookhaven and SLAC.

It finally put to rest discussions of the reality of quarks; it became all too obvious that *they do exist but are confined inside hadrons.*

Even prior to this discovery Appelquist and Politzer wrote a paper [132] suggested that if heavy quarks did exist, they would form *heavy quarkonia*, which can be treated non-relativistically, like the positronium in QED. Subsequent studies of the charmonium ($\bar{c}c$) and the bottonium ($\bar{b}b$) spectra[2] revealed that a combination of the Coulomb and linear confining potentials

$$V(r) = -\frac{4}{3}\frac{\alpha_s(r)}{r} + Kr \tag{2.3}$$

with the string tension $K \approx (440\,\text{MeV})^2 \approx 1\,\text{GeV/fm}$ can actually describe the levels rather well.

Let me remind the reader that in atoms and nuclei the velocities are of the order $v_{\text{atomic}} \sim e^2/\hbar c = 1/137$; $v_{\text{nuclear}} \sim 1/5$. In heavy quarkonia the role of the fine structure constant is played by α_s, which is not very small. Furthermore, the Coulomb energy scale for levels is $E \sim \alpha_s^2 M_Q$, so rotation frequencies of heavy quarkonia are parametrically growing with the quark mass. For charmonia and bottonia, however, due to a number of corrections, the frequencies (deduced from level differences) are actually nearly identical.

If one attempts the usual non-relativistic expansion, including the usual *spin–spin* interaction

$$H_{\text{spin-spin}} = \frac{4\alpha_s(r)}{3m_im_j}\left[\frac{8\pi}{3}\vec{S}_i\vec{S}_j\delta^3(\vec{r}_{ij}) + \frac{1}{r_{ij}^3}\left(\frac{3(\vec{S}_i\vec{r}_{ij})(\vec{S}_j\vec{r}_{ij})}{r_{ij}^2} - (\vec{S}_i\vec{S}_j)\right)\right], \tag{2.4}$$

one also finds a reasonable description of quarkonia. Note that it is basically the same Hamiltonian which was so successful in atomic physics. The situation with the *spin–orbit* term is more complicated; it includes derivatives of the radial potential which should of course contain both the one-gluon exchange and the confining (string) part. Because it was confirmed that heavy quarkonia are very much atom-like non-relativistic systems, we will not discuss them in this book any further.

These findings have led to what I would call the "minimal QCD-based model" according to which *the one-gluon exchange plus confinement* are the only building blocks one needs to understand hadronic physics, including all hadrons made of light quarks. According to this ideology, which started with the early influential paper by DeRujula, Georgi and Glashow [94], the chiral symmetry breaking is doing nothing else but supplying the effective "constituent quark mass", large enough to treat them non-relativistically. Perhaps the most detailed studies of the kind have been made by Karl and Isgur [95] who worked out spectra for many excited baryons and mesons in such a model.

---

[2]The last top quark t is so heavy that its weak decay time is shorter than the time needed to form the topponium states.

The main point of most of the subsequent discussion is that actually this ideology is far from the truth. Fortunately for us, even at rather small distances there exist much more interesting phenomena in QCD than just one gluon exchange, and so the QCD is far from being a dull repetition of QED, at a somewhat larger (and running) coupling.

2.1.2. The "usual" hadrons

Let me start with what I would call *the "usual mesons"*, for which the "minimal non-relativistic ideology" outlined above is actually fine. The lowest SU(3) multiplets for the total angular momentum J and parity P such that they form a "vector" ($J^P = 1^-$), an "axial" ($J^P = 1^+$), a "tensor" ($J^P = 2^+$), etc. all appear in complete flavor nonets, a singlet plus octet ($3 * 3 = 9 = 1 + 8$).

In the case of equal quark masses $m_u = m_d = m_s$ the SU(3) symmetry would demand that those 9 states be a degenerate octet plus a separate SU(3)-singlet state. In reality in "normal" cases the value of m_s is large enough to separate strange from the non-strange hadrons.

We will often discuss vector mesons below, so let me use this nonet as an example. (There are many others; see Ref. [91].) There are 4 non-strange vectors; a triplet of $\rho(771)$ mesons[3] with the isospin $I = 1$, plus isosinglet $\omega(782)$ $I = 0$, $(1/\sqrt{2})(\bar{u}u + \bar{d}d)$. The semi-strange vectors $K^*(892)$ $I = 1/2$ form 4 states, since there are separate strange and anti-strange states. Finally the last 9th vector is the $\bar{s}s$ state $\phi(1020)$.

Note that strange states are systematically heavier. Furthermore, extra mass due to strangeness appears to be simply additive, so the non-relativistic picture of additive non-strange and strange effective masses (plus small "binding") seems to be adequate.

The mixing between these vector states is found to be very small, which is known as a "Zweig rule". In principle, strong processes may transform $\bar{u}u \leftrightarrow \bar{d}d \leftrightarrow \bar{s}s$; but experimentally it is not the case for vector mesons. For example, ω, ϕ practically do not mix; so e.g. the $\phi \to \pi\pi$ decay matrix element[4] is very small compared to that for $\phi \to KK$. The perturbative explanation of that is that in the vector channel one has to go to 3 gluons to make a flavor transition. Non-perturbatively, the explanation is that the instantons do not operate in the vector channel.

The SU(3) representations for baryons (3 quarks) are $3 * 3 * 3 = 10 + 8 + 8 + 1$. As an example of *the "usual baryons"* let me mention the spin 3/2 decuplet first. [E] It has the simplest wave function since the spin and coordinate wave functions are trivially symmetric (all quarks in the same state), and so it is the antisymmetric ϵ_{ijk} color tensor which makes the total wave function antisymmetric, as Fermi statistics

---

[3]The number near the symbol here and below is the rounded particle mass in MeV.

[4]Not the partial widths since these two modes have very different phase space.

demands. The particles are

$$\Delta(1232, I = 3/2), \quad \Sigma^*(1385, I = 1), \quad \Xi^*(1530, I = 1/2), \quad \Omega(1672, I = 0). \quad (2.5)$$

The last one has no isospin because it is purely strange sss state. The additivity of the masses approximately holds; e.g. $M_\Delta \approx \frac{3}{2} m_\rho$, $M_\Omega \approx \frac{3}{2} m_\phi$, etc. The SU(3) breaking rules in first order are working very well.

Exercise 2.2. If the fourth c-quark were light, how many states would there be in such a multiplet? Answer: 20.

The other famous baryon multiplet is the lowest octet with spin $= 1/2$;

$$N(939, I = 1/2), \quad \Lambda(1115, I = 0), \quad \Sigma(1189, I = 1), \quad \Xi(1315, I = 1/2). \quad (2.6)$$

Here simple additivity would not work; for example M_N is not close to $\frac{3}{2} m_\rho$; the difference has led to a discussion of spin–spin forces, which we will treat in Section 2.2.5 below.

2.1.3. *The "unusual" mesons*

Let us now have a look at what I would call *unusual mesons*, starting from the lowest — pseudoscalar — states. The lowest states in quantum mechanics are always at zero angular momentum or s-wave states, which usually have positive parity because $P = (-)^L$. However the fermion–antifermion pair has the so called intrinsic negative parity; thus for mesons $P = (-)^{L+1}$ and the lowest s-wave mesons are pseudoscalars.

$\boxed{\text{E}}$

Like vector mesons they form a nonet of states

$$\pi(139, I = 1), \quad K(493, I = 1/2), \quad \eta(547, I = 0), \quad \eta'(958, I = 0), \quad (2.7)$$

which however looks very different. As we already discussed in Chapter 1, the octet is light because in the chiral limit these mesons are massless Goldstone particles. The fact that η' is very heavy — the so called Weinberg U(1) problem — is due to explicit breaking of chiral U(1) symmetry by the chiral anomaly, see Section 1.5.2.

The mixing pattern in this nonet is very different from the "usual" mesons; η' is close[5] to a completely symmetric $(\bar{u}u + \bar{d}d + \bar{s}s)/\sqrt{3}$ state, the SU(3) singlet. It means that the pseudoscalars seem to ignore the strange quark mass which can only be the case if the forces which act in this case are *much stronger* than the SU(3) violating terms, proportional to m_s.

---

[5]The best descriptions have a rotation with a small angle about $10°$ between physical η, η' states and ideal octet and singlet.

The case of the light-quark scalars is the most complicated one in hadronic spectroscopy, not yet clarified completely in spite of several decades of hard work by experimentalists and theorists.

Already the very first discovered state — known as σ or $f_0(600)$ — was very puzzling. First observed as a broad enhancement in $\pi\pi$ scattering, it (together with pions) became the basis of the classical "sigma model" by Gell–Mann and Levy [28]. It obviously does not fit into $\bar{q}q$ systematics of non-relativistic quark models; the scalars must non-relativistically be the p-wave states, which are placed well above $1\,\text{GeV}$. Due to its light mass and relatively large width, the sigma meson has been thrown out of the particle tables. After a long period of absence, it is however back there, due to recent very nice observations using heavy quark systems. A summary of the σ parameters, as well as those for its strange brother κ, is given in the Table 2.1 from Ishida [110].

There are also flavored scalars such as (old name δ) $a_0(1400)$ $(I = 1)$ and $K^*(1430)$ $(I = 1/2)$, which more or less correspond to the usual quark model place natural for the $\bar{q}q$ scalar nonet. The unflavored scalars in the same region are $f_0(1400, 1525, 1590, 1710)$, so even assuming that one of them is the scalar glueball we have 3 states and only 2 places left in the $\bar{q}q$ nonet.

On top of this, there are scalars which are even lighter; those include 4 states $f_0(975)$ $I = 0$ and $a_0(980)$ $I = 1$, plus the σ, κ already mentioned. A possible interpretation for them is that they are the 4-quark $(\bar{q}\bar{q}qq)$ states, see e.g. Jaffe [99]. We have borrowed a nice Fig. 2.1 from Jaffe's lecture, which shows that the mass structure of such a nonet should be inverted compared to the usual one because there are more states containing the heavier strange quarks. And indeed, $f_0(975)$ and $a_0(980)$ decay into kaons rather than pions, in agreement with such assignment. σ in this model is understood as a $\pi\pi$ state.

Table 2.1. The mass and width of the σ and κ mesons, from various sources.

Processes	m_σ (MeV)	Γ_σ (MeV)
$\pi\pi \to \pi\pi$	535~675	385 ± 70
$J/\psi \to \omega\pi\pi$(DM2)	482	325
$J/\psi \to \omega\pi\pi$(BES)	390^{+60}_{-36}	282^{+77}_{-50}
$\Upsilon(mS) \to \Upsilon(nS)\pi\pi$	526^{+48}_{-37}	301^{+145}_{-100}
$p\bar{p} \to 3\pi^0$	540^{+36}_{-29}	385^{+64}_{-80}
$D^+ \to \pi^-\pi^+\pi^+$	$478^{+24}_{-23} \pm 17$	$324^{+42}_{-40} \pm 21$
$\tau^- \to \pi^-\pi^0\pi^0\nu_\tau$	555	540
Processes	m_κ (MeV)	Γ_κ (MeV)
$K\pi \to K\pi$	905^{+65}_{-30}	545^{+235}_{-110}
$D^+ \to K^-\pi^+\pi^+$	$797 \pm 19 \pm 42$	$410 \pm 43 \pm 85$
$J/\psi \to K^{*0}K^-\pi^+$	~800	~300

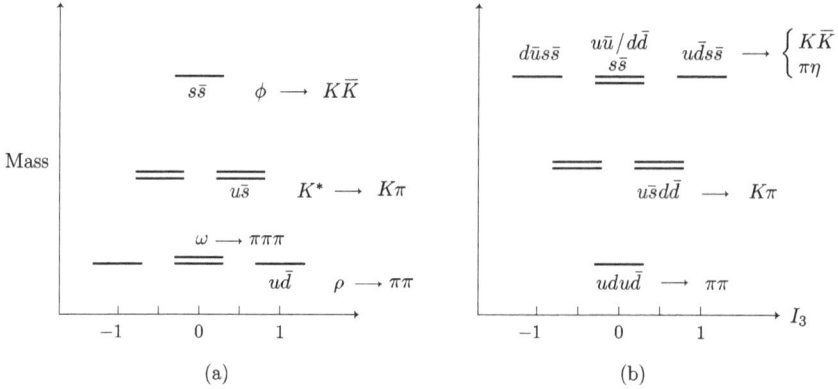

Fig. 2.1. The mass pattern, quark content and natural decay couplings of (a) a $\bar{q}q$ nonet and (b) a $\bar{q}\bar{q}qq$ nonet, according to Ref. [107].

At the time of this writing a newly discovered narrow resonance $D_s(2320)$ was reported by the BABAR and CLEO collaborations [108], in $D_s\pi$ mode. Since the decay violates isospin, a small width is expected. It is the low mass of this — presumably scalar — resonance, which attracted attention, only about 0.35 GeV above the D_s ground state. Again, it is too low for a P-wave state of the potential model, and is thus the 4-quark state suspect. Naive ideas about a weakly bound molecular state of a $D_s + \eta$ type would also put it 200 MeV higher.

Why are all those 4-quark states so light, much lower than the 2-quark ones?

One obvious answer is that four quarks can couple to the 0^{++} channel without a unit of orbital excitation. (Now the relative $P = -1$ parity of the fermion–antifermion enters squared and does not matter.)

My tentative answer (which, admittedly, remains to be investigated) is that the non-diagonal matrix elements between $\bar{q}q$ and $(\bar{q}q)^2$ are very large due to 't Hooft interaction, and thus the level repulsion between those is much stronger than anticipated. For example, in the new D_s state such mixing can be $s \leftrightarrow s\bar{q}q$.

The mechanism may be revealed by the wave function, which may resemble one of the following alternatives: (i) a 4-quark bag, (ii) a meson–meson molecule, well separated in space or (iii) a diquark–diquark molecule; (iv) about 50–50 mix of a 2 and 4-quark state, in case of the non-diagonal mixing element dominating.

The issue can in principle be settled by studies of the decay patterns, and also theoretically, e.g. by considering couplings of these states to operators of different structure on the lattice. Recent work by Alford and Jaffe [107] is the first attempt of the kind, with only the meson–meson-type sources, but much more should be done.

Exercise 2.3.   Take massive quark and antiquark as plane wave spinors and show that when both are at rest, $\vec{k}_1, \vec{k}_2 \to 0$ the pseudoscalar combination $\bar{q}\gamma_5 q$ is non-zero but the scalar one vanishes, $\bar{q}q = 0$.

2.1.4. *The exotic hadrons*

Another set of unusual hadrons are *glueballs*, which have no quarks at all. The simplest theoretical way to learn something about them is to study the theory without quarks — the gluodynamics — on the lattice, see Section 5.3.1. Let me only mention here one striking observation; the mass scale of glueballs seems to be much higher than that of normal mesons. The question "Why are glueballs so heavy?" will be repeatedly discussed below.

In real world glueballs can mix with other $\bar{q}q$ states with the same quantum numbers. Still it is believed that the experimental resonances $f_0(1710), f_0(1500)$ have the largest share of the lowest scalar glueball. The observed 2^+ tensor state $f_2(2300)$ is the leading candidate for the role of the tensor glueball. The next — the pseudoscalar, according to lattice data — believed to be still heavier and (to my knowledge) has not yet been associated with any experimental candidate resonance.

All hadrons other than $\bar{q}q$ mesons and qqq baryons can be called exotic. However the standard terminology uses the word *exotic* for those states which have quantum numbers impossible for $\bar{q}q$ mesons and qqq baryons. In this strict sense, the $\bar{q}\bar{q}qq$ states considered above and the glueballs are not truly exotic.

Particle tables now include two vector meson candidates which have truly exotic quantum numbers $J^{PC} = 1^{-+}$. They are $\pi_1(1400)$ and $\pi_1(1600)$, seen experimentally in the late 1990s. If confirmed by further studies, they are the first examples of the hybrid $\bar{q}qg$-type mesons. (A need for an extra gluon is seen from flipped charge conjugation parity, different from that of the usual vector mesons.) Their masses are however somewhat lower than lattice-based expectations for a hybrid which were in the 1.7–1.9 GeV range.

In July 2003[6] the field of exotic hadron searches was shaken by a discovery of truly exotic baryon $\Theta^+(1540)$ with a small $\Gamma < 9\,\mathrm{MeV}$ width. The discovery made in Japan (T. Nakano *et al.*, *Phys. Rev. Lett.* **91**, 012002 (2003)) was in a matter of months confirmed by about 8 other experiments. The observed angular distribution suggests a likely spin $1/2$ state, with so far unknown parity. Its minimal quark content is a pentaquark, i.e. $(ud)^2\bar{s}$. The anti-decuplet flavor assignment was further strengthened by an observation by the NA49 collaboration (C. Alt *et al.*, hep-ex/0310014.) of a family of exotic Ξ baryons, with a mass of 1.86 GeV and the width smaller than the experimental resolution of 18 MeV. The exotic ones have quantum numbers of $(us)^2\bar{d}$ and $(ds)^2\bar{u}$.

Although existence of anti-decuplet of exotic hadrons was anticipated in the framework of the Skyrmion model, with *positive* parity for these states, the very small width is a problem for this description. It is obviously a sign that internal wave function of this state, whatever it is, is very different from $N + K$. More traditional "shell model ideology" (e.g. the MIT bag model or nonrelativistic constituent quark

---

[6]After the book was submitted to the publisher, so this part is a note added in proofs.

models) tends to put as many quarks as possible in the lowest shell, and thus predict negative parity for the lowest state.

We see these finding as new confirmation of the picture advocated in this book, in which the key element are the *instanton-induced*[7] diquarks. Due to Pauli principle at the level of instanton zero modes, two quarks of the same flavor cannot interact with the same instanton. This forces diquarks to be flavor antisymmetric, equivalent to antiquarks, and forcing all multiquark states to be in the *lowest possible* flavor representation, avoiding many other possible exotic states, both in the meson and baryon sectors. In particular, even these newly discovered states, although truly exotic, still are in a way analogous to the decuplet baryons.

A simple model of pentaquarks based on these ideas have been proposed by Jaffe and Wilczek, hep-ph/0307341 and by I. Zahed and myself hep-ph/0310270. The following shorthand notation for diquark flavors in $SU(3)_f$ is used in the latter work:

$$\underline{S} = (u^T C \gamma_5 d); \qquad \underline{U} = (s^T C \gamma_5 d); \qquad \underline{D} = (u^T C \gamma_5 s).$$

The model treats diquarks (scalar or tensor) on equal footing with constituent quarks. Because of their similar mass and quantum numbers, certain approximate symmetries appear between states with the same numbers of "bodies". The $\bar{q}q$ mesons (a) are a well known example of the 2-body objects, as well as the quark–diquark states (b) (the octet baryons qq). Furthermore, the diquark–antidiquark states (c) are in this model the 2-body objects as well. So, to *zeroth* order, both non-strange mesons (like ρ, ω), the nucleon, and some 4-quark states all have the same mass $2\Sigma \approx 840\,\text{MeV}$. To *first* order, including one-gluon-exchange Coulomb and confinement, the degeneracy should still hold, as color charges and masses of quarks and diquarks are the same. Only in *second* order, when the spin–spin and other residual forces are included, they split. Note that this new symmetry between N and ρ is actually more accurate than the old SU(6) symmetry, relating the octet and the decuplet baryons such as N and Δ.

Pentaquarks in this model are treated as 3-body objects, with two correlated diquarks plus an antiquark, and thus there are simple relations between masses of various "3-body objects" similar to the decuplet baryons. If both diquarks are scalar they are identical bosons and symmetry demands p-wave ($l = 1$): otherwise they are scalar and tensor diquarks. In both cases the total parity is predicted to be *negative*.

From the color point of view, all 3-body states involve the same ϵ_{ijk} wave function, just like the ordinary color singlet baryons. From the flavor point of view, one can also write it similarly, fully utilizing the notations we introduced above. For example, $\Theta^+(1540) = (ud)(ud)\bar{s} = \underline{SS}\bar{s}$ is an analogue of anti-Ω, and is thus the top of the antidecuplet. New exotic $\Xi(1860)$ are $\underline{UU}\bar{u}$ and $\underline{DD}\bar{d}$ make two

---

[7]Although scalar diquarks are also attractive channel for a single-gluon exchange, such forces do not lead to the structure we discuss as they are flavor blind.

remaining ends of the triangle. The remaining 7 members can mix with the octet of pentaquarks, which make flavor $(8_f \oplus \overline{10}_f)$ multiplets. Using our flavor notations one can easily construct all states in complete analogy to antibaryons, changing from bar to underline where needed. If however a scalar and a tensor diquarks are used, the flavor representation is different, including 27-plet. The θ^+ (1540) is then the member of isotriplet.

In Chapter 11 we will discuss that dense baryonic matter as a very strong color superconductor, with scalar diquarks as Cooper pairs. It is gratifying to see that states of 5 quarks seem to be already made of one antiquark and 2 Cooper pairs. On the other hand, shell-model ideology, so successful for atoms and nuclei, is unlikely to be adequate for multiquark systems. It predicts many states in flavor-symmetric multiplets which were not observed.

An exotic state which was sought a lot but never found is the dibaryon H (also called dilambda) which is supposed to be the analog of the helium atom or the alpha particle of nuclear physics; the first closed spin-flavor shell, $(uds)^2$. For a review of theoretical/experimental information on it see Ref. [100]; we will discuss why it is so important in connection with strange quark matter in Chapter 11.

2.1.5. *Remarks about highly excited states*

One famous regularity is observed for hadrons with large angular momentum J. All such particles follow linear Regge trajectories[8]

$$m_J^2 = m_0^2 + \alpha' J, \qquad (2.8)$$

where the slope α' is the universal slope of Regge trajectories. My version of such a traditional plot is shown in Fig. 2.2. I arbitrarily selected two sets of well-studied mesons; of course there are many more linear trajectories for baryons. Two features of Regge trajectories are clearly seen: (i) their linear behavior fits the masses very well, and the ρ and the f_2 seem to be on the same trajectory; (ii) subleading resonances seem to be on parallel "daughter" trajectories. For comparison I have also plotted the Pomeron trajectory (which is only measured for negative t of course, and its extrapolation far from the measured region not really justified). It is supposed to hit the $f_2(2300)$ resonance which is the tensor glueball, and it may do so if it continues linearly into the large mass region.

This behavior is expected from the known features of confining forces. It can be shown that just this behavior is expected from two light charges (quarks for mesonic resonances, quark and diquark for baryonic ones) connected by a "string" and rotating relativistically. In this simple model the string tension is related to the

---

[8]Recall that in the usual quantum mechanics of the rotating solid rotor is quite the opposite; $E = J(J+1)/2I$, where I is the momentum of inertia. The difference apparently is that the hadrons deform with J rather than have a fixed I.

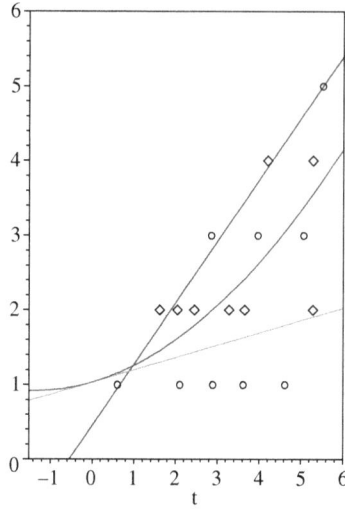

Fig. 2.2. On the plot of J (spin of the meson) versus t (mass squared) the masses of all the known ρ and f_2 resonances are shown as circles and diamonds, respectively. The lines are the traditional linear Regge trajectory and the Pomeron trajectory, shown both as a non-linear and linear extrapolations from the measured region (7.61) at negative $t = [-1, 0]\,\mathrm{GeV}^2$.

Regge slope by

$$K = \frac{1}{2\pi\alpha'} \approx (420\,\mathrm{MeV})^2 \tag{2.9}$$

and K is about the same as observed from spectra of the quarkonia.

Furthermore, one may expect that the probability of string breaking, by production of a new $\bar{q}q$ pair, is also a universal constant. If so, both the width of such hadrons and their mass are proportional to the string length, or $\Gamma(J)/M(J)$ should be independent of J. This is indeed the case phenomenologically.

The situation is not so nice in many other cases, however. To illustrate a problem, the very first excitation of the nucleon, $N^*(1440)$ or Ropper resonance, has the same quantum numbers as the ground state. (A similar excitation is known for other baryons as well.)

Following the usual logic of quantum mechanics one would assign it to the radial excitation of the nucleon. The problem is, using the non-relativistic quark model and solving the Schrödinger equation with an inter-particle potential like (2.3) one finds that it should be much heavier. This nasty problem persisted in lattice calculations as well, which consistently found that the next excitation is N^* with the negative parity and not the Ropper resonance with the positive parity. The experiment tells us however that the order should be inverted, as the $P = -1$ state has the mass 1535 and the $P = 1$ Ropper is only 1440. Only recently have I seen a lattice calculation [312] with chiral fermions and small enough quark masses, which has observed that these two levels cross at $m_\pi^2 \sim 0.1\,\mathrm{GeV}^2$ because the Ropper mass starts suddenly decreasing with the unusually strong slope as small quark masses get smaller.

What can the reason be for this behavior? One of the possibilities which I think is likely to be the case is that the Ropper $N^*(1440)$ has a $\bar{q}qqqq$ structure (or component), most probably a molecular state of the type $N + \sigma$. It is certainly not compatible with the radially excited qqq state.

Finally, in striking difference with the lowest states, the excited states of both baryons and mesons show a rather puzzling "parity doubling" phenomenon, discussed by Glozman [109]. For example, there is quite a significant gap between lowest vector and axial mesons, to say nothing about scalar and pseudoscalars. The first opposite parity state with nucleon quantum number is $N^*(1535)$, separated by quite a gap of 600 MeV from the nucleon, etc. However concerning the excited nucleon states starting at the mass $M \sim 1.7$ GeV, one observes three almost perfect parity doublets with spins $J = 1/2, 3/2, 5/2$, see Fig. 2.3 from Glozman [109].

According to the results of the partial wave analysis of mesonic resonances from 1.8 GeV to 2.4 GeV obtained in $p\bar{p}$ annihilation at LEAR, among mesonic states there are even possible scalar–pseudoscalar degenerate quartets [109]:

$$\pi(1801 \pm 13) - f_0(1770 \pm 12) - a_0(?) - \eta(1760 \pm 11),$$
$$\pi(2070 \pm 35) - f_0(2040 \pm 40) - a_0(2025) - \eta(2010^{+35}_{-60}),$$
$$\pi(2360 \pm 25) - f_0(2337 \pm 14) - a_0(?) - \eta(2285 \pm 20).$$

(Two question marks for a_0 means that they are still missing.)

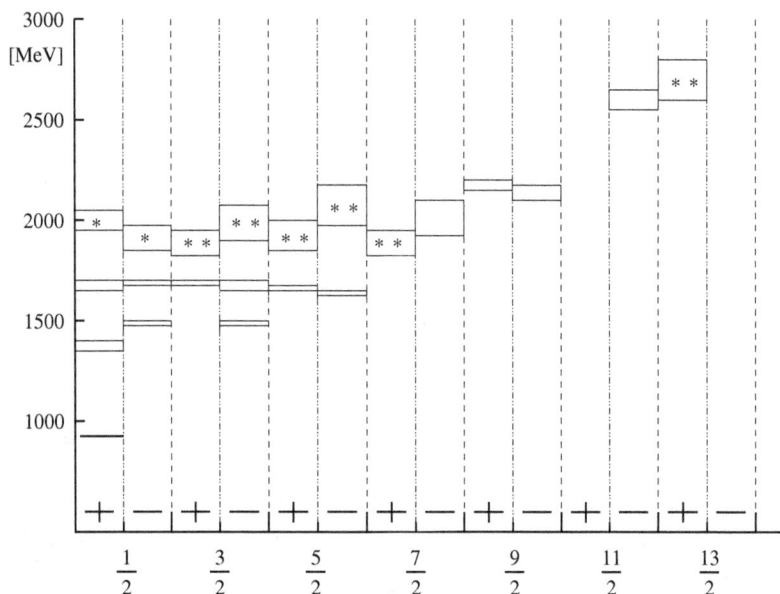

Fig. 2.3. Excitation spectrum of the nucleon. The real part of the pole position is shown. Boxes represent experimental uncertainties. Those resonances which are not yet established are marked by two or one stars according to the PDG classification. The one-star resonances with $J = 1/2$ around 2 GeV are given according to the recent Bonn (SAPHIR) results.

From the point of view of the "minimal picture" of the non-relativistic quark model this phenomenon looks very unnatural. The opposite parity states correspond to different shells because they have different orbital momentum, and it is difficult to imagine why spin forces should be fine-tuned in order to generate these close pairs.

The picture makes more sense if one looks at it in a framework of massless quarks. Chiral symmetry is broken by specific interaction of the quark pair in a meson via the instantons, which is a quasi-local operator in terms of the quark and the antiquark. The lowest hadrons have relatively small sizes so that their role is large; but as the excitation grows the wave function at the origin decreases and the role of instanton effects gets smaller and smaller. Effectively, states can follow the pattern of the original chiral symmetry, in which $U(N_f)_L \times U(N_f)_R$ multiplets start to emerge.

2.2.   Models of hadronic structure

2.2.1.   *Generalities*

A brief introduction to a few widely used models of the nucleon structure is provided by[9] Fig. 2.4. The first picture (a) shows the essence of the *non-relativistic quark model* suggested in the 1960s. It shows a family of three rather massive "constituent quarks" (with $M_{\text{eff}} = 350$–$400\,\text{MeV}$) kept together by their mutual attraction.

Figure 2.4(b) represents *the MIT bag* [113], suggested in the mid 1970s, in the early days of QCD. It shows a completely different picture; the main objects are nearly massless "current" quarks, bouncing back and forth between the confining walls with the speed of light. The object exists because colored quarks are unable to get out of a "bag"; they are simply not admitted in the "physical vacuum" outside it. They are no more attracted to each other than inmates which happen to be in the same prison cell. Forces between them are perturbative ones, based on one-gluon exchange and are mainly magnetic spin–spin interactions.

Figure 2.4(c) is a completely different picture; it is the so called *Skyrmion* model. Although originally proposed in the early 1960s by Skyrme [116] it became fashionable only in the 1980s, after several puzzling questions about its properties had been

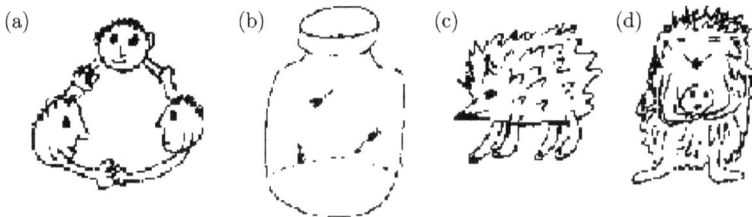

Fig. 2.4. Four pictures of the nucleon structure: (a) the non-relativistic quark model, (b) the MIT bag, (c) the Skyrmion and (d) the "chiral bag".

---

[9]I beg the reader to forgive the poor quality of the sketch.

clarified. The "hedgehog" shown is supposed to be a classical configuration made out of the pion field. Its non-trivial topology is represented by the pins (direction of the pion field in $3d$ pion isospin space) which go radially in the usual space $\vec{\pi} \sim \vec{r}$. If the number of colors is large, $N_c \gg 1$, the baryons become parametrically heavy objects. Like molecules or heavy nuclei, those rotate slowly and produce a tower of rotational states, of which the nucleon (with the spin and isospin $S = I = 1/2$) and the Δ ($S = I = 3/2$) are supposed to be the first two rotational excitations.

The last picture (d) shows a combination of the previous two; it is the so called *little bag* surrounded by a hedgehog-shaped pion cloud. Gerry Brown and collaborators, who did the surgery combining two beasts together, took care to ensure that at the boundary the pressure and other important quantities like "chiral current" are continuous. A smile of the hedgehog should remind us about the so called "Cheshire Cat Principle", according to which the location of the "scar" separating two descriptions should be invisible and irrelevant.

Even a brief look at these pictures shows that there is no way all of those models can be at least partially true at the same time. Even trying to ignore differences in description and language used, one cannot fail to see the completely different physics involved. For example, according to the MIT bag model, all hadronic properties directly follow from confinement physics, with masses (and other dimensional quantities) simply related to the bag constant B. It ignores chiral symmetry and related phenomena. The non-relativistic quark model acknowledges their existence, but limits their entire effect to one simple quantity — a "constituent quark" mass — making the main part of the hadronic mass. The Skyrmion picture has no quarks at all. One may say that this model views the $\bar{q}q$ attraction and the resulting binding as so strong that quarks and antiquarks can only travel together, in a pion form. If so, there is simply no place left for the confinement effect, and indeed the model has nothing colored in it and no need for confinement.

We will discuss all these models a bit more below, emphasizing their successes and problems. Let me point out here one generic problem common to all of them; they do not follow the general wisdom that emerged from solutions of multiple quantum mechanics problems; try to understand *the ground state* first. If it is understood then properties of the excitations will follow naturally. The lowest hadronic modes we care most about like pions or nucleons are the low-lying collective excitations of the QCD vacuum, like phonons in solids or liquid He or heavy nuclei. One of the main points of this book is that an analogous process takes place in QCD; recent progress in understanding the vacuum state immediately explained many puzzles in hadronic spectroscopy.

At this point it is fair to recall one single early-day model which had actually followed this strategy closely. It is the Nambu–Jona–Lasinio (NJL) model[10] [37] we will discuss in Section 2.4.1. Although suggested in 1961, long before QCD and even

---

[10]There was also a simultaneous but nearly forgotten independent paper by Vaks and Larkin [38] with basically the same idea.

before quarks, it outlived many popular models suggested much later and is still widely used in the literature today.

The NJL model was based on a hypothetical attractive four-fermion interaction (inspired by the BCS theory of superconductivity). It was shown that if the interaction is strong enough, it can rearrange the vacuum into a chirally asymmetric (kind of "superconducting") phase, with light pions and other mesons, in analogy with the Cooper pairs with reasonably large effective masses. We will discuss in detail in this book that we now have quite convincing evidence that such a kind of 4- or even 6-fermion interaction actually exists in QCD, known as the 't Hooft interaction, and its exact form follows directly from the semiclassical theory based on instantons.

Another general remark I would like to make at the outset is related to often asked questions: what are the limits of Particle Tables? Although atomic physics is limited to states of known elements, as chemistry teaches us, the number of all possible molecules is truly infinite. In fact the same should be true of QCD since, as suggested in my paper [789], at least a matter made entirely of heavy quarks is driven by the Coulomb-like perturbative interaction similar to that for electrons and positrons. Their analogs — "excitons" — well known in solid state physics, are indeed known to produce bound states with three and more particles, up to a macroscopically populated condensed "exiton liquid". Thus, it is quite plausible that for the matter made entirely of b quarks and antiquarks the same phenomena should exist as well.

One can then ask whether large clusters of "quark matter" may still be stable according to strong interactions if some fraction of b quarks are substituted by charmed or even strange quarks. Of course, the heavy quarks can still decay due to weak interaction, but in the terminology of the Review of Particle Properties they would all still be called "stable particles".

Finally, logically speaking it can be that some macroscopically large piece of matter made of strange and light quarks is stable even for weak decays, because in principle strong binding can compensate for strange quark mass (just as nuclear binding makes neutrons stable inside the nuclei). We will return to this question in Chapter 11.

2.2.2.   *MIT bag*

A very elegant model, suggested by Chodos, Jaffe, Johnson, Thorn and Weisskkopf [113], treats all hadrons like small drops of another — *perturbative* — phase of QCD. All non-perturbative physics is included in one universal quantity, the bag constant B which describes the difference in energy density between the perturbative and physical vacua. The Hamiltonian of the model is a sum of three terms

$$H = H_{\text{kinetic}} + H_{\text{spin–spin}} + BV, \tag{2.10}$$

where the first term describes the kinetic energy of the quarks confined in the bag, the second describes the spin–spin interaction, and the last "bag" term contains the bag constant B times the volume of the bag interior.

If quarks are (nearly) massless, it is clear for dimensional reasons that the only answer for the first term can be $E_{\text{kinetic}} \sim 1/R$ where R is the bag radius; it tries to expand the bag. Solving the Dirac equation in a cavity with boundary conditions $in_\mu\gamma_\mu q = q$ one gets $E_{\text{quark}} = 2.04/R$. More elaborate versions of the bag model included the kinetic energy not only of the confined valence quarks, but also of the virtual quark and gluon colored fields confined in a bag, the so called Casimir energy; this too must be $E_{\text{Casimir}} \sim 1/R$ for the same dimensional reasons. The bag term is $B(4\pi/3)R^3$, it tries to contract the bag; as a result the equilibrium is reached. This bag constant — the only dimensional quantity of the model — determines the masses and sizes of all the hadrons.

The fit to masses of the "usual" hadrons[11] made in the original MIT bag model was rather successful, producing the following value of the bag constant:

$$B_{\text{MIT}} = 56\,\text{MeV}/\text{fm}^3. \tag{2.11}$$

Theoretically speaking, the MIT bag model has a number of problems which subsequent development has tried to cure. One serious flaw is that chiral symmetry is badly broken, which is seen from the fact that the axial current is *not* conserved at the bag boundary. Indeed, when a quark is reflected from the bag wall, its momentum is flipped but the spin is not; thus a left-handed quark can become right-handed. We will discuss what people tried to do about it in the next section.

The other problem is that the universality of B, however beautiful it may be as a concept, is not supported by observations. In particular, the MIT bag model predicts the same scale of masses for glueballs as for ordinary mesons. Solutions of the Yang–Mills equations in a cavity brings in about the same kinetic energies as the Dirac equation, modulo numerical coefficients of the order (and really not far from) unity.

In my paper [216] it has been pointed out that the MIT value (2.11) is at least an order of magnitude smaller than the value suggested by the QCD sum rules. A different argument related with quarkonia, but to the same effect, has been simultaneously proposed by P. Hazenfratz and Kuti [114]. As we will discuss in Chapter 8, QCD phase transitions at high T further constrain the real bag constant, defined as the difference between the vacuum and the QGP ground energy, and have shown it to be indeed around a much larger value,

$$B = 500\text{–}1000\,\text{MeV}/\text{fm}^3 \gg B_{\text{MIT}}, \tag{2.12}$$

which is consistent with the QCD sum rules.

Furthermore, the universality of the MIT bag constant is in contradiction with the large N_c arguments, as pointed out by Bardeen and Zakharov [80]. Indeed, the standard N_c counting of the non-perturbative vacuum energy is $\epsilon = O(N_c^2)$, while the bag constant B (like mesonic masses) is assumed to be independent of N_c.

---

[11]The "unusual" ones like pions, η' or scalars are not of course reproduced.

The smallness of the MIT bag constant thus reflects an important generic point: all known hadrons are not at all the small drops of the new phase, but rather relatively small perturbations of the QCD vacuum.

Let me mention one more aspect of the bag model, much discussed in the literature. If hadrons are small bubbles of a new phase of matter, one may expect them to behave as all other bubbles do and coalesce if they meet each other. If the non-zero surface tension is introduced it becomes even more obvious, but it is true even with only the volume bag term. The nucleons in the nuclei do not coalesce like that. If one puts a pair of nucleons on top of each other a rather strong repulsive core is observed, as directly seen by the scattering phases of NN scattering. Spin forces in the MIT bag model have fixed this problem for the NN case, but not in general. The most symmetric closed shell 6-quark state — the famous dibaryon H — still happens to be lower than twice the mass of the Λ [99]. This state has been never seen, in spite of multiple experimental efforts.

$\boxed{\text{E}}$

Exercise 2.4.   Show that in the bag model without spin forces the energy of the 6-quark bag is less than that for two 3-quark bags.

2.2.3. *Skyrmions*

The idea to use a classical pion field as the basis for the description of baryons may seem strange at first sight. Its basic justification follows from the large N_c arguments for baryons put forward by Witten. The main idea is that the baryon mass is large $M = O(N_c)$ while its size remains constant $R = O(N_c^0)$, and as a consequence we have a source of pion field with increasing intensity. For a relatively detailed pedagogical description of Skyrmions and chiral bags I recommend the book by Nowak, Rho and Zahed [117]; here I will only briefly mention a few main points.

Can a soliton made of a classical pion field be a fermion? Has it exactly one unit of the baryon charge? The answer to both questions is "yes" although to see that one has to go to SU(3) flavor and add the Wess–Zumino–Witten term (2.51) to the Skyrme action. We will not go into this fascinating but technically complex story, but refer the reader to the original literature [118,120] or the book mentioned above [117].

A specific model for such an object was in fact proposed long before all these considerations by Skyrme [116], using the following Lagrangian:

$$L = \left(\frac{1}{4}f_\pi^2\right) \text{Tr}(\partial_\mu U^+ \partial_\mu U) + \left(\frac{1}{4}\epsilon^2\right) \text{Tr}([U^+\partial_\mu U, U^+\partial_\nu U]^2). \qquad (2.13)$$

The first term is the usual Weinberg term of the chiral effective Lagrangian, and the second — the so called Skyrme term — was suggested on the basis of simplicity, without a derivation. As will be clear from what follows, any positively-defined term with four derivatives (and of course correct symmetries) would be as good as this one. The trace is over the flavor indices which the exponentiated pion operator $U = \exp[i\pi^a(x)\tau^a/f_\pi]$ has.

Let us now see if there is a stable soliton or not. If the exponentiated pion field U changes by $O(1)$ over the soliton radius, the volume integral over the first term is $\sim R^{3-2} \sim R$ while it is $\sim R^{3-4} \sim 1/R$ for the second term with four derivatives. Since these two terms have thus opposite trends, they indeed can balance each other and produce a stable soliton.

The field itself is found from the "hedgehog ansatz", $m = 1, 2, 3$,

$$\pi^a = \frac{x^a}{r}F(r), \quad r = \sqrt{x^2}, \tag{2.14}$$

with the explicit form of the radial function found from the minimization of the energy. The Skyrme model has only one free parameter — the dimensionless coefficient of the Skyrme term ϵ. It can for example be fixed from the condition that the soliton has correct coupling to the axial current (and pions) at large distances, namely $g_A = 1.25$. Then one can calculate the mass which turns out to be 1.4 GeV, which is rather large.

The baryon number is identified with the topological charge

$$B = \frac{1}{24\pi^2} \int d^3x \, \epsilon_{ijk} \, \mathrm{Tr}(L_i L_j L_k), \tag{2.15}$$

where $L_i = U^+\partial_i U$. It is normalized in such a way that the minimal soliton gives $B = 1$. People have studied of course the minima with $B = 2$; those however happen to be not the spherically symmetric hedgehogs but strange toroidal objects with a hole in the middle instead. I have never seen any paper trying to explain what that may mean.

Atoms and spherical nuclei, in their ground states, are examples of symmetric objects which cannot be "rotated". In contrast to that, molecules and deformed nuclei can be rotated, except around the symmetry axes. The Skyrmion has a particular orientation in space and therefore belongs to the latter category. Quantization of its rotation is done in the adiabatic approximation, which means that the back reaction of centrifugal forces on its shape is ignored. The nucleon, Δ etc. are the beginning of a rotational band of states in which $I = J = 1/2, 3/2, \ldots$. The spectrum of excitations contains the moment of inertia I of the soliton, which can be evaluated. The resulting $m_\Delta - m_N = 3/2I \approx 290$ MeV is very close to experiment, although it is not really very small compared to the nucleon mass. Unfortunately, no $I = J = 5/2$ state of the kind has been seen; the argument given here is that centrifugal forces become too strong and rip the Skyrmion apart by this excitation.

The Skyrmion is an important testing ground for many relations, which actually need only the large N_c limit. For a review of the modern theory of large-N_c baryons see Manohar [79]. However the Skyrmion is not a especially good model from a practical point of view. It is hardly surprising, as neither has its Lagrangian been actually derived, nor is the used expansion parameter $1/N_c$ small enough to guarantee high accuracy. Another problem is that the hedgehog ansatz is quite natural for the SU(2) flavor group, but not for SU(3).

However some works were really impressive; e.g. Mattis and Karliner [123] considered the π–*baryon* scattering in this model and achieved a remarkably good description of most lowest baryonic resonances seen experimentally. It seems to confirm that those states may be more meson + baryon states rather than true excitations of qqq configurations.

2.2.4. Chiral bags

Under this name I actually mean several models united by the idea that it is very important to keep chiral symmetry and its consequences correctly represented in any models of baryons. The chiral bags can be divided into two classes, depending on whether they do or not care to enforce confinement.

An example of the former class is the so called "*little bag model*" suggested by Brown and collaborators, for a review of the model see Brown and Zahed [124] and somewhat later a broader view on the issues involved can be found in the book [117]. One of the reasons the bag is small is that a larger bag constant is used, but it is not the main one.

This model is a combination of the two models discussed above; it literally attempts a very radical change of the degrees of freedom; (i) inside the bag there is a perturbative phase, based on massless quarks and gluons, with their perturbative interactions, (ii) outside the bag there is a "pion cloud" described by a Skyrmion.

Usually the boundary is selected in such a way that the baryon number is divided between two components 50–50; but it is neither necessary nor even important, according to the so called *Cheshire Cat Principle* which states that change of language does not mean any change of physics.

An important example of continuity is that the axial current is among other physical observables which are continuous across the bag boundary. We saw above that a quark bouncing from the boundary of the MIT bag violated chiral symmetry; now it no longer does so since the result is a radiation of a pion outside, to which the chirality is passed.

However the following problem of the model still remains unsolved; the "scar" between the two descriptions does not go away at the quantum level; for example Casimir energies inside (due to gauge and quark fields) and outside (due to pions) are difficult to match because of the different and large color factor involved in the former case.

In the chiral bag models of the second class the "bag" no longer represents confinement, and it is also made of the pion field. Probably the most developed version of such a model is known as the *Chiral Quark–Soliton Model*. It has been proposed by Diakonov *et al.* [131]. It includes quarks interacting with the pion field via the second term in the Lagrangian

$$L = \bar{q}[i\partial_\mu\gamma_\mu - MU_5(\pi)]q, \qquad (2.16)$$

containing the constituent quark mass (which is a flavor matrix) M times the (exponentiated with γ_5) pion field $U_5 = \exp(i\pi^a\tau^a\gamma_5)$. The pion field is treated as a classical background in a Skyrmion-like hedgehog configuration, which provides the "bag" and binds the quarks.

The baryon number in this model is still carried by three valence quarks, however, the bound state of the quarks in the pion field is not deep enough to plunge them into the negative Dirac continuum with energies $E < -M$. (It is in this case that the valence quarks disappear and the topologically nontrivial pion field has to be responsible for nucleon quantum numbers.)

In this model the quantum corrections — the "sea" quarks produced by the pion field — have also been calculated, by evaluating the Dirac operator determinant in the pion background. This allows a comparison to the structure functions not only for valence quarks (as many other models did) but for the "sea" as well. I have found it impressive that the model can successfully describe even flavor asymmetry of the sea, a different number and x-distribution of the $\bar{u}$ and the $\bar{d}$ in the nucleon.

2.2.5. *Evolving views on the nature of the spin forces*

There are in general three different mechanisms suggested in the literature to explain the spin–spin forces in hadrons;

(i) perturbative spin–spin interaction due to "gluo-magnetic moments" [94];
(ii) the instanton-induced quasi-local interaction originating from 't Hooft interaction [96–98, 103]; and
(iii) pion-mediated forces [111].

Broadly speaking, these three interactions care about

(i) color-spin;
(ii) chirality and flavor; and
(iii) spin-flavor combinations of the quantum numbers;

and can thus be in principle distinguished.

The first was introduced by DeRujula, Georgi and Glashow [94] based on one-gluon exchange (still a new topic in 1975, boosted by the discovery of charmonium a year before!) and the first relativistic corrections,

$$\frac{V}{(\text{color factor})\alpha_s(r)} = \frac{1}{r} - \frac{1}{2m_i m_j}\left(\frac{\vec{p}_i\vec{p}_j}{r} + \frac{(\vec{p}_i\vec{r})(\vec{p}_j\vec{r})}{r^3}\right) - \frac{1}{2}\pi\delta^3(\vec{r})\left(\frac{1}{m_i^2} + \frac{1}{m_j^2}\right)$$

$$- \frac{1}{m_i m_j}\left(\frac{8\pi}{3}(\vec{s}_i\vec{s}_j)\delta^3(\vec{r}) + \frac{1}{r^3}[3(\vec{s}_i\hat{r})(\vec{s}_j\hat{r}) - (\vec{s}_i\vec{s}_j)]\right)$$

$$- \frac{1}{2r^3}\left(\frac{1}{m_i^2}\vec{r}\times\vec{p}_i\vec{s}_i - \frac{1}{m_j^2}\vec{r}\times\vec{p}_j\vec{s}_j\right.$$

$$\left. + \frac{1}{m_i m_j}[2\vec{r}\times\vec{p}_i\vec{s}_j - 2\vec{r}\times\vec{p}_j\vec{s}_i]\right). \tag{2.17}$$

The first line includes the Coulomb and Darwin terms, then come the so called Zitterbewegung, spin–spin and spin–orbit terms.

The color factors include the product of the color matrices $t_i^a t_j^a$ of quarks ij involved, averaged over the color wave function; they are $-2/3$ for baryons and $-4/3$ for mesons. It follows from this color factor that all spin–spin forces in mesons are twice as strong as in baryons, since in the latter the "color spins" make an angle $120°$ to each other, not $180°$. This factor 2 indeed roughly corresponds to reality.

Another feature of the non-relativistic model is that spin–spin forces are due to magnetic moments and thus are *inversely* proportional to quark masses. Let us see if this is true. Let us define

$$\Delta M = \frac{\vec{S}_i \vec{S}_j}{\gamma_i \gamma_j} M_0, \quad M_0 = \frac{-8\pi}{3} \frac{\alpha_s}{m^2} |\psi(0)|^2.$$

The simplest case is light quark mesons, $\langle(\vec{S}_i\vec{S}_j)\rangle = -3/4$ if $S = 0$ and $1/4$ if $S = 1$. The average mass of the non-strange meson $\bar{M} = (3m_\rho + m_\pi)/4 = 617\,\text{MeV}$ is a benchmark. The splitting is then $M_0 = (3/4)(m_\rho - m_\pi) = 480\,\text{MeV}$. Simple scaling by mass taking crudely $M = 300$, $M_s = 500\,\text{MeV}$ then give $378\,\text{MeV}$ for K^*-K mass difference (experimental value $394\,\text{MeV}$), $M_c = 1500$, $M_b = 5000\,\text{MeV}$ give $126\,\text{MeV}$ for $D^* - D$ (experimental value $143\,\text{MeV}$), and 38 (experimental value 52) MeV for $B^* - B$. It seems to work!

Now let us discuss the lowest baryons. For the $N - \Delta$ mass difference one needs to know that

$$\langle \vec{S}_1\vec{S}_2 + \vec{S}_1\vec{S}_3 + \vec{S}_2\vec{S}_3 \rangle = S(S+1)/2 - 9/8,$$

from which, with the color factor $2/3$, one gets $m_\Delta(S = 3/2) - m_N(S = 1/2) = M_0$. Of course, the experimental mass difference is not $480\,\text{MeV}$, as for mesons, but only about $300\,\text{MeV}$. This is however not surprising since baryons have larger size compared to mesons, and thus the wave function at the origin is smaller. In what follows we take a new value of this matrix element as an input and evaluate splittings for other baryons.

Now let us look at hyperons (baryons with strangeness). The projection $\vec{s}_u\vec{s}_d = 1/4$ in Σ is positive, but it is $-3/4$ in Λ. From the total spin $1/2$ of both particles it is easy to do other spin projections and calculate the total,

$$\boxed{E}$$

$$\Delta M_\Sigma = (2/3)(1/4 - 1/\gamma), \quad \Delta M_\Sigma^* = (2/3)(1/4 + 1/2\gamma). \quad (2.18)$$

Their difference leads to the following relation

$$M_\Sigma^* - M_\Sigma = M_0/\gamma = (m_\Delta - m_N)/\gamma,$$

the l.h.s. is $191\,\text{MeV}$, and we get the strangeness suppression factor $\gamma = 1.5$. The value is close to the ratio of the constituent quark masses M_s/M_u.

The Λ–Σ mass difference is proportional to $1/M_u - 1/M_s$ and indeed it has the right sign! Furthermore, one can derive

E

$$m_\Sigma - m_\Lambda = (2/3)(1 - 1/\gamma)(m_\Delta - m_N), \qquad (2.19)$$

which also works well.

The story continues with magnetic moments, which happen to be also well described with the same strangeness suppression factor. The spin-1/2 wave function for mixed symmetry octet should now be written. We will use superscript $+, -$ for spin projections, so a proton with spin up is

$$p^+ = \tfrac{1}{\sqrt{6}}(2u^+u^+d^- - u^+u^-d^+ - u^-u^+d^+),$$

and calculate the magnetic moment. Let $\mu \equiv e/2m$ be the *quark* magneton, times the quark electric charges $2/3$ for u, $-1/3$ for d, s, which we will write explicitly below. We just take the average over that wave function and get

$$\mu_p = \tfrac{1}{6}\mu\left[4\left(\tfrac{2}{3} + \tfrac{2}{3} + \tfrac{1}{3}\right) + \left(\tfrac{2}{3} - \tfrac{2}{3} - \tfrac{1}{3}\right) + \left(-\tfrac{2}{3} + \tfrac{2}{3} - \tfrac{1}{3}\right)\right] = \mu,$$

while for neutron (interchanging all u and d)

$$\mu_n = -(2/3)\mu_p.$$

Here is the great triumph of the non-relativistic quark model; $\mu_p = -2.79(e/2m_n)$ while $\mu_n = 1.91$. Not only does their ratio work very well, but the magnitude itself suggests that the quark mass is indeed about $1/3$ of the nucleon mass.

Other baryons with known magnetic moment can also be compared

$$\mu_\Lambda = -0.61 \ \text{versus} \ -1/3\gamma,$$
$$\mu_{\Sigma+} = 2.33 \ \text{versus} \ (1/9)(4/3 + 1/\gamma),$$
$$\mu_{\Sigma-} = -1.41 \ \text{versus} \ (4/9)(-1 + 1/3\gamma),$$
$$\mu_{\Xi0} = -1.25 \ \text{versus} \ -(4/9)(2/3 - 1/\gamma),$$
$$\mu_{\Xi-} = -0.69 \ \text{versus} \ (4/9)(1 - 1/\gamma),$$
$$\mu_{\Omega-} = -2.0 \ \text{versus} \ -1/\gamma.$$

Once again, they all fit well, with only one strangeness suppression parameter $\gamma = M_s/M_u = 1.52$.

There is another way to measure quark magnetic moments, now in mesons, via radiative magnetic M1 transitions. The best examples are spin flips like $\omega \to \pi\gamma$. The isospin wave functions are

$$\omega = (\bar{u}u + \bar{d}d)/\sqrt{2}, \qquad \rho, \pi = (-\bar{u}u + \bar{d}d)/\sqrt{2},$$

and the spin ones are obvious, so the matrix elements read

$$\langle\omega|Q\sigma_-|\pi\rangle = (1/2)[-(2/3)+(-1/3)] = -1/2,$$
$$\langle\rho|Q\sigma_-|\pi\rangle = (1/2)[(2/3)+(-1/3)] = 1/6,$$

and the partial width should then be related by

$$\frac{\Gamma(\omega\to\pi\gamma)}{\Gamma(\rho\to\pi\gamma)} = 9 \text{ versus experimental } \frac{890\,\text{keV}}{70\,\text{keV}}.$$

There are many other reactions like that (vectors can be also ϕ, K^*, pseudo-scalars $\eta\eta'$, and similar M1 transitions are also known in baryons like $\Delta\to N\gamma$) and they also are described very well with $\gamma = M_s/M_u \approx 1.5$.

The *spin–orbit part of the one-gluon forces* was, in contrast, a complete fiasco. Experimentally it is nearly absent; e.g. compare close pairs of states with different l such as $f_1(1285), f_2(1270)$ or $a_3(2050), a_4(2040)$. Isgur and Karl [95] basically ignored the spin–orbit part assuming that non-perturbative terms (confining? others?) cancel it somehow.[12]

After the simplest spin–spin interaction in a non-relativistic quark model seemed to be such a complete success, one may think that other ideas have no chance to succeed. But let us look closer.

Let us go to the second item, the *instanton-induced* interaction. Rosner and myself [98] argued that the dependence on the quark mass for instanton-induced forces is actually exactly the same as in the non-relativistic quark model, although for a completely different reason. The original 't Hooft Lagrangian is the 6-fermion operator of the $\bar{u}u\bar{d}d\bar{s}s$ structure, see Section 4.2.2. So when u and d quarks interact, there is an s-quark loop, and if say d and s quarks interact the loop is a u-quark one, see Fig. 2.5. Although such a loop is zero perturbatively for massless quarks, because chirality must be flipped, it is possible non-perturbatively. One may describe this loop in the mean field approximation by an effective mass M^* (proportional to the chirality-flip vacuum amplitude or the quark condensate), which appears as a factor in the fermionic determinant and the effective interaction. For the strange quark

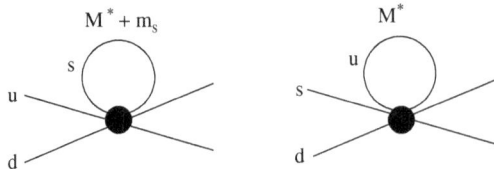

Fig. 2.5. Two effective instanton-induced operators, for the interaction of ud and ds quark pairs.

[12]This situation is in striking contrast to nuclear physics, in which vector and scalar exchanges nearly cancel each other in radial potential but add up for the spin–orbit, making it very strong.

Table 2.2. Expressions and masses for octet and decuplet baryons in the Rosner–Shuryak model.

Particle	Expression	Value	Exp.
N	$3U - \frac{3}{2}E_{ud}$	938.0	938.0
Λ	$S + 2U - E_{ud} - E_{us}/2$	1116	1116
Σ	$S + 2U - \frac{3}{2}E_{us}$	1184	1195
Ξ	$2S + U - \frac{3}{2}E_{us}$	1329	1318
Δ	$3U$	1239	1232
Σ^*	$S + 2U$	1383	1385
Ξ^*	$2S + U$	1528	1533
Ω	$3S$	1672	1672

loop it is enhanced $M_s^* = M_u^* + m_s$. So the ratio of the two forces, for ud and us quarks, is actually the same $\gamma = (M_u^* + m_s)/M^*$ as above.

Of course, there remain other significant differences between the instanton-induced spin forces and the perturbative ones. There are terms $(t_1^a t_2^a)(\vec{\sigma}_1 \vec{\sigma}_2)$ in the 't Hooft Lagrangian (4.14), but there are also other terms. In fact *there are no instanton-induced forces* in the vector channel at all. The splitting with the pion is entirely due to interaction in the pion channel, not the ratio 3-to-1 as the usual spin–spin operator suggests. And indeed, as we will see in Chapter 6, there are no strong deviations from the pQCD behavior in vector correlation functions, while there are such for the pion.

Another difference is that the instanton-induced forces care about flavor; there are no such forces between say uu or ss quarks at all. Can a model based on them describe all octet and decuplet masses? As shown in Ref. [98], a very simple model *a la* non-relativistic quark model, in which light and strange quarks have some effective energies E_q, E_s[13] *plus* an additional negative energy for spin-zero diquarks made of different flavors (with and without strange quark) $\delta E_{ud}, \delta E_{us}$. All masses of the baryon octet and decuplet can be well described by this primitive model very well, see Table 2.2, with the following values of the parameters (U, S stand for constituent quark masses and the other two for binding energies):

$$U = 412.9\,\text{MeV}, \quad S = 557.5\,\text{MeV}, \quad \delta E_{ud} = 200.5\,\text{MeV}, \quad \delta E_{us} = 132.7\,\text{MeV}.$$

$$(2.20)$$

Let me add the following remarks.

(i) E_q is also consistent with the magnitude of the light quark energy in the light–heavy hadrons ($D, B, \ldots$).

---

[13] We speak about effective energies in order not to go into a discussion of which part of it is the "constituent quark mass", and which is its kinetic or potential energy.

(ii) δE is also close to the binary interaction found from the splitting in the light–heavy baryons.

Also the strange sector leads to no surprises.

(iii) The ratio $S/U \approx 1.35$ is not far from the ratio of the binding energies $E_{\rm su}/E_{\rm ud} \approx 1.5 \approx \gamma$, and it is also close to the "strangeness suppression" factor in the magnetic moments.

Let us summarize our findings; they suggest the idea that the nucleon is light because it contains spin-zero qq pairs, which are subject to *attractive* instanton-induced interaction. We will show below in the book that one can get much more detailed confirmation on this fact from lattice and instanton liquid calculations.

Just for comparison, let us use this very simple model for the H–dibaryon problem. If it is made of ud, us, ds scalar diquarks — the most attractive channels — with negligible kinetic energy (as above), the H mass is simply $2S + 4U - 2E_{us} - E_{ud} \approx 2300\,{\rm MeV}$, which is about 65 MeV *below* the $2M_\Lambda$ of this additive model. This calculation of course assumes that the size of H is the same as for octet and decuplet baryons. Note that Pauli principle forced us to break (ud) diquark and couple these light quarks with strange ones, thus near zero binding. The result would be very different if s quark would be lighter.

Some recent papers [104, 105] have argued that the H can be very compact, with much stronger instanton-induced attraction and therefore a strong binding. However it is clear that a single instanton cannot bind 3 (or even 2 such as us, ud) diquarks at the same time since there is only one incoming line per flavor in the 't Hooft Lagrangian. This correlates with an earlier observation by Takeuchi and Oka [103] who pointed out that 3-body forces induced by an instanton are in fact repulsive.

Let us now comment about the third contender, the pion exchange forces of Glozman and Riska [111]. Without going into a discussion as to whether the pion is compact enough to fit inside the nucleon, let us look at the results. Unlike one-gluon exchange, they now care not about the color but flavor matrices, since they are proportional to $(\lambda_i^a \lambda_j^a)(\vec{S}_i \vec{S}_j)$. Obviously they lead to a different structure of the excitations. In particular, the Ropper resonance $N^*(1440)$, the $\Delta(1600)$ and the $\Lambda(1405)$ which were a big problem for the non-relativistic model are now in the right place. The relation to pion and chiral physics also explains why the lattice calculations had problems reproducing the positions of these states as well, if too large quark masses were used.

One should keep in mind that even for a light pion these forces are short-range because pions interact with a derivatives at the vertex. The one-massless-pion exchange potential is not $1/r$ but $1/r^3$. On the other hand, a pion is basically a sequence of $\bar{q}q$ going through the instanton, see Fig. 2.7, and one may wonder to what extent pion exchange and instanton forces are the same thing.

Exercise 2.5.   Derive the relations (2.18) and (2.19).

2.3. Models of the QCD vacuum: an overview

2.3.1. *Condensates and scales*

One of the main objectives of this book is to explain our current understanding of the structure of the ground state of QCD. It is a very dense state of matter, composed of gauge fields and quarks that interact in a complicated way. Unlike in free theory (or QED, which is close enough to it) one cannot decompose it into separate harmonic oscillators, with simple Gaussian ground state wave functions and independent zero point oscillations. All oscillators we have to deal with are nonlinear and all coupled together.

More precisely, for sufficiently large Euclidean momenta $p_\mu^2 \gg \Lambda_{\text{QCD}}^2$ (very virtual harmonics of the field) the effective coupling $g(p)$ is relatively small and the perturbative picture of the vacuum, with the Gaussian weight $W \sim \exp(-S)$ of harmonics with approximately quadratic action $S \sim \int d^4p\, p^2 A(p)^2$ is reasonable. The correlator of such free fields — the usual gluon propagator $\langle A(x)A(0) \rangle \sim \int d^4x\, e^{ipx}(1/p^2) \sim 1/x^2$ — can be used to generate perturbative diagrams. But when the field virtuality goes down to momenta of the order of 1 GeV or less the coupling gets strong and there is no use for the Gaussian representation any more.

One way to proceed is to evaluate the non-Gaussian functional integral numerically on the lattice; we return to it in Chapter 5. Another is to develop some phenomenology of the non-perturbative effects. It can be rather systematically developed using the *operator product expansion* (OPE), which relates the short-distance behavior of various current correlation functions to the so called *condensates*, the vacuum expectation values (VEVs) of a set of lowest-dimension quark and gluon operators. This idea was suggested by Shifman, Vainshtain and Zakharov (SVZ) [352] and has been used in a huge amount of subsequent papers known as "QCD sum rules". Two most important condensates,

$$\langle \bar{q}q \rangle = -(240\,\text{MeV})^3, \qquad \langle g^2 G^2 \rangle = (850\,\text{MeV})^4, \qquad (2.21)$$

we have already met earlier in this book.[14]

The significance of the quark condensate is due to the fact that it is an order parameter for the spontaneously broken chiral symmetry; we will return to it in the next section in depth.

The gluon condensate is important because via the QCD trace anomaly,

$$T_{\mu\mu} = \sum_f m_f \langle \bar{q}_f q_f \rangle - \frac{b}{32\pi^2} \langle g^2 G^2 \rangle, \qquad (2.22)$$

---

[14]By definition, "condensates" are defined for a particular normalization scale μ, as they include all field harmonics with virtual momenta $p_\mu^2 < \mu^2$. The numbers above correspond to $\mu = 1$ GeV.

it is related to the energy density $\epsilon_0 \simeq -500\,\mathrm{MeV/fm}^3$ of the QCD vacuum. Here, $T_{\mu\nu}$ is the energy momentum tensor and $b = 11N_c/3 - 2N_f/3$ is the first coefficient of the beta function.

More generally, the gluon condensate determines the normalization of the non-perturbative fields in the vacuum, and so any model of it should be appropriately normalized to its value.

However, it is not the only important scale; there are various other "non-perturbative scales" in QCD. In the early days of QCD, in the 1970s, that was not clear. Somewhat surprisingly some naive simplistic ideas are still alive today. I would argue below against the picture of non-perturbative objects as some structure-less fields with typical momenta of the order of $p \sim \Lambda_{\mathrm{QCD}} \sim (1\,\mathrm{fm})^{-1}$. In the mid 1970s it fitted well to a view of hadrons as structure-less "bags" filled with near-massless perturbative quarks, with mild non-perturbative effects appearing at its boundaries and confining them at the scale of $1\,\mathrm{fm}$.

One logical consequence of this picture would be applicability of the derivative expansion of the non-perturbative fields or Operator Product Expansion (OPE), the basis of the QCD sum rules. However, after the first successful applications of the method [352] rather serious problems [396] have surfaced. All spin-zero channels (as we will see, those are the ones directly coupled to instantons) related with quark or gluon-based operators alike, indicate unexpectedly large non-perturbative effects and deviate from the OPE predictions at very small distances.

It provided a very important lesson: *the non-perturbative fields form structures with sizes significantly smaller than 1 fm* and local field strength much larger than Λ^2. Instantons are one of them; in order to describe many of these phenomena in a consistent way one needs instantons of small size, as I suggested in Ref. [226], $\rho \sim 1/3\,\mathrm{fm}$. We have now a direct confirmation of it from the lattice, but not yet a real understanding of why there are no large-size instantons.

Furthermore, the instanton is not the only such small-scale gluonic object. We also learned from the lattice-based works that QCD flux tubes (or confining strings) also have small radii, only about $r_{\mathrm{string}} \approx 1/5\,\mathrm{fm}$. So, all hadrons (and clearly the QCD vacuum itself) have a *substructure*, with "constituent quarks" generated by instantons connected by such flux tubes.

Clearly this substructure should play an important role in hadronic physics. We would like to know why the usual quark model has been so successful in spectroscopy, and why so little of exotic states has been seen. Also, high energy hadronic collisions must tell us a lot about substructure, since the related objects should be readily produced.

2.3.2.   *Condensate factorization and stochastic vacuum model*

Most of the early attempts to understand the QCD ground state were based on the idea that the vacuum is dominated by very soft gauge fields which are nearly constant fields [137], or regions of constants fields patched together, as in the so called "spaghetti vacuum" introduced by the Copenhagen group [138]. One reason these attempts were unsuccessful, however, was because constant fields were found to be unstable against quantum perturbations. (This instability we discussed in Section 1.4.1 in the discussion of quantum gluons moving in a background magnetic field.)

Without specifying the fields explicitly, Shifman *et al.* [352] have proposed the so called *factorization hypothesis* or *vacuum dominance* used in order to evaluate the VEVs of more complicated operators. It assumes that if one can split an operator into two parts, so that the vacuum expectation value of each exists, it should be the dominant one. For example, it was assumed that

$$\langle (G^a_{\mu\nu})^4 \rangle \approx \langle (G^a_{\mu\nu})^2 \rangle^2. \tag{2.23}$$

Using such a prescription systematically, for all gluonic operators, Shifman [139] tried to re-sum the OPE series and see whether the results do or do not agree with phenomenology. An example he followed in detail is the VEV of a small-size Wilson loop. We expect from the Wilson criterion of confinement, that it behaves as the exponent of its area. A good news item was that Shiman's result had indeed there an exponential of the loop area. However instead of a decreasing function he got a Bessel function with unphysical oscillations. One can conclude that this hypothesis cannot be a good approximation.

More generally, while trying to model the vacuum fields one has to make a choice on the following general dilemma: (i) are they more or less a kind of a quantum noise, or (ii) are these fields formed out of some specific objects. We will discuss two models, one of each category, and it will become clear what I mean by this.

The first of them is known as *the stochastic vacuum model*, and it was proposed in the late 1980s by Dosch and Simonov [140]. Their main idea was to model the field strength distribution as Gaussian, but in terms of the field strength, not A_μ:

$$\int_{p_\nu^2 < \mu^2} DA\, e^{-S[A]} \rightarrow \int DG\, e^{-G_{\mu\nu} K_{\mu\nu\kappa\lambda} G_{\kappa\lambda}}, \tag{2.24}$$

where the inverse to $K_{\mu\nu\kappa\lambda}$ defines the "non-perturbative propagator", the main ingredient of the model. It is parameterized as follows:

$$K^{-1}_{\mu\nu\delta\lambda} = \langle g^2 G^2_{\mu\nu} \rangle \Big(\frac{\kappa}{12} (\delta_{\mu\delta}\delta_{\nu\lambda} - \delta_{\mu\lambda}\delta_{\nu\delta}) D(x^2)$$

$$+ \frac{1-\kappa}{2} [\partial_\mu(x_\delta\delta_{\nu\lambda}) + \partial_\nu(x_\lambda\delta_{\mu\delta})] D_1(x^2) \Big), \tag{2.25}$$

where D, D_1 are two functions and the κ in the numerator is a parameter of the model. In the Abelian theory one gets the field correlator with only the second structure, so one may say in this case $\kappa = 0$.

The first non-Abelian structure can be separately normalized in a nontrivial way; it was later shown that, following the rules of the model, one can calculate the Wilson loop and derive the area law, with the coefficient — the string tension — being the integral

$$K = \langle g^2 G_{\mu\nu}^2 \rangle \frac{\kappa}{72} \int_0^\infty d\rho \int_{-\infty}^\infty d\tau D(\rho^2 + \tau^2), \qquad (2.26)$$

which constrains the function further. The function itself has been fit to lattice data on the correlator $\langle G_{\mu\nu}(x)UG_{\kappa\lambda}(0)U^+ \rangle$ (where the U are color transporters needed for gauge invariance) determined on the lattice. The model has been rather successfully applied to high energy scattering [419]. However, it only addresses color exchange processes, not production, because the model has only "non-perturbative propagators" without vertices.

The main assumption of the model is that higher order cumulants, absent in the Gaussian ensemble, are negligibly small. There is no reason for this to be true in general, and this condition is badly violated in the model which we will discuss in the next section.

2.3.3. An example of a highly inhomogeneous model: the instanton vacuum

The fields in the stochastic vacuum model are just quantum noise, they do not obey any equations, and there is no procedure to improve the model. It is obvious that there is no way to define "quantum corrections" to this model. Another alternative is that vacuum contains some semiclassical objects with strong gauge fields, which are not random but rather classical solutions of the nonlinear Yang–Mills equations. In this case smallness of quantum corrections is guaranteed and one can make loop improvements consistently.

A well known example of this kind is the "instanton" solution which we will discuss in detail in the next two chapters. We will also discuss phenomenological arguments which suggest that those are the strongest field fluctuations. The following Fig. 2.6 shows a sketch of what the vacuum of the model looks like. Although the field strength in the instantons is much stronger than in the stochastic vacuum model, it is localized in rather dilute instanton gas, so that the overall field strength normalization is still given by the same value of the gluon condensate.

We will study instantons in detail in the next two chapters; here we only provide a pictorial overview. The main distinction of this model from the ones just discussed in the preceding section is that the quark and especially the gauge fields are not at all homogeneous, but assembled into small-size *structures* with plenty of "empty" space around them. As one of the coordinates is the (Euclidean) time, in one of my early papers on the subject it was called a *twinkling* vacuum.

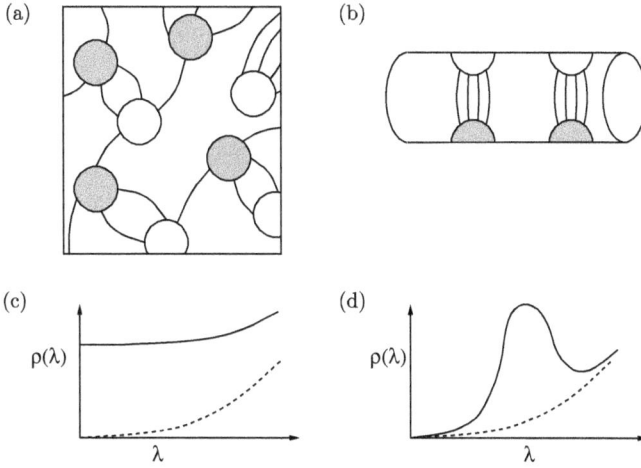

Fig. 2.6. (a) Schematic picture of the instanton liquid at zero temperature and (b) above the chiral phase transition. Instantons and anti-instantons are shown as open and shaded circles. The lines correspond to fermion exchanges. Figures (c) and (d) show the schematic form of the Dirac spectrum in the configurations (a) and (b), respectively.

Other objects are present in the QCD vacuum as well, and the debate on which exactly role each of them plays is still going on. Those are $3d$ monopoles or $2d$ vortices or QCD flux tubes. We return to those in Section 2.5.

Instantons are classical solutions to the Euclidean equations of motion. They are characterized by a topological quantum number and correspond to tunneling events between degenerate classical vacua in Minkowski space. As in quantum mechanics, tunneling lowers the ground state energy. Therefore, instantons provide a simple understanding of the negative non-perturbative vacuum energy density. In the presence of light fermions, instantons are associated with fermionic zero modes. Zero modes not only play a crucial role in understanding the axial anomaly, they are also intimately connected with spontaneous chiral symmetry breaking. When instantons interact through fermion exchanges, zero modes can become delocalized, forming a collective quark condensate.

A crude picture of quark motion in vacuum can then be formulated as follows [see Fig. 2.6(a)]. Instantons act as a potential well, in which light quarks can form bound states (the zero modes). If instantons form an interacting liquid, quarks can travel over large distances by hopping from one instanton to another, in a similar way as electrons do in a conductor. Just as the conductivity is determined by the density of states near the Fermi surface, the quark condensate is given by the density of eigenstates of the Dirac operator near zero virtuality. A schematic plot of the distribution of eigenvalues of the Dirac operator is shown in [Fig. 2.6(c)]. For comparison, the spectrum of the Dirac operator for non-interacting quarks is depicted by the dashed line. If the distribution of instantons in the QCD vacuum is sufficiently random, there is a non-zero density of eigenvalues near zero, and chiral symmetry is broken.

The quantum numbers of the zero modes produce very specific correlations between quarks. First, since there is exactly one zero mode per flavor, quarks with different flavor (say u and d) can travel together, but quarks with the same flavor cannot. Furthermore, since zero modes have a definite chirality (left-handed for instantons, right-handed for anti-instantons), quarks flip their chirality as they pass through an instanton. This is very important phenomenologically, because it distinguishes instanton effects from perturbative interactions, in which the chirality of a massless quark does not change. It also implies that quarks can only be exchanged between instantons of the opposite charge.

Based on this picture, we can also understand the formation of hadronic bound states. Bound states correspond to poles in hadronic correlation functions. As an example, let us consider the pion, which has the quantum numbers of the current $j_\pi = \bar{u}\gamma_5 d$. The correlation function $\Pi(x) = \langle j_\pi(x) j_\pi(0) \rangle$ is the amplitude for an up quark and a down anti-quark with opposite chiralities created by a source at point 0 to meet again at the point x. In a disordered instanton liquid, this amplitude is large, because the two quarks can propagate by the process shown in [Fig. 2.7(a)]. As a result, there is a light pion state. For the ρ meson, on the other hand, we need the amplitude for the propagation of two quarks with the same chirality. This means that the quarks have to be absorbed by different instantons (or propagate in non-zero mode states), see [Fig. 2.7(c)]. The amplitude is smaller, and the meson state is much less tightly bound.

Using this picture, we can also understand the formation of a bound nucleon. Part of the proton wave function is a scalar ud diquark coupled to another u quark. This means that the nucleon can propagate as shown in [Fig. 2.7(b)]. The vertex in

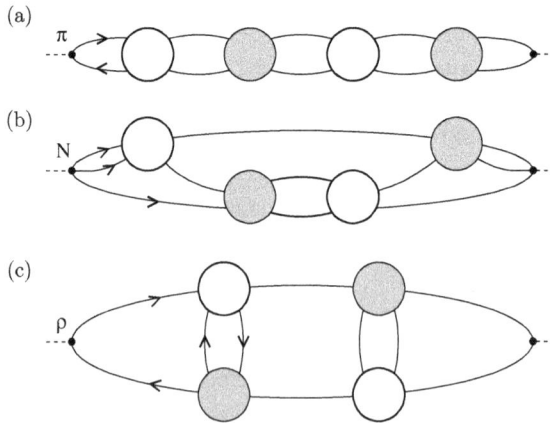

Fig. 2.7. Instanton contribution to hadronic correlation functions: (a) shows the pion, (b) the nucleon and (c) the ρ meson correlator. The solid lines correspond to zero mode contributions to the quark propagator.

the scalar diquark channel is identical to the one in the pion channel with one of the quark lines reversed.[15] The Δ resonance has the quantum numbers of a vector diquark coupled to a third quark. Just as in the case of the ρ meson, there is no first order instanton induced interaction, and we expect the Δ to be less bound than the nucleon.

2.4. Chiral symmetry breaking and effective low energy theory

2.4.1. *Spontaneous breaking of the chiral symmetry*

In Section 1.5.2 we saw that U(1) chiral symmetry is killed by the anomaly, and so it is not a symmetry at the quantum level. However $SU(N_f)$ chiral symmetry has no anomaly and thus remains a true symmetry in QCD with massless quarks.

In the QCD vacuum the latter chiral symmetry is nevertheless *broken spontaneously*. Although the action is symmetric under such transformations, *the ground state of the theory is not*. In this section we will discuss why this happens and what the consequences are. In Section 8.5.4 we will show that at high temperature this symmetry is restored.

Let me start with a *paradox*, pointing out what at first glance looks like terrible disagreement between the chiral symmetry and the real world. On one hand, chiral rotations include γ_5 and therefore they look different if reflected in a mirror interchanging left and right. In a more technical language, these rotations are P-odd (P is reflection of the space coordinates) and mix states of different parity. On the other hand they are symmetry transformations, thus they commute with the Hamiltonian and can only mix states of the same energy. Parity is respected by the strong interaction, so P is a good quantum number of a state. The conclusion is that each state of the QCD (in the massless limit) must have a degenerate partner(s) of the opposite parity, which can be mixed by the chiral transformations. On the other hand, we observe that excitations of QCD — individual hadrons — all have a well-defined P-parity, and *no such parity-partners are observed!*[16] For example, a pion is a pseudoscalar, a $P = -1$ state, and there is no scalars ($P = 1$) states of the same mass in the particle table.

The paradox is resolved by the spontaneous breakdown of the chiral symmetry. This phenomenon (SBCS for short), reminds us once more that the symmetry of the action does not automatically imply the symmetry of the ground state. There are plenty of examples of the kind, let me mention two most popular ones.

---

[15] For more than three flavors the color structure of the two vertices is different, so there is no massless diquark in SU(3) color.

[16] As we discussed in Section 2.1.5, higher excitations seem to converge to this regime.

Example 1.   In spite of exact translational invariance in quantum mechanics, the ground state of the systems made of most (but He) atoms are solids, or periodic crystals.

Example 2.   In spite of exact rotational invariance the ground state of many spin systems is a ferromagnetic state, in which all of them have the same direction.

It is difficult to move an atom from its equilibrium position in a crystal potential well. However if one moves the crystal as a whole, translational invariance guarantees that it costs no energy. The same is true for rotation of all spins as a whole in the second example. More formally, we see that if the ground state is not a singlet under the symmetry, then there must exist other vacua with the same energy, making together a complete representation of the symmetry group.

Let us now ask the following question: what happens if our asymmetry parameters are weakly changing in space, say oscillate with a small amplitude and large wavelength? If an infinitely long-wavelength transformation costs no energy, one may think that finite but long-wave ones would be rather cheap as well. And indeed, crystals have massless long-wave phonons, ferromagnetics have massless magnetic waves, etc.

The most general formulation of this phenomenon is the subject of the famous Goldstone theorem [29], which states that any spontaneously broken continuous symmetry must generate a massless mode, so that at $k \to 0$ its energy vanishes $\omega_k \to 0$. In the QCD vacuum this long-wave Goldstone mode is identified with the pseudoscalar mesons, pions for $N_f = 2$ plus kaons and eta if $N_f = 3$. In the massless quark limit all of them must therefore be massless.

The existence of such massless pseudoscalar excitations naturally explains the paradox mentioned above; all states of this theory are P-degenerate. Indeed, an infinitely long-wave pion can be added to any state, say a nucleon, changing its parity without any change in energy.

2.4.2.   *The Goldstone modes: oscillations of the quark condensate*

For all spontaneously broken symmetries one can define the so called *order parameter*. Say, for our Example 1 from the previous subsection, a set of $N \gg 1$ interacting atoms. It is clear that as a function of inter-atomic distances the energy has a minimum and some crystal state can be formed. The question however is about motion as a whole; one cannot select a particular crystal without fixing *all* atoms. It is the position of their center of mass which is the order parameter in this case.

One may argue that the stationary states should be characterized by a momentum $\vec{p}_{\text{tot}}$ and the minimal energy state is obviously at zero total momentum. This means that the probability to find atoms would be homogeneously distributed all over the available volume. This seems to contradict everyday experience. The resolution of this simple paradox is that for macroscopic bodies the energy differences of such states $\vec{p}_{tot}{}^2/2M_{tot}$ become so small, that in the thermodynamic limit $N \to \infty$ a tiny external potential V can localize the center of mass with any needed precision.

In Example 2 from the previous subsection, a set of $N \gg 1$ interacting spins, one finds the same phenomenon. The Hamiltonian is spherically symmetric, so the total angular momentum is conserved. Thus one should characterize stationary states by the value of the total spin and its projection S^{tot}, S_z^{tot}. Suppose the minimum of the energy is at some non-zero value of total spin, its value per particle $M = S^{tot}/N$ one may call a magnetization. The direction of the total spin is still arbitrary because of the high degeneracy of such a state, and one may think that the real system would constantly move among those, following any arbitrarily small external perturbation. In order to measure magnetization one may switch on in tiny external magnetic field $\vec{h}$ which would violate rotational symmetry, split the states with different S_z^{tot} and select the one in which magnetization is directed along $\vec{h}$.[17] What is important to recognize is that there is a double limit here, to be taken in the following order: (i) $N \to \infty$ first, and (ii) $h \to 0$ second. The reverse order does not produce any non-zero magnetization.

The order parameter which quantifies a "chiral asymmetry" of the QCD vacuum, the quark condensate, can be defined as a vacuum expectation value of a local scalar operator

$$\langle \bar{q}_i q_j \rangle = \langle \bar{q}_i^L q_j^R + \bar{q}_i^R q_j^L \rangle, \tag{2.27}$$

where $i, j = 1-N_f$ are flavor indices which are at this point arbitrary. It is an amplitude in a vacuum to emit the left-handed quark and absorb the right-handed one, plus vice versa.

If the vacuum state is chirally symmetric, its vacuum expectation should be zero. Indeed, under vector and axial flavor rotations R and L components transform differently; in other words, R and L can be rotated separately, by arbitrary flavor matrices, and would be changed in an arbitrary way, with zero average.

Like the magnetization direction of the ferromagnetic, in the real world the quark condensate is actually nonzero, and the phenomenological value of its vacuum expectation value (VEV)[18] is

$$\langle 0|\bar{q}q|0\rangle \approx -(240\,\text{MeV})^3. \tag{2.28}$$

---

[17] A macroscopic piece of iron ore is indeed naturally magnetized, with frozen magnetization in the direction of the Earth's magnetic field at the time of its formation. The so called paleomagnetizm is based on this fact.

[18] We mention a standard value normalized at momentum scale $\mu = 1\,\text{GeV}$. Although the condensate itself has an anomalous dimension $\langle \bar{q}q \rangle \sim [\log(\mu/\Lambda_{QCD})]^{4/b}$, the same and the anomalous dimension with the opposite sign hold for the quark mass, so that the combination $m\langle \bar{q}q \rangle$ does not dependent on the normalization scale μ.

Chiral rotations change it, as explained above; the corresponding states with different values of the condensate are *other possible vacua* of the massless QCD, with the
same energy as ours. (Note again the analogy to states of the ferromagnetic with
rotated direction of magnetization.)

Now one can understand the properties of the Goldstone modes; those are nothing else but small amplitude waves of the condensate in the direction of symmetry
transformations. In order to understand that we have to make an infinitely small
symmetry transformation of the condensate,

$$\psi \to (1 + i\beta^a \lambda^a \gamma_5)\psi, \tag{2.29}$$

and put it into the condensate. Its variation is then $\bar{\psi}(\beta\gamma_5 + i\beta^a\lambda^a\gamma_5)\psi$. Nine angles
β are analogs of S_y, which is oscillating in the magnons. They indicate how many
degenerate vacua there are close to ours. If their amplitude is slowly oscillating in
space and time, they also describe 8 pseudoscalar mesons known as π, K, η. In the
chiral limit all of them would be massless. Their existence explains the paradox we
started with in this section.

2.4.3.   *Quark condensate and Dirac eigenvalue spectrum*

Let us now give a slightly deeper analysis of the quark condensate, trying to understand its underlying dynamics in terms of quark motion in background gauge fields.
Again, there are two non-commuting limits, singular for exactly the same reason as
the limit $h \to 0$ is singular for ferromagnetics; a non-zero h however small selects
one vacuum out of many degenerate ones.

Suppose we consider a *single* gauge field configuration, and the quark motion
in it, as is described by the Dirac operator

$$i\hat{D} = (i\partial_\mu + gt^a A_\mu^a(x)/2)\gamma_\mu.$$

A formal definition of the quark condensate is

$$\langle \bar{\psi}\psi \rangle = \text{Tr}\, S(x, x), \tag{2.30}$$

where $S(x, y)$ is the fermion propagator.

(Trivial example: free massless quarks have a propagator $S(x, 0) = x_\mu \gamma_\mu/(2\pi^2 x^4)$
and their trace is zero. So there is no condensate without inter-quark interaction.)

A standard way to approach a propagator is by inversion of the Dirac operator.
In any field A one can diagonalize as follows:

$$i\hat{D}\psi_\lambda = \lambda\psi_\lambda, \tag{2.31}$$

and then express the inverse operator as

$$S(x, y) = \sum_\lambda \frac{\psi_\lambda^+(x)\psi_\lambda(y)^+}{(\lambda + im)}. \tag{2.32}$$

Note the following:

(i) All eigenvalues should be real, as eigenvalues of the Hermitian operator.

(ii) Using chiral symmetry it is easy to prove that the Dirac eigenvalue spectrum is symmetric around zero. Indeed, if ψ_λ is found, one may multiply it with γ_5 and get another solution with eigenvalue $-\lambda$.

(iii) The fermionic determinant can be considered positive; combining two paired values we get $(im + \lambda)(im - \lambda) = -(m^2 + \lambda^2)$.

(iv) Only the $\lambda = 0$ state can be unpaired, say only left-handed.

Let us take $x = y$ and average (integrate) over the whole 4-dimensional volume V_4 of the Universe (or lattice). Then, using the normalization of the eigenvectors, we get the following expression for the quark condensate:

$$\langle \bar{q}q \rangle = \frac{1}{V_4} \int d^4x\, S(x,x) = \frac{1}{V_4} \sum_\lambda \frac{1}{\lambda + im}. \tag{2.33}$$

Now we may use property (ii) to sum in the propagator only over positive eigenvalues or "virtualities",

$$\langle \bar{q}q \rangle = \frac{1}{V_4} \int d^4x\, S(x,x) = \frac{1}{V_4} \sum_{\lambda > 0} \frac{2im}{\lambda^2 + m^2}. \tag{2.34}$$

One can now see how tricky the limit $m \to 0$ can be. If it is taken *first* and all λ values are finite, the expression above shows that $\bar{\psi}\psi$ is zero. It is a particular case of a more general statement; no finite system can spontaneously break a continuous symmetry.

(Note that a lattice *is* a finite system; thus some sufficiently large quark mass is needed. This restriction is rather important and still limits the predictive power of lattice QCD. In Section 5.2 we will return to this non-trivial issue.)

However, in the "thermodynamic limit" of a very large system, the λ spectrum becomes continuous and there are eigenvalues arbitrarily close to zero. Their contribution to the above sum, for finite m, is like $1/m$, not like m. Therefore, *the quark condensate is connected with quark states possessing arbitrarily small "virtualities"* λ.

If the thermodynamic limit is taken first, and the $m \to 0$ limit is taken second, one gets the correct non-zero condensate. Quite an elegant general relation for it (known as the Casher–Banks formula) is

$$\langle \bar{\psi}\psi \rangle = \frac{1}{\pi} \left\langle \frac{dN}{d\lambda}\Big|_{\lambda=0} \right\rangle, \tag{2.35}$$

where the r.h.s. is a "spectral density" of states at $\lambda = 0$. Note that we also put angular brackets in the r.h.s.; those denote averaging over an ensemble of appropriately weighted gauge field configurations describing the QCD vacuum. This relation is often used in different approaches (including the lattice one) to evaluate the quark condensate.

(One more comment on a trivial example mentioned above, the *free fermion* case. The eigenvectors are plane waves, so λ is just a momentum length — remember, we are in Euclidean space–time. The spectrum is simply given by $\sum_\lambda \cdots = \int d^4 p \cdots$, so at $\lambda = 0$ its density is indeed zero, so the condensate vanishes as well.)

2.4.4. *Elements of chiral perturbation theory*

Let us now introduce small but finite quark masses as a perturbation which breaks chiral symmetry, fixes certain directions in the flavor space and selects one unique vacuum we live in. One may expect that physical quantities can in principle be expanded in powers of quark masses around their values at the massless point. The calculation of those coefficients is the subject matter of chiral perturbation theory. We will not go deeply into this vast subject, and only discuss one example of such an approach, to the masses of the Goldstone modes.

As a result of chiral symmetry violating masses, the poles in the propagators of the pseudoscalar mesons are shifted from the zero mass value, and they become massive. If the effect is small, one may expect it to be simply linear in the respective perturbations, e.g. for π, K mesons,

$$m_\pi^2 = B(m_u + m_d) + O(m^2), \qquad m_K^2 = B(m_u + m_s) + O(m^2), \qquad (2.36)$$

where as perturbations we have used masses of valence quarks for those particles.

Here the unknown constant B is assumed to be flavor independent in the chiral limit. Eliminating it one can get the following quark mass ratios:

$$\frac{m_u + m_d}{m_u + m_s} = \frac{m_\pi^2}{m_K^2} \approx \frac{1}{13}. \qquad (2.37)$$

Thus the u, d quarks should be much lighter than the strange one. The order-of-magnitude value for the mass of the strange quark is known, e.g. from the splitting of strange and non-strange baryons, and so it had been known already in the 1960s that $m_s = 100$–150 MeV. This value, together with the above ratio, leads to conclusion that m_u and m_d are very small, only few MeV. (Additional evidence comes of course from the isotopic mass differences, once the electromagnetic effects are subtracted.)

So far we have assumed that the linear expansion in masses works. Fortunately we can check its validity for the two remaining pseudoscalar mesons, η and η'. We assume the former to be an ideal SU(3) octet state $\eta = (\bar{u}u + \bar{d}d - 2\bar{s}s)/\sqrt{6}$ and the latter an SU(3) singlet $\eta' = (\bar{u}u + \bar{d}d + \bar{s}s)/\sqrt{3}$. For η one can eliminate the unknown constant and obtain the so called Gell–Mann–Okubo formula

$$m_\eta^2 = m_K^2 \tfrac{4}{3} - m_\pi^2 \tfrac{1}{3} \approx (566\,\mathrm{MeV})^2, \tag{2.38}$$

which compares well with the experimental value $m_\eta = 549\,\mathrm{MeV}$ and is thus our first example of a successful prediction by chiral perturbation theory.

However, for the η' meson the prediction turns out to be a disaster; omitting small light quark masses one gets

$$m_\eta = m_K (2/3)^{1/2} \approx 400\,\mathrm{MeV}, \tag{2.39}$$

to be compared with the experimental value of $m_{\eta'} = 958\,\mathrm{MeV}$. This problem was pointed out by S. Weinberg [36] and it is known as the $U(1)_A$ problem. Its general resolution follows from the chiral anomaly that we discussed in Section 1.5.2. However to get the solution *in practice* one should be able to calculate the right mass for the η'; we will return to it in Section 6.2.4.

The Gell–Mann–Oakes–Renner relation [30] is an immediate consequence. For nonzero quark masses, the divergence of the axial current is related to the pseudoscalar density by $\partial_\mu(\bar{u}\gamma^\mu\gamma_5 d) = (m_u + m_d)\bar{u}i\gamma_5 d$. The vacuum-to-pion matrix element of this equality can be written as a relation $M_\pi^2 F_\pi = (m_u + m_d)G_\pi$, where F_π, G_π are constants defined by the l.h.s. and r.h.s.

$$\langle \pi(p)|\bar{q}\gamma_\mu\gamma_5 q|0\rangle = -ip_\mu f_\pi, \qquad \langle \pi(p)|\bar{q}\gamma_5 q|0\rangle = G_\pi. \tag{2.40}$$

The coefficient of the leading term in the expansion of the pion mass in powers of the quark masses, $M_\pi^2 = (m_u + m_d)B + \cdots$, is therefore determined by the condensate and by the pion decay constant; $B = G/F = C/F^2$. The above relations involve two independent low energy constants, F and B. In particular, the value of the quark condensate in the massless theory may be expressed in terms of these: $\langle 0|\,\bar{u}u\,|0\rangle = -F^2 B$.

Finally, we found the numerical value of the constant B defining the magnitude of the pseudoscalar masses,

$$B = |\langle\bar{u}u\rangle|/F_\pi^2 \approx 1700\,\mathrm{MeV} \tag{2.41}$$

(the numerical value for $1\,\mathrm{GeV}$ normalization, as condensate). Unlike the case of the quark masses, B is a parameter internal to QCD and on general grounds it is expected to be $B = \mathrm{const} \cdot \Lambda_{\mathrm{QCD}}$ with the constant of $O(1)$. In reality it happens to be rather $O(10)$; it is an example of an unusually large scale related to chiral dynamics. However strange it may sound, the lightest hadron — the pion — is in fact *surprisingly heavy*.

2.4.5. *Effective chiral Lagrangian*

In the previous subsection we saw that the hadronic states in QCD with light quarks possess a certain hierarchy; the Goldstone modes are lighter than all others. They are massless in the chiral limit and are very light in the real world. As a result, Goldstone modes dominate strong interaction at large distances. The subject of this subsection is the properties of such interactions, which are described by the so called *effective chiral Lagrangian* $\mathcal{L}_\chi$.

There are three main reasons why this subject is worth studying. First of all, it supersedes earlier developments known as current algebra (early 1960s) and partial conservation of axial current (PCAC, late 1960s). Second, it is an extremely nice example of how a simple general principle — the chiral symmetry — can be very powerful, relating many properties of such interactions, and even fixing some uniquely. The third reason: $\mathcal{L}_\chi$ is interesting because it helped us to understand how a renormalization prescription for a non-renormalizable interaction[19] should be done.

Let the quark condensate in vacuum be slightly deformed, namely, let the orientation of the flavor axis be rotated according to its standard position by some (space–time dependent) unitary matrix $\langle \bar{q}q \rangle' = U \langle \bar{q}q \rangle$ such as

$$U(x) = \exp\left(\frac{i\pi^a(x)\lambda^a}{F_\pi} \right), \tag{2.42}$$

where λ^a are flavor Gell–Mann matrices. The rotation angles are identified with the pion field $\pi^a(x)$, traditionally normalized by the pion decay constant[20] $F_\pi \approx 92.4\,\mathrm{MeV}$.

Unlike the case of global rotation, in which π^a is space–time independent and the chiral symmetry guarantees that we get another vacuum with the same energy, the oscillating pion field costs a certain amount of energy. This is basically the integral of the chiral effective Lagrangian we are looking for.

First of all, since an x-independent U just rotates one vacuum into another and cannot have any energy, all the terms in this Lagrangian should be proportional to some derivatives of the phase U. The first term is the so called "kinetic" term, suggested by Weinberg[21] [31]

$$\mathcal{L}_\chi = \frac{1}{4} F_\pi^2 \operatorname{Tr}(\partial_\mu U \partial_\mu U^+). \tag{2.43}$$

---

[19] There is nothing wrong with a low energy effective interaction behaving badly at large momenta, in UV. Also, renormalization is of course always possible, but non-renormalizable theories need infinitely many constants. It is OK, provided there exist enough experimental data to fix them.

[20] The reason for that will become clear later in this subsection.

[21] In fact he used another parametrization of the chiral phase $U = \sigma + \pi^a(x)\lambda^a$ in terms of the sigma model. Unitarity of this matrix restricts σ, π^a to be on a unit sphere, which is less convenient in practice than the standard parameterization with the angles we use here.

We put "kinetic" within quotation marks for the following reason: although it indeed starts with the ordinary kinetic $(1/2)(\pi^a(x))^2$ term, it also contains an infinite set of nonlinear terms in the pionic field. Those can be identified with $\pi\pi$ binary scattering and even rather complicated multiple scattering processes. It looks like a miracle: all those terms have prefixed coefficients and so Weinberg was able to derive nontrivial predictions for low-energy processes, valid to the *second* power of momenta. An example is the following simple expression for the $\pi\pi$ scattering amplitude

$$T_{ABCD} = \frac{4}{F_\pi^2}(s\delta_{AB}\delta_{CD} + t\delta_{AC}\delta_{BD} + u\delta_{AD}\delta_{CB}), \qquad (2.44)$$

where s, t, u are the usual kinematical variables and $A, B, C, D = 1$–3 are pion flavor indices. When these results were compared with the data the agreement was found to be very good.

Let me now explain how the normalization constant F_π was fixed. At the quark level we know how rotation in flavor space of the left- and right-handed quark fields is generated by the vector and axial transformations. The phase U is changed under left and right rotations with some matrices $U \geq A_L U B_R$. By making some small transformations of the fields one can find the variation of the Weinberg Lagrangian and, by a standard procedure, determine the corresponding left- and right-handed currents,

$$j_\mu^L = \tfrac{1}{2}iF_\pi \, \mathrm{Tr}(\lambda^a U \partial_\mu U^+) \approx -F_\pi \partial_\mu \pi^a + O(\pi^2)\cdots, \qquad (2.45)$$

$$j_\mu^R = \tfrac{1}{2}iF_\pi \, \mathrm{Tr}(\lambda^a U^+ \partial_\mu U) \approx F_\pi \partial_\mu \pi^a + O(\pi^2)\cdots, \qquad (2.46)$$

where expressions on the r.h.s. above are in the $O(\pi)$ approximation. They lead to the definition of F_π given above. The left-hand current is the main element in the theory of weak interactions, and so one can by standard rules relate F_π to the lifetime of the charged pion decay $\pi \to \mu\nu_\mu$

$$1/\tau_{\pi \to \mu\nu_\mu} = F_\pi^2 G_F^2 \frac{\cos^2\theta_c}{4\pi} m_\pi m_\mu^2 \left(1 - \frac{m_\mu^2}{m_\pi^2}\right), \qquad (2.47)$$

where G_F, θ_c are the Fermi constant and the Cabbibo angle, respectively. Comparing this expression with the lifetime of the pion decay, one can finally fix F_π to the numerical value mentioned above.

The mass term, violating the chiral symmetry, is of first order in the (3×3) mass matrix m

$$\mathcal{L}_m = \tfrac{1}{2}F_\pi^2 B \, \mathrm{Tr}(m\,U^\dagger + U m^\dagger). \qquad (2.48)$$

It gives the pion a nonzero mass as discussed above. Next terms of mass corrections to order p^4 and further [34] are

$$
\begin{aligned}
\mathcal{L}_{np}^{(4)} = {} & L_1 \langle D_\mu U^\dagger D^\mu U \rangle^2 + L_2 \langle D_\mu U^\dagger D_\nu U \rangle \langle D^\mu U^\dagger D^\nu U \rangle \\
& + L_3 \langle D_\mu U^\dagger D^\mu U D_\nu U^\dagger D^\nu U \rangle + L_4 \langle D_\mu U^\dagger D^\mu U \rangle \langle U^\dagger \chi + \chi^\dagger U \rangle \\
& + L_5 \langle D_\mu U^\dagger D^\mu U (U^\dagger \chi + \chi^\dagger U) \rangle + L_6 \langle U^\dagger \chi + \chi^\dagger U \rangle^2 \\
& + L_7 \langle U^\dagger \chi - \chi^\dagger U \rangle^2 + L_8 \langle U^\dagger \chi U^\dagger \chi + \chi^\dagger U \chi^\dagger U \rangle \\
& - i L_9 \langle F_{\mu\nu}^R D^\mu U D^\nu U^\dagger + F_{\mu\nu}^L D^\mu U^\dagger D^\nu U \rangle \\
& + L_{10} \langle F_{\mu\nu}^R U F^{L\,\mu\nu} U^\dagger \rangle - i L_{11} D_\mu \theta \langle U^\dagger D^\mu U D_\nu U^\dagger D^\nu U \rangle \\
& + L_{12} D_\mu \theta D^\mu \theta \langle D_\nu U^\dagger D^\nu U \rangle + L_{13} D_\mu \theta D_\nu \theta \langle D^\mu U^\dagger D^\nu U \rangle \\
& + L_{14} D_\mu \theta D^\mu \theta \langle U^\dagger \chi + \chi^\dagger U \rangle - i L_{15} D_\mu \theta \langle D^\mu U^\dagger \chi - D^\mu U \chi^\dagger \rangle \\
& + i L_{16} \partial_\mu D^\mu \theta \langle U^\dagger \chi - \chi^\dagger U + \cdots \rangle,
\end{aligned}
\tag{2.49}
$$

where the covariant derivatives D include external vector and axial fields, $\chi \equiv 2Bm$ can even be slowly x-dependent. Some of the couplings involve the field strengths of the external fields. The traceless matrices $F_{\mu\nu}^R$, $F_{\mu\nu}^L$ collect the octet components of the field strengths belonging to the right- and left-handed fields $r_\mu = v_\mu + a_\mu$, $l_\mu = v_\mu - a_\mu$, respectively, while $v_{\mu\nu}^0$, $a_{\mu\nu}^0$ are the Abelian field strengths of the singlets v_μ^0, a_μ^0.

Needless to say, each term can contribute to many different processes. As an example of a concrete use of such a term, consider higher order corrections to the pion mass m_π [34]

$$
m_\pi^2 = \mathring{M}_\pi^2 \left(1 + \frac{\mathring{M}_\pi^2}{32\pi^2 F^2} \ln \frac{\mathring{M}_\pi^2}{\Lambda_A^2} - \frac{\mathring{M}_\eta^2}{96\pi^2 F^2} \ln \frac{\mathring{M}_\eta^2}{\Lambda_B^2} \right) + O(m^3),
\tag{2.50}
$$

where $\mathring{M}$ is of course the lowest order result and the two scales Λ_A, Λ_B are completely determined by the coupling constants L_4, L_5, L_6, L_8.

Finally we mention one more term in the chiral effective Lagrangian which is known on very general grounds. It is the so called Wess–Zumino term [35] which we present in the beautiful 5-dimensional form suggested by E. Witten:

$$
S_{WZ} = \frac{iN_c}{240\pi^2} \int d^5x \, \epsilon_{\alpha\beta\delta\gamma\delta} \operatorname{Tr}(L_\alpha L_\beta L_\gamma L_\delta L_\epsilon).
\tag{2.51}
$$

The reader is probably puzzled why the integral here is taken over a 5-dimensional space. This is an example of a topological action which is in fact a divergence of some vector; it can be rewritten as a surface integral over a $d = 4$ boundary, assumed to be our 4-dimensional space–time. Then the Wess–Zumino original form is recovered, which is however less symmetric.

Note that this term has a dimensionless constant in front, which is completely determined. More than that, it contains the number of colors N_c, which should puzzle an attentive reader because chiral properties and related flavor rotations have nothing to do with color. Let me just provide a hint: anomalies are at work here. In Witten's paper mentioned above it is shown that it is this term which makes a Skyrmion a boson or a fermion, depending whether N_c is even or odd.

2.4.6. *Nambu–Jona–Lasinio model*

In our discussion of chiral symmetry breaking we have not yet discussed *why* it actually happens. Ideas about the existence of chiral symmetry and what it may take to break it spontaneously appeared in 1961, in two papers [37, 38] inspired by the BCS theory of superconductivity.

The NJL model [37] was the first microscopic model which attempted to derive dynamically the properties of chiral symmetry breaking and pions, starting from some *hypothetical 4-fermion interaction*.

$$\mathcal{L}_{\text{NJL}} = \bar{\psi} i \not{\partial} \psi + G[(\bar{\psi}\psi)^2 + (\bar{\psi} i \gamma_5 \vec{\tau} \psi)^2], \tag{2.52}$$

where G is some unknown coupling constant. From this viewpoint, the interaction is a sum of pseudoscalar and scalar currents. Rewriting it as $L \pm R$ one can see that chirally non-invariant terms cancel out, and thus this interaction is consistent with chiral symmetry. This interaction is also a point-like local attraction between quarks, and as we will see below, if G is strong enough it can indeed make the normal perturbative vacuum unstable, and create the quark–antiquark condensation.

What can the quantum numbers of the condensate be? If it is analogous to that in superconductors, it is a scalar. For the $\bar{q}q$ channel, however, it is not a $L = 0$ or *s*-wave. The parity in this case is $P = (-1)^{L+1} = 1$ and the charge parity is $C = (-1)^{L+S} = 1$. The only solution is $S = L = 1$. (This resembles the special phase of He^3 with spin-1 Cooper pairs.) The simplest diagram is the mass operator shown in Fig. 2.8. For generality, we will show the result for the non-zero temperature T and the chemical potential

$$\Sigma = 4GN_cN_fm \int \frac{d^3p}{(2\pi)^3} \frac{1}{E_p} [1 - f^+(\vec{p}, \mu) - f^-(\vec{p}, \mu)] \tag{2.53}$$

with the Fermi distribution functions

$$f_p^\pm \equiv 1/[e^{\beta(E_p \pm \mu)} + 1] = f^\pm(\vec{p}, \mu), \quad E_p^2 = p^2 + m^2.$$

In vacuum the expression is the same without the f-terms.

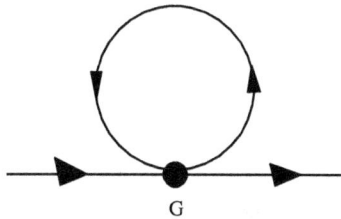

Fig. 2.8. The mass operator for the NJL model.

The l.h.s. is the mass operator, while the r.h.s. has the mass m. What if both are the same thing? This indeed leads to the so called *gap equation*, solving which one can find the generated mass.

In order to derive this equation properly out we should write down the total thermodynamical potential,

$$\Omega(m) = \frac{m^2}{4G} - \gamma \int \frac{d^3p}{(2\pi)^3} E_p - \gamma T \int \frac{d^3p}{(2\pi)^3} \log\left[1 + e^{-\beta(E_p+\mu)}\right]\left[1 + e^{-\beta(E_p-\mu)}\right],$$
(2.54)

where $\beta = 1/T$ and $\gamma = 2N_cN_f$ is the degeneracy factor. The first term is the "potential energy", while the latter two are just a sum over all energies of all fermionic levels, in negative states of the Dirac sea (the second term) and in the heat bath (the third). The diagrams are not divergent because the NJL interaction is assumed to exist only in some finite momentum interval; all three momentum integrals are understood to be regulated by a cutoff Λ.

The value of the spontaneously generated mass m should be chosen so as to minimize the thermodynamic potential (or maximize the statistical sum Z).

Exercise 2.6.   Show that by differentiating Ω with respect to m one finds the gap equation given above.

Of course not all solutions are to be accepted, but only those associated with a potential lower than in the trivial vacuum. If such a potential exists, the chiral symmetry will be spontaneously broken. In the $T = 0$ limit, and omitting m from both sides, the gap equation reads

$$\frac{\pi^2}{2GN_cN_f} = \int_{p_F}^{\Lambda} dp \frac{p^2}{\sqrt{p^2 + m^2}},$$
(2.55)

and at zero density Fermi momentum $p_F = 0$. No solution exists for small G, but it appears above a critical value.

The standard set of NJL parameters is $\Lambda = 0.65\,\text{GeV}$ and $G = 5.01\,\text{GeV}^{-2}$, which are fixed from the observed values of F_π and the value of the quark condensate.

Let me make a few parting comments about the NJL model.

(i) It was the first bridge between the BCS theory of superconductivity and quantum field theory, leading the way to the Standard Model. It first showed that the vacuum can be truly nontrivial, a superconductor of a kind, with the mass gap $\Delta = 330$–$400\,\text{MeV}$, known as "constituent quark mass".

(ii) The symmetries of the NJL model are not quite right; there is no $U(1)_A$ breaking and η (the analog of η' for the $N_f = 2$ version we are considering) is as massless as the pion.

(iii) The NJL model has two parameters; the strength of its 4-fermion interaction G and the cutoff $\Lambda \sim 0.8$–$1\,\text{GeV}$. The latter regulates the loops (the model is non-renormalizable, which is OK for an effective theory) and is known as the "chiral scale". For example, the size of the constituent quark is of this scale. We will return to the question whether the NJL-type interaction exists in QCD and why its parameters have such values in Chapter 4. We will relate Λ to the typical instanton size ρ, and G to a combination $n\rho^2$ of the size and density of instantons.

(iv) One non-trivial prediction of the NJL model was that the mass of the scalar is $m_\sigma \approx 2m_{\text{const.quark}} \approx m_\rho$. Because this state is the P-wave in non-relativistic language, it means that there is strong attraction which is able to compensate for rotational kinetic energy. Experimentally $m_\sigma \approx 500\,\text{MeV}$ is even lighter. As for other hadrons, such as a nucleon, there have been conflicting papers on this issue all along. It is not even clear whether or not it is a resonance. The NJL model is non-renormalizable and results depend on how exactly the UV cutoffs are implemented.

2.5. Color confinement

2.5.1. *Static potential*

The colored objects — quarks and gluons — are not observed as physical excited states, only *colorless* hadrons are observed. This fact, commonly attributed to the famous *color confinement*, is probably the most striking feature of QCD. A closer look however raises many questions and a need for a more precise definition of confinement; for a nice pedagogical description in many QCD-like theories see Ref. [75].

Suppose one takes two static quarks and adiabatically slowly pulls them apart. What is then observed, in experiment or on the lattice, is that at some distance a pair of light quarks would appear out of the vacuum and screen the static charges, resulting with a pair of heavy–light mesons. Does this prove confinement? Quite the contrary, it only shows that the 2-meson state is the lowest of all possible ones. In fact, this pair production process makes it impossible to argue that the real world — QCD with light quarks — is a confining theory.

Let us now repeat the same experiment, but in gluodynamics *without* light dynamical quarks.[22] Now nothing can prevent us from discussing the linearly growing potential between two fundamental charges; the only dynamical particles, the gluons, have an adjoint color representation and cannot screen the fundamental charge. The "triality" or center of the SU(3) group is a quantum number which can be assigned to the electric field flux corresponding to a fundamental charge.

---

[22] We still need a pair of quarks for this experiment, but those are infinitely heavy ones, which is irrelevant for the gluon dynamics. It is nothing but an external probe.

On the lattice one can do it in great detail, and observe that although at small distances around a charge the electric field lines go radially from the charge, as in QED, at larger distances these field lines converge into the so called flux tubes or "QCD strings", with its properties stabilized far from the charge. It means that the potential energy V is growing linearly with its length r, $V(r) = Kr$.

Returning to the real world with light quarks, we find that in fact strings of finite lengths actually exist. We know about those from two independent sources, already discussed at the beginning of this chapter; (i) the levels of charmonium and bottonium states; and (ii) Regge trajectories.

Unfortunately, turning to the theoretical side of the problem, we are in a much worse position. In spite of multiple attempts resulting in spectacular progress in understanding many QCD-like theories, we are not able to understand why it is so in pure gluodynamics.

Historically, one of the first important papers on the subject was a very influential paper by K. Wilson [251] which we already discussed in connection with the lattice formulation. First of all, Wilson formulated a practical criterion for confinement, related to a confining potential for two static quarks. Consider some elongated rectangular loop with length T in the time direction, with L some spatial direction, and with $T \gg L$. It can be considered as the following physical process: instantaneous creation of a quark–antiquark pair, being then separated by the distance L for a time period T, with subsequent annihilation. Its action on QCD vacuum is associated with such a loop-type external current corresponding to the following factor to the statistical sum,[23] known as the *Wilson loop*:

$$W(C) = \operatorname{Re} \operatorname{Tr} P \exp \left(ig \int_C A_\mu \, dx_\mu \right). \tag{2.56}$$

Wilson's confinement criterion means that the vacuum expectation value of the "Wilson loop" W satisfies the *area law*

$$\langle W(C) \rangle = \exp[-T * V(L)] = \exp[-T * L * K], \tag{2.57}$$

where K is the string tension.

The very intriguing property of the lattice theory shown by K. Wilson in the same classical work was a demonstration that confinement is natural in *the strong coupling expansion*. Although the same is true for Abelian theory, and although the physically relevant regime of the lattice theory is the opposite limit of the weak coupling, it was a very important step.

Further excitement was raised by the first numerical simulation of the SU(2) gauge theory done by M. Creutz in 1980 [270], which indicated that this indeed seems to be the case; we return to those lattice results in Chapter 5.

---

[23] The definition of the path-ordered exponent of x-dependent matrices was discussed in Section 1.1.

2.5.2. *Dual superconductivity*

Duality these days is a term with a multitude of meanings in field theory; we adopt here the oldest meaning of all, the Maxwellian (or Dirac) electric–magnetic duality of the Abelian gauge theory.

Nambu, 't Hooft and Mandelstam [60] have argued that *electric* confinement in gauge theory can be analogous to *magnetic* confinement in superconductors. In all quantum field theories in which confinement has been proven, namely in compact U(1) gauge theory in the Villain formulation, the Georgi–Glashow model and especially the SUSY Yang–Mills theories [886], this scenario is indeed realized.

It is well known that superconductors expel the magnetic field (the so called Meissner effect). There are two different ways in which it happens. In the so called superconductors of the second type the magnetic field can survive inside the so called Abrikosov–Nielsson–Olesen (ANO) flux tubes [61] with a quantized flux. Repulsive interaction between the tubes stabilizes the system in the form of a triangular lattice. In superconductors of the first type the interaction between flux tubes is attractive, and thus the tubes are glued into macroscopic regions with a non-zero field and broken superconductivity.

Suppose one gets a pair of a magnetic monopole and an antimonopole[24] and embeds it into a superconductor. The monopoles would be connected by a flux tube, which would confine them with a linear potential. (Note that since we only have one flux tube, the issue of interaction between tubes is irrelevant. Both kinds of superconductors will display linearly confined magnetic charges.)

Following this analogy further, one may think that the condensation of Cooper pairs in a superconductor may correspond to Bose-condensation of magnetic monopoles in the vacuum.

As a warning, the analogy to an ordinary superconductor (even after Abelian projection) is not complete. Even ignoring the difference between Abelian and non-Abelian charges, note that electrons that form Cooper pairs in BCS theory are all negatively charged. However, the monopoles that are thought to condense in QCD can carry both negative and positive magnetic charges. Because the Cooper pairs and monopoles are quite different objects, the interaction between them is different as well. It is also by no means clear what a small parameter (to allow one to use simple BCS theory) can be in the QCD vacuum. The idea here is that since the magnetic coupling is inverse to the electric one, the former can be small at large distances since the latter is large.

The main applications of the dual-superconductivity-based-models are to model the QCD string by analogy to the ANO vortex, as was done in a series of papers [62]. Some of the definite successes of this approach are: (i) the prediction of weak string–string interaction, putting it close to the boundary of types I

---

[24]They do not really exist, but are only needed for a thought experiment, to demonstrate the point.

and II superconductivity; (ii) the prediction of a whole set of potentials other than a central potential. Both predictions agree well with available lattice data, for a review see Ref. [65].

The effective Lagrangian used in Ref. [62] is

$$L = \tfrac{4}{3}\left[\tfrac{1}{4}(\partial_\mu C_\nu - \partial_\nu C_\mu)^2 + \tfrac{1}{2}|(\partial_\mu - ig_m C_\mu)\phi|^2 + \tfrac{1}{4}\lambda(|\phi|^2 - |\phi_0|^2)^2\right], \quad (2.58)$$

where we have omitted interaction with quarks at the ends. C_μ is a dual color potential coupled to the Higgs with magnetic coupling $g_m = 2\pi/g$. Assuming that we are exactly at the boundary of types I and II of superconductivity, the masses of the Higgs and the "dual photon" are equal $M_\phi = M_C = g_m\phi_0$. The (classical) string tension is directly related to Higgs VEV

$$\sigma = \tfrac{4}{3}\pi|\phi_0|^2. \quad (2.59)$$

2.5.3.   *Structure of flux tubes*

Using lattice data as a source of empirical information about confining flux tubes (strings), one may ask many different questions. Some of them, with increasing level of sophistication, we will consider in this subsection.

The first question: *how short should a string be in order to exist?*

Before we discuss the data, consider first simple theoretical expectations. A charge–anti-charge pair at a small distance r from each other is nothing but an electric *dipole*, expected to interact directly with the vacuum electric fields. Since their non-zero average can only exist for the field squared, the prediction of the second order in dipole approximation [64] for this interaction is

$$V(r) - V_{\text{Coulomb}}(r) = r^2 \int d\tau \exp\left(-\frac{3\alpha_s(r)\tau}{2r}\right)\langle 0|G_{\mu\nu}(\tau)U_\tau G_{\mu\nu}(0)U_\tau^+|0\rangle, \quad (2.60)$$

where the field strengths are separated by the time delay τ, with U_τ being the appropriate parallel transports. The exponent originates from the difference between attractive and repulsive Coulomb potential between the $\bar{q}q$ in the octet representation, which is the case during the time between the absorption of the first and emission of the second gluon. The bottom line is that the interaction is expected to be $O(r^2)$.

Several other cases, including this one, have been discussed in my paper [63]; and in all of them without exception the expected short-distance behavior from similar considerations is in disagreement with actual observations.

The lattice data by G. Bali *et al.* [65] for the short-range part of the potential obtained in quenched simulation are shown in Fig. 2.9. Although uncertainties related with the exact definition of what the perturbation theory really predicts, remain, the authors nevertheless concluded from these data that oscillator-like potential (2.60), "it clearly fails to describe the short range static potential as obtained from quenched lattice simulations, at least for source separations bigger

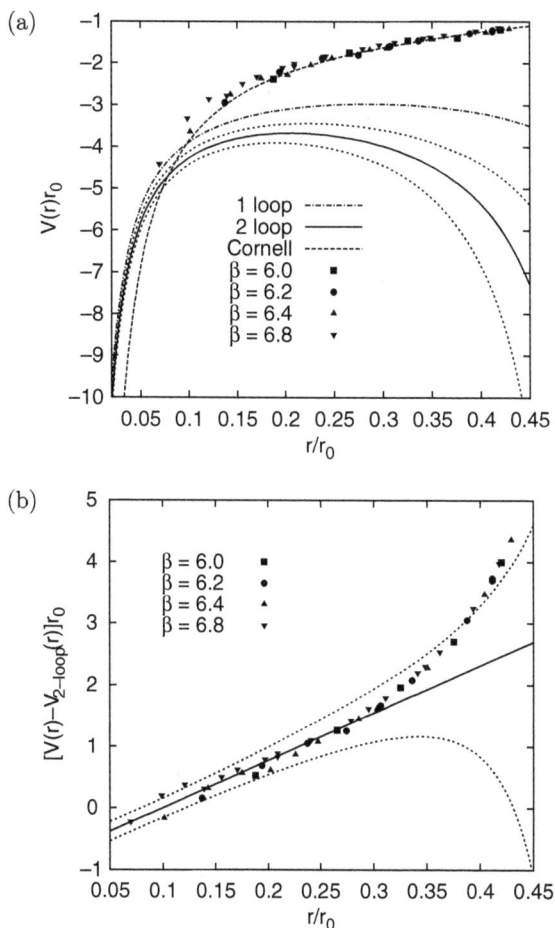

Fig. 2.9. (a) The quenched lattice potential versus perturbation theory in units of $r_0 \approx 0.5$ fm. The only free parameter is the self-energy of the static sources. The lattice data have been normalized such that $V(r_0) = 0$. (b) The small-distance part with the perturbative 2-loop potential subtracted, in comparison with a linear curve with slope $7.7 r_0^{-2}$.

than $(6\,\mathrm{GeV})^{-1}$. The difference between the non-perturbatively determined potential and perturbation theory at short distance is well parameterized by a linear term. We quote the two-loop slope, $(1.20 \pm 0.36)\,\mathrm{GeV}^2$ as our final result, which is significantly bigger than the string tension $K \approx 0.21\,\mathrm{GeV}^2$."

The only comment on the situation I can make is that an appearance of a linear-looking potential at such small r may imply that the string radius is rather small. Indeed: *how thick is a confining string?*

The answer is somewhat puzzling, as discussed in Ref. [65]. While Abelian projected distributions of the action and the energy densities show thin tubes with a radius of only about 0.2 fm, the full SU(2) action shows a size about twice as large.

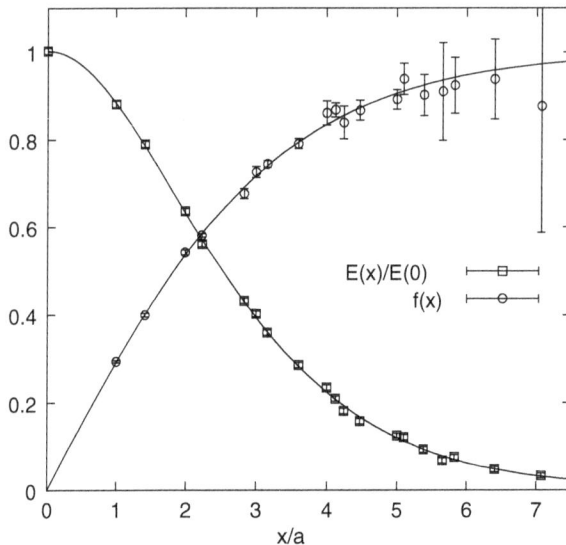

Fig. 2.10. Comparison of the profiles of the electric field and the amplitude of the Ginzburg–Landau wave function as a function of the distance from the center of the ANO vortex.

The third question is: *is the radial distribution of the electric field in the string, as well as of the effective Higgs field, the same as in an ANO vortex?* Quite an impressive fit of the lattice data to the distribution expected from an effective Abelian model is shown in Fig. 2.10, from Ref. [65]. Let me simply quote the conclusion of this paper, "*The dual Maxwell equations have been verified in Abelian projected SU(2) and the fields are adequately described by the dual Ginzburg–Landau equations with the values* $\lambda = 0.15(2)$ *fm and* $\xi = 0.25(3)$ *fm for penetration and coherence length, respectively. These values correspond to a (dual) photon mass* $m_\gamma \approx 1.3$ *GeV* $\approx 3\sigma$ *and a Higgs mass of* $m_H \approx 0.8$ *GeV* $\approx 2\sigma$, *the ratio of which,* $\kappa = \lambda/\xi = 0.59(13)$, *indicates the vacuum of SU(2) gauge theory to be a (weak) type I superconductor.*"

Here is the fourth and last question of this section: *what is the microscopic mechanism of confinement?* We already discussed the dual superconductivity idea; before we comment further on it let us mention some alternatives.

Apart from the Abelian Projection and Monopoles we mentioned, there is model [67] which claims that it is not (3*d*) monopoles but (2*d*) center vortices which are crucial for the Wilson area law. By eliminating all non-topological or "photon" part of the lattice fields and preserving only center vortices, one can reproduce the string tension. These two ideas do not necessarily exclude each other since both monopoles and vortices are found to be correlated with each other.

There was a suggestion [66] that the instanton liquid model can generate a significant static potential due to a tail of instanton size distribution at large sizes, without contradiction to the usual assumption that the dominant instantons have small sizes $\rho \sim 1/3$ fm (see Chapter 3). Since the large-size instanton is basically a constant field, this idea is similar to the anti-ferromagnetic or domain vacuum going back to the Savvidi papers [137]. The situation with respect to such long-range fields was addressed on the lattice and the status is still somewhat inconclusive [69]. On the other hand, it is difficult to see how such models can generate small-size confining strings, which phenomenology and lattice seem to demand.

One more old idea is related with the so called *merons* [214], the hypothetical objects with the topological charge 1/2 into which instantons can dissociate at large sizes, when the coupling and quantum fluctuations get big enough. Unlike instantons, which interact at large distances weakly, as dipoles, and cannot effectively disorder the Wilson loop, the merons interact like charges and can do so. For a recent study of this idea see Ref. [76].

Finally, let me mention an ancient idea by Callan, Dashen and Gross [217] and myself [216], suggesting that the color fields are expelled from the QCD vacuum because they suppress instantons, and thus prevent lowering the ground state energy due to tunneling. A strong negative correlation between flux tubes and instantons has indeed been demonstrated on the lattice, but to my knowledge no serious study of its magnitude has ever been made.

2.5.4. *Interaction of flux tubes*

We have already mentioned that the flux tubes can be classified by the so called center group $Z(N_c)$, which is a transformation changing quark fields $q \to e(i2\pi k/N_c)$, where k is some integer. The gluons do not transform under it.

In SU(3) the corresponding quantum number is called *triality*. This group does not allow for a large choice of representations, the triality can be 1 or 2. Diquarks exist in two representations; $3 * 3 = 6 + 3$. The 3-diquark is antisymmetric, and writing it as $\epsilon_{ijk} q_j q_k$ one see that it is the same representation as the antiquark. (That is why high spin mesons and baryons have the same string.) The string for the symmetric 6 representation should however have the same tension, due to charge conjugation,

$$\sigma(k) = \sigma(N_c - k). \tag{2.61}$$

In summary, there is only one string with only one standard tension K in the SU(3) gauge theory.

Higher color groups allow for the so called k-strings, which are caused by sources which transform by a factor of z^k under a global gauge transformation. There are various conjectures in the literature of what the tension of such k-strings is.

One of them is the (M-theory inspired) conjecture [70]

$$\frac{\sigma_k}{\sigma_1} = \frac{\sin(k\pi/N_c)}{\sin(\pi/N_c)}, \tag{2.62}$$

where σ_k is the tension of the k-string. Note that in the large N_c limit those are k non-interacting strings, with $O(1/N_c^2)$ residual interaction.

Another possible choice is the so called *Casimir scaling* conjecture, originating from the scaling of the pQCD and potentials at small distances [71],

$$\frac{\sigma_k}{\sigma_1} = \frac{k(N_c - k)}{N_c - 1}, \qquad (2.63)$$

which for large N_c would again mean k non-interacting strings plus parametrically larger pairwise attraction,

$$\frac{\sigma_k}{\sigma_1} = k - \frac{k(k - 1)}{N_c} + \cdots \qquad (2.64)$$

Recent lattice calculations have addressed the issue and found that the string tensions of $k = 2$-strings are consistent with both predictions mentioned above, although they are slightly closer to the M-theory conjecture than to "Casimir scaling". Teper *et al.* [72] have also brought forward an interesting argument that closed k-strings provide a natural way for non-perturbative effects to introduce $O(1/N_c)$ corrections into the pure gauge theory, in contradiction to the conventional diagrammatic expectation.

Performing simulations in pure gauge SU(4) and SU(5) theories, one indeed definitely finds that strings are *bound*, namely $\sigma_{k=2}/\sigma_1 < 2$ beyond any doubt. Moreover the continuum-extrapolated value lies between the two conjectures which are numerically very similar for $k = 2$. The results of Teper *et al.* [72] are shown in Fig. 2.11, while a more recent calculation [73] for SU(4) and especially SU(6) gauge

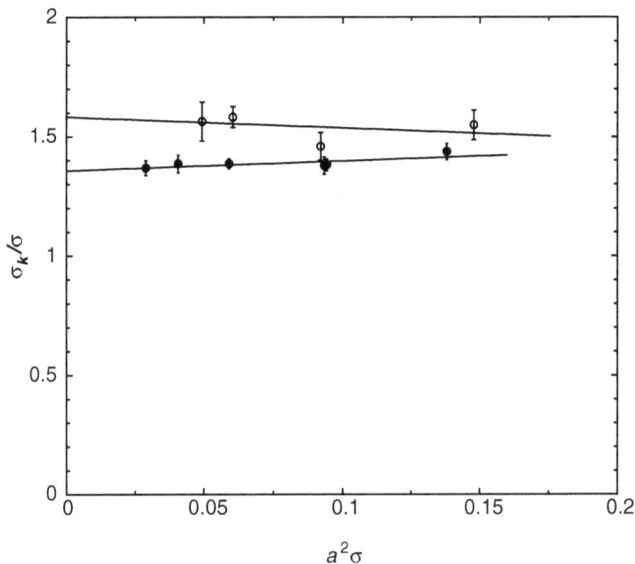

Fig. 2.11. $k = 2$ string tensions in SU(4) ($\bullet$) and SU(5) ($\circ$).

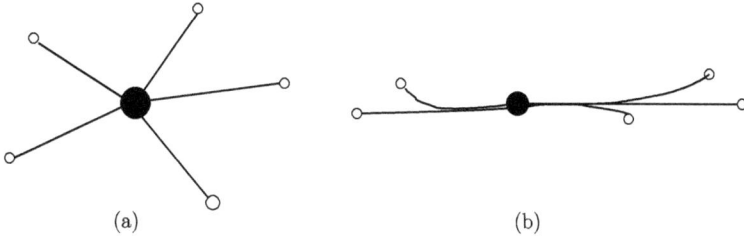

Fig. 2.12. A schematic picture of two possible structures of the large N_c baryons; (a) a conventional quasi-independent quark versus (b) a pencil-like structure created by attractive string–string interaction. The black circle at the center is the string junction, small open circles are quarks. (Although the figure is meant for large N_c, it shows an SU(5) color.)

group (where $k = 3$ is also a non-trivial case) seem to favor the MQCD conjecture more.

Why is it important to know which result is the case? Although both predict that in the large N_c limit strings are weakly attractive, the magnitude of the interaction is qualitatively different, $O(1/N_c^2)$ and $O(1/N_c)$ respectively.

Quite independently of these developments G. Carter and myself [74] put forward the following general argument. If the residual interaction between color objects (like instantons and color strings) is $O(1/N_c)$ at large N_c, this smallness can be compensated by the presence of N_c objects interacting collectively. Although there would be no few-body bound states, their multi-body binding is possible.[25] In Fig. 2.12 we schematically indicate that, should such be the case, it may make large-N_c baryons of pencil-like (b) rather than conventional[26] round shape (a). If the effect is not strong enough for the ground state baryons, it is definitely important for excited states, especially rotated (large orbital momentum $l \sim N_c$) excitations. For large enough N_c a transition between the two images of Fig. 2.12 should have the form of a phase transition. The situation is reminiscent of a transition from spherical to deformed nuclei in the nuclear shell model.

---

[25] Example: in weak interaction theory W, Z are not bound to each other, but clusters of $O(1/\alpha_w)$ of them can be bound; the electroweak sphaleron solution is an example.

[26] For clarity; speaking about baryons we have quarks, of course, which however are not assumed to be light and lead to chiral symmetry breaking and Skyrmion-like baryons.

CHAPTER 3

Euclidean Theory of Tunneling: From Quantum Mechanics to Gauge Theories

3.1. Tunneling in quantum mechanics

3.1.1. *Brief history of tunneling*

Tunneling through a classically impenetrable barrier is the most amazing consequence of quantum mechanics, discovered in 1927–1928 when it was just few months old.[1] First came the paper by Hund [143], devoted to energy levels of a double-well potential, exactly of the type we will discuss below in detail. The context in which it came about for the first time was vibrational states of diatomic molecules.

But by far the most famous paper[2] is that by George Gamow [144], who demystified the alpha-decay of heavy nuclei. The apparent puzzle was why the Rutherford scattering experiments with alpha particles demonstrated the existence of the repulsive barrier at a height of 10 MeV or more, while in the decay of the same nuclei alpha particles emerged with smaller energies, of only a few MeV. The tunneling probability is the square of the famous Gamow factor in the amplitude, which is $\boxed{E}$

$$P_{\text{Gamow}} \sim \exp\left(-\frac{2\pi(2e^2)(Z-2)}{hv}\right), \qquad (3.1)$$

where $2(Z-2)$ is the product of the charges of the alpha particle and the remaining nucleus, and v is the velocity of the outgoing alpha particle. This expression has a characteristic small parameter v in denominator of the exponent, explaining why it takes up to a billion years to decay, in spite of the fact that would-be alpha-clusters beat against the wall about 10^{22} times per second.[3] The resulting semiclassical WKB approximation, which we all learned in quantum mechanical

---

[1] More details can be found e.g. in Ref. [142].

[2] The paper [145] with a similar idea (but without the analytic Gamow factor) was submitted the next *day* after Gamow's paper. My advice to the reader is obvious; never delay a paper, even for a day.

[3] Another advice to the reader; anything may happen once somebody is persistent enough, few or even lots of failures never prove that something is impossible.

courses, was developed in 1930s. A new interesting development along the WKB-line is a convergent semiclassical iterative procedure recently proposed by Friedberg, Lee *et al.* [146].

Many years later, when one might have thought that the subject is completely worked out, Feynman introduced his formulation of quantum mechanics in terms of path integrals. We have already introduced path integrals in an imaginary Euclidean time in Section 1.3. Feynman himself emphasized the role of the classical paths in the classically allowed region, as the ones at which phase is stationary and the contribution of nearby paths do not cancel each other. Further use of Euclidean time led to classical paths describing motion under the barrier. In this case there are no phases and one simply looks for the most probable path. The weight of any path is $\exp(-S[x(\tau)])$, and thus we look for minima of the action. Those satisfy the equations of motion, and thus literally prescribe how tunneling is happening, by *classical* equations in the *classically forbidden* region. These methods are much more powerful than WKB since they can be easily used in multi-dimensional problems. This section is essentially a detailed discussion of two important examples of its use, for the double well potential and for pure gauge theory.

Exercise 3.1.   Derive the Gamow factor (3.1) using the classical under-the-barrier path in Euclidean time.

Exercise 3.2.   Derive a semiclassical expression for the production of a pair of massive scalar particles in a homogeneous electric field (as was done already in 1931, by F. Sauter, *Z. Phys.* **69** (1931) 742, in the 1950s solved exactly by Schwinger). Consider a pair produced at point $x = 0$ and tunneling to the distance where the energy gain from the field can pay for the produced mass $2m$. The energy of a particle is $\sqrt{p^2 + m^2} - eEx$, it is zero if the initial point is $x = 0, p = im$ and the final point is $p = 0, x = m/eE$. Find the Euclidean path between them and calculate its action.

3.1.2.   *Double-well problem and instantons*

The word "tunneling" hints that one can pass the mountain (the barrier due to a repulsive potential) not by climbing and then descending from it, but by going through it, *as if* there were a path through the tunnel. The point of this section is to show that not only *can one imagine* a path through the mountain, it is very easy and useful to find the *best* one.

Before we do it in more detail, let me first give a hint why we have to use unphysical Euclidean time for this to happen. Since the energy is conserved, let us read the Schrödinger equation in this way $E = E_{\text{kinetic}} + V(x)$ where $E_{\text{kinetic}} = \hat{p}^2/2m = -\partial_x^2\psi/(2m\psi)$. In a classically allowed region, $E_{\text{kinetic}} > 0$ and the wave function resemble a wave $\psi \sim \exp(ipx)$ with *real* p. However, if we are in a classically *forbidden* region, in which $E_{\text{kinetic}} < 0$, the momentum is *imaginary* because $\psi \sim \exp(-|p|x)$.

Here comes a formal trick; if p is imaginary, why do we not try to interpret it as a motion in *imaginary time*? Changing t to $\tau = it$ one finds a new classical equation of motion

$$m\frac{d^2x}{d\tau^2} = +\frac{dV}{dx}.$$ (3.2)

The different sign comes of course from i^2 in the l.h.s., but this is the same as flipping the potential upside down! The instantons we are looking for would be classical paths satisfying such a flipped equation.

Let us start with the simplest question, the probability to find a particle at a given point x_0 in the ground state. In the Schrödinger formulation it is given by the square of the wave function $|\psi_0(x_0)|^2$. For example, for a harmonic oscillator

$$S = (1/2m)\int d\tau\,(\dot{x}^2 + \omega_0^2 x^2),$$ (3.3)

the equation with a flipped potential is easy to solve. The resulting path looks as follows:

$$x_{\text{cl}}(\tau) = x_0\exp(-\omega_0|\tau|).$$ (3.4)

It reaches the prescribed point x_0 from both sides, with a break at the derivative, but relaxes to zero — the classical vacuum — both at large negative and positive τ. Such paths, considered in my work [157], I called "fluctons". The corresponding action is

$$S_{\text{cl}} = m\omega_0 x_0^2,$$ (3.5)

and so the probability we are looking for is $P \sim \exp(-m\omega_0 x_0^2)$, the same as in the textbooks.

Although this procedure reminds us of the WKB, there are important differences between them. The WKB can be used for any state, not just the ground state. However, our Euclidean path approach can easily be used for multidimensional barriers[4] or even barriers with an infinite number of dimensions, as is the case for quantum field theories.

Now is the time to introduce the double-well potential, which is traditionally used for the demonstration of tunneling phenomena,

$$S = \int d\tau\left(m\frac{\dot{x}^2}{2} + \lambda(x^2 - f^2)^2\right)$$ (3.6)

---

[4]Say, in case of complicated molecules making chemical reactions, with many atoms moving simultaneously. One cannot solve the Schrödinger equations, but still can do it for the Newton ones! (Quantum field theory is the complicated case, with infinitely many coordinates, so this is important.)

(where dot means the derivative with respect to τ). It has minima at $\pm f$, the two "classical vacua".

Following Hund, let us consider the energy levels of this quantum system, on three levels of sophistication:

(i) If one first ignores tunneling, the energy levels are given by zero point oscillations in each well, with the energy

$$E_0 = \omega/2, \quad \omega = f\sqrt{8\lambda}. \tag{3.7}$$

In order to have a better contact with the gauge theory later in the chapter, we will eliminate f from all expressions substituting it by the ω just defined. The maximal hight of the barrier, for example, is then $V_{\max} = \omega^4/64\lambda$, etc.

(ii) At a low value of the only dimensionless parameter of the model $\lambda/\omega^3 \ll 1$ (the high barrier limit), one can further calculate a whole series of *perturbative corrections* which go as powers of this parameter,

$$E_0 = \tfrac{1}{2}\omega\left[1 + \Sigma_n C_n(\lambda/\omega^3)^n\right]. \tag{3.8}$$

We have already discussed in Chapter 1 how to use Feynman diagrams to calculate these corrections.

(iii) Finally one takes into account tunneling. The left–right degeneracy is lifted and one has close pairs of levels, corresponding to symmetric and antisymmetric wave functions under $x \leftrightarrow -x$, with energies

$$E_{\pm} = \frac{\omega}{2}\left(1 \mp \sqrt{\frac{2\omega^3}{\pi\lambda}}e^{-\omega^3/12\lambda}\right), \tag{3.9}$$

which are separated by an exponentially small gap.

Note the following characteristic feature: the tunneling correction cannot be expanded in series in λ. All Teylor terms for this famous function, $\exp(1/\lambda)$ are *zeros!* It simply means that the Feynman diagrams of the perturbation series "do not know" about the existence of the other well. We mentioned that, since in the QFT context there are multiple attempts to re-sum all the diagrams, in the hope that this will give the ultimate solution of the theory. Alas! All of these diagrams, no matter how many, also do not know about other minima, if such exist.

The tunneling probability is given by the exponent of the minimal necessary action for passing the barrier. Again, to find it one needs only to solve the classical equation of motion with flipped sign of the potential. The solution leading from one maximum of the flipped potential to the other is

$$x_{\mathrm{cl}}(\tau) = f\tanh[\omega(\tau - \tau_0)/2]. \tag{3.10}$$

This path we will call the "*instanton*". The corresponding action is easy to calculate without even differentiation; the action is twice the average potential energy, so

$$S_{\text{cl}} = 2 \int d\tau \lambda (x_{\text{cl}}^2 - f^2)^2 = \frac{\omega^3}{12\lambda} \tag{3.11}$$

and of course the tunneling probability is $P \sim \exp(-S_{\text{clas}})$, so we have already reproduced the exponent in (3.9).

3.1.3. *Pre-exponent and zero modes*

Now we would like to make one more step, which is much more difficult, but it will explain a lot about what we should do later for quantum field theories. Let us write the general tunneling path as small quantum deviations from the classical path

$$x(\tau) = x_{\text{cl}}(\tau) + \delta x(\tau), \tag{3.12}$$

and expand the action in powers of quantum corrections $\delta x(\tau)$. The linear term is absent, because the classical path is an extremum, and we start with the quadratic one

$$S = S_{\text{cl}} + \frac{1}{2} \int d\tau \, \delta x(\tau) \hat{O} \delta x(\tau) \tag{3.13}$$

where there is the following differential operator:

$$\hat{O} = -\frac{m}{2} \frac{d^2}{d\tau^2} + \frac{\delta^2 V}{\delta x^2}\bigg|_{x=x_{\text{cl}}} = -\frac{m}{2} \frac{d^2}{d\tau^2} + 4\lambda x^2 (3x^2 - f^2). \tag{3.14}$$

At times distant from the tunneling moment τ_0 (also called the position of the instanton), the last term is just constant, $8\lambda f^2$. But around τ_0 this last term is strongly changing and it is negative.

We would like to calculate the Gaussian integral over the quantum deviations $\delta x(\tau)$, but the problem is that the coefficient in the quadratic form is time-dependent. Since it is an example of a more general case, let me make a small digression on general multi-dimensional Gaussian integrals of the type

$$I = \int \Pi_n \, dx_n \, \exp\left(-\frac{1}{2} \sum_{i,j} x_i M_{ij} x_j\right). \tag{3.15}$$

If one can diagonalize the matrix, by finding new coordinates y_n such that the exponent looks as a single sum $\exp(-\sum_n \epsilon_n y_n^2/2)$, the integral splits into a product of Gaussian integrals and is easily evaluated

$$I = \Pi_n \sqrt{2\pi/\epsilon_n} = (2\pi)^{N/2} [\det M]^{-1/2}. \tag{3.16}$$

To complete the argument, note that the determinant is invariant under linear rotations, so there is no need to find the diagonal basis y_n.

This idea is generalized to functional integrals. Now instead of the determinant of the matrix we have the determinant of a differential operator. Again, its calculations can go as follows: we have to *diagonalize the operator* by solving the equation

$$\hat{O}x_n(\tau) = \epsilon_n x_n(\tau). \tag{3.17}$$

It is like solving the one-dimensional Schrödinger equation, with τ interpreted as a coordinate

$$-\frac{d^2}{d\tau^2}x_n(\tau) + \omega^2\left(1 - \frac{3}{2\cosh^2(\omega\tau/2)}\right)x_n(\tau) = \epsilon_n x_n(\tau), \tag{3.18}$$

so it is quite doable. In fact just for this potential everything is done in Quantum Mechanics of Landau and Lifshitz (and many other texts).

It turns out that there are exactly two localized solutions, plus of course, the scattering states. The lowest eigenvalue is zero, $\epsilon_0 = 0$, and the next is positive $\epsilon_1 = \frac{3}{4}\omega^2$, and the entire scattering energy is of course above the potential far from the well, which is $\epsilon_k > \omega^2$. The wave function for the zero mode is

$$x_0(\tau) \sim \frac{1}{\cosh^2(\omega\tau/2)}, \tag{3.19}$$

and one can find the normalization constant from the usual normalization condition $\int d\tau x_n^2 = 1$ to be const $= \sqrt{3\omega/8}$.

Every zero should have a simple explanation. And indeed, there is a simple reason for the zero eigenvalue. There is translational symmetry in the problem, so the action does not depend on the instanton displacement. Furthermore, the last formula for the zero mode can be obtained very simply, by differentiation of the instanton solution over τ_0.

$$x_0(\tau) = S_0^{-1/2}\frac{d}{d\tau_0}x_{cl}(\tau - \tau_0). \tag{3.20}$$

Now, if there is a zero eigenvalue, the determinant is zero too, and the tunneling probability under discussion is in fact infinite! Indeed, one of the integrals is actually non-Gaussian; $\epsilon_0 = 0$ and nothing prevents large amplitudes of fluctuations in this direction. This is the simplest case of the so called "valleys" in the functional space.

The solution to this problem is however quite simple; one may not take this integral at all! All we have to do is to rewrite the integral over dC_0 as the integral over the collective coordinate τ_0. Consider a modification of C_0 by dC_0. The path changes by $dx = x(\tau)\, dC_0$ [remember the definition $x(\tau) = \Sigma_n c_n x_n(\tau)$]. At the same time we have another definition of the zero mode from which follows

$$dx = \frac{dx_{cl}}{d\tau_0}d\tau_0 = -\sqrt{S_0}x_0(\tau)\, d\tau_0. \tag{3.21}$$

Equalizing two expressions for dx, we have

$$dC_0 = \sqrt{S_0}\, d\tau_0.\tag{3.22}$$

Returning to our functional integral over the quantum fluctuations, we now have the following form:

$$\int D\delta x(\tau) e^{-S} = e^{-S_{\rm cl}} \prod_{n>0} \sqrt{\frac{2\pi}{\epsilon_n}} \sqrt{S_0} \int d\tau_0.\tag{3.23}$$

The product includes the determinant without the zero mode, it is often called $\det'(\hat{O})$. Its calculation is tedious but straightforward; it just gives a finite factor.

We have to explain however what to do with the divergent integral over the τ_0. Suppose the whole path integral is taken over some finite time, from 0 to $t_{\max}$; then the contribution of tunneling — described by the instanton-type path — grows linearly with $t_{\max}$. One may say that the finite quantity is the *transition probability per unit time*, but it is not a satisfactory solution in the long run. If $t_{\max}$ is very large, it overcomes the smallness of the exponent; at this point we will have to think about an ensemble of *many* instantons.

3.1.4. *Instanton gas*

What happens if one follows the calculation outlined above for a very long time? Certainly, at some point the expression for the probability we derived becomes larger than 1, which makes no sense. The solution would be that there are in fact more than one tunneling events.

Suppose one has a path with n instantons (anti-instantons), placed at $\tau_1 < \cdots < \tau_n < \tau_0$, If they are all separated sufficiently far from each other (as one says, the instanton gas is *dilute*) the total action is the sum of the individual actions. Furthermore, determinants become factorized, and the expression for the transition amplitude in this case reads

$$G(f, -f, \tau_0) = \left(\sqrt{\omega/\pi}\, \exp^{-\omega\tau_0/2}\right)\left(\sqrt{6S_0/\pi}\, e^{-S_0}\right)^n$$
$$\times \int_0^{\tau_0} \omega\, d\tau_n \cdots \int_0^{\tau_2} \omega\, d\tau_1,\tag{3.24}$$

where the integrals are carried out over ordered positions. The factor which repeatedly appears here,

$$d = \sqrt{\frac{6S_0}{\pi}} e^{-S_0} \omega,\tag{3.25}$$

we will call the *instanton density* (per unit time). One can relax the nesting condition on the tunneling moments by integrating over all intervals, but then dividing the result by n factorial. It looks then like the exponential series, but since n is actually odd, one gets the following final expression for the Green function:

$$G(f, -f, \tau_0) = \left(\sqrt{\frac{\omega}{\pi}} \exp^{-\omega\tau_0/2} \right) \sinh\left(\sqrt{\frac{6S_0}{\pi}} e^{-S_0} \omega\tau_0 \right). \qquad (3.26)$$

Now, it may be used at any time. If it is very large, one can utilize another asymptotic form of the sinh, the exponential one. Notice that the total dependence on τ_0 is now again exponential, with the *corrected ground state energy*,

$$E_0 = \frac{\omega}{2} - \sqrt{\frac{6S_0}{\pi}} e^{-S_0} \omega, \qquad (3.27)$$

which is precisely what one also gets for a level shift from the Schrödinger equation in the semiclassical approximation.

Even if the instanton gas is dilute, it is still interesting to ask what happens if two instantons are close, $\tau_k - \tau_{k-1} \sim 1/\omega$. Then they certainly do interact, because the total action is actually *less* than $2S_0$. Numerical studies indeed show that there is a strong positive correlation between the instanton positions at smaller time intervals. We will return to interacting instantons later.

Finally, let me introduce the last issue of this section, related with the so called correlation functions of the coordinates. Those characterize properties of the ground state, and in particular show the crucial role of the dilute instanton gas. The simplest observable we can think of is just the particle coordinate, so let us define the correlator of the coordinates as

$$K(\tau_1 - \tau_2) = \langle x(\tau_1) x(\tau_2) \rangle, \qquad (3.28)$$

where the averaging is supposed to be done by appropriately weighted quantum paths. We imagine now the length of all paths τ_0 to be infinite, so the correlator depends only on the time *difference*.

Recalling the "old fashioned" quantum mechanical expressions, one can use matrix elements of the coordinate operator, and write this function as a sum over stationary states which can be excited by the operator of the coordinate,

$$K(\tau) = \Sigma_n |\langle 0|x|n\rangle|^2 \exp(-E_n\tau). \qquad (3.29)$$

This function is not only *positive* but it is a *monotonically decreasing* one. Although the sum runs over all states, only *parity odd* ones can be excited by the coordinate operator. For example, the symmetric ground state is absent in this sum, because $\langle 0|x|0 \rangle = 0$.

Thus, if one knows the correlation function at large times, one knows E_1, the lowest parity odd state as well. This is analogous to what we will do in Chapter 6 to calculate the lowest excitations in QCD, mesons and baryons.

Now, let us try to understand what the correlation function looks like, if we are not specifically interested in large times. Clearly, the problem has two time scales: (i) the perturbative scale $\tau_{\text{pert}} = 1/\omega$ and (ii) the tunneling scale $\tau_{\text{tunneling}}$, the inverse tunneling rate.

If one imagines the particle sitting in the same well all the time, the correlator is $K(\tau) \approx f^2$. However, the tunneling kills the correlation, and eventually $K(\tau) \to 0$. One can write approximately the paths as sequences of steps or kinks,

$$x(\tau) = f \prod_i \text{sign}(\tau - \tau_i), \tag{3.30}$$

and calculate the correlation function

$\boxed{\text{E}}$

$$K = f^2 \frac{\Sigma_n \int \Pi_n \, d\tau_n \, d^n \, \text{sign}(\tau - \tau_i) \, \text{sign}(-\tau_i)}{\Sigma_n \int \Pi_n \, d\tau_n \, d^n} \tag{3.31}$$

where $d = \sqrt{6S_0/\pi} \, e^{-S_0} \omega$. Its large-time limit is

$\boxed{\text{E}}$

$$K(\tau) \sim \exp(-2\tau d). \tag{3.32}$$

Comparing it to the expression above, one gets the *gap*, the energy splitting between the vacuum and the first excited state, $E_1 - E_0 = 2d$. Intuitively factor 2 appears because of the instanton plus the anti-instantons.

Summarizing this discussion: if the classical action is large $S_0 \gg 1$ (actually, much larger than the Plank constant), the ensemble of paths is an exponentially dilute gas of instantons and anti-instantons. All such paths together lead to "exponentiation" of the tunneling correction, and correspond to the negative shift of the ground state energy. They also randomize the sign of the coordinate and create the gap between the vacuum and the excited states. As we will see later in the chapter, exactly the same thing happens in gauge theories.

Exercise 3.3. Derive expression (3.32) for the correlator of coordinates in the dilute instanton gas approximation.

3.1.5. *Two-loop quantum corrections*

Another advantage of the Euclidean path approach to tunneling is that it allows to develop the semiclassical methods into a systematic procedure, with improvements on its accuracy to any desired order in the expansion parameter $\lambda/\omega^3 \sim 1/S_0$. In this subsection we show how to formulate and use the perturbation theory around the instanton solution.

The one-loop calculation of the determinant of the operator is a bit special, so we use as an example the two-loop calculation. More specifically, we will consider the $1/S_0$ correction to the level splitting in the double well, following the work by my student Wohler and myself [151], which resolved some mistakes in the earlier literature [149, 150]. This quantity has also been evaluated by the numerical simulations I reported in Ref. [157]. The result itself was first obtained using the

Schrödinger equation (rather than diagrams) by Zinn–Justin [152], which was later found out by us.

To next order in $1/S_0$, the tunneling amplitude can be written as

$$\langle -f|e^{-H\tau}|f\rangle = |\psi_0(f)|^2 \left(1 + \frac{2A}{S_0} + \cdots \right)$$

$$\times \exp\left[-\frac{\omega\tau}{2}\left(1 + \frac{B}{S_0} + \cdots\right)\right] 2d\left(1 + \frac{C}{S_0} + \cdots\right)\tau, \qquad (3.33)$$

where the corrections A and B are those to the wave function at its minimum, and to the energy due to the anharmonicity of the oscillations. These two corrections are unrelated to tunneling and we can get rid of them by dividing the amplitude by $\langle f| \exp(-H\tau)|f\rangle$, in which they appear in the same way. We are interested in the coefficient C, the next order correction to the tunneling amplitude (instanton density d) and eventually to the level splitting.

In order to calculate the next order correction to the instanton result, we have to expand the action beyond order $(\delta x)^2$. The result can be interpreted in terms of a new set of Feynman rules in the presence of an instanton (see Fig. 3.1). The triple and quartic coupling constants are $\alpha = -\sqrt{2\lambda} + 4x_{\mathrm{cl}}(t)$ and $\beta = \lambda$ (compared to $\alpha = -\sqrt{2\lambda}$ and $\beta = \lambda$ for the anharmonic oscillator). The propagator is the Green function of the differential operator $\hat{O}$ we defined above. There is one complication due to the fact that this operator has a zero mode and cannot be inverted. The propagator in such cases is uniquely defined in a Hilbert subspace orthogonal to the (translational) zero mode. The analytic form for it found in Ref. [150] is

$$G(x,y) = g_0(x,y)\left[2 - xy + \tfrac{1}{4}|x-y|(11 - 3xy) + (x-y)^2\right]$$
$$+ \tfrac{3}{8}\left(1 - x^2\right)\left(1 - y^2\right)\left[\log\left(g_0(x,y)\right) - \tfrac{11}{3}\right], \qquad (3.34)$$

$$g_0(x,y) = \frac{1 - |x-y| - xy}{1 + |x-y| - xy},$$

where $x = \tanh(t/2)$, $y = \tanh(t'/2)$. There are four diagrams at two-loop order, see Fig. 3.1. The first three diagrams are of the same form as the anharmonic oscillator diagrams. Subtracting these contributions, we get rid of anharmonic effects far from

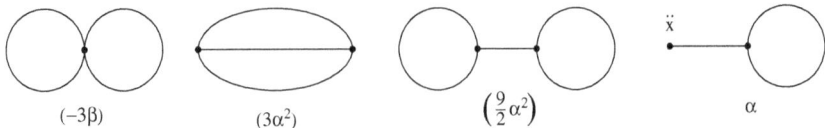

$$(-3\beta) \qquad (3\alpha^2) \qquad \left(\tfrac{9}{2}\alpha^2\right) \qquad \alpha$$

Fig. 3.1. The diagrams for the two-loop corrections to tunneling. The first three come from anharmonic terms in the Lagrangian in the standard way, while the fourth originates from the Jacobian of the zero mode.

the tunneling event. The resulting final contributions are

$$a_1 = -3\beta \int_{-\infty}^{\infty} dt \left[G^2(t,t) - G_0^2(t,t) \right] = -\frac{97}{1680},$$

$$b_{11} = 3\alpha^2 \int_{-\infty}^{\infty} \int_{-\infty}^{\infty} dt\, dt' \left[\tanh(t/2)\tanh(t'/2)\, G^3(t,t') - G_0^3(t,t') \right] = -\frac{53}{1260},$$

$$b_{12} = \frac{9}{2}\alpha^2 \int_{-\infty}^{\infty} \int_{-\infty}^{\infty} dt\, dt' \left[\tanh(t/2)\tanh(t'/2)\, G(t,t)G(t,t')G(t',t') \right.$$

$$\left. - G_0(t,t)G_0(t,t')G_0(t',t') \right] = -\frac{39}{560}.$$

(3.35)

The last diagram comes from expanding the Jacobian in δx. This leads to a tadpole graph proportional to $\ddot{x}_{cl}$, which has no counterpart in the anharmonic oscillator case. We get

$$c_1 = -9\beta \int_{-\infty}^{\infty} \int_{-\infty}^{\infty} dt\, dt' \, \frac{\tanh(t/2)}{\cosh^2(t/2)}$$

$$\times \tanh(t'/2)\, G(t,t')G(t',t') = -\frac{49}{60}.$$

(3.36)

The sum of the four diagrams is $C = a_1 + b_{11} + b_{12} + c_1 = -71/72$, in agreement the result obtained in Ref. [152] and with my numerical studies [157]. The fact that the next order correction is of order one and negative is significant. This implies that quantum fluctuations work against tunneling.

We have presented this calculation in detail in order to show that the instanton method can be systematically extended to higher orders in $1/S$. For gauge theory the same four diagrams appear, but so far a similar calculation has not been performed.

3.2. A digression: tunneling versus perturbative series

3.2.1. *Convergence of perturbative series*

3.2.1.1. *Dyson instability*

Divergent series are the invention of the devil, and it is shameful to base on them any demonstration whatsoever wrote N.H. Abel in 1828. But modern physicists tend to be "non-Abelian", and they use the perturbation theory widely, its divergent series notwithstanding.

The main idea which proved that the perturbative series diverge was suggested by F. Dyson [164] in 1952. In short, the argument went as follows. In QED one makes an expansion in e^2, order by order. Let us think a bit what would happen if $e^2 < 0$. Then protons and electrons would no longer attract each other, so atoms would dissolve. On the other hand, electrons would start attracting each other and congregate in large numbers, and so would protons or positrons. A bit of thought shows that the binding energy in this case can be made arbitrarily large; there would come a complete collapse of the theory and no stable ground state would be possible. Now, if the positive and negative vicinities of the $e^2 = 0$ point are that different, the series expansion around this point obviously cannot be nice and convergent.

In spite of the fact that Dyson's paper was well known to all theorists of the day, and although QED was then on the top of the theory list, nobody had picked up this idea in the 1950s. That happened in the 1960s, by three bright students,[5] A. Vainshtain in 1964 [165] and C. Bender and T.T. Wu in 1969 [163] independently pointed out that a calculation of the rate of the Dyson instability provides the asymptotics of the perturbative series. Both papers used quantum mechanics but had field theories in mind; this connection was made explicit a bit later by Lipatov [166] and then many others, see e.g. a book of reprints [168].

Vainshtein used a cubic potential $V = x^2/2 + gx^3$ which is always unstable, but we will use the quartic one used in the Bender–Wu work,

$$V = \tfrac{1}{2}x^2 + \lambda x^4. \tag{3.37}$$

In this case $\lambda > 0$ leads to a stable single-well problem while $\lambda < 0$ has only a metastable region at $x \sim 0$ with the catastrophic negative potentials at $x \to \pm\infty$.

3.2.1.2.   *Perturbative series in high orders*

The perturbative expansion of the ground state energy,

$$E^0(\lambda) - 1/2 = \sum E_n^0 \lambda^n, \tag{3.38}$$

can be calculated either by the Feynman diagrams or by the recursive relation derived in Ref. [163]. In high orders it has the following behavior:

$$E_n^0 = -(-3)^n \frac{6^{1/2}}{\pi^{3/2}} \Gamma(n + 1/2) \left(1 - \frac{95}{72n} + O(1/n^2)\right). \tag{3.39}$$

Needless to say, the gamma function is a factorial and the series thus badly diverges. The factorial basically originates from the number of diagrams. So, are such series of any use, or was Abel's pessimism correct?

This depends on the value of λ. A simple investigation, by plotting a partial sum till some n displays a typical pattern of *asymptotic* series, shows that they first converge and then diverge. The critical order at which the minimum is reached is $1/3\lambda$ and the value of the error is $\sim\exp(-1/3\lambda)$.

Can these divergent series be actually re-summed into some function, and if so, is such re-summation unique? Before we try to answer those questions, let us recall

---

[5]I emphasize this point to the reader: it was a fruit of independent thinking of very young people. I know for fact that Vainshtein's adviser was not interested and did not even advise him to publish the work. I do not know the setting in which Bender and Wu started their work, but there is also no sign of any senior advisers.

the so called Borel improvement of series, which provides a useful language for that purpose. The so called *Borel transform* of a series with coefficients f_n is defined by

$$B(t) = \sum_n f_n \frac{t^n}{n!}.$$ (3.40)

If this is a nice and unique function, the original series is given by the inverse Borel transform in its integral form,

$$f(\lambda) = \frac{1}{\lambda} \int dt\, e^{t/\lambda} B(t).$$ (3.41)

So let us apply this tool to our perturbative series. The factorial is canceled and we get the answer by a geometric series. It turns out to be different for different signs of λ, namely,

$$B(t) \sim \frac{1}{1 + \text{sign}(\lambda)3t}.$$ (3.42)

So for positive λ there is a unique function, while in the opposite case the pole at $t = 1/3$ is on the integration path in (3.41) and the integral should be better defined by some further prescription. This means that the contribution of the order of $\exp(-1/3\lambda)$ remains basically undefined by the perturbative series. This is also what we have concluded above, from estimates of the series accuracy.

3.2.1.3. *Semiclassical evaluation of Dyson's instability*

Now the main idea of this section is finally going to be applied. Since the Dyson instability determines the singularity at $\lambda = 0$, let us *evaluate its rate*. It is given by the tunneling rate from the potential well around $x = 0$ to the large x region where the negative λx^4 term overtakes the quadratic one. The turning point is obviously at zero energy, so $x_0^2/2 + \lambda x_0^4 = 0$ or $x_0 = 1/(-2\lambda)^{1/2}$. Solving the classical equation of motion in Euclidean time with zero energy, one gets the instanton solution we discussed previously in this chapter. In the notation we now use it has the form and the action

$$x_{\text{inst}}(\tau) = \frac{x_0}{\cosh(\tau - \tau_0)}, \qquad S_{\text{inst}} = \frac{1}{3\lambda}.$$ (3.43)

Thus the value of the instanton classical action finally explains why this particular number, $1/3$, appeared in the high-order coefficients and in the position of the pole in the Borel transform.

Furthermore, performing the calculation of the quantum fluctuations around the classical solution one gets the instability rate in absolute units,

$$\text{Im}\, E^0(\lambda) = \exp\left(-\frac{1}{3\lambda}\right) \left(\frac{2}{-\pi\lambda}\right)^{1/2}.$$ (3.44)

Once the magnitude of the jump on the cut is known, a standard dispersion relation gives the real part of the function which is nothing else but the leading term in (3.39).

3.2.1.4. *High orders of perturbative series in field theories*

The first generalization of the above ideas to field theory has been made by Lipatov [166], who considered a $\lambda\phi^4$ scalar theory. The corresponding instanton, or *lipaton* as it was sometimes called, is easily found using the $4d$ Euclidean spherical ansatz $\phi(x^2)$.

$\boxed{\text{E}}$

However, in the context of quantum field theories the story does not end with the tunneling and the instantons. One can show that the corresponding Borel transform $B(t)$ has also *other singularities*. Nobody knows how many of them exist, but at least one other example (besides instantons) is the so called *renormalons*, related to the Landau pole due to running of the coupling. They generate poles at the value of the Borel parameter $t = n/b$ where n is an integer and b is the first coefficient of the beta function. Physically they correspond to power OPE correction we will discuss in Chapter 6 on correlation functions. Those terms are therefore also undefined by the perturbative series, and should be determined non-perturbatively.

Exercise 3.4.   For the $4d$ scalar field theory $\lambda\phi^4$ with negative λ find the spherical instanton-like solution and its action. Determine the corresponding constant in the asymptotics of the perturbative series in high orders in λ.

3.2.2.   *Instanton–anti-instanton interaction and one more correction to the ground state energy*

We now switch back from the digression we made in the preceding section, which focused on the role of tunneling in the *single*-well potential, when it becomes metastable. Now we go back to the *double*-well potential, with positive quartic x^4 but negative x^2 terms, which has tunneling at the physical value of the coupling constant $\lambda > 0$. Clearly this example is closer to QCD, in which we are eventually interested.

We stopped at the determination of the $1/S_0$ correction to the tunneling rate (instanton density d, or the gap $E_1 - E_0 = 2d$ between the ground state and the first excited state). In this section we discuss how the semiclassical theory can be used to calculate not the gap but the shift of the mean position of the two levels, $E_{\text{ctr}} = (E_0 + E_1)/2$.

In this context, it is customary to define the double well potential by $V = (x^2/2)(1 - gx)^2$. The coupling constant g is related to the coupling λ used above by $g^2 = 2\lambda$. Unlike the splitting, the mean energy is related to topologically *trivial* paths, the simplest of which is an instanton–anti-instanton pair.

If we take the interaction among instantons into account, the contribution from instanton–anti-instanton pairs is given by

$$\langle f|e^{-H\tau_m}|f\rangle = \tau_m \int \frac{d\tau}{\pi g^2} \exp\left(S_{IA}(\tau)\right), \tag{3.45}$$

where $S_{IA}(\tau)$ is the action of an instanton–anti-instanton pair with separation τ and the factor $(\pi g^2)^{-1}$ is the square of the single instanton density. The action of an instanton–anti-instanton (IA) pair can be calculated given an ansatz for the path that goes from one minimum of the potential to the other and back. An example for such a path is the "sum ansatz"

$$q_{\text{sum}}(\tau) = \frac{1}{g}\left(\frac{1}{\tanh(\tau - \tau_I)} - \frac{1}{\tanh(\tau - \tau_A)} - 1\right). \tag{3.46}$$

This path has the action

$$S_{IA}(T) = (1/g^2)(1/3 - 2e^{-T} + O(e^{-2T})), \quad T = |\tau_I - \tau_A|.$$

It is qualitatively clear that if the two instantons are separated by a large time interval $T \gg 1$, the action $S_{IA}(T)$ is close to $2S_0$. In the opposite limit $T \to 0$, the instanton and the anti-instanton annihilate and the action $S_{IA}(T)$ should tend to zero. In that limit, however, the IA pair is at best an approximate solution of the classical equations of motion and it is not clear how the path should be chosen.

The best way to deal with this problem is the "streamline" or "valley" method [211]. In this case one starts with a well separated IA pair and lets the system evolve using the method of steepest descents. A sequence of paths obtained numerically can be found in my paper [157]. Analytically, one can obtain the following result at large T [158]:

$$S(T) = \frac{1}{g^2} \left[\tfrac{1}{3} - 2e^{-T} - 12Te^{-2T} + O(e^{-2T})\right]$$
$$\times \left[1 - 24Te^{-T} + O(e^{-T})\right]\left(1 - \tfrac{71}{6}g^2\right), \tag{3.47}$$

where the first term is the classical streamline interaction up to next-to-leading order, the second term is the quantum correction (one loop) to the leading order interaction, and the last term is the two-loop correction to the single instanton interaction.

If one tries to use the instanton result (3.45) in order to calculate corrections to E_{ctr} one encounters two problems. First, the integral diverges at large T. This is simply related to the fact that IA pairs with large separation should not be counted as pairs, but as independent instantons. This problem is easily solved, just subtract the square of the single instanton contribution. Second, once regulated, the integral is dominated by the region of small T, where the action is not reliably calculable. This problem is real, it is related to the fact that E_{ctr} is not determined by tunneling, but given by the sum of ordinary perturbation theory. Perturbation theory in g is however not convergent (not even Borel summable), so the calculation of the IA contribution requires a suitable definition of perturbation theory.

This problem can be dealt with by the analytic continuation in g. For g imaginary (g^2 negative), the IA contribution is well defined (the integral over T is dominated by $T \sim -\log(-g^2)$). The IA contribution to E_{ctr} is [159]

$$E_{\text{ctr}}^{(2)} = \frac{e^{-1/3g^2}}{\pi g^2} \left[\log\left(-\frac{2}{g^2}\right) + \gamma + O(g^2 \log(g^2)) \right], \qquad (3.48)$$

where $\gamma = 0.577\ldots$ is the Euler constant. When we now continue back to positive g^2, we get both real and imaginary contributions to E_{ctr}. Since the sum of all contributions to E_{ctr} is certainly real, the imaginary part has to cancel against a small $O(e^{-1/3g^2})$ imaginary part in the perturbative part. This allows us to determine the imaginary part $\text{Im} E_{\text{ctr}}^{(0)}$ of the analytically continued perturbative sum.[6]

From the knowledge of the imaginary part one can determine the large order behavior of the perturbation series $E_{\text{ctr}}^{(0)} = \sum_k g^{2k} E_{\text{ctr},k}^{(0)}$ [166, 167]. The coefficients are given by the dispersion integrals

$$E_{\text{ctr},k}^{(0)} = \frac{1}{\pi} \int_0^\infty \text{Im}\left[E_{\text{ctr},k}^{(0)}(g^2) \right] \frac{dg^2}{g^{2k+2}}. \qquad (3.49)$$

Since the semiclassical results are reliable for small g, we can calculate the large order coefficients. Including the corrections calculated in Ref. [158], we have

$$E_{\text{ctr},k}^{(0)} = \frac{3^{k+1}k}{\pi} \left(1 - \frac{53}{18k} + \cdots \right). \qquad (3.50)$$

The result can be compared with the exact coefficients [167]. For small k the result is completely wrong, but already for $k = 5, 6, 7, 8$ the ratio of the asymptotic result to the exact coefficients is close to 1, namely 1.04, 1.11, 1.12, 1.11.

An interesting way to use this formula was suggested in Ref. [158]. In order to improve the estimate of E_{ctr}, we use the first N terms of the perturbation series and add the IA contribution. The best accuracy occurs when $|N - 1/3g^2| \sim 1$. In this case, the best estimate is given by[7]

$$E_{\text{ctr}} = \sum_{n=0}^{N} g^{2n} E_{\text{ctr},n}^{(0)} + \frac{3Ne^{-N}}{\pi} \left(\log 6N + \gamma + \frac{1}{3}\sqrt{\frac{2\pi}{N}} \right) \left(1 - \frac{53}{18N} \right). \qquad (3.51)$$

In summary: we learned in this section that a common shift of both levels E_{ctr} is related to configurations with zero net topology, and the calculation of non-perturbative effects requires better definition of the perturbation sum. This can

---

[6] How can the perturbative result develop an imaginary part? After analytic continuation, the perturbative sum is Borel summable, because the coefficients alternate in sign. If we define $E_{\text{ctr}}^{(0)}$ by analytic continuation of the Borel sum, it will have an imaginary part for positive g^2.

[7] The fermionic determinant and related issues we will discuss in the next chapter in detail.

be accomplished using analytic continuation, and in this case we can perform a reliable interacting instanton calculation. The result not only provides an estimate of the large order behavior of perturbation theory, but also an accurate formula for E_{ctr}.

3.3. Fermions coupled to the double-well potential

In this section we will consider one fermionic degree of freedom ψ_α ($\alpha = 1, 2$) coupled to the double-well potential. This model provides additional insight into the vacuum structure not only of quantum mechanics, but also of gauge theories; we will see that fermions are intimately related to tunneling, and that the fermion-induced interaction between instantons leads to strong instanton–anti-instanton correlations. Another motivation for studying fermions coupled to the double-well potential is that for a particular choice of the coupling constant, the theory is *supersymmetric*. Therefore all perturbative corrections to the vacuum energy cancel out, and the instanton contribution (the only shift there remains) is more easily defined.

The model is defined by the action

$$S = \frac{1}{2} \int dt \, (\dot{x}^2 + W'^2 + \psi\dot{\psi} + cW''\psi\sigma_2\psi), \tag{3.52}$$

where dots denote time and primes spatial derivatives, and $W' = x(1 - gx)$. We will see that the vacuum structure depends crucially on the Yukawa coupling c. For $c = 0$ fermions decouple and we recover the double-well potential studied in the previous sections, while for $c = 1$ the classical action is supersymmetric. The supersymmetry transformation is given by

$$\delta x = \zeta\sigma_2\psi, \quad \delta\psi = \sigma_2\zeta\dot{x} - W'\zeta, \tag{3.53}$$

where ζ is a Grassmann variable. W is usually referred to as the super potential. The action (3.52) can be rewritten in terms of two bosonic partner potentials [160,161]. Nevertheless, it is instructive to keep the fermionic degree of freedom, because the model has many interesting properties that also apply to QCD, where the action cannot be bosonized.

As before, the potential $V = \frac{1}{2}W'^2$ has degenerate minima connected by the instanton solution. The tunneling amplitude is given by

$$\text{Tr} \left(e^{-\beta H} \right) = \int d\tau \, J \, \frac{1}{\sqrt{\det \mathcal{O}'_B}} \det \mathcal{O}_F \, e^{-S_{\text{cl}}}, \tag{3.54}$$

where S_{cl} is the classical action, $\mathcal{O}_B$ is the bosonic operator and $\mathcal{O}_F$ is the fermionic Dirac operator

$$\mathcal{O}_F = \frac{d}{dt} + c\sigma_2 W''(x_{\text{cl}}). \tag{3.55}$$

As explained in Section 3.1.3, $\mathcal{O}_B$ has a zero mode, related to translational invariance. This mode has to be treated separately, which leads to a Jacobian J and an integral over the corresponding collective coordinate τ. The fermion determinant also has a zero mode,[8] given by

$$\chi^{I,A} = N \exp\left(\mp \int_{-\infty}^{t} dt'\, cW''(x_{\text{cl}})\right) \frac{1}{\sqrt{2}} \begin{pmatrix} 1 \\ \pm i \end{pmatrix}. \tag{3.56}$$

Since the fermion determinant appears in the numerator of the tunneling probability, the presence of a zero mode implies that the tunneling rate is zero!

The reason for this is as follows: the two vacua have different fermion number, so they cannot be connected by a bosonic operator. The tunneling amplitude is non-zero only if a fermion is created during the process, $\langle 0, +|\psi_+|0, -\rangle$, where $\psi_\pm = (1/\sqrt{2})(\psi_1 \pm i\psi_2)$ and $|0, \pm\rangle$ denote the corresponding eigenstates. Formally, we get a finite result because the fermion creation operator absorbs the zero mode in the fermion determinant. As we will see later, this mechanism is completely analogous to the axial $U(1)_A$ anomaly in QCD and baryon number violation in electroweak theory. For $c = 1$, the tunneling rate is given by [160]

$$\langle 0, +|\psi_+|0, -\rangle = \sqrt{\frac{2}{\pi g^2}} e^{-1/6g}. \tag{3.57}$$

Let us now return to the calculation of the ground state energy. For $c = 0$, [E] the vacuum energy is the sum of perturbative contributions and a negative non-perturbative shift $O(e^{-1/6g})$ due to individual instantons. For $c \neq 0$, the tunneling amplitude (3.57) will only enter squared, so one needs to consider instanton–anti-instanton pairs. Between the two tunneling events, the system has an excited fermionic state, which causes a new interaction between the instantons. For $c = 1$, supersymmetry implies that all perturbative contributions (including the zero-point oscillation) to the vacuum energy cancel. Using supersymmetry, one can calculate the vacuum energy from the tunneling rate[9] (3.57) [160]. The result is $O(e^{-1/3g})$ and positive, which implies that supersymmetry is broken.[10] While the dependence on g is what we would expect for a gas of instanton–anti-instanton molecules, understanding the sign in the context of an instanton calculation is more subtle (see below).

---

[8]In the supersymmetric case, the fermion zero mode is the super partner of the translational zero mode.

[9]The reason is that for SUSY theories, the Hamiltonian is the square of the SUSY generators Q_α, $H = \frac{1}{2}\{Q_+, Q_-\}$. Since the tunneling amplitude $\langle 0, +|\psi_+|0, -\rangle$ is proportional to the matrix element of Q_+ between the two different vacua, the ground state energy is determined by the square of the tunneling amplitude.

[10]This was indeed the first known example of the non-perturbative breaking of supersymmetry, due to Witten [162].

Exercise 3.5. The quantum mechanical formulation of a double-well system coupled to fermions can be directly solved numerically; we deal with just two coupled Schrödinger equations. For $c = 1$ it is instructive to see that the vacuum energy shift indeed becomes exponentially small.

For $c \neq 1$ the instanton–anti-instanton contribution to the vacuum energy has to be calculated directly. Also, even for $c = 1$, where the result can be determined indirectly, this is a very instructive calculation. For an instanton–anti-instanton path, there is no fermionic zero mode. Writing the fermion determinant in the basis spanned by the original zero modes of the individual pseudo-particles, we have

$$\det\left(\mathcal{O}_F\right)_{ZMZ} = \begin{pmatrix} 0 & T_{IA} \\ T_{AI} & 0 \end{pmatrix}, \tag{3.58}$$

where T_{IA} is the overlap matrix element

$$T_{IA} = \int_{-\infty}^{\infty} dt\, \chi_A \left[\partial_t + c\sigma_2 W''(x_{IA}(t))\right] \chi_I. \tag{3.59}$$

Clearly, mixing between the two zero modes shifts the eigenvalues away from zero and the determinant is non-zero. As before, we have to choose the correct instanton–anti-instanton path $x_{IA}(t)$ in order to evaluate T_{IA}. Using the valley method introduced in the last section the ground state energy is given by [211]

$$E = \frac{1}{2}\left(1 - c + O(g)\right) - \frac{1}{2\pi} e^{-1/3g} \left(\frac{g}{2}\right)^{c-1} 2c^c$$
$$\times \int_0^\infty d\tau \, \exp\left(-2c(\tau - \tau_0) + \frac{2}{g} e^{-2\tau}\right), \tag{3.60}$$

where τ is the instanton–anti-instanton separation and $\exp(-2\tau_0) = cg/2$. The two terms in the exponent inside the integral correspond to the fermionic and bosonic interaction between instantons. One can see that fermions cut off the integral at large τ. There is an attractive interaction which grows with distance and forces instantons and anti-instantons to be correlated. Therefore, for $c \neq 0$ the vacuum is no longer an ensemble of random tunneling events, but consists of correlated instanton–anti-instanton molecules.

The fact that both the bosonic and fermionic interactions are attractive means that the integral (3.60), just like (3.45), is dominated by small τ where the integrand is not reliable. This problem can be solved as outlined in the last section, by analytic continuation in the coupling constant. As an alternative, Balitsky and Yung suggested to shift the integration contour in the complex τ-plane, $\tau \to \tau + i\pi/2$. On this path, the sign of the bosonic interaction is reversed and the fermionic interaction picks up a phase factor $\exp(ic\pi)$. This means that there is a stable saddle point, but the instanton contribution to the ground state energy is in general complex.

The imaginary part cancels against the imaginary part of the perturbation series, and only the sum of the two contributions is well defined.

A special case is the supersymmetric point $c = 1$. In this case, perturbation theory vanishes and the contribution from instanton–anti-instanton molecules is real,

$$E = \frac{1}{\pi} e^{-1/3g} \left[1 + O(g) \right]. \tag{3.61}$$

This implies that at the SUSY point $c = 1$, there is a well defined instanton–anti-instanton contribution. The result agrees with what one finds from the supersymmetric Hamiltonian $H = \frac{1}{2}\{Q_+, Q_-\}$ or directly from the Schrödinger equation.

In summary: in the presence of light fermions, the tunneling is possible only if the fermion number changes during the transition. Fermions create a long-range attractive interaction between instantons and anti-instantons and the vacuum is dominated by instanton–anti-instanton "molecules". It is non-trivial to calculate the contribution of these configurations to the ground state energy, because topologically trivial paths can mix with perturbative corrections. The contribution of molecules is most easily defined if one allows the collective coordinate (time separation) to be complex. In this case, there exists a saddle point where the repulsive bosonic interaction balances the attractive fermionic interaction and molecules are stable. These objects give a non-perturbative contribution to the ground state energy, which is in general complex, except in the supersymmetric case where it is real and positive.

3.4.   Instantons in gauge theories

3.4.1.   *Topologically nontrivial objects*

Before we go to specifics related to $4d$ instantons, let me remind the reader that those belong to long-studied family of topological solitons in different dimensions.

The simplest $1d$ topological object is a *domain wall*, named after a wall interpolating between domains with different magnetization of a ferromagnet. Their topology is related to discrete groups, with several distinct vacua. An example of that is the double-well potential we studied above, and quantum-mechanical instanton interpolating between those is also an example of the $1d$ domain wall topology. In $1d$ there are indeed two infinities, $\tau = \pm\infty$, which can be mapped on the field non-trivially.

But perhaps much more familiar are vortices, $2d$ objects describing specific rotational motion of a liquid or gas. Cyclones and hurricanes are routinely used in weather forecasts, allowing one to make sense of quite complicated pictures of local flow velocities and temperatures. Its long-time stability is known empirically and is easily understood e.g. in ideal hydrodynamics. Quantum liquid helium or superconductors have also been studied in great detail. In contrast to classical ones, their vorticity is quantized.

Let us recall why this happens. Consider the simplest theory with one scalar field ψ, describing the collective wave function of *condensed* He^4 atoms. Let us think of it as a modulus and the phase $\psi = |\psi| \exp(i\phi)$. For one vortex with its axis in the z direction, those depend on $2d$ coordinates, say the distances from the axis r and the azimuthal angle α. The simplest vortex solution has a phase proportional to the angle, in order to have velocity — the gradient of the phase — to be rotational. The main point here is the following: the coefficient can only be an integer n,

$$\phi = n\alpha, \tag{3.62}$$

simply because the original wave function ψ must be a unique function of its coordinate. What we have described here is the simplest example of a mapping of one circle to another, which is described by the so called *winding number* n. Note that in this (and all other examples to follow) one *maps the boundary of the space* (in this case $2d$ plane) described by angle α *on the field variables* (the angle ϕ).

The stability of topological quantum numbers like the winding number n with respect to any smooth perturbations is quite obvious in this example: if the phase winds n times while we go around a large circle, it cannot be smoothly changed to another value. This explains why the lifetime of metastable quantized vortices in liquid He^4 can be very long, up to an hour.

The vorticity implies that velocity decays slowly with distance from the vortex core, namely $v_\phi \sim 1/r$, and therefore their energy $E_{\text{vortex}} \sim \int v_\phi^2 r \, dr \sim \log(r_{\text{max}}/r_{\text{min}})$ is IR and UV divergent. This is not the case if a gauge field is also present as is the situation in superconductors. The gauge field can compensate for rotational velocity and lead to exponential decay of rotational velocity at large r and finite-mass-per-length objects.

A similar thing happens with the $3d$ topological object, such as the 't Hooft–Polyakov monopole, namely its mass becomes finite and the magnetic flux quantized because of interplay of the gauge and scalar fields. Another $3d$ object is the Skyrmion described in the preceding chapter.

Any of these objects can serve both as soliton and as instanton, depending on whether all of the coordinates are spatial or one of them is interpreted as an Euclidean time. In this book we would not describe any of these objects in detail, nor discuss mathematical aspects of topological objects. Our interests would be focused on a $4d$ objects, gauge field instantons. A reader interested in topology can use for an introduction e.g. a book [169]. Good introductory texts on instantons are Coleman's lecture [170] and "ABC of instantons" [171] by Novikov *et al.*

3.4.2. *Topologically distinct pure gauge configurations*

Before we study tunneling phenomena in Yang–Mills theory, we have to become more familiar with the classical vacuum of the theory. In the Hamiltonian formulation, it is convenient to use the temporal gauge $A_0 = 0$[11] so that the conjugate momentum to the field variables $A_i(x)$ is just the electric field $E_i = \partial_0 A_i$. The Hamiltonian is given by

$$H = \frac{1}{2g^2} \int d^3x \, (E_i^2 + B_i^2), \tag{3.63}$$

where E_i^2 is the kinetic and B_i^2 the potential energy term. The classical vacuum corresponds to configurations with zero field strength. For non-Abelian gauge fields this does not imply that the potential has to be constant, but limits the gauge fields to be "pure gauge"

$$A_i = iU(\vec{x})\partial_i U(\vec{x})^\dagger. \tag{3.64}$$

In order to enumerate the classical vacua we have to classify all possible gauge transformations $U(\vec{x})$. This means that we have to study equivalence classes of maps from 3-space R^3 into the gauge group $SU(N)$. In practice, we can restrict ourselves to matrices satisfying $U(\vec{x}) \xrightarrow{x \to \infty} 1$ [214]. Such mappings can be classified using an integer called the winding (or Pontryagin) number, which counts *how many times the group manifold is covered by mapping*

$$n_W = \frac{1}{24\pi^2} \int d^3x \, \epsilon^{ijk} \, \mathrm{Tr} \left[(U^\dagger \partial_i U)(U^\dagger \partial_j U)(U^\dagger \partial_k U) \right]. \tag{3.65}$$

Because of its topological meaning, continuous deformations of the gauge fields do not change n_W. In the case of $SU(2)$, an example of a mapping with winding number n can be found from the "hedgehog" ansatz

$$U(\vec{x}) = \exp(if(r)\tau^a \hat{x}^a), \tag{3.66}$$

where $r = |\vec{x}|$. For this mapping, we find

$$n_W = \frac{2}{\pi} \int dr \, \sin^2(f) \frac{df}{dr} = \frac{1}{\pi} \left[f(r) - \frac{\sin(2f(r))}{2} \right]_0^\infty. \tag{3.67}$$

In order for $U(\vec{x})$ to be uniquely defined, $f(r)$ has to be a multiple of π at both zero and infinity, so that n_W is indeed an integer. Any smooth function with $f(r \to \infty) = 0$ and $f(0) = n\pi$ provides an example for a function with winding number n.

We conclude that there is an infinite set of classical vacua enumerated by an integer n. Since they are topologically different, one cannot go from one vacuum to another by means of a continuous gauge transformation. Therefore, there is no path from one vacuum to another, such that the energy remains zero all the way.

---

[11] Here we use matrix notation $A_i = A_i^a t^a / 2$, where the $SU(N)$ generators satisfy $[t^a, t^b] = 2if^{abc}t^c$ and are normalized according to $\mathrm{Tr}(t^a t^b) = 2\delta^{ab}$.

3.4.3. *Digression: spherically symmetric Yang–Mills fields*

After we have discussed pure gauge fields and identified classical vacua of the theory, the next logical step is to consider static magnetic fields which provide the barrier separating those vacua. For this purpose, as well for others to follow later in Chapter 7, we will need to consider spherically symmetric Yang–Mills fields. We will do so in this subsection, following Witten's paper [202].

For the SU(2) color subgroup we are mainly interested in, for spherically symmetric field configurations four components of the field A_μ can be expressed through four functions $A(t,r) \cdots D(t,r)$ by using the following space ($j = 1{-}3$) and color ($a = 1{-}3$) structures

$$\mathcal{A}_j^a = A(r,t)\Theta_j^a + B(r,t)\Pi_j^a + C(r,t)\Sigma_j^a,$$

$$\mathcal{A}_0^a = D(r,t)\frac{x^a}{r}, \tag{3.68}$$

with

$$\Theta_j^a = \frac{\epsilon_{jam}x^m}{r}, \qquad \Pi_j^a = \delta_{aj} - \frac{x_a x_j}{r^2}, \qquad \Sigma_j^a = \frac{x_a x_j}{r^2}. \tag{3.69}$$

It is convenient to re-express functions $A, B, C,$ and D through the new set of r, t dependent parameters, which are related to the Abelian gauge ($A_{\mu=0,1}$) and the Higgs (ϕ, α) field on a hyperboloid [202]

$$A = \frac{1 + \phi\sin\alpha}{r}, \qquad B = \frac{\phi\cos\alpha}{r}, \qquad C = A_1, \qquad D = A_0. \tag{3.70}$$

One can express the field strength in terms of those as

$$\mathcal{E}_j^a = \mathcal{G}_{0j}^a = \frac{1}{r}[\partial_0\phi\sin\alpha + \phi\cos\alpha(\partial_0\alpha - A_0)]\Theta_j^a$$

$$+ \frac{1}{r}[\partial_0\phi\cos\alpha - \phi\sin\alpha(\partial_0\alpha - A_0)]\Pi_j^a + (\partial_0 A_1 - \partial_1 A_0)\Sigma_j^a, \tag{3.71}$$

$$\mathcal{B}_j^a = \frac{1}{2}\epsilon_{jkl}\mathcal{G}_{kl}^a$$

$$= \frac{1}{r}[-\partial_1\phi\cos\alpha + \phi\sin\alpha(\partial_1\alpha - A_1)]\Theta_j^a$$

$$+ \frac{1}{r}[\partial_1\phi\sin\alpha + \phi\cos\alpha(\partial_1\alpha - A_1)]\Pi_j^a + \frac{1 - \phi^2}{r^2}\Sigma_j^a, \tag{3.72}$$

where $\partial_0 \equiv \partial_t$ and $\partial_1 \equiv \partial_r$.

Action in 4d Minkowski space then looks as follows:

$$S = \frac{1}{4g^2}\int d^3x\, dt\, [(\mathcal{B}_j^a)^2 - (\mathcal{E}_j^a)^2]$$

$$= 4\pi\int dr\, dt\left((\partial_\mu\phi)^2 + \phi^2(\partial_\mu - a_\mu)^2 + \frac{(1 - \phi^2)^2}{2r^2} - \frac{r^2}{2}(\partial_0 A_1 - \partial_1 A_0)^2\right), \tag{3.73}$$

with the summation over 2-dimensional Minkowski $(-,+)$ metric.

The spherical ansatz is consistent with remaining special gauge transformations generated by a unitary matrix of the type

$$U(r, t) = \exp(i\beta(r, t)\tau^a x^a/(2r)).$$ (3.74)

These transformations coincide with the gauge symmetry of the corresponding 2-d Abelian Higgs model,

$$\phi' = \phi, \qquad \alpha' = \alpha + \beta, \qquad A'_\mu = A_\mu + \partial_\mu\beta.$$ (3.75)

The symmetry can be used to gauge out, for example, the $\mathcal{A}_0$ field.

For completeness, we also give expressions for the topological current (3.81) and charge to be explained in detail in the next subsections. In the spherical ansatz and the $A_0 = 0$ gauge the topological current is

$$K^0 = \frac{1}{8\pi^2 r^2}[(1 - \phi^2)(\partial_1\alpha - A_1) - \partial_1(\alpha - \phi\cos\alpha)],$$

$$K^i = \frac{x^i}{8\pi^2 r^3}[(1 - \phi^2)\partial_0\alpha - \partial_0(\alpha - \phi\cos\alpha)],$$ (3.76)

and the topological charge is

$$Q = \partial_\mu K^\mu$$

$$= \frac{1}{8\pi^2 r^2}[-\partial_0((1 - \phi^2)(\partial_1\alpha - A_1)) + \partial_1((1 - \phi^2)(\partial_0\alpha - A_0))].$$ (3.77)

The Chern–Simons number is

$$N_{CS} = \int d^3x\, K_0$$

$$= -\frac{1}{2\pi}\int dr\,(1 - \phi^2)(\partial_1\alpha - A_1) + \frac{1}{2\pi}(\alpha - \cos\alpha)|_{r=0}^{r=\infty},$$ (3.78)

where the first term is gauge invariant and is called often the "corrected" or true Chern–Simons number $\tilde{N}_{CS}$.

At any given time it is possible to use special gauge transformation (3.74) with time-independent angle β to gauge out $A_1(r)$ not affecting the general $A_0 = 0$ gauge. At $t = 0$ the energy of the field is thus expressed in terms of ϕ, α only

$$E = \frac{4\pi}{g^2}\int dr\left((\partial_r\phi)^2 + \phi^2(\partial_r\alpha)^2 + \frac{(1 - \phi^2)^2}{2r^2}\right).$$ (3.79)

3.4.4. *Static magnetic configurations and their minimal energy*

Now we are ready to make the first step toward the understanding of the gauge field configurations which are *not* pure gauge. They will be however still restricted by several simplifying assumptions.

First, let us start with *static* (time-independent) fields; before understanding the dynamical effects and motion, it is a good idea to know the potential energy itself. Since we are still in the $A_0 = 0$ gauge it means that *no electric field is present* and the non-zero gauge field strength is purely *magnetic*.

The first physics point we will address is that there exists a generalization of the topological winding number formula which can be used for non-pure-gauge configurations; *the Chern–Simons number*

$$n_{CS} = \frac{1}{16\pi^2} \int d^3x\, \epsilon^{ijk} \left(A_i^a \partial_j A_k^a + \tfrac{1}{3} f^{abc} A_i^a A_j^b A_k^c \right).$$ (3.80)

It can also be interpreted as some charge, since it is related to the integral $N_{CS} = \int d^3x K_0$ of the zeroth component of the *topological current*

$$K_\mu = -\frac{1}{32\pi^2} \epsilon^{\mu\nu\rho\sigma} (\mathcal{G}_{\nu\rho}^a \mathcal{A}_\sigma^a - \tfrac{1}{3} g \epsilon^{abc} \mathcal{A}_\nu^a \mathcal{A}_\rho^b \mathcal{A}_\sigma^c).$$ (3.81)

Although this current is not gauge invariant, its divergence is related to the (gauge invariant) local topological charge

$$\partial_\mu K^\mu = -\frac{1}{32\pi^2} \mathcal{G}_{\mu\nu}^a \tilde{\mathcal{G}}_{\mu\nu}^a = Q.$$ (3.82)

Exercise 3.6. The reader is encouraged to check that (i) the Chern–Simons number indeed coincides with the winding number n_W (3.65) for pure gauge field; and (ii) to derive relation (3.82), in order to see that this central formula is just an identity.

The Chern–Simons number is the "topological coordinate" for gauge fields; as we will see shortly the topological barrier and tunneling are described in its terms. The meaning of all of that will be clarified in subsequent sections.

The next question is: can one find the gauge fields with a given N_{CS} with the *minimal possible energy*? If so, those would serve as the best root to go from one (zero energy) classical vacuum to another. (As we learned in the previous section, for integer $N_W = N_{CS}$ pure gauge without any field strength and thus with zero energy would do.) We will indeed be able to find the shape of the topological barrier, following in this section the work by Ostrovsky, Carter and myself [175].

As an additional simplification, we will consider the O(3) spherically symmetric fields; convenient notations have just been discussed in the preceding subsection 3.4.3. The motivation for that is simple; perhaps the field configurations with the minimal energy have the maximal possible symmetry.

Before we can do so we have to break the dilatation (or scale) symmetry of the problem, which prevents any configuration of finite size to be the minimum of energy. This can be achieved by setting a requirement that the ratio

$$\langle r^2 \rangle = \frac{\int d^3x\, r^2 \mathcal{B}^2}{\int d^3x\, \mathcal{B}^2},$$ (3.83)

has some particular value on the static solution we are looking for. To keep both Chern–Symons number and mean radius constant we introduce two Lagrange multipliers $1/\rho^2, \eta$ and search for the minimum of the following combination:

$$\tilde{E} = \frac{4\pi}{g^2} \int dr \left(1 + \frac{r^2}{\rho^2}\right) \left((\partial_r \phi)^2 + \phi^2 (\partial_r \alpha)^2 + \frac{(1 - \phi^2)^2}{2r^2}\right)$$

$$+ \frac{\eta}{2\pi} \int dr \, (1 - \phi^2) \partial_r \alpha. \tag{3.84}$$

It is convenient to introduce the new variable $\xi = 2 \arctan(r/\rho) - \pi/2$. Then

$$\tilde{E} = \frac{8\pi}{g^2} \left[\int_{-\pi/2}^{\pi/2} d\xi \left((\partial_\xi \phi)^2 + \phi^2 (\partial_\xi \alpha)^2 + \frac{(1 - \phi^2)^2}{2 \cos^2 \xi} + \kappa(1 - \phi^2)\partial_\xi \alpha\right)\right], \tag{3.85}$$

where $\kappa = \eta \rho g^2/(32\pi^2)$. The Euler–Lagrange equations are

$$\partial_\xi^2 \phi - \phi(\partial_\xi \alpha)^2 + \frac{(1 - \phi)^2 \phi}{\cos^2 \xi} + 2\kappa \phi \partial_\xi \alpha = 0,$$

$$\partial_\xi(\phi^2 \partial_\xi \alpha) + \kappa \partial_\xi(1 - \phi^2) = 0. \tag{3.86}$$

Finiteness of the energy demands the boundary conditions $\phi^2(\xi = -\pi/2) = \phi^2(\pi/2) = 1$.
The second of Eqs. (3.86) gives

$$\partial_\xi \alpha = -\kappa \frac{1 - \phi^2}{\phi^2}, \tag{3.87}$$

with the integration constant equal to 0 as it follows from the form of energy. After substitution of $\partial_\xi \alpha$ in Eq. (3.86) one has

$$\partial_\xi^2 \phi + \frac{(1 - \phi^2)\phi}{\cos^2 \xi} = \kappa^2 \frac{1 - \phi^4}{\phi^3}. \tag{3.88}$$

The solution to this equation exists for $-1 < \kappa < 1$; it is $\phi^2 = 1 - (1 - \kappa^2)\cos^2 \xi$. Assuming ϕ to be positive one finds finally

$$\phi(r) = \left(1 - (1 - \kappa^2)\frac{4\rho^2 r^2}{(r^2 + \rho^2)^2}\right)^{1/2}, \tag{3.89}$$

$$\partial_r \alpha(r) = -2\kappa \frac{1 - \phi^2}{\phi^2} \frac{\rho}{r^2 + \rho^2}. \tag{3.90}$$

For any κ, the mean radius of the solution is the same $\langle r^2 \rangle = \rho^2$, and the energy density, the total energy, and the (corrected) Chern–Symons number are, respectively,

$$B^2/2 = 24(1 - \kappa^2)^2 \rho^4 / (r^2 + \rho^2)^4,$$
$$E_{\text{stat}} = 3\pi^2 (1 - \kappa^2)^2 / (g^2 \rho),$$
$$\tilde{N}_{CS} = \text{sign}(\kappa)\,(1 - |\kappa|)^2 (2 + |\kappa|)/4.$$

(3.91)

The last two equations define the parametric form of the potential, see Fig. 3.2. More precisely, the same profile continues as a periodic potential with zeros at all integer values of N_{CS}, as a chain of mountains separated by valleys. In fact there are mountains of any height, but the tall ones are narrow. If the energy is expressed in units of $1/g\rho$, it becomes unique. Note that the maximum is about parabolic but near zero energy the behavior is linear; so valleys are actually more like deep canyons. There are alternative ways to derive the potential, in a dynamical path across those mountains; see Ref. [175] for more details.

The maximum has a special meaning, it is the so called *sphaleron solution* corresponding to $\kappa = 0$ and $N_{CS} = 1/2$,

$$\phi = \frac{|r^2 - \rho^2|}{r^2 + \rho^2}, \qquad \alpha = \pi\theta(r - \rho).$$

(3.92)

The name was suggested originally in the context of electroweak theory by Manton [173], it means in Greek "ready to fall". Indeed, the maximum, although a solution to the equation of motion, is unstable. In electroweak theory the Higgs

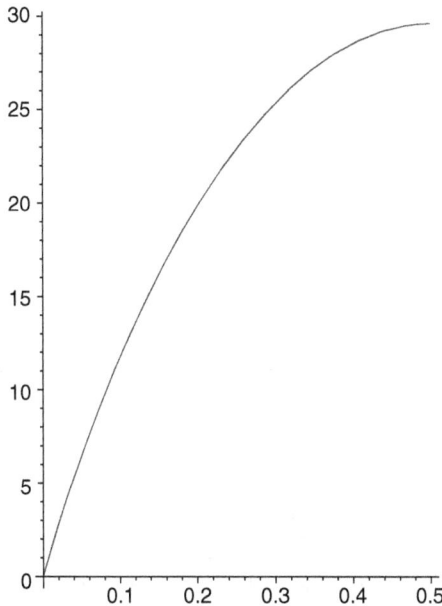

Fig. 3.2. The potential energy E in units of $1/\rho$ versus the Chern–Simons number $\tilde{N}_{CS}$, for the solution (3.91).

VEV sets a scale and there is no need also for ρ and the size constraint either; so it is just a solution to the field equations found first by Manton and Klinkhamer [173] by a variational method. The presence of the Higgs field makes equations more complicated and the solution is only known numerically. Sphalerons will play a significant role in Chapter 7, where we will discuss their production in experiment.

3.5.   Tunneling and BPST instanton

3.5.1.   *Instanton solution*

In this section we are going to look for a tunneling path in gauge theory, which connects topologically different classical vacua, following the original work by Belavin, Polyakov, Schwartz and Tyupkin [176]. From the quantum mechanical example we know that we have to look for classical solutions of the Euclidean equations of motion. The best tunneling path is the solution with minimal Euclidean action connecting vacua with different Chern–Simons numbers. To find these solutions, it is convenient to exploit the following identity:

$$
\begin{aligned}
S &= \frac{1}{4g^2} \int d^4x \, G^a_{\mu\nu} G^a_{\mu\nu} \\
&= \frac{1}{4g^2} \int d^4x \left[\pm G^a_{\mu\nu} \tilde{G}^a_{\mu\nu} + \tfrac{1}{2} \left(G^a_{\mu\nu} \mp \tilde{G}^a_{\mu\nu} \right)^2 \right],
\end{aligned}
\tag{3.93}
$$

where $\tilde{G}_{\mu\nu} = 1/2\epsilon_{\mu\nu\rho\sigma} G_{\rho\sigma}$ is the dual field strength tensor (the field tensor in which the roles of electric and magnetic fields are interchanged). Since the first term is a topological invariant (see below) and the last term is always positive, it is clear that the action is minimal if the field is (anti)*self-dual*[12]

$$
G^a_{\mu\nu} = \pm \tilde{G}^a_{\mu\nu}.
\tag{3.94}
$$

In simpler language, it means that Euclidean electric and magnetic fields are the same. One can also show directly that the self-duality condition implies the equations of motion,[13] $D_\mu G_{\mu\nu} = 0$. This is a useful observation, because in contrast to the equation of motion, the self-duality equation (3.94) is a first order differential equation. In addition to that, one can show that the energy momentum tensor vanishes for self-dual fields.

E

> **Exercise 3.7.**   Prove that any of the self-dual fields has zero energy density and even zero classical stress tensor.

---

[12]This condition is written in Euclidean notation; in Minkowski space an extra i appears in the electric field.

[13]The reverse is not true, but one can show that non-self-dual solutions of the equations of motion are saddle points, not local minima of the action.

The action of a self-dual field configuration is determined by the *topological charge* (or 4d Pontryagin index)

$$Q = \frac{1}{32\pi^2} \int d^4x \, G^a_{\mu\nu} \tilde{G}^a_{\mu\nu}. \tag{3.95}$$

From (3.93), we have $S = (8\pi^2|Q|)/g^2$ for self-dual fields. For finite action field configurations, Q has to be an integer. This can be derived as follows. We already mentioned above that the integrand is a total derivative of the topological current

$$Q = \frac{1}{32\pi^2} \int d^4x \, G^a_{\mu\nu} \tilde{G}^a_{\mu\nu}$$

$$= \int d^4x \, \partial_\mu K_\mu$$

$$= \int d\sigma_\mu K_\mu. \tag{3.96}$$

For finite action configurations, the gauge potential has to be pure gauge at infinity $A_\mu \to iU\partial_\mu U^\dagger$. Similarly to the arguments given in the last section, all maps from the three sphere S_3 (corresponding to $|x| \to \infty$) into the gauge group can be classified by a winding number n. Inserting $A_\mu = U\partial_\mu U^\dagger$ into (3.96) one finds that $Q = n$.

Furthermore, if the gauge potential falls off sufficiently rapidly at spatial infinity,

$$Q = \int dt \, \frac{d}{dt} \int d^3x \, K_0$$

$$= n_{CS}(t = \infty) - n_{CS}(t = -\infty), \tag{3.97}$$

which shows that field configurations with $Q \neq 0$ connect different topological vacua. In order to find an explicit solution with $Q = 1$, it is useful to start from the simplest winding number $n = 1$ configuration. Similarly to (3.66), we can take $A_\mu = iU\partial_\mu U^\dagger$ with $U = i\hat{x}_\mu \tau^+_\mu$, where $\tau^\pm_\mu = (\vec{\tau}, \mp i)$. Then $A^a_\mu = 2\eta_{a\mu\nu}x_\nu/x^2$, where we have introduced the 't Hooft symbol $\eta_{a\mu\nu}$. It is defined by

$$\eta_{a\mu\nu} = \begin{cases} \epsilon_{a\mu\nu}, & \mu, \nu = 1, 2, 3, \\ \delta_{a\mu}, & \nu = 4, \\ -\delta_{a\nu}, & \mu = 4. \end{cases} \tag{3.98}$$

We also define $\bar{\eta}_{a\mu\nu}$ by changing the sign of the last two equations. Further properties of $\eta_{a\mu\nu}$ are summarized in Appendix A.4.

One can look for a solution by using the O(4) symmetric ansatz

$$A^a_\mu = (2/g)\eta_{a\mu\nu}x_\nu f(x^2)/x^2, \tag{3.99}$$

where the unknown function $f(x^2)$ has to satisfy the "radial" equations. The boundary condition for it is $f \to 1$ as $x^2 \to \infty$; otherwise the commutator is not canceled by the derivative, and the field strength becomes non-zero. Substituting it into the definition of the field we get, after some algebra,

$$G_{\mu\nu}^a = -\frac{4}{g}\left(\eta_{a\mu\nu}\frac{f(1-f)}{x^2} + \frac{f(1-f) - x^2 f'}{x^4}(x_\mu \eta_{a\nu\gamma}x_\gamma - x_\nu \eta_{a\mu\gamma}x_\gamma)\right). \quad (3.100)$$

Its square is a Lagrangian, we propose it as the following excercise: $\boxed{\text{E}}$

Exercise 3.8.   (i) Derive the effective Lagrangian in terms of f and its derivatives and show that it is can be reduced to a variant of a double-well potential we studied earlier in this section.
(ii) Find the instanton solution by solving classical equations of motion following from it.

Instead of doing so we will use a shortcut following the original BPST work [176] using the self-duality. Indeed, the expression for the dual field is

$$\tilde{G}_{\mu\nu}^a = -\frac{4}{g}\left(\eta_{a\mu\nu}f' + \frac{f(1-f) - x^2 f'}{x^4}(x_\mu \eta_{a\nu\gamma}x_\gamma - x_\nu \eta_{a\mu\gamma}x_\gamma)\right), \quad (3.101)$$

and from self-duality (3.94) one gets the simpler *first* order equation

$$f(1-f) - x^2 f' = 0. \quad (3.102)$$

This equation with the boundary condition is solved by $f = x^2/(x^2 + \rho^2)$, which gives the famous BPST instanton solution [176]

$$A_\mu^a(x) = \frac{2}{g}\frac{\eta_{a\mu\nu}x_\nu}{x^2 + \rho^2}. \quad (3.103)$$

Here ρ is an arbitrary parameter characterizing the size of the instanton. As in the potential we discussed in the preceding section, its appearance is dictated by the scale invariance of classical Yang–Mills equations.

Note that all fields are $\sim 1/g$, much stronger than ordinary, perturbative, fields which do not have such a factor. It is convenient to change field normalization so that the coupling constant only appears as a factor in front of the action. This convention is very convenient in dealing with classical solutions. We rescale the fields by putting $1/g$ inside the instanton gauge potential.

A solution with the topological charge $Q = -1$, the *anti-instanton*, can be obtained by replacing $\eta_{a\mu\nu} \to \bar{\eta}_{a\mu\nu}$. The corresponding field strength is distributed in space as

$$(G_{\mu\nu}^a)^2 = \frac{192\rho^4}{(x^2 + \rho^2)^4} \quad (3.104)$$

and one can see that it is very well localized. On the other hand the gauge potential is long-range, $A_\mu \sim 1/x$. The invariance of the Yang–Mills equations under coordinate inversion implies that the singularity of the potential can be shifted from infinity to the origin by means of a (singular) gauge transformation $U = i\hat{x}_\mu \tau^+$. The gauge potential in singular gauge is given by

$$A_\mu^a(x) = \frac{2}{g} \frac{x_\nu}{x^2} \frac{\bar{\eta}_{a\mu\nu}\rho^2}{x^2 + \rho^2}. \tag{3.105}$$

This singularity at the origin is unphysical, pure gauge, but now in order to calculate the topological charge from a surface integral of the topological current K_μ one finds that it originates from a small sphere around the singularity. The topology of this configuration is therefore located at the origin,[14] not at infinity.

In significant share of literature the instantons in a singular gauge have been considered as a kind of curiosity, and emphasized that the topology actually comes from a *big* sphere near infinity. However I think such an attitude is absolutely wrong, and in fact the instanton topology comes actually from the *small* spheres near their centers. The difference becomes evident if one considers not a single instanton but multi-instanton configurations; there is only one large sphere but very many small ones.

Having found one solution, one can find many others simply by acting on it by symmetry transformations of the classical YM theory. In particular, the solution can be shifted in space, or rotated both in the usual and color space. As a result, a set of instanton solutions has a number of degrees of freedom, known as collective coordinates. In the case of SU(2), the solution is characterized by the instanton size ρ, the instanton position z_μ, and three parameters which determine the color orientation of the instanton. The group orientation can be specified in terms of the SU(2) matrix U, $A_\mu \to UA_\mu U^\dagger$, or the corresponding rotation matrix $R^{ab} = \frac{1}{2}\mathrm{tr}(U\tau^a U^\dagger \tau^b)$, such that $A_\mu^a \to R^{ab} A_\mu^b$. Due to the symmetries of the instanton configuration, ordinary rotations do not generate new solutions different from what one gets by color rotations.

The SU(3) instantons can be constructed by embedding the SU(2) solution. For $|Q| = 1$, there are no genuine SU(3) solutions. The number of parameters characterizing the color orientation is seven, not eight, because one of the SU(3) generators leaves the instanton invariant. For SU(N), the number of collective coordinates (including position and size) is $4N$.

There exist exact n-instanton solutions with $4nN$ parameters, all of them with the same action. A simple solution where the relative color orientations are fixed was given by 't Hooft (unpublished), see Refs. [201, 202]. A general solution has been found in Ref. [203], but in a complicated form which is difficult to use. Only decades later, P. van Baal and collaborators [205] have plotted numerically the action distribution for the $Q = 2$ solution of Ref. [203] and found with surprise that when two instantons approach each other instead of a double-strength maximum

---

[14]It is useful to think about a small sphere as an empty space inside a small hypercube, in a lattice definition of gauge fields. It is another way to learn that some singularities of the gauge fields are actually allowed/included in a lattice-based path integral.

they produce a toroid (like Skyrmions!) with a hole in the middle. Although the action stays the same for all $Q = 2$ configurations, the measure in the collective space clearly has something like a hard core, as solutions with two instantons on top of each other simply do not exist.

Let me summarize this section. We have explicitly constructed the tunneling path that connects different topological vacua. The instanton action is given by $S = (8\pi^2|Q|)/g^2$, implying that the tunneling probability is

$$P_{\text{tunneling}} \sim \exp(-8\pi^2/g^2). \tag{3.106}$$

More details on this to follow.

3.5.2. *Theta vacua*

We have seen that non-Abelian gauge theory has a periodic potential, and that instantons connect the different vacua. This means that the ground state of QCD cannot be described by any of the topological vacuum states, it has to be a superposition of all vacua. This problem is similar to the motion of an electron in the periodic potential of a crystal. It is well known that the solutions form a band ψ_θ, characterized by a phase $\theta \in [0, 2\pi]$ (sometimes referred to as quasi-momentum). The wave functions are Bloch waves satisfying the periodicity condition $\psi_\theta(x+n) = e^{i\theta n}\psi_\theta(x)$.

Let us see how this band arises from tunneling events. If instantons are sufficiently dilute, then the amplitude to go from one topological vacuum $|i\rangle$ to another $|j\rangle$ is given by

$$\langle j| \exp(-H\tau)|i\rangle = \sum_{N_+}\sum_{N_-} \frac{\delta_{N_+-N_--j+i}}{N_+!N_-!} \left(K\tau e^{-S}\right)^{N_++N_-}, \tag{3.107}$$

where K is the pre-exponential factor in the tunneling amplitude and $N_\pm$ are the numbers of instantons and anti-instantons. Using the identity

$$\delta_{ab} = \frac{1}{2\pi} \int_0^{2\pi} d\theta\, e^{i\theta(a-b)}, \tag{3.108}$$

the sum over instantons and anti-instantons can rewritten as

$$\langle j| \exp(-H\tau)|i\rangle = \frac{1}{2\pi} \int_0^{2\pi} d\theta\, e^{i\theta(i-j)} \exp\left[2K\tau \cos(\theta) \exp(-S)\right]. \tag{3.109}$$

This result shows that the true eigenstates are the theta vacua $|\theta\rangle = \sum_n e^{in\theta}|n\rangle$. Their energy is

$$E(\theta) = -2K\cos(\theta)\exp(-S). \tag{3.110}$$

The width of the zone is on the order of the tunneling rate. The lowest state corresponds to $\theta = 0$ and has negative energy. This is as it should be, tunneling lowers the ground state energy.

Does this result imply that in QCD there is a continuum of states, without a mass gap? Not at all; although one can construct stationary states for any value of θ, they are not excitations of the $\theta = 0$ vacuum, because in QCD the value of θ cannot be changed. As far as the strong interaction is concerned, different values of θ correspond to different worlds. Indeed, we can fix the value of θ by adding an additional term

$$\mathcal{L} = \frac{i\theta}{32\pi^2} G_{\mu\nu}^a \tilde{G}_{\mu\nu}^a \tag{3.111}$$

to the QCD Lagrangian.

Does physics depend on the value of θ? Naively, the interaction (3.111) violates both T and CP invariance. On the other hand, (3.111) is a surface term and one might suspect that confinement somehow screens the effects of the θ-term. A similar phenomenon is known to occur in 3-dimensional compact electrodynamics [147]. In QCD, however, one can show that if the $U(1)_A$ problem is solved (there is no massless η' state in the chiral limit) and none of the quarks is massless, a non-zero value of θ implies that CP is broken.

Consider the expectation value of the CP violating observable $\langle G\tilde{G} \rangle$. Expanding the partition function in powers of θ, we have $\langle G\tilde{G} \rangle = \theta(32\pi^2)\chi_{\text{top}}$. Furthermore, a low energy theorem determines the topological susceptibility for small quark masses. Using these results, one has

$$\langle G\tilde{G} \rangle = -\theta(32\pi^2) f_\pi^2 m_\pi^2 \frac{m_u m_d}{(m_u + m_d)^2} \tag{3.112}$$

for two light flavors to leading order in θ and the quark masses. Note that the pion mass squared is proportional to $m_u + m_d$ and so the r.h.s. vanishes in the chiral limit. In fact, it vanishes even if *any* quark mass is zero; this can be understood from the fact that in such a case the topological charge cannot fluctuate at all.

Similar estimates can be obtained for CP violating observables that are directly accessible to experiment. The most severe limits on CP violation in the strong interaction coming from the electric dipole of the neutron. These limits were worked out in Refs. [193–195]

$$|\theta| < 10^{-9}. \tag{3.113}$$

The question *why* is θ so small is known as the strong CP problem. The status of this problem is unclear. As long as we do not understand the source of CP violation in nature, it is not clear whether the strong CP problem is expected to have a solution within the standard model, or whether there is some mechanism outside the standard model that adjusts θ to become small.

One possibility is provided by the fact that the state with $\theta = 0$ is the bottom of the zone and thus has *the lowest* energy. This means that if θ becomes a dynamical variable, the vacuum can relax to the $\theta = 0$ state (just like electrons can drop to the bottom of the conduction band by emitting phonons). This is the basis of the axion mechanism [190–192]. The axion is a hypothetical pseudo-scalar particle, which couples to $G\tilde{G}$. The equations of motion for the axion field automatically remove the effective θ term, which is now a combination of $\theta_{\rm QCD}$ and the axion expectation value. Experimental limits on the axion coupling are very severe, but an "invisible axion" might still exist [196–199].

The simplest way to resolve the strong CP problem is to assume that the mass of one of the quarks (say u-quark) vanishes (presumably because of some unknown symmetry not manifest in the standard model). Unfortunately, this possibility appears to be ruled out phenomenologically, but there is no way to know for sure before this scenario is explored in more detail on the lattice. More recently, it was suggested that QCD might undergo a phase transition near $\theta = 0, \pi$. In the former case some support for this idea has come from lattice simulations [184], but the instanton model and lattice measurements of the topological susceptibility etc. do not suggest any singularity around $\theta = 0$. The latter limit $\theta = \pi$ also conserves CP and has a number of interesting properties (see e.g. Ref. [183]); in this world all instanton-induced effects would have a sign opposite to that at $\theta = 0$ or in the world we live in. Clearly, it is important to understand the properties of QCD with a non-zero θ-angle in more detail.

3.5.3.   *Tunneling amplitude*

The next natural step is the one-loop calculation of the pre-exponent in the tunneling amplitude. In gauge theory, this is a rather tedious calculation which was done in the classic paper by 't Hooft [177]. Basically, the procedure is completely analogous to what we did in the context of quantum mechanics. The field is expanded around the classical solution, $A_\mu = A_\mu^{\rm cl} + \delta A_\mu$. In QCD, we have to make a gauge choice. In this case, it is most convenient to work in a background field gauge $D_\mu(A_\nu^{\rm cl})\delta A_\mu = 0$.

We have to calculate the one-loop determinants for gauge fields, ghosts and possible matter fields (which we will deal with later). The determinants are divergent both in the ultraviolet, like any other one-loop graph, and in the infrared, due to the presence of zero modes. As we will see below, the two are actually related. In fact, the QCD beta function is partly determined by zero modes (while in certain supersymmetric theories, the beta function is completely determined by zero modes, see Section 12.3.1).

First one has to deal with the $4N_c$ zero modes of the system. The integral over the zero mode is traded for an integral over the corresponding collective variable. For each zero mode, we get one factor of the Jacobian $\sqrt{S_0}$. The group integration is compact, so it just gives a factor, but the integral over size and position we have to keep. As a result, we get a differential tunneling rate

$$dn_I \sim \left(\frac{8\pi^2}{g^2}\right)^{2N_c} \exp\left(-\frac{8\pi^2}{g^2}\right) \rho^{-5} d\rho\, d^4 z, \qquad (3.114)$$

where the power of ρ can be determined from dimensional considerations. Note that we for the first time meet here $5d$ Anti-de-Sitter space with the 5th coordinate being a scale; it is the same as in the famous Maldacena duality we will discuss in Chapter 12.

The ultraviolet divergence is regulated using the Pauli–Vilars scheme, the only known method to perform instanton calculations (the final result can be converted into any other scheme using a perturbative calculation). This means that the determinant $\det O$ of the differential operator O is divided in $\det(O + M^2)$, where M is the regulator mass. Since we have to extract $4N_c$ zero modes from $\det O$, this gives a factor M^{4N_c} in the numerator of the tunneling probability.

In addition to that, there will be a logarithmic dependence on M coming from the ultraviolet divergence. To one-loop order, it is just the logarithmic part of the polarization operator. For any classical field A_μ^{cl} the result can be written as a contribution to the effective action [180]

$$\delta S_{NZM} = \frac{2}{3}\frac{g^2}{8\pi^2} \log(M\rho) S(A^{cl}). \qquad (3.115)$$

In the background field of an instanton the classical action cancels the prefactor $g^2/8\pi^2$, and $\exp(-\delta S_{NZM}) \sim (M\rho)^{-2/3}$. Now, we can collect all terms in the exponent of the tunneling rate

$$dn_I \sim \exp\left(-\frac{8\pi^2}{g^2} + 4N_c \log(M\rho) - \frac{N_c}{3}\log(M\rho)\right)\rho^{-5} d\rho\, dz_\mu$$

$$\equiv \exp\left(-\frac{8\pi^2}{g^2(\rho)}\right)\rho^{-5} d\rho\, dz_\mu, \qquad (3.116)$$

where we have recovered the running coupling constant

$$\frac{8\pi^2}{g^2(\rho)} = \frac{8\pi^2}{g^2} - \frac{11N_c}{3}\log(M\rho).$$

Thus, the infrared and ultraviolet divergent terms combine to give the coefficient of the one-loop beta function, $b = 11N_c/3$, and the bare charge and the regulator mass M can be combined into a running coupling constant. At two-loop order, the renormalization group requires the miracle to happen once again, and the non-zero mode determinant can be combined with the bare charge to give the two-loop beta function in the exponent, and the one-loop running coupling in the pre-exponent.

The remaining constant from the determinant of the non-zero modes was calculated in Refs. [177, 178]. The result is

$$dn_I = \frac{0.466 \exp(-1.679 N_c)}{(N_c - 1)!(N_c - 2)!} \left(\frac{8\pi^2}{g^2}\right)^{2N_c} \exp\left(-\frac{8\pi^2}{g^2(\rho)}\right) \frac{d^4 z \, d\rho}{\rho^5}. \qquad (3.117)$$

The tunneling rate dn_A for anti-instantons is of course identical. Using the one-loop beta function the result can also be written as

$$\frac{dn_I}{d^4 z} \sim \frac{d\rho}{\rho^5} (\rho \Lambda)^b, \qquad (3.118)$$

and because of the large coefficient $b = 11 N_c/3 = 11$, the exponent overcomes the Jacobian and small size instantons are strongly suppressed. On the other hand, there appears to be a divergence at large ρ, although the perturbative beta function is not applicable in this regime.

Although the setting of the next order calculations is in principle quite analogous to those we have discussed in quantum mechanical context in Section 3.1.5, and even the diagrams are the same, in all the years passed since 1976 the two-loop correction to the 't Hooft formula has not been calculated.

CHAPTER 4

Instanton Ensemble in QCD

4.1. Brief history of instantons

There is a widespread trend to teach physics theories presenting them as logical constructions, omitting "for simplicity" their history (which looks very illogical). However, it is very important to know the internal logics of the development, and what exactly the major breakthroughs were. Very rarely do we manage to transfer to our students that each of these theories was created by the efforts of a few enthusiasts outnumbered by surrounding skeptics. If nothing else, it will teach them not be surprised when they find themselves in a similar situation as well.

As one can see from our brief presentation, the history of instantons also had its high points separated by years of frustration. I have split it into three very different periods, which roughly correspond to the 1970s, 1980s and 1990s.

4.1.1. *Discovery and early applications*

The famous classical solution of the Yang–Mills equations was discovered by Polyakov and collaborators [176], driven by the quest for nontrivial 1-d topology in analogy with the 3-d Polyakov-'t Hooft monopole. Its physical relation with tunneling was clarified only in the subsequent year [147]. Existence of many classical vacua with different Chern–Simons number, as well as the periodic potential separating them, was discussed in Refs. [181,182]. Polyakov has found [147] that instantons in some 3-dimensional models can explain confinement: and the main hopes at that time were that it may also be the case in 4-dimensional theories as well; but without success.

Another direction was set by the classic paper by 't Hooft [177].[1] Apart from a technical advance (explicit one-loop calculation of the tunneling amplitude), 't Hooft has discover fermionic zero modes. He realized that tunneling is intimately

---

[1] By the way, the word "instanton" itself originates from this paper; Polyakov *et al.* called it a *"pseudoparticle"*.

related with fermionic effects: ultimately these studies have "de-mystified" the long-standing theoretical puzzle, the chiral anomalies. This work has changed quantum field theory forever; in particular, it has explained *how* the baryon charge is violated in the Standard Model.

In QCD the chiral U(1) symmetry is violated by the 't Hooft effective interaction between light quarks, which explains why η' is not a Goldstone mode. Furthermore, as Witten [188] and Veneziano [189] suggested, the η' mass can be related with the so called topological susceptibility. It was very important historically and also led later to significant efforts in lattice studies.

Unfortunately, this issue was immediately related with a difficult question of the large-N_c limit: this resulted into significant confusion. Both Witten and Veneziano expressed serious doubts about the relation between the topological susceptibility and instantons.[2]

Early attempts to relate instantons with practical applications to QCD were summarized by an important paper by Callan, Dashen and Gross [214]. Incorporating the dipole forces [208] they tried to create a self-consistent theory of interacting instantons. With "merons" they have attempted to explain confinement. Using the 't Hooft interaction they studied a possibility to break spontaneously the SU(N_f) chiral symmetry. Later it became clear that the dilute gas approximation they used could not really be a consistent description of the vacuum; then Callan, Dashen and Gross tried to apply it to the hadron *interior* [217] where the instanton ensemble is believed to be somewhat more dilute. In my paper [216] and also in Ref. [217] it was suggested that suppression of instantons by a color field can lead to a bag-type picture and be related to confinement.

Although many of those efforts were not immediately successful, they definitely affected the thinking of many later workers in the field. The very idea that an effective theory describing QCD and similar theories may not necessarily be quantum field theories, but rather a kind of statistical mechanics with a *finite* number of effective degrees of freedom, the collective coordinates, was not forgotten.

By the end of the 1970s the pessimism was at its maximum. There was no solid confirmation of any instanton-induced effect in experiment, and no general reason was found for applicability of the semi-classical methods in QCD. Most of the major players of the early days (like Polyakov, 't Hooft, Callan, Dashen and Gross) were strongly disappointed and left the field.

4.1.2. *Phenomenology leads to a qualitative picture*

When it is difficult to build the theory from first principles, one may always look at the phenomena under consideration from an empirical point of view, searching for hints. It is precisely what happened in the early 1980s.

---

[2]Even the title of Veneziano's paper ended by "... without instantons". Today lattice studies have shown that this quantity is indeed explained by instantons without doubt, see Section 5.4.

By that time, the QCD sum rules [352] has been widely used, and they provided some understanding of the behavior of the QCD correlation functions at small distances. Furthermore, it was pointed out in Ref. [396] that "all hadrons are not alike": the analysis based on the Operator Product Expansion (OPE) does not actually work for all channels. Large deviations from OPE predictions found in all scalar and pseudoscalar channels (and even more so, for gluonic operators) needed explanation. Spin-flavor properties of the 't Hooft interaction has successfully explained why those effects are large in some cases (e.g. for the pion [395]) and small in others (e.g. for vectors and axials). At sufficiently small distances the effects are sufficiently small, and therefore one can apply corrections in the single instanton approximation.

In order to explain the available "phenomenology of the vacuum", a qualitative model was proposed in my work [226], which suggested the so called "instanton liquid" model. This model provided a quantitative description of $SU(N_f)$ and $U(1)_A$ chiral symmetry breaking, and reproduced the pseudoscalar correlators at small distances [227], both for attractive channels (like the pion) and repulsive (η') ones. Even the gluonic correlators evaluated in this work were reasonable, with the scalar glueball mass predicted then to be around 1.5 GeV. It became instantly clear that it can cure the most acute problems the OPE-based sum rules have encountered.

4.1.3. *Technical development during 1980s*

What was needed was a consistent many-body theory of the instanton ensemble. The first step was a simple hard-core model by Ilgenfritz and Mueller–Preussker [219]. The second was a variational approach by Diakonov and Petrov [220]. For a simple "sum ansatz" (the gauge potential equal to a sum of the potentials, in a singular gauge, for individual instantons and anti-instantons) the classical interaction was indeed found to be rather repulsive, and the mean-field treatment of statistical mechanics indeed reproduced the diluteness of the equilibrium ensemble. Inclusion of light quarks followed [229], which led to a picture of collectivized quasi-zero fermionic modes. Random phase approximation and bosonization methods [229, 231] were applied to some mesonic correlation functions. All of those are variants of the *mean field approximation* (MFA) used to calculate the quark condensate and the quark effective mass.

It was nevertheless clear from the very beginning, that MFA is not really justified numerically in this problem, and so more direct and accurate methods had to be developed. Numerical methods to treat the ensemble were developed in a series of my papers [234]. It was eventually possible to include the 't Hooft effective interaction to *all oders*. Experimentally known correlators were reproduced by the model at small distances in a number of channels, now at a quantitative level. The model was definitely working there, and it was tempting to test it further and further.

Nevertheless, its theoretical foundations remained unclear. The repulsion derived in Ref. [220] from the sum ansatz was criticized, and a number of works [210, 211, 213] devoted to the "valley" of instanton–anti-instanton configurations, concluded that (for most attractive relative orientation) it smoothly connected them to the zero field one. It was realized that there was no wall separating instanton-induced and perturbative effects. Although the interaction corresponding to the "streamline" (the bottom of the valley) was found, its practical application in fact became possible only in combination with the repulsive core introduced *ad hoc*.

4.1.4.   *Recent progress*

The early 1990s was the time of mostly phenomenological development. The random instanton liquid model (RILM) was used for direct calculation of correlators, for many mesonic channels [237, 238]. Unlike previous attempts, significant numerical efforts were made, allowing one to calculate the relevant correlation functions up to larger distances (about 1.5 fm), following their decay by a few decades. As a result, the predictive power of the model was explored in substantial depth. Many of the coupling constants and even hadronic masses were calculated with surprising accuracy for such a crude model as the RILM. In most cases the agreement with experiment was in fact within the numerical error bars. Subsequent calculations of baryonic correlators [239] revealed further surprising facts. In the instanton vacuum the nucleon was shown to be a deeply bound state of constituent quarks, while Δ was weakly bound.

This analysis has recently been reinforced [244]; RILM happens to reproduce rather accurately the ALEPH data on τ decay for both vector and axial correlators. A comparison between the correlators calculated in RILM and on the lattice [399] has also shown good agreement, including the channels unreachable by usual phenomenology (e.g. baryonic ones). Studies of the "wave functions" [240] and even glueball ones [241] has followed, again with results very close to the lattice ones.

The "instanton liquid" itself started emerging from the "fog" of quantum fluctuations, when lattice configurations were subjected to "cooling". In Ref. [399] both the instanton separation and the mean size were evaluated; within the accuracy involved ($\sim$10%) the results coincided with the parameters suggested a decade earlier [226]. And moreover: the correlators calculated *after cooling* remained stubbornly about the same, indicating that perturbative phenomena (killed by cooling) and even confinement (strongly reduced by it) were not in fact very important for hadronic properties!

Further lattice studies of the instanton size distribution [400] have clearly shown that suppression of the large-size instantons does exist. Why this is the case is still not understood, we will discuss those issues in Section 4.3.1.

Technical progress in numerical simulation of the *interacting* ensemble of instantons [243, 245] has resulted in self-consistent determination of all the parameters, provided the interaction is given. It is important, that the 't Hooft interaction is included in all orders, so this step is analogous to transition from "quenched" (no fermions) to unquenched lattice calculations. Significant improvement in two "repulsive" channels (η', δ) relative to the random model was demonstrated.

In the late 1990s a breakthrough enabled one to use good chiral fermions on the lattice. This allowed in the last couple of years a rather extensive lattice test of a general claim that the lowest Dirac eigenstates of lattice configurations actually are superpositions of instantons zero modes. And indeed, that was confirmed and their "local chirality" property became well established. We will discuss these results in Section 5.4.

4.1.5. *Instantons at finite temperatures and chiral restoration*

The instanton solution can be generalized to a non-zero temperatures [547].[3] Because instantons include the electric field, in a plasma phase at high T they should be suppressed [216, 548] by the ordinary Debye screening. It was therefore assumed by many authors, that "melting" of a quark condensate (chiral symmetry restoration) is a direct consequence of the instanton suppression.

However, in the low-T limit worked out by Velkovsky and myself [556] only a very weak T-dependence of the instanton density was found. This prediction was confirmed by direct lattice studies, e.g. there is very little change in the topological susceptibility for $T < T_c$.

A different mechanism for the chiral restoration phase transition was suggested by Ilgenfritz and myself [557]. According to it, at T close to T_c instantons are not suppressed but rearranged into pairs, forming the so called instanton–anti-instanton molecules. Later this suggestion was studied in some detail. Not only was it confirmed by direct simulations [245] and analytic studies [558], but it was also found to be relevant at zero T, however in QCD with larger number of light quarks [559]. These findings provide a completely new perspective on the properties of the quark–gluon plasma at intermediate temperatures $3T_c > T > T_c$ in which free quarks and gluons coexist with large non-perturbative effects.

4.1.6. *Instantons and color superconductivity at high densities*

The second most attractive channel in QCD is the interaction of two quarks in the scalar $S = I = 0$ channel. It also follows from the 't Hooft effective Lagrangian, and suppressed by a factor $1/(N_c - 1)$ relative to the most attractive scalar $\bar{q}q$ channel. It was pointed out in two simultaneous papers[4] [827, 828] in 1997 that the same interaction led to a very robust Cooper pairing in high density QCD. Within a few years this field boomed and has now a bibliography of about 500 papers. We will discuss this in Sections 11.3.1 and 11.3.5.

---

[3]This solution is also referred to as "calorons".

[4]They were submitted on the web on the same day.

4.2. Tunneling and light quarks

4.2.1. *Relating gauge field topology to the axial charge*

When we discussed the topologically different classical vacua, we treated them as distinct states. But why are those different classical vacua really physically different, and not a different copy of the same one? In more precise language, why is the Chern–Simons number treated as a periodic coordinate and not as an angle?

In fact there is a physical quantity which allows us to distinguish them: it is the *axial charge*, or the difference between the number of left and right quarks. In order to see that, first one has to recall the chiral anomaly,

$$\partial_\mu j_\mu^5 = \frac{N_c}{16\pi^2} G_{\mu\nu}^a \tilde{G}_{\mu\nu}^a, \qquad j_\mu^5 = \bar{q}\gamma_\mu\gamma_5 q, \quad q = u, d, s, \dots . \tag{4.1}$$

The second ingredient of the argument is the observation that the r.h.s., the topological charge, is the divergence of the topological current, $\partial_\mu K_\mu$. So it means that in QCD one can still define a conserved axial current, a combination of the two. The corresponding charges, 3-dimensional integrals of the 0th components, are the axial charge and Chern–Simons number. One can now conclude that the anomaly equation simply means that each time N_{CS} changes by 1 unit, the axial charge changes by $2N_f$. (A subtlety in the argument is the fact that N_{CS} is not by itself gauge invariant: only its variation is.)

In order to see how instantons can lead to the non-conservation of axial charge, let us calculate the change in axial charge

$$\Delta Q_5 = Q_5(t = +\infty) - Q_5(t = -\infty) = \int d^4x \, \partial_\mu j_\mu^5. \tag{4.2}$$

In terms of the fermion propagator, ΔQ_5 is given by

$$\Delta Q_5 = \int d^4x \, N_f \partial_\mu \operatorname{tr}(S(x, x)\gamma_\mu\gamma_5). \tag{4.3}$$

The fermion propagator is the inverse of the Dirac operator, $S(x, y) = \langle x|(i\slashed{D})^{-1}|y\rangle$. For any given gauge field, we can determine the propagator in terms of the eigenfunctions $\slashed{D}\psi_\lambda = \lambda\psi_\lambda$ of the Dirac operator

$$S(x, y) = \sum_\lambda \frac{\psi_\lambda(x)\psi_\lambda^\dagger(y)}{\lambda}. \tag{4.4}$$

Using the eigenvalue equation, we can now evaluate ΔQ_5

$$\Delta Q_5 = N_f \int d^4x \operatorname{tr}\left(\sum_\lambda \frac{\psi_\lambda(x)\psi_\lambda^\dagger(x)}{\lambda} 2\lambda\gamma_5\right). \tag{4.5}$$

For every non-zero λ, $\gamma_5 \psi_\lambda$ is an eigenvector with eigenvalue $-\lambda$. But this means, that ψ_λ and $\gamma_5 \psi_\lambda$ are orthogonal, so only *zero modes* can contribute to (4.5)

$$\Delta Q_5 = 2N_f(n_L - n_R), \tag{4.6}$$

where $N_{L,R}$ is the number of left (right)-handed zero modes and we have used the fact that the eigenstates are normalized.

4.2.2. *Fermionic zero modes*

After such an extensive introduction, the reader should not be surprised to see that indeed a very important property of instantons, originally discovered by 't Hooft [177], is that the Dirac operator has a zero mode $i\not{D}\psi_0(x) = 0$ in the instanton field. For an instanton in the singular gauge, the zero mode wave function is

$$\psi_0(x) = \frac{\rho}{\pi}\frac{1}{(x^2+\rho^2)^{3/2}}\frac{\gamma \cdot x}{\sqrt{x^2}}\frac{1+\gamma_5}{2}\phi, \tag{4.7}$$

where $\phi^{\alpha m} = \epsilon^{\alpha m}/\sqrt{2}$ is a constant spinor[5] in which the SU(2) color index α is coupled to the spin index $m = 1, 2$. Let us briefly digress in order to show that (4.7) is indeed a solution of the Dirac equation. First observe that[6]

$$(i\not{D})^2 = \left(-D^2 + \tfrac{1}{2}\sigma_{\mu\nu}G_{\mu\nu}\right). \tag{4.8}$$

We can now use the fact that $\sigma_{\mu\nu}G^{(\pm)}_{\mu\nu} = \mp\gamma_5\sigma_{\mu\nu}G^{(\pm)}_{\mu\nu}$ for (anti) self-dual fields $G^{(\pm)}_{\mu\nu}$. In the case of a self-dual gauge potential the Dirac equation $i\not{D}\psi = 0$ then implies $(\psi = \chi_L + \chi_R)$

$$\left(-D^2 + \tfrac{1}{2}\sigma_{\mu\nu}G^{(+)}_{\mu\nu}\right)\chi_L = 0, \quad -D^2\chi_R = 0, \tag{4.9}$$

and vice versa $(+ \leftrightarrow -, L \leftrightarrow R)$ for anti-self-dual fields. Since $-D^2$ is a positive operator, χ_R has to vanish and the zero mode in the background field of an instanton has to be left-handed, while it is right-handed in the case of an anti-instanton. This result is not an accident. Indeed, there is a mathematical theorem (the Aliyah–Singer index theorem), that requires that $Q = n_L - n_R$ for every species of chiral fermions. In the case of instantons, this relation was proven in Ref. [222]. A general analysis of the solutions of (4.9) was given in Ref. [177]. For (multi) instanton gauge

---

[5]The norm is such that this mode is normalized to $\int d^4x\, \bar{\psi}_0\psi_0 = 1$.

[6]We use Euclidean Dirac matrices that satisfy $\{\gamma_\mu, \gamma_\nu\} = 2\delta_{\mu\nu}$. We also will use the combinations of gamma matrices $\sigma_{\mu\nu} = i/2[\gamma_\mu, \gamma_\nu]$ and $\gamma_5 = \gamma_1\gamma_2\gamma_3\gamma_4$.

potentials of the form $A_\mu^a = \bar\eta_{\mu\nu}^a \partial_\nu \log \Pi(x)$ the solution is of the form [223]

$$\chi_\alpha^m = \sqrt{\Pi(x)}\,\partial_\mu \left(\frac{\Phi(x)}{\Pi(x)}\right) (\tau_\mu^{(+)})^{\alpha m}. \tag{4.10}$$

The Dirac equation requires $\Phi(x)$ to be a harmonic function, $\Box\Phi(x) = 0$. Using this result, it is straightforward to verify (4.7). Again, we can obtain an SU(3) solution by embedding the SU(2) result.

It is also important that there is no chirality partner for zero modes: the "pairing" theorem mentioned in Section 1.4.3 holds only for non-zero modes. So what was wrong with the proof? Of course the assumption that $\gamma_5\psi_\lambda$ can be used as *another* eigenvector: it would not work for purely chiral solutions.

So, what does the existence of the zero mode mean for the tunneling rate? At first glance, the zero mode is a problem since it causes the fermionic determinant to vanish in the partition function. Indeed, the determinant is of the operator $i\slashed{D}+im$, and since the former term gives zero on a zero mode, one has to conclude that for massless fermions the tunneling probability vanishes. Not necessarily, argued 't Hooft, since the mass term can be supplemented by an external scalar current. What it all means is that there is no tunneling *unless* a $\bar q_R q_L$ pair for each massless flavor is produced. We have already seen how it works for a much simpler problem, of a quantum mechanical system coupled to a fermion in Section 3.3.

Still the whole process looks very mysterious. The final "demystification" of the anomaly was made by V.N. Gribov around 1980. It has explained why a strong and topologically non-trivial field can do what any finite number of emitted or absorbed gluons cannot do: to change the axial charge, the number of left-handed minus the number of right-handed quarks.

So let me provide another view on the issue. One can follow the tunneling configurations adiabatically, changing time very slowly, and for each value of time we are looking for static energy levels of the Dirac particle, while ignoring all time derivatives. One then finds that the levels move in such a way, that all left-handed states make one step down, to the next level, and all right-handed ones make one step up. A hint that this is the case can be explained as follows: in the adiabatic approximation (slow change in time) the time-dependent solution is

$$\psi(t,x) = \psi_{\text{static}}(t,x)\exp\left(-\int_0^t dt'\epsilon_{\text{static}}(t')\right). \tag{4.11}$$

If the energy is positive for large t and negative for $t \to \infty$, the corresponding time-dependent wave function is 4-dimensionally normalizable. The explicit 't Hooft zero mode is such a normalizable solution. Thus, if only one such solution exists, it means that only *one* state has passed the zero energy mark.

So, when tunneling is finished, the spectrum is of course the same, but it is the *level occupation* which is different! This conclusion, known as the "infinite hotel story", complements the discussion of the same phenomenon in a different setting from the one we had in Section 1.5.3.

4.2.3. The 't Hooft effective interaction

Instead of the tunneling amplitude, let us calculate a $2N_f$-quark Green function $\langle \prod_f \bar{\psi}_f(x_f)\Gamma\psi_f(y_f)\rangle$, containing one quark and anti-quark of each flavor. Contracting all the quark fields, the Green function is given by the tunneling amplitude multiplied by N_f fermion propagators. Every propagator has a zero mode contribution with one power of the fermion mass in the denominator,

$$S(x,y) = \frac{\psi_0(x)\psi_0^+(y)}{im} + \sum_{\lambda \neq 0} \frac{\psi_\lambda(x)\psi_\lambda^+(y)}{\lambda + im}, \tag{4.12}$$

where I have written the zero mode contribution separately. Note that if both points x,y are far from the instanton center (relative to ρ), one can use the asymptotic expression for ψ_0 which at large arguments behaves as a constant spinor times $1/x^3$. Since this behavior is nothing else but just the free propagator for a massless fermion, one sees that in this limit the first term can be interpreted as two free propagators, from x to z and from y to z, times some constant vertex. The procedure we have described is in fact standard "amputation of external legs" of the Green functions, used when one would like to derive the effective vertex or Lagrangian.

Let us now look at the dependence on the light quark mass. Suppose there are N_f light quark flavors, so that the instanton amplitude is proportional to m^{N_f} (or, more generally, to $\prod_f m_f$) due to the fermionic determinant in the weight.

As a result, the zero mode contribution to the Green function is finite in the chiral limit.[7]

The result can be written in terms of an effective Lagrangian [177]. (We will return to it in Section 4.3.3, where we give it a different form.) It is a non-local $2N_f$-fermion interaction, where the quarks are emitted or absorbed in zero mode wave functions. In general, it has a fairly complicated structure, but under certain assumptions, it can be significantly simplified. First, if we limit ourselves to low momenta, the interaction is effectively local. Second, if instantons are uncorrelated, one can average over their orientation in color space. For SU(3) color and $N_f = 1$ the result is [218]

$$\mathcal{L}_{N_f=1} = \int d\rho\, n_0(\rho)\left(m\rho - \tfrac{4}{3}\pi^2\rho^3 \bar{q}_R q_L\right), \tag{4.13}$$

where $n_0(\rho)$ is the tunneling rate without fermions. Note that the zero mode contribution acts like a mass term. This is quite natural, because for $N_f = 1$, there is

---

[7]Note that Green functions involving more than $2N_f$ legs are not singular as $m \to 0$. The Pauli principle always ensures that no more than $2N_f$ quarks can propagate in zero mode states.

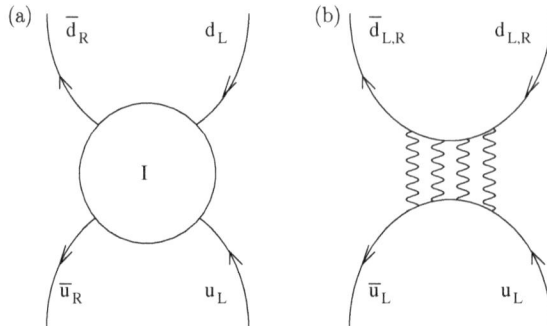

Fig. 4.1. The instanton-induced 't Hooft vertex (a) for 2 flavor QCD versus the ordinary gluon exchange diagrams (b). Note a very different chiral structure of the two: the latter does not violate any chiral symmetry because chirality is conserved along each line.

only one chiral $U(1)_A$ symmetry, which is anomalous. Unlike the case $N_f > 0$, the anomaly can therefore generate a fermion mass term.

For $N_f = 2$, the result is

$$\mathcal{L}_{N_f=2} = \int d\rho\, n_0(\rho) \left[\prod_f \left(m\rho - \tfrac{4}{3}\pi^2\rho^3 \bar{q}_{f,R}q_{f,L} \right) \right.$$

$$\left. + \frac{3}{32} \left(\tfrac{4}{3}\pi^2\rho^3 \right)^2 \left(\bar{u}_R\lambda^a u_L \bar{d}_R\lambda^a d_L - \tfrac{3}{4}\bar{u}_R\sigma_{\mu\nu}\lambda^a u_L \bar{d}_R\sigma_{\mu\nu}\lambda^a d_L \right) \right], \quad (4.14)$$

where the λ^a are color Gell–Mann matrices. One can easily check that the interaction is $SU(2) \times SU(2)$ invariant, but $U(1)_A$ is broken. This means that the 't Hooft Lagrangian provides another derivation of the $U(1)_A$ anomaly. Furthermore, we will argue below that the importance of this interaction goes much beyond the anomaly, and that it explains the physics of chiral symmetry breaking and the spectrum of light hadrons.

One may view this process as a kind of non-local vertex, in which a pair of each light quark flavor is produced, the so called *'t Hooft effective interaction*. Chirality of quark and anti-quark lines is different, so an attempt to close these lines in loops gives zero for the massless quarks.

We will show later in this chapter and especially in Chapters 6 and 7 how this interaction leads to many important non-perturbative phenomena. Let me provide here one example (which we would not discuss in detail). It was noticed by Bjorken [247] that decays of η_c has three large 3-body modes, about 5% each of the total width: $KK\pi; \pi\pi, \eta; \pi, \pi, \eta'$. Note that there is no $\pi\pi\pi$ or many other decays one may think of: in fact the average multiplicity of these decays is rather large and 3-body decays is an anomaly by itself. He pointed out that the vertex seems to be $\bar{u}u\bar{d}d\bar{s}s$ and suggested that these decays exist due to the 't Hooft interaction. I tried the calculation of those a few times but got stuck till finally Schafer and Zetocha [248] did it. The results reasonably well reproduced the ratios between the channels and even the absolute width. Furthermore, these calculations provide about the most accurate evaluation of the average instanton size available, see Fig. 4.2.

Finally, in order to complete the effects of light quarks on the tunneling, we need to include the effects of non-zero modes. One effect is that the coefficient in the beta function is changed to $b = 11N_c/3 - 2N_f/3$. In addition to that, there is an overall constant that was calculated in Refs. [177, 215]

$$n(\rho) \sim (1.34m\rho)^{N_f}(1 + N_f(m\rho)^2 \log(m\rho) + \cdots), \tag{4.15}$$

where we have also specified the next order correction in the quark mass. Note that at two-loop order, one not only gets the *two-loop beta function* in the running coupling, but also the *one-loop anomalous dimensions* of the quark masses.

4.2.4. *Baryon number violation in the standard model*

In this section we deviate briefly from our discussion of QCD and consider instantons made of W, Z fields. Very little changes in terms of formulae are need, while the numbers involved are drastically different. The Higgs VEV sets a scale and therefore instanton size also gets fixed [177]. The charge at electroweak scale is small, and therefore the probability of tunneling is now extremely small

$$P \sim \exp\left(\frac{-16\pi^2}{g_w^2}\right) \sim 10^{-169}, \tag{4.16}$$

so it seems out of question that one can observe any manifestations of it.

At the same time, interesting phenomena happen with fermions. As we have seen, each light fermion gets a zero mode. It means that there must be anomalous production of 12 fermions, say for instanton with flavor orientation "up" those will be $e, \mu, \tau, 3u, 3c, 3t$. (The quarks go in three copies since the electroweak instanton does not interact with color and each is just as an extra index.) Of course, orientation in flavor space can be arbitrary, so instead of an electron one may get an electron neutrino, and so on. Furthermore, the corresponding effective interaction breaks a lot of conservation laws, including the *baryon charge*! This happens because only left-handed fermions interact with the W, Z fields, and there is no difference between vector and axial currents. (Recall that in QCD vector currents are conserved because L, R loops cancel each other in the anomaly.)

In nuclear physics people were able to observe the "induced radioactive decay" with excited states of nuclei: then the tunneling rate is much larger. Can we help to penetrate the barrier by using high energy collision of hadrons? The barrier height is known from the "sphaleron mass": substituting weak charge instead of strong one gets about 15 TeV (the exact number depends on the so far unknown Higgs mass). As was shown in [441] and will be discussed in Chapter 7, making high energy collision really helps: exactly *half* of the instanton action disappears. This is about 80 orders of magnitude enhancement! However, it is not enough for observation even then, and so instanton effects in weak interactions remain essentially a theorist's dream.

4.3.   Instanton ensemble

4.3.1.   *Qualitative discussion of the instanton ensembles*

The semiclassical analysis we have made so far is only valid for sufficiently small instantons, which indeed form a simple dilute gas in purely gauge theory. (Let me remind the reader that the corresponding condition is that the action $S_{\text{instanton}} = 8\pi^2/g^2(\rho)$ should be large compared to 1 (or, rather, the Planck constant $\hbar$)). However, the semiclassical formula for the instanton density is strongly divergent at large ρ, and therefore one cannot estimate its effect. As we will see below, instantons interact with each other in a rather nontrivial manner, in particular by exchanging light fermions. On top of that there are unknown interactions with confining forces and maybe other effects. For all these reasons it is not easy to calculate the properties of the instanton ensemble from first principles.

For small enough instantons one can hope to use the OPE and describe their interaction with vacuum using the so called "condensates". The first correction,

$$dn = dn_{\text{semiclassical}} \left(1 + \frac{\pi^3 \rho^2}{2g^4} \langle (gG_{\mu\nu}^a)^2 \rangle \right), \tag{4.17}$$

predicts positive correction, making the growth with size ρ even stronger. One can estimate from this expression what the size is and at which correction it is still small, and the result is that such is the case at $\rho < 2\,\text{fm}$. However empirically and also from the lattice (see Section 5.4) one finds instead that the semiclassical expression is modified by a cutoff which is well described by $\exp(-\text{const} * \rho^2)$ [63].

In 1982 I proposed to use a simple model [226] with fixed instanton size and density, used as parameters to be determined phenomenologically. It used the instanton size distribution in the simplified form

$$\frac{dn}{d\rho} = n_0 \delta(\rho - \rho_0). \tag{4.18}$$

A rough idea what the instanton density in the QCD vacuum can be was known at the time: it should be bounded by the value of the gluon condensate

$$n_0 < n_c = \langle (gG_{\mu\nu}^a)^2 \rangle / 32\pi^2 \sim (1\,\text{fm})^{-4} \tag{4.19}$$

(where the density means both pseudoparticles together, I and $\bar{I}$). Another observable known is the topological susceptibility

$$\chi_{\text{top}} = \frac{1}{V_4} \langle Q^2 \rangle \approx (180\,\text{MeV})^4, \tag{4.20}$$

which is equal to the density if the instantons are uncorrelated and the distribution is Poissonian. The magnitude of the instanton size was unclear. I found that

$$\rho_0 \sim \tfrac{1}{3}\,\text{fm} = 1/600\,\text{MeV} \tag{4.21}$$

can generate a reasonable quark condensate and other properties of chiral breaking. I then proceeded calculating corrections to different quantities, in particular correlation functions of the pseudoscalar currents [227] and found that the results were in nice agreement with phenomenology. The Instanton Liquid Model (ILM for short) was born.

Over the years, dozens of observables sensitive to the instanton size have been proposed and their determination always resulted in the value mentioned above. In Section 6.6.2 we will for example discuss the pion and nucleon form factors, which are sensitive to it, see Fig. 6.24. Let me show here the most recent examples of such plots, see Fig. 4.2. They are from the paper by Zetocha and Schafer [248], who considered instanton-induced decays of charmonium states η_c, χ, among others. Note that in this case the sensitivity to the mean instanton size is so strong because it is a simultaneous form factor of 2 or 3 mesons, times the basic 't Hooft vertex squared, with its many powers of the size ρ.

Let me summarize some important qualitative features of this model:

1. The *diluteness parameter* is small

$$n_0 \rho_0^4 = (\rho/R)^4 \sim (1/3)^4, \tag{4.22}$$

   where R is the typical distance between the pseudoparticles. So only a few percent of the space–time is occupied by the strong field. The factorization hypothesis is thus violated, by the inverse diluteness.

2. *Semiclassical formulae are applicable.* The action of a typical instanton is large enough

$$S_0 = 8\pi^2/g(\rho)^2 \sim 10 \gg 1. \tag{4.23}$$

   Quantum corrections go as $1/S_0$ and are presumably small enough.

3. *The interaction does not destroy instantons.* Estimated by the dipole formula, the interaction was found to be typically

$$|\delta S_{\text{int}}| \sim (2-3) \ll S_0. \tag{4.24}$$

4. *It is a liquid, not a gas.* The interaction is not small in the statistical mechanics of instantons, on the contrary correlations are strong

$$\exp |\delta S_{\text{int}}| \sim 20 \gg 1. \tag{4.25}$$

Now let me provide a qualitative discussion of the fermionic effects. If instead of one instanton one considers many, it is necessary to come back to the *spectrum of eigenvalues λ of the Dirac operator*. Its discussion can be made using the analogy to atomic physics, because we have very good intuition for different kinds of matter made of atoms.

(a)

(b)

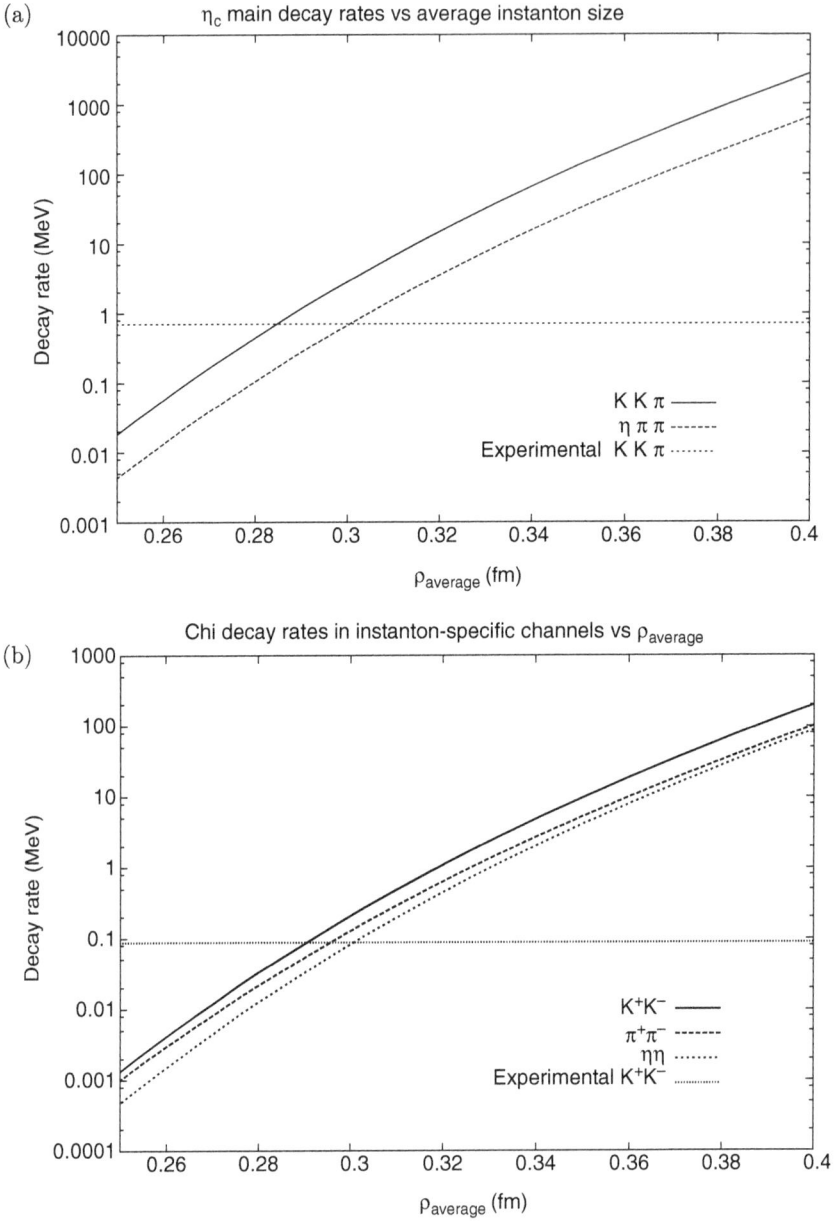

Fig. 4.2. (a) Decay widths $\eta_c \rightarrow KK\pi$ and $\eta_c \rightarrow \eta\pi\pi$ as a function of the average instanton size ρ. The short horizontal dashed line shows the experimental $KK\pi$ width. (b) Decay widths $\chi_c \rightarrow K^+K^-, \pi^+\pi^-$ and $\eta\eta$ as a function of the average instanton size ρ. The short horizontal dashed line shows the experimental K^+K^- width.

Let us make the following simple vocabulary, explaining what I mean:

instantons → atomic potential well
fermionic zero modes → electrons bound in the well
non-zero modes → unbound electron states
4-dimensional Dirac operator → atomic Hamiltonian
spectrum of "virtualities" λ → spectra of electron energies
quark condensate → conductivity

Indeed, if one has many instantons and anti-instantons, one has to study statistical mechanics of this system and find out what phase is realized in nature as the ground state of QCD. In principle, there is a wide selection of possibilities, e.g.

1. The ordered system, or "instantonic crystal".
2. Disordered "liquid" of still strongly interacting instantons.
3. Nearly ideal "gas" of instanton–anti-instanton pairs ("molecules").
4. More complicated dilute instanton matter, including "polymeric chains" of instantons connected by through-going quark lines.

The first possibility implies spontaneous *breaking of translational*[8] *invariance*, and, as turns out, also of the global color symmetry (instantons tend to have not only fixed relative positions in the crystal, but fixed relative orientations as well). It does *not* however break the chiral symmetry. This is certainly not the world we live in.

Possibility No. 2, the "liquid", is obviously the best one, it provides the chiral symmetry breaking we need. As we will show in the next section, possibility No. 3, the molecular gas, does not provide spontaneous breaking of the chiral symmetry either, and describes the QGP phase of QCD at high temperatures. Finally, the exotic possibility No. 4 is what happens in high density QCD, and the polymers are the world lines of the Cooper pairs.

But the main point now is that one cannot just select the phase one prefers. As is also true for the atomic system, it is really impossible to say what phase is actually the case under given conditions before some realistic calculation is made. All phases are logically possible, so we need quantitative evaluation of the partition function in order to tell which phase is to be preferred. It means, a hard job lies ahead.

Avoiding this, one can *assume* randomness of the ensemble and calculate what the quark propagator in this field looks like. If the spectrum and eigenvectors of the Dirac operators are known, it is easy to do. It turns out that the "random instanton liquid model" (RILM) works quite well, it reproduces correlators, mesonic and baryonic masses. It fails however exactly where the "quenched lattice approximation" also fails, in describing effects crucially depending on the fermionic determinant.

---

[8]Note that after time translation is broken, the energy is no longer conserved like the quasi-momentum in crystals, where the energy can be obtained from the system as a whole. It is very unfortunate that we do not know of any case in which such a phase would be the ground state.

4.3.2.   *Mean field approximation: pure glue*

The method described in this subsection is widely used in many branches of physics. Let us start with the partition function for a system of instantons in pure gauge theory

$$Z = \frac{1}{N_+! N_-!} \prod_i^{N_+ + N_-} \int [d\Omega_i \, n(\rho_i)] \exp(-S_{\text{int}}). \qquad (4.26)$$

Here, $N_\pm$ are the numbers of instantons and anti-instantons, $\Omega_i = (z_i, \rho_i, U_i)$ are the collective coordinates of the instanton i, $n(\rho)$ is the semi-classical instanton distribution function (3.117) and S_{int} is the bosonic instanton interaction.

In general, before doing full scale simulations of this partition function, it is wise to use some simple approximation. The mean field approximation is one of them, expected to be reliable if correlations between individual instantons are weak. The first such attempt was made by Callan, Dashen and Gross [214] using the dipole interaction, which cannot stabilize the system. In order to deal with this problem, Ilgenfritz and Müller–Preußker [219] introduced an *ad hoc* "hard core" in the instanton interaction which excludes large instantons and leads to a well behaved partition function. It is important to note that one cannot simply cut the size integration at some ρ_c, but has to introduce the cutoff in a way that does not violate the conformal invariance of the classical equations of motion. This guarantees that the results do not spoil the renormalization properties of QCD, most notably the trace anomaly. They have proposed to scale the core to instanton sizes

$$S_{\text{int}} = \begin{cases} \infty & |z_I - z_A| < (a\rho_I^2\rho_A^2)^{1/4} \\ 0 & |z_I - z_A| > (a\rho_I^2\rho_A^2)^{1/4} \end{cases}, \qquad (4.27)$$

which leads to an excluded volume in the partition function, controlled by the *dimensionless* parameter a, and all scale properties are preserved. The next development, by Diakonov and Petrov [220], replaced the core by the repulsive interaction following from the sum ansatz.

In MFA the partition function is evaluated using a trial distribution function chosen to be the product of single instanton distributions $\mu(\rho)$

$$Z_1 = \frac{1}{N_+! N_-!} \prod_i^{N_+ + N_-} \int d\Omega_i \, \mu(\rho_i) = \frac{1}{N_+! N_-!} (V\mu_0)^{N_+ + N_-}, \qquad (4.28)$$

where $\mu_0 = \int d\rho \, \mu(\rho)$. The distribution $\mu(\rho)$ is then determined from the variational principle, or maximization of $Z \, \delta \log Z_1/\delta\mu = 0$. In quantum mechanics a variational wave functions always provides an upper bound on the true ground state energy. The analogous statement in statistical mechanics is known as Feynman's variational principle. Using convexity,

$$Z = Z_1 \langle \exp(-(S - S_1)) \rangle \geq Z_1 \exp(-\langle S - S_1 \rangle), \tag{4.29}$$

where S_1 is the variational action, one can see that the variational vacuum energy is always higher than the true one.

In our case, the single instanton action is given by $S_1 = \log \mu(\rho)$ while $\langle S \rangle$ is the average action calculated from the variational distribution (4.28). Since the variational ansatz does not include any correlations, only the average interaction enters. For the sum ansatz

$$\langle S_{\text{int}} \rangle = \frac{8\pi^2}{g^2} \gamma^2 \rho_1^2 \rho_2^2, \qquad \gamma^2 = \frac{27}{4} \frac{N_c}{N_c^2 - 1} \pi^2, \tag{4.30}$$

the same for both IA and II pairs. Clearly, (4.30) is of the same form as the hard core (4.27) discussed above, with a different dimensionless parameter γ^2. Applying the variational principle, one finds [220]

$$\mu(\rho) = n(\rho) \exp\left[-\frac{\beta\gamma^2}{V} N \bar{\rho}^2 \rho^2\right], \tag{4.31}$$

where $\beta = \beta(\bar{\rho})$ is the average instanton action and $\bar{\rho}^2$ is the average size. We observe that the single instanton distribution is cut off at large sizes by the average instanton repulsion. The average size $\bar{\rho}^2$ is determined by the self-consistency condition $\bar{\rho}^2 = \mu_0^{-1} \int d\rho\, \mu(\rho) \rho^2$. The result is

$$\bar{\rho}^2 = \left(\frac{\nu V}{\beta\gamma^2 N}\right)^{1/2}, \qquad \nu = \frac{b-4}{2}, \tag{4.32}$$

which determines the dimensionless diluteness of the ensemble, $\rho^4(N/V) = \nu/(\beta\gamma^2)$. Using the pure gauge beta function $b = 11$, $\gamma^2 \simeq 25$ from above and $\beta \simeq 15$, we get the diluteness $\rho^4(N/V) = 0.01$, even more dilute than phenomenology requires. The instanton density can be fixed from the second self-consistency requirement, $(N/V) = 2\mu_0$ (the factor 2 comes from instantons and anti-instantons). One gets

$$\frac{N}{V} = \Lambda_{PV}^4 \left[C_{N_c} \beta^{2N_c} \Gamma(\nu)(\beta\nu\gamma^2)^{-\nu/2}\right]^{2/(2+\nu)}, \tag{4.33}$$

$$\chi_{\text{top}} \simeq \frac{N}{V} = (0.65\Lambda_{PV})^4, \qquad (\bar{\rho}^2)^{1/2} = 0.47\Lambda_{PV}^{-1} \simeq \tfrac{1}{3}R, \qquad \beta = S_0 \simeq 15. \tag{4.34}$$

It may be consistent with the phenomenological values if $\Lambda_{PV} \simeq 300\,\text{MeV}$. It is instructive to calculate the free energy as a function of the instanton density.

Using $F = -1/V \cdot \log Z$, one has

$$F = \frac{N}{V} \left\{ \log \left(\frac{N}{2V\mu_0} \right) - \left(1 + \frac{\nu}{2} \right) \right\}. \tag{4.35}$$

The instanton density is determined by minimizing the free energy, $\partial F / \partial (N/V) = 0$. The vacuum energy density is given by the value of the free energy at the minimum, $\epsilon = F_0$. We find $N/V = 2\mu_0$ as above and

$$\epsilon = -\frac{b}{4} \left(\frac{N}{V} \right). \tag{4.36}$$

Estimating the value of the gluon condensate in a dilute instanton gas, $\langle g^2 G^2 \rangle = 32\pi^2 (N/V)$, we see that (4.36) is consistent with the trace anomaly.

The second derivative of the free energy with respect to the instanton density, the compressibility of the instanton liquid, is given by

$$\left. \frac{\partial^2 F}{\partial (N/V)^2} \right|_{n_0} = \frac{4}{b} \left(\frac{N}{V} \right), \tag{4.37}$$

where n_0 is the equilibrium density. This observable is also determined by a low energy theorem based on broken scale invariance [396]

$$\int d^4x \, \langle g^2 G^2(0) g^2 G^2(x) \rangle = (32\pi^2) \frac{4}{b} \langle g^2 G^2 \rangle. \tag{4.38}$$

Here, the left hand side is given by an integral over the field strength correlator, suitably regularized and with the constant disconnected term $\langle g^2 G^2 \rangle^2$ subtracted. For a dilute system of instantons, the low energy theorem gives

$$\langle N^2 \rangle - \langle N \rangle^2 = \frac{4}{b} \langle N \rangle. \tag{4.39}$$

Here, $\langle N \rangle$ is the average number of instantons in a volume V. The result (4.39) shows that density fluctuations in the instanton liquid are not Poissonian. Using the general relation between fluctuations and the compressibility gives the result (4.37). This means that the form of the free energy near the minimum is determined by the renormalization properties of the theory. Therefore, the functional form (4.35) is more general than the mean field approximation used to derive it.

How reliable are these results? The accuracy of the MFA can be checked by doing statistical simulations of the full partition function, see below. The accuracy of the sum ansatz can be checked explicitly by calculating the induced current $j_\mu = D_\mu G^{cl}_{\mu\nu}$ in the classical gauge configurations [210]. This current measures the failure of the gauge potential to be a true saddle point. The sum ansatz indeed is not a good starting point for a self-consistent solution, which is not surprising since it was chosen without any justification other than simplicity. One looked at the streamline method which provided a better idea about the $\bar{I}I$ interaction. However, applying the variational method to the streamline configurations [213] is also not satisfactory, because the ensemble contains too many close pairs and too many large instantons. Stabilization of the density cannot really be reached without a repulsive core of some kind.

4.3.3. *Quark condensate in the mean field approximation*

Proceeding from pure glue theory to QCD with light quarks, one has to deal with the much more complicated problem of quark-induced interactions. In the chiral limit the very presence of instantons cannot be understood for a single instanton: the legs of the 't Hooft vertex have to go somewhere! In the QCD vacuum, however, the chiral symmetry is spontaneously broken and the quark condensate $\langle \bar{q}q \rangle$ is non-zero. The quark condensate is the amplitude for a quark to flip its chirality, and the vacuum thus can absorb these legs. In other words, the effective mass can substitute the fundamental one.

Given the importance of chiral symmetry breaking, we will discuss this phenomenon on three different levels:

(i) In this section, we would like to give a simple qualitative derivation following Ref. [226], which in turn has followed some ideas from earlier works on chiral symmetry breaking by instantons [214, 215].
(ii) The next level would be the approach by Diakonov and Petrov [229], who followed the development of the NJL model.
(iii) Finally we will return to the direct diagonalization of the Dirac operator and discuss its eigenvalue spectrum: a qualitative introduction to it has already been given in Section 4.3.1.

The simplest case is the QCD with just one light flavor, $N_f = 1$, with the only chiral symmetry being the axial $U(1)_A$ broken by the anomaly. This means that there is no spontaneous symmetry breaking, and the quark condensate appears at the level of a single instanton. The condensate is given by

$$\langle \bar{q}q \rangle = i \int d^4x \, \text{tr}(S(x,x)). \tag{4.40}$$

In the chiral limit, non-zero modes do not contribute to the quark condensate. Using the zero mode propagator $S(x,y) = -\psi_0(x)\psi_0^\dagger(y)/(im)$ the contribution of a single instanton to the quark condensate is given by $-1/m$. Multiplying this result by the density of instantons, we have $\langle \bar{q}q \rangle = -(N/V)/m$. Since the instanton density is proportional to the quark mass m, the quark condensate is finite in the chiral limit.

The situation is different for $N_f > 1$ due to a chiral $SU(N_f)_L \times SU(N_f)_R$ symmetry which is spontaneously broken to $SU(N_f)_V$. *This effect cannot be understood at the level of a single instanton*: the contribution to $\langle \bar{q}q \rangle$ is still $(N/V)/m$, but the density of instantons is proportional to $(N/V) \sim m^{N_f}$ and the product vanishes as $m \to 0$.

Spontaneous symmetry breaking has to be a collective effect involving infinitely many instantons. This effect is most easily understood in the context of the MFA, in which one average the $2N_f$-fermion operator appearing in the 't Hooft Lagrangian over the physical vacuum. If its VEV is estimated using the "vacuum dominance" (or factorization) approximation

$$\langle \bar{\psi}\Gamma_1\psi\bar{\psi}\Gamma_2\psi\rangle = \frac{1}{N^2}\left(\text{Tr}\,[\Gamma_1]\,\text{Tr}\,[\Gamma_2] - \text{Tr}\,[\Gamma_1\Gamma_2]\right)\langle\bar{q}q\rangle^2, \tag{4.41}$$

where $\Gamma_{1,2}$ is a spin, isospin, color matrix and $N = 4N_fN_c$ is the corresponding degeneracy factor. Using this approximation, we find that the factor $\prod_f m_f$ in the instanton density should be replaced by $\prod_f m_f^*$, where the effective quark mass is given by[9]

$$m_f^* = m_f - \tfrac{2}{3}\pi^2\rho^2\langle\bar{q}_f q_f\rangle. \tag{4.42}$$

Thus, if chiral symmetry is broken, the instanton density is $O((m^*)^{N_f})$, finite in the chiral limit.

The main idea is to make this estimate self-consistent: to that end we have to calculate the quark condensate from instantons. If we replace the current mass by the effective mass also in the quark propagator,[10] the contribution of a single instanton to the quark condensate is given by $1/m^*$ and, for a finite density of instantons, we expect

$$\langle\bar{q}q\rangle = -\frac{(N/V)}{m^*}. \tag{4.43}$$

This equation, taken together with (4.42), gives a self-consistent value of the quark condensate [226]

$$\langle\bar{q}q\rangle = -\frac{1}{\pi\rho}\sqrt{\frac{3N}{2V}}. \tag{4.44}$$

Using the phenomenological values $(N/V) = 1\,\text{fm}^{-4}$ and $\rho = 0.33\,\text{fm}$, we get $\langle\bar{q}q\rangle \simeq -(215\,\text{MeV})^3$, quite consistent with the experimental value $\langle\bar{q}q\rangle \simeq -(230\,\text{MeV})^3$. The effective quark mass is given by $m^* = \pi\rho(2/3)^{1/2}(N/V)^{1/2} \simeq 170\,\text{MeV}$. The self-consistent pair of equations (4.42), (4.43) has the general form of a gap equation.

An alternative road to MFA for the quark condensate [229] was to formulate an effective fermionic theory that describes the effective interaction between quarks generated by instantons, similar to NJL model. For this purpose, one can integrate over the gauge field first and obtain the 't Hooft Lagrangian. For this purpose we

---

[9]The overall sign of the second term is positive since the condensate is negative.

[10]We will give a more detailed explanation for this approximation in Section 4.3.4.

rewrite the partition function of the instanton liquid,

$$Z = \frac{1}{N_+! N_-!} \prod_i^{N_+ + N_-} \int [d\Omega_i \, n(\rho_i)] \exp(-S_{\text{int}}) \det(\slashed{D} + m)^{N_f}, \qquad (4.45)$$

in terms of a fermionic effective action

$$Z = \int d\psi \, d\psi^\dagger \exp\left(\int d^4 x \psi^\dagger (i\slashed{\partial} + im)\psi \right) \left\langle \prod_{I,f} (\Theta_I - im_f) \prod_{A,f} (\Theta_A - im_f) \right\rangle,$$
$$(4.46)$$

$$\Theta_{I,A} = \int d^4 x \left(\psi^\dagger(x) i\slashed{\partial} \phi_{I,A}(x - z_{I,A}) \right) \int d^4 y \left(\phi^\dagger_{I,A}(y - z_{I,A}) i\slashed{\partial}\psi(y) \right), \qquad (4.47)$$

which describes quarks interacting via the 't Hooft vertices $\Theta_{I,A}$. The expectation value $\langle \cdot \rangle$ corresponds to an average over the distribution of instanton collective coordinates. Formally, (4.46) can be derived by "fermionizing" the original action. In practice, it is easier to check the result by performing the integration over the quark fields and verifying that one recovers the fermion determinant in the zero mode basis.

Here, however, we want to use a different strategy and exponentiate the 't Hooft vertices $\Theta_{I,A}$ in order to derive the effective quark interaction. For this purpose we calculate the average in Eq. (4.46) with respect to the variational single instanton distribution (4.31). There are no correlations, so only the average interaction induced by a single instanton enters. For simplicity, we only average over the position and color orientation and keep the average size $\rho = \bar{\rho}$ fixed

$$Y_\pm = \int d^4 z \int dU \prod_f \Theta_{I,A}. \qquad (4.48)$$

In order to exponentiate $Y_\pm$ we insert factors of unity,

$$\int d\Gamma_\pm \int d\beta_\pm / (2\pi) \exp(i\beta_\pm(Y_\pm - \Gamma_\pm)),$$

and integrate over $\Gamma_\pm$ using the saddle point method

$$Z = \int d\psi \, d\psi^\dagger \exp\left(\int d^4 x \, \psi^\dagger i\slashed{\partial}\psi \right) \int \frac{d\beta_\pm}{2\pi} \exp(i\beta_\pm Y_\pm)$$
$$\times \exp\left[N_+ \left(\log\left(\frac{N_+}{i\beta_+ V} \right) - 1 \right) + (+ \leftrightarrow -) \right], \qquad (4.49)$$

where we have neglected the current quark mass. In this partition function, the saddle point parameters $\beta_\pm$ play the role of an activity for instantons and anti-instantons.

The form of the saddle point equations for $\beta_\pm$ depends on the number of flavors. The simplest case is $N_f = 1$, where the Grassmann integration is quadratic. The average over the 't Hooft vertex is most easily performed in momentum space,

$$Y_\pm = \int \frac{d^4k}{(2\pi)^4} \int dU \psi^\dagger(k) \left[\slashed{k} \phi_{I,A}(k) \phi^\dagger_{I,A}(k) \slashed{k} \right] \psi(k), \qquad (4.50)$$

where $\phi(k)$ is the Fourier transform of the zero mode profile (see the Appendix). Performing the average over the color orientation, we get

$$Y_\pm = \int \frac{d^4k}{(2\pi)^4} \frac{1}{N_c} k^2 \varphi'^2(k) \psi^\dagger(k) \gamma_\pm \psi(k), \qquad (4.51)$$

where $\gamma_\pm = (1 \pm \gamma_5)/2$ and $\varphi'(k)$ is defined in the Appendix. Clearly, the saddle point equations are symmetric in $\beta_\pm$, so that the average interaction is given by $Y_+ + Y_-$, which acts like a mass term. This can be seen explicitly by first performing the Grassmann integration

$$Z = \int \frac{d\beta_\pm}{2\pi} \exp \left[N_\pm \left(\log \frac{N_\pm}{i\beta_\pm V} - 1 \right) \right.$$
$$\left. + N_c V \int \frac{d^4k}{(2\pi)^4} \operatorname{tr} \log \left(\slashed{k} + \gamma_\mp \beta_\pm \frac{k^2 \varphi'^2(k)}{N_c} \right) \right], \qquad (4.52)$$

and then doing the saddle point integral. Varying with respect to $\beta = \beta_\pm$ gives the gap equation [229]

$$\int \frac{d^4k}{(2\pi)^4} \frac{M^2(k)}{k^2 + M^2(k)} = \frac{N}{4N_c V}, \qquad (4.53)$$

where $M(k) = \beta k^2 \varphi'^2(k)/N_c$ is the momentum-dependent effective quark mass. The gap equation determines the effective constituent mass $M(0)$ in terms of the instanton density N/V. For the parameters (4.34), the effective mass is $M \simeq 350$ MeV. We can expand the gap equation in the instanton density [207]. For small N/V, one finds $M(0) \sim \rho(N/2VN_c)^{1/2}$, which parametrically behaves like the effective mass m^* introduced above. Note that a dynamical mass is generated for arbitrarily small values of the instanton density. This is expected for $N_f = 1$, since there is no spontaneous symmetry breaking and the effective mass is generated by the anomaly at the level of an individual instanton.

In the context of QCD, the more interesting case is the one of two or more flavors. For $N_f = 2$, the effective 't Hooft vertex is a four-fermion interaction

$$Y_\pm = \left[\prod_{i=1,4} \int \frac{d^4k_i}{(2\pi)^4} k_i \varphi'(k_i) \right] (2\pi)^4 \delta^4 \left(\sum_i k_i \right)$$
$$\times \frac{1}{4(N_c^2 - 1)} \left(\frac{2N_c - 1}{2N_c} (\psi^\dagger \gamma_+ \tau_a^- \psi)^2 + \frac{1}{8N_c} (\psi^\dagger \gamma_+ \tau_a^- \psi)^2 \right), \qquad (4.54)$$

where $\tau_a^- = (\vec{\tau}, i)$ is an isospin matrix and we have suppressed the momentum labels on the quark fields. In the long wavelength limit $k \to 0$, the 't Hooft vertex (4.54) corresponds to a local four-quark interaction[11]

$$\mathcal{L} = \beta(2\pi\rho)^4 \frac{1}{4(N_c^2 - 1)}$$

$$\times \left(\frac{2N_c - 1}{2N_c} \left[(\psi^\dagger \tau_a^- \psi)^2 + (\psi^\dagger \gamma_5 \tau_a^- \psi)^2 \right] + \frac{1}{4N_c} (\psi^\dagger \gamma_+ \tau_a^- \psi)^2 \right). \tag{4.55}$$

This Lagrangian is of the type first studied by Nambu and Jona–Lasinio [37] and widely used as a model for chiral symmetry breaking and as an effective description for low energy chiral dynamics, see e.g. a nice review by Klevansky [39].

Unlike the NJL model, however, the interaction has a natural cut-off parameter $\Lambda \sim \rho^{-1}$, and the coupling constants in (4.55) are determined in terms of a physical parameter, the instanton density (N/V). The interaction is attractive for quark–anti-quark pairs with the quantum numbers of the π and σ mesons. If the interaction is sufficiently strong, the vacuum is rearranged, quarks condense and a light (Goldstone) pion is formed. The interaction is repulsive in the pseudoscalar–isoscalar (the SU(2) singlet η') and the scalar–isovector δ channels, showing the effect of the U(1)$_A$ anomaly. Note that to first order in the instanton density, there is no interaction in the vector ρ, ω, a_1, f_1 channels. We will explore the consequences of this interaction in much more detail below.

In the case of two (or more) flavors the Grassmann integration cannot be done exactly, since the effective action is more than quadratic in the fermion fields. Instead, we perform the integration over the quark fields in the mean field approximation. This procedure is consistent with the approximations used to derive the effective interaction (4.48). The MFA is most easily derived by decomposing the fermion bilinears into a constant and a fluctuating part. The integral over the fluctuations is quadratic and can be done exactly. Technically, this can be achieved by introducing auxiliary scalar fields L_a, R_a into the path integral[12] and then shifting the integration variables in order to linearize the interaction. Using this method, the four-fermion interaction becomes

$$\begin{aligned}
(\psi^\dagger \tau_a^- \gamma_- \psi)^2 &\to 2(\psi^\dagger \tau_a^- \gamma_- \psi)L_a - L_a L_a, \\
(\psi^\dagger \tau_a^- \gamma_+ \psi)^2 &\to 2(\psi^\dagger \tau_a^- \gamma_+ \psi)R_a - R_a R_a.
\end{aligned} \tag{4.56}$$

In the mean field approximation, the L_a, R_a integration can be done using the saddle point method. Since isospin and parity are not broken, only $\sigma = L_4 = R_4$ can have

---

[11]The structure of this interaction is identical to the one given in Eq. (4.14), as one can check using Fierz identities. The only new ingredient is that the overall constant β is determined self-consistently from a gap equation.

[12]In the MFA, we do not need to introduce auxiliary fields $T_{\mu\nu}$ in order to linearize the tensor part of the interaction, since $T_{\mu\nu}$ cannot have a vacuum expectation value.

a non-zero value. At the saddle point, the free energy $F = -1/V \cdot \log Z$ is given by

$$F = 4N_c \int \frac{d^4k}{(2\pi)^4} \log \left(k^2 + \beta\sigma k^2 \varphi'^2(k) \right)$$
$$-2\frac{2N_c(N_c^2 - 1)}{2N_c - 1}\sigma^2 - \frac{N}{V}\log\left(\frac{\beta V}{N}\right). \qquad (4.57)$$

Varying with respect to $\beta\sigma$ gives the same gap equation as in the $N_f = 1$ case, now with $M(k) = \beta\sigma k^2 \varphi'^2(k)$. We also find $(N/V) = 2f\sigma^2\beta$ where $f = 2N_c(N_c^2 - 1)/(2N_c - 1)$. Expanding everything in (N/V) one can show that $M(0) \sim (N/V)^{1/2}$, $\sigma \sim (N/V)^{1/2}$ and $\beta \sim$ const.

The fact that the gap equation is independent of N_f is a consequence of the mean field approximation. This implies that even for $N_f = 2$, chiral symmetry is spontaneously broken for arbitrarily small values of the instanton density. As we will see in the next chapter, this is not correct if the full fermion determinant is included. If the instanton density is too small, the instanton ensemble forms a molecular gas and chiral symmetry is unbroken.

The quark condensate is given by

$$\langle \bar{q}q \rangle = -4N_c \int \frac{d^4k}{(2\pi)^4} \frac{M(k)}{M^2(k) + k^2}. \qquad (4.58)$$

Solving the gap equation numerically, we get $\langle \bar{q}q \rangle \simeq -(255\,\mathrm{MeV})^3$. It is easy to check that $\langle \bar{q}q \rangle \sim (N/V)^{1/2}\rho^{-1}$, in agreement with the results obtained above.

The procedure for three flavors is very similar, so we do not need to go into detail here. Let us simply quote the effective Lagrangian for $N_f = 3$ [231],

$$\mathcal{L} = \beta(2\pi\rho)^6 \frac{1}{6N_c(N_c^2 - 1)} \epsilon_{f_1 f_2 f_3} \epsilon_{g_1 g_2 g_3}$$
$$\times \left(\frac{2N_c + 1}{2N_c + 4} (\psi_{f_1}^\dagger \gamma_+ \psi_{g_1})(\psi_{f_2}^\dagger \gamma_+ \psi_{g_2})(\psi_{f_3}^\dagger \gamma_+ \psi_{g_3}) \right.$$
$$+ \frac{3}{8(N_c + 2)} (\psi_{f_1}^\dagger \gamma_+ \psi_{g_1})(\psi_{f_2}^\dagger \gamma_+ \sigma_{\mu\nu}\psi_{g_2})(\psi_{f_3}^\dagger \gamma_+ \sigma_{\mu\nu}\psi_{g_3})$$
$$+ (+ \leftrightarrow -)), \qquad (4.59)$$

which was first derived in a slightly different form in Ref. [218]. So far, we have neglected the current quark mass dependence and considered the SU(N_f) symmetric limit. Real QCD is intermediate between the $N_f = 2$ and $N_f = 3$ cases. Flavor mixing in the instanton liquid with realistic values of quark masses was studied in Ref. [231] to which we refer the reader for more details.

As for NJL, the pseudo-scalar spectrum is most easily obtained by bosonizing the effective action. In order to correctly describe the U(1)$_A$ anomaly and the scalar σ channel, one has to allow for fluctuations of the number of instantons. The fluctuation properties of the instanton ensemble can be described by the following "coarse grained" partition function [231]:

$$
S_{\text{eff}} = \frac{b}{4} \int d^4 z \, (n^+(z) + n^-(z)) \left(\log \left(\frac{n^+(z) + n^-(z)}{n_0} \right) - 1 \right)
$$
$$
+ \frac{1}{2n_0} \int d^4 z \, (n^+(z) - n^-(z))^2
$$
$$
+ \int d^4 z \, (\psi^\dagger (i\partial\!\!\!/ + im)\psi - n^+(z)\overline{\Theta}_I(z) - n^-(z)\overline{\Theta}_A(z)), \qquad (4.60)
$$

where $n^\pm(z)$ is the local density of instantons/anti-instantons and $\overline{\Theta}_{I,A}(z)$ is the 't Hooft interaction (4.47) averaged over the instanton orientation. This partition function reproduces the low energy theorem (4.39) as well as the relation $\chi_{\text{top}} = N/V$ expected for a dilute system in the quenched approximation. In addition to that, the divergence of the flavor-singlet axial current is given by $\partial_\mu j^5_\mu = 2N_f(n^+(z) - n^-(z))$, consistent with the axial U(1)$_A$ anomaly.

Again, the partition function can be bosonized by introducing auxiliary fields, linearizing the fermion interaction and performing the integration over the quarks. In addition to that, we expand the fermion determinant in derivatives of the meson fields in order to recover kinetic terms for the meson fields. This procedure gives the standard nonlinear σ-model Lagrangian. To leading order in the quark masses, the pion and kaon masses satisfy the Gell–Mann, Oakes, Renner relations and all other chiral identities. The value of the pion decay constant is

$$
f_\pi^2 = 4N_c \int \frac{d^4 k}{(2\pi)^4} \frac{M^2(k)}{(k^2 + M^2(k))^2} \simeq (100 \, \text{MeV})^2. \qquad (4.61)
$$

There is the anomalous contribution to the η' mass. It agrees with the effective Lagrangian originally proposed in Ref. [189] and leads to the Witten–Veneziano relation

$$
f_\pi^2 (m_{\eta'}^2 + m_\eta^2 - 2m_K^2) = N_f(N/V). \qquad (4.62)
$$

Diagonalizing the mass matrix for $m = 5$ MeV and $m_s = 120$ MeV, we find $m_\eta = 527$ MeV, $m_{\eta'} = 1172$ MeV and a mixing angle $\theta = -11.5°$. The η' mass is somewhat too heavy, but given the crude assumptions, the result is certainly not bad. One should not forget that the result corresponds to MFA, an ensemble of uncorrelated instantons. In full QCD, however, the topological susceptibility is zero and correlations between instantons have to be present, see Section 4.4.1.

4.3.4. *The single instanton approximation*

In Section 4.2.2 we derived the 't Hooft Lagrangian using a configuration with a *single instanton*. We found a finite answer as a result of the exact compensation between (i) small $\sim \Pi_f m_f$ (or vanishing, in the chiral limit) probability to have one instanton and (ii) large (or infinite) $\sim 1/\Pi_f m_f$ contribution of diagrams with

N_f fermion lines. However, in a real vacuum there are many instantons and anti-instantons, and so one may wonder how one should change the argumentation in order to account for the effect of one instanton *belonging to the ensemble*. A very short answer to that is that in both the fermionic determinant and the fermionic propagators some *effective quark masses* appear. (Both remain finite in the chiral limit and so no compensation of infinitely large factors is needed.) However, in contrast to the mean field approximation discussed above, their values are *not* the same. One can strictly define [172, 233] both effective masses, and relate them to the *spectrum of Dirac eigenvalues* λ.

Let us focus at an ensemble with $Q = 0$, equal number of instanton and anti-instantons, without exact zero modes. The propagator in such a background field can be written as follows:

$$S(x, y) = \sum_{IJ} \psi_I(x) \left(\frac{1}{T + im} \right)_{IJ} \psi_J^\dagger(y) + S_{NZM}, \qquad (4.63)$$

and we would in this section ignore the last term representing the contribution of all non-zero modes. The capital letters number all instantons and anti-instantons, $\psi_I(x)$ is the zero mode of the Ith instanton and T_{IJ} denotes the so called *overlap matrix in the zero-modes subspace*

$$T_{IJ} = \int d^4z \, \psi^\dagger(z)_I (i\slashed{D}) \psi(z)_J. \qquad (4.64)$$

In (4.63), two zero-modes are amplitudes to pick up and emit quarks at initial and final points by some instantons. The inverse "hopping matrix" T_{IJ} is the amplitude to jump from one instanton to another.

Here comes the main idea. Suppose we calculate some diagram with several propagators, and all their arguments are either at the same point, or at a few points which are sufficiently close to one another, for all $x, y, \ldots |x - y| \ll R$ ($R = n^{-1/4} \sim 1$ fm is the typical instanton separation). It is clear that the biggest term in the propagator is associated to the closest instanton, I^*, while all the terms in the sum representing instantons far away from the points x and y will be sub-leading.

$$S(x, y) \simeq \psi_{I_*}(x) \left(\frac{1}{T + im} \right)_{I_* I_*} \psi_{I_*}^\dagger(y) \approx \frac{\psi_{I_*}(x) \psi_{I_*}^\dagger(y)}{m^*}, \qquad (4.65)$$

where we have approximated the diagonal matrix element by its average,

$$(T + im)_{I_* I_*}^{-1} \simeq N^{-1} \sum_I (T + im)_{II}^{-1}.$$

So we now see that a proper definition of the effective mass involves the inverse matrix containing all the information about the particular configuration of the instanton ensemble. Furthermore, in order to evaluate various correlation functions, containing different numbers of propagators, one needs to average over all possible configurations.

For example, the quark condensate is

$$\langle 0|\bar{u}(x)u(x)|0\rangle = \int d^4 z \int d\rho\, n_0\, d(\rho) \left[\frac{-2\rho^2}{[(z-x)^2 + \rho^2]^3 \pi^2\, m_{uu}} \right]$$

$$= -\frac{n_0}{m_{uu}}, \tag{4.66}$$

where, for reasons that will become clear shortly, we have denoted with m_{uu} the quark effective mass and $n_0 := n_0^I + n_0^{\bar{I}}$.

Now, repeating the same calculation in the ensemble:[13]

$$\chi_{uu} = \left\langle \text{Tr}\left[\sum_{I,J} \psi_{0\,I}(x) \left(\frac{1}{T}\right)_{IJ} \psi_{0\,J}^\dagger(x) \right] \right\rangle, \tag{4.67}$$

where the average is over all configurations of the ensemble. A comparison provides another definition of an effective mass

$$m_{uu} := -\frac{n_0}{\left\langle \text{Tr}\left[\sum_{I,J} \psi_{0\,I}(x) \left(\frac{1}{T}\right)_{IJ} \psi_{0\,J}^\dagger(x) \right] \right\rangle}. \tag{4.68}$$

Let us now consider another VEV, of a scalar quark density *squared*, namely:

$$\chi_{uudd} := \langle 0|\text{Tr}[\bar{u}(x)u(x)] \cdot \text{Tr}[\bar{d}(x)d(x)]|0\rangle = \langle [\text{Tr}S(x,x)]^2 \rangle. \tag{4.69}$$

Such condensate receives a double contribution from zero-modes. For a single instanton one obtains:

$$\chi_{uudd} = \int d\rho\, d(\rho) \frac{n_0}{5\pi^2 \rho^4 m_{uudd}^2}, \tag{4.70}$$

where we have now denoted with m_{uudd} one more "quark effective mass".

Comparing, as before, with the result of ensemble calculations leads to still another effective mass

$$m_{uudd}^2 = \left(\int d\rho\, d(\rho) \frac{n_0}{5\pi^2 \rho^4} \right)$$

$$\times \left\langle \left[\text{Tr} \sum_{I,J} \psi_{0\,I}(x) \left(\frac{1}{T}\right)_{IJ} \psi_{0\,J}^\dagger(x) \right]^2 \right\rangle^{-1}. \tag{4.71}$$

---

[13] Here, we have neglected all small current quark mass terms in $1/T$.

Now, are these two masses the same, namely $(m_{uu})^2 = m^2_{uudd}$? Generally speaking there is no reason it would be true, because if so it would imply a relation

$$\frac{\left\langle \text{Tr} \left[\sum_{I,J} \psi_{0\,I}(x) \left(\frac{1}{T}\right)_{I\,J} \psi^\dagger_{0\,J}(x) \right] \right\rangle^2}{\left\langle \left[\text{Tr} \sum_{I,J} \psi_{0\,I}(x) \left(\frac{1}{T}\right)_{I\,J} \psi^\dagger_{0\,J}(x) \right]^2 \right\rangle} = \frac{5\pi^2 n_0}{\int d\rho\, d(\rho) 1/\rho^4}$$

$$\simeq 5\pi^2 n_0 \bar\rho^4 \sim \frac{5}{8}, \qquad (4.72)$$

for which there is apparently no reason. Moreover, since different configurations and even points have different leading instanton, the corresponding value $T_{I^*I^*}$ fluctuates, and the average of its different powers in general leads to different effective masses.

The propagators are not the only place where the Dirac spectrum appears: they form the fermionic determinant as well. The very first effective mass we defined previously was defined in this way. After averaging over the appropriate ensemble, one can define the so-called "determinantal masses":

$$m^i_{\text{det}} := \frac{\langle (\det[\slashed{D}])^{i/N} \rho^i \rangle}{\langle \rho^i \rangle}, \quad i = 1, 2, \ldots \qquad (4.73)$$

where the index i refers to the number of flavors and N denotes the number of instantons. Their values tell us how much the presence of fermions reduces the instanton density, compared to the same ensemble without them.

In principle, there is no reason why the values of m_{det} and m^2_{det} should agree with other effective masses defined in the previous section; the average of positive rather than negative powers of the overlap matrix may be quite different. And indeed they do not; see the comparison calculated numerically for two ensembles called the RILM and in the IILM (to be defined below): the results are reported in the following Table 4.1.

The general reason why these masses are rather small is the following. Instantons have fluctuating strength of interaction with others in the ensemble: some of them are "hermits" and have small matrix elements in the corresponding entries of the overlap matrix T. As in all expressions we average the *inverse* of this matrix, the contribution of such "hermits" is enhanced. This lowers the value of the effective

Table 4.1. Determinantal masses evaluated in the RILM and in the IILM as compared to m_1 and m^2_2 defined in the text.

Mass	RILM calculation	IILM calculation
m_1	120 MeV	177 MeV
m_{det}	63 MeV	102 MeV
m^2_2	$(65\,\text{MeV})^2$	$(103\,\text{MeV})^2$
m^2_{det}	$(64\,\text{MeV})^2$	$(103\,\text{MeV})^2$

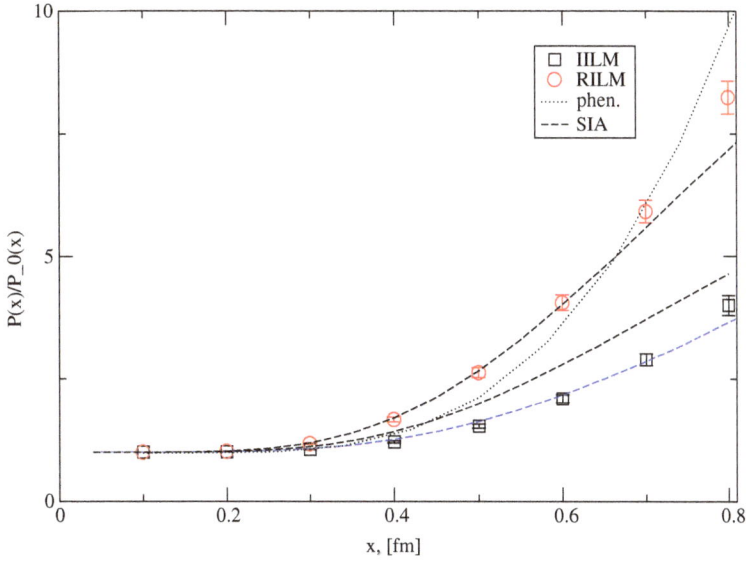

Fig. 4.3. Pion pseudo-scalar two-point function normalized to the same correlation function in the free theory. The open circles (squares) represent RILM (IILM) points, the dashed lines represent SIA calculations with masses given in Table 4.1 and the dotted line is the phenomenological curve obtained from the spectral decomposition.

masses. Furthermore, because a random ensemble of RILM has more such "hermits", as compared to IILM (where the fermionic determinant in the statistical weight suppresses them), these masses are smaller in RILM as compared to IILM. Such discrepancy reflects the fact that the two ensembles give actually quite different correlation functions [172].

Are these effective masses useful for any practical applications? If so, how good is SIA? In general, reliability of the SIA depends on the vacuum diluteness and the quantity one evaluates. The more propagators, the better it works: we found that SIA is not good for the quark condensate but quite accurate for any observable of dimension 6 such as the condensate squared or correlation functions.

One can find many comparisons in the original work [233], let me give one example, the pseudoscalar correlation function. For the definitions and explanations see Chapter 6. The correlation function of two pseudoscalar currents is plotted versus the distance between them. The normalization is to free quark propagation, so convergence of all results to 1 for $x < 3\,\mathrm{fm}$ is just due to asymptotic freedom. Points are full (multi-instanton) simulations for two ensembles, those are to be compared to the upper and the lower dashed lines, representing SIA with the masses from the Table 4.1. Note that except for the last points at $x \sim 0.8\,\mathrm{fm}$, the agreement is very good.

The dotted line is the parameterization of the experiment: the middle dashed line — tangent to the dotted one — is the SIA with the phenomenologically tuned effective mass. If one would like to consider this effective mass to be the third parameter of the instanton liquid (after n_0, ρ_0), a lot of applications to short-distance observables — correlators, formfactors etc. — become possible in the SIA framework.

4.4.   The interacting instanton liquid model

We have discussed above two approximate methods, the MFA and SIA, but in order to understand the ensemble of interacting instantons one can also perform direct simulations of the appropriate partition function.

$$Z = \int \Pi_{I,A} d\Gamma(\alpha_{I,A}) \exp(-S_{\text{int}}) \det{}^{N_f} [i\hat{D} + im]. \tag{4.74}$$

In the SU(3) color group there are 12 collective variables per instanton: 4 coordinates of the center, the size ρ and 7 color rotation angles. Recall that in general it has the following three main ingredients:

1. Semiclassical tunneling probability, which I will denote $d\Gamma(\alpha)$, depending on the collective coordinates of the pseudoparticle.
2. Gluonic interaction between pseudoparticles S_{int}.
3. Quark-induced interaction, in the form of the fermionic determinant.

Let me subsequently comment on them. The measure in $d\Gamma(\alpha)$, e.g. for the multi-instanton solutions, has quite nontrivial geometry of the manifold of collective coordinates, but very few studies (see Ref. [888]) of it have so far been done. In practice it is usually understood as a product of measures for independent instantons, with the difference in principle to be included into the gluonic interaction.

For $\bar{I}I$ pairs there is freedom in choosing an ansatz for the gauge field, corresponding to finding a convenient parametrization of the field configurations in the vicinity of an approximate saddle point. In general, we have to integrate over all field configurations anyway, but the ansatz determines the way we split coordinates into approximate zero modes and non-zero modes.

For well separated IA pairs, the fields are not strongly distorted and the interaction is well defined. For very close pairs, on the other hand, the fields are strongly modified and the interaction is not well determined. In addition to that, if the instanton and anti-instanton begin to annihilate, the gauge fields become perturbative and should not be included in semi-classical approximations.

The interaction between instantons at large distances was derived by Callan, Dashen and Gross [214] (see also Ref. [208]). They began by studying the interaction of an instanton with a weak, slowly varying external field $(G^a_{\mu\nu})_{\text{ext}}$. In order ensure that the gauge field is localized, the instanton is put in the singular gauge.

One then finds

$$S_{\text{int}} = -\frac{2\pi^2}{g^2} \rho^2 \bar{\eta}^a_{\mu\nu} (G^a_{\mu\nu})_{\text{ext}}. \tag{4.75}$$

This result can be interpreted as the external field interacting with the color magnetic "dipole moment" $(2\pi^2/g^2)\rho^2\bar{\eta}^a_{\mu\nu}$ of the instanton. Note that the interaction vanishes if the external field is self-dual. If the external field is taken to be an anti-instanton at distance R, the interaction is

$$S_{\text{int}} = -\frac{32\pi^2}{g^2} \rho_I^2 \rho_A^2 \eta^a_{\mu\rho} \eta^b_{\nu\rho} R^{ab} \frac{\hat{R}_\mu \hat{R}_\nu}{R^4}, \tag{4.76}$$

where R^{ab} is the relative color orientation matrix $R^{ab} = \frac{1}{2}\text{tr}(U\tau^a U^\dagger \tau^b)$ and $\hat{R}$ is the unit vector connecting the centers of the instanton and anti-instanton. The dipole–dipole form of the interaction is quite general, the interaction of topological objects in other theories, such as Skyrmions or $O(3)$ instantons, is of a similar type. In order to study this interaction in more detail, it is useful to introduce some additional notation. We can characterize the relative color orientation in terms of the four vector $u_\mu = (1/2i)\text{tr}(U\tau^+_\mu)$, where $\tau^+_\mu = (\vec{\tau}, -i)$. Note that for the gauge group SU(2), u_μ is a real vector with $u^2 = 1$, whereas for SU($N > 2$), u_μ is complex and $|u|^2 \leq 1$. Also note that for SU($N > 2$), the interaction is completely determined by the upper 2×2 block of the SU(N) matrix U. We can define a relative color orientation angle θ by

$$\cos^2\theta = |u\hat{R}|^2/|u|^2. \tag{4.77}$$

In terms of this angle, the dipole interaction is given by

$$S_{\text{int}} = -\frac{32\pi^2}{g^2} \frac{\rho_1^2 \rho_2^2}{R^4} |u|^2 (1 - 4\cos^2\theta). \tag{4.78}$$

The orientational factor $d = 1 - 4\cos^2\theta$ varies between 1 and -3. We also observe that the dipole interaction vanishes upon averaging over all angles θ.

The interactions has been studied for the sum ansatz (a simple sum of A_μ for instantons in the singular gauge, in which the decrease with distance is maximal). It however creates singular $G_{\mu\nu}$ fields which can be cured by the "ratio" ansatz [234]

$$A^a_\mu = \left(2R^{ab}_I \bar{\eta}^b_{\mu\nu} \frac{\rho_I^2 (x - z_I)_\nu}{(x - z_I)^4} + 2R^{ab}_A \eta^b_{\mu\nu} \frac{\rho_A^2 (x - z_A)_\nu}{(x - z_A)^4} \right)$$
$$\times \left(1 + \frac{\rho_I^2}{(x - z_I)^2} + \frac{\rho_A^2}{(x - z_A)^2} \right)^{-1}, \tag{4.79}$$

whose form was inspired by the 't Hooft's exact multi-instanton solution. This ansatz ensures that the field strength is regular everywhere and that (at least if they have the same orientation) there is no interaction between pseudo-particles of the same charge. Both the sum and the ratio ansatz lead to the dipole interaction at large distances, but the amount of repulsion at short distance is significantly reduced in the ratio ansatz. Yung [212] has proposed an even better ansatz,

$$A_\mu^a = 2\eta_{\mu\nu}^a x_\nu \frac{1}{x^2 + \rho^2 \lambda} + 2R^{ab}\eta_{\mu\nu}^b x_\nu \frac{\rho^2}{\lambda} \frac{1}{x^2(x^2 + \rho^2/\lambda)}, \tag{4.80}$$

where $\rho = \sqrt{\rho_1 \rho_2}$ is the geometric mean of the two instanton radii and λ is the conformal parameter

$$\lambda = \frac{R^2 + \rho_1^2 + \rho_2^2}{2\rho_1\rho_2} + \left(\frac{(R^2 + \rho_1^2 + \rho_2^2)^2}{4\rho_1^2\rho_2^2} - 1\right)^{1/2}, \tag{4.81}$$

which approximately satisfies the streamline equation at all distances [213]. The interaction for this ansatz is given by [213]

$$S_{IA} = \frac{8\pi^2}{g^2} \frac{1}{(\lambda^2 - 1)^3} \left\{-4\left[1 - \lambda^4 + 4\lambda^2 \log(\lambda)\right]\left[|u|^2 - 4|u \cdot \hat{R}|^2\right]\right.$$
$$+ 2\left[1 - \lambda^2 + (1 + \lambda^2)\log(\lambda)\right]$$
$$\left. \times \left[(|u|^2 - 4|u \cdot \hat{R}|^2)^2 + |u|^4 + 2(u)^2(u^*)^2\right]\right\}. \tag{4.82}$$

The fermionic interaction is usually treated by separating the subspace of the zero mode of all instantons and calculating the determinant exactly in this subspace,

$$i\displaystyle{\not}D = \begin{pmatrix} 0 & T_{IA} \\ T_{AI} & 0 \end{pmatrix}, \tag{4.83}$$

while using factorization for the non-zero modes into the product for each instanton. The matrix elements have the meaning of a hopping amplitude for a quark from one pseudo-particle to another. Indeed, the amplitude for an instanton to emit a quark is given by the amputated zero mode wave function $\displaystyle{\not}D\psi_{0,I}$. This shows that the matrix element can be written as two quark-instanton vertices connected by a propagator, $\psi_{0,I}^\dagger i\displaystyle{\not}D(i\displaystyle{\not}D)^{-1}i\displaystyle{\not}D\psi_{0,A}$. At large distance, the overlap matrix element decreases as $T_{IA} \sim 1/R^3$, which corresponds to the exchange of a massless quark. The determinant of the matrix (4.83) can be interpreted as the sum of all closed diagrams consisting of (zero mode) quark exchanges between pseudo-particles. In other words, the logarithm of the Dirac operator (4.83) is the contributions of the 't Hooft effective interaction to the vacuum energy to all orders. It is easy to see explicitly how writing the determinant as a product of one element from each column and row leads to the sum of all loop diagrams with the 't Hooft effective vertices. Due to the chirality of the zero modes, the matrix elements T_{II} and T_{AA} between instantons of the same charge vanish. In the sum ansatz, we can use the equations of motion to replace the covariant derivative in T_{IA} by an ordinary one. The dependence on the relative orientation is then given by $T_{IA} = (u \cdot \hat{R})f(R)$. This means that, like the gluonic dipole interaction, the fermion overlap is maximal if

$\cos\theta = 1$. The matrix element can be parameterized by

$$T_{IA} = i(u \cdot R)\frac{1}{\rho_I \rho_A}\frac{4.0}{(2.0 + R^2/(\rho_I \rho_A))^2}. \qquad (4.84)$$

A parametrization of the overlap matrix element for the streamline gauge configuration can be found in Ref. [237].

As an example, the contribution of a single instanton–anti-instanton pair to the partition function is

$$Z_{IA} = V_4 \int d^4 z \, dU \, \exp(N_f \log |T_{IA}(U, z)|^2 - S_{\text{int}}(U, z)). \qquad (4.85)$$

Here, z is the distance between the centers and U is the relative orientation of the pair. The fermionic part is attractive, while the bosonic part is either attractive or repulsive, depending on the orientation. If the interaction is repulsive, there is a real saddle point for an z-integral, whereas for an attractive orientation there is only a saddle point in the complex z plane (as in Section 3.3). The calculation of the partition function (4.85) in the saddle point approximation was recently attempted in Ref. [558]. There it was found that for a large number of flavors, $N_f > 5$, the ground state energy oscillates as a function of N_f. The period of the oscillations is 4, and the real part of the energy shift vanishes for even $N_f = 6, 8, \ldots$. The reason for these oscillations is exactly the same as in the case of SUSY quantum mechanics: the saddle point gives a complex contribution with a phase that is proportional to the number of flavors.

In order to go beyond the mean field approximation and study the instanton liquid with the 't Hooft interaction included to all orders, we have performed numerical simulations of the interacting instanton liquid [231, 234, 235]. In these simulations, we make use of the fact that the quantum field theory problem is analogous to the statistical mechanics of a system of pseudo-particles in 4 dimensions. The distribution of the $4N_cN$ collective coordinates associated with a system of N pseudo-particles can be studied using standard Monte Carlo techniques (e.g. the Metropolis algorithm), originally developed for simulations of statistical systems containing atoms or molecules.

These simulations have a number of similarities to lattice simulations of QCD, see the review [250]. Like lattice gauge theory, we consider systems in a finite 4-dimensional volume, subject to periodic boundary conditions. This means that both approaches share finite size problems, especially the difficulty to work with realistic quark masses. Also, both methods are formulated in Euclidean space which means that it is difficult to extract real time (in particular non-equilibrium) information. However, in contrast to the lattice, space–time is continuous, so we have no problems with chiral fermions. Furthermore, the number of degrees is drastically reduced and meaningful (unquenched!) simulations of the instanton ensemble can be performed in a few minutes on an average workstation. Finally, using the analogy with an interacting liquid, it is easier to develop an intuitive understanding of the numerical simulations.

The gauge interaction between instantons is approximated by a sum of pure two-body interactions $S_{int} = \frac{1}{2}\sum_{I\neq J} S_{int}(\Omega_{IJ})$. Genuine three-body effects in the instanton interaction are not important as long as the ensemble is reasonably dilute. This part of the interaction is fairly easy to deal with. The computational effort is similar to that of a statistical system with long range forces.

The fermion determinant, however, introduces non-local interactions among many instantons. Changing the coordinates of a single instanton requires the calculation of the full N-instanton determinant, not just N two-body interactions. In practice, it is simpler to study the instanton ensemble for a fixed particle number $N = N_+ + N_-$. This means that instead of the grand canonical partition function we consider the canonical ensemble

$$Z_N = \frac{1}{N_+!N_-!} \prod_i^{N_++N_-} [d\Omega_i\, n(\rho_i)] \exp(-S_{int}) \prod_f^{N_f} \det(\slashed{D} + m_f), \qquad (4.86)$$

for different densities and determine the ground state by minimizing the free energy $F = -1/V \log Z_N$ of the system. Furthermore, we will only consider ensembles with no net topology. The two constraints $N/V = \text{const}$ and $Q = N_+ - N_- = 0$ do not affect the results in the thermodynamic limit. The only exceptions are of course fluctuations of Q and N (the topological susceptibility and the compressibility of the instanton liquid). In order to study these quantities one has to consider appropriately chosen subsystems, see Section 4.4.1.

Using the sequence of configurations generated by the Metropolis algorithm, it is straightforward to determine expectation values by averaging measurements in many configurations

$$\langle \mathcal{O} \rangle = \lim_{N\to\infty} \frac{1}{N} \sum_{j=1}^{N} \mathcal{O}(\{\Omega_i\}). \qquad (4.87)$$

This is how the quark and gluon condensates, as well as the hadronic correlation functions discussed in this and the following sections have been determined. However, more work is required to determine the partition function, which gives the overall normalization of the instanton distribution. Knowledge of the partition function is necessary in order to calculate the free energy and the thermodynamics of the system. In practice, the partition function is most easily evaluated using the thermodynamic integration method. For this purpose we write the total action as

$$S(\alpha) = S_0 + \alpha\Delta S, \qquad (4.88)$$

which interpolates between a solvable action S_0 and the full action $S(\alpha = 1) = S_0 + \Delta S$. If the partition function for the system governed by the action S_0 is known, the full partition function can be determined from

$$\log Z(\alpha = 1) = \log Z(\alpha = 0) - \int_0^1 d\alpha' \langle 0|\Delta S|0\rangle_{\alpha'}, \qquad (4.89)$$

where the expectation value $\langle 0| \cdot |0\rangle_\alpha$ depends on the coupling constant α. The obvious choice for decomposing the action of the instanton liquid would be to use the single-instanton action, $S_0 = \sum_i \log n(\rho_i)$, but this does not work since the ρ integration in the free partition function is not convergent. We therefore consider the decomposition

$$S(\alpha) = \sum_{i=1}^{N_+ + N_-} \left(-\log n(\rho_i) + (1 - \alpha)\nu \frac{\rho_i^2}{\bar\rho^2} \right) + \alpha \left(S_{\text{int}} + \text{tr}\log(\not{D} + m_f) \right), \qquad (4.90)$$

where $\nu = (b - 4)/2$ and $\bar\rho^2$ is the average size squared of the instantons with the full interaction included. The ρ_i^2 term serves to regularize the size integration for $\alpha = 0$. It does not affect the final result for $\alpha = 1$. The specific form of this term is irrelevant, our choice here is motivated by the fact that $S(\alpha = 0)$ gives a single instanton distribution with the correct average size $\bar\rho^2$. The $\alpha = 0$ partition function corresponds to the variational single instanton distribution

$$Z_0 = \frac{1}{N_+! N_-!} (V\mu_0)^{N_+ + N_-}, \qquad \mu_0 = \int_0^\infty d\rho\, n(\rho) \exp\left(-\nu \frac{\rho^2}{\bar\rho^2} \right), \qquad (4.91)$$

where μ_0 is the normalization of the one-body distribution. The full partition function obtained from integrating over the coupling α is

$$\log Z = \log(Z_0) + N \int_0^1 d\alpha' \langle 0|\nu \frac{\rho^2}{\bar\rho^2}$$
$$- \frac{1}{N} \left(S_{\text{int}} + \text{tr}\log(\not{D} + m_f) \right) |0\rangle_{\alpha'}, \qquad (4.92)$$

where $N = N_+ + N_-$. The free energy density is finally given by $F = -1/V \cdot \log Z$ where V is the four-volume of the system. The pressure and the energy density are related to F by $p = -F$ and $\epsilon = T(dp/dT) - p$.

A general problem in the interacting instanton model is the treatment of very close instanton–anti-instanton pairs. In practice we have decided to deal with this difficulty by introducing a phenomenological short range repulsive core

$$S_{\text{core}} = \frac{8\pi^2}{g^2} \frac{A}{\lambda^4} |u|^2, \qquad \lambda = \frac{R^2 + \rho_I^2 + \rho_A^2}{2\rho_I \rho_A} + \left(\frac{(R^2 + \rho_I^2 + \rho_A^2)^2}{4\rho_I^2 \rho_A^2} - 1 \right)^{1/2}, \qquad (4.93)$$

into the streamline interaction. Here, λ is the conformal parameter (4.81) and A controls the strength of the core. This parameter essentially governs the dimensionless diluteness $f = \rho^4 (N/V)$ of the ensemble. The second parameter of the instanton liquid is the scale $\Lambda_{\rm QCD}$ in the instanton size distribution, which fixes the absolute units.

We have defined the scale parameter by fixing the instanton density to be $N/V = 1\,{\rm fm}^{-4}$. This means that in our units, the average distance between instantons is $1\,{\rm fm}$ by definition. Alternatively, one can proceed as in lattice gauge simulations, and use an observable such as the ρ meson mass to set the scale. Using N/V is very convenient and, as we will see in the next section, using the ρ or nucleon mass would not make much of a difference. We use the same scale setting procedure for all QCD-like theories, independently of N_c and N_f. This provides a simple prescription to compare dimensional quantities in theories with different matter content.

The remaining free parameter is the (dimensionless) strength of the core A, which determines the (dimensionless) diluteness of the ensemble. In Ref. [559], we chose $A = 128$ which gives $f = \bar{\rho}^4(N/V) = 0.12$ and $\bar{\rho} = 0.43$ fm. As a result, the ensemble is not quite as dilute as phenomenology seems to demand $((N/V) = 1\,{\rm fm}^{-4}$ and $\bar{\rho} = 0.33$ fm), but comparable to the lattice result $(N/V) = (1.4\text{–}1.6)\,{\rm fm}^{-4}$ and $\bar{\rho} = 0.35$ fm [399]. The average instanton action is $S \simeq 6.4$ while the average interaction is $S_{\rm int}/N \simeq 1.0$, showing that the system is still semi-classical and that interactions among instantons are important, but not dominant. Detailed simulations of the instanton ensemble in QCD are discussed in Ref. [559]. As an example, we show the free energy versus the instanton density in pure gauge theory (without fermions) in Fig. 4.4. At low densities the free energy is roughly proportional to the instanton density, but at larger densities repulsive interactions become important, leading to a well-defined minimum.

We also show the average action per instanton as a function of density. The average action controls the probability $\exp(-S)$ to find an instanton, but has no minimum in the range of densities studied. This shows that the minimum in the free energy is a compromise between maximum entropy and minimum action. Fixing the units such that $N/V = 1\,{\rm fm}^{-4}$, we have $\Lambda = 270\,{\rm MeV}$ and the vacuum energy density generated by instantons is $\epsilon = -526\,{\rm MeV}/{\rm fm}^3$. We have already stressed that the vacuum energy is related to the gluon condensate by the trace anomaly. Estimating the gluon condensate from the instanton density, we have $\epsilon = -b/4(N/V) = -565\,{\rm MeV}/{\rm fm}^3$, which is in good agreement with the direct determination of the energy density.

Not only the depth of the free energy, but also its curvature (the instanton compressibility) is fixed from the low energy theorem (4.37). The compressibility determined from Fig. 4.4 is $3.2(N/V)^{-1}$, to be compared with $2.75(N/V)^{-1}$ from the low energy theorem. At the minimum of the free energy we can also determine the quark condensate see [Fig. 4.4(c)]. In quenched QCD, we have $\langle \bar{q}q \rangle = -(251\,{\rm MeV})^3$, while a similar simulation in full QCD gives $\langle \bar{q}q \rangle = -(216\,{\rm MeV})^3$, in good agreement with the phenomenological value.

In Fig. 4.5 we show the spectrum of the Dirac operator in the instanton liquid for $N_f = 0, 1, 2, 3$ light flavors [236]. Clearly, the results are qualitatively consistent with

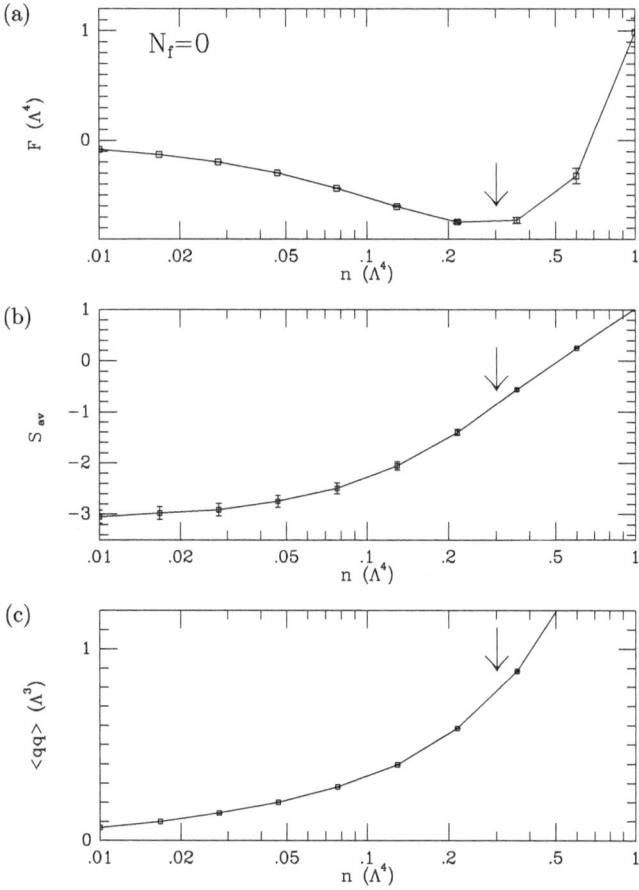

Fig. 4.4. Free energy, average instanton size and quark condensate as a function of the instanton density in the pure gauge theory, from Ref. [559]. All dimensional quantities are given in units of the scale parameter $\Lambda_{\rm QCD}$.

the Smilga–Stern theorem[14] for $N_f \geq 2$. In addition to that, the trend continues for $N_f < 2$, where the result is not applicable. We also note that for $N_f = 3$ (massless!) flavors, a gap starts to open up in the spectrum. In order to check whether this gap indicates chiral symmetry restoration in the infinite volume limit, one has to investigate finite size scaling. The problem was studied in more detail in Ref. [242], where it was concluded that chiral symmetry is restored in the instanton liquid between $N_f = 4$ and $N_f = 5$. Another interesting problem is the dependence on the dynamical quark mass in the chirally restored phase $N_f > N_f^{\rm crit}$. If the quark mass is increased, the influence of the fermion determinant is reduced, and eventually "spontaneous" symmetry breaking is recovered. As a consequence, QCD has an

---

[14]Numerically, the slope in the $N_f = 3$ spectrum appears to be too large, but it is not clear how small λ has to be for the theorem to be applicable.

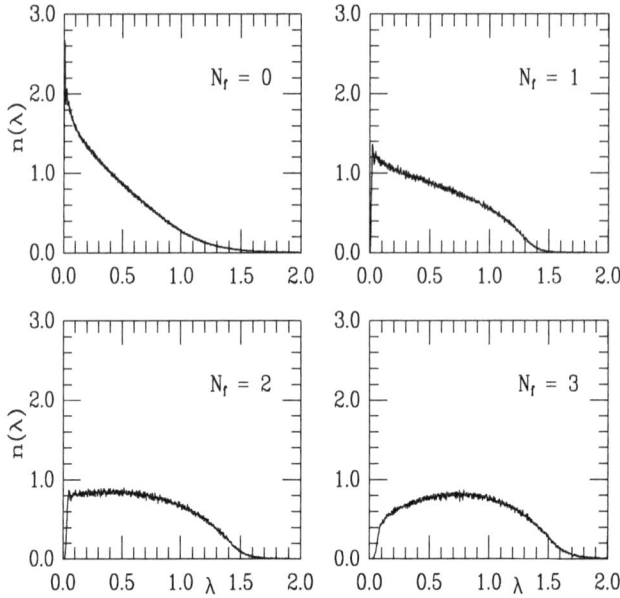

Fig. 4.5. Spectrum of the Dirac operator for different values of the number of flavors N_f, from Ref. [236]. The eigenvalue is given in units of the scale parameter $\Lambda_{\rm QCD}$ and the distribution function is normalized to one.

interesting phase structure as a function of the number of flavors and their masses, even at zero temperature.

4.4.1. *Screening of the topological charge*

Out of all aspects of this nontrivial statistical system let us discuss one phenomenon associated with dynamical quarks: the *screening* of the topological charge. It sheds some light on properties of the η' meson, strong CP violation, and the structure of QCD at finite θ angle.

The topological susceptibility is related to a correlation function

$$\chi_{\rm top} = \int d^4x \, \langle Q(x)Q(0)\rangle$$

$$= -\frac{m\langle \bar{q}q\rangle}{2N_f} + \frac{m^2}{4N_f^2} \int d^4x \, \langle \bar{q}\gamma_5 q(x)\bar{q}\gamma_5 q(0)\rangle. \tag{4.94}$$

Since the correlation function on the r.h.s. does not have any massless poles in the chiral limit, the topological susceptibility $\chi_{\rm top} \sim m$ as $m \to 0$. More generally, $\chi_{\rm top}$ vanishes if there is at least one massless quark flavor.

Alternatively, we can use the fact that the topological susceptibility is the second derivative of the vacuum energy with respect to the θ angle. Writing the QCD partition function as a sum over all topological sectors and extracting the zero modes from the fermion determinant, we have

$$Z = \sum_\nu e^{i\theta\nu} \int_\nu dA_\mu e^{-S} \det{}_f(\not{D} + m_f)$$

$$= \sum_\nu \left(e^{i\theta} \det \mathcal{M}\right)^\nu \int_\nu dA_\mu e^{-S} \prod_{n,f}{}' \left(\lambda_n^2 + m_f^2\right). \qquad (4.95)$$

Here, ν is the winding number of the configuration, $\mathcal{M}$ is the mass matrix and Π' denotes the product of all eigenvalues with the zero modes excluded. The result shows that the partition function depends on θ only through the combination $e^{i\theta} \det \mathcal{M}$, so the vacuum energy is independent of θ if one of the quark masses vanishes.

The fact that χ_{top} vanishes implies that all fluctuations in the topological charge are suppressed, so instantons and anti-instantons have to be correlated. Every instanton is surrounded by a cloud of anti-instantons which completely screen its topological charge, analogous to Debye screening in ordinary plasmas. The screening length is nothing else but the mass of the η'.

Detailed numerical studies of topological charge screening in the interacting instanton model were performed by Verbaarschot and myself [243]. Indeed, the complete screening takes place if one of the quark masses goes to zero and the screening length is consistent with the η' mass. We suggested a new way to extract the η' mass from topological charge fluctuations. The idea is not to study the limiting value of $\langle Q^2 \rangle / V$ for large volumes, but instead to determine its dependence on V for small volumes $V < 1\,\text{fm}^4$. In this case, one has to worry about possible surface effects. It is therefore best to consider the topological charge in a segment $H(l_4) = l_4 \times L^3$ of the torus L^4 (a hypercube with periodic boundary conditions). This construction ensures that the surface area of $H(l_4)$ is independent of its volume. Using the effective meson action introduced above, we expect (in the chiral limit)

$$K_P(l_4) \equiv \langle Q(l_4)^2 \rangle = \frac{1}{m_{\eta'}}(1 - e^{-m_{\eta'} l_4}). \qquad (4.96)$$

Numerical results for $K_P(l_4)$ are shown in Fig. 4.6. The full line shows the result for a random system of instantons with a finite topological susceptibility $\chi_{\text{top}} \simeq (N/V)$ and the dashed curve is a fit using the parametrization (4.96). Again, we clearly observe topological charge screening. Furthermore, the η' mass extracted from the fit is $m_{\eta'} = 756$ MeV (for $N_f = 2$), quite consistent with what one would expect. The figure also shows the behavior of the scalar correlation function, related to the compressibility of the instanton liquid. The instanton number $N = N_+ + N_-$ is of course not screened, but the fluctuations in N are reduced by a factor $4/b$ due to the interactions, see Eq. (4.37). For a more detailed analysis of the correlation function, see Ref. [243].

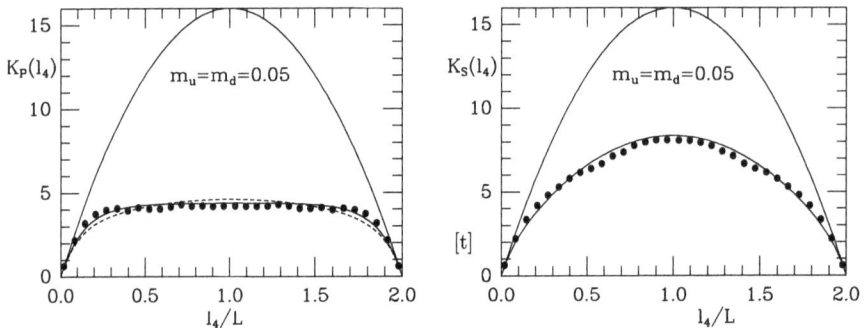

Fig. 4.6. Pseudoscalar correlator $K_P(l_4)$ (left panel) and scalar gluonic correlator $K_S(l_4)$ (right panel) as a function of the length l_4 of the sub-volume $l_4 \times L^3$, from Ref. [243] for QCD with $m_u = m_d = 10\,\mathrm{MeV}$ and $m_s = 150\,\mathrm{MeV}$. The upper solid lines correspond to a random system of instantons, while the other lines show some parameterizations. Note the qualitative difference between the data for topological and number fluctuations. The topological charge correlator is flat, corresponding to charge screening, while the number fluctuations are only reduced in size as compared to the random ensemble.

4.5.   Instantons for larger number of colors

A general discussion of the large N_c limit has already been given in Section 1.7.1, and here we will discuss what all of that means for instantons. In the one-loop order the instanton action is given by $S_0 = 8\pi^2/g^2 = -b\log(\rho\Lambda)$ where $b = 11N_c/3$ is the first coefficient of the beta function in pure gauge QCD. In the 't Hooft limit $N_c \to \infty$ with $g^2 N_c = $ const we expect $S_0 = O(N_c)$ and $\rho = O(1)$. This leads to Witten's suggestion that instantons do not survive the large-N_c limit. However the argument has loopholes: one should keep in mind that (i) instantons in QCD come in all sizes, and (ii) their interaction as well as entropy of configurations is $O(N_c)$ as well.

We will discuss the effects of both subsequently, closely following a detailed study by Schafer [246], who managed to show that few known paradoxes of the dilute gas approximation disappear in the interacting instanton liquid. In fact, a self-consistent picture of the ensemble emerges, which agrees well with the pre-existing theoretical expectations, including the Witten's conjectures about the size of the topological susceptibility and the η' mass. Another remarkable feature of this regime is that what we know about the instanton ensemble in QCD with $N_c = 2, 3$ is not really drastically changed, even in the large N_c limit.

In brief, main features of this regime are as follows. The density of instantons is predicted to grow as N_c, whereas the typical instanton size remains finite. The effective diluteness (accounting for the fact that instantons not overlapping in color do not interact) remains constant. Interactions between instantons are important and suppress fluctuations of the topological charge. As a result the $U(1)_A$ anomaly is effectively restored even though the number of instantons increases. Using mean field approximation and then numerical IILM simulations one finds that this scenario does not require fine tuning but arises naturally if the instanton ensemble is stabilized by a classical repulsive core. Although the total instanton density is large, the instanton liquid remains effectively dilute because instantons are not strongly overlapping in color space.

4.5.1. *Naive counting and expectations*

Since the *total instanton density* is related to the non-perturbative gluon condensate

$$\frac{N}{V} = \frac{1}{32\pi^2} \langle g^2 G^a_{\mu\nu} G^a_{\mu\nu} \rangle, \tag{4.97}$$

the N_c counting suggests that $\langle g^2 G^2 \rangle = O(N_c)$ and we are led to the conclusion that $(N/V) = O(N_c)$. This is also consistent with the expected scaling of the vacuum energy. Using Eq. (4.97) and the trace anomaly relation

$$\langle T_{\mu\mu} \rangle = -\frac{b}{32\pi^2} \langle g^2 G^a_{\mu\nu} G^a_{\mu\nu} \rangle, \tag{4.98}$$

the vacuum energy density is given by

$$\epsilon = -\frac{b}{4} \left(\frac{N}{V} \right). \tag{4.99}$$

Using $N/V = O(N_c)$ we find that the vacuum energy scales as $\epsilon = O(N_c^2)$ which agrees with our expectations for a system with N_c^2 gluonic degrees of freedom.

Note that $N/V = O(N_c)$ implies that the effective diluteness of instantons remains constant in the large N_c limit. Indeed, in spite of the large density most instantons do not see each other: the number of mutually commuting SU(2) subgroups of SU(N_c) scales as N_c.

If instantons are distributed randomly then fluctuations in the number of instantons and anti-instantons are expected to be Poissonian. This leads to the predictions

$$\langle N^2 \rangle - \langle N \rangle^2 = \langle N \rangle, \tag{4.100}$$
$$\langle Q^2 \rangle = \langle N \rangle, \tag{4.101}$$

where $N = N_I + N_A$ is the total number of instantons and $Q = N_I - N_A$ is the topological charge. Equation (4.101) implies that

$$\chi_{\text{top}} = \frac{\langle Q^2 \rangle}{V} = \frac{N}{V}. \tag{4.102}$$

Using $N/V = O(N_c)$ we observe that $\chi_{\text{top}} = O(N_c)$ which is *in contradiction to Witten's conjecture* $\chi_{\text{top}} = O(1)$. However, as we shall see, the interactions between instantons suppress the fluctuations and invalidate Eqs. (4.100), (4.101).

4.5.2. *Mean field arguments and the chiral condensate*

We now include the fermion-related dynamics, and ask how the chiral condensate scales with N_c, using first analytic MFA.[15] After averaging over the color orientation of the instanton the effective Lagrangian becomes

$$\mathcal{L} = \int n(\rho) d\rho \frac{2(2\pi\rho)^4 \rho^2}{4(N_c^2 - 1)} \epsilon_{f_1 f_2} \epsilon_{g_1 g_2} \left(\frac{2N_c - 1}{2N_c} (\bar{\psi}_{L,f_1} \psi_{R,g_1})(\bar{\psi}_{L,f_2} \psi_{R,g_2}) \right.$$
$$\left. - \frac{1}{8N_c} (\bar{\psi}_{L,f_1} \sigma_{\mu\nu} \psi_{R,g_1})(\bar{\psi}_{L,f_2} \sigma_{\mu\nu} \psi_{R,g_2}) + (L \leftrightarrow R) \right). \tag{4.103}$$

We observe from it that the explicit N_c dependence is given by $1/N_c^2$. This is again related to the fact that instantons are $SU(2)$ objects. Quarks can only interact via instanton zero modes if they overlap with the color wave function of the instanton. As a result, the probability that two quarks with arbitrary color propagating in the background field of an instanton interact is $O(1/N_c^2)$.

The MFA gap equation for the spontaneously generated constituent quark mass is

$$M = GN_c \int \frac{d^4 k}{(2\pi)^4} \frac{M}{M^2 + k^2}, \tag{4.104}$$

where M is the constituent mass and G is the effective coupling constant in Eq. (4.103). The factor N_c comes from doing the trace over the quark propagator. The coupling constant G scales as $1/N_c$ because the density of instantons is $O(N_c)$ and the effective Lagrangian contains an explicit factor $1/N_c^2$. We conclude that the coefficient in the gap equation is $O(1)$ and that the dynamically generated quark mass is $O(1)$ also. This also implies that the quark condensate, which involves an extra sum over color, is $O(N_c)$.

The results in the mean field approximation are summarized in Fig. 4.7(a) which shows that for $N_c > 4$ the average instanton size is essentially constant while the instanton density grows linearly with N_c. This can also be verified by expanding $\log(N/V)$ in powers of N_c and $\log(N_c)$: one observes that independently of the details of the interaction, the instanton density scales at most as a power, not an exponential, in N_c.

Another way to see why the instanton density scales as the number of colors is as follows: the size distribution is regularized by the interaction between instantons. This means that there has to be a balance between the average single instanton action and the average interaction between instantons. If the average instanton action satisfies $S_0 = O(N_c)$ we expect that $\langle S_{\text{int}}^{\text{tot}} \rangle = O(N_c)$ also. Using $\langle S_{\text{int}}^{\text{tot}} \rangle = (N/V)\langle S_{\text{int}} \rangle$ and the fact that the average interaction between any two instantons satisfies $\langle S_{\text{int}} \rangle = O(1)$, we expect the density to grow as N_c.

---

[15] For definiteness, we will consider the case $N_f = 2$ but the conclusions are of course independent of the number of flavors.

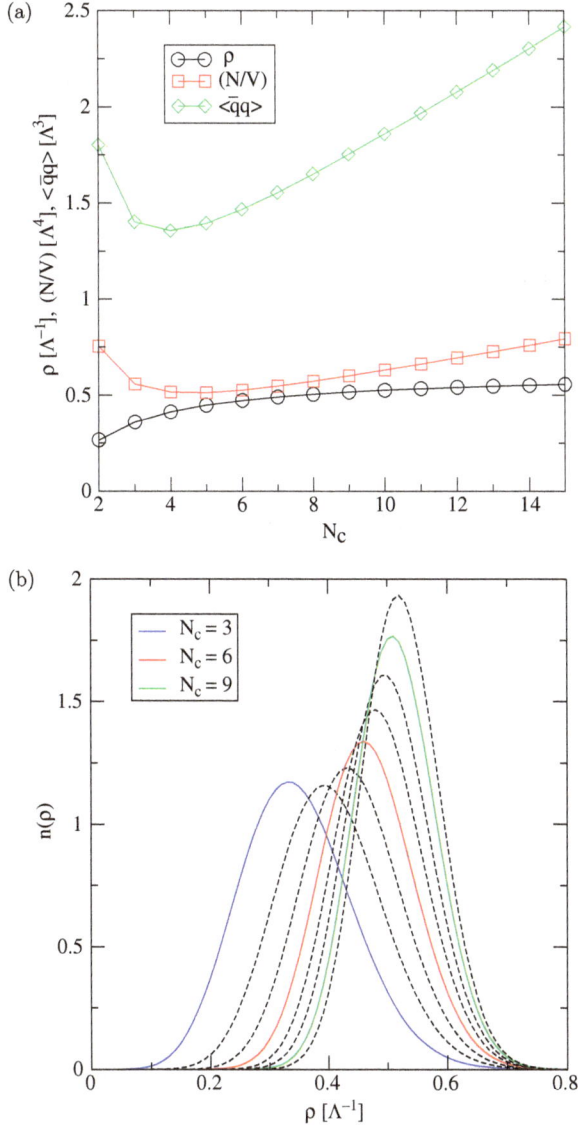

Fig. 4.7. Average instanton size ρ, instanton density N/V and quark condensate $\langle \bar{q}q \rangle$ for different numbers of colors N_c. The results shown in this figure were obtained using the mean field approximation. All quantities are given in units of the QCD scale parameter Λ. Instanton size distribution $n(\rho)$ for different numbers of colors $N_c = 3, \ldots, 10$.

Figure 4.7(b) shows the instanton size distribution for different numbers of colors. We observe that the number of small instantons is strongly suppressed as $N_c \to \infty$ but the average size stabilizes at a finite value $\bar{\rho} < \Lambda^{-1}$. We also note that there is the critical size ρ^* for which the number of instantons does not change as $N_c \to \infty$. The value of ρ^* is easy to determine analytically. We write

$n(\rho) = \exp(N_c F(\rho))$ with $F(\rho) = a \log(\rho) + b\rho^2 + c$ where the coefficients a, b, c are independent of N_c in the large N_c limit. The critical value of ρ is given by the zero of $F(\rho)$. We find $\rho^* = 0.49\Lambda^{-1}$. The existence of a critical instanton size for which $n(\rho^*)$ is independent of N_c was discussed in Refs. [83–85]. It was found on the lattice by Lucini and Teper [72], see Section 5.4.4.

4.5.3.   *Fluctuations in the interacting instanton liquid*

Fluctuations in the net instanton number are related to the second derivative of the free energy with respect to N (4.39). This result is in agreement with a low energy theorem (4.38) based on broken scale invariance [396], based solely on the renormalization group equations. The left hand side is given by an integral over the field strength correlator, suitably regularized and with the constant term $\langle G^2 \rangle^2$ subtracted. For a dilute system of instantons Eq. (4.38) reduces to Eq. (4.39). The result (4.39) shows that fluctuations of the instanton ensemble are suppressed by $1/N_c$. This is in agreement with general arguments showing that fluctuations are suppressed in the large N_c limit. We also note that the result (4.39) clearly shows that even if instantons are semi-classical, interactions between instantons are crucial in the large N_c limit.

Fluctuation in the topological charge can be studied by adding a θ-term to the partition function. The mean square is just the average instanton number

$$\langle Q^2 \rangle = \langle N \rangle, \tag{4.105}$$

which is identical to the result in the random instanton liquid and disagrees with Witten's hypothesis $\chi_{\rm top} = O(1)$. However, Diakonov *et al.* noticed that Eq. (4.105) is a consequence of the fact that in the sum ansatz the average interaction between instantons of the same charge is identical to the average interaction between instantons of opposite charge. In general there is no reason for this to be the case and more sophisticated instanton interactions do not have this feature [213, 234]. If r denotes the ratio of the average interaction between instantons of opposite charge and instantons of the same charge, $r = \langle S_{IA} \rangle / \langle S_{II} \rangle$, then

$$\langle Q^2 \rangle = \frac{4}{b - r(b - 4)} \langle N \rangle. \tag{4.106}$$

This result shows that for any value of $r \neq 1$ fluctuations in the topological charge are suppressed as $N_c \to \infty$. We also note that $\chi_{\rm top} = O(1)$, in agreement with Witten's hypothesis.

The free energy of the instanton liquid as a function of the instanton density for $N_c = 3, \ldots, 6$ is shown in [Fig. 4.8(a)]. The equilibrium density (at the minimum of the function $F(N/V)$) is shown in [Fig. 4.8(b)]. We observe that N/V increases linearly with N_c whereas the free energy is quadratic. The slope of N/V as a function of N_c is small, in agreement with the mean field result shown in Fig. 4.7(a).

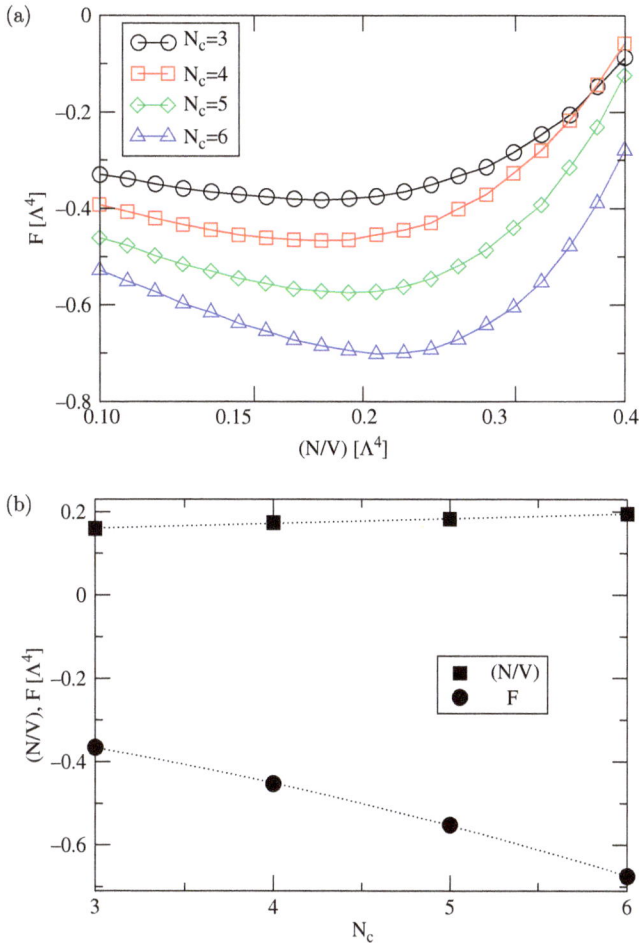

Fig. 4.8. (a) Free energy F of the quenched instanton liquid as a function of the instanton density N/V for $N_c = 3, \ldots, 6$ colors. Both N/V and F are given in units of the QCD scale parameter. The results shown in this figure were obtained using numerical simulations with $N = 32$ instantons. (b) Instanton density N/V and free energy F in a pure gauge instanton ensemble for $N_c = 3, \ldots, 6$ colors. Both N/V and F are given in units of Λ^4 where Λ is the QCD scale parameter. The dashed lines show fits of the form $a_1 N_c^2 + a_2 N_c + a_3$ (for the free energy F) and $a_2 N_c + a_3$ (for the instanton density N/V).

In contrast to the mean field result the linear behavior already sets in at small $N_c \simeq 3$. These results are stable with respect to changing the parameters.[16]

In Ref. [246] the instanton size distribution, the topological susceptibility and the spectrum of the Dirac operator for different numbers of colors have been determined in IILM numerically. The instanton size distribution is shown in Fig. 4.9. As

---

[16] All results are however dependent on the assumption that the parameter A in Eq. (4.27) is *not* a function of N_c. As that parameter exists in SU(2), we see no reason why it should depend on it, but since its nature is not yet clear this warning is perhaps justified.

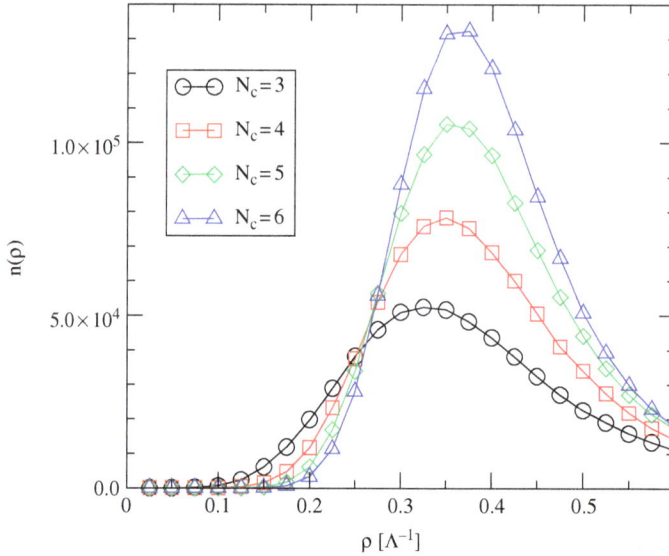

Fig. 4.9. Instanton size distribution in a pure gauge instanton ensemble for different numbers of colors. The results were obtained using numerical simulations with $N = 128$ instantons.

expected, small instantons are strongly suppressed as the number of colors increases. We observe a clear fixed point in the size distribution at $\rho^* \Lambda \simeq 0.27$. The simulations were carried out in the total topological charge $Q_{\text{top}} = 0$ sector of the theory. One can nevertheless determine the topological susceptibility by measuring the average Q_{top}^2 in a sub-volume $V_3 \times l_4$ of the Euclidean box $V_3 \times L_4$ [238]. The topological susceptibilities for $N_c = 3$ agree well with the expectation based on Poissonian statistics, $\chi_{\text{top}} \simeq N/V$. For $N_c > 3$, however, fluctuations are significantly suppressed and the topological susceptibility increases more slowly than the density of instantons, consistent with a Witten scenario in which χ_{top} remains finite as $N_c \to \infty$.

The chiral condensate for $m_q = 0.1\Lambda$ and topological susceptibility are shown in Fig. 4.10. We clearly see that $\langle \bar{q}q \rangle$ is linear in N_c while χ_{top} approach a constant.

4.5.4. *Do instantons cluster at large N_c?*

At the end of Section 12.5 we will discuss a remarkable finding by Dorey, Khoze and Mattis [888] that in the $\mathcal{N} = 4$ SUSY theory a complete clustering of many instantons occurs. In clusters many instantons share a common location and size. The reason for that is the exchange of adjoint fermions which have zero modes of all colors.

Such adjoint fermions are not present in QCD, and we have just argued in the previous subsection that instantons of non-overlapping colors do not interact. Despite that, Carter and myself [74] argued that a similar clustering is possible in QCD-like theories at large N_c as well. A generic argument is that even a rather weak attractive force can lead to binding provided sufficiently many bosonic objects

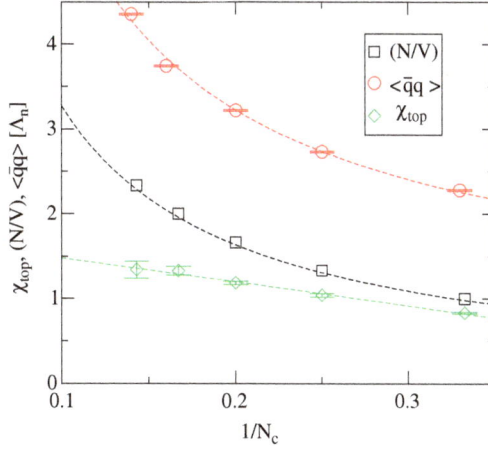

Fig. 4.10. Dependence of the chiral condensate $\langle \bar{\psi}\psi \rangle$ and the topological susceptibility χ_{top} on the number of colors. The instanton density N/V was assumed to scale as $N/V \sim N_c$. The dashed lines show fits of the form $a_1 N_c + a_2$ (for $\langle \bar{\psi}\psi \rangle$ and N/V) and $a_2 + a_3/N_c$ (for χ_{top}).

pile up and act coherently. For example, relativistic W, Z, H bosons interact weakly with each other, $\alpha_w \sim 1/100$, but if there are about a 100 of them, a bound state — the sphaleron — can exist.

At the quantum level there appear generic Van der Waals type forces even between instantons of non-overlapping colors. Although the corresponding coupling must inevitably be weak at large N_c, this weakness can be overcome by the large $O(N_c)$ number of members acting coherently, so that clustering ensues.

The corresponding effective interaction will be described in a model-dependent way, by exchange of a colorless scalar field interacting with the scalar combination of the gauge field $(G^a_{\mu\nu})^2$. This scalar field can be interpreted as that of the 0^{++} glueball. Apart of the simplicity of this toy model, it is supported by the following observations: (i) the scalar glueball is by far the lightest. For this reason it may be reasonable to consider it as an elementary dynamical field, analogous to pion-mediated dynamics in nuclei; (ii) the scalar glueball is also believed to be rather small, with a size of about 0.2 fm which is comparable to the dominant instanton size; (iii) the scalar glueball has been shown to be strongly coupled to instantons, unlike e.g. the second $J^{PC} = 2^{++}$ tensor glueball state. Although we expect scalar to be dominant, nevertheless there is no theoretical parameter involved, and it remains a model-dependent assumption.

The Euclidean action of the model has the form

$$S = \int d^4x \left(\frac{1}{4g^2} G^a_{\mu\nu}(x) G^a_{\mu\nu}(x) + \frac{1}{2} \partial_\mu \phi(x) \partial_\mu \phi(x) \right.$$

$$\left. + \lambda \phi(x) G^a_{\mu\nu}(x) G^a_{\mu\nu}(x) \right). \tag{4.107}$$

Here the gauge field strength tensor is $G_{\mu\nu}^a = \partial_\mu A_\nu^a - \partial_\nu A_\mu^a + f^{abc} A_\mu^a A_\nu^a$, the parameter λ is a coupling constant with the dimension of length, and as usual the factor g has been absorbed in the gauge field. While one could naturally include a potential for the field ϕ we see no motivation to do so in this case. Minimizing Eq. (4.107) generates the equations of motion:

$$\left(\frac{1}{4g^2} + \lambda\phi(x)\right) D_\mu G_{\mu\nu}^a(x) = 0, \tag{4.108}$$

$$\partial^2 \phi(x) - \lambda\sqrt{Z_\phi} G_{\mu\nu}^a(x) G_{\mu\nu}^a(x) = 0. \tag{4.109}$$

The self-dual instanton solution, unchanged by the presence of ϕ, is

$$A_\mu^a(x) = \frac{2\eta_{\mu\nu}^a x_\nu}{(x-z)^2 + \rho^2}, \tag{4.110}$$

where $\eta_{\mu\nu}^a$ is the usual 't Hooft symbol, with ρ and z the instanton size and position.

The parameter λ can be determined by the scalar glueball mass, $m_\phi \simeq 1.5\,\text{GeV}$. After squaring the interaction term and Fourier-transforming the nonlocal result, the leading contribution to the glueball self-energy is

$$\Pi(q^2) = (32\pi^2)^2\lambda^2 \int d\rho\, \nu(\rho)\, (q\rho)^4 K_2(q\rho)^2, \tag{4.111}$$

where $K_2(x)$ is a modified Bessel function. From the static limit the mass is obtained,

$$m_\phi^2 = \Pi(0) = \frac{4}{b}(32\pi^2)^2\lambda^2 n. \tag{4.112}$$

Note that since $n \sim N_c$ and $b \sim N_c$, in the large N_c limit we have finite mass and coupling. With this result, one finds the coupling in order to fit a glueball mass of 1.5 GeV. The glueball mass is assumed to be stable with respect to the number of colors [396], whereas the instanton density is assumed to scale as $n \sim N_c$, as above.

The attraction between any pair of pseudo-particles is quantified by the change in the action, parametrized as usual by the instanton sizes and the separation between their centers, R. For two isolated instantons, this is

$$\Delta S_{\text{pair}}(R, \rho_1, \rho_2) = -\frac{\lambda^2}{2} \int d^4x\, d^4y\, [G_{\mu\nu}(x - R, \rho_1)]^2$$
$$\times\ D(x - y)[G_{\mu\nu}(y, \rho_1)]^2, \tag{4.113}$$

where the scalar propagator in Euclidean space is

$$D(x) = \frac{m}{4\pi^2|x|} K_1(m|x|), \tag{4.114}$$

and the sources are the field strength tensor evaluated over an instanton action distribution. This will describe the extent of interactions between two instantons in completely distinct SU(2) color subgroups, as well as the interaction between an instanton and anti-instanton.

Even if the attractive force between any pair of pseudo-particles is small, it is relevant, and becomes in fact dominant, when one considers a cluster of instantons with a number of constituents on the order of the number of colors. Specifically, the total change in the action from such a cluster is in this case parametrically large and will eventually overcome the entropy of random placements, so that the large-N_c ensemble will be one of strongly bound instanton clusters with $\sim N_c$ constituents.

Can such a clustering process continue indefinitely, until all of them collapse into the same mega-cluster? The answer is negative for two reasons. The first is that instantons and anti-instantons in the same SU(2) subgroup would obviously annihilate one another. The second, less trivial, effect which would limit clustering is the rather interesting non-linear deformation of the two-instanton solution. Although this is, strictly speaking, only known for instanton pairs in the same color subgroup, we think it is more general and leads to a kind of a shell model situation (without any fermions) in which instantons and anti-instantons tend to occupy only *non-overlapping* SU(2) subgroups. We now consider the latter effects in detail.

Any group of pseudo-particles will likely include as many anti-instantons as instantons, for the scalar exchange knows no difference.[17] The inevitable annihilation between instantons and anti-instantons within a given cluster will leave some fraction of the original number in a cluster of instantons or anti-instantons alone. Statistical fluctuations in $N_I - N_{\bar{I}}$ will however guarantee that these clusters persist, still with a number of constituents on the order of N_c.

Instantons in the same SU(2) subgroup may be analyzed using an exact multi-instanton solutions known as the ADHM construction [203]. We will show that it leads to effective repulsion, in two different ways.

First, when two such pseudo-particles converge, their distinct peaks in the action density are deformed continuously into a toroidal cloud in two of the four spatial directions, see Ref. [205]. There is a minimal separation, $R_{\min} = \sqrt{\rho_1 \rho_2 / 2}$, after which a certain coordinate transformation reveals re-separation into a direction orthogonal to the original R_μ. Although the classical action is the same for all these configurations, effectively a hard-core repulsion between instantons of the same color projection appears. It is generated by the non-trivial measure of the 2-instanton collective coordinate space, and thus enter at the quantum level only (in the pre-exponent). Although geometrically the 2-instanton manifold is smooth, in order not to describe the same configuration twice, a coordinate discontinuity

---

[17]Consequently, in this ensemble the instantons and anti-instantons of different colors are statistically independent and therefore its topological susceptibility is *not suppressed* relative to an unclustered ensemble, $\chi \sim N_c \Lambda^4$. A more refined model with more complicated interactions is therefore needed to provide for Witten's suppression, $\chi \sim \Lambda^4$, such as one that restricts the topological charge of each cluster to $Q = -1, 0, 1$.

appears at the toroidal configuration in question. (To the best of our knowledge, this observation has not been made in the literature before.)

The second reason for effective repulsion appears due to clustering. Note that the action density becomes less concentrated as the instantons are deformed into a toroidal-like configuration. Consider now the interaction between a toroidal 2-instantons in one subgroup with a cluster of instantons of other colors, which have an unmodified shape. The attraction between all of them is reduced. Indeed, instead of Eq. (4.113) we now would have

$$\Delta S_{\text{pair}}(R, \rho_1, \rho_2) = -\frac{\lambda^2}{2} \int d^4x \, d^4y \, [G_{\mu\nu}(x - R, \rho_1)]^2$$

$$\times D(x - y) \frac{1}{2} s(y, \rho_2, \rho_2), \tag{4.115}$$

where

$$\tfrac{1}{4} s(x, \rho_1, \rho_2) = -\tfrac{1}{2} \partial^2 \partial^2 \ln \det [\Delta(x)^\dagger \Delta(x)] \tag{4.116}$$

is the action density for the ADHM solution.

The matrix $\Delta(x)$ is a rather complicated function and we refer the interested reader to Ref. [205] for a thorough analysis. Since we are only interested in quantitative scalar exchanges between an instanton and such a solution, we consider two instantons in the same SU(2) subgroup and with a color group angle of $\pi/2$ between the two (the case of maximal deformation). In that case, we have

$$\det [\Delta(x)^\dagger \Delta(x)] = \left[(x - R)^2 + \rho_1^2 + \frac{(\rho_1 \rho_2)^2}{R^2} \right]$$

$$\times \left[x^2 + \rho_2^2 + \frac{(\rho_1 \rho_2)^2}{R^2} \right] - \frac{(2\rho_1 \rho_2)^2}{R^2} x_1^2. \tag{4.117}$$

Here we have chosen $R_\mu = (R, 0, 0, 0)$ and color group elements σ_0 and σ_1 for the two pseudo-particles, which leads to a torus in the plane of coordinates x_0 and x_1.

With this specification we have computed via Monte Carlo the "binding action" generated by scalar exchange between an instanton and half an ADHM pair, which is compared to that for two more prosaic instantons. From this we can see that the cluster formation will prefer filling each SU(2) subgroup once to allow for maximal attraction. This generates a medium of isolated clusters with the order of N_c individual instantons. We now depart from pure glue theory and consider the effects of light fermions on the clustering process. Specifically, we discuss QCD-like fermions with fundamental color charges rather than the "gluino"-like adjoint fermions present in supersymmetric theories.[18]

The clustering of instantons as described effectively reduces the instanton density, n, by a factor of N_c. This will have severe consequences for the spontaneous

---

[18]The difference is crucial, since the former will have $O(N_f)$ and the latter $O(N_c)$ fermionic zero modes.

breaking of chiral symmetry. In a background of unclustered instantons, the chiral condensate and quark effective mass scale as [229]:

$$\langle \bar{\psi}\psi \rangle \sim N_c, \quad M_q \sim N_c^0. \tag{4.118}$$

In a clustered environment we will instead have

$$\langle \bar{\psi}\psi \rangle \sim \sqrt{N_c}, \quad M_q \sim 1/\sqrt{N_c}. \tag{4.119}$$

Spontaneous chiral symmetry breaking is therefore more robust with the unclustered instantons, and a large number of quark flavors might keep the system in a phase of strongly broken chiral symmetry rather than the clustered medium. To estimate how many flavors would be necessary, we note that the fermion determinant built of zero-mode overlap matrices would raise the action, possibly overcoming the reduction from scalar exchange (4.113) which will appear in the exponent of the partition function. Quantitatively, for clustering to persist we must have

$$\det \left(\mathcal{T} \right)^{2N_f} e^{-N_c \Delta S_{\text{pair}}} \equiv e^{-\Delta S_{\text{eff}}} > 1, \tag{4.120}$$

where $\mathcal{T}$ is the matrix of overlaps. To extract the N_c dependence from the determinant, we note that the overlap matrices involve two zero-mode propagators and hence

$$\ln \det \left(\mathcal{T} \right) \sim \ln \left(\frac{R^3}{R_0^3} \right) \sim \frac{3}{4} \ln \left(\frac{n_0}{n} \right) \sim \frac{3}{4} \ln N_c. \tag{4.121}$$

The total effective change in the action is thus

$$\Delta S_{\text{eff}} = \frac{3}{2} N_f \ln N_c + N_c \Delta S_{\text{pair}}. \tag{4.122}$$

This becomes positive — and clustering becomes unstable — for

$$N_f > -\frac{2}{3} \Delta S_{\text{pair}} \frac{N_c}{\ln N_c} \gtrsim \frac{2N_c}{\ln N_c}, \tag{4.123}$$

where a typical value for ΔS_{pair} has been inserted. Thus we see that the clustering phenomena will be inhibited by a rather small N_f/N_c value.

CHAPTER 5

Lattice QCD

5.1. Generalities

5.1.1. *Brief history*

Any treatment of quantum field theories starts with their "regularization", a selection of some finite subset of degrees of freedom out of the infinite set. Unfortunately, the simplest UV regularization methods (such as the familiar Pauli–Villars method, a subtraction of diagrams with a large "regulator mass") have some unwanted features: they do not respect symmetries of the theory, in particular, the gauge invariance. So, the original motivation for putting the gauge fields on the space–time lattice was to obtain a gauge invariant regularization. How to do so we have already discussed in detail in the introductory Chapter 1.

However, as it often happens, any good theory has its own logic which drives us much further than it was originally expected. A transition from the momentum representation, natural for Feynman rules of the perturbation theory, back to coordinate space has made the theory much more transparent. Interesting connections with statistical spin systems have led to the idea to apply powerful methods previously developed in this field, both analytic (e.g. the high temperature expansion) and numerical.

A feasibility of "numerical experiments" on the lattice has caused a revolution in the theory of quantum fields. In the 1970s, when the lattice formulation was proposed, it was considered evident, that such problems are far too complicated for any imaginable computer. Even for quantum mechanical systems with few (non-separable) variables the practical limitations were very severe. However, if instead of looking for the wave function in multidimensional space one generated ensemble of configurations corresponding to the (Euclidean) Feynman integrals the problem would become treatable.[1]

---

[1]A "retrospective" description of the early days of lattice gauge theory, with a more detailed reference list, can be found in Ref. [250].

The pioneering numerical simulations, such as by M. Creutz [270] for the SU(2) gauge theory, has opened the way. Quite characteristically, he immediately attacked the most difficult problem, that of confinement. Using Wilson criteria he found the first hints suggesting that perhaps the transition from strong to weak coupling is quite rapid and therefore the latter can be reachable with much more modest resources than previously thought.

After the early 80s, when many pioneering works had been done, there was a long period of technical development. Experimentation with "improved" and then "perfect" actions, first for pure gauge and then with fermions took a lot of time and efforts. We will not be able to go into these issues in detail: in this chapter I will only try to explain the dramatic story of the "chiral fermions" formulation on the lattice. This issue is so crucial since the chiral symmetries are so important for QCD applications. A difficulty of its implementation has been noticed already in the pioneering work by Wilson [281], but it took the efforts of many people and more than 20 years to solve it.[2]

Among the main points in the chapter is the story of the so called *quenched* approximation, which neglects the fermionic determinant[3] while weighting the gauge field configurations. (In other words, one ignores all fermionic vacuum loop diagrams.) In the 80s, based on "large-N_c" ideology, people thought that it may give a relatively accurate approximation to hadronic masses. In the 90s, when the first signs of trouble had surfaced, they were related with a wrong treatment of $U(1)_A$ symmetry breaking producing fake massless η'. Gradually, it was demonstrated that in quenched QCD not all correlation functions satisfy even the spectral positivity. It means that it not only is a bad approximation to masses, strictly speaking the quenched QCD does not even resemble a normal quantum field theory and does not even have *any* mass spectrum. Let me add that all these troubling phenomena were first found in the much smaller and cheaper simulations of the instanton liquid: but very few of lattice practitioners cared to pay any attention, until the moment they had found the same thing themselves.

The remaining challenges to lattice gauge theory are still quite significant. To illustrate it by an example, let me refer to a discussion session at the last lattice conference (Lattice-02 at MIT) on the issue of the so called *chiral extrapolation* [308]. For reasons we will discuss below, the actual calculations still have to be done with quark masses significantly exceeding the physical ones. In the region where calculations are made (when a light quark has a mass like a physical strange quark) the mass dependence looks smooth and linear: but it took a lot of work

---

[2] For about 15 of those years the whole thing looked as total "no-go", and I recall asking people involved in that whether there was any progress, with the answer, "not yet, but we understand much better what the problem is". Then suddenly all was clarified and several different solutions were found nearly at the same time.

[3] Except for the charge renormalization, or units adjustments, which of course is not a good substitution for a determinant and can only roughly account for it on average.

to demonstrate clearly that the continuation of this linear dependence does *not* produce correct results, with sometimes deviations from experimental data by as much as 50 percent or so. It may look very strange, but although most of the quark modes on the lattice have high momenta, of the order of $1/a = 2$–$4\,\text{GeV}$, nevertheless somehow the results are still sensitive to quark mass as small as, say, changing between 20 and 40 MeV. Most people think that the remedy is a correct account for the contribution of the "pion cloud". For some quantities perhaps, but the deeper reason is again the existence of narrow "zero mode zone" generated by instantons, with its small eigenvalues. More studies are still needed to understand these effects and to work out a reliable chiral extrapolation.

Another current challenge to lattice gauge theory is to put one more fundamental symmetry on the lattice, namely the *supersymmetry*. If possible, it would be really nice to bring together the efforts of two different communities, and to compare or complement the numerical and the analytic studies.

And finally, this overview cannot be concluded without mentioning really heroic efforts of many lattice practitioners in pushing various computer projects to completion. In spite of the existence of a powerful computer industry, at the high end of the computer spectrum one still finds today several "home-made", machines designed and built by lattice practitioners themselves. Those provide more "flops-per-dollar" than any commercial machine can deliver. It is really fascinating that such amateuristic-looking projects succeed again and again.

5.1.2. *Lattice limitations*

Let us start with recalling the main setting of lattice calculations. In the introductory chapter we have already discussed the lattice formulation for the gauge fields, in terms of which the partition function is written. The first task then is to generate a *well equilibrated* ensemble of gauge fields, corresponding to the weights prescribed by the partition function. How to do it we will discuss in the next section.

The second task is to "perturb the vacuum" by several $(1, 2, 3, \dots)$ external probes and do the measurements of their correlation functions. As we had already discussed in Chapter 2, the VEVs of some local operators, also called the "vacuum condensates", tell us about the *average* properties of the vacuum state. Green functions with 2, 3 and more (multi-point) correlators give us further details, about physical excitation modes propagating in the QCD vacuum and their properties.

Formally speaking, the necessary conditions for both tasks are the same, namely that the distance scale we want to explore (let us call it $1/m$) is safely separated from both the UV and IR cutoffs provided by the lattice spacing a and the lattice size L, respectively,

$$a \ll 1/m \ll L. \tag{5.1}$$

The expression "much larger" is however too loose and in practical applications one would like to make it more precise, defining the kind of accuracy (100%, 10% or 1%)

one may expect in real numerical studies, in order to optimize given computational resources. The answer depends very much on a specific quantity one would like to measure.

Consider first the problem of limited space–time volume of the QCD vacuum, for which the calculations can be made. The correlation length scale $\xi = 1/m$ is typically the geometric mean ($\sqrt{aL}$) of macro and micro scales L, a. So, if one wants to increase the ratio of the measured mass to UV or IR cutoffs by a factor ξ, the number of lattice points would grow as $\xi^{1/2d} \sim \xi^{1/8}$. Even ignoring further slowing down related with fermions, one can see that even given an impressive exponential growth of the memory and the power of computers, the progress in parameters like ma is rather slow. In practice "much larger" still means 3–4 times larger at best, with little hope for significant improvements. So, in order to obtain accurate lattice results one has to understand the "finite-size effects" well and learn how to correct for them.

The size of the "box" in physical units depends on the value of the input coupling constant, which is strongly limited by the necessity to be in the "scaling window" (5.1). In practice $L \approx 1$ fm in the pioneer papers and now it is up to about 3 fm in the latest works.

If one puts a hadron into such a box, the periodic boundary conditions used are not that harmless. If one thinks of many copies of the box taken together, it becomes obvious that instead of the problem concerning one single hadron in a closed box one in fact is dealing with a dense "hadronic crystal", periodically repeated in $4d$ with period L.

As always in the problems with periodic potentials, the spectrum is given by Brillouin zones, with widths proportional to the tunneling probability between the neighboring cells. The question is whether such zones are wide or narrow. A very instructive analysis of this problem was made by Hasenfratz and Montvay [402], who just obtained numerical solution of the Schrödinger equation for some toy quark models of hadrons in a similar box. Their conclusion was rather disappointing: light quarks easily penetrate through the barriers separating a hadron from its "mirror image", generating an artificial interaction between them. This of course reduces the masses, etc.

Even if one successfully takes the box size L large enough, so that the quarks are prevented from tunneling to the next cells, there exists a similar exchange of the lightest hadrons, especially of the pions. We return to this problem in Section 5.1.3.

In order to get some qualitative idea about the magnitude of the corrections involved, consider the following analogy. In the center of heavy nuclei one finds the denser matter existing under normal conditions. In it a specific volume per nucleon is about $6 \, \text{fm}^3$. The 3-volume of a lattice, L^3 is either comparable or a few times larger than this volume. So, when we put (by an appropriate operator) a single nucleon into the lattice box, we should remember that the periodic boundary conditions imply existence of its mirror images. On the other hand, experiment tells us what the modifications are of the nucleon properties in nuclei in detail: e.g. for an effective nucleon mass and magnetic moment, the modifications are as large as 20%. A scale of modification by finite volume lattice effects can be crudely normalized to this number, as the effect is expected to scale with the density.

Summarizing the point I would like to make, let me say that putting a hadron into a box is not enough: one should also take care that enough room around is left for the unperturbed vacuum!

The same phenomenon can be cast into somewhat different terms with respect to the time direction. Usually this is the direction in which the decaying correlation signal is measured: and the periodic boundary conditions make it look like $\cosh[m(\tau_1 - \tau_2)]$ rather than the exponent. It means that not more than a half of the box size L can be used for hadron propagation: otherwise the stronger signal comes from another direction on the torus. In practice one tends to make the 4th dimension of the box about twice as large as the other three.

As we will discuss in detail in Chapter 8, one can interpret the finite size effects in the 4th direction differently. The periodicity in Euclidean time corresponds to a nonzero Matsubara temperature $T = 1/L_4$. For example, in the pioneer spectroscopic calculations (e.g. those made by Hamber and Parisi [305]) the vacuum was in fact heated up to a temperature of about $T = 100$ MeV, and even in the modern calculations T is still above 50 MeV. Ironically enough, hot hadronic matter (being the aim of experimentalist's efforts) is in fact the best approximation to the vacuum state which is technically feasible, in the most modern computers!

So far we have discussed only the IR "finite size" effects related to the box size L in physical units. There are also UV "finite size" effects related to the micro-scale a, the lattice spacing. Naive thinking is that it should be small in a "hadronic size scale" of about 1 fm. However, as we have repeatedly emphasized above, there is no single unique scale at which all non-perturbative phenomena take place.

In particular, in order to correctly account for the instantons or the flux tubes one needs $a \ll 0.2$–0.3 fm: otherwise a significant part of it would be missed.[4]

Finally, imagine that some day one is able to deal with lattices large enough to take care of all finite size effects, both UV and IR. Lattice spectroscopy would still be confronted with a different type of problem, a limitation on the ability to measure sufficiently weak signals because of statistical noise. Indeed, the measured signal behaves as $\exp(-m\tau)$ and at large time it is very weak. The statistical noise decreases as $1/\sqrt{N}$, where N is the number of measurements. It is difficult to measure a signal smaller than, say for the sake or argument, 0.01. It means, that even at infinite lattice at distance $\tau > 6/m$ one would not find any correlation function,

---

[4]Some part of it is always missed, and the question is how to correct for it. For example, small-size instantons which did not fit into the lattice can be accounted for by extra 't Hooft $2N_f$-fermion interaction. To my knowledge this "lattice improvement" has never be implemented: the reader may do it as an exercise.

just noise. If one measures the hadronic masses by the logarithmic derivative of the correlation function, in such a setting, it is not possible to do better than achieve an accuracy of about 10%! Quite similarly, one may argue that it is not possible to prove confinement on the lattice simply because one cannot test the area law, considering

$$W \sim \exp(-KL^2) = \exp\left[-\left(\frac{L}{0.5\,\text{fm}}\right)^2\right] \qquad (5.2)$$

is masked by the statistical noise already for Wilson loops of several square fm.

5.1.3.   *Mesoscopic regime and the random matrix theory*

We start this section with one outstanding question which has received a lot of attention. It is related with the lightest hadron of all — the pion — which is obviously most difficult to force into the "safe window" (5.1). Here is the question: *If the pion does not fit into the box, is it really a problem?* As you will see from discussion below the answer is a definite "no".

The basic reason for that is that soft pions interact weakly, due to their Goldstone nature, and the resulting corrections can be accounted for. A rather straightforward way to do so, using pion–hadron scattering lengths, was proposed by Luscher [309].

Let us however proceed directly to the so called mesoscopic regime of QCD. In condensed matter physics, a mesoscopic system is a system of which the linear size is larger than the elastic mean free path of the electrons but smaller than the phase coherence length, which is essentially the inelastic mean free path. A typical size is about $1\,\mu$m. The conductance fluctuations of mesoscopic samples is closely related to their spectral properties.

In lattice studies it is precisely the regime in which *the pion does not fit into the box but the other hadrons do*, namely

$$\frac{1}{\Lambda_{\text{QCD}}} \sim \frac{1}{m} \ll L \ll \frac{1}{m_\pi}. \qquad (5.3)$$

As the quark mass goes to zero into the chiral limit, the r.h.s. diverges as $1/\sqrt{m_q}$, and so this limit can always be reached for any L and small enough quark mass. This is the "mesoscopic" limit of lattice theory.

Now, if the pion field does not vary appreciably over the size of the box, then the derivative terms are small. For constant fields, the space–time integral in the action can be replaced by the four-volume, and, effectively, we only have to deal with a much simpler zero-dimensional or matrix theory, the integral over constant pion modes.

An alternative way of thinking about these issues is in terms of Dirac eigenvalues. The lowest excitations of the lattice must be related with the lowest part of their spectrum. In fact in Section 2.4.3 we learned the so called Casher–Banks formula, which tells us that the density of eigenvalues at small values is nothing else but the quark condensate. One may wonder if more detailed general results can be obtained about their properties.

It is indeed so, and many nice theoretical results have been obtained in fact in the framework of the so called "Chiral Random Matrix Theory" (or RMT for short). In general, RMT is a powerful way to describe universal correlations of eigenvalues of complex systems. In general, the statistical properties of these eigenvalues are universal and follow from the assumption of randomness (or complexity, or disorder), with rather simple modifications taking care of the specific global symmetries of the QCD partition function.

This theory originated from two different sources. One was pure theoretical studies of the chiral Lagrangians and related issues in the mesoscopic limit by Smilga and Leutwyler [275] who derived some sum rules for eigenvalues etc.

The idea of randomness for the Dirac matrix was first used by Diakonov and Petrov [229] who argued on that ground that the "zero-mode-zone" should look like a semi-circle. While doing dynamical studies of the instanton liquid (described in Chapter 4) Verbaarschot and I looked at many Dirac matrices and did not fail to notice that although they look rather random, the spectrum did not look like a semi-circle at all. Even more relevant, we saw a deep dip of the density near the zero eigenvalue due to the finite-box effects. We concluded [276] that RMT should be modified by adding the determinants in the appropriate powers, and chiral symmetry, so that the chRMT was born.

This led to many analytic results and lattice conformations, for a review see e.g. Ref. [278] (which we follow below to some extent). Let me go into one example only. As was first explained by Dyson, the eigenvalues of a matrix are like electric charges on a circle: they repel each other and thus make something resembling the Wigner crystal. In ordinary RMT one can only see this by looking at the 2-eigenvalue correlation functions. In the chiral chRMT each eigenvalue should appear twice, at positive and negative positions: in terms of the effective charges one may say that each of them has a mirror image. Collective repulsion from those images is the reason for the dip near $\lambda = 0$. Furthermore, as now some point on the λ axis is special, one can observe a crystalline distribution in a single-λ density shown in Fig. 5.1(b).

To explain the notation, recall first that there is a θ angle in the QCD partition function, which however can be absorbed by the phase of the quark masses. To this end, we introduce complex masses m_f for the right-handed quarks and the complex conjugate masses m_f^* for the left-handed quarks. In general, one can consider separately sectors with different global topological number $Q = \nu$.

The chiral RMT is defined by a partition function [277]

$$
Z_{N_f,\nu}^\beta(m_1, \ldots, m_{N_f})
$$

$$
= \int DW \prod_{f=1}^{N_f} \det(\mathcal{D} + m_f) \exp\left(-\frac{N\beta}{4} \, \mathbf{Tr}\, v(W^+ W)\right), \qquad (5.4)
$$

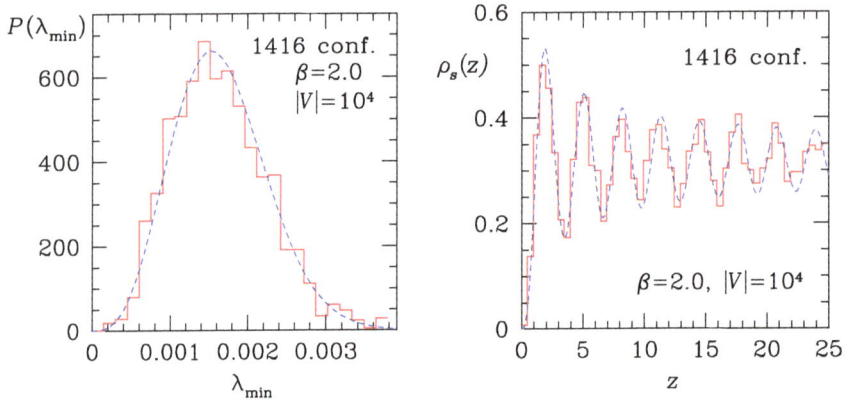

Fig. 5.1. Distribution of the smallest eigenvalue (left) and microscopic spectral density (right) of the staggered Dirac operator in quenched SU(2). The dashed lines are the predictions of the chSE for $N_f = 0$ and $\nu = 0$. (From Ref. [279].)

where β is the Dyson index,

$$\mathcal{D} = \begin{pmatrix} 0 & iW \\ iW^\dagger & 0 \end{pmatrix}, \tag{5.5}$$

and W is an $n \times m$ matrix with $\nu = m - n$ and $N = n + m$. The interpretation of this model is that N low-lying modes interact via a random interaction. We assume that ν does not exceed $\sqrt{N}$ so that, to a good approximation, $n = N/2$ for large N. The parameter N is proportional to the volume of space–time and is considered large. The potential v can be any polynomial — universal results do not depend on that — but the Gaussian one $v(\phi) \sim \phi^2$ is usually used. The matrix elements of W are either real [$\beta = 1$, chiral Gaussian Orthogonal Ensemble (chGOE)], complex [$\beta = 2$, chiral Gaussian Unitary Ensemble (chGUE)], or quaternion real [$\beta = 4$, chiral Gaussian Symplectic Ensemble (chGSE)]. In the latter case, the eigenvalues of $\mathcal{D}$ are doubly degenerate, and the use of Majorana fermions is implemented by replacing the determinant by its square root.

One can calculate many properties of the eigenvalues based on this definition, see in Ref. [278]. As two examples let me show two distributions in Fig. 5.1, the distribution of the smallest eigenvalue (left) and the microscopic spectral density (right). The results of the numerical lattice studies agree with the analytical expressions following from chRMT completely.

The lesson learned from that is as follows: very small Dirac eigenvalues $\lambda \sim 1/(V_4|\langle \bar{q}q \rangle|)$ do not carry any dynamical information other than the magnitude of the chiral condensate (which determines their density). The "instanton liquid" or complete lattice simulation shows true disorder at very large spatial (or very small momentum) scale.

5.1.4. *Art of numerical simulation of multi-dimensional integrals*

This section is intended for non-specialists, a mini-lecture about the basic ideas used in simulations. Lattice practitioners and other numerical experts should skip it.

Confronting a multi-dimensional integral for the first time, everybody starts with two obvious ideas:

1. Use a *simple grid*, say with N points per coordinate, and take the sum of the values at all sites. It is very easy to write a program, and for $d = 1$–3 it usually works. However, since the number of points needed is $N_{points} \sim N^d$, it is clearly impossible to use it for say $d \sim 10$.

2. *Naive Monte Carlo* method. One puts N_{points} random points into the d-dimensional box and average the function. A naive expectation is that the accuracy depends on the number of points as $1/\sqrt{N_{points}}$ for any d. In fact, it would be true only if the function has $O(1)$ values for all points, which is not true for most functions in multidimensional space. In reality the efficiency of this algorithm is also exponentially small because only a small fraction of the volume $\sim [(\text{useful volume})/(\text{total volume})]^d$ contains the important region of space, while most of the random points are simply wasted. Playing with the box size or changing variables helps sometimes, provided one understands the function well. Nevertheless, a lot of people have wasted with this method an enormous amount of computer and personal time, sometimes reaching wrong answers without realizing it.

Let us see how many variables we need for a simulation of some statistical/ quantum mechanical model on the lattice. The simplest case is the path integrals for $1d$ quantum mechanics, for which one needs typically 100–1000 points. The $2d$ spin models can be reasonably well studied on a 100*100 lattice, so there are 10^4 variables to sum over. Even the simplest statistical mechanics problem, the Ising model, already has 2^{10000} configurations on such lattice, so any attempt to do a statistical sum "by definition" is doomed. A bit more complex models with "continuous spin" such as the Heisenberg model for ferromagnetics or the more general Landau functionals with scalar fields need about a $d = 10000$-dimensional integral. Gauge field theory, with all its matrix variables and fermions, needs many more. Clearly any application of the "naive Monte Carlo" to any of those problems is completely hopeless.

At this point people usually start looking into numerical libraries, but (for reasons I never could understand) do not find much useful information there.[5] However, very powerful methods to deal with d as large as a million or so variables do exist and are widely used. True, in order to understand why or how they work one has to know a bit of math. (Fortunately, most of it can also be explained in physics terms.) I recommend the reader to play a bit with some examples and see how they work

---

[5]Otherwise very useful and popular books, e.g. the "Numerical recipes", only mention the Metropolis algorithm vaguely.

for themselves. These methods are universal and can be used in a wide variety of problems. Even if you have a not-so-multi-dimensional integral to deal with, they may save sometimes time by avoiding potential problems: e.g. no box needs to be defined etc.

The main difference between the "naive" methods mentioned above and the sophisticated ones to be described next is that the trial points are not put independently of each other. Instead there is the so called Markowian process, generating a set of points in the "configuration space" (as we will call our set of variables). It can be: (i) purely stochastic (Monte Carlo methods); (ii) a solution of some differential equation (molecular dynamics); (iii) a solution to stochastic equations (Langevin equation); (iv) a combination of the two (hybrid algorithms).

The procedure should accomplish two main goals. First, the set of points should sample mostly the region where the weight function is the largest. On the other hand, it is not just going to reach for the maximum: the multiplicity of configurations (*"entropy"*) should also be a player in the process, balancing the weight (*"energy"*) in a correct way. Finally, it should allow one to move away from the maxima with some probability, and not being stuck near one of them forever (this problem is most obvious in molecular dynamics). Finally, any algorithm of course goes with the mathematical proof (into which we would not go) that the random walk it generates asymptotically converges to the right ensemble prescribed by the given weight function. Usually what is shown is that the ratio of probabilities for a pair of points, x and y, is given by the ratio of the prescribed weight functions. In practice the "asymptotic convergence" is not enough and one would like to use the method which converges it in the shortest computer time possible.

To outline a map of the territory to be explored, let me first enumerate the primary methods used, with brief comments:

1. *The Heat Bath Method*: One variable at a time is "updated" with correct distribution (is put in contact with the heat bath). It is very powerful but correct updates can be implemented only in special cases.

2. *The Metropolis algorithm*: in this case one also updates one variable at a time. But the change from the previous value is usually relatively small, so that the weight is not changed by more than factor 2 or so. It also may or may not be updated. It is very general and has no error except a statistical one: this is the one I will use below as an example.

3. *The Langevin-based methods* are based on solutions of the equations of motion which have in the r.h.s. a stochastic force normalized to the "temperature" of the ensemble. In a few cases an equation of motion allows for analytic treatment, which is nice. However, as a differential equation, it requires small modifications per one time step. As for all numerically solved differential equations, that should be handled with care. More important in practice is however a simple fact that a necessity to go by small steps make the Langevin methods less efficient.

4. "Molecular dynamics": the name is historical and comes from the simulations of gases and liquids with inter-atomic forces and classical equations of motion. In the context we are interested in, lattice gauge theories and the like, it is used as follows. One treats the action (e.g. with the plaquettes $UUUU$) as a potential energy, then imagines a fictitious "computer time" t (which has nothing to do with Euclidean $\tau = x_4$) and a fictitious "kinetic term" (e.g. $\sim \text{Tr}\, \dot{U}\dot{U}^+$). If those are in place, one derives the equation of motion and just follows the classical path. This method is also "microcanonical" rather than canonical, in the sense that the energy is conserved and so its value for the input configuration (per variable) is what is conjugate to the "temperature" or coupling constant in other approaches. In this case the trajectories of classical mechanics (or classical field theory) appear in place of random Brownian-like motion in stochastic algorithms, so the method often samples different places quicker and get a good convergence rate. However it relies on special properties of the system and does not work in some simple case is when the system is *not* chaotic by itself, such as free fields or exactly solvable models with many integrals of motion.

5. *The Hybrid algorithms* may be the fastest of all, if tuned well. They alternate in computer time, usually between the Metropolis and the Molecular Dynamics algorithms. The path then consists of Brownian periods and classical trajectories, combining the randomization of the former with the quick motion of the latter. It is especially useful for complicated non-local actions when evaluation of the action is very costly.

The update of a single variable (the ith coordinate of the path) at a time is the basic step of the program. A subsequent update of all sites is called one *iteration* of the system. Making many subsequent iterations leads to the *relaxation* process, monitored by measuring some average value over computer time, say $\langle x^2 \rangle$. Although those are different for differently chosen initial conditions, they converge to the *same* values characteristic of the *statistical equilibrium*. Usually people compare the "cold start" $x_i = 0$ versus the "hot start" with random x_i. After the equilibrium is believed to have been reached with the desired accuracy, a set of *measurements* is performed. The process continues until the needed statistical accuracy is reached. The *observables* include average values of some operators like $\langle x^2 \rangle, \langle x^4 \rangle$, or the distribution over x, or non-local measurements like that of the coordinate correlator $G(\tau) = \langle x(t + \tau)x(t) \rangle$.

Exercise 5.1. Perform a set of simulations of a quantum-mechanical path integral, or the $1d$ set of about 100 "spins", the coordinate values x_i at the sites. An example of such program can be found in Appendix B.3. In its core is a small universal subroutine for the Metropolis algorithm update. (Provided in Fortran, but hopefully self-explained form.) For a double-well potential calculate (i) the mean potential energy; and (ii) the correlator of two coordinates $\langle x(\tau_1)x(\tau_2) \rangle$.

5.2.   Fermions on the lattice

5.2.1.   *Fermionic doublers*

Putting fermions into the lattice gauge theory, we imagine that we consider a single gauge field configuration defined by the link variables U_μ which we discussed in Section 1.1.2. The fermions are "matter fields", so they are naturally defined on lattice cites. Their gauge transformation is thus just gauge rotations, in general independent and arbitrary at any cite. We have already given a preliminary discussion of Grassmanian path integral and Euclidean notation for quarks and gamma matrices in Section 1.4.3.

Let me recall that in the continuum (Minkowski) space–time the fermion action is

$$S_f^{\text{Minkowski}} = \int d^4x\, \bar\psi(i\hat D - m)\psi, \tag{5.6}$$

with the covariant derivative $i\hat D = (i\partial_\mu + gt^a A_\mu^a(x)/2)\gamma_\mu$ containing the field. Apart from the color matrices, the action and long derivative are of course analogous to the familiar QED ones. The long derivative convoluted with gamma matrices is called the Dirac operator. On the lattice, the derivative like the gauge field itself are both naturally defined on links.

In general, one would like to have the following conditions to be fulfilled by the lattice fermions:

- No doublers, only N_f fermions with the desired masses.
- Locality of the operator, $|\langle\vec n|D|\vec n'\rangle| < \exp(-\text{const}|\vec n - \vec n'|)$ with a finite constant.
- Correct continuum limit $a \to 0$.
- The chiral symmetry is exact, protecting the masslessness of quarks and pions.

Unfortunately, as follows from the Nielsen–Ninomiya "no-go" theorem [283], such a Dirac operator on the lattice simply does not exist. The resolution of the problem took a long time; but it was solved in the late 1990s, and even in several independent ways.

Let us start at the beginning and see what happens if one naively approximates the derivative by a difference, on both sides of the cite n,

$$\partial_\mu\psi \approx \frac{1}{2a}(\psi_{n+1} - \psi_{n-1}). \tag{5.7}$$

For the one-dimensional lattice the Hamiltonian,

$$H = \sum_n iK[\bar\psi_n\gamma_1\psi_{n+1} - \bar\psi_{n+1}\gamma_1\psi_n] \tag{5.8}$$

is diagonalized by the plane waves with momentum q and the energy spectrum is

$$E(q) = 2K \sin(qa). \tag{5.9}$$

At small q it is linear and vanishes at $q = 0$, OK, but at the edge of the zone (which corresponds to $q = \pm\pi/a$) the energy of the fermion is zero again. This is the "doubler", or a "staggered mode" with $\psi_{n+k} = (-)^k \psi_n$. Adding the mass term to the Hamiltonian of the form $M \sum \bar{\psi}_n \psi_n$ leads to a modified spectrum

$$E(q) = (4K^2 \sin^2(qa) + M^2)^{1/2}, \tag{5.10}$$

which does not solve the problem: both $q = 0$ and the doubler modes have the same gap.

5.2.2. *Wilson fermions*

Wilson's main idea [281] was to introduce an additional term to the fermionic action which would kill all the unwanted doublers and preserve only one massless mode. This is achieved with the Hamiltonian

$$H = \sum iK[\bar{\psi}_n(\gamma_1 - r)\psi_{n+1} - \bar{\psi}_{n+1}(\gamma_1 + r)\psi_n], \tag{5.11}$$

where r is a new (the so called "hopping") parameter. The spectrum now is

$$E(q) = \{4K^2 \sin^2(qa) + [M - 2Kr\cos(qa)]^2\}^{1/2} \tag{5.12}$$

and the gap at $q = 0$ is $M - 2Kr$ while at $q = \pi/a$ it is $M + 2Kr$. So by tuning r one can cause the former value to be small as compared to the latter, and the unwanted doubler is finally killed, as $a \to 0$.

However this simple solution of the doubler problem goes with a heavy price: now the chiral symmetry at finite lattice spacing a is gone, only to be recovered in the limit of $a \to 0$.

Furthermore, quantum renormalization of the quark masses is not protected by any symmetry and should be tuned to desired values in each new lattice parameter. This is usually done by monitoring the dependence of the pion correlator/mass on the hopping parameter, and tuning it to the desired value.

Another early formulation of lattice fermions — called staggered or Kogut–Susskind fermions [282] — tried to make use of the doublers. More specifically, they have interpreted four species of lattice light quarks as four flavors. It was shown that even some remnant chiral symmetry survives in this formulation, which keeps some component of the pion multiplet massless. We shall not discuss it in detail, but jump directly to a newer development which we will discuss in the next section.

5.2.3.   *Ginsparg–Wilson relation and lattice chiral symmetry*

In a paper [284] which remained basically unknown for long time, Ginsparg and Wilson had proposed that the lattice Dirac operator should better satisfy the relation

$$\gamma_5 D + D\gamma_5 = aD\gamma_5 RD, \tag{5.13}$$

where R is some so far arbitrary local operator. In the continuum limit $a \rightarrow 0$ its r.h.s. is zero and the usual commutation relation between D and γ_5 is recovered.

Only much later was it shown by Luscher [289] that this relation actually implies the existence of the *modified* non-local lattice chiral symmetry at non-zero a, with the transformation

$$\delta\psi = i\alpha\gamma_5 \left[1 - \tfrac{1}{2}aD\right]\psi. \tag{5.14}$$

Note that D is not a local operator, but only approximately local. Still, the existence of such symmetry with all its consequences is extremely important and would be very much welcome by the lattice practitioners.

5.2.4.   *Known solutions to GW relation*

The first solution, known as *overlap* fermions, was proposed by Neuberger [285] in 1997. It has the following form

$$D_{\rm N} = \frac{1}{a}\left(1 - \frac{A}{\sqrt{A^+A}}\right), \quad A = 1 - aD_{\rm W}, \tag{5.15}$$

where $D_{\rm W}$ is the Wilson operator. One can show that it indeed satisfies the GW relation (5.13) with $R = 1/2$.

Exercise 5.2.   In order to understand what the overlap operator (5.15) does it is convenient to see which eigenvalues this operator prescribes for free fermions (no gauge field). The lattice solutions for naive fermions moving in any direction with momentum p_μ (changing between $-\pi/a, \pi/a$) yield the $\sin(p_\mu a)$ solutions for Wilson fermions. Prove that overlap fermions have eigenvalues which span the circle with radius $1/a$ and center at $1/a$.

The main point of the transformation from Wilson to overlap fermions is to divide out the norm and put all Dirac eigenvalues onto a circle touching the imaginary axis. For small eigenvalues these eigenvalues can be simply mapped to a straight line. In contrast, the previous W and KS fermions provide a cloud of eigenvalues with various real and imaginary parts, converging into a line only for small eigenvalues and in the $a \rightarrow 0$ limit.

From the form of the projection itself it is easy to see that the eigenvalues should lie on this circle for any configuration with any field and input operator D_W, since we divide A by its norm. This by itself would clarify the issue of the zero eigenvalues and modes.

One more practical solution (see e.g. Ref. [562] and references therein) is now used. One writes the Dirac operator as an ordinary link difference plus more complicated link-like paths with undetermined coefficients. Then, by plugging this expression into the GW relation, one gets multiple quadratic equations for those coefficients because the r.h.s. has the operator squared. By solving them numerically, with some truncation of high enough terms, one can get rather accurate (and still not too expensive) solutions of the GW equation.

A more satisfactory solution however comes from a long line of investigations aimed at improvement of the lattice action, related with the so called *perfect* action.[6] Hasenfratz and Niedermayer [287] realized that in the classical limit one can numerically construct the so called *classical perfect lattice actions*, which are completely free of finite lattice spacing artifacts. Those classically perfect actions were then used for quantum (non-smooth) configurations and turned out to be rather good also at the quantum level, as was shown in Ref. [291] and subsequent works.

In particular, classically perfect actions have reproduced correct values for the instanton action, even for rather small-size lattice instantons with size $\rho \sim a$. I was present at P. Hasenfratz's talk in Cambridge in 1997 on that subject, and asked him: "If you can get the action that accurately, does in also mean that the fermionic zero mode would also be exactly correct?" Peter replied with the question "Why do you think the zero mode is a classical object?" But when he thought about it, he managed to prove that a *classically perfect Dirac operator* would actually not only do this, but also satisfy the GW relation [288]. Thus following the renormalization group, people found one more way toward nice chiral lattice fermions.

5.2.5. *Domain wall fermions*

In this formulation, suggested by D. Kaplan [286], the fermion is kept massless by the topological index theorem we discussed in Section 3.3. Small perturbations cannot affect this mass of the topological origin, and so these fermions do not need any fine tuning.

The setting includes introduction of the 5th coordinate x_5, on which gauge fields does not depend. The fermions have however a mass which is x_5-dependent. More specifically, this coordinate is periodic and the mass term jumps twice, from $+m$ to $-m$ and then back, see Fig. 5.2. As will be clear from the following, the exact shape of $m(x_5)$ is unimportant. All that is needed is two domains, with positive and negative mass, with two 4d "domain walls" separating them.

---

[6]We will briefly discuss it for the quantum mechanical rotor in Section 5.4.

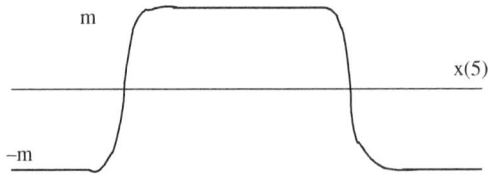

Fig. 5.2. Schematic dependence of the fermion mass on the 5th coordinate, in domain wall formulation.

The main idea is based on the phenomenon we already discussed in Section 3.3, although in a somewhat different language. We recall that there the background field was a quantum mechanical instanton, or kinks in Euclidean time τ. Now we have kinks in what one thought of as the $x(5)$ spatial coordinates.

As is the case for all topologically non-trivial backgrounds, the Dirac equation has the zero mode solutions localized around it. If the coordinate is Euclidean time, its interpretation is the same as for all other instantons we discuss extensively above. If the coordinate is spatial, like a domain wall, this zero mode is interpreted it as the static zero energy bound state.

Furthermore, an infinitely separated kink has the zero mode which is "chiral". In the kink–anti-kink setting, as for instanton–anti-instanton molecules, we have that one wave function is *nearly* purely left- and the other right-handed. The accuracy to which this is true depends on the interaction between the two walls, which decays exponentially $\sim \exp(-mL_5/2)$ with the size of the lattice in the 5th direction L_5, and so it can be made very small. In practice people were able to reduce the "residual quark mass" to quite small values, less than one MeV, which is quite sufficient.

Let us for a moment recall the instanton liquid picture and think what it would look like in this formulation. The lowest Dirac eigenstates were described in it as a superposition of instanton and antiinstanton zero modes. The former and latter are now separated in x_5 because they live on L- and R-handed walls. The vacuum quarks, jumping from instanton to antiinstantons and back, now have to not only move in 4 ordinary coordinates, but also ping-pong between the two walls.

For all QCD applications we have of course operators which do not depend on x_5 and thus averaging leads to an equal share of L- and R-handed quarks. However the domain wall fermions allow us in principle to simulate *chiral theories*, such as the electroweak sector of the Standard Model or supersymmetric theories. For example, if only L-handed fermions are needed, all one has to do is to localize all the measured operators on the L-wall.

5.3.   Hadronic spectroscopy on the lattice

In principle, in lattice QCD one can determine all hadronic parameters directly from first principles. This has not happened yet, although the quality of the results is incomparable to what was available when the first edition of this book was being prepared. A large variety of opinions can be found even among people working in the field (as it always happens in rapidly developing branches of science). Therefore it hardly makes sense to present a detailed description of current data: just in a couple of years even their authors themselves would consider them as "some old stuff". Instead, we will start with glueballs, then discuss the status of the quenched approximation, and then the "remaining problems" which restrict the accuracy of the results.

5.3.1. *Glueballs in gluodynamics*

Lattice gauge theories without quarks are much simpler, so they were studied first. Even more important is the fact that there are no very light glueballs, and also the separation of the lowest and excited states is very significant. This significantly relaxes the conditions on the minimal lattice box size. On the other hand, as a scalar glueball is a rather small object, it increases demands on the UV side, forcing us to have very fine lattices.

There exists extensive literature on this subject, including the SU(2) and the SU(3) color groups, various actions and lattices. We are not going to present all these data and discuss them in detail, our aim is only to show the tendencies and the present state of this art. Historically studies of "mass gap" by plaquette–plaquette correlation started in the early 1980s. For some period the lowest glueball mass was suggested to be around $700\,\mathrm{MeV}$, where sigma or $f_0(600)$ is located. With finer lattices the scale was changing upward, and in mid-80s (when the first edition of this book was written) the results were closer to the $f_0(1300)$ resonance. And as the reader will see shortly, they increased again later.

Among methodical questions that are discussed are (i) the "universality" (action independence) of the results; (ii) quality of restoration of rotational (Lorentz) symmetry, and (iii) extrapolation to continuum $a \to 0$ limit.

As a representative example we selected one paper [304] which provided rather extensive simulations. In Fig. 5.3a one can see how different states of the cubic lattice converge to the same mass of 2^{++}, 3^{++} glueballs as a gets small. One can also see that the two scalars show a variation for rather small a, unlike 2^{++}, 3^{++}. The latter of course have a centrifugal barrier and the wave function is zero at the origin: so they do not interact with small-size objects. It seems clear that two scalars do have some short-distance interaction, presumably with small-size objects. As we will discuss in Section 6.5.5 those seem once again to be instantons.

The extrapolated continuum spectrum of other glueballs is shown in Fig. 5.3b, with the width of the line corresponding to uncertainty. To convert the glueball masses into physical units, the value of the hadronic scale r_0 must be specified: $r_0^{-1} = 410(20)\,\mathrm{MeV}$ was used to obtain the scale shown on the right-hand vertical axis. This estimate was obtained by combining Wilson action calculations of a/r_0 with values of the lattice spacing a determined using quenched simulation

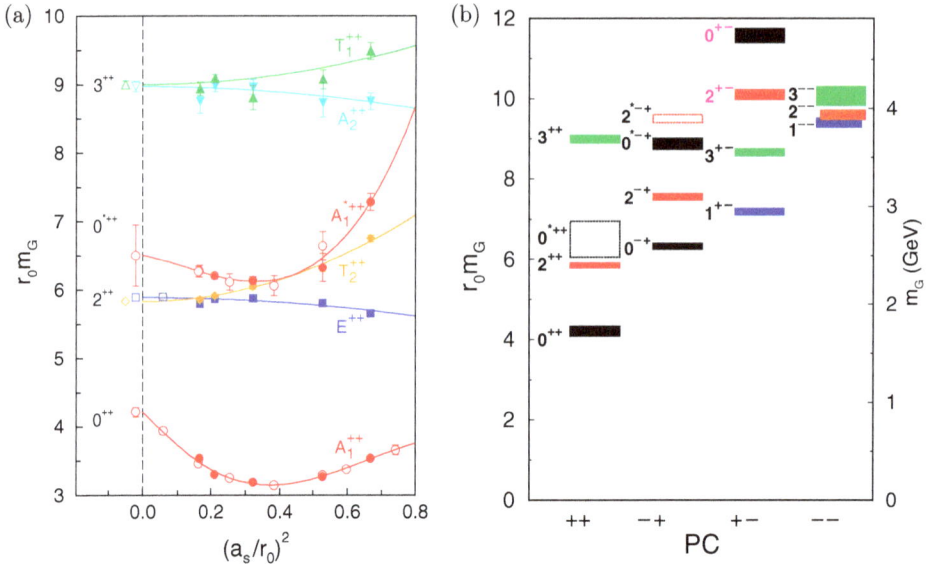

Fig. 5.3. From Ref. [304]. (a) Mass estimates of the $PC = ++$ glueballs in terms of r_0 against the lattice spacing $(a_s/r_0)^2$. Various symbols indicate results from different actions, the solid curves are best fits. (b) The mass spectrum of glueballs in the pure SU(3) gauge theory extrapolated to the $a = 0$ continuum limit. The masses are given in terms of the hadronic scale r_0 along the left vertical axis and in terms of GeV along the right vertical axis (assuming $r_0^{-1} = 410\,\text{MeV}$). The mass uncertainties indicated by the vertical extents of the boxes do *not* include the uncertainty in setting r_0. The locations of states of which the interpretation requires further study are indicated by the dashed hollow boxes.

results of various physical quantities, such as the masses of the ρ and ϕ mesons, the decay constant f_π, and the 1P–1S splittings in charmonium and bottomonium. For the lowest-lying glueballs, the masses are $m(0^{++}) = 1730(50)(80)\,\text{MeV}$ and $m(2^{++}) = 2400(25)(120)\,\text{MeV}$, where the first error comes from the uncertainty in $r_0 m_G$ and the second error comes from the uncertainty in r_0^{-1}.

Note that this figure plots the masses, not masses squared as we did in Section 2.1 when we discussed the π, ρ, η' trio. If we did so, we would see a similar splitting pattern for the $0^{++}, 2^{++}, 0^{-+}$ trio, with the tensor 2^{++} state being the unperturbed "anchor". Such a pattern is expected from the first order instanton-induced effects, which lead to similar attraction-nothing-repulsion in those three states.

5.3.2. *Light quark spectroscopy in quenched approximation*

Let me start with a general warning: light fermions go together with chiral symmetry breaking and light pion states, which makes the demand on the box size quite severe. In principle, however, once can correct for the fact that "pions do not quite fit into the box" in a relatively simple way, by accounting for the scattering amplitude for missing pions, see Ref. [309].

First works on the hadronic spectroscopy such as Ref. [305, 306] have used the so called "quenched" approximation, in which the fermionic determinant in the weight is ignored. It means that there are only the "valence" quarks created by the operators involved, which move in the purely gluonic vacuum. There are no virtual quark loops, as those belong to the neglected determinant.

The motivation for this approximation can only be made in the limit of a large number of colors, where indeed the back reaction of all quark loops on gauge fields can be ignored in the leading order. In the 1980s people hoped that the real world with $N_c = 3$ is actually very close to this limit. Perturbatively, one may indeed think so. For example, consider the error one would make if one neglected virtual quark loops in the charge renormalization: in the leading beta function term it would be about 2/11 only. Furthermore, part of it can be absorbed in the scale difference between quenched and true QCD.

On the other hand, one may ask when the fermion determinant may really provide quite different weights to specific gauge field configurations. The simplest example is the configurations with the non-zero net topological charge: in the chiral limit the fermion determinant simply demands the weight for such configurations to vanish. More importantly, as we have seen in Chapter 4, the determinant creates a specific interaction between the instantons and radically changes their ensemble.

In practice, for more than a decade quenched results looked rather promising or even quite accurate, with the first signs of trouble showing up only in early the 1990s. On the theory side the first studies of wrong properties of quenched QCD were those by Sharpe [307] who pointed out that this approximation led to a massless η' which not only shows in this particular channel, but also in all others in the form of "false chiral logs" of the kind $\log m_q$, to be contrasted to true chiral logs which can only appear as $m_q \log m_q$. This qualitatively wrong feature of quenched QCD is seen in many variables, even in such a global parameter of the chiral symmetry breaking as the quark condensate. First in RILM and then in quenched QCD it was demonstrated that the spectrum of the Dirac eigenvalues at small λ does not follow the smooth behavior expected of the Casher–Banks formula. Instead, it is singular $dN/d\lambda \sim 1/\lambda^p$ with some small index p. From the Banks–Casher relation it follows then that in the chiral limit $\langle \bar{q}q \rangle \to \infty$, while it should be finite in true QCD. A recent confirmation of that, in simulations with the domain wall fermions from the Brookhaven group [321] is shown in Fig. 5.4. One can see from it that there is a bizarre behavior of the condensate at small masses instead of a smooth limit to a constant.

Another observable in which wrongness of the quenching (and thus of the large-N_c ideology) is seen in the most bizarre way: consider the correlator of two $I = 1$ scalar (δ or a_0) local currents. As we will show in Section 6.5.2, it was found in the 1993 work by Verbaarschot and myself [237] that the quenched version of the instanton liquid, RILM, led to a very good description of most correlators, except the η' and a_0 ones, which very rapidly decreased with distance and *changed sign*

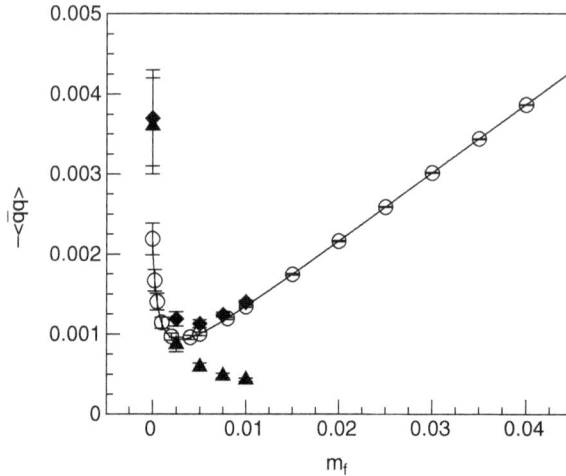

Fig. 5.4. $\langle \bar{q}q \rangle$ for quenched simulations done on $16^3 \times 32$ lattices at $\beta = 6.0$ for $L_s = 16$. In addition to the direct results for the quark condensate, we also plot the values obtained from the 18 eigenvectors studied here (filled triangles) and the sum of those values with the contribution from far off-shell states (filled diamonds).

at $x \sim 0.3$ fm see Section 6.1. This contradicts positivity and spectral decomposition. This unphysical behavior was first attributed to defects of the instanton model, because lattice calculations done at the time did not have such features. However, with time computer power allowed the use of smaller fermion masses, and very similar negative a_0 correlators were also observed on the lattice [319]. It was rather shocking, although in principle such features as the positivity and spectral decomposition should only be present in legitimate field theories, while the quenched QCD is obviously not one of them.

5.3.3.   *Spectroscopy with dynamical quarks*

So, in order to reproduce QCD one has to take explicitly into account light dynamical quarks, which requires very significant calculational efforts. In order to exemplify the present status of such calculations let me mention a recent work by the UKQCD collaboration [310]. The results of Fig. 5.5a show vector quark masses, obtained from the logarithmic derivative of the correlation functions. They are plotted versus the pseudoscalar meson mass *squared* since those are proportional to the first power of the perturbations (such as e.g. quark masses). One can see from various simulations that all results seem to agree very well with those in quenched approximation. Furthermore, both seem to indicate a simple linear relation between the vector meson and quark mass. However the phenomenological extrapolation of ϕ and K^* into this plot is *not* on this straight line.

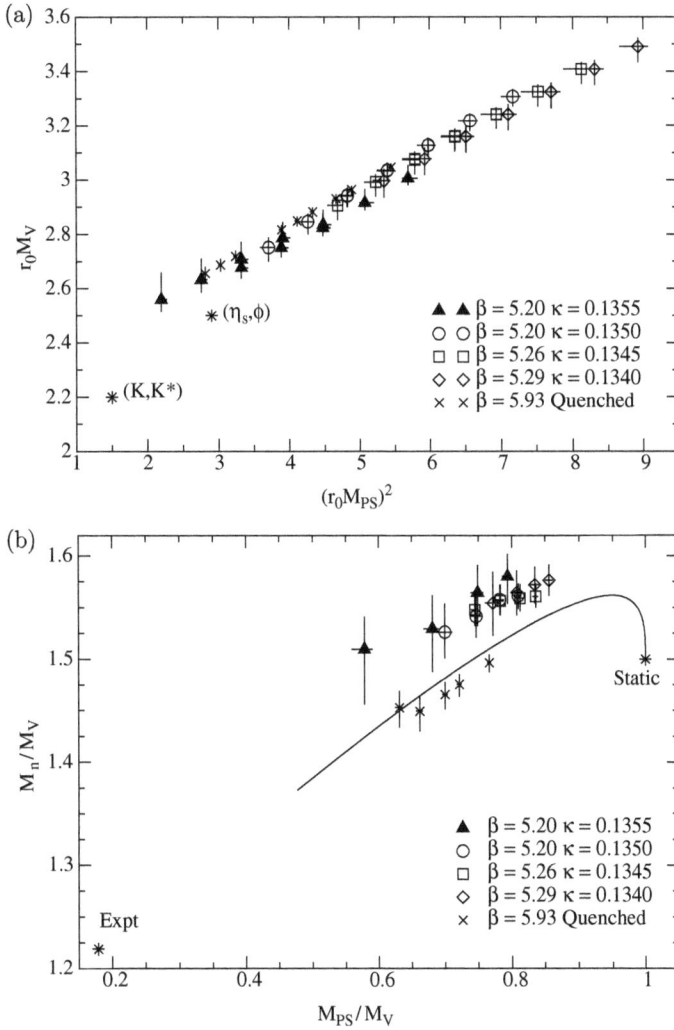

Fig. 5.5. From Ref. [310]. (a) Vector mass plotted against pseudoscalar mass squared in units of r_0, together with the experimental data points. (b) The so called Edinburgh plot for all the data sets. The phenomenological curve (from Ref. [311]) has been included as a guide to the eye.

The results shown in Fig. 5.5b show a similar plot for the nucleon-to-vector mass ratio versus the pseudoscalar-to-vector mass ratio, known as the "Edinburgh plot". Now the difference with the quenched results is much more visible. Also it is quite clear that the expected dependence on the quark mass has to be non-linear, as the linear fit to the data points obviously would not point to the experimental nucleon-to-ρ mass ratio (the lone star in the left lower corner).

The conclusion a reader may draw out of these plots is that available computer power still does not allow the use of physical quark masses. As a result, the pions are not so light as they should be, and uncertainties related with chiral extrapolation dominate all other errors.

To make the conclusion drawn in the last subsection even more convincing, let me show one more set of lattice results, the moments of the structure functions calculated in Ref. [313]. Again, a nonlinear dependence on the quark mass is inevitable, as the experimental values tell us. Similar problems have been encountered for nuclear magnetic moments and g_a, where deviations between calculation with currently available quark masses and data are at the level of 50 percent.

5.4.   Topology on the lattice

5.4.1.   *Quantum-mechanical topology and perfect actions*

Let us start with the simplest example we so extensively discussed in Chapters 1 and 3, that of tunneling events in quantum mechanics. We will start with a double well and then turn to the quantum rotator.

At the classical level the topology of a double-well problem is very simple: there are three topologically different classical paths: the instanton $[x(-\infty) = -f,$ $x(\infty) = f]$, the anti-instanton $[x(-\infty) = f, x(\infty) = -f]$, and the trivial one $x(\tau) = 0$. So naively one may think it is rather straightforward to look at a quantum path, such as shown in Fig. 5.7, and decide which of these three it resembles most (the instanton, for the path shown in this figure).

However, in order for the procedure to be meaningful, two basic conditions should be fulfilled:

- The topological objects should be well separated.
- The discretization should be good enough to interpolate.

Note that the former condition is a basic requirement that a small semiclassical parameter should be actually available, while the latter is a condition on discretization.

The *interpolation* mentioned above is the key word, it is a specific way to reconstruct the continuous field out of its lattice version. For the paths one can either use set of segments (obtaining the path shown by the thin solid line in Fig. 5.7) or use the more complicated "improved" version of a spline, as shown by the thick dashed line. Either way, it seems clear that the *global* topological charge of this configuration,

$$Q = \frac{1}{2f} \int d\tau \frac{dx}{d\tau} = \frac{x(L_\tau) - x(0)}{2f}, \tag{5.16}$$

is the same as that for one instanton.

It is much less clear what the *local* topological charge is, or the number of topological objects $N_{\mathrm{I}} + N_{\bar{\mathrm{I}}}$. Look at the right upper side of the path in

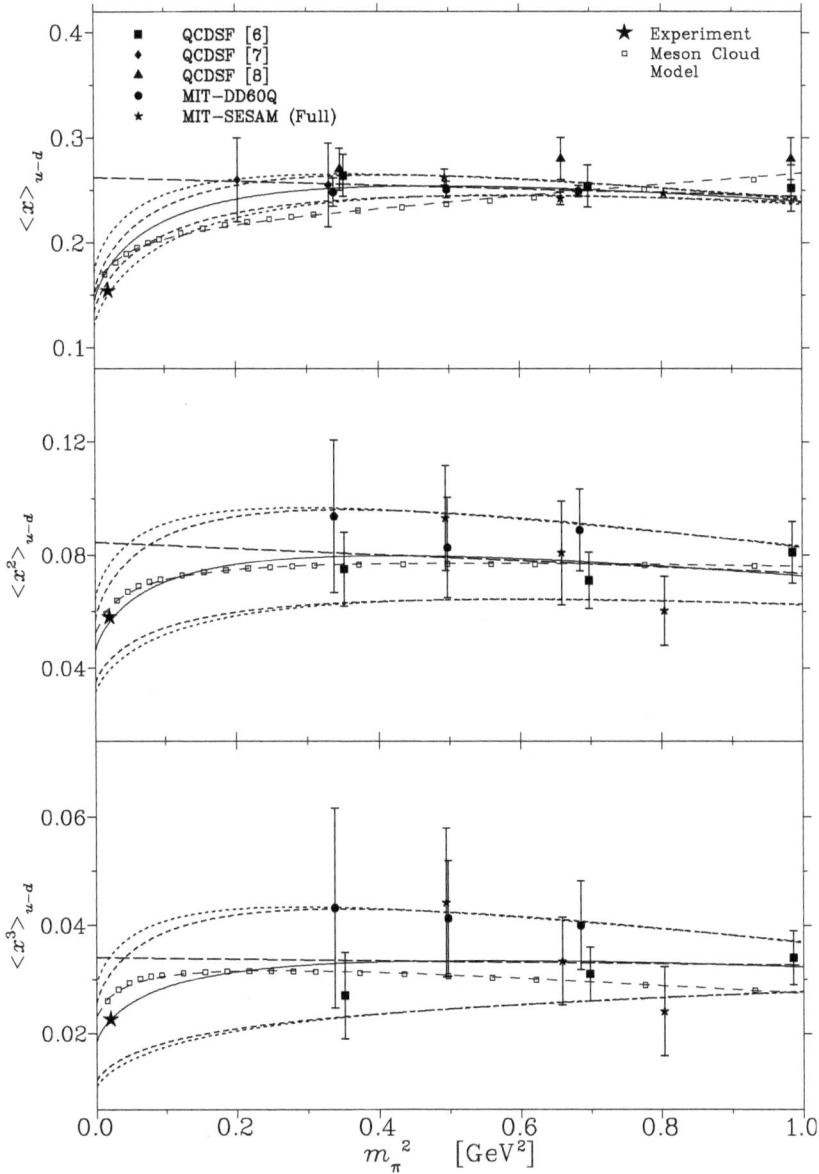

Fig. 5.6. Moments of the unpolarized structure function, versus the pseudoscalar quark mass squared. More precisely, it is the difference between the average x, x^2, x^3 for up and down quarks in a nucleon. The curves are phenomenological interpolation to experimentally observed values (left side of the figures).

Fig. 5.7: it contains a fluctuation with large $dx/d\tau$ which even crosses the barrier center. Is it a close $\bar{I}I$ pair? Or a "flucton" discussed in Section 3.1.2? Or just a perturbative fluctuation? It is clear that those questions are not really meaningful as the answer depends on rather arbitrary definitions separating those objects. The

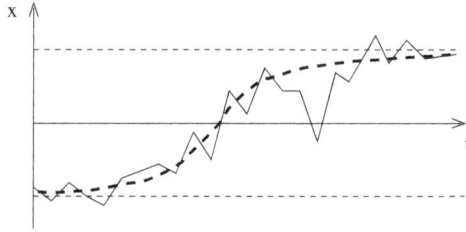

Fig. 5.7. A quantum path with a propagation from one well to another. The thicker dashed line shows schematically its interpolated (or smoothened, or cooled) version, corresponding to the classical instanton.

meaningful question are: What is the probability of such a fluctuation to happen? Which role do fluctuations play in particular observables?

The situation is even more subtle for our second example, the *quantum rotator* we discussed already in Section 1.1.5. In this case the coordinate variable is an angle, defined in principle on a compact manifold with natural periodicity. Say we have a quantum rotator and its latticized configurations are described by a set of positions $\phi_i, i = 1\text{–}N$ on a circle. We want to interpolate those to continuum $\phi(\tau)$ and classify all paths by their winding number. The interpolation procedure is unique if jumps between the subsequent points are all small $|\phi_{i+1} - \phi_i|/2\pi \ll 1$. However if even one of them is large, say a jump comparable to π, one has to choose on which side of the circle to interpolate. Say, interpolation along the smallest one seems most reasonable: it suppresses "defects" or topological objects of a size comparable to lattice spacing. As soon as the interpolation algorithm is chosen, there is a continuous path, associated with an *integer* winding number.

The quantum rotator was extensively studied from many points of view, and some of those elucidate many important developments in lattice theory. One of them unquestionably is the issue of improvements of the lattice action, eventually leading to so called "perfect" actions which have *no* corrections in any power of a, either at the classical or at quantum level. For the rest of this subsection we follow the paper of Ref. [314] on the quantum rotator. We derived its spectrum and the topological angle theta in Section 1.1.5. From that one can get a finite temperature partition function

$$Z(\theta) = \text{Tr} \exp\{-\beta H(\theta)\} = \sum_m \exp\left[-\frac{\beta\hbar^2}{2I}\left(m - \frac{\theta}{2\pi}\right)^2\right]. \tag{5.17}$$

From the partition function one derives the topological charge distribution,

$$p(Q) = \frac{1}{2\pi}\int_{-\pi}^{\pi} d\theta\, Z(\theta)\exp\{-i\theta Q\} = \sqrt{\frac{2\pi I}{\beta\hbar^2}}\exp\left(-\frac{2\pi^2 I}{\beta\hbar^2}Q^2\right), \tag{5.18}$$

and from that the topological susceptibility

$$\chi_t = \frac{1}{\beta} \frac{\sum_Q Q^2 p(Q)}{\sum_Q p(Q)} = \frac{1}{\beta} \frac{\sum_Q Q^2 \exp\{-2\pi^2 I Q^2 / \beta\hbar^2\}}{\sum_Q \exp\{-2\pi^2 I Q^2 / \beta\hbar^2\}}. \tag{5.19}$$

In the zero temperature limit $\beta \to \infty$ one obtains

$$\chi_t = \left.\frac{d^2 E_0(\theta)}{d\theta^2}\right|_{\theta=0} = \frac{\hbar^2}{4\pi^2 I}. \tag{5.20}$$

To formulate the path integral for the quantum rotor on a Euclidean time lattice with lattice spacing a, we first construct the transfer matrix,

$$\langle \varphi_{t+a} | T | \varphi_t \rangle = \left\langle \varphi_{t+a} \left| \exp\left(-\frac{a}{\hbar} H(\theta)\right) \right| \varphi_t \right\rangle$$

$$= \sum_m \langle \varphi_{t+a} | m \rangle \exp\left[-\frac{\hbar a}{2I}\left(m - \frac{\theta}{2\pi}\right)^2\right] \langle m | \varphi_t \rangle$$

$$= \frac{1}{2\pi} \sum_m \exp\left[-\frac{\hbar a}{2I}\left(m - \frac{\theta}{2\pi}\right)^2 + im(\varphi_{t+a} - \varphi_t)\right]. \tag{5.21}$$

Hence, the Fourier transform of the transfer matrix with respect to $\varphi = \varphi_{t+a} - \varphi_t$ is $\exp[-\hbar a (m - \theta/2\pi)^2/2I]/2\pi$. Using the Poisson re-summation formula, the transfer matrix can also be written as,

$$\langle \varphi_{t+a} | T | \varphi_t \rangle = \sqrt{\frac{I}{2\pi\hbar a}} \sum_n \exp\left[-\frac{I}{2\hbar a}(\varphi_{t+a} - \varphi_t + 2\pi n)^2 + i\frac{\theta}{2\pi}(\varphi_{t+a} - \varphi_t + 2\pi n)\right]. \tag{5.22}$$

This follows when we consider the Fourier transform of this expression with respect to φ, that

$$\frac{1}{2\pi} \int_{-\pi}^{\pi} d\varphi \sqrt{\frac{I}{2\pi\hbar a}} \sum_n \exp\left(-\frac{I}{2\hbar a}(\varphi + 2\pi n)^2 + i\frac{\theta}{2\pi}(\varphi + 2\pi n) - im\varphi\right)$$

$$= \frac{1}{2\pi} \exp\left[-\frac{\hbar a}{2I}\left(m - \frac{\theta}{2\pi}\right)^2\right], \tag{5.23}$$

which agrees with the Fourier transform of the expression in Eq. (5.21).

The exact partition function can be written as a path integral on a Euclidean time lattice with $N = \hbar\beta/a$ points. Starting from

$$Z = \mathrm{Tr}\exp\{-\beta H\} = \mathrm{Tr}\, T^N, \tag{5.24}$$

one inserts complete sets of states $|\varphi_t\rangle$ at each time step (Chapman–Kilogram equation) and one obtains

$$Z = \int \mathcal{D}\varphi \mathcal{D}n \exp\left(-\frac{1}{\hbar}S[\varphi, n] + i\theta Q[\varphi, n]\right). \tag{5.25}$$

The measure of the path integral is given by

$$\int \mathcal{D}\varphi \mathcal{D}n = \prod_{t=0}^{(N-1)a} \sqrt{\frac{I}{2\pi\hbar a}} \int_{-\pi}^{\pi} d\varphi_t \sum_{n_{t+a/2}}, \tag{5.26}$$

with periodic boundary conditions $\varphi_{Na} = \varphi_0$. The action takes the form

$$S[\varphi, n] = a \sum_{t=0}^{(N-1)a} \frac{I}{2}\left(\frac{\varphi_{t+a} - \varphi_t + 2\pi n_{t+a/2}}{a}\right)^2. \tag{5.27}$$

This is the Villain action of the XY model, which — by construction — turns out to be a *quantum perfect action*. In the continuum limit $a \to 0$, it converges to the continuum action

$$S[\varphi] = \int_0^{\hbar\beta} dt\, \frac{I}{2}\dot{\varphi}(t)^2. \tag{5.28}$$

Similarly, the perfect topological charge is given by

$$Q[\varphi, n] = \frac{a}{2\pi} \sum_{t=0}^{(N-1)a} \frac{\varphi_{t+a} - \varphi_t + 2\pi n_{t+a/2}}{a}, \tag{5.29}$$

which — due to periodicity in time — is equivalent to $Q[\varphi, n] = \sum_{t=0}^{(N-1)a} n_{t+a/2}$. In the continuum limit it turns into

$$Q[\varphi] = \frac{1}{2\pi} \int_0^{\hbar\beta} dt\, \dot{\varphi}(t), \tag{5.30}$$

as it should. The interested reader can further follow many details of this paper: while we shift our attention to gauge fields.

5.4.2. *Naive and geometric methods for gauge fields*

Let me first remind the reader that although the SU(3) gauge group of QCD, with its 8 generators and angles (gauge fields) does not generate a non-trivial topology by itself, its smaller subgroups Z_3, U(1), SU(2) do. The U(1) has one angle and it generates (i) *vortexes* with nontrivial flux $\oint_C A_\mu dx_\mu$ for all contours C around them and (ii) *monopoles* with the flux $\oint_\sigma G_{\mu\nu} d\sigma_{\mu\nu}$ over a 2d surface around it. In 4d space–time the paths of those objects make 2d surfaces and 1d paths, respectively. The SU(2) generates instantons, with the topological charge $\oint_\sigma K_\mu d\sigma_\mu$ integrated over 3d surfaces, such as the small and large spheres we discussed in Chapter 4. Of course, in all cases one can use a generalized Gauss theorem and take the integral over a higher dimension over the full derivative of the form, say in the last case one may integrate $Q = \partial_\mu K_\mu$ over the $4d$ volume instead. All of those topological objects were identified and studied in significant details on the lattice, but we will not go into details.

The method we have called naive is the direct replication of the continuum expression for the topological charge (3.95), namely

$$Q = \frac{1}{32\pi^2} \int d^4x\, G^a_{\mu\nu} \tilde{G}^a_{\mu\nu}, \qquad (5.31)$$

in which the field strength is calculated from a plaquette as in (1.12) (but without the trace) and then at each site the sum is taken over all mutually orthogonal 2-planes in $4d$, with finally a color trace and a sum over all sites. Although for large-size instantons, with a size much larger than the lattice spacing $\rho \gg a$, it should work, in practice this definition is rather inconvenient. Part of the problem is that the result is not even an integer, but also lattice artifacts are notoriously large.

The situation has improved somewhat owing to the invention of various "geometric" definitions, starting from early important works in the $2d$ sigma model [315] and gauge theory [316]. The general idea is to relate a given lattice configuration with some continuous mapping, from space to the gauge group, so that the result is an integer. Generically, all SU(N) groups can be viewed as a combination of spheres, and whatever the specific lattice may be, one can always think of "triangulation" of space (segments in $1d$, triangles in $2d$, tethrahedrons in $3d$ etc.). So the problem can always be an interpolation of mapping when only discrete points are given. We have discussed the ambiguity in $1d$ mapping of angle into angle in the preceding subsection. A $2d$ triangle mapped on a $2d$ sphere with three corners fixed can still be interpolated in two ways: and the smoother one is achieved if that of smaller area is chosen. Other geometric algorithms work also like this. And still, no method is really free from ambiguities related with the so called "defects", the topological objects with size $\rho \sim a$.

Jumping to practical applications, we first note that modern lattice simulations have lattice size in the range $a = 0.1$–0.05 fm which is sufficiently fine for covering typical instantons with $\rho \sim 1/3$ fm. It is very important that the true instanton density at $\rho \sim a$ should be very small and unimportant.[7]

An example of cooled configurations which reveal the instantons are shown in Fig. 5.8.

Moreover, one can test now not only the magnitude of the action and the topological charge of such objects, but even local details. As an illustration, let me ask the following question: *are the topological clusters locally selfdual?* The results for one configuration from Ref. [290] are shown in Fig. 5.9. But it can also be studied

---

[7]This can be checked by comparing observables calculated with a different lattice action which treats them very differently. Some (e.g. the so called Iwasaki action) basically kill of topology at small scale, and some may encourage more defects.

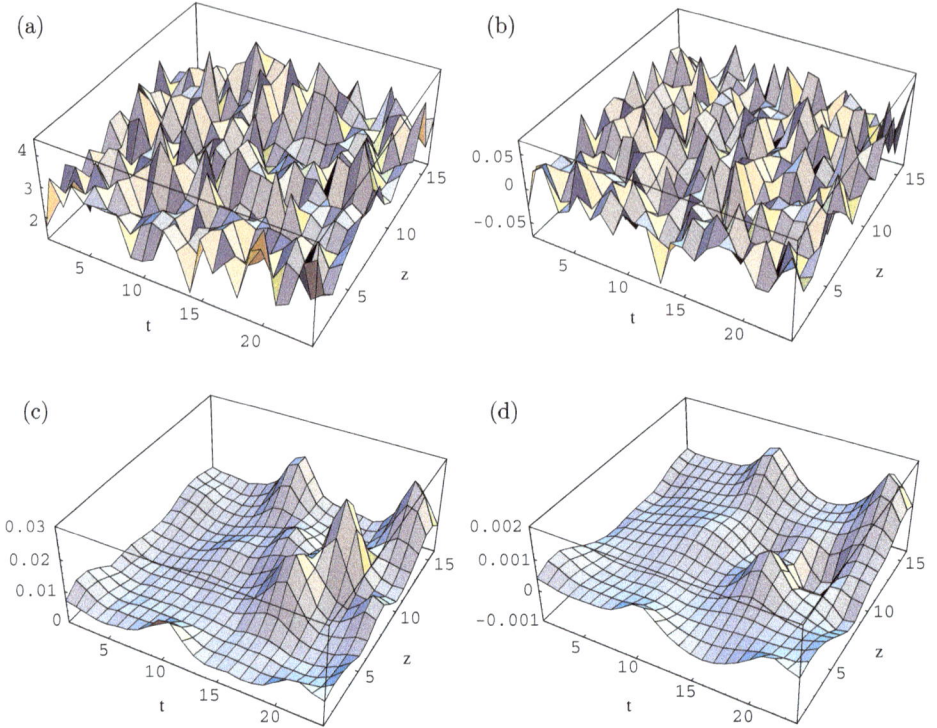

Fig. 5.8. Examples of cooling of lattice configurations. (a) and (b) are "cooled" into (c) and (d). The left (a) and (c) show the action, while the right (b) and (d) the topological charge.

statistically. If one defines the ratio

$$r(x) = \frac{\text{Tr}\, F_{\mu\nu}(x)' F_{\mu\nu}(x)' - \text{Tr}\, F_{\mu\nu}(x)' \widetilde{F_{\mu\nu}(x)}'}{\text{Tr}\, F_{\mu\nu}(x)' F_{\mu\nu}(x)' + \text{Tr}\, F_{\mu\nu}(x)' \widetilde{F_{\mu\nu}(x)}'}. \tag{5.32}$$

For a space–time point x where the gauge field is self-dual the numerator will vanish while the denominator is finite and $r(x)$ equals 0. Conversely for an x where the gauge field is anti self-dual the role of numerator and denominator are interchanged and $r(x) = \infty$. For space–time points without definite duality properties, $r(x)$ assumes some finite value between 0 and ∞. The transformation $R(x) = 4/\pi \arctan r(x) - 1$ maps the interval $[0, \infty)$ into the interval $[-1, 1]$. For configurations which are dominated by (anti) self-dual lumps one expects values near ± 1. As for the local chirality variable of Ref. [319], it is interesting to study different selections for the lattice points x in $R(x)$. In the figure we use all lattice points (top curve), the subset of 50% of the lattice points supporting the highest peaks of $|\text{Tr}\, F_{\mu\nu}(x)' \widetilde{F_{\mu\nu}(x)}'|$ (middle curve) and also a cut of 10% (bottom curve). The histograms were computed by averaging over 50 configurations.

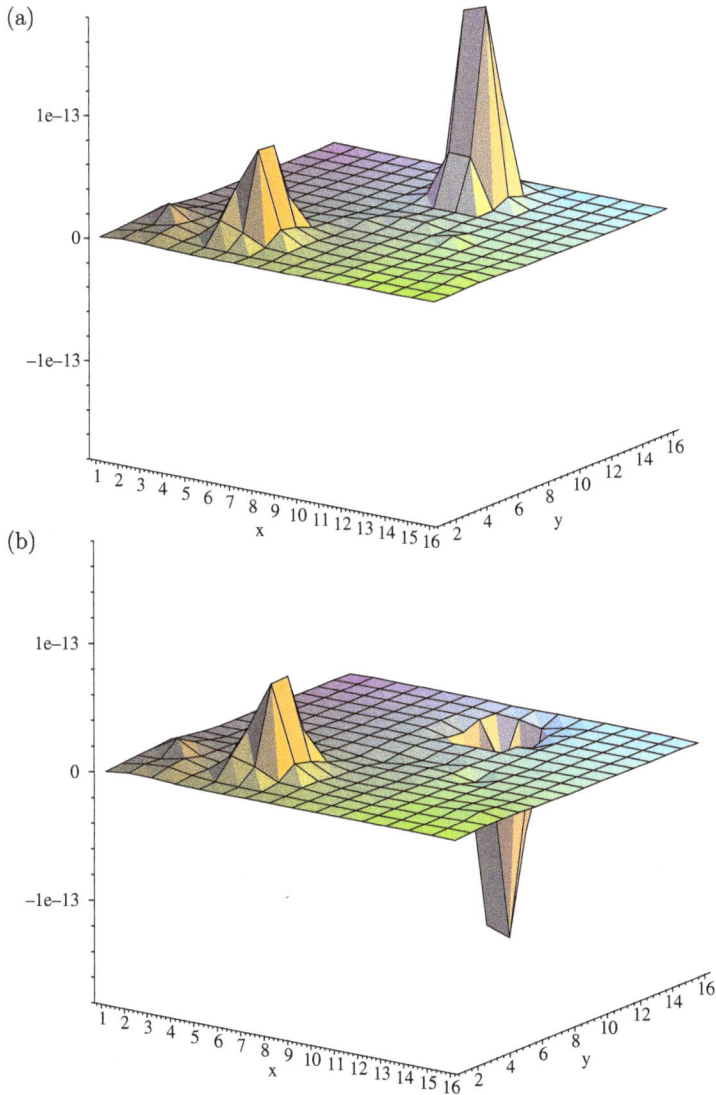

Fig. 5.9. Local $\mathrm{Tr}\, F_{\mu\nu}(x)' F_{\mu\nu}(x)'$ (a) and $\mathrm{Tr}\, F_{\mu\nu}(x)' \, \widetilde{F_{\mu\nu}}(x)'$ (b). The primes indicate that one takes into account only the contributions of the lowest 6 near zero-modes. From Ref. [290].

5.4.3. *Are the lowest Dirac eigenstates locally chiral?*

As we have argued many times in this book, for the QCD applications the topology of the gauge field is so important because it affects the light quarks in a dramatic way. Therefore it is crucial to see that this connection is indeed working properly on the lattice.

These studies start with the so called *fermionic* definition of the topological charge, based on the index theorem which tells us that the configuration with the

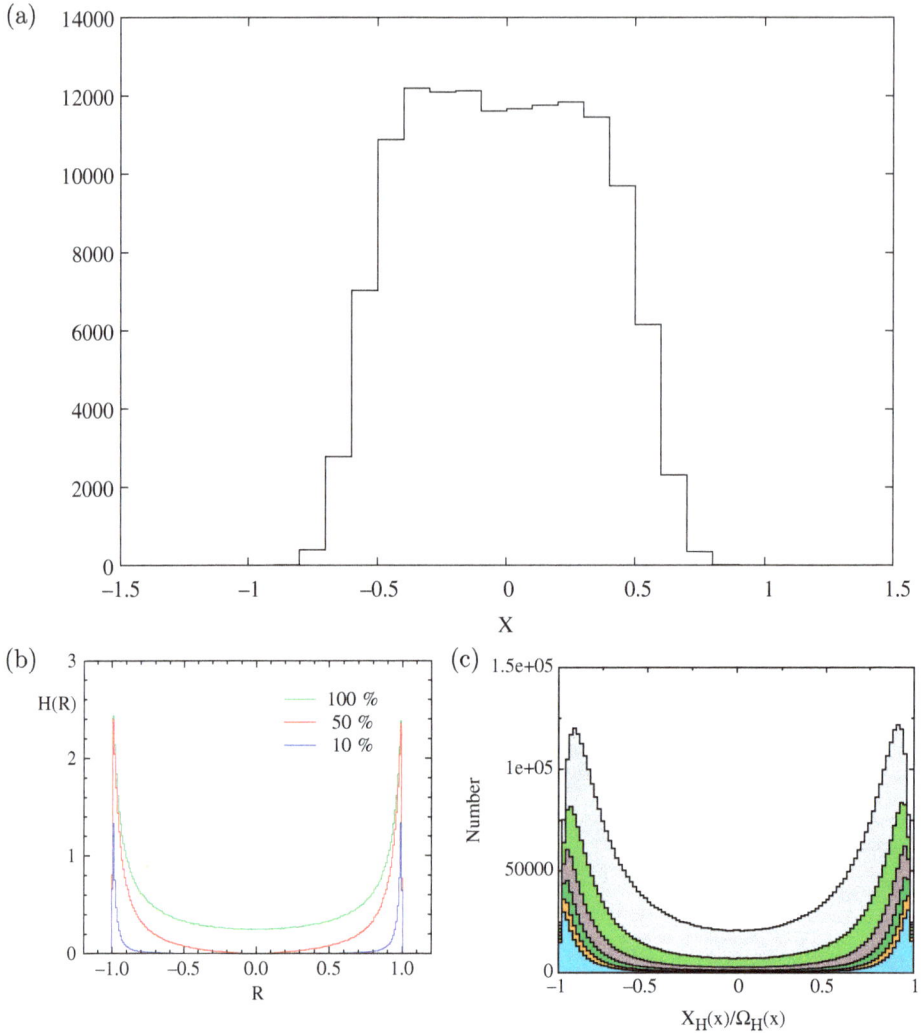

Fig. 5.10. (a) Local chirality distribution, from Ref. [320]. (b) Same from Ref. [290]. We display our data without a cut (top curve), a cut of 50% (middle curve) and a cut of 10% on the number of lattice points supporting the highest peaks of the gauge topological charge $|\text{Tr}\, F_{\mu\nu}(x)' \widetilde{F_{\mu\nu}}(x)'|$. (c) Same from Blum *et al.* [321].

topological charge Q should be accompanied by at least Q *exactly zero modes* of the Dirac operator. With good enough fermions one indeed can identify very small or zero modes for each configuration. From their average number one can get the value of the topological susceptibility, and for quenched simulations it is close to the "phenomenological"[8] one. It worked in fact even with Wilson or Kogut–Susskind

---

[8] We recall that we put it in parentheses since although this value is determined from a real η' mass, it refers not to the real but to quenched or large-N_c ensemble without the fermionic determinant.

fermions. We would not provide details of that, and proceed directly to more recent studies for which the "old" fermions failed.

We now come to the question put in the title of this section. The exact zero modes of course have to be chiral. However even all quasi-zero modes — according to the instanton liquid model we described in Chapter 4 — should be made of zero modes of many instantons and anti-instantons. Since the ensemble is dilute, one expects that (i) all low-lying modes go in approximately spherical lumps of size $\rho_0 \sim 0.3$ fm; and (ii) are "locally chiral".

Is it so on the lattice? Horvath *et al.* [320] have studied the local chirality distribution for near-real eigenvalues and found them to be peaked far from ± 1, see Fig. 5.10a. They used Wilson fermions and $12^3 \times 24$, and the gauge coupling is $\beta = 5.7$ (corresponding to a lattice spacing of $a^{-1} \simeq 1.18$ GeV). If that would be true, it would be the end of the instanton liquid model. However, such fermions failed to see complete chirality even for exact zero modes, which was a warning sign.

Soon after that result was announced, several other papers [290, 321, 322] very quickly showed these results to be wrong. Better chiral fermions did made a difference, with them one finds a completely different picture! For example, the domain wall fermions and larger lattices used by the BNL/Columbia group [321] see perfect 100% chirality for exact zero modes and very nice double-peak structure for near-zero states, see Fig. 5.10b. Needless to say, this is in perfect agreement with the instanton liquid model.

As one more test of the ideas inspired by the ILM, one may ask how important the lowest Dirac eigenstates are for hadronic observables.

In order to test further instanton-motivated ideas, lattice practitioners have looked whether the lowest Dirac eigenstates do or do not make good approximation to quark propagators and hadronic correlation functions at large distances. The procedure has been named *truncated eigenmode approach* (TEA).

Let me show an example of the results from Ref. [323] from the work by Neff *et al.* [323] for the pion and the two-loop η'-correlators. As shown in Fig. 5.11, the

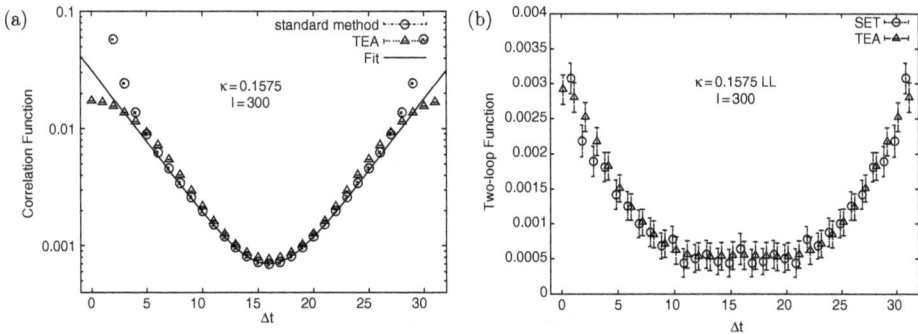

Fig. 5.11. (a) Comparison of the π correlation function as provided by TEA with $l = 300$, with the one obtained from the standard method (solving linear systems) on local sinks and sources. (b) Same but for two-loop pseudoscalar (η' or SU(3) singlet) correlator.

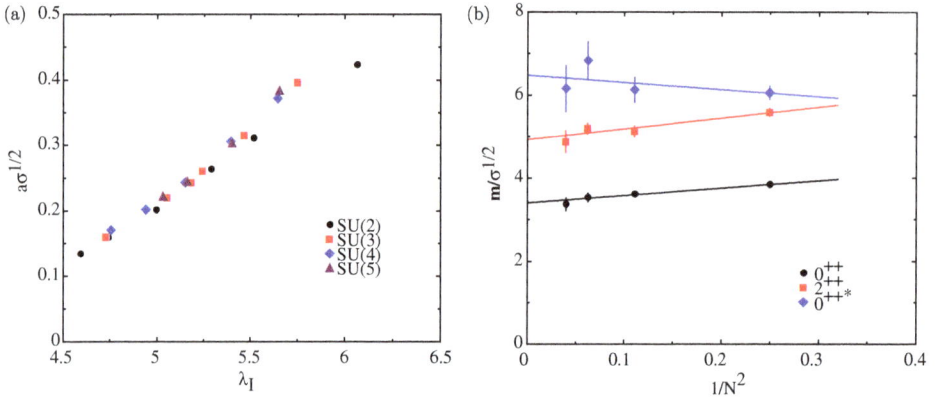

Fig. 5.12. (a) The universality of the string tension in terms of the 't Hooft coupling. (b) The dependence of the lowest glueball masses.

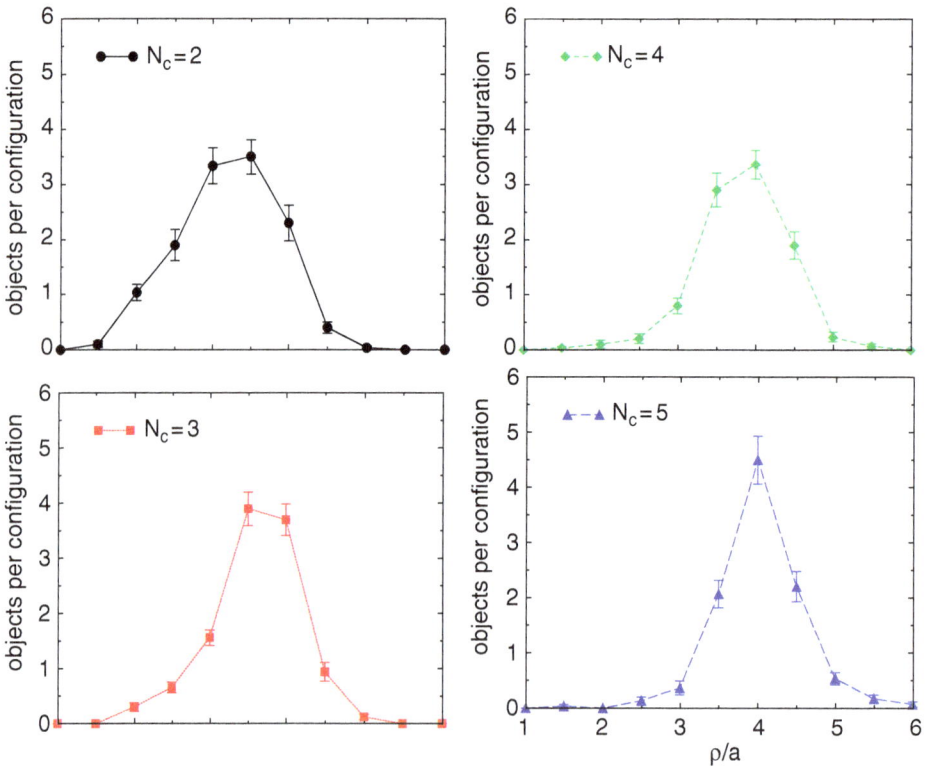

Fig. 5.13. The size distribution of instanton-like objects as seen in the topological charge density after 10 cooling sweeps. The radius is calculated from the shape around the peak value.

approximation (triangles) works as well as standard much more costly calculations (open circles).

5.4.4. *Testing the large N_c limit on the lattice*

This section is mostly based on works by M. Teper *et al.* [72] which simulated pure gauge lattice theory with the number of colors $N_c = 2$–6. In Fig. 5.12a one can see that the string tension seems to depend only on a particular combination of the bare lattice coupling g and the number of colors, called the 't Hooft coupling $\lambda = g^2 N_c$.

Masses of the lowest glueballs are shown in Fig. 5.12b, they seem to show a simple linear behavior as a function of $1/N_c^2$, with a smooth and finite limit. Moreover, even corrections for $N_c = 2$, 3 are not really large as compared to the large N_c limit, and so one may think that in the gluonic sector the limit is nearly reached at any N_c.

Topology has also been studied for those N_c, both with (slightly cooled) gluonic and fermionic definitions. Local "clusters" of topology are seen at all N_c. One way of studying their local structures is to examine correlation functions, e.g. the auto-correlation function of the pseudoscalar (chiral) density $w(x) = \psi^\dagger(x)\gamma_5\psi(x)$, or of the topological charge $Q(x)$.

Extracting the instanton size distribution is more tricky, as it includes fitting of individual topological or chiral fermion clusters to a certain size. The lattice results are very close to what instanton liquid predicts, see e.g. Section 4.5. In particular, there is a clear shift in the typical sizes of the bulk of instanton-like objects from about 0.36 fm to 0.48 fm as we increase N_c from 2 to 5 and a suppression of small instantons at large N_c.

One more interesting feature observed in these simulations is the apparent disappearance of the local chirality of the quasi-zero modes. It is however unclear whether it really signals a change in the chiral symmetry breaking mechanism, or a simple overlap of instantons with antiinstantons of many different SU(2) subgroups.

CHAPTER 6

QCD Correlation Functions

6.1.   Generalities

Correlation functions are the main tools used in studies of structure of the QCD vacuum. They can be obtained in several ways. First, they can in many cases be deduced phenomenologically, using the vast set of data accumulated in hadronic physics. Second, they can be directly calculated *ab initio* using quantum field theory methods, such as lattice gauge theory, or semiclassical methods. A significant amount of work has also been done in order to understand their small-distance behavior, based on the Operator Product Expansion (OPE). The large distance limit can also be understood using effective hadronic approaches or various quark models of hadronic structure. In this section we focus on the available *phenomenological information* about the correlation functions, emphasizing the most important observations, which are then compared with *predictions of various theoretical approaches*; lattice numerical simulations, the operator product expansion and the interacting instantons approximation. As a "common denominator" for our discussion we have chosen *the point-to-point correlation functions in coordinate representation*.

6.1.1.   *Why the correlation functions?*

We will discuss two types of operators, mesonic ones of the type

$$O_{\mathrm{mes}}(x) = \bar{\psi}_i \psi_j \delta_{ij}, \tag{6.1}$$

and baryonic ones

$$O_{\mathrm{bar}}(x) = \psi_i \psi_j \psi_k \epsilon^{ijk}, \tag{6.2}$$

where the color indices are explicitly shown (but they will be suppressed below). Other indices like spin and flavor ones are not yet specified, this we will do later. As all color indices are properly contracted and all quark fields are taken at the same point x, these operators are manifestly gauge invariant.

The correlation function is the vacuum expectation value (VEV) of the product of two (or more) of them taken at different points, such as

$$K(x - y) = \langle 0|O(x)O(y)|0\rangle. \tag{6.3}$$

Since the vacuum is homogeneous, the correlator depends on the relative distance only. This one of the points can always be at the origin of the coordinate system, and we will below put $y = 0$.

Throughout this chapter the distance between the points x–y is assumed to be *space-like*. The reason is that we prefer to deal with *virtual* (instead of real) propagation of quarks or hadrons from one point to another, so one deals with simple decaying (instead of complicated oscillating) correlation functions (see below).

At this point I was inevitably asked the old question: how can any correlation between the fields be *outside of the light cone*? It was essentially answered by Feynman long ago, who had to defend his propagator and $i\epsilon$ prescription following from Euclidean definitions. In the path integral the particles can propagate along *any* path going from x to y. Depending on the reference frame, an observer may possibly view the path to be a sequence of spontaneous pair creation and annihilation events. These correlations in vacuum *do not contradict causality* because one cannot use it for the signal transfer.

One can look at a pair of points separated by a Euclidean distance in terms of two distinct limits. Either they can be two different points in space, taken at the same moment in time, or be two events at the same spatial point separated by a non-zero interval of the *imaginary (Euclidean) time*; $ix_0 - iy_0 = \tau$. Below we will use both, depending on which is more convenient at the moment. Due to Lorentz invariance (rotational O(4) in Euclid) the answers are of course the same.

At small distances x (remember, $y = 0$) the "asymptotic freedom" of QCD tells us that (up to small and calculable radiative corrections) quarks and gluons propagate freely. Therefore[1] $K(x) \approx K_{\text{free}}(x)$, where the free quark correlator in the mesonic (baryonic) case is essentially the square (cube) of the free quark propagator, $S(x) = \langle \bar{q}(x)q(0)\rangle = \gamma_\mu x_\mu)/2\pi^2 x^4$, where I have ignored the small quark mass. From the dimension of the free correlators they can only be for the mesonic (baryonic) channels $K_{\text{free}}(x) \sim x^{-6}$ ($\sim x^{-9}$). Of course, QCD does have a dimensional parameter Λ_{QCD}, which shows up in physical (non-free) correlators $K(x)$ and causes deviations from $K_{\text{free}}(x)$. However, in perturbation theory it only comes in via the radiative corrections. Therefore, at small $x\Lambda \ll 1$, those produce corrections to our estimates above containing powers of $\alpha_s(x) \sim 1/\log(1/x\Lambda)$. If quarks are allowed to propagate to larger distances, they start interacting with the non-perturbative vacuum fields. If corrections are not too large, one can take these effects into account using the operator product expansion (OPE) formalism. At intermediate distances the description of the correlation functions becomes in general very complicated,

---

[1]Strictly speaking, if small x includes $x = 0$, the correlators may have local terms of the kind of $\delta(x)$ and its derivatives, which we will also ignore unless the integral over x is carried out.

and one may only evaluate them either using lattice numerical simulations or some "vacuum models" (e.g. the instanton liquid).

At *large distances* one can again understand the behavior of the correlation functions in simple terms using now a completely different kind of arguments. Instead of thinking in terms of fundamental fields, one may just use the formal relation for the time evolution of an operator $O(t) = e^{iHt}O(0)e^{-iHt}$ where H is a Hamiltonian, and then insert a complete set of *physical intermediate states* between the two operators. In Minkowski time this means

$$K(t) = \sum_n |\langle 0|O(0)|n\rangle|^2 e^{-itE_n}. \tag{6.4}$$

Now one can analytically continue the correlation function into the Euclidean domain $\tau = it$, and get a sum over decreasing exponents.

Physically, the application of such a relation in QCD means that one considers propagation of physical excitations, or *hadrons* between two Euclidean points, so $K(x) \sim \exp(-mx)$ at large x, where m is the mass of the lightest particle with the corresponding quantum numbers. Note, that this is essentially the idea of Yukawa, who had related the range of the nuclear forces to the (then hypothetical) meson mass.

Concluding this discussion of limits we emphasize that *the same* function can be considered on many different levels: in terms of the fundamental QCD fields, quark and gluons; or in terms of the physical hadrons.

After we have recollected these general facts, let us try to explain *why the correlation functions are so important in non-perturbative QCD and hadronic physics.* The answer is: it is the most effective way to study the *inter-quark effective interaction*. Applications of hadronic models (we discussed in Chapter 2) to hadronic spectroscopy strongly resemble the state of the nuclear physics in its early days, when only limited information about the nuclear forces was known from the properties of the simplest bound states (e.g. the deuteron). In order to reproduce those merely qualitative features of the interaction (some attractive potential well with appropriately tuned parameters) was enough. Only later extensive studies of the NN scattering phases in different channels and energies had eventually shown all the details of nuclear forces, with their complicated spin–isospin structure.

Quite similarly, applications of quark models are mostly averaged over few lowest states, and the precise dependence of the inter-quark interaction on distance and momenta remains unknown. A confining potential with few additions (like spin forces) fits the spectrum.

Since, due to confinement, the qq or the $\bar{q}q$ scattering is experimentally impossible, a set of correlation functions $K(x)$ per channel are substitutes for the phase shifts in nuclear physics. Roughly speaking, the correlator tells us something about *virtual $\bar{q}q$ or qq scattering*, using instead of physical hadrons the *wave packets of a variable size*.

6.1.2.   *Different representations of the correlation functions*

The correlation functions in Euclidean space–time $K(\vec{x})$ (or $K(\tau)$) defined above are the main object of our discussion in what follows. As their argument x is the distance between the two points in (Euclidean) space–time, we call them *point-to-point correlation functions*.

We need a special name because in various applications people have used other representations of correlators, related to the above mentioned ones by some integral transformations. In this section we compare their definitions and briefly comment on their advantages and disadvantages.[2]

If one makes a Fourier transform of $K(x)$, the resulting function $K_{\text{mom}}(q^2)$ depends on the momentum transfer q flowing from one operator to another. For clarity we use the following notation, introducing momentum squared with negative sign $Q^2 = -q^2$, so for the *virtual* space-like momenta $q^2 < 0$ we are interested in (like in scattering experiments) $Q^2 > 0$.

Due to causality, it satisfies standard dispersion relations

$$K_{\text{mom}}(q^2) = \int \frac{ds}{\pi} \frac{\text{Im}\, K_{\text{mom}}(s)}{(s - q^2)}, \tag{6.5}$$

where the r.h.s. contains the so called *physical spectral density* $\text{Im}\, K_{\text{mom}}(s)$. It describes the squared matrix elements of the operator in question between the vacuum and all hadronic states with the invariant mass $s^{1/2}$, and certainly is non-zero only for *positive* s. Note, that because we will only consider *negative* q^2, or the semi-plane without singularities, we never come across a vanishing denominator and therefore ignore $i\epsilon$ which is usually put in the denominator. This simplification is possible because our discussion is restricted to *virtual* processes (although in the r.h.s. we do use information coming from the *real* experiments of annihilation type).[3]

The dispersion relation is the basis of the so called *sum rules*. The general idea is as follows: suppose one knows the l.h.s. $K_{\text{mom}}(q^2)$ in some region of the argument; it means that some integral in the r.h.s. of the physical spectral density is known. It can be used to relate a set of physical parameters of different excitations, etc.

Unfortunately, the so called *finite energy sum rules* using directly momentum space are not very productive; most of the dispersion integrals are divergent, leading to usable sum rules only after some subtractions. This introduces extra parameters and significantly undermines their prediction power.

For example, we have mentioned above that at small x mesonic correlators are just given by $K_{\text{free}} \sim 1/x^6$, the free quark propagator squared. It is not difficult to perform the Fourier transform and see (by dimension) that $K_{\text{mom}}(q^2) \sim q^2 \log q^2$

---

[2] As we actually do not use any of them in what follows, some readers may well skip this section.

[3] In principle, virtual processes contain complete information, but of course in practice it is much more difficult to go in the opposite direction, and reproduce the physical spectral density from the correlators considered.

and the imaginary part of the log gives for the spectral density in the r.h.s. $\operatorname{Im} K_{\text{mom}}(s) \sim s$. Therefore, putting it into the dispersion relation given above one finds an ultraviolet divergent integral, which signals that in the last argument something is missing. In this particular example it is clear what is it; the constant under the log is lost.

A simple way to go around this is to consider the second derivative over Q^2 of both sides of (6.5); then one finds a convergent dispersion relation. However, while going back to the original function $K_{\text{mom}}(s)$ from its derivative, one has to fix the integration constant.[4] However, in applications of the finite energy sum rules these undefined constants should also be determined from the data.

Several ideas were suggested to improve these sum rules. First, after taking a sufficient number of derivatives, one may take $Q = 0$ and arrive at the so called *moments*

$$M_n = \int \frac{ds}{\pi} \frac{\operatorname{Im} K_{\text{mom}}(s)}{s^{n+1}}. \tag{6.6}$$

Following the original paper [352], this method is traditionally used in the discussion of "charmonium sum rules" related to correlators of $\bar{c}c$ currents.

Another idea also suggested in Ref. [352] is to introduce the so called *Borel transform* of the function $K_{\text{mom}}(Q)$ defined as follows

$$K_{\text{bor}}(m) = \lim_{n,Q^2 \to \infty, m^2 = Q^2/n = \text{fixed}} \frac{Q^{2n}}{(n-1)!} \left(-\frac{d}{dQ^2}\right)^n K_{\text{mom}}(Q^2). \tag{6.7}$$

Applying it to the dispersion relation one gets the sum rules in the Borel-transformed representation,

$$K_{\text{bor}}(m) = \int \frac{ds}{\pi} \operatorname{Im} K_{\text{mom}}(s) \exp\left(-\frac{s}{m^2}\right). \tag{6.8}$$

Now the integral is cut off at large s by the exponent. Since usually we know the contributions of the lowest states better, the exponential cut off "hides" our ignorance and is therefore welcome. Such a form for the sum rules has been used in multiple papers based on the OPE, see e.g. references in Refs. [355–357, 360].

However, one may avoid the Borel transformation, as well as a Fourier transform, by using the sum rules directly in the coordinate space. By applying Fourier transformation to (6.5) one obtains the following nearly self-explanatory form [324]:

$$K(x) = \int \frac{ds}{\pi} \operatorname{Im} K_{\text{mom}}(s) D(s^{1/2}, x), \tag{6.9}$$

the former function describes the amplitude of production of all intermediate states of mass $\sqrt{s}$, while the latter function,

$$D(m, x) = (m/4\pi^2 x) K_1(mx), \tag{6.10}$$

---

[4]Note that polynomials in s generate local terms in $K(x)$ we ignore, assuming that x is never zero.

is nothing else but the Euclidean propagation amplitude of these states up to distance x. The difference between this expression and Borel sum rules is not really very significant; at large x the propagator goes as $\exp(-mx)$, therefore one also has an exponential cutoff, only $\exp(-\sqrt{s}x)$ replaces $\exp(-s/m^2)$. The space–time one can be calculated numerically on the lattice or in instanton liquid, and also analytic formulae are simpler to derive.

For completeness, let me also mention one more type of correlation function, the one traditionally used in LGT. This is the so called *plane-to-plane* correlation function, obtained from $K(x)$ by an integration over the 3-dimensional plane:

$$K_{\text{plane-to-plane}}(\tau) = \left\langle \int d^3x \, O(x, \tau) \, O(0,0) \right\rangle. \tag{6.11}$$

In other terms, spatial integration causes the momentum of intermediate states to be zero, so the dispersion relations are done in energy only. This function, in turn, can be related to the physical spectral density by

$$K_{\text{plane-to-plane}}(\tau) = \int \frac{dm}{\pi} \, \text{Im} \, K_{\text{mom}}(m) \exp(-\tau m). \tag{6.12}$$

If this function is known, the mass of a lowest hadron can be obtained directly from its logarithmic derivative at large τ. However, its essential disadvantage is that it *mixes contributions of small and large distances*, and therefore it is really not useful for OPE.

(i) The original point-to-point function $K(x)$ in coordinate space.
(ii) The Fourier transform $K_{\text{mom}}(q^2)$.
(iii) The moments of the spectral density M_n.
(iv) The Borel-transform function $K_{\text{mom}}(m)$.
(v) The plane-to-plane correlation function $K_{\text{plane-to-plane}}(\tau)$.

Although for various reasons different people use all of them, we still suggest that for the understanding of the underlying physics it is still better to use the original point-to-point function $K(x)$, and we will do so in what follows.

6.1.3.   *Quantum numbers and inequalities*

The *flavored* and *unflavored* currents have different types of quark diagrams; as shown in Fig. 6.1 the former ones have only the one-loop contributions (a), while the latter have the two-loop diagrams (b) as well. For example, we will discuss the flavored ($I = 1$) channels like π^+, ρ^+ etc. which have operators of the type $\bar{u}\Gamma d$ (where Γ is the appropriate Dirac matrix). Since two quark lines are in this case of different flavor, $\bar{u}$ and d respectively, obviously one cannot have a double-loop. On the other hand, considering similar unflavored $I = 0$ channels like $\eta, \omega \ldots$ one has the $\bar{u}u, \bar{d}d, \bar{s}s$ terms which can be "looped". Thus, if one would be interested in

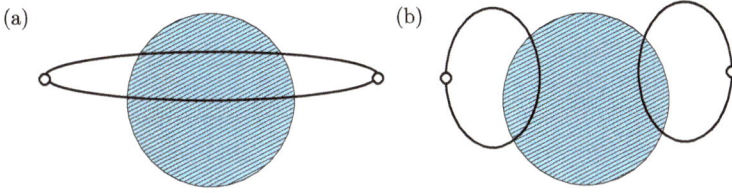

Fig. 6.1. Two diagrams for mesonic correlators in a some gluonic background (indicated by the dashed circle). The flavored currents have only the single loop diagram (a) while the unflavored currents have also the two-loop contributions of the type (b).

a *difference* between say ρ and ω channels, that would be entirely due to the two-loop diagram (b). Furthermore, one can be interested in the so called *non-diagonal* correlators, for example with different flavors such as $\langle \bar{u}(x)\Gamma u(x)\bar{d}(0)\Gamma d(0)\rangle$. Again, one is restricted to the two-loop diagrams only.

The main focus of lattice work deals with the one-loop diagrams and therefore with the flavored channels; the reason is technical into which we would not go.

What I would like to point out here is that for one-loop diagrams there exist some important general inequalities. In order to derive those following Weingarten [328], one uses the following relation for the propagator in the backward direction:

$$S(x, y) = -\gamma_5 S^+(y, x)\gamma_5. \tag{6.13}$$

Second, one can decompose it into a sum over all 16 Dirac matrices $S = \Sigma a_i \Gamma_i$ where $\Gamma_i = 1, \gamma_5, \gamma_\mu, i\gamma_5\gamma_\mu, i\gamma_\mu\gamma_\nu$ where the last term is anti-symmetric only and $\mu \neq \nu$. The third step: write all one-loop correlators of the type $\Pi = \text{Tr}(S(x, y)\Gamma_i S(y, x)\Gamma_i)$, perform the traces, and write them explicitly in terms of the coefficients.

Exercise 6.1. Prove the Weingarten expression for the inverse propagator.

Exercise 6.2. Derive the expressions for all the diagonal correlators with $\Gamma_i = 1, \gamma_5, \gamma_\mu, i\gamma_5\gamma_\mu, i\gamma_\mu\gamma_\nu$ in terms of the propagator decomposition of the same type. Use the Weingarten expression for the inverse propagator.

The resulting expression for the $I = 1$ pseudoscalar (pion) correlator contains a simple sum of all coefficients squared;

$$\Pi_{\text{PS}} \sim |a_1|^2 + |a_5|^2 + |a_\mu|^2 + |a_{\mu 5}|^2 + |a_{\mu\nu}|^2, \tag{6.14}$$

while others have some negative signs, e.g. the scalar one is instead

$$\Pi_{\text{S}} \sim -|a_1|^2 - |a_5|^2 + |a_\mu|^2 + |a_{\mu 5}|^2 - |a_{\mu\nu}|^2. \tag{6.15}$$

As a result, the *Weingarten inequality* follows:

$$\Pi_{\text{PS}}(x) > \Pi_{\text{S}}(x), \tag{6.16}$$

and the pseudoscalar correlator should exceed the scalar one at all distances. The non-trivial consequence is that the masses of the lowest states should also then satisfy the inequality, $m_{PS} < m_S$. This of course is satisfied in the real world, as the physical pion is indeed lighter than any scalars. Note however that we did not say a word about chiral symmetry breaking and the Goldstone theorem here; the result is more general. Note also that at large x the scalar correlator must be much much smaller than the pseudoscalar one, since the different lowest masses are in the exponent. It means that there must be a very delicate cancellation between different components of the quark propagator in all channels except the pseudoscalar one (in which all terms appear as squares with positive coefficients).

More information is provided by similar relations for the vector (ρ) and axial (A_1) channels,

$$\Pi_V \sim 2|a_1|^2 - 2|a_5|^2 + |a_\mu|^2 - |a_{\mu 5}|^2, \tag{6.17}$$
$$\Pi_A \sim -2|a_1|^2 + 2|a_5|^2 + |a_\mu|^2 - |a_{\mu 5}|^2, \tag{6.18}$$

and the Verbaarschot inequalities follow:

$$\Pi_{PS}/\Pi_{PS}^{\text{free}} > \left(\frac{1}{2}\right)\left(\Pi_V/\Pi_V^{\text{free}} + \Pi_A/\Pi_A^{\text{free}}\right), \tag{6.19}$$

$$\Pi_{PS}/\Pi_{PS}^{\text{free}} > \left(\frac{1}{4}\right)\left(\Pi_V/\Pi_V^{\text{free}} - \Pi_A/\Pi_A^{\text{free}}\right). \tag{6.20}$$

More information about other general inequalities can be found in the review by Nussinov [329].

Note that these inequalities are identities, to be satisfied for *any configuration* of the gauge field, and therefore theoretically they are not very restrictive. Other features of the correlators are however less obvious and for example do not hold for quenched QCD. For example, spectral decomposition demands that the (diagonal) correlators themselves are (i) positive, and (ii) even monotonically decreasing functions. Some field configurations do produce *negative* contributions to correlators; so their weight in the ensemble of vacuum fields should not be too large.

6.1.4. *Correlators with chirality flips*

In many cases it is instructive to use a specific combination of correlation functions, which focus on the phenomena we would like to study. In particular, one would like to understand how breaking of the $SU(N_f)$ and $U(1)_A$ manifest themselves in the correlation functions. In the next section we will perform a systematic study of several phenomenologically important channels; now we would point out their combinations which a theorist would like to pinpoint.

Let me give two examples of such choices. The first example concerns the linear combinations $\Pi_{V-A} = \Pi_V - \Pi_A$ of the $I = 1$ vector and axial correlation functions. All effects of the interaction in which quark chirality is preserved throughout the loop cancels out, because two γ_5 are produced $(\pm)^2 = 1$. Furthermore, Π_{V-A} is non-zero only due to the effects of chiral symmetry breaking. This can be most clearly expressed if one writes the two currents in terms of left and right handed currents, as $\bar{q}_L \gamma_\mu q_L \pm \bar{q}_R \gamma_\mu q_R$; then this correlator is manifestly non-diagonal in chirality

$$\Pi_{\mu\nu}^{V-A} = 4 \langle \bar{q}_L \gamma_\mu(x) q_L \bar{q}_R \gamma_\mu q_R(y) \rangle. \tag{6.21}$$

It is obviously zero to any order of the perturbation theory. The experimental data for this V–A combination is available from τ decays; we will discuss those in the next section.

The second example is a similar difference, but between the $I = 1$ scalar (called a_0 or δ) and the $I = 1$ pseudoscalar (the pion π).

$$R^{NS}(\tau) = \frac{A_{flip}^{NS}(\tau)}{A_{non\text{-}flip}^{NS}(\tau)} = \frac{\Pi_\pi(\tau) - \Pi_\delta(\tau)}{\Pi_\pi(\tau) + \Pi_\delta(\tau)}, \tag{6.22}$$

where $\Pi_\pi(\tau)$ and $\Pi_\delta(\tau)$ are pseudo-scalar and scalar NS two-point correlators related to the currents $J_\pi(\tau); = \bar{u}(\tau) i \gamma_5 d(\tau)$ and $J_\delta(\tau); = \bar{u}(\tau) d(\tau)$. If the propagation is chosen along the (Euclidean) time direction, $A_{flip(non\text{-}flip)}^{NS}(\tau)$ represents the probability amplitude for a $|q\bar{q}\rangle$ pair with iso-spin 1 to be found after a time interval τ in a state in which the chirality of the quark and anti-quark is interchanged (not interchanged) Notice that the ratio $R^{NS}(\tau)$ must vanish as $\tau \to 0$ (no chirality flips), and must approach 1 as $\tau \to \infty$ (infinitely many chirality flips). Again, this amplitude receives no leading perturbative contributions. The difference with Example 1 is that this ratio is proportional to the chirality structure $\bar{R}L\bar{R}L + (R \leftrightarrow L)$, same as for the 't Hooft vertex, while in Example 1 it was $\bar{L}L\bar{R}R$.

The fact we would like to emphasize is that this second combination is growing surprisingly rapidly already at small distances, ~ 0.3–0.6 fm. At large distance it is obviously 1 since the pion is about massless while a_0 is very heavy, it has a mass of about 1.4 GeV. The reason (to be discussed below in detail) is that this correlator admits a direct instanton-induced 't Hooft vertex, while the V–A does not. This combination has received some special attention, see references in the paper by Faccioli and DeGrand [330] we follow in this subsection.

In Fig. 6.2a one finds a calculation of this ratio in two instanton liquid ensembles, the random one RILM which has no fermionic determinant, and the interacting IILM, in which it is included. Note that in RILM the chirality flip ratio rises rapidly and exceeds 1, while the unquenched IILM follows the unitarity requirement, $R^{NS}(\tau) \leq 1$. The result of the quenched lattice calculation shown in Fig. 6.2b follows very accurately the same "overshooting" of 1 as the RILM results. (The unquenched lattice ones are not yet available.) On the other hand, the naive constituent quark models, which only include chirality flips due to the constituent quark mass, show a very slow rise illustrated by the lowest curve in Fig. 6.2b.

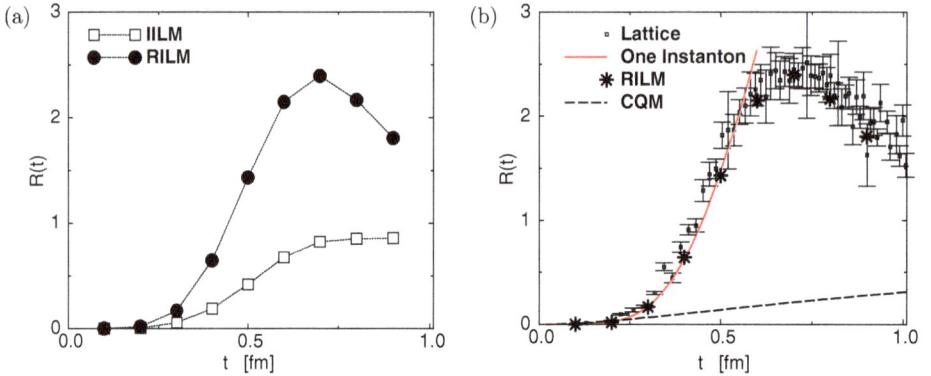

Fig. 6.2. The chirality-flip ratio, $R^{NS}(\tau)$, in lattice and in two phenomenological models. (a) Circles are RILM (quenched) results, squares are IILM (unquenched) results. (b) Squares are lattice points of previous DeGrand simulation. Stars are RILM points obtained numerically from an ensemble of 100 instantons of $1/3$ fm size in a 5.3×2.65^3 fm^4 box. The solid line is the contribution of a single-instanton, calculated analytically. The dashed curve was obtained from two free "constituent" quarks with a mass of 400 MeV. Such a curve qualitatively resembles the prediction of a model in which chiral symmetry is broken through a vector coupling (as in the present DSE approaches).

6.2. Phenomenology of mesonic correlation functions

6.2.1. *Vector and axial correlators*

We start the discussion of the correlation functions with the vector and axial currents for obvious reasons; in contrast to the many other operators to be used below, these currents really exist in nature, as the *electromagnetic* ones coupled to photons and (the parts of) the *weak* current coupled to the W, Z bosons. Therefore, we know a lot about such correlators. In fact, quite complete spectral densities have been experimentally measured, in e^+e^- annihilation into hadrons and in weak decays of the heavy lepton τ.

The currents and their correlation functions will be denoted by the name of the lightest meson in the corresponding channel, in particular

$$j_\mu^{\rho 0} = \frac{1}{\sqrt{2}}[\bar{u}\gamma_\mu u - \bar{d}\gamma_\mu d], \qquad j_\mu^{\rho-} = \bar{u}\gamma_\mu d, \tag{6.23}$$

$$j_\mu^{\omega} = \frac{1}{\sqrt{2}}[\bar{u}\gamma_\mu u + \bar{d}\gamma_\mu d], \qquad j_\mu^{\phi} = \bar{s}\gamma_\mu s, \tag{6.24}$$

(see more definitions in Table 6.1).

The electromagnetic current is the following combination of these currents:

$$j_\mu^{em} = \frac{2}{3}\bar{u}\gamma_\mu u - \frac{1}{3}\bar{d}\gamma_\mu d + \cdots$$
$$= (1/2^{1/2})j_\mu^{\rho} - (1/2^{1/2}3)j_\mu^{\omega} + \cdots. \tag{6.25}$$

Table 6.1. Definition of various currents and hadronic matrix elements referred to in this work.

Channel	Current	Matrix element	Experimental value
π	$j_\pi^a = \bar{q}\gamma_5\tau^a q$	$\langle 0\|j_\pi^a\|\pi^b\rangle = \delta^{ab}\lambda_\pi$	$\lambda_\pi \simeq (480\,\text{MeV})^3$
	$j_{\mu 5}^a = \bar{q}\gamma_\mu\gamma_5\frac{1}{2}\tau^a q$	$\langle 0\|j_{\mu 5}^a\|\pi^b\rangle = \delta^{ab}q_\mu f_\pi$	$f_\pi = 93\,\text{MeV}$
δ	$j_\delta^a = \bar{q}\tau^a q$	$\langle 0\|j_\delta^a\|\delta^b\rangle = \delta^{ab}\lambda_\delta$	
σ	$j_\sigma = \bar{q}q$	$\langle 0\|j_\sigma\|\sigma\rangle = \lambda_\sigma$	
η_{ns}	$j_{\eta_{ns}} = \bar{q}\gamma_5 q$	$\langle 0\|j_{\eta_{ns}}\|\eta_{ns}\rangle = \lambda_{\eta_{ns}}$	
ρ	$j_\mu^a = \bar{q}\gamma_\mu\frac{1}{2}\tau^a q$	$\langle 0\|j_\mu^a\|\rho^b\rangle = \delta^{ab}\epsilon_\mu(m_\rho^2/g_\rho)$	$g_\rho = 5.3$
a_1	$j_{\mu 5}^a = \bar{q}\gamma_\mu\gamma_5\frac{1}{2}\tau^a q$	$\langle 0\|j_{\mu 5}^a\|a_1^b\rangle = \delta^{ab}\epsilon_\mu(m_{a_1}^2/g_{a_1})$	$g_{a_1} = 9.1$
N	$\eta_1 = \epsilon^{abc}(u^a C\gamma_\mu u^b)\gamma_5\gamma_\mu d^c$	$\langle 0\|\eta_1\|N(p,s)\rangle = \lambda_1^N u(p,s)$	
N	$\eta_2 = \epsilon^{abc}(u^a C\sigma_{\mu\nu} u^b)\gamma_5\sigma_{\mu\nu} d^c$	$\langle 0\|\eta_2\|N(p,s)\rangle = \lambda_2^N u(p,s)$	
Δ	$\eta_\mu = \epsilon^{abc}(u^a C\gamma_\mu u^b)u^c$	$\langle 0\|\eta_\mu\|N(p,s)\rangle = \lambda^\Delta u_\mu(p,s)$	

The correlation functions are defined as

$$\Pi_{\mu\nu}(x) = \langle 0|j_\mu(x)j_\nu(0)|0\rangle \tag{6.26}$$

and the Fourier transform (in Minkowski space–time) is traditionally written as

$$i\int d^4x\, e^{iqx}\Pi_{\mu\nu}(x) = \Pi(q^2)(q_\mu q_\nu - q^2 g_{\mu\nu}). \tag{6.27}$$

The r.h.s. is explicitly "transverse" (it vanishes if multiplied by momentum q), because all vector currents are conserved.

The dispersion relation for the scalar functions $\Pi(q^2)$ has the usual form

$$\Pi(Q^2 = -q^2) = \int \frac{ds}{\pi}\frac{\text{Im}\,\Pi(s)}{(s+Q^2)}, \tag{6.28}$$

where the physical spectral density $\text{Im}\,\Pi_i(s)$ is directly related with the cross section of e^+e^- annihilation into hadrons. As this quantity is dimensionless, it is proportional to the normalized cross section

$$R_i(s) = \frac{\sigma_{e^+e^- \to i}(s)}{\sigma_{e^+e^- \to \mu^+\mu^-}(s)}, \tag{6.29}$$

where the denominator includes the cross section of the muon pair production[5] $\sigma_{e^+e^- \to \mu^+\mu^-} = (4\pi\alpha^2/3s)$ and α is the fine structure constant. If the current considered has only one type of quark (like e.g. the ϕ quark) one gets

$$\text{Im}\,\Pi_s(s) = \frac{R_s(s)}{12\pi e_s^2} = \frac{R_s(s)}{12\pi(1/9)}, \tag{6.30}$$

where e_s is the s-quark electric charge. Generalization to ρ, ω channels is straightforward; instead of the charge, there appear the corresponding coefficients in the

[5]The muon mass is neglected in this expression.

expression for the electromagnetic current (6.25);

$$\text{Im}\,\Pi_\rho(s) = \frac{1}{6\pi} R_\rho(s), \qquad \text{Im}\,\Pi_\omega(s) = \frac{3}{2\pi} R_\omega(s). \tag{6.31}$$

(The reader may wonder how the experimental selection of the channels is actually made. It is clear enough for heavy flavors (c and b); if the final state has a pair of such quarks, there are many more chances that they were directly produced in the electromagnetic current rather than by strong "final state interaction". Below we use this idea for the strange quark as well, although some corrections should, in principle, be applied in this case. It is also possible to separate light quark ρ, ω channels; they have a different isospin $I = 1, 0$, which is conserved by any strong final state interaction. As is well known, C-parity plus isotopic invariance leads to the so called G-parity conservation, and pions have *negative* G-parity. Therefore, strong interactions do not mix states with even and odd number of pions. The currents j_ρ, j_ω have fixed G-parity as well, and therefore pionic states created by them can have only even or odd number of pions, respectively.)

Let us start the simple and well known predictions of QCD; all the ratios $R_i(s)$ have very simple limit at high energies s. It is conjugate to the small-distance limit $x \to 0$ in which $\Pi \to \Pi_{\text{free}}$ because quarks and anti-quarks propagate there as free particles. For currents containing only one quark flavor q the only difference with the muon is a different electric charge and a color factor:

$$\lim_{s\to\infty} R_q(s) = e_q^2 N_c, \tag{6.32}$$

which for ϕ case give $\lim_{s\to\infty} R_\phi(s) = 1/3$. For the ρ and ω cases one may use the following decomposition of the electromagnetic current;

$$\lim_{s\to\infty} R_\rho(s) = 3/2; \qquad \lim_{s\to\infty} R_\omega(s) = 1/6. \tag{6.33}$$

As we will show shortly, these relations are well satisfied experimentally (being historically one of the first and simplest justification for QCD).

Coming back to coordinate representation of the dispersion relation one obtains

$$\Pi_{i,\mu\nu}(x) = (\partial^2 g_{\mu\nu} - \partial_\mu \partial_\nu)\frac{1}{12\pi^2} \int_0^\infty ds\, R_i(s) D(s^{1/2}, x), \tag{6.34}$$

where, we recall, $D(m, x)$ is just the propagator of a scalar mass-m particle to distance x. Convoluting indices and using the equation $-\partial^2 D(m, x) = m^2 D(m, x) +$ contact term (which we disregard), one finally obtains the following for the dispersion relation:

$$\Pi_{i,\mu\mu}(x) = \frac{1}{4\pi^2} \int_0^\infty ds\, s R_i(s) D(s^{1/2}, x). \tag{6.35}$$

Since the r.h.s. is experimentally available, this equation would serve as our "experimental definition" of the l.h.s., the vector correlation functions in Euclidean space–time.

As the correlators are very strongly decreasing functions of x, it is more convenient to plot them normalized to free massless quark ones $\Pi_{\mu\mu}(x)/\Pi_{\mu\mu}^{\text{free}}(x)$ where $\Pi_{\mu\mu}^{\text{free}}(x)$ corresponds to the simple loop diagram, describing free quark propagation. Such ratios directly tell us something about the inter-quark interaction, with values above 1 corresponding to attraction and below 1 to repulsion. In addition all inessential normalization factors (such as the electromagnetic charges) drop out.

Figure 6.3a from my review [324] shows a sample of experimental data on $R_\rho(s)$ at sufficiently low energies.[6] One can see that this function consists of two quite

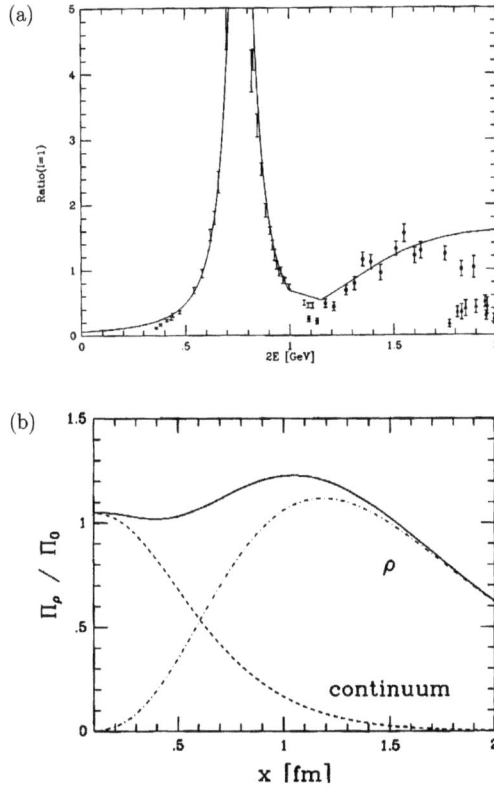

Fig. 6.3. (a) The ratio of $\sigma(e^+e^- \rightarrow n\pi, \text{even } n)/\sigma(e^+e^- \rightarrow \mu^+\mu^-)$ versus the total invariant mass of the hadronic system. The data points correspond to the following final states: the error bars without points to 2π, stars to 4π and triangles to 6π states. We have not shown all data points available near the top of the ρ peak, nor the region $2E > 2\,\text{GeV}$, where the agreement between multiple data points and our fitted curve (the solid line) is very good. This line should be compared with the total cross section in $I = 1$ channel, or to the sum of all contributions with even number of pions. (b) The ratio of the $I = 1$ vector correlation function to that corresponding to free quark propagation $R(x) = \Pi_{\mu\mu}/\Pi_{\mu\mu}^{\text{free}}$ versus the distance x (in fm). The dash-dotted curve is the ρ meson contribution (technically defined as the contribution to the integral of the region below a total energy of $1\,\text{GeV}$), the dashed one called the "continuum" is the complementary contribution of all hadronic states above $1\,\text{GeV}$. The solid curve is their sum.

[6]For a set of references to original experimental papers see Ref. [324].

different parts: (i) *the prominent rho-meson resonance*, seen in 2π channel; and (ii) a mixture of multi-pion states, which starts with (at least) two "primed" resonances, $\rho'(1450)$ and $\rho'(1700)$, seen mainly in *the 4-pion channel*. Furthermore, taken together with (iii) *the 6-pion channel*, they add up to some rather smooth non-resonance "continuum", and already at energies about 1.5 GeV this spectral density follows the asymptotic QCD prediction $R_\rho = 3/2$ made above.

Before we go any further let me comment on the accuracy of the curve. Although we do not put error bars on the curves in Fig. 6.3b, the experimental uncertainties are certainly present. Their magnitude is seen in Fig. 6.3a. In the ρ region the data from Novosibirsk[7] are quite accurate, and the resulting error at large distances $x >$ 0.6 fm is about 5 percent. The high energy domain is covered with rather accurate SPEAR data from SLAC, which fix the correlator in the small-x region, also to a few percent accuracy. In the region of medium energies 1–3 GeV there are now data from Beijing BES, which appeared after the review [324] and are not included. The e^+e^- data suffer from the problem of the luminosity normalization, different at each collider. It has been avoided however in the τ decay data such as those from ALEPH (to which we turn below); those provide the whole spectral density with one common normalization. Unfortunately there remain some unresolved experimental problems here, since the summary of e^+e^- and τ data differ by more than the error bars.[8]

The solid line corresponds to the following parametrization of $R_\rho(E)$:

$$R_\rho(E) = \frac{9}{1 + 4(E - m_\rho)^2/\Gamma_\rho^2} + \frac{3}{2}(1 + \alpha_s(E)/\pi)\frac{1}{1 + \exp[(E_0 - E)/\delta]}, \qquad (6.36)$$

where

$$E_0 = 1.3 \, \text{GeV}, \qquad \delta = 0.2 \, \text{GeV}, \qquad \alpha_s(E) = \frac{0.7}{\ln(E/0.2 \, \text{GeV})h}.$$

The physical meaning of the parameter E_0, called the duality interval, is the energy above which the asymptotic freedom is restored and simple asymptotic parton model estimates for the cross section become approximately valid.

Using this parametrization and the dispersion relation (6.35) one calculates the correlation function, taking the integral over all energies. The resulting curve is shown in Fig. 6.16b, where the contributions of the resonance and continuum components of the spectral density mentioned above are also shown separately. Note that, starting with a rather complicated spectral density, containing high peaks and low dips, one arrives at a very smooth function of the distance. Clearly, the way back, from the Euclidean space to the physical spectral density would be a next to impossible task! (See more on that in Section 6.2.5.)

---

[7]In relation with the so called $g - 2$ muon experiment in Brookhaven, new measurements were done, which improved accuracy at the lowest energies. Those are not included; but it will make very little difference for the correlators we discuss.

[8]This difference apparently is also larger than the QED-originated difference due to different radiative corrections for $\rho^0, \rho^\pm$ channels.

A quite striking observation made in Ref. [324] (which is specific to vector currents only, as we will see later in this section) is that the resonance and continuum contributions complement each other very accurately. As a result the ratio $\Pi(x)/\Pi_{\text{free}}(x)$ remains close to unity *up to distances as large as 1.5 fm*(!), while each function falls off by orders of magnitude. This "fine tuning" was called in Ref. [324] a *superduality*; let me explain why this is indeed a remarkable (and so far unexplained) fact.

At small distances it is very natural to expect the so called "hadron–parton duality" between the sum over hadronic states and the pQCD quark-based description; basically it is a simple consequence of the "asymptotic freedom" up to about $x < 1\,\text{GeV}^{-1} = 0.2\,\text{fm}$. In the interval $x = 0.2$–$1.5\,\text{fm}$ the correlator itself drops by more than 4 orders of magnitude, while its ratio to the free loop (free quark propagation) remains close to 1 within 10–15 percent! What this remarkable phenomenon means is that in this channel all kind of interactions — perturbative, instanton and confinement-related ones — cancel out in a wide range of distances. In lectures I usually present this fact by saying that in the vector channel the effective $\bar{q}q$ interaction is, like for a dog on a leash, apparent only at distances exceeding its length.

Let us continue our discussion of vector correlators, looking now at the isoscalar ω current made of light quarks, and then turning to strange ϕ channel. The corresponding data for the cross section of e^+e^- annihilation into an *odd* number of pions, now summed over all channels, are shown in Fig. 6.4a. The corresponding data with a kaon pair produced are shown in Fig. 6.4b. Although the tops of the peaks are not shown, the Breit–Wigner curves (with the width value taken from the Particle Data) is of course just perfect there. One can also see a trace of the ϕ peak due to the ω–ϕ mixing (which is disregarded). Note the change of scale and larger error bars, compared to the $I = 1$ channel; but inside the uncertainties the continuum magnitude still fits well to its asymptotic value $R_\omega = 1/6$ and is reached at about the same energy. For narrow resonances we use the resonance contribution in the simple form

$$\Pi_{\mu\mu}(\tau) = 3f_{\text{res}}^2 m_{\text{res}}^2 D(m_{\text{res}}, \tau) + \text{continuum}, \qquad (6.37)$$

where the coupling constants of the currents to mesons are defined by

$$\langle 0|j_\mu|\text{resonance}\rangle = f_{\text{res}} m_{\text{res}} \epsilon_\mu, \qquad (6.38)$$

the ϵ_μ being the polarization vector of the vector meson. These couplings and the partial widths to the e^+e^- channel are related by

$$f_{\text{res}}^2 = \frac{3m_{\text{res}}\Gamma(\text{res} \to e^+e^-)}{4\pi\alpha^2}. \qquad (6.39)$$

For the record let us list them: $f_\rho \approx 152\,\text{MeV},^9 \; f_\omega = 46\,\text{MeV}, \; f_\phi \approx 79\,\text{MeV}$.

---

[9] Note that for the ρ the narrow resonance approximation is not really accurate.

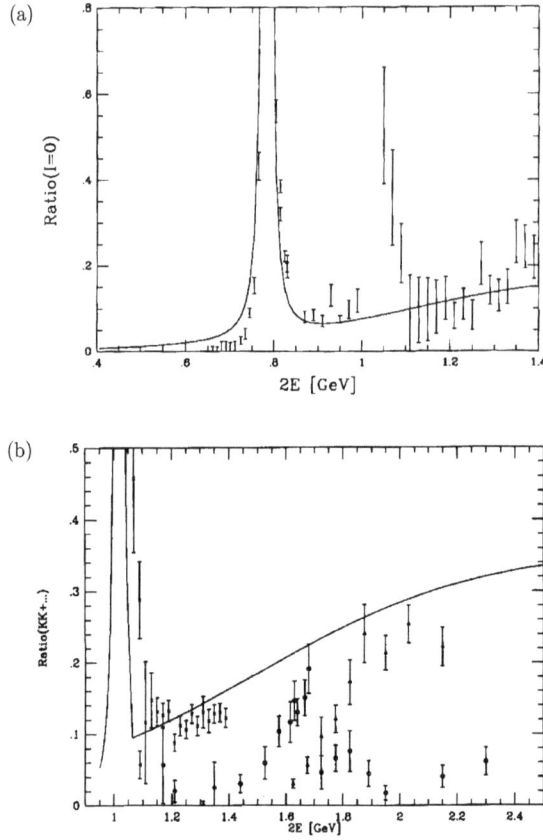

Fig. 6.4. (a) The ratio of $\sigma(e^+e^- \to n\pi, n = \text{odd})/\sigma(e^+e^- \to \mu^+\mu^-)$ versus the total invariant mass of the hadronic system. (b) The cross section of e^+e^- annihilation into all channels containing a pair of K mesons. The points marked by crosses, closed dots, open dots and triangles correspond to the following final states: K^+K^-, $K_S K_L$, $K_S^0 K^- \pi^+ + K_S^0 K^+ \pi^-$ and $K^+K^-\pi^+\pi^-$, respectively. The solid curve is our fit to their sum.

For completeness, let us also add the last (strange) vector channel of the nonet, with the current $j_{K^*} = \bar{u}\gamma_\mu s$ and the lowest meson called $K^*(892)$.

Phenomenological analysis in these cases is based on the vector part of the weak currents, from the weak decay process $\tau \to \nu_\tau + \text{hadrons}$. Comparing (Cabbibo suppressed) production of K^* to (Cabbibo allowed) production ρ in this decay, one can obtain the relation

$$\frac{\text{Br}(\tau \to \nu_\tau + K^*)}{\text{Br}(\tau \to \nu_\tau + \rho)} = \tan^2\theta_c \left(\frac{f_{K^*}}{f_\rho}\right)^2 \frac{(1 - m_{K^*}^2/m_\tau^2)^2}{(1 - m_\rho^2/m_\tau^2)^2} \frac{(1 + 2m_{K^*}^2/m_\tau^2)}{(1 + 2m_\rho^2/m_\tau^2)}, \quad (6.40)$$

where θ_c is Cabbibo angle. Using for the l.h.s. the experimental ratio 0.0143 ± 0.0031 one then finds the following ratio of the coupling constants:

$$f_{K^*}/f_\rho = 1.1 \pm 0.1. \quad (6.41)$$

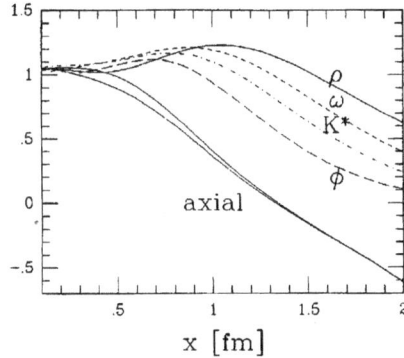

Fig. 6.5. All vector correlators together, plus the a_1 axial correlator for comparison.

So, inside the error bars the resonance coupling in both channels is the same, and the difference thus comes from the mass difference.

Now is the time to see what all four vector correlators look like. Figure 6.5 shows the resulting Euclidean correlators, again in the normalized form $\Pi(x)/\Pi_{\text{free}}(x)$. Comparing all the correlators, one can see that, in spite of completely different widths of the resonances and different even and odd-pion continuum states, all four vector correlators are *remarkably similar*. This demonstrates the deep consistency between four independent sets of data which is rather impressive.

To understand what this phenomenon implies theoretically, let us look at the difference between the ρ^0 and ω correlators. The former current has the $\bar{u}u - \bar{d}d$ flavor structure, while the latter is $\bar{u}u + \bar{d}d$. Thus the difference between them is the *flavor-changing* correlator

$$K^{\omega-\rho}(x) = 2(\Pi_{\omega,\mu\mu} - \Pi_{\rho,\mu\mu}) = \langle \bar{u}\gamma_\mu u(x)\bar{d}\gamma_\mu d(0)\rangle \tag{6.42}$$

and the data presented above tell us that this amplitude is basically zero, inside the error bars. Only at distances as large as 2 fm does the difference between the omega and rho correlators become clearly observable. It means that now the flavor-changing correlation function becomes comparable to the flavor-diagonal ones[10] only when the latter drops by several orders of magnitude. A more traditional way of looking at the famous "Zweig rule", forbidding the flavor-changing transitions, is provided by the following facts. (i) The rho–omega mass difference is only 12 MeV. (ii) The omega–phi mixing angle is only 1–3 degrees.

What can the reason be for such strong suppression on flavor-changing transitions in vector channels? In pQCD in the vector channel one needs not two but at least three gluons. However, this argument is good only at small x, and should not work at $x > 1$ fm. Non-perturbatively, the argument (to be discussed in the next

---

[10]We show that for pseudoscalar correlators such deviation happens at much smaller distances, where the correlation function is about 4 orders of magnitude larger!

chapter) is that in vector channels there are no direct instanton contributions in the first order in the 't Hooft interaction, and the effects of the second order tend to cancel.

6.2.2.   *Comparing axial and vector channels*

The axial $I = 1$ channel with the A_1 quantum numbers is associated with the current

$$j_\mu^{A_1} = \bar{u}\gamma_\mu\gamma_5 d, \tag{6.43}$$

which is also a part of the weak current. The corresponding data are available from the $\tau \to \nu_\tau + $ hadrons decay into the corresponding neutrino and hadrons, a sample of those is shown in Fig. 6.6. It is useful to combine vector and axial into sum and difference, as the weak current has both components. Such approach has been followed in the paper [244] which will be discussed also in Section 6.5.2 below.

Since we deal with the charged current due to the W exchange, there is no $I = 0$ component (as for photons), so the production of an *odd* number of pions is now entirely due to the axial part of the current, while an *even* number of pions corresponds to the vector $I = 1$ current discussed above.

The coupling constant f_{A_1} is defined similarly to those of vector resonances,

$$\langle 0|\bar{d}\gamma_\mu\gamma_5 u|A_1\rangle = f_{A_1} m_{A_1}\epsilon_\mu. \tag{6.44}$$

From the experimental branching ratios and the theoretical expression

$$\frac{\mathrm{Br}(\tau \to \nu_\tau + A_1)}{\mathrm{Br}(\tau \to \nu_\tau + \rho)} = \left(\frac{f_{A_1}}{f_\rho}\right)^2 \frac{\left(1 - m_{A_1}^2/m_\tau^2\right)^2}{\left(1 - m_\rho^2/m_\tau^2\right)^2}\frac{\left(1 + 2m_{A_1}^2/m_\tau^2\right)}{\left(1 + 2m_\rho^2/m_\tau^2\right)}, \tag{6.45}$$

one finds that $f_{A_1}/f_\rho = 1.0 \pm 0.07$. Again, withing the errors, there is the same coupling for both resonances.

Fixing the resonance contribution, we proceed to that of "continuum" states at larger invariant masses. Unfortunately, the tau lepton is not heavy enough, and direct observation of the axial spectral density is limited by it mass, 1784 MeV. Therefore, at high energies the spectral density is approximated by its asymptotic limit using the so called *Weinberg sum rules* which must be satisfied exactly.

Let me add a bit of general discussion of the axial correlators. In the case of *exact* chiral symmetry (all quark masses are zero), the *non-singlet axial currents are conserved*. So, one might think that, like the vectors, their correlators have only the transverse part $\Pi_{\mu\nu}^A(q) \sim (q_\mu q_\nu - g_{\mu\nu}q^2)$ but this is not the case. The massless Goldstone mode, the pion, is coupled to the axial current and contributes to the longitudinal term

$$\Pi_{\mu\nu}^A(q) = \Pi_t(q^2)(q_\mu q_\nu - g_{\mu\nu}q^2) + f_\pi^2\frac{q_\mu q_\nu}{q^2}. \tag{6.46}$$

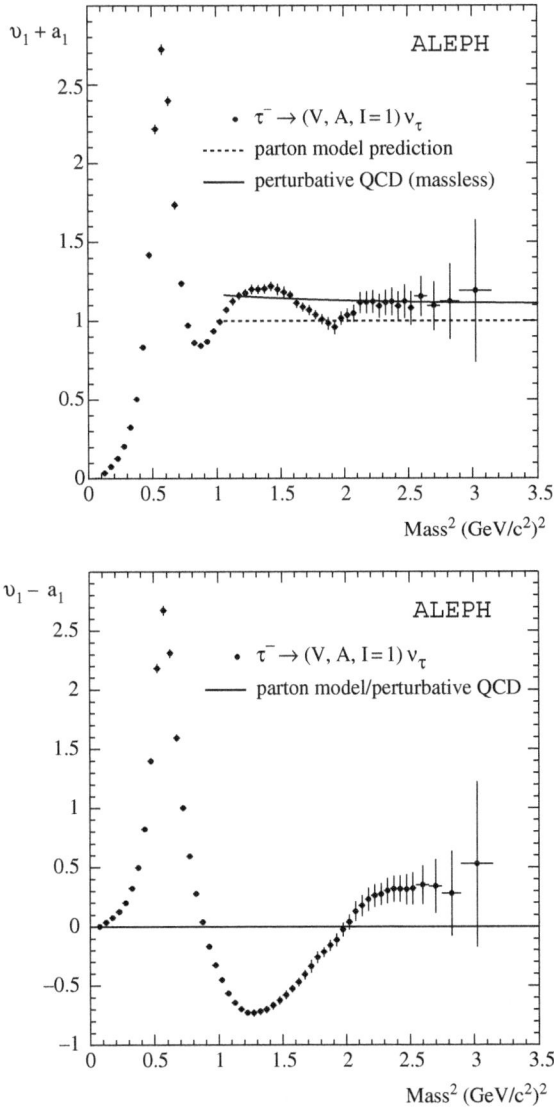

Fig. 6.6. Spectral functions $v(s) \pm a(s) = 4\pi^2(\rho_V(s) \pm \rho_A(s))$ extracted by the ALEPH collaboration from τ lepton decays.

This term does not spoil current conservation; if we multiply this expression by q_μ the denominator cancels, so we will only have a term linear in q. Thus there is neither propagator nor singularity; in the x-space it is just a contact term, which we can ignore or remove. Note also, that if one simply convolutes indices the pole is also canceled.

In the real world quark masses are non-zero. Still we have a longitudinal part due to a pion contribution, now depending on x as $\partial_\mu \partial_\nu D(m_\pi, x)$. (Let me recall that the derivatives come from the pion coupling to the axial current.) Now, taking a divergence ∂_μ (or convoluting indices $\mu\nu$) we obtain the expression $\partial^2 D(m_\pi, x) = -m_\pi^2 D(m_\pi, x)$+ contact term. Now we do have a longitudinal contribution with a propagator, non-trivially depending on distance. It is however small because of the factor m_π^2, or to the quark mass. This result is not unexpected; the axial current is not conserved in this case, and its divergence is $O(m_q) = O(m_\pi^2)$.

Completing this short excursion into the theory, we conclude that the form for the axial correlator one may use is

$$\Pi_{\mu\mu}^A(\tau) = 3f_A^2 m_A^2 D(m_A, \tau) - f_\pi^2 m_\pi^2 D(m_\pi, \tau)$$
$$+ \frac{3}{4\pi} \int dE\, E^3 D(E, \tau) \frac{1 + \alpha_s(E)/\pi}{1 + \exp[(E_0 - E)/\delta]}, \qquad (6.47)$$

where we have the A_1, π parts, together with the non-resonant continuum. The latter is here written in our standard way, with a perturbative contribution starting at some E_0, taken to be 1.5 and 1.7 GeV (as above, we took $\delta = 0.2$ GeV). This contribution is also shown in Fig. 6.5; one can see that there is a significant splitting between vector and axial channels.

6.2.3. *Pseudoscalar* SU(3) *octet* (π, K, η) *channels*

The definition of the octet pseudoscalar operators we use is

$$j_\pi = \frac{i}{\sqrt{2}}(\bar{u}\gamma^5 u - \bar{d}\gamma^5 d), \qquad j_K = i\bar{u}\gamma^5 s, \qquad j_\eta = \frac{i}{\sqrt{6}}(\bar{u}\gamma^5 u + \bar{d}\gamma^5 d - 2\bar{s}\gamma^5 s).$$
$$(6.48)$$

We have already discussed in Chapter 2 that the pseudoscalar (and scalar) mesons, are in many respects rather exceptional members of the family of hadrons. There are some surprisingly large numbers attached to them, in particular the pion constant B (2.41) in its mass and in the pion coupling constants to the pseudoscalar current (2.40). Therefore, the contributions of these particles to correlators *at small* x are important as well, telling us a lot about the *short range* vacuum structure.

For the K meson we know from its weak decays that

$$f_K \approx 1.24 f_\pi, \qquad (6.49)$$

and we then extrapolate from the π, K cases to the η one,

$$f_\eta \approx (4/3)f_K - (1/3)f_\pi \approx 1.32 f_\pi. \qquad (6.50)$$

(The idea is that the deviations from the SU(3) symmetry are due to strangeness, and η is 2/3 "strange", while K is only 1/2 "strange".) Only information about π, K, η couplings to the weak axial currents is accurately known; in order to evaluate

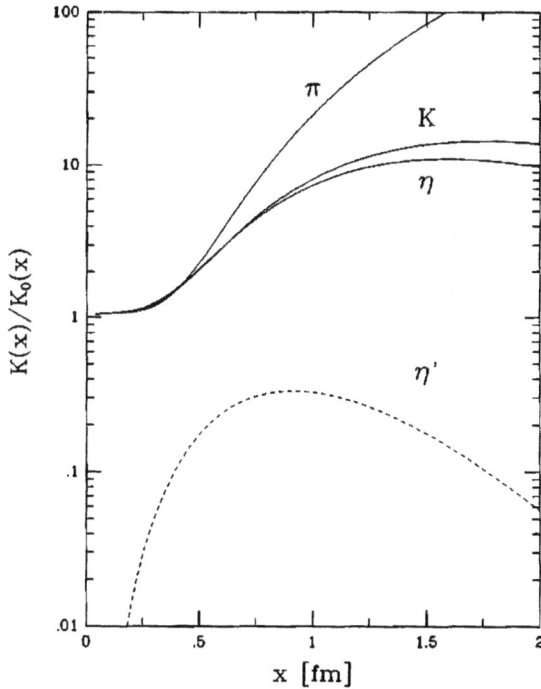

Fig. 6.7. (a) The correlation functions for the pseudoscalar nonet. Note that in contrast to the preceding figures it is now shown in the logarithmic scale.

their contributions to the *pseudoscalar* ones we will *assume* that[11]

$$\lambda_K/\lambda_\pi \sim f_K/f_\pi, \qquad \lambda_\eta/\lambda_\pi \sim f_\eta/f_\pi. \tag{6.51}$$

We will also assume (although it makes very little difference) that the non-resonance continuum with $E_0 \approx 1.6\,\mathrm{GeV}$.

The resulting π, K, η pseudoscalar correlators, in the form $K(x)/K_{\mathrm{free}}(x)$, are shown in Fig. 6.7. Note the marked difference compared to the vector correlators considered above; instead of changes within 10–20% (in the interval of distances $x \sim 1\,\mathrm{fm}$ considered), for pseudoscalars the ratio K/K_{free} has changed by up to two orders of magnitude!

One reason for that behavior is of course the small masses of the pseudoscalar mesons. In terms of the $\bar{q}q$ interaction, this implies a very strong attraction. Note also, that up to distances of the order of 0.5 fm there is *no marked difference between the three curves*, which means that all effects proportional to the strange quark mass are irrelevant in this region. Actually heavier mesons have slightly larger couplings, making the curves for different channels even more similar.

[11] In any case, the 10–20 percent level of accuracy is good enough for most of our conclusions below, and at this level all couplings can just be considered as equal.

What is however surprising, due to the contributions of the lowest mesons alone, is that the *asymptotic freedom is violated at very small distances*, about $1/5$ fm. This fact, noticed first by Novikov *et al.* [396], shows once again that the pseudoscalar channels differ substantially from the vector and axial ones, already at very small distances.

6.2.4.   SU(3) *singlet pseudoscalars*

The SU(3) singlet channel (to be called the η' one for brevity) can be associated with three important operators. One is the singlet axial current

$$j_{\eta'}^{\mu} = \frac{i}{\sqrt{3}} \left(\bar{u}\gamma^5\gamma_\mu u + \bar{d}\gamma^5\gamma_\mu d + \bar{s}\gamma^5\gamma_\mu s \right). \tag{6.52}$$

The second is the corresponding pseudoscalar operator (without γ_μ) and the third is the gluonic $G\tilde{G}$.

The relation between the matrix elements of these three operators is given by sandwiching (between vacuum and the η' state) of the famous Adler–Bell–Jackiw anomaly relation we discussed in Chapter 1,

$$\partial_\mu j_{\eta'}^{\mu} = \sqrt{3} \left(2im_s\bar{s}\gamma_5 s + \frac{3g^2}{16\pi^2}G\tilde{G} \right), \tag{6.53}$$

where the contributions proportional to the light quark masses are ignored.

The matrix element of the axial current is connected with $f_{\eta'}$ in the usual way, defining the coupling constant,

$$\langle 0|j_{\eta'}^{\mu}|\eta'\rangle = if_{\eta'}k_\mu. \tag{6.54}$$

Various estimates of $f_{\eta'}$ made by Novikov *et al.* [396] all suggest it to be somewhat smaller than f_π,

$$f_{\eta'} \approx 0.7f_\pi. \tag{6.55}$$

Let me outline the simplest way to estimate the relevant matrix elements from the radiative decays $J/\psi \to \gamma+$ pseudoscalar meson. If the charmed quarks are considered as sufficiently heavy, one can describe the $\bar{c}c \to \gamma gg$ annihilation in terms of *local* operators. Two gluons forming the pseudoscalar operator of the lowest dimension naturally lead to the operator of interest $G\tilde{G}$. The exact coefficient of this operator in the effective Lagrangian describing the decays is irrelevant for relative decay probabilities into different states,

$$\frac{\Gamma(\psi \to \gamma\eta')}{\Gamma(\psi \to \gamma\eta)} = \left| \frac{\langle 0|G\tilde{G}|\eta'\rangle}{\langle 0|G\tilde{G}|\eta\rangle} \right|^2 \left(\frac{p_{\eta'}}{p_\eta} \right)^3, \tag{6.56}$$

where the last factor contains the momenta ratio *cubed* because of the P-wave decay. Experimentally the l.h.s. ratio is 4.9 ± 0.5,[12] from which one finds the ratio of the matrix elements to be

$$\frac{\langle 0|G\tilde{G}|\eta'\rangle}{\langle 0|G\tilde{G}|\eta\rangle} = 2.46 \pm 0.1. \tag{6.57}$$

Since Novikov *et al.*'s work was published, another large contribution in the radiative decay of ψ was found, into higher state $\eta(1430)$ (originally called ι). Repeating the same argument, one obtains an even larger[13] matrix element for this particle:

$$\frac{\langle 0|G\tilde{G}|\eta(1430)\rangle}{\langle 0|G\tilde{G}|\eta'\rangle} = 1.12 \pm 0.2. \tag{6.58}$$

What we see from these results is that, unlike all previous channels we have discussed, the coupling of the $G\tilde{G}$ to subsequent states $\eta, \eta', \eta(1440)$ is not decreasing but growing. This is in fact what one should expect, since all these matrix elements are due to glue–$\bar{q}q$ mixing. The true pseudoscalar glueball, which should dominate the correlator, is heavy, with its mass 2.5–3 GeV according to lattice. Presumably, that (so far unknown state[14]) would eventually dominate the correlator of two $G\tilde{G}$.

In order to evaluate the *absolute* magnitude of all these matrix elements, Novikov *et al.* sandwiched anomaly relation between the vacuum and the η state (rather than η'). Of course, if η is ideal member of the SU(3) octet[15] the l.h.s. should vanish because the current is the SU(3) singlet. Thus, one should have an exact cancellation in the r.h.s., between the $O(m_s)$ and anomalous parts, so that

$$\frac{3g^2}{16\pi^2}\langle 0|G\tilde{G}|\eta\rangle = -2im_s\langle 0|\bar{s}\gamma_5 s|\eta\rangle = \sqrt{(3/2)}f_\eta m_\eta^2, \tag{6.59}$$

which fixes the absolute scale of the l.h.s. matrix elements, since the r.h.s. is known. Furthermore, using the ratios derived above, one obtains other couplings such as (6.55) and

$$\langle 0|G\tilde{G}|\eta\rangle \approx 0.9 \, \text{GeV}^3, \qquad \langle 0|G\tilde{G}|\eta'\rangle \approx 2.2 \, \text{GeV}^3, \qquad \langle 0|G\tilde{G}|\eta(1440)\rangle \approx 2.9 \, \text{GeV}^3. \tag{6.60}$$

For the sake of comparison with other pseudoscalar correlators, we have plotted in Fig. 6.7 also the η' contribution, making an "educated guess" (based on the ratio of $f_{\eta'}/f_\pi$ just derived), namely assuming the same ratio $\lambda_{\eta'} \approx 0.74\lambda_\pi$. Whatever the uncertainties in these couplings, may be, the reader is invited to be impressed

---

[12]The numbers, we recall, are from Ref. [324], which used data as they were in 1993.

[13]Actually this is even the *lower limit* of the matrix element, since the decay branching ratio into $\gamma\eta(1440)$ we use actually contains the branching ratio of $\eta(1440) \to \bar{K}K\pi$.

[14]The upsilon decay $\Upsilon \to \gamma + \text{PS glueball}$ is the natural place to look.

[15]In fact, the η–η' mixing angle $\theta_{\text{mixing}} \approx 10$–$20°$ and corrections to this statement are very small, $O(\theta_{\text{mixing}}^2)$.

by the quite striking difference between the SU(3) octet and the singlet correlators
that obvious from this figure. Unlike octet channels, the η' one clearly shows the
repulsive interaction between quarks. The splitting of the octet from the free correlator happens already at distances as small as $x_{\text{splitting}} \sim 0.3\,\text{fm}$. We will return
to the origin of this difference below in this chapter.

To complete the discussion of this section let us now do some preliminary estimates of the contribution of three η states to the pseudoscalar gluonic correlator.
As in all other channels, at small distances the correlator is dominated by a perturbative "asymptotically free" gluonic contribution, which is equal to

$$K(x) = \langle 0|G\tilde{G}(x)G\tilde{G}(0)|0\rangle, \qquad K_{\text{free}}(x) = \frac{48(N_c^2 - 1)}{\pi^4 x^8}. \qquad (6.61)$$

It is thus instructive to ask a simple question: at which x is the contribution of all
$\eta, \eta', \eta(1440)$ together to $K(x)$ equal to K_{free}? The answer is, at $x \approx 0.2\,\text{fm}$, to be
compared with about 1 fm for all vector resonances. And the reader should keep in
mind that we still have not seen the contribution of "the *true pseudoscalar glueball*"
yet, which is expected to overshadow the contribution of η.

The main conclusion one can draw from this discussion (the main point of
Ref. [396]) is that scales of the boundary between pQCD and non-perturbative
physics should in the ps gluonic channel be unusually high, at several GeV or so.

6.2.5. *Hadron–parton duality*

The philosophy of the previous sections is that of the original "QCD sum rules",
namely: *in order to compare experimental data with the theory one had better
translate the data into Euclidean correlators, which are easier to calculate*. In the
rest of this chapter we will show that this indeed works well, and the Euclidean
approaches — the lattice and its "smaller brother", the instanton liquid — do a
good job on the Euclidean correlators. The reason for it is the so called *quark–hadron
duality* which states that somewhat smoothened or averaged spectral densities correspond to a simple description in terms of quarks and gluons [331], while the observed
oscillations and fluctuations related to final state interaction and some "hadronic
details" do not survive the averaging.

In this section however we will discuss a more ambitious program aimed at
reproducing directly the experimentally accessible (Minkowskian) spectral densities.
We follow in this section Shifman's lecture [332], which summarizes attempts to
understand the *deviations* from the hadron–parton duality.

If one could calculate $\Pi(Q^2)$ in the Euclidean domain *exactly*, all one should
then do is to analytically continue the result to the Minkowski domain, and then
take the imaginary part. *There would be no need for duality.* We can calculate it
approximately, as will be shown later in this chapter, but while attempting analytic
continuation one quickly realizes that uncertainty gets exponentially amplified.

The OPE expansion is an example of a systematic theoretical approach, basically fixing the coefficients of all terms singular at zero distances $x^2 = 0$. There is no question however that there can be many other singularities, at some finite x^2 or at infinity.

As an example, consider an expression (familiar from instantons); $1/(x^2+\rho^2)$ has a pole at the (Euclidean distance squared) $x^2 = -\rho^2$ but it is regular at the origin and is missing from the usual OPE series. Its Fourier transform, upon analytic continuation, contains $E^{-\kappa}\sin(-E\rho)$, where E stands for the total energy, $E = \sqrt{q^2}$, and κ is some positive power which damps the oscillations. Needless to say that determining κ is one of the most important tasks in the issue of the quark–hadron duality.

To be more specific, let us look at a contribution of an instanton of a specific size ρ; to my knowledge, the observation that they generate an oscillating component was first made in Ref. [334]. The only formula we will need is

$$\int d^4x \frac{1}{(x^2 + \rho^2)^\nu} e^{iqx} = \frac{2\pi^2}{\Gamma(\nu)} \left(\frac{Q\rho}{2}\right)^{\nu-2} \frac{K_{2-\nu}(Q\rho)}{\rho^{2\nu-4}}, \tag{6.62}$$

in the Euclidean domain, which implies that in the Minkowski domain, at large q^2,

$$\mathrm{Im} \int d^4x \frac{1}{(x^2 + \rho^2)^\nu} e^{iqx} \propto s^{(\nu/2)-(5/4)} \sin(\sqrt{s}\rho - \delta), \quad s = q^2. \tag{6.63}$$

Here K is the McDonald function, and δ is a constant phase which is of no concern to us here.

To calculate some process in the instanton background is what we called a single-instanton approximation (SIA) in Section 4.3.4. The polarization operator is therefore the product of two Green functions, at large Euclidean momenta,

$$\Pi_{\mu\nu} \propto \int d^4x e^{iq(x-y)} d^4z \frac{1}{[(x - z)^2 + \rho^2][(y - z)^2 + \rho^2]^3}$$

$$\propto \int d^4x\, e^{iqx} \frac{1}{x^2 + \rho^2} \times \int d^4z\, e^{iqz} \frac{1}{(z^2 + \rho^2)^3}$$

$$\propto K_1(Q\rho)K_{-1}(Q\rho) \propto \frac{1}{Q} \exp(-2Q\rho), \tag{6.64}$$

which implies, in turn, that at large $E = \sqrt{s}$ the oscillating component of the spectral density is

$$\Delta\rho(s) \propto \frac{1}{E^3} \sin(2E\rho). \tag{6.65}$$

Equation (6.65) reproduces the high-energy asymptotics of the exact result for the polarization operator in the one-instanton approximation.

Does the $s^{-3/2}$ fall off of the oscillating (duality violating) component make sense? In Fig. 6.8 one can see a comparison to the ALEPH data on τ decays. There exist some data above $3\,\mathrm{GeV}^2$ from e^+e^- annihilation, but the error bars are

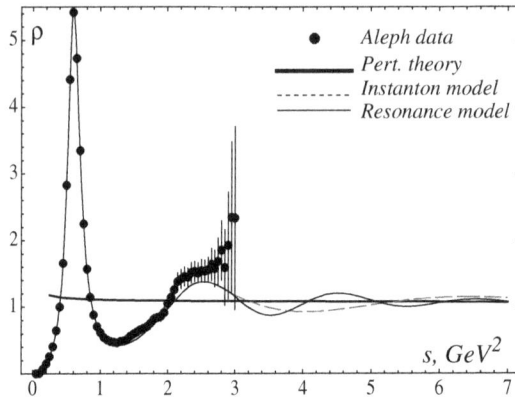

Fig. 6.8. Experimental data (ALEPH) on the spectral density in the vector channel and the duality violation models, from Ref. [332]. The dashed line is the instanton model described in the text and the solid line is a set of equidistant resonances with a fixed width. Both normalized to the maximum of the second oscillation.

so large that plotting these points would just obscure the picture. We know that their mean value is in agreement with the zeroth order parton model prediction; as for oscillations, they are not good enough to sort out systematics of different experiments from the real effect.

It is clear from the figure that these data indicate the occurrence of two oscillations, known as ρ and ρ' resonances. Although Shifman's discussion ends here, let me add that the Particle Tables list several more rho resonances, namely $\rho(1450), \rho(1700), \rho(1900), \rho(2150)$ (see e.g. our Fig. 2.2). If those are the next oscillations of this cross section, the period in mass seems to be approaching a constant. Clearly more accurate experimental measurements will most certainly provide us with the material needed for construction of a reliable and well calibrated model of the duality violations.

For the time being, theorists have studied some solvable models like the 't Hooft $1+1d$ model. One may find in Ref. [333], e.g. a picture where the widths of the first 500 resonances are plotted and compared to the duality predictions.

6.3. Operator product expansion and QCD sum rules

6.3.1. *Brief history and overview*

The aim of the Operator Product Expansion (OPE) is to develop a systematic description of the correlation functions *at small distances*. The main idea, originating from K. Wilson [335], is the *separation of scales*, for pedagogical discussion of the theoretical issues involved see Ref. [336].

After the scales are separated, the phenomena at "hard" momenta larger than a chosen *normalization scale* μ are treated explicitly (usually perturbatively), while the contributions of "soft" momenta are parameterized by VEVs of some operators. Even if one cannot calculate these VEVs, they can be fixed from phenomenology of some correlators and then used for prediction of many others.

The coefficients — calculated perturbatively — predict a specific dependence on the normalization (or the separation) scale; usually in the form of calculable powers of the $\log(\mu/\Lambda_{QCD})$, known as anomalous dimensions. In the early days of QCD the OPE was widely used for analysis of deep inelastic scattering data, see e.g. [337,338], precisely because of such anomalous dimensions. (Later the DGLAP evolution equation used directly for structure functions has substituted it.)

The OPE way of thinking was very important for the so called *higher twist* effects, suppressed by powers of the hard scale Q^2. The coefficients describe the correlations between partons in a hadron, see Refs. [339–341]. The OPE ideology was also instrumental in the formulation of the theory of hard exclusive reactions, from the power counting rules [342–346] to the definition of the so called hadronic wave functions in Refs. [347,348]. A discussion of these issues however would take us too far away from the correlation functions.

Returning to the OPE-based sum rules for vacuum correlation functions, we note that it has been developed mostly by the ITEP group — Shifman, Vainshtein, Zakharov, Ioffe and many others — [352] in late 70s, and it now has hundreds of papers and reviews [355–357, 360] devoted to it. In the first edition of this book it was discussed in more detail, while now I would be more brief, especially in technical points. The reason for that is that, as a practical tool aimed at calculation of the non-perturbative effects and hadronic properties, it unfortunately proved to be much less useful than it was expected two decades ago. As we show below, deviations from the OPE-based predictions often occur at so small distances that it is difficult to connect them to hadronic physics or data.[16]

The first signs of trouble have been detected by the authors themselves. In a very influential paper [396], entitled "Are all hadrons alike?", they have pointed out that both light quark and gluonic correlators of *spin zero* (both scalars and pseudo-scalars) display significantly larger non-perturbative effects as compared to other channels. (We have just followed part of their argument for the $\tilde{G}G$ channel in the preceding section.) Furthermore, they found that the standard OPE expressions are unable to reproduce these large scales.

Large corrections at small distances are obviously a sign of very strong gauge fields in the QCD vacuum. By learning which correlators do have and which do not have such corrections, we learn about the structure of those fields. As we will show later in this chapter, subsequent studies pointed toward the *instantons* as the origin of OPE violation in scalar and pseudoscalar channels.

---

[16]Perhaps the best example illustrating the situation is to note that the gluon condensate, introduced decades ago, remains numerically uncertain up to factor 2 or so because even in the best experimentally studied cases, the vector and axial correlators, the uncertainties are still too large to identify this term.

6.3.2. *Separation of scales*

Suppose one would like to estimate the correlators at small distances x (or large momentum transfer Q), smaller than any relevant scale for strong interactions. Those are presumably set by some typical correlation lengths of the fluctuations in the QCD vacuum. Naively it is expected to be of the order of $1\,\mathrm{fm} \sim 1/\Lambda_{\mathrm{QCD}}$, so the OPE is thought to be just an expansion in powers of $x\Lambda_{\mathrm{QCD}}$ or Λ_{QCD}/Q, a kind of Taylor expansion of the correlators. As for any Taylor series, the coefficients are defined locally at one point, while the expansion is supposed to reproduce the function behavior in some range.

The word "operator" in OPE means that the short-distance coefficients are the same, independently of which physical states the relation would be sandwiched over. Its generic form is

$$\hat{J}(x)\hat{J}(0) = \sum_i C_i(x,\mu)\hat{O}^i_{(\mu)}, \qquad (6.66)$$

which includes (x-depending) coefficients and some local operators $\hat{O}$ composed out of the fundamental fields.

We are interested in vacuum correlators, so we will average expressions like that over the vacuum state; so *only the scalar operators contribute*. No specific assumptions about the vacuum structure are made, and their VEVs are treated as some parameters. In applications to deep inelastic scattering mentioned in the introduction one averages such relations over the target hadron, the nucleon. In applications to exclusive hard processes this is done between the vacuum and the hadron. Clearly, in all these cases the quantum numbers of the operators should be appropriate.

Already at this generic level one finds that the OPE has a certain predictive power. The correlation functions are physical — that is gauge invariant — quantities, and should so the relevant operators be. The scalar gauge invariant gluonic operator with the lowest dimension[17] is $G^2_{\mu\nu}$, and thus one expects corrections at small x to be $\sim x^4 \langle G^2_{\mu\nu}\rangle$. So, by looking at deviations from the perturbative behavior one can estimate the gluon condensate. It was first done from an analysis of the charmonium correlators and the so called "standard value" was obtained

$$\langle 0|(G^a_{\mu\nu})^2|0\rangle \simeq 0.5\,\mathrm{GeV}^4, \qquad (6.67)$$

which sets a scale of non-perturbative fields.

The logic we so far used is however rather naive, appropriate at a *classical* level when the vacuum fields are some smooth background, perturbing a bit the propagators and correlators. At the *quantum* level the results are not so simple because quantum fluctuations exist at all possible scales. In particular, in any loop diagram one finds integrals over virtual momenta. So one should be more specific

---

[17]The lower-dimension operator A^2_μ for example is not allowed, as it is not gauge invariant.

and prescribe which fluctuations are included in the coefficients and which ones in the operators.

Thus the key element of the OPE is the introduction of the so called "normalization point" μ, typically 1 or 2 GeV. By definition, all fluctuations with momenta $p < \mu$ are included in the operators, while those of the momenta $p > \mu$ are included in the coefficients. The QCD radiative corrections, renormalized by standard methods with logarithmic accuracy, produce $\log(x\mu)$ in coefficients and $\log(\mu/\Lambda_{\mathrm{QCD}})$ in the operator VEVs. In sum, for *physical* quantities,[18] the dependence on μ should disappear. This statement formally means that the renormalization group equations for the coefficients and operators have opposite signs.

One more general comment about OPE is in order here. One can either (i) expand *point-to-point* correlators at small distance x, or (ii) expand its Fourier transfer in inverse powers of large momentum transfer Q. Although so far we have made no distinctions, these two forms of the OPE are not quite analogous. Each of them sees very well terms which are regular in this representation, but those become singular in the *other* representation and are easily missed. In particular, an expansion in $1/Q$ corresponds to singular terms at $x = 0$. For example, $x^n \log x$ are retained but not a constant or positive powers of x. Only by using both can one get a complete picture.[19]

So far, the OPE is only a convention, thus it is meaningless to ask whether it is right or wrong. In practice however some particular assumptions are made. A standard one is that the "small scale physics" can be approximately accounted for by perturbation theory, while the nonperturbative phenomena are connected with the "large scale physics" and attributed to the operators. If so, μ is a scale separating the two, and as such becomes an important physical quantity.

One may ask what the accuracy of this assumption is and whether it is appropriate to use $\mu = 1\text{–}2\,\mathrm{GeV}$. On one hand, there exist nonperturbative fluctuations (e.g. instantons) of arbitrarily small size; if need be, those can be included in the coefficients.

6.3.3. *OPE in a background field*

Perturbative evaluation of the OPE coefficients can in principle be made with the help of the ordinary Feynman diagrams. However, it is a rather inadequate way of making such calculations in the case when the operator contains the gluonic field strength. First of all, separate diagrams are not even gauge invariant, only their sum is. Second, covariant operators contain terms of different order in the coupling constant (e.g. covariant derivatives), and in order to reconstruct them one has to

---

[18]Examples of "physical" unrenormalized quantities are e.g. the electromagnetic or weak currents or their correlators. Their anomalous dimensions are zero because of the current conservation.

[19]V. Zakharov had coined a good way to describe their relations, saying that the x and p representations are like two eyes, each observing well the deficiencies of the other but missing its own.

perform calculations in several different orders consecutively. It is more convenient to use a background field formalism due to Schwinger [362] which allows one to keep the gauge invariance of the calculations at all stages and the possibility to work directly in the coordinate representation.

Consider a particle propagating in an arbitrary background field.[20] Introduce the formal basis of coordinate states $|x\rangle$ normalized by $\langle x|y\rangle = \delta(x - y)$, being the eigenvectors of the coordinate operator

$$X_\mu |x\rangle = x_\mu |x\rangle. \tag{6.68}$$

The momentum operator is acting in this basis as a covariant derivative

$$P_\mu = -i\partial_\mu + g A_\mu^a(x) T^a. \tag{6.69}$$

(As usual, the color generator appears in the group representation corresponding to the particle under consideration.) These operators satisfy the commutation rules[21]

$$[P_\mu X_\nu] = i g_{\mu\nu} \quad [P_\mu P_\nu] = i g T^a G_{\mu\nu}^a. \tag{6.70}$$

In such formalism the quark propagator can be written as

$$S(x,y) = \langle x|\frac{1}{P_\mu \gamma_\mu - m}|y\rangle, \tag{6.71}$$

while its Fourier transform can be represented as

$$S(q) = \int d^4x \, e^{iqx} \langle x|\frac{1}{P_\mu \gamma_\mu - m}|y\rangle = \int d^4x \langle x|\frac{1}{(P_\mu + q_\mu)\gamma_\mu - m}|y\rangle, \tag{6.72}$$

where the following identities were used: $\exp(iqX)P = (P + q)\exp(iqX)$ and $\exp(iqX)|0\rangle = |0\rangle$. This expression is very convenient for the OPE, one may say that there is large numerical part of the momentum q and a small (but operator) one P, so it is tempting to expand in P/q,

$$\frac{1}{\hat{P} + \hat{q} - m} = \frac{1}{\hat{q} - m} - \frac{1}{\hat{q} - m}\hat{P}\frac{1}{\hat{q} - m} + \cdots. \tag{6.73}$$

Note that because P is just a differential operator, we have indeed expanded the nonlocal object — the propagator — in series over the local derivatives and fields. All that is left to do are some algebraic manipulations.

Coming to physical examples, let me recall the saying (ascribed to Fermi): "What is the hydrogen atom of this problem?" Taking it literally, we follow the paper [55] and consider the propagation of a meson made of a very heavy quark and the light antiquark. The virtue of the heavy quark is that its large mass significantly simplifies

---

[20] In particular, it is not assumed to be weak.

[21] We use in this section Minkowski notation, in Euclidean space-time the metrics are just delta symbols etc.

its dynamics, in particular, it allows the use of more familiar non-relativistic language. The light quark is needed because in this case one may ignore the Coulomb-type interaction with the heavy one, so important for states with two heavy quarks. Therefore, the "probe" we are going to use is $j = \bar{\Psi}\Gamma\psi$ where Ψ, ψ are the fields of the light and heavy quarks. At sufficiently small distances the polarization operator is given by the loop made of free propagators,

$$K_0 = \mathrm{Tr}\left[\Gamma S_0^{\mathrm{heavy}}(x,0)\Gamma S_0^{\mathrm{light}}(0,x)\right]. \tag{6.74}$$

The propagation direction x may be taken along the (Euclidean) time axis, $x^2 = -\tau^2$, and the relevant free fermion propagators are

$$S_0^{\mathrm{heavy}} = \left(\frac{1+\gamma_0}{2}\right)e^{-m\tau}\left(\frac{m}{2\pi\tau}\right)^{3/2}, \tag{6.75}$$

$$S_0^{\mathrm{light}} = \gamma_0\frac{1}{2\pi^2\tau^3} + \frac{1}{4N_c}\langle\bar{q}q\rangle, \tag{6.76}$$

whereas in the first nonperturbative correction, we included the light quark condensate, as the gauge invariant scalar $\bar{q}q$ operator of the lowest dimension. Substituting this propagator to the correlation function, one obtains the first OPE term for heavy–light correlators in any channels [55]. $\boxed{\mathrm{E}}$

Exercise 6.3. Using the form for the light quark propagator (6.76), calculate the $O(\langle\bar{q}q\rangle)$ corrections for four practically important heavy–light cases, with $\Gamma = i\gamma_5, 1, \gamma_\mu, \gamma_\mu\gamma_5$.

Exercise 6.4. Using the same form for the propagator and the same four channels, find $O((\langle\bar{q}q\rangle)^2)$ corrections for light–light correlators.

Let me now show how, including interaction of propagating quarks with the field in the operator formalism, one can calculate further terms in the energy representation. The correlator then takes the form

$$\Pi_{AB} = \langle x|\bar{q}\Gamma_A\frac{1}{(iD+q)_\mu\gamma - \mu - M}\Gamma_A q|0\rangle, \tag{6.77}$$

where in the heavy quark propagator D is a covariant derivative and M is the heavy quark mass. To simplify things, we consider the heavy quark mass very large and make the non-relativistic reduction for the heavy quark propagator,

$$\frac{1}{\hat{P}+\hat{q}-M} \approx \frac{1+\gamma_0}{2}\frac{1}{\hat{P}_0-E}, \tag{6.78}$$

where $E = q_0 - M$ is the non-relativistic energy (counted from the mass). Expanding it in P/E we observe that the zero order term is the quark condensate one, while the first correction vanishes $\langle \bar{q}D_0 q \rangle = 0$ because only Lorentz scalars have the non-zero vacuum average values. The second correction contains momentum squared which can be rewritten as

$$\langle \bar{q}P_0^2 q \rangle = \frac{1}{4}\langle \bar{q}P_\mu^2 q \rangle = -\frac{1}{16}\langle \bar{q}ig\sigma_{\mu\nu}G_{\mu\nu}^a t^a q \rangle, \tag{6.79}$$

where at the last step we have used the Dirac equation in its squared form. The reason we can do so is that the "outgoing" spinors $\bar{q}, q$ are emitted "on shell" in the background field, which means one can use $\gamma_\mu D_\mu q = 0$. Two first corrections in the E-representation are

$$\delta\Pi(E) = \mp\frac{\langle \bar{q}q \rangle}{2E} + \frac{\langle \bar{q}ig\sigma_{\mu\nu}G_{\mu\nu}^a t^a q \rangle}{32E^2} + \cdots. \tag{6.80}$$

Note also, that although we seemingly expanded the propagator of the heavy quark, the operator acts on light ones, and the correction is actually due to the magnetic moment of light quarks, and not the heavy ones (which is $1/M$ and is neglected).

Further refinement of the operator formalism appears in the so called *fixed point gauge*

$$x_\mu A_\mu(x) = 0, \tag{6.81}$$

which was independently suggested by Fock, Schwinger and many others [361, 362]. In this gauge the following convenient expression holds:

$$A_\mu(x) = x_\nu \int_0^1 d\alpha\, \alpha G\mu\nu(\alpha x)$$

$$= \sum_{k=0}^{\infty} \frac{1}{k!(k+2)} x_\nu x_{\mu_1} \cdots x_{\mu_k} (D_{\mu_1} \cdots D_{\mu_k}) G_{\mu\nu}|_{x=0}, \tag{6.82}$$

expressing the potential in terms of the field and its local covariant derivatives. In particular, from it follows that $A_\mu(0) = 0$, and this fact explains the name of this gauge.

A useful application of expression (6.82) is the OPE expansion of the VEV of the small-size Wilson loop

$$W(C) = \left\langle \frac{1}{N_c} \mathrm{Tr}\, \mathrm{Re}\, P \exp\left(ig \int_C A_\mu\, dx_\mu \right) \right\rangle. \tag{6.83}$$

Following Shifman [139] for any small-size contour C is

$$W(C) = 1 - \frac{(gG_{\mu\nu}^a)^2}{192N_c^2} \int\int (dx\, dy)(xy) - (xdx)(ydy) + \cdots. \tag{6.84}$$

For the plane contour the latter integral is proportional to A^2 where A is its area. Further corrections also mostly depend on the area rather than on the contour shape, say the next correction is as follows:

$$\cdots + \frac{g^3 f^{abc} G^a_{\mu\nu} G^b_{\nu\sigma} G^c_{\sigma\mu}}{768 N_c} \int\int (dx\,dy)[(xy)(x^2 + y^2) - 2x^2 y^2] + O(a^8). \qquad (6.85)$$

In Ref. [139] the next terms of the OPE were calculated. However an attempt to evaluate the sum using the factorization hypothesis for the vacuum fields has produced a nonsensical result which we discussed in Section 2.3.

Finally, a more difficult (but useful) example: the propagator of a scalar massless particle in the external field [339]. In our symbolic formalism it looks as follows:

$$D(q) = \int d^4 x \left\langle x \left| \frac{1}{(P+q)^2} \right| 0 \right\rangle. \qquad (6.86)$$

Expanding as above in powers of P/q we obtain some polynomials in P. The utility of (6.82) is seen when one tries to rewrite it in a more familiar way, in terms of the field G and its covariant derivatives. Note that each covariant derivative consists of two parts, the ordinary derivative and the term with A. The former gives zero acting to the left because $\int dx \langle x|\partial_\mu \cdots |0\rangle = 0$, as an integral over the derivative of the delta function, while A gives zero acting to the right because $A(x)|0\rangle = A(0)|0\rangle = 0$. All one has to do is to move all derivatives to the left and all potentials to the right. As a result, only the commutators (derivatives of A) survive. For their evaluation a relation is of great use, giving covariant expressions for these derivatives.

Let me give the first terms we would need for scalar and spinor propagators, $\boxed{E}$ (for higher order terms see e.g. Ref. [339]),

$$D(q) = \frac{1}{q^2} - \frac{g}{3q^6} D_\alpha G^a_{\alpha\beta} t^a q_\beta$$

$$- \frac{g^2}{2q^8} \left(q_\alpha q_\gamma G^a_{\alpha\beta} G^b_{\beta\gamma} + \tfrac{1}{4} G^a_{\alpha\beta} G^b_{\alpha\beta} q^2 \right) t^a t^b + \cdots \qquad (6.87)$$

$$S(q) = \frac{1}{\hat{q}} - \frac{g}{2q^4} (q_\alpha \tilde{G}_{\alpha\beta} \gamma_\beta) \gamma_5 + \cdots \qquad (6.88)$$

We have seen above that the non-zero quark condensate produces directly some non-zero correction to the propagator of the light quark. Let us now see what the effect of the gluon condensate is. To this end, let us try to average the expressions for the propagators over the vacuum fields, looking only at the G^2 combination. It turns out [334] that there are *no such corrections* in scalar and spinor cases. (In the gluon propagator it appears due to the effect of its "magnetic moment".)

Exercise 6.5. Derive the Adler–Bardeen–Jackiw "triangular anomaly" relation for the divergence of the axial current, by writing the current with separated points (1.90) and the P-exponent between them. Then use the fixed point gauge expressions given here to first order in A_μ, both in the P-exponent and in the propagator.

6.3.4. *Sum rules for heavy–light mesons*

We have already considered OPE for the heavy–light correlator, as the simplest example in which the first order correction $O(\langle\bar{q}q\rangle)$ appears. Let us now see how this can be related to the hadronic world.

Before addressing more complicated issues of inter-quark interaction in $\bar{q}q$ or qqq systems, it is logical to start with the propagation in the QCD vacuum of *a single light quark*. This is the line of reasoning which has led us to studies of such mesons. In a sense, it is the simplest system with light quarks, so-to-say, a *hydrogen atom of hadronic physics.*

By combining a light quark with a heavy static one we have the so called gauge invariant version of the propagator $\tilde{S}$,

$$\tilde{S}(x) = \left\langle \bar{q}(x) P \exp\left(\frac{ig}{2}\int_0^x A_\mu^a t^a d\tau\right) q(0)\right\rangle, \tag{6.89}$$

where the path-ordered exponent contains an integral to be taken over the straight line, going from 0 to x.

Returning to our main object, the two-point correlators, we are going to discuss the correlation functions of the type

$$K_i(\tau) = \langle \bar{Q}(\tau)\Gamma_i q(\tau)\bar{q}(0)\Gamma_i^+ Q(0)\rangle, \tag{6.90}$$

where Γ_i is some gamma matrix and τ is the "Euclidean time difference" between the two points, $Q(q)$ stands for the field of heavy (light) quarks. We assume that the heavy anti-quark just stands at $\vec{x} = 0$ all the time. All energies are naturally measured relative to the heavy quark mass M_Q. An additional simplification due to the large M_Q limit is the absence of a spin splitting; the spin direction of the super-heavy quark cannot be of any importance. Thus, the pseudoscalar and the vector mesons are degenerate, as well as the scalar and the axial ones. Therefore, we have to consider only the splitting in parity P, "$-$" for the two former cases and "$+$" for the two latter ones, denoting these correlators by $K_-(\tau)$ and $K_+(\tau)$, respectively. These functions were evaluated numerically both on the lattice and in the instanton liquid. The small-τ behavior of the corresponding correlation function is given trivially by the product of free propagators

$$K^{\text{free}}(\tau) = \text{Tr}\left[S_q^{\text{free}}(\tau)\Gamma S_Q^{\text{free}}(\tau)\Gamma\right], \tag{6.91}$$

$$S_q^{\text{free}}(\tau) = -\gamma_0/(2\pi^2\tau^3), \quad S_Q^{\text{free}}(\tau) \sim (1+\gamma_0). \tag{6.92}$$

(Here we have dropped all unimportant factors in the heavy quark propagator.) The next correction discussed in the preceding section comes from the light quark propagator, which is modified due to the presence of the non-zero quark condensate as follows:

$$S_q(\tau) = -\gamma_0/(2\pi^2\tau^3) + \langle\bar{q}q\rangle/12 + \cdots \tag{6.93}$$

and therefore the ratio of the result corrected to zeroth order is simply

$$R_\pm(\tau) = 1 - (\pm)\frac{1}{6}\pi^2\tau^3|\langle \bar{q}q \rangle|. \tag{6.94}$$

(We recall that $\pm$ stands for the parity of the state considered, and we use the modulus of the quark condensate to avoid confusion related with its sign.) Thus the non-zero quark condensate (manifesting a breaking of the chiral symmetry) naturally produces the splitting of the correlation functions with the opposite parity. Then the $P = -1$ one curves up (which means attraction, the PS mesons are lighter), the $P = 1$ one curves down (which means repulsion, the scalar mesons are heavier).

Following Ref. [55] further, let us use the parametrization of the physical spectral density

$$\text{Im}\,\Pi(E) = 6n\pi\delta(E - E_{\text{res}}) + (3E^2/\pi)\theta(E - E_0), \tag{6.95}$$

with three parameters, the density of the light quark at the origin n, the position of the resonance E_r, and the perturbative threshold E_0. After this the spectral density is put into the dispersion relation, and the correlation function is calculated,

$$\frac{\Pi(\tau)}{\Pi_{\text{free}}(\tau)} = \frac{\pi^2\tau^3}{3}\left[6ne^{-E_{\text{res}}\tau} + \frac{3}{\pi^2}\int_{E_0}^\infty dE\, E^2 \exp(-E\tau)\right]. \tag{6.96}$$

Then the parameters can be selected in order to get the best agreement between the OPE and the parameterization. In Fig. 6.9 we show curves with the following parameters for $P = \pm$

$$E_{\text{res}}^- = 0.45\,\text{GeV}, \qquad E_c^- = 0.7\,\text{GeV}, \qquad n^- = 0.005\,\text{GeV}^3, \tag{6.97}$$

$$E_{\text{res}}^+ = 1.3\,\text{GeV}, \qquad E_c^+ = 1.9\,\text{GeV}, \qquad n^+ = 0.05\,\text{GeV}^3. \tag{6.98}$$

Note only that we have predicted the scalars to be much heavier, but their volume (the inverse of n) is about 10 times larger. The density n or the wave function at the origin, for D, B mesons estimated above is in reasonable agreement with lattice and experimental data.

The experimentally known masses of relevant "charmed" D and "beautiful" B are

$$M(D(0^-)) \approx 1865\,\text{MeV},$$
$$M(D(1^-)) \approx 2007\,\text{MeV}, \qquad M(D(1^+)) \approx 2420\,\text{MeV}, \tag{6.99}$$
$$M(B(0^-)) \approx 5279\,\text{MeV}, \qquad M(B(1^-)) \approx 5325\,\text{MeV},$$

and so we do not really know the parity splitting in B mesons.

Using the values of s, c, b quark masses from the QCD sum rule analysis $m_c \approx 1250\,\text{MeV}$ and $m_b \approx 4800\,\text{MeV}$, and assuming that the spin splitting is caused by the usual $\vec{S}_1\vec{S}_2$ gluon-exchange-type interaction,[22] we have the following

---

[22] Note that one of the quarks is heavy, and so they do not participate in the 't Hooft interaction.

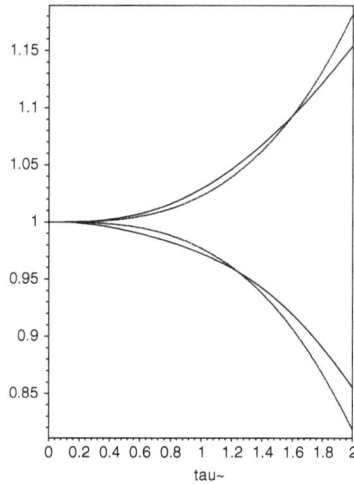

Fig. 6.9. Corrections to free propagation of quarks for heavy–light systems, for scalar (lower curves) and pseudoscalar (upper curves). The solid lines represent the theoretical expression to the first order in the quark condensate (6.94), while the dashed ones are fits described in the text.

expression for the "spin-averaged" masses of the negative $(P = -1)$ and the positive $(P = 1)$ parity states

$$E^{\pm} = \frac{3}{4}E(1^{\pm}) + \frac{1}{4}E(0^{\pm}), \tag{6.100}$$

where 1, 0 are spins. Using the experimental numbers, one concludes that both $E_{\text{res}}^{-} \approx 450\,\text{MeV}$ and the splitting of the two parity states $E^{+} - E^{-} \approx 450\,\text{MeV}$. We conclude from that, that for the heavy mesons the energy of the light quarks is, for PS and S channels

$$E_{\text{res}}^{-} \approx 450\,\text{MeV}, \qquad E_{\text{res}}^{+} \approx 900\,\text{MeV}. \tag{6.101}$$

Comparing with the results from the sum rules one can see, that the lowest energy of the light quark due to quark condensate E^{-} was evaluated correctly, while the splitting $E^{+} - E^{-}$ is strongly overestimated.

Summarizing this section, we see that some splitting patterns for correlators at small distances can be related with hadronic properties. Note that a very simple DPE-based splitting — due to first order of the quark condensate — has a complicated explanation in quark models. Non-relativistically, the splitting is related with a centrifugal barrier, present in $L = 1$ states and absent in $L = 0$ ones. No obvious reason for the $\pm$ symmetry of the splitting follows from any such models, although it is quite obvious in the OPE context.

6.3.5. *OPE for light quark baryons*

In this section we will further show how the simplest effect of the quark condensate for the correlators corresponds to phenomenology of hadrons made of light quarks.

Fig. 6.10. The diagrams for the baryonic correlation function for (a) free quarks and (b) including the effects of four-quark operators.

Unlike the heavy–light correlators just considered above, in this case the chirality constraints do not allow corrections of the *first* order in $\langle \bar{q}q \rangle$, and the chiral symmetry breaking appears starting from the 4-fermionic operators, or $O(\langle \bar{q}q \rangle^2)$.

For the reasons which will become clear later, we jump over the light mesons considered in the original SVZ paper [352] and proceed directly to the baryonic currents. The OPE and the corresponding sum rules for them have been worked out by Ioffe and others [374–376]. Since those have three quarks in a current and two of them we will send to vacuum, as shown in Fig. 6.10, the correction we would like to calculate would be schematically $1/x^9 + C\langle \bar{q}q\bar{q}q \rangle/x^3 + \cdots$. It is subleading to the gluon condensate correction $\sim \langle G^2 \rangle/x^5$ at small x. However much smaller coefficients of the former operator makes it unimportant and, following the original Ioffe paper, we will ignore them.

The proton current can be constructed by coupling a d-quark to a uu-diquark. The diquark has the structure $\epsilon_{abc}u_b C\Gamma u_c$ which requires that the matrix $C\Gamma$ be symmetric. This condition is satisfied for the V and T gamma matrix structures. The two possible currents (with no derivatives and the minimum number of quark fields) with positive parity and spin $1/2$ are given by [374]

$$\eta_1 = \epsilon_{abc}\left(u^a C\gamma_\mu u^b\right)\gamma_5\gamma_\mu d^c, \qquad \eta_2 = \epsilon_{abc}\left(u^a C\sigma_{\mu\nu} u^b\right)\gamma_5\sigma_{\mu\nu}d^c. \qquad (6.102)$$

It is sometimes useful to rewrite these currents in terms of scalar and pseudo-scalar diquarks

$$\eta_{1,2} = (2,4)\left\{\epsilon_{abc}(u^a C d^b)\gamma_5 u^c \mp \epsilon_{abc}(u^a C\gamma_5 d^b)u^c\right\}. \qquad (6.103)$$

Nucleon correlation functions are defined by $\Pi^N_{\alpha\beta}(x) = \langle \eta_\alpha(0)\bar{\eta}_\beta(x) \rangle$, where α, β are the Dirac indices of the nucleon currents. In total, there are six different nucleon correlators; the diagonal $\eta_1\bar{\eta}_1, \eta_2\bar{\eta}_2$ and off-diagonal $\eta_1\bar{\eta}_2$ correlators, each contracted with either the identity or $\gamma \cdot x$, see Table 6.2. Let us focus on the first two of these correlation functions. For more detail, we refer the reader to Ref. [239] and references therein.

In this section we will not give the OPE expression for all possible baryonic correlators, but simply follow the original selection of currents and sum rules. For the normalization of both functions we do the same as for the mesons, dividing out

Table 6.2. Definition of nucleon and delta correlation functions.

Correlator	Definition	Correlator	Definition
$\Pi_1^N(x)$	$\langle\mathrm{tr}(\eta_1(x)\bar{\eta}_1(0))\rangle$	$\Pi_1^\Delta(x)$	$\langle\mathrm{tr}(\eta_\mu(x)\bar{\eta}_\mu(0))\rangle$
$\Pi_2^N(x)$	$\langle\mathrm{tr}(\gamma\cdot\hat{x}\eta_1(x)\bar{\eta}_1(0))\rangle$	$\Pi_2^\Delta(x)$	$\langle\mathrm{tr}(\gamma\cdot\hat{x}\eta_\mu(x)\bar{\eta}_\mu(0))\rangle$
$\Pi_3^N(x)$	$\langle\mathrm{tr}(\eta_2(x)\bar{\eta}_2(0))\rangle$	$\Pi_3^\Delta(x)$	$\langle\mathrm{tr}(\hat{x}_\mu\hat{x}_\nu\eta_\mu(x)\bar{\eta}_\nu(0))\rangle$
$\Pi_4^N(x)$	$\langle\mathrm{tr}(\gamma\cdot\hat{x}\eta_2(x)\bar{\eta}_2(0))\rangle$	$\Pi_4^\Delta(x)$	$\langle\mathrm{tr}(\gamma\cdot\hat{x}\hat{x}_\mu\hat{x}_\nu\eta_\mu(x)\bar{\eta}_\nu(0))\rangle$
$\Pi_5^N(x)$	$\langle\mathrm{tr}(\eta_1(x)\bar{\eta}_2(0))\rangle$		
$\Pi_6^N(x)$	$\langle\mathrm{tr}(\gamma\cdot\hat{x}\eta_1(x)\bar{\eta}_2(0))\rangle$		

Table 6.3. Definition of various baryon correlation functions. We also give the form of the resonance contribution (with the propagators suppressed) and the invariance properties under chiral $SU(2)_A$ and $U(1)_A$ transformations.

Correlator	Definition	Resonance contribution	$SU(2)_A$	$U(1)_A$
$\Pi_1^N(x)$	$\langle\mathrm{tr}(\eta_1(x)\bar{\eta}_1(0))\rangle$	$\|\lambda_1^{N^+}\|^2 m_{N+} - \|\lambda_1^{N^-}\|^2 m_{N-}$	no	no
$\Pi_2^N(x)$	$\langle\mathrm{tr}(\gamma\cdot\hat{x}\eta_1(x)\bar{\eta}_1(0))\rangle$	$\|\lambda_1^{N^+}\|^2 + \|\lambda_1^{N^-}\|^2$	yes	yes
$\Pi_3^N(x)$	$\langle\mathrm{tr}(\eta_2(x)\bar{\eta}_2(0))\rangle$	$\|\lambda_2^{N^+}\|^2 m_{N+} - \|\lambda_2^{N^-}\|^2 m_{N-}$	no	no
$\Pi_4^N(x)$	$\langle\mathrm{tr}(\gamma\cdot\hat{x}\eta_2(x)\bar{\eta}_2(0))\rangle$	$\|\lambda_2^{N^+}\|^2 + \|\lambda_2^{N^-}\|^2$	yes	yes
$\Pi_5^N(x)$	$\langle\mathrm{tr}(\eta_1(x)\bar{\eta}_2(0))\rangle$	$\lambda_1^{N^+}(\lambda_2^{N^+})^* m_{N+} - \lambda_1^{N^-}(\lambda_2^{N^-})^* m_{N-}$	no	no
$\Pi_6^N(x)$	$\langle\mathrm{tr}(\gamma\cdot\hat{x}\eta_1(x)\bar{\eta}_2(0))\rangle$	$\lambda_1^{N^+}(\lambda_2^{N^+})^* + \lambda_1^{N^-}(\lambda_2^{N^-})^*$	yes	no

Table 6.4. Definition of delta correlation functions; the form of the resonance contribution (with the propagators suppressed) and the invariance properties under chiral $SU(2)_A$ and $U(1)_A$ transformations.

Correlator	Definition
$\Pi_1^\Delta(x)$	$\langle\mathrm{tr}(\eta_\mu(x)\bar{\eta}_\mu(0))\rangle$
$\Pi_2^\Delta(x)$	$\langle\mathrm{tr}(\gamma\cdot\hat{x}\eta_\mu(x)\bar{\eta}_\mu(0))\rangle$
$\Pi_3^\Delta(x)$	$\langle\mathrm{tr}(\hat{x}_\mu\hat{x}_\nu\eta_\mu(x)\bar{\eta}_\nu(0))\rangle$
$\Pi_4^\Delta(x)$	$\langle\mathrm{tr}(\gamma\cdot\hat{x}\hat{x}_\mu\hat{x}_\nu\eta_\mu(x)\bar{\eta}_\nu(0))\rangle$

the correlator which is non-zero in the free quark approximation, namely

$$\Pi_N^{\mathrm{free}}(\tau) = \frac{6}{\pi^6\tau^9}, \quad \Pi_\Delta^{\mathrm{free}}(\tau) = \frac{18}{\pi^6\tau^9}. \tag{6.104}$$

The dimension-6 OPE predicts [374]

$$\Pi_1^N/\Pi_2^{N0} = \frac{1}{12}\pi^2|\langle\bar{q}q\rangle|\tau^3, \tag{6.105}$$

$$\Pi_2^N/\Pi_2^{N0} = 1 + \frac{1}{768}\langle G^2\rangle\tau^4 + \frac{1}{72}\pi^4|\langle\bar{q}q\rangle|^2\tau^6. \tag{6.106}$$

In the case of the delta resonance, there exists only one independent current, given by (for the Δ^{++})

$$\eta_\mu^\Delta = \epsilon_{abc}\left(u^a C\gamma_\mu u^b\right)u^c. \tag{6.107}$$

However, the spin structure of the correlator $\Pi_{\mu\nu;\alpha\beta}^\Delta(x) = \langle\eta_{\mu\alpha}^\Delta(0)\bar\eta_{\nu\beta}^\Delta(x)\rangle$ is much richer. In general, there are ten independent tensor structures, but the Rarita–Schwinger constraint $\gamma^\mu\eta_\mu^\Delta = 0$ reduces this number to four, see Table 6.2. The OPE predicts

$$\Pi_1^\Delta/\Pi_2^{\Delta 0} = \frac{4}{12}\pi^2|\langle\bar qq\rangle|\tau^3, \tag{6.108}$$

$$\Pi_2^\Delta/\Pi_2^{\Delta 0} = 1 - \frac{25}{18}\frac{1}{768}\langle G^2\rangle\tau^4 + \frac{6}{72}\pi^4|\langle\bar qq\rangle|^2\tau^6. \tag{6.109}$$

6.3.6. *OPE for mesons made of light quarks*

Now let us proceed to practically and historically important $\bar qq$ currents. We will be focusing mainly on the simplest flavored currents, in which the correlation function is basically just a trace of the propagator squared. A simple dimensional analysis shows that (unlike the cases considered above) the 4-fermion operators and the correlator itself have the same dimensions. So there can be corrections of two types:

$$\Pi(x) \sim 1/x^6 + C_G/x^2\langle G^2\rangle + \langle\bar qq\bar qq\rangle(\alpha_s C_1\log x + C_2). \tag{6.110}$$

The Taylor-expanded propagators in the background field do not have logarithms and thus the C_1 corrections come from a virtual gluon exchange, so $C_1 \sim \alpha_s$. The second term is a part of a regular expansion, which are invisible in the $1/Q$ expansion used by SVZ. Alternatively one may say that the singularity does not fix the constant under the log.

Although the gluon condensate corrections are small, let me still point out how one can calculate it. There can be two potential sources of contributions of the gluon condensate $\langle G^2\rangle/x^2$: (i) the $O(\langle G^2\rangle)$ terms in one of the propagators; (ii) the $O(G)$ term in each of the propagators. We have seen in Section 6.3.3 that the first term is *absent* in the fixed-point gauge; therefore in this gauge one has only the square of the first order term in (6.87), which leads to

$$\Pi_{\mu\nu} = \frac{3}{\pi^4 x^8}\left(2x_\mu x_\nu - g_{\mu\nu}x^2\right) + \frac{2\langle g^2 G^2\rangle}{3(16\pi^2)^2 x^4}\left(2x_\mu x_\nu + g_{\mu\nu}x^2\right) + \cdots. \tag{6.111}$$

(Note that the polarization operator is transverse $\partial_\mu\Pi_{\mu\nu} = 0$, as it should be due to current conservation.) This result is the same as originally obtained by SVZ [352], in a much more complicated diagrammatic calculation.

For clarity we omit some terms which are in practice unimportant, like the $m_q\bar qq$ operators. We also do not discuss lengthy expressions for higher dimension operators because they are not actually used in applications.

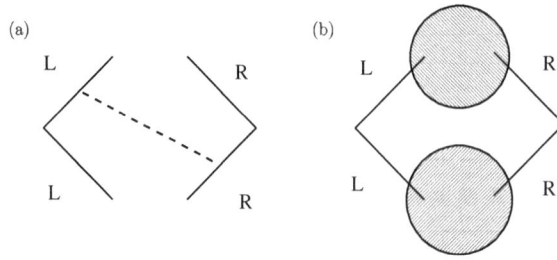

Fig. 6.11. Two contributions to the chirality-flipping correlator: (a) perturbative version, with the one-gluon exchange, and (b) non-perturbative, with instanton and anti-instanton pair.

In Euclidean time τ (which is the same as the spatial distance x used above) the normalized correlation functions for the ρ and A_1 channels are given by [352],

$$\frac{\Pi_{\mu\mu}^{\rho,A_1}(\tau)}{\Pi_{\mu\mu}^{\text{free}}(\tau)} = 1 + \frac{\alpha_s(\tau)}{\pi} - \frac{\langle(gG_{\mu\nu}^a)^2\rangle\tau^4}{3\cdot 2^7} + \frac{\pi^2\tau^6}{16}\log\frac{1}{\tau\mu}\langle O_{\rho,A_1}\rangle. \qquad (6.112)$$

The four-fermionic operators O_{ρ,A_1} are different for the vector and axial channels,

$$O_\rho = \frac{1}{2}\pi\alpha_s(\bar{u}\gamma_\mu\gamma_5 t^a u - \bar{d}\gamma_\mu\gamma_5 t^a d)^2 + \frac{1}{9}\pi\alpha_s(\bar{u}\gamma_\mu t^a u + \bar{d}\gamma_\mu t^a d)(\Sigma_q\bar{q}\gamma_\mu t^a q), \quad (6.113)$$

$$O_{A_1} = O_\rho + 2\pi\alpha_s(\bar{u}_L\gamma_\mu t^a u_L - \bar{d}_L\gamma_\mu t^a d_L)(\bar{u}_R\gamma_\mu t^a u_R - \bar{d}_R\gamma_\mu t^a d_R), \qquad (6.114)$$

where R, L mean right and left-hand polarized quarks. The difference has a specific chirality-flip structure we discussed at the beginning of this section; they come from diagrams shown in Fig. 6.11a.

Estimates of their vacuum expectation values were made using the so called "vacuum dominance" hypothesis [352].[23] The recipe is as follows: one should make all kinds of transformations of this operator, trying to write it as a product of two scalars, so one can sandwich the vacuum state there. In other words, one essentially tries to factorize the four-fermion operators into the product of two quark condensates. For the operators mentioned above the answer is [352]

$$\langle O_\rho\rangle \approx \frac{224\pi}{81}\alpha_s\langle\bar{\psi}\psi\rangle^2, \qquad \langle O_{A_1}\rangle \approx -\frac{352\pi}{81}\alpha_s\langle\bar{\psi}\psi\rangle^2. \qquad (6.115)$$

Note the *opposite sign* of these VEVs, which will be important shortly.

How these expressions work is demonstrated in Fig. 6.12, where the solid curves are phenomenological ones (derived from experimental data in Section 6.2.1, the short-dashed line corresponds to perturbative and gluon condensate corrections,

---

[23] The accuracy of this estimate should of course be questioned, and lattice data, instanton liquid and other vacuum models can be used for that. See the instanton based discussion in Ref. [226]. In short, it was found that "instanton liquid" strongly contradicts this hypothesis for gluonic operators (deviation is about an order of magnitude), it contradicts by a factor 1.5–2 scalar-type operators like $(\bar{q}q)^2$, and more or less agrees with it for the operators currently considered.

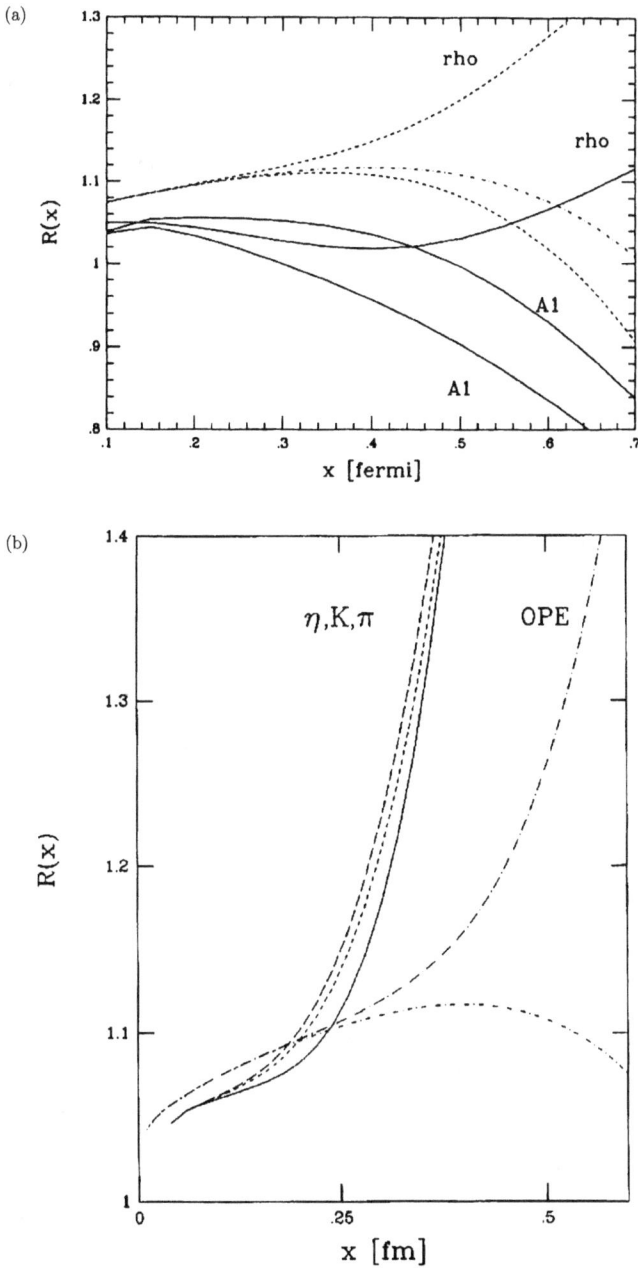

Fig. 6.12. (a) The correlation functions for the vector and axial $I = 1$ channel compared to OPE prediction. Note the extended scale. (b) The octet correlators at small distances compared with the OPE prediction.

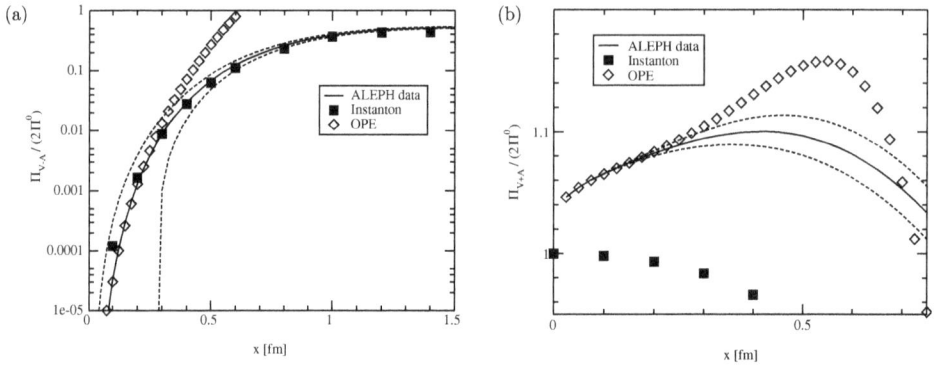

Fig. 6.13. (a) The $\Pi_V(x) - \Pi_A(x)$ correlator plotted on a logarithmic scale and (b) $\Pi_V(x) + \Pi_A(x)$ correlator shown in more detail. In both the experimental data are shown by a solid line plus an error corridor, the OPE is shown by open points following the SVZ estimates, and the closed points correspond to RILM.

while those marked ρ, SVZ and A_1, SVZ correspond to four-fermion operators included. One can see, that the general behavior of these correlation functions is indeed reproduced well, up to distances of about $1/2$ fm. In particular:

(i) the splitting between the vector and axial $I = 1$ channels happens exactly in the right place;

(ii) the magnitude of the splitting is also correct[24];

(iii) in the vector case the quark and gluon corrections nearly compensate each other, so that at least the *beginning of "superduality"* is reproduced.

A different analysis by Schafer and myself [244] used different data, from the ALEPH collaboration on τ decay, and it also organized the correlators differently, focused on $V \pm A$ combinations. Fig. 6.13 shows how the experimentally determined correlator (with the allowed region shown by two dashed lines around it) corresponds to OPE (diamonds). In the $V - A$ combination it starts with the four-fermion operator, of which the magnitude is tuned to small-x behavior, and the OPE described the data well for $x < 0.3$ fm. In the $V + A$ combination the main deviation from 1 is actually the perturbative $\alpha_s(x)/\pi$ correction, and (with running coupling included[25]) it also describes the data for $x < 0.3$ fm.

Although OPE looks great at this point, there is the following puzzling question. It was noticed at the beginning of this section that the OPE as used by SVZ in the momentum space, ignores the regular (non-singular at $x = 0$) part of the correlation functions such as the one shown in Fig. 6.11b. Let us estimate what corrections can

---

[24] A discrepancy between OPE curves and experimental ones, about 10 percent in absolute normalization, is probably due to higher order radiative corrections. They are also comparable to the experimental uncertainties in the axial channel.

[25] Note that one can deduce the Λ_{QCD} value from these data, as has been done by a number of authors prior to our paper. Needless to say, our extracted value agrees with theirs.

arise from them. In fact, the theory of the regular terms is even much simpler than that of the singular ones; it is just the ordinary Taylor expansion. In particular, the constant term at $x \to 0$ is nothing else but just the original current *squared* and averaged; $K_{regular}(x \to 0) = \langle j_\mu^2(0) \rangle$. One can alternatively evaluate their vacuum expectation values in the same "vacuum dominance" approximation, gaining some insight into the effect of those regular terms. Moreover, as all currents considered in this section have simple flavor structure $\bar{u}d$, two quark lines do not mix and one can simply get all $O(\langle \bar{\psi}\psi \rangle^2)$ corrections by using the quark propagator in the form $S = S_0 + \frac{1}{12}\langle \bar{\psi}\psi \rangle$ as we did for heavy-light systems. Simple results which follow

$$\Pi_{\mu\mu}^{\rho,A1}(\tau)/\Pi_{\mu\mu}^{free}(\tau) = 1 \pm \frac{\pi^4 \tau^6}{18}\langle \bar{\psi}\psi \rangle^2, \tag{6.116}$$

$$\Pi_{\pi,a0}(\tau)/\Pi_{free}(\tau) = 1 \pm \frac{\pi^4 \tau^6}{36}\langle \bar{\psi}\psi \rangle^2. \tag{6.117}$$

Those terms have behavior and magnitude similar to the original OPE ones. The clarity is provided by lattice studies with "cooled" quantum configurations, which are supposed to eliminate quantum gluons and the diagrams (a). As one can see from the right panel of Fig. 6.19, the "cooled" (classical) vector correlator is the same as the uncooled (quantum); one can conclude from this that the perturbative effect cannot be the dominant one.

Let us now look at other channels for which there is experimental information, in particular at the pseudoscalar ones. The OPE expression used were obtained already in the first paper [352]. In coordinate the representation we use, they are

$$\frac{\Pi_\pi(\tau)}{\Pi_{free}(\tau)} = 1 + \frac{11\alpha_s(\tau)}{3\pi} + \langle (gG_{\mu\nu}^a)^2 \rangle \frac{\tau^4}{384} + \frac{\pi^2}{48}\tau^6 \log\frac{1}{\tau\lambda}\langle O_1 + O_2 \rangle, \tag{6.118}$$

where the four-fermion operators are defined as

$$O_1 = -\pi\alpha_s(\bar{u}\sigma_{\mu\nu}t^a u)(\bar{d}\sigma_{\mu\nu}t^a d), \tag{6.119}$$

$$O_2 = (\pi\alpha_s/2)\left[(\bar{u}\sigma_{\mu\nu}t^a u)^2 + (\bar{d}\sigma_{\mu\nu}t^a d)^2\right] + (\pi\alpha_s/3)\left[(\bar{u}\gamma_\mu t^a u)(\Sigma_q \bar{q}\gamma_\mu t^a q)\right]. \tag{6.120}$$

Their average values are estimated according to the "vacuum dominance hypothesis", and the main one is

$$\langle O_2 \rangle \approx \frac{56\pi}{27}\alpha_s\langle \bar{\psi}\psi \rangle^2. \tag{6.121}$$

The resulting curves are compared to phenomenological ones for π, K, η mesons in Fig. 6.12b. Although the OPE prediction has "good intentions" in terms of the sign, but numerically these effects are way too small to reproduce the experimental trend. This failure was discussed by Novikov *et al.* [396] as part of a general trend; for all spin-zero correlation functions some large effects show up which are "not seen" in the OPE framework. The regular terms discussed above are also too small to help.

We will see in the next section that these puzzles are resolved once the instanton-induced effects are included.

6.4.   Instantons and the correlators: analytic results

6.4.1.   *Propagator in the field of a single instanton*

The propagators of massless scalar bosons, gauge fields and fermions in the background field of a single instanton can be determined analytically[26] [206]. We do not go into details of the construction, which is quite technical, but only provide the main results.

We have already seen that the quark propagator in the field of an instanton is ill behaved because of the presence of a zero mode. The remaining non-zero mode part of the propagator satisfies the equation

$$i\slashed{D} S^{nz}(x,y) = \delta(x-y) - \psi_0(x)\psi_0^\dagger(y), \tag{6.122}$$

which ensures that all modes in S^{nz} are orthogonal to the zero mode. Formally, this equation is solved by

$$S^{nz}(x,y) = \overrightarrow{\slashed{D}}_x \, \Delta(x,y) \left(\frac{1+\gamma_5}{2}\right) + \Delta(x,y) \, \overleftarrow{\slashed{D}}_y \left(\frac{1-\gamma_5}{2}\right), \tag{6.123}$$

where $\Delta(x,y)$ is the propagator of a scalar quark in the fundamental representation, $-D^2\Delta(x,y) = \delta(x,y)$. Equation (6.123) is easily checked using the methods we employed in order to construct the zero mode solution, see Eq. (4.8). The scalar propagator does not have any zero modes, so it can be constructed using standard techniques. The result (in singular gauge) is Ref. [206]

$$\Delta(x,y) = \frac{1}{4\pi^2(x-y)^2} \frac{1}{\sqrt{1+\rho^2/x^2}} \frac{1}{\sqrt{1+\rho^2/y^2}} \left(1 + \frac{\rho^2\tau^- \cdot x\tau^+ \cdot y}{x^2 y^2}\right). \tag{6.124}$$

For an instanton located at z, one has to make the obvious replacements $x \to (x-z)$ and $y \to (y-z)$. The propagator in the field of an anti-instanton is obtained by interchanging τ^+ and τ^-. If the instanton is rotated by the color matrix R^{ab}, then $\tau^\pm$ have to be replaced by $(R^{ab}\tau^b, \mp i)$.

Using the result for the scalar quark propagator and the representation (6.123) of the spinor propagator introduced above, the non-zero mode propagator is given by

$$\begin{aligned}
S^{nz}(x,y) = {}& \frac{1}{\sqrt{1+\rho^2/x^2}} \frac{1}{\sqrt{1+\rho^2/y^2}} \left[S_0(x,y)\left(1 + \frac{\rho^2\tau^- \cdot x\tau^+ \cdot y}{x^2 y^2}\right) \right. \\
& - D_0(x,y)\frac{\rho^2}{x^2 y^2} \left(\frac{\tau^- \cdot x\tau^+ \cdot \gamma\tau^- \cdot (x-y)\tau^+ \cdot y}{\rho^2 + x^2} \gamma_+ \right. \\
& \left. \left. + \frac{\tau^- \cdot x\tau^+ \cdot (x-y)\tau^- \cdot \gamma\tau^+ \cdot y}{\rho^2 + x^2} \gamma_- \right) \right],
\end{aligned} \tag{6.125}$$

---

[26] The result is easily generalized to 't Hooft's exact multi-instanton solution, but much more effort is required to construct the quark propagator in the most general (ADHM) instanton background [204].

where $\gamma_\pm = (1 \pm \gamma_5)/2$. The propagator can be generalized to arbitrary instanton positions and color orientations in the same way as the scalar quark propagator discussed above.

At short distances, as well as far away from the instanton, the propagator reduces to the free one. At intermediate distances, the propagator is modified due to gluon exchanges with the instanton field,

$$S^{nz}(x,y) = -\frac{\gamma \cdot (x-y)}{2\pi^2 (x-y)^4} - \frac{1}{16\pi^2 (x-y)^2}(x-y)_\mu \gamma_\nu \gamma_5 \tilde{G}_{\mu\nu} + \cdots. \qquad (6.126)$$

This result is consistent with the OPE of the quark propagator in a general background field. It is interesting to note that all the remaining terms are regular as $(x-y)^2 \to 0$. This has important consequences for the OPE of hadronic correlators in a general self-dual background field [334].

Finally, we need the quark propagator in the instanton field for small but non-vanishing quark mass. Expanding the quark propagator for small m, we get

$$S(x,y) = \frac{1}{i\not{D} + im}$$
$$= \frac{\psi_0(x)\psi_0^\dagger(y)}{m} + S^{nz}(x,y) + m\Delta(x,y) + \cdots. \qquad (6.127)$$

6.4.2. *First order in the 't Hooft effective vertex*

We have seen in Section 4.2.2 that the zero modes of the Dirac operator in the instanton field play a special role in the chiral limit $m_q \to 0$. For flavored currents the single-loop diagram in which only the zero mode parts of both propagators is used leads to an expression of the type

$$K(x-y) = \frac{n}{m_u m_d} \int d^4 z \, \bar\psi_0(x-z)\Gamma\psi_0(x-z)\bar\psi_0(y-z)\Gamma\psi_0(y-z), \qquad (6.128)$$

where for the time being $N_f = 2$ and we ignore the strange quark entirely, n is the instanton density and the matrix Γ is the gamma matrix in the vertex.

Now recall that for the instanton background field the quark zero mode is right-handed, while the antiquark one is left-handed only. (It is flipped L $\leftrightarrow$ R for the anti-instanton.) This leads to the following general conclusions.

- Such contribution in the vector or the axial channels, when $\Gamma = \gamma_\mu, \gamma_\mu\gamma_5$, is *zero* since these currents are non-zero only if both spinors have the same chirality.
- For scalar and pseudoscalar[27] channels $\Gamma = 1, i\gamma_5$, it is non-zero and has the *opposite sign*.

---

[27] Note that we put i into the current, in order to keep the zeroth order free quark loop the same in both cases.

- For flavored current one can see that the pseudoscalar π gets a positive relative correction while the scalar δ or a_0 gets a negative one.
- An analogous calculation for flavor-singlet scalar σ (or f_0) and pseudoscalar we would still consider that η' gets split as well, with the former getting a positive and the latter a negative contribution.
- The absolute magnitude of these corrections for all four cases considered is the same.

So, all the signs are in perfect agreement with phenomenology, which (as discussed in the previous chapter) does indeed suggest light π, σ and heavy a_0, η'.

Now, as we are satisfied that the signs are correct, what about the absolute magnitude of these corrections? The product of small quark masses appears in the denominator, which however can be tamed in a single-instanton background; its density n also has the product of these masses due to the fermion determinant.

However, a consistent evaluation of any effect cannot proceed without accounting for broken chiral symmetry in the QCD vacuum and condensates. In the single instanton approximation (SIA) discussed in Chapter 4, instead of the bare quark masses one should substitute properly defined effective masses. In the first paper on the subject [227] it was done more crudely, in the MFA, for the π, K, η, η' correlators.

Let us return to a single instanton background and proceed to the vector channel. We have shown above that the zero mode term does not contribute in this case, so the correlator is actually finite in the chiral limit without $m_u * m_d$ in the fermionic determinant. The calculation itself was done by Andrei and Gross [389], and this paper created at the time significant controversy.

Non-vanishing contributions come from the non-zero mode propagator (6.125) and from interference between the zero mode part and the leading mass correction in (6.127). The latter term survives even in the chiral limit, because the factor m in the mass correction is canceled by the $1/m$ from the zero mode.

$$\Pi_\rho^{AG}(x,y) = \text{Tr}\left[\gamma_\mu S^{nz}(x,y)\gamma_\mu S^{nz}(y,x)\right]$$
$$+ 2\,\text{Tr}\left[\gamma_\mu \psi_0(x)\psi_0^\dagger(y)\gamma_\mu \Delta(y,x)\right]. \tag{6.129}$$

After averaging over the instanton coordinates, the result is[28]

$$\Pi_\rho^{SIA}(x) = \Pi_\rho^0 + \int d\rho\, n(\rho)\frac{12}{\pi^2}\frac{\rho^4}{x^2}\frac{\partial}{\partial(x^2)}\left\{\frac{\xi}{x^2}\log\frac{1+\xi}{1-\xi}\right\}, \tag{6.130}$$

where $\xi^2 = x^2/(x^2 + 4\rho^2)$.

---

[28]There is a mistake by an overall factor 3/2 in the original work, originating from color traces. In other words, the result is correct in SU(2).

The reason we discuss this result is its relations to the OPE. Expanding (6.130), we get

$$\Pi_\rho^{\text{SIA}}(x) = \Pi_\rho^0(x) \left(1 + \frac{\pi^2 x^4}{6} \int d\rho\, n(\rho) \right). \qquad (6.131)$$

This agrees exactly with the OPE expression, provided we use the average values of the operators in the dilute gas approximation

$$\langle g^2 G^2 \rangle = 32\pi^2 \int d\rho\, n(\rho), \qquad m\langle \bar{q}q \rangle = -\int d\rho\, n(\rho). \qquad (6.132)$$

Note, that the value of $m\langle \bar{q}q \rangle$ is "anomalously" large in the dilute gas limit. This means that the contribution from dimension-4 operators is attractive, in contradiction with the OPE prediction based on the canonical values of the condensates.

An interesting observation is the fact that (6.131) is the only singular term in the DIGA correlation function. In fact, the OPE of *any* mesonic correlator in *any* self-dual field contains only dimension-4 operators [334]. This means that for all higher order operators either the Wilson coefficient vanishes (as it does, for example, for the triple gluon condensate $\langle f^{abc} G^a_{\mu\nu} G^b_{\nu\rho} G^c_{\rho\mu} \rangle$) or the matrix elements of various operators of the same dimension cancel each other.[29]

This is a very remarkable result, because it helps to explain the success of QCD sum rules based on the OPE in many channels. In the instanton model, the gluon fields are very inhomogeneous, so one would expect that the OPE fails for $x > \rho$. The Dubovikov–Smilga result shows that quarks can propagate through very strong gauge fields (as long as they are self-dual) without suffering strong interactions.

6.4.3. *Propagator in the instanton ensemble*

In this section we generalize the results of the last section to the more general case of an ensemble consisting of many pseudo-particles. The quark propagator in an arbitrary gauge field can always be expanded as

$$S = S_0 + S_0 \mathcal{A} S_0 + S_0 \mathcal{A} S_0 \mathcal{A} S_0 + \cdots, \qquad (6.133)$$

where the individual terms have an obvious interpretation as arising from multiple gluon exchanges with the background field. If the gauge field is a sum of instanton

---

[29] Here, we do not consider radiative corrections like $\alpha_s \langle \bar{\psi}\psi \rangle^2$. Technically, this is because we evaluate the OPE in a fixed (classical) background field without taking into account radiative corrections to the background field.

contributions, $A_\mu = \sum_I A_{I\mu}$, then (6.133) becomes

$$S = S_0 + \sum_I S_0 \mathcal{A}_I S_0 + \sum_{I,J} S_0 \mathcal{A}_I S_0 \mathcal{A}_J S_0 + \cdots \tag{6.134}$$

$$= S_0 + \sum_I (S_I - S_0) + \sum_{I \neq J} (S_I - S_0) S_0^{-1} (S_J - S_0)$$

$$+ \sum_{I \neq J, J \neq K} (S_I - S_0) S_0^{-1} (S_J - S_0) S_0^{-1} (S_K - S_0) + \cdots . \tag{6.135}$$

Here, $I, J, K, \ldots$ refer to both instantons and anti-instantons. In the second line, we have re-summed the contributions corresponding to an individual instanton. S_I refers to the sum of zero and non-zero mode components. At large distance from the center of the instanton, S_I approaches the free propagator S_0. Thus Eq. (6.135) has a nice physical interpretation; Quarks propagate by jumping from one instanton to the other. If $|x - z_I| \ll \rho_I, |y - z_I| \ll \rho_I$ for all I, the free propagator dominates. At large distances, terms involving more and more instantons become important.

In the QCD ground state, chiral symmetry is broken. The presence of a condensate implies that quarks can propagate over large distances. Therefore, we cannot expect that truncating the series (6.135) will provide a useful approximation to the propagator at low momenta. Furthermore, we know that spontaneous symmetry breaking is related to small eigenvalues of the Dirac operator. A good approximation to the propagator is obtained by assuming that $S_I - S_0$ is dominated by fermion zero modes

$$(S_I - S_0)(x, y) \simeq \frac{\psi_I(x)\psi_I^\dagger(y)}{im}. \tag{6.136}$$

In this case, the expansion (6.135) becomes

$$S(x, y) \simeq S_0(x, y) + \sum_I \frac{\psi_I(x)\psi_I^\dagger(y)}{im}$$

$$+ \sum_{I \neq J} \frac{\psi_I(x)}{im} \left(\int d^4 r \psi_I^\dagger(r)(-i\slashed{\partial} - im)\psi_J(r) \right) \frac{\psi_J^\dagger(y)}{im} + \cdots , \tag{6.137}$$

which contains the overlap integrals T_{IJ} defined in Eq. (4.64). This expansion can easily be summed to give

$$S(x, y) \simeq S_0(x, y) + \sum_{I,J} \psi_I(x) \frac{1}{T_{IJ} + im\, D_{IJ} - im\, \delta_{IJ}} \psi_J^\dagger(y). \tag{6.138}$$

Here, $D_{IJ} = \int d^4 r \psi_I^\dagger(r)\psi_J(r) - \delta_{IJ}$ arises from the restriction $I \neq J$ in the expansion (6.135). The quantity mD_{IJ} is small in both the chiral expansion and in the packing fraction of the instanton liquid and will be neglected in what follows. Comparing the re-summed propagator (6.138) with the single instanton propagator (6.136) shows the importance of chiral symmetry breaking. While (6.136) is proportional to $1/m$, the diagonal part of the full propagator is proportional to $(T^{-1})_{II} = 1/m^*$.

The result (6.138) can also be derived by inverting the Dirac operator in the basis spanned by the zero modes of the individual instantons

$$S(x, y) \simeq S_0(x, y) + \sum_{I,J} |I\rangle \left\langle I \left| \frac{1}{i\slashed{D} + im} \right| J \right\rangle \langle J|. \tag{6.139}$$

The equivalence of (6.138) and (6.139) is easily seen using the fact that in the sum ansatz, the derivative in the overlap matrix element T_{IJ} can be replaced by a covariant derivative.

The propagator (6.138) can be calculated either numerically or using the mean field approximation introduced in Section 6.4.1. We will discuss the mean field propagator in the following section. For our numerical calculations, we have improved the zero mode propagator by adding the contributions from non-zero modes to first order in the expansion (6.135). The result is

$$S(x, y) = S_0(x, y) + S^{\mathrm{ZMZ}}(x, y) + \sum_I \left(S_I^{\mathrm{NZM}}(x, y) - S_0(x, y) \right). \tag{6.140}$$

How accurate is this propagator? We have seen that the propagator agrees with the general OPE result at short distances. We also know that it accounts for chiral symmetry breaking and spontaneous mass generation at large distances. In addition to that, we have performed a number of checks on the correlation functions that are sensitive to the degree to which (6.140) satisfies the equations of motion, for example by testing whether the vector correlator is transverse (the vector current is conserved).

6.4.4. *Propagator in the mean field approximation*

In order to understand the propagation of quarks in the "zero mode zone" it is very instructive to construct the propagator in the mean field approximation. The mean field propagator can be obtained in several ways. Most easily, we can read off the propagator directly from the effective partition function (4.52). We find

$$S(p) = \frac{\slashed{p} + M(p)}{p^2 + M^2(p)}, \tag{6.141}$$

with the momentum dependent effective quark mass

$$M(p) = \frac{\beta}{2N_c} \frac{N}{V} p^4 \varphi'^2(p). \tag{6.142}$$

Here, β is the solution of the gap equation (4.53). Originally, the result (6.141) was obtained by Diakonov and Petrov from the Dyson–Schwinger equation for the quark propagator in the large N_c limit [229]. At small momenta, chiral symmetry breaking generates an effective mass $M(0) = (\beta/2N_c)(N/V)(2\pi\rho)^2$. The quark condensate was already given in Eq. (4.58). At large momenta, we have $M(p) \sim 1/p^6$ and constituent quarks become free current quarks.

For comparison with our numerical results, it is useful to determine the mean field propagator in coordinate space. Fourier transforming the result (6.141) gives

$$S_V(x) = \frac{1}{4\pi^2 x} \int dp \frac{p^3}{p^2 + M^2(p)} J_2(px), \tag{6.143}$$

$$S_S(x) = \frac{1}{4\pi^2 x} \int dp \frac{p^2 M(p)}{p^2 + M^2(p)} J_1(px). \tag{6.144}$$

The result is shown in Fig. 6.15. The scalar and vector components of the propagator are normalized to the free propagator. The vector component of the propagator is exponentially suppressed at large distance, showing the formation of a constituent mass. The scalar component again shows the breaking of chiral symmetry. At short distance, the propagator is consistent with the coordinate dependent quark mass $m = (\pi^2/3) \cdot x^2 \langle \bar{\psi}\psi \rangle$ inferred from the OPE expression. At large distance the exponential decay is governed by the constituent mass $M(0)$.

For comparison, we also show the quark propagator in different instanton ensembles. The result is very similar to the mean field approximation, the differences are mainly due to different values of the quark condensate. The quark propagator is not very sensitive to correlations in the instanton liquid. In Refs. [237, 393], the quark propagator was also used to study heavy–light mesons in the $1/M_Q$ expansion. The results are very encouraging, and we refer the reader to the original literature for details.

6.4.5. *Correlators in the random phase approximation*

The random phase approximation (RPA) is a standard way to find collective excitations in many-body problems. In our context it corresponds to iteration of the average 't Hooft vertex in the s-channel via the Bethe–Salpeter equation [229, 394]

$$\Pi_\pi^{\text{RPA}}(x) = \Pi_\pi^{\text{MFA}}(x) + \Pi_\pi^{\text{int}}, \qquad \Pi_\rho^{\text{RPA}}(x) = \Pi_\rho^{\text{MFA}}(x). \tag{6.145}$$

Here Π_Γ^{MFA}, denotes the mean field (non-interacting) part of the correlation functions

$$\Pi_\Gamma^{\text{MFA}}(x) = \text{Tr}\left[\bar{S}(x)\Gamma\bar{S}(-x)\Gamma\right], \tag{6.146}$$

where $\bar{S}(x)$ is the mean field propagator discussed in Section 6.4.4. In the ρ meson channel, the 't Hooft vertex vanishes and the correlator is given by the mean field contribution only. The interacting part of the π, η' correlation functions is given by

$$\Pi_{\pi,\eta'}^{\text{int}}(x) = N_c(N_cV/N) \int d^4q\, e^{iq\cdot x}\, \Gamma_5(q) \frac{\pm 1}{1 \mp C_5(q)} \Gamma_5(q) \tag{6.147}$$

where the elementary loop function C_5 and the vertex function Γ_5 are given by

$$C_5(q) = 4N_c \left(\frac{V}{N}\right) \int \frac{d^4p}{(2\pi)^4} \frac{M_1 M_2 (M_1 M_2 - p_1 \cdot p_2)}{(M_1^2 + p_1^2)(M_2^2 + p_2^2)}, \tag{6.148}$$

$$\Gamma_5(q) = 4 \int \frac{d^4p}{(2\pi)^4} \frac{\sqrt{M_1 M_2}(M_1 M_2 - p_1 \cdot p_2)}{(m_1^2 + p_1^2)(M_2^2 + p_2^2)}, \tag{6.149}$$

where $p_1 = p + q/2$, $p_2 = p - q/2$ and $M_{1,2} = M(p_{1,2})$ are the momentum dependent effective quark masses. We have already stressed that in the long wavelength limit, the effective interaction between quarks is of NJL type. Indeed, the pion correlator in the RPA approximation to the NJL model is given by

$$\Pi_\pi^{\mathrm{NJL}}(x) = \int d^4q \, e^{iq \cdot x} \frac{J_5(q)}{1 - G J_5(q)}, \tag{6.150}$$

where $J_5(q)$ is the pseudo-scalar loop function

$$J_5(q) = 4N_c \int^\Lambda \frac{d^4p}{(2\pi)^4} \frac{(M^2 - p_1 \cdot p_2)}{(M^2 + p_1^2)(M^2 + p_2^2)}, \tag{6.151}$$

G is the four fermion coupling constant and the (momentum independent) constituent mass M is the solution of the NJL gap equation. The loop function (6.151) is divergent and is regularized using a cutoff Λ.

Clearly, the RPA correlation functions (6.147) and (6.150) are very similar, the only difference is that in the instanton liquid both the quark mass and the vertex are momentum dependent. The momentum dependence of the vertex ensures that no regulator is required, the loop integral is automatically cut off at $\Lambda \sim \rho^{-1}$. We

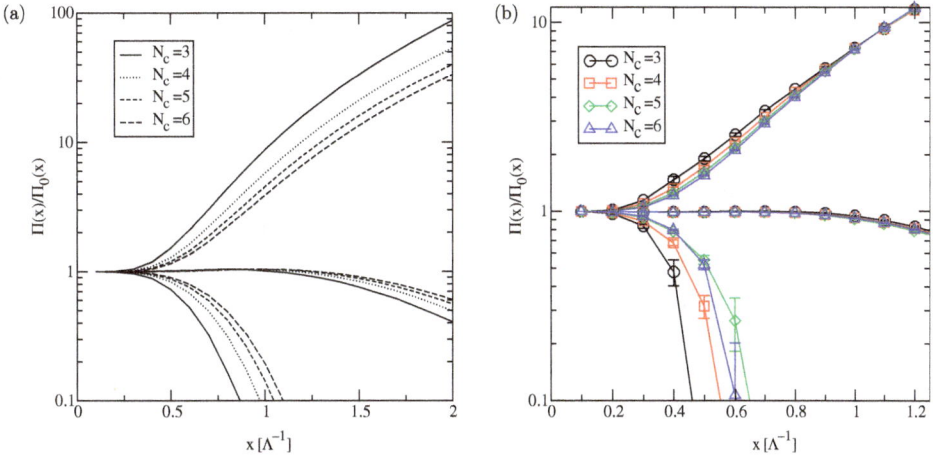

Fig. 6.14. Correlation functions in the pion, rho meson, and η' meson channel. The correlators are shown as a function of the distance in units of the inverse scale parameter. The correlation functions are normalized to free field behavior, $\Pi_0(x) \sim N_c/x^6$. The results shown in Fig. (a) were obtained using the mean field approximation for different number of colors. In Fig. (b) the same correlators are shown calculated numerically in pure gauge (quenched) instanton liquid.

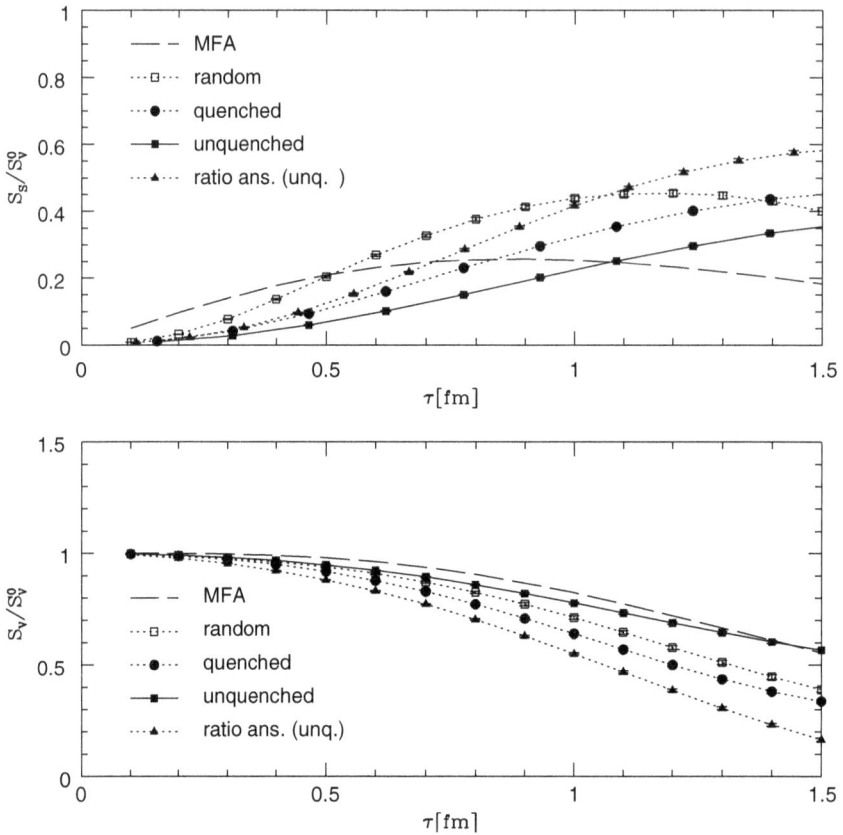

Fig. 6.15. Scalar and vector component of the quark propagator, normalized to the free vector propagator. The dashed lines show the result of the mean field approximation while the data points were obtained in different instanton ensembles.

also note that the momentum dependence of the effective interaction at the mean field level is very simple, the vertex function is completely separable.

The RPA and numerical (quenched) results [246] for the pion, rho meson, and η' are shown in the following figure (Fig. 6.14) for several values of $N_{\rm c}$. One can see that the qualitative behavior is as phenomenology demands. One can also observe that the η' correlator gets less repulsive as $N_{\rm c}$ grows.

What is the quality and the range of validity of the RPA? The RPA is usually motivated by the large $N_{\rm c}$ approximation, and the dimensionless parameter that controls the expansion is $\rho^4(N/V)/(4N_c)$. This parameter is indeed very small, but in practice there are additional parameters that determine the size of corrections to the RPA. First of all, $\rho^4(N/V)$ is a useful expansion parameter only if the instanton liquid is random and the role of correlations is small. As discussed in Section 4.3.1, if the density is very small the instanton liquid is in a molecular phase, while it is in a crystalline phase if the density is large. Clearly, the RPA is expected to fail in both of these limits. In general, the RPA corresponds to using linearized equations for the fluctuations around a mean field solution. In our case, low lying meson states are collective fluctuations of the chiral order parameter. For isovector scalar mesons, the RPA is expected to be good, because the scalar mean field (the condensate) is large and the masses are light. For isosinglet mesons, the fluctuations are much larger, and the RPA is likely to be less useful.

6.5. Correlators in the instanton liquid

6.5.1. *Quark propagator in the instanton liquid*

As we have emphasized above, the complete information about mesonic and baryonic correlation functions is encoded in the quark propagator in a given gauge field configuration. Interactions among quarks are represented by the failure of expectation values to factorize, e.g. $\langle S(\tau)^2 \rangle \neq \langle S(\tau) \rangle^2$. In the following, we will construct the quark propagator in the instanton ensemble, starting from the propagator in the background field of a single instanton.

From the quark propagator, we calculate the ensemble averaged meson and baryon correlation functions. However, it is also interesting to study the vacuum expectation value of the propagator[30]

$$\langle S(x) \rangle = S_S(x) + \gamma \cdot x S_V(x), \qquad (6.152)$$

From the definition of the quark condensate, we have $\langle \bar{q}q \rangle = S_S(0)$, which means that the scalar component of the quark propagator provides an order parameter for chiral symmetry breaking. To obtain more information, we can define a gauge invariant propagator by adding a Wilson line

$$S_{\mathrm{inv}}(x) = \langle \psi(x) P \exp \left(\int_0^x A_\mu(x') \, dx'_\mu \right) \bar{\psi}(0) \rangle. \qquad (6.153)$$

This object has a direct physical interpretation, because it describes the propagation of a light quark coupled to an infinitely heavy, static, source [226, 237, 393]. It therefore determines the spectrum of heavy–light mesons (with the mass of the heavy quark subtracted) in the limit where the mass of the heavy quark goes to infinity.

---

[30]The quark propagator is of course not a gauge invariant object. Here, we imply that a gauge has been chosen or the propagator is multiplied by a gauge string. Also note that before averaging, the quark propagator has a more general Dirac structure $S(x) = E + P\gamma_5 + V_\mu\gamma_\mu + A_\mu\gamma_\mu\gamma_5 + T_{\mu\nu}\sigma_{\mu\nu}$. This decomposition, together with positivity, is the basis of a number of exact results about correlation functions [328, 329].

6.5.2. *Mesonic correlators*

In the following we will therefore discuss results from numerical calculations of hadronic correlators in the instanton liquid. These calculations go beyond the RPA in two ways: (i) the propagator includes genuine many instanton effects and non-zero mode contributions; (ii) the ensemble is determined using the full (fermionic and bosonic) weight function, so it includes correlations among instantons. In addition to that, we will also consider baryonic correlators and three-point functions that are difficult to handle in the RPA.

We will discuss correlation functions in three different ensembles, the random ensemble (RILM), the quenched (QILM) and fully interacting (IILM) instanton ensembles. In the random model, the underlying ensemble is the same as in the mean field approximation, only the propagator is more sophisticated. In the quenched approximation, the ensemble includes correlations due to the bosonic action, while the fully interacting ensemble also includes correlations induced by the fermion determinant. In order to check the dependence of the results on the instanton interaction, we study correlation functions in two different unquenched ensembles, one based on the streamline interaction (with a short-range core) and one based on the ratio ansatz interaction. The bulk parameters of these ensembles are compared in Table 6.5.

Correlation functions in the different instanton ensembles were calculated in [238, 239, 242] to which we refer the reader for more details. The results are shown in Fig. 6.16 and summarized in Table 6.6. The pion correlation functions in the different ensembles are qualitatively very similar. The differences are mostly due to different values of the quark condensate (and the physical quark mass) in the different ensembles. Using the Gell-Mann, Oaks, Renner relation, one can extrapolate the pion mass to the physical value of the quark masses, see Table 6.6. The results are consistent with the experimental value in the streamline ensemble (both quenched and unquenched), but clearly too small in the ratio ansatz ensemble. This is a reflection of the fact that the ratio ansatz ensemble is not sufficiently dilute.

In Fig. 6.16 we also show the results in the ρ channel. The ρ meson correlator is not affected by instanton zero modes to first order in the instanton density. The

Table 6.5. Bulk parameters of different instanton ensembles.

	Streamline	Quenched	Ratio ansatz	RILM
n	$0.174\Lambda^4$	$0.303\Lambda^4$	$0.659\Lambda^4$	$1.0\,\mathrm{fm}^4$
$\bar{\rho}$	$0.64\Lambda^{-1}$	$0.58\Lambda^{-1}$	$0.66\Lambda^{-1}$	$0.33\,\mathrm{fm}$
	$(0.42\,\mathrm{fm})$	$(0.43\,\mathrm{fm})$	$(0.59\,\mathrm{fm})$	
$\bar{\rho}^4 n$	0.029	0.034	0.125	0.012
$\langle \bar{q}q \rangle$	$0.359\Lambda^3$	$0.825\Lambda^3$	$0.882\Lambda^3$	$(264\,\mathrm{MeV})^3$
	$(219\,\mathrm{MeV})^3$	$(253\,\mathrm{MeV})^3$	$(213\,\mathrm{MeV})^3$	
Λ	$306\,\mathrm{MeV}$	$270\,\mathrm{MeV}$	$222\,\mathrm{MeV}$	—

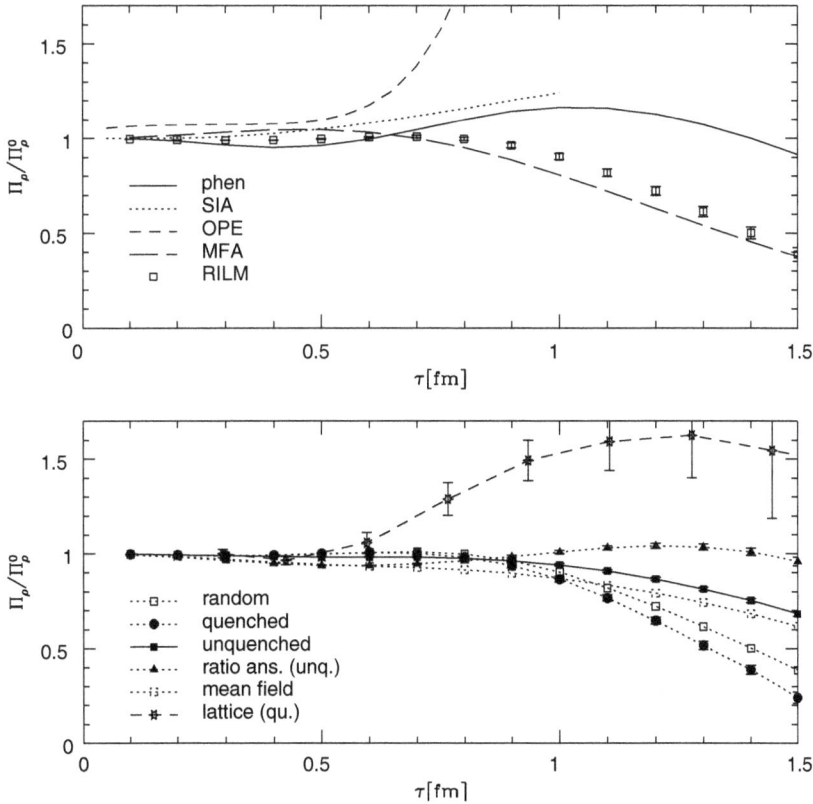

Fig. 6.16. Rho meson correlation functions. The dashed squares show the non-interacting part of the rho meson correlator in the interacting ensemble.

results in the different ensembles are fairly similar to each other and all fall somewhat short of the phenomenological result at intermediate distances $x \simeq 1$ fm. We have determined the ρ meson mass and coupling constant from a fit, the results are given in Table 6.6. The ρ meson mass is somewhat too heavy in the random and quenched ensembles, but in good agreement with the experimental value $m_\rho = 770$ MeV in the unquenched ensemble.

Since there are no interactions in the ρ meson channel to first order in the instanton density, it is important to study whether the instanton liquid provides any significant binding. In the instanton model, there is no confinement, and m_ρ is close to the two-(constituent)-quark threshold. In QCD, the ρ meson is also not a true bound state, but a resonance in the 2π continuum. In order to determine whether the continuum contribution in the instanton liquid is predominantly from 2-π or 2-quark states would require the determination of the corresponding three-point functions, which has not been done yet. Instead, we have compared the full correlation function with the non-interacting (mean field) correlator (6.146), where we use the average (constituent quark) propagator determined in the same ensemble,

Table 6.6. Meson parameters in the different instanton ensembles. All quantities are given in units of GeV. The current quark mass is $m_u = m_d = 0.1\Lambda$.
Except for the pion mass, no attempt has been made to extrapolate the parameters to physical values of the quark mass.

	Unquenched	Quenched	RILM	Ratio ansatz (unqu.)
m_π	0.265	0.268	0.284	0.128
m_π (extr.)	0.117	0.126	0.155	0.067
λ_π	0.214	0.268	0.369	0.156
f_π	0.071	0.091	0.091	0.183
m_ρ	0.795	0.951	1.000	0.654
g_ρ	6.491	6.006	6.130	5.827
m_{a_1}	1.265	1.479	1.353	1.624
g_{a_1}	7.582	6.908	7.816	6.668
m_σ	0.579	0.631	0.865	0.450
m_δ	2.049	3.353	4.032	1.110
$m_{\eta_{ns}}$	1.570	3.195	3.683	0.520

see Fig. 6.16. This comparison provides a measure of the strength of interaction. We
observe that there is an attractive interaction generated in the interacting liquid due
to correlated instanton–anti-instanton pairs. This is consistent with the fact that the
interaction is considerably smaller in the random ensemble. In the random model,
the strength of the interaction grows as the ensemble becomes more dense. However,
the interaction in the full ensemble is significantly larger than in the random model
at the same diluteness. Therefore, most of the interaction is due to dynamically
generated pairs.

We have already discussed ALEPH τ-decay data in Section 6.3.5, and have
shown in Fig. 6.13 the data compared to OPE and the calculation in the random
instanton liquid model (RILM) by Schafer and myself [244]. Another figure from
this work, Fig. 6.17, shown here shows the larger-x part of the correlator studied.
As one can see, RILM works for the whole $V - A$ curve, and, with 10% radiative
correction α_s/π, it works very well for $V + A$ as well. There is no fit of any parameter
here, and in fact the calculation preceded the experiment by a few years.

The situation is drastically different in the η' channel. Among the $\sim$40 correlation functions calculated in the random ensemble, only the η' (and the isovector-
scalar δ discussed in the next section) are completely unacceptable. The correlation
function decreases very rapidly and becomes negative at $x \sim 0.4$ fm. This behavior is incompatible with the positivity of the spectral function. The interaction in
the random ensemble is too repulsive, and the model "over-explains" the $U(1)_A$
anomaly.

The results in the unquenched ensembles (closed and open points) significantly improve the situation. This is related to dynamical correlations between
instantons and anti-instantons (topological charge screening). The single instanton

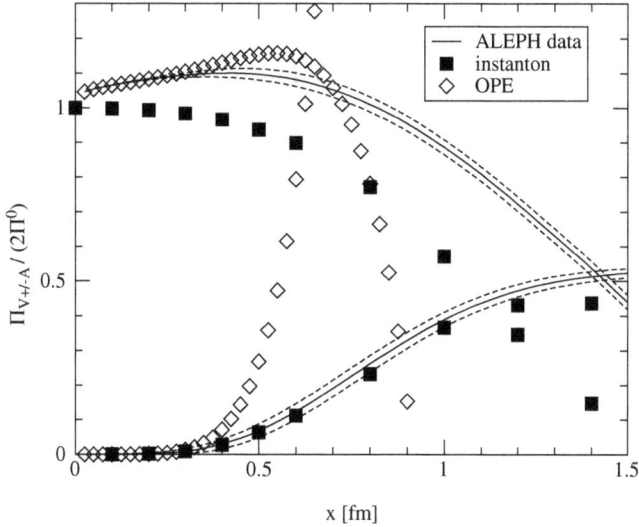

Fig. 6.17. Euclidean coordinate space correlation functions $\Pi_V(x) \pm \Pi_A(x)$ normalized to free field behavior. The solid lines show the correlation functions reconstructed from the ALEPH spectral functions and the dotted lines are the corresponding error band. The squares show the result of a random instanton liquid model and the diamonds the OPE fit described in the text.

contribution is repulsive, but the contribution from pairs is attractive [245]. Only if correlations among instantons and anti-instantons are sufficiently strong, are the correlators prevented from becoming negative. Quantitatively, the δ and η_{ns} masses in the streamline ensemble are still too heavy as compared to their experimental values. In the ratio ansatz, on the other hand, the correlation functions even shows an enhancement at distances on the order of 1 fm, and the fitted masses is too light. This shows that the η' channel is very sensitive to the strength of correlations among instantons.

In summary, pion properties are mostly sensitive to global properties of the instanton ensemble, in particular its diluteness. Good phenomenology demands $\bar{\rho}^4 n \simeq 0.03$, as originally suggested in Ref. [226]. The properties of the ρ meson are essentially independent of the diluteness, but show sensitivity to IA correlations. These correlations become crucial in the η' channel.

Let us comment on the dependence of the correlators on the number of colors N_c, studied in Ref. [246]. The correlation functions measured in numerical simulations of the instanton liquid for $N_c = 3, \ldots, 6$ are shown in Fig. 6.14. The results were obtained from simulations with $N = 128$ instantons in a Euclidean volume $V\Lambda^4 = V_3 \times 5.76$. V_3 was adjusted such that $N/V = (N_c/3)\Lambda^4$. In order to avoid finite volume artifacts the current quark mass was taken to be rather large, $m_q = 0.2\Lambda$. We observe that the rho meson correlation function exhibits almost perfect scaling with N_c and as a result the rho meson mass is practically independent of N_c. The scaling is not as good in the case of the pion. As a consequence there is some variation in the pion mass. However, this effect is consistent with $1/N_c$ corrections that amount to about 40% of the pion mass for $N_c = 3$. Finally, we study the behavior of the η' correlation function. There is a clear tendency toward $U(1)_A$ restoration, but the correlation function is still very repulsive for $N_c = 6$. It was also found that the η' correlation function only approaches the pion correlation for fairly large values of N_c. For example, the η' correlation function does not show intermediate range attraction unless $N_c > 15$.

After discussing the π, ρ, η' in some detail we only briefly comment on other correlation functions. The remaining scalar states are the isoscalar σ and the isovector δ (the f_0 and a_0 according to the notation of the particle data group). The sigma correlator has a disconnected contribution, which is proportional to $\langle \bar{q}q \rangle^2$ at large distance. In order to determine the lowest resonance in this channel, the constant contribution has to be subtracted, which makes it difficult to obtain reliable results. Nevertheless, we find that the instanton liquid favors a (presumably broad) resonance around 500–600 MeV. The isovector channel is in many ways similar to the η'. In the random ensemble, the interaction is too repulsive and the correlator becomes unphysical. This problem is solved in the interacting ensemble, but the δ is still very heavy, $m_\delta > 1\,\text{GeV}$.

The remaining non-strange vectors are the a_1, ω and f_1. The a_1 mixes with the pion, which allows a determination of the pion decay constant f_π (as does a direct measurement of the π–a_1 mixing correlator). In the instanton liquid, disconnected contributions in the vector channels are small. This is consistent with the fact that the ρ and the ω, as well as the a_1 and the f_1 are almost degenerate.

Finally, we can also include strange quarks. SU(3) flavor breaking in the 't Hooft interaction nicely accounts for the masses of the K and the η. More difficult is a correct description of η–η' mixing, which can only be achieved in the full ensemble. The random ensemble also has a problem with the mass splittings among the vectors ρ, K^* and ϕ [238]. This is related to the fact that flavor symmetry breaking in the random ensemble is so strong that the strange and non-strange constituent quark masses are almost degenerate. This problem is improved (but not fully solved) in the interacting ensemble.

6.5.3. *Baryonic correlation functions*

As emphasized a few times above, the existence of a strongly attractive interaction in the pseudo-scalar quark–anti-quark (pion) channel also implies an attractive interaction in the scalar quark–quark (diquark) channel. This interaction is phenomenologically very desirable, because it not only explains why the spin 1/2 nucleon is lighter than the spin 3/2 Delta, but also why Lambda is lighter than Sigma.

The vector components of the diagonal correlators receive perturbative quark-loop contributions, which are dominant at short distance. The scalar components of the diagonal correlators, as well as the off-diagonal correlation functions,

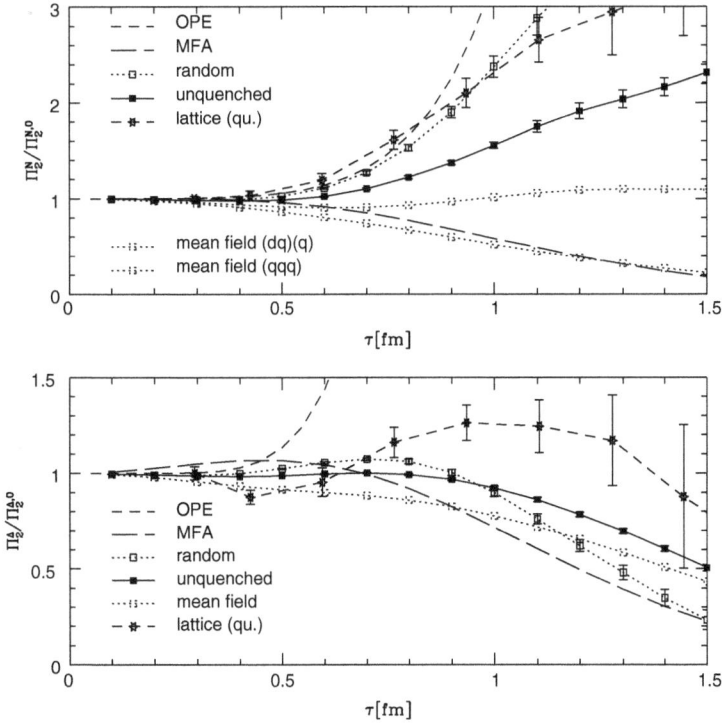

Fig. 6.18. Nucleon and Delta correlation functions Π_2^N and Π_2^Δ.

are sensitive to chiral symmetry breaking, and the OPE starts at order $\langle \bar{q}q \rangle$ or higher. Single instanton corrections to the correlation functions were calculated in Refs. [397, 398].[31] Instantons introduce additional, regular, contributions in the scalar channel and violate the factorization assumption for the four-quark condensates. Similarly to the pion case, both of these effects increase the amount of attraction already seen in the OPE.

The correlation function Π_2^N in the interacting ensemble is shown in Fig. 6.18. There is a significant enhancement over the perturbative contribution which corresponds to a tightly bound nucleon state with a large coupling constant. Numerically, we find[32] $m_N = 1.019 \, \text{GeV}$ (see Table 6.7). In the random ensemble, we have measured the nucleon mass at smaller quark masses and found $m_N = 0.960 \pm 0.30 \, \text{GeV}$. The nucleon mass is fairly insensitive to the instanton ensemble. However, the strength of the correlation function depends on the instanton ensemble. This is reflected by the value of the nucleon coupling constant, which is smaller in the interacting model.

---

[31] The latter paper corrects a few mistakes in the original work by Dorokhov and Kochelev.

[32] Note that this value corresponds to a relatively large current quark mass $m = 30 \, \text{MeV}$.

Table 6.7. Nucleon and delta parameters in the different instanton ensembles. All quantities are given in units of GeV. The current quark mass is $m_u = m_d = 0.1\Lambda$.

	Unquenched	Quenched	RILM	Ratio ansatz (unqu.)
m_N	1.019	1.013	1.040	0.983
λ_N^1	0.026	0.029	0.037	0.021
λ_N^2	0.061	0.074	0.093	0.048
m_Δ	1.428	1.628	1.584	1.372
λ_Δ	0.027	0.040	0.036	0.026

Figure 6.18 also shows the nucleon correlation function measured in a quenched lattice simulation [399]. The agreement with the instanton liquid results is quite impressive, especially given the fact that before the lattice calculations were performed, there was no phenomenological information on the value of the nucleon coupling constant and the behavior of the correlation function at intermediate and large distances.

The fitted position of the threshold is $E_0 \simeq 1.8\,\mathrm{GeV}$, larger than the mass of the first nucleon resonance, the Roper N*(1440), and above the $\pi\Delta$ threshold $E_0 = 1.37\,\mathrm{GeV}$. This might indicate that the coupling of the nucleon current to the Roper resonance is small. In the case of the $\pi\Delta$ continuum, this can be checked directly using the phenomenologically known coupling constants. The large value of the threshold energy also implies that there is little strength in the (unphysical) three-quark continuum. The fact that the nucleon is deeply bound can also be demonstrated by comparing the full nucleon correlation function with that of three non-interacting quarks, see Fig. 6.18. The full correlator is significantly larger than the non-interacting (mean field) result, indicating the presence of a strong, attractive interaction.

Some of this attraction is due to the scalar diquark content of the nucleon current. This raises the question whether the nucleon (in our model) is a strongly bound diquark very loosely coupled to a third quark. In order to check this, we have decomposed the nucleon correlation function into quark and diquark components. Using the mean field approximation, that means treating the nucleon as a non-interacting quark–diquark system, we get the correlation function labeled (diq) in Fig. 6.18. We observe that the quark–diquark model explains some of the attraction seen in Π_2^N, but falls short of the numerical results. This means that while diquarks may play some role in making the nucleon bound, there are substantial interactions in the quark–diquark system. Another hint for the qualitative role of diquarks is provided by the values of the nucleon coupling constants $\lambda_N^{1,2}$. Using (6.103), we can translate these results into the coupling constants $\lambda_N^{s,p}$ of nucleon currents built from scalar or pseudo-scalar diquarks. We find that the coupling to the scalar diquark current $\eta_s = \epsilon_{abc}(u^a C \gamma_5 d^b) u^c$ is an order of magnitude bigger than the coupling to the pseudo-scalar current $\eta_p = \epsilon_{abc}(u^a C d^b)\gamma_5 u^c$. This is in agreement with the idea that the scalar diquark channel is very attractive and that these configurations play an important role in the nucleon wave function.

The Delta correlation function in the instanton liquid is shown in Fig. 6.18. The result is qualitatively different from the nucleon channel, the correlator at intermediate distance $x \simeq 1$ fm is significantly smaller and close to perturbation theory. This is in agreement with the results of the lattice calculation [399]. Note that, again, this is a quenched result which should be compared to the predictions of the random instanton model.

The mass of the delta resonance is too large in the random model, but closer to experiment in the unquenched ensemble. Note that similarly to the nucleon, part of this discrepancy is due to the value of the current mass. Nevertheless, the Delta–nucleon mass splitting in the unquenched ensemble is $m_\Delta - m_N = 409$ MeV, still too large as compared to the experimental value 297 MeV. Similarly to the ρ meson, there is no interaction in the Delta channel to first order in the instanton density. However, if we compare the correlation function with the mean field approximation based on the full propagator, see Fig. 6.18, we find evidence for substantial attraction between the quarks. Again, more detailed checks, for example concerning the coupling to the πN continuum, are necessary.

6.5.4. *Comparison to correlators on the lattice*

The study of hadronic (point-to-point) correlation functions on the lattice was pioneered by the MIT group [399] which measured correlation functions of the $\pi, \delta, \rho, a_1, N$ and Δ in quenched QCD. The correlation functions were calculated on a $16^3 \times 24$ lattice at $6/g^2 = 5.7$, corresponding to a lattice spacing of $a \simeq 0.17$ fm. A more detailed investigation of baryonic correlation functions on the lattice can be found in Ref. [401]. We have already shown some of the results of the MIT group in Figs. 6.16–6.18. The correlators were measured for distances up to $\sim$1.5 fm. Using the parametrization introduced above, they extracted ground state masses and coupling constants and found good agreement with phenomenological results. What is even more important, they found the *full correlation functions* to agree with the predictions of the instanton liquid, even in channels (like the nucleon and Delta) where no phenomenological information is available.

In order to check this result in more detail, they also studied the behavior of the correlation functions under cooling [399]. The cooling procedure was monitored by studying a number of gluonic observables, like the total action, the topological charge and the Wilson loop. From these observables, the authors conclude that the configurations are dominated by interacting instantons after $\sim$25 cooling sweeps. Instanton–anti-instanton pairs are continually lost during cooling, and after $\sim$50 sweeps, the topological charge fluctuations are consistent with a dilute gas. The characteristics of the instanton liquid were already discussed in Section 5.4. After

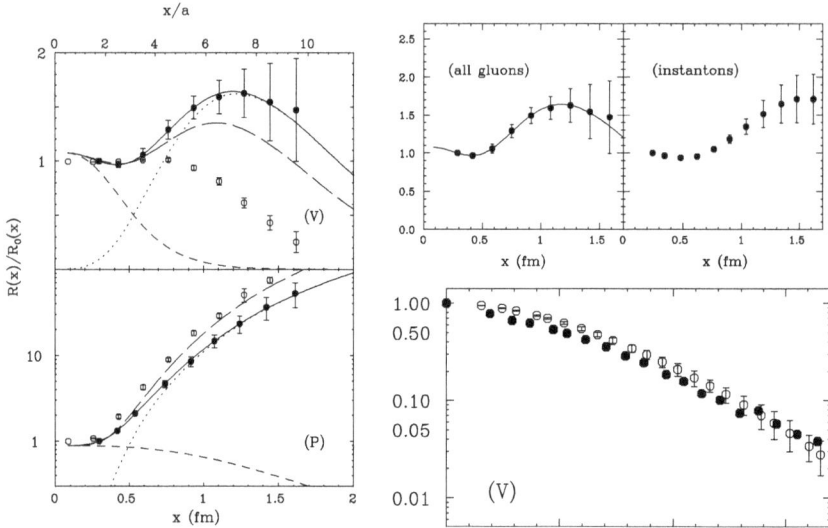

Fig. 6.19. The left panel shows the correlation functions for the vector (marked (V)) ρ channel and the pseudoscalar (marked (P)) π channel. The long-dashed lines are phenomenological ones, open and closed circles stand for RILM [237] and lattice calculation [399], respectively. The upper right panel compares vector correlators before and after "cooling". The lower part shows the same comparison for the ρ wave function; the closed and open points here correspond to "quantum" and "semiclassical" vacua, respectively.

50 sweeps the action is reduced by a factor $\sim$300 while the string tension (measured from 7×4 Wilson loops) has dropped by a factor 6.

The first comparison made between the instanton liquid results and those obtained on the lattice [399] are shown in Fig. 6.19a, for ρ vector (V) and π pseudoscalar (P) channels.

Even more direct was a comparison between the correlator calculated on the "quantum" configurations, as compared to "cooled" or "semiclassical" lattice configurations [399]. The behavior of the pion and nucleon correlation functions under cooling is shown in Fig. 6.20. The behavior of the ρ and Δ correlators (not shown) was quite similar. During the cooling process the scale was readjusted by keeping the nucleon mass fixed.[33]

6.5.5. Gluonic correlation functions

One of the most interesting problems in hadronic spectroscopy is whether one can identify glueballs, bound states of pure glue, among the spectrum of observed

---

[33]This introduces only a small uncertainty, the change in scale is $\sim$16%. We observe that the correlation functions are *stable under cooling*, they agree almost within error bars. This is also seen from the extracted masses and coupling constants. While m_N and m_π are stable by definition, m_ρ and g_ρ change by less than 2%, λ_π by 7% and λ_N by 1%. Only the delta mass is too small after cooling, it changes by 27%.

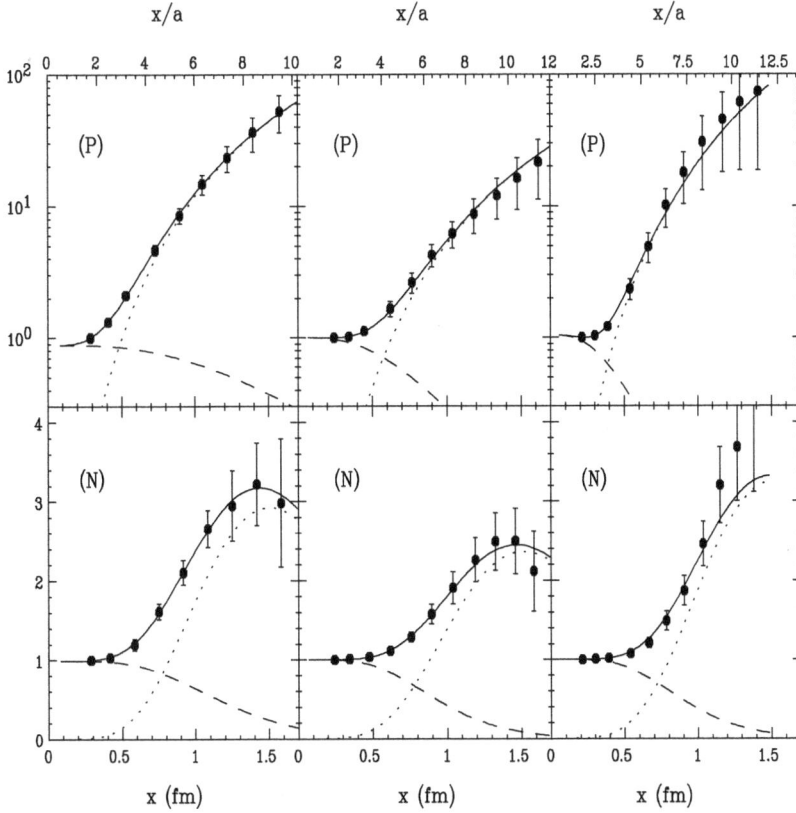

Fig. 6.20. Behavior of pion and proton correlation functions under cooling, from Ref. [399]. The left, center, and right panels show the results in the original ensemble, and after 25 and 50 cooling sweeps. The solid lines show fits to the data based on a pole plus continuum model for the spectral function. The dotted and dashed lines show the individual contributions from the pole and the continuum part.

hadrons. This question has two aspects. In pure glue theory, stable glueball states exist and they have been studied for a number of years in lattice simulations. In full QCD, glueballs mix with quark states, making it difficult to unambiguously identify glueball candidates.

Even in pure gauge theory, lattice simulations still require large numerical efforts. Nevertheless, a few results appear to be firmly established [300]. (i) The lightest glueball is the scalar 0^{++}, with a mass in the 1.5–1.8 GeV range. (ii) The tensor glueball is significantly heavier $m_{2++}/m_{0++} \simeq 1.4$, and the pseudo-scalar is heavier still, $m_{0-+}/m_{0++} = 1.5$–1.8 [302]. (iii) The scalar glueball is much smaller than other glueballs. The size of the scalar is $r_{0++} \simeq 0.2$ fm, while $r_{2++} \simeq 0.8$ fm [303]. For comparison, a similar measurement for the π and ρ mesons gives 0.32 fm and 0.45 fm, indicating that spin-dependent forces between gluons are stronger than between quarks.

Gluonic currents with the quantum numbers of the lowest glueball states are the field strength squared (S $= 0^{++}$), the topological charge density (P $= 0^{-+}$), and the energy momentum tensors (T $= 2^{++}$);

$$j_S = (G^a_{\mu\nu})^2, \qquad j_P = \frac{1}{2}\epsilon_{\mu\nu\rho\sigma}G^a_{\mu\nu}G^a_{\rho\sigma}, \qquad j_T = \frac{1}{4}(G^a_{\mu\nu})^2 - G^a_{0\alpha}G^a_{0\alpha}. \quad (6.154)$$

The short distance behavior of the corresponding correlation functions is determined by the OPE [396]

$$\Pi_{S,P}(x) = \Pi^0_{S,P}\left(1 \pm \frac{\pi^2}{192g^2}\langle f^{abc}G^a_{\mu\nu}G^b_{\nu\beta}G^c_{\beta\mu}\rangle x^6 + \cdots\right), \qquad (6.155)$$

$$\Pi_T(x) = \Pi^0_T\left(1 + \frac{25\pi^2}{9216g^2}\langle 2\mathcal{O}_1 - \mathcal{O}_2\rangle \log(x^2)x^8 + \cdots\right), \qquad (6.156)$$

where we have defined the operators $\mathcal{O}_1 = (f^{abc}G^b_{\mu\alpha}G^c_{\nu\alpha})^2$, $\mathcal{O}_2 = (f^{abc}G^b_{\mu\nu}G^c_{\alpha\beta})^2$ and the free correlation functions are given by

$$\Pi_{S,P}(x) = (\pm)\frac{384g^4}{\pi^4 x^8}, \qquad \Pi_T(x) = \frac{24g^4}{\pi^4 x^8}. \qquad (6.157)$$

Power corrections in the glueball channels are remarkably small. The leading-order power correction $O(\langle G^2_{\mu\nu}\rangle/x^4)$ vanishes,[34] while radiative corrections of the form $\alpha_s \log(x^2)\langle G^2_{\mu\nu}\rangle/x^4$ (not included in (6.155)), or higher order power corrections like $\langle f^{abc}G^a_{\mu\nu}G^b_{\nu\rho}G^c_{\rho\mu}\rangle/x^2$ are very small.

On the other hand, there is an important low energy theorem that controls the large distance behavior of the scalar correlation function [396]

$$\int d^4x \, \Pi_S(x) = \frac{128\pi^2}{b}\langle G^2\rangle, \qquad (6.158)$$

where b denotes the first coefficient of the beta function. In order to make the integral well defined, we have to subtract the constant term $\sim\langle G^2\rangle^2$ as well as singular (perturbative) contributions to the correlation function. Analogously, the integral over the pseudo-scalar correlation functions is given by the topological susceptibility $\int d^4x\Pi_P(x) = \chi_{\rm top}$. In pure gauge theory $\chi_{\rm top} \simeq (32\pi^2)\langle G^2\rangle$, while in unquenched QCD $\chi_{\rm top} = O(m)$, see Section 4.4.1. These low energy theorems indicate the presence of rather large non-perturbative corrections in the scalar glueball channels. This can be seen as follows. We can incorporate the low energy theorem into the sum rules by using a subtracted dispersion relation

$$\frac{\Pi(Q^2) - \Pi(0)}{Q^2} = \frac{1}{\pi}\int ds \frac{\mathrm{Im}\,\Pi(s)}{s(s + Q^2)}. \qquad (6.159)$$

In this case, the subtraction constant acts like a power correction. In practice, however, the subtraction constant totally dominates over ordinary power corrections.

---

[34]There is a $\langle G^2_{\mu\nu}\rangle\delta^4(x)$ contact term in the scalar glueball correlators which, depending on the choice of sum rule, may enter momentum space correlation functions.

For example, using pole dominance, the scalar glueball coupling $\lambda_S = \langle 0|j_S|0^{++}\rangle$ is completely determined by the subtraction, $\lambda_S^2/m_S^2 \simeq (128\pi^2/b)\langle G^2\rangle$.

For this reason, we expect instantons to give a large contribution to scalar glueball correlation functions. Expanding the gluon operators around the classical fields, we have

$$
\Pi_S(x,y) = \langle 0|G^{2\,\mathrm{cl}}(x)G^{2\,\mathrm{cl}}(y)|0\rangle
$$
$$
+ \langle 0|G_{\mu\nu}^{a,\mathrm{cl}}(x)\left[D_\mu^x D_\alpha^y D_{\nu\beta}(x,y)\right]^{ab} G_{\alpha\beta}^{b,\mathrm{cl}}(y)|0\rangle + \cdots, \qquad (6.160)
$$

where $D_{\mu\nu}^{ab}(x,y)$ is the gluon propagator in the classical background field. If we insert the classical field of an instanton, we find [226, 241, 396]

$$
\Pi_{S,P}^{SIA}(x) = \int \rho^4 dn(\rho) \frac{12288\pi^2 \rho^{-8}}{y^6(y^2+4)^5}\Big[y^8 + 28y^6 - 94y^4 - 160y^2 - 120
$$
$$
+ \frac{240}{y\sqrt{y^2+4}}(y^6 + 2y^4 + 3y^2 + 2)\,\mathrm{asinh}\left(\frac{1}{2}y\right)\Big], \qquad (6.161)
$$

with $y = x/\rho$.

There is no classical contribution in the tensor channel, since the stress tensor in the self-dual field of an instanton is zero. Note that the perturbative contributions in the scalar and pseudo-scalar channels have opposite sign, while the classical contributions have the same sign. To first order in the instanton density, we therefore find the three scenarios discussed in Section 6.5.2: *attraction* in the scalar channel, *repulsion* in the pseudo-scalar and *no* effect in the tensor channel. The single-instanton prediction is compared with the OPE in Fig. 6.21. We clearly see that classical fields are much more important than power corrections.

Quantum corrections to this result can be calculated from the second term in (6.160) using the gluon propagator in the instanton field [206]. The singular contributions correspond to the OPE in the instanton field. There is an analog of the Dubovikov–Smilga result for glueball correlators; In a general self-dual background field, there are no power corrections to the tensor correlator [396]. This is consistent with the result (6.156), since the combination $\langle 2\mathcal{O}_1 - \mathcal{O}_2\rangle$ vanishes in a self-dual field. Also, the sum of the scalar and pseudo-scalar glueball correlators does not receive any power corrections (while the difference does, starting at $O(G^3)$).

Numerical calculations of glueball correlators in different instanton ensembles were performed in Ref. [241]. At short distances, the results are consistent with the single instanton approximation. At larger distances, the scalar correlator is modified due to the presence of the gluon condensate. This means that (like the σ meson), the correlator has to be subtracted and the determination of the mass is difficult. In the pure gauge theory we find $m_{0^{++}} \simeq 1.5\,\mathrm{GeV}$ and $\lambda_{0^{++}} = 16 \pm 2\,\mathrm{GeV}^3$. While the mass is consistent with QCD sum rule predictions, the coupling is much larger than expected from calculations that do not enforce the low energy theorem [358, 359].

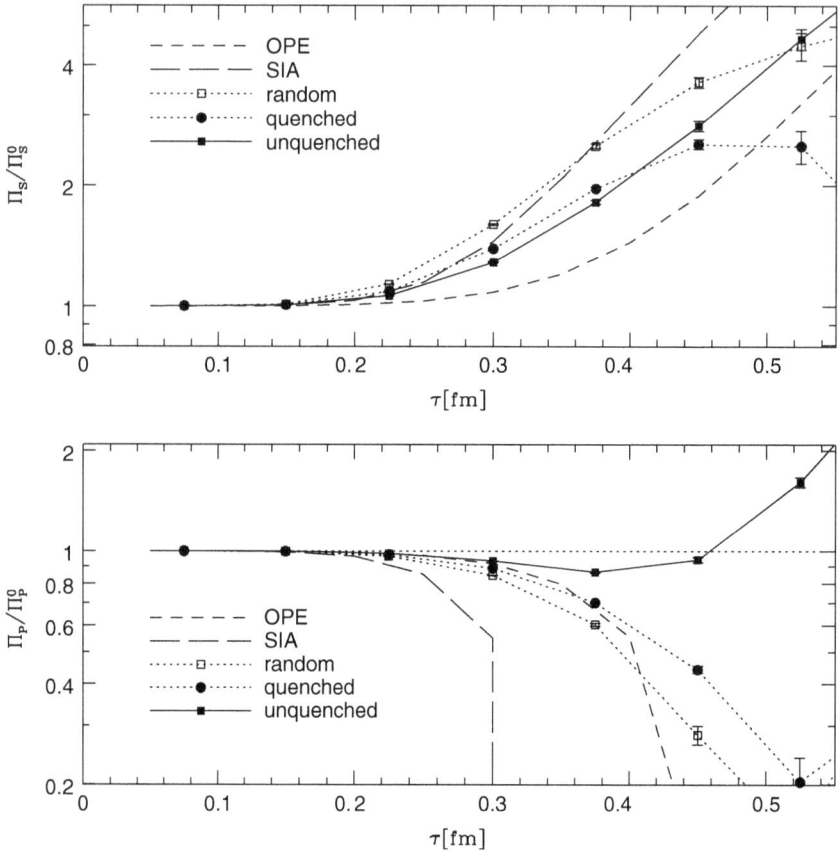

Fig. 6.21. Scalar and pseudo-scalar glueball correlation functions. Curves labeled as in Fig. 6.16.

Table 6.8. Scalar glueball parameters in different instanton ensembles.

	Random	Quenched	Unquenched
$m_{0^{++}}$ [GeV]	1.4	1.75	1.25
$\lambda_{0^{++}}$ [GeV^3]	17.2	16.5	15.6

    In the pseudo-scalar channel the correlator is very repulsive and there is no clear indication of a glueball state. In the full theory (with quarks) the correlator is modified due to topological charge screening. The non-perturbative correction changes sign and a light (on the glueball mass scale) state, the η' appears. Non-perturbative corrections in the tensor channel are very small. Isolated instantons and anti-instantons have a vanishing energy–momentum tensor, so the result is entirely due to interactions.

In [241] we also measured glueball wave functions. The most important result is that the scalar glueball is indeed small, $r_{0++} = 0.2$ fm, while the tensor is much bigger, $r_{2++} = 0.6$ fm. The size of the scalar is determined by the size of an instanton, whereas in the case of the tensor the scale is set by the average distance between instantons. This number is comparable to the confinement scale, so the tensor wave function is probably not very reliable. On the other hand, the scalar is much smaller than the confinement scale, so the wave function of the 0^{++} glueball may provide an important indication for the importance of instantons in pure gauge theory.

6.6. Hadronic structure and n-point correlators

When we introduced hadronic phenomenology in Chapter 2 we discussed mostly masses and quantum numbers of hadrons, rather than their sizes, form factors and wave functions. The reason is partly that such phenomenological information is not yet as systematic as we would like it to be. Even general questions cannot be answered, such as: *are the sizes and form factors of the "unusual"* $(\pi, \eta'$, *glueballs,"molecular" states etc.) hadrons any different from those of "ordinary" ones?*

The most studied hadron is of course the proton, and we have learned over the years many details about its form factors. Its electric and magnetic form factors are measured, and we discuss it below. Let me add here a comment that unfortunately we do not know directly form factors in channels other than vector and axial directly, in lack of probes other than γ, W, Z.

Still scattering experiments with protons reveal that other operators probing the proton see its size to be very different from the e.m. radius. The so called scalar form factor (interacting with the operator $\bar{q}q$) sees *larger* size, exceeding 1 fm, indicating that valence quarks affect the quark condensate (nearly eliminating it) in a large sphere around a nucleon. For more about this phenomenon read Refs. [409, 410].

On the other hand, scalar processes with the Pomeron, which in the lowest order are considered to be the interaction with the scalar operator $G_{a\mu\nu}^2$, indicate that glue is distributed inside a proton in a spot with radius significantly smaller than the e.m. radius, of only about 0.4 fm in m.s.r. We will discuss this issue in Chapter 7.

Further insight into hadronic structure can be obtained via *multi-point* correlators, which can provide calculations of resonance decay widths, form factors, structure functions etc. We cannot provide a systematic description of this vast field, but only discuss two examples of exploratory attempts. All of them focus on the issues related with the *shapes* and *sizes* of various hadrons.

In non-relativistic quantum mechanics we are used to calculating wave functions of the bound states which are related to the experimentally measurable form factors by the Fourier transform. Unfortunately, this familiar language is not literally adequate for hadrons.

But still, before we go to technical details, it is useful to recapitulate the issue to the readers, in the old-fashioned way first. Basically, there are two kind of hadrons: some of them are bound by confinement and some "from within", by short-range instanton-induced forces. As we repeatedly emphasized above, ρ is an example of the former, and π of the latter variety. If one would like to connect their form factors and wave functions to some Schrödinger-like equation, the effective potential would be a rising potential in the former case and something like a delta function at the origin (like for deuteron) in the latter. For convenience, taking a quadratic (rather than linear) potential in the former case, one ends up with a Gaussian wave function and a form factor. In the latter case a cusp at the origin would lead to a form factor decreasing only as a power of the momentum transfer. This leads us to the main idea: in order to tell experimentally one kind of hadrons from another, one has to *study form factors at intermediate momentum transfer $Q \sim$ few GeV.*[35]

6.6.1.   *Wave functions*

One quantity which is close to a *hadronic wave function* is the so called Bethe–Salpeter amplitude (which is different from the so called light-cone one). Such Bethe–Salpeter amplitudes have been measured in a number of lattice gauge simulations, both at zero [403–406] and at finite temperature [407, 408].

In the pion case this quantity is defined by

$$\psi_\pi(y) = \int d^4x \, \langle 0|\bar{d}(x) P \exp\left(i \int_x^{x+y} A(x') \, dx'\right) \gamma_5 u(x+y)|\pi\rangle. \quad (6.162)$$

In practice, it is extracted from the three point correlator

$$\Pi_\pi(x, y) = \langle 0|T(\bar{d}(x) P \exp\left(i \int_x^{x+y} A(x') \, dx'\right) \gamma_5 u(x+y) \bar{d}(0) \gamma_5 u(0))|0\rangle$$

$$\sim \psi(y) e^{-m_\pi x}, \quad (6.163)$$

where x has to be a large space-like separation in order to ensure that the correlation function is dominated by the ground state and that y is the separation of the two quarks in the transverse direction $((x\dot{y}) = 0)$. In practice it is convenient to divide the 3-point by the 2-point function, canceling the x-dependent part and the coupling constants.

Like the two point correlation functions, the Bethe–Salpeter amplitudes are calculated from the light quark propagator.[36]

---

[35] Unfortunately one cannot go to too large Q either, because pQCD gluon exchange leads to some universal power dependences for all hadrons. We mean here Q above the region where the pQCD asymptotics works.

[36] In general, the inclusion of the Schwinger P exp factor is expected to give an important contribution to the measured wave functions, since it corresponds to an additional string type potential, but not in the instanton model.

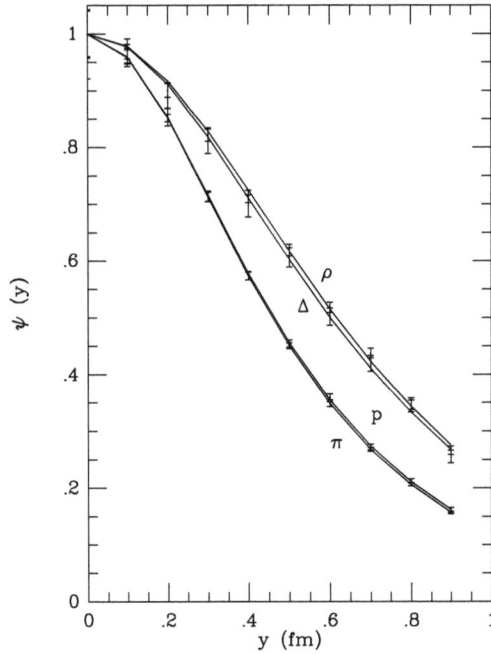

Fig. 6.22. Hadronic wave functions of the pion, rho meson, proton and Delta resonance in the random instanton ensemble.

A qualitative understanding of the wave functions can be obtained using the single-instanton approximation. For small transverse separations y and $x \to \infty$ we get a very simple result

$$\psi_\pi(y) = 1 - y^2/(2\rho)^2 + \cdots , \qquad (6.164)$$

indicating that a pion radius (as determined by Bethe–Salpeter amplitude) is directly related to the instanton radius.

The wave functions in the random ensemble were calculated in Ref. [241]. Those for π, ρ, N and Δ are shown in Fig. 6.22. We observe that the pion and the proton as well as the rho meson and the Delta resonance have very similar wave functions, but the sizes for the pion and the proton are *smaller* than for the rho meson and the Delta resonance. We have already argued that a small-size scalar diquark in the nucleon is linked with the instanton-induced attraction.

6.6.2. *Form factors*

Let us now switch to the second application of the 3-point correlator, related with the form factors. Unlike Bethe–Salpeter amplitudes, those are directly measurable in experiments. We consider only the simplest example, namely the *pion form factor*.

It is a very important quantity, for which well-defined pQCD asymptotics is known, and a comparison of the asymptotic behavior and the experiment is crucial

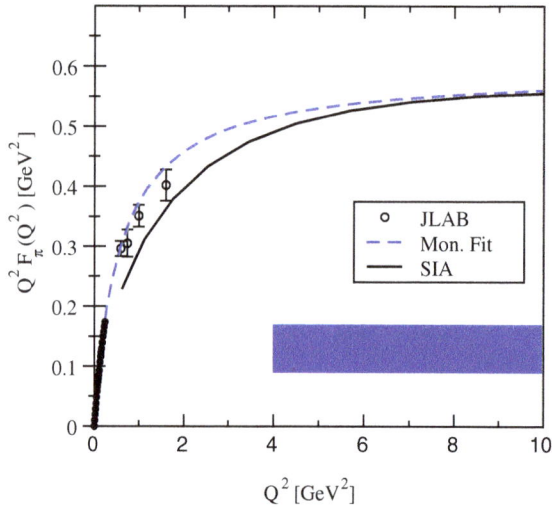

Fig. 6.23. The JLAB data for $Q^2 F_\pi(Q^2)$ in comparison with the asymptotic pQCD prediction (thick black bar, for a typical $\alpha_s \approx 0.2$–0.4, the monopole fit (dashed line), and our SIA calculation (solid line). The SIA calculation is not reliable below $Q^2 \sim 1\,\text{GeV}^2$. The solid circles denote older SLAC data.

for understanding the boundary between the perturbative and non-perturbative regimes of QCD.

Recently, the charged pion form factor has been measured very accurately at momentum transfers $0.6\,\text{GeV}^2 < Q^2 < 1.6\,\text{GeV}^2$ by the so called F_π collaboration in the Jefferson Laboratory (JLAB) [411]. Not only are the data at the highest experimentally accessible momenta still very far from the asymptotic limit, but the trend is still away from the pQCD prediction shown by black area in Fig. 6.23.

In the single-instanton approximation the problem was addressed in Ref. [413]. Unlike the previous works (done in the QCD sum rule framework), these authors have started with a 3-point correlator including electromagnetic and two *pseudoscalar* (rather than axial) currents. In such a correlator (as well as other scalar and pseudoscalar channels) the instanton contribution with maximal number of the zero-mode terms in the quark propagators is possible, resulting in enhanced (relative to e.g. the vector or axial channels) contribution, by a factor $1/(m^\star\rho)^2$, where $m^\star$ denotes the effective quark mass defined and discussed in detail in Section 4.3.4. This feature, however, is not generically related to the pion itself and depends on the particular three-point function under investigation. For example, the enhancement is absent, when one considers the pion contribution to the axial correlator. Similarly, there is no such enhancement of the $\gamma\gamma^\star\pi^0$ neutral pion transition form factor. The relevant instanton effects for this process are not due to (enhanced) zero modes, but are either related to non-zero mode propagators in the instanton background or to multi-instanton effects, which are suppressed by the instanton diluteness. This conclusion is nicely supported by recent CLEO measurements of this form factor, which indeed show that the asymptotic pQCD regime is reached much earlier, at $Q^2 \sim 2\,\text{GeV}$ [412].

Let us select points as follows:

$$\Gamma_\mu(x,y) = \langle j_5^+(-x/2)j_\mu(y)j_5(x/2)\rangle, \tag{6.165}$$

and, for simplicity, think of a charged pion, so that axial currents are made of quarks of different flavor and therefore only the "triangular" diagram has to be considered. This diagram for small x and y was calculated in Ref. [413], and (after plugging in standard RILM parameters) performing a relatively complicated numerical analysis (removing the non-pion contribution), the pion form factor for $Q^2 \sim 1\,\text{GeV}$ was obtained, in good agreement with data.

Blotz and myself [414] extended the calculation of the 3-point correlator to multi-instanton background (RILM), which has allowed us to take $x, y \sim 1\,\text{fm}$ and therefore to ensure dominance of the pion pole. Furthermore, one can obtain the pion mass and coupling constant from 2-point correlators for the same ensemble and check its consistency with the measured 3-point one. After that, the pion form factor can be rather accurately determined. Its standard parametrization is a "monopole form" $F_\pi(Q^2) = M^2/(Q^2 + M^2)$ was found to work well.

Furthermore, the dependence of the parameter M on the instanton size was specially studied by Blotz and myself; see Fig. 6.24. The results confirmed that

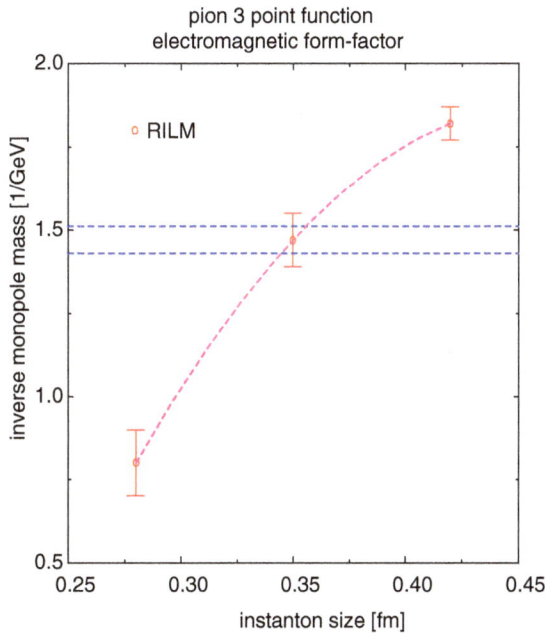

Fig. 6.24. The dependence of the parameter in the pion form factor fit on the instanton size ρ used in the ensemble. The region between two horizontal lines corresponds to experimental uncertainty.

the pion size is directly related with the instanton size, and the experimentally observed value is indeed obtained for $\rho \approx 0.35$ fm. So, in a way, *the pion form factor is basically the instanton form factor.* See also similar plots in Fig. 4.2 on p. 152 of Ref. [248] for a discussion of the instanton-induced decays of the charmonium states into 2 and 3 pseudoscalar mesons.

One further conclusion of this work is worth mentioning. A question of whether this mass M is or is not the mass of the ρ meson is an old one. What we found is that the pion size parameter M has strong dependence on the instanton size, while m_ρ has not. Comparing random and interacting ensembles we also found that m_ρ is sensitive to instanton correlations while M is not. So one can conclude that although they are numerically close, it is probably mere coincidence.

Another step has been made in Ref. [415], where the pion formfactor was calculated at intermediate momentum transfers, $2\,\mathrm{GeV}^2 < Q^2 \lesssim 10\,\mathrm{GeV}^2$. The central prediction following from analytic SIA calculation is shown in Fig. 6.23 in comparison to the recent JLAB measurements. We find the very intriguing result that the instanton contribution to the formfactor is completely consistent with the monopole fit at intermediate momentum transfers, even where the vector dominance model has no justification.

For larger momenta transfer, $Q^2 > 20\,\mathrm{GeV}^2$, the SIA breaks down, as it is necessary to increase the distances in order to isolate the pion ground state. At these needed distances, however, the correlation functions will become sensitive to multi-instanton effects.

Let me briefly explain the main points of this calculation. We consider the spatial Fourier transforms of the Euclidean three-point function and two-point function,

$$G_\mu(t, \mathbf{p} + \mathbf{q}; -t, \mathbf{p}) = \int d^3x\, d^3y\, e^{-i\mathbf{p}\cdot\mathbf{x} + i(\mathbf{p}+\mathbf{q})\cdot\mathbf{y}}$$

$$\times \langle 0| j_5(t, \mathbf{y}) J_\mu(0, \mathbf{0}) j_5^\dagger(-t, \mathbf{x}) |0\rangle, \qquad (6.166)$$

$$G(2t, \mathbf{p}) = \int d^3x\, e^{i\mathbf{p}\cdot\mathbf{x}} \langle 0| j_5(t, \mathbf{x}) j_5^\dagger(-t, \mathbf{0}) |0\rangle, \qquad (6.167)$$

where the pseudo-scalar current $j_5(x) = \bar{u}(x)\gamma_5 d(x)$ excites states with the quantum numbers of the pion, and $J_\mu(0)$ denotes the electro-magnetic current operator. In the large t limit (at fixed momenta), both correlation functions are dominated by the pion pole contribution and the ratio of the three-point function to the two-point function becomes proportional to the pion form factor [417]. In the Breit frame, $\mathbf{p} = -\mathbf{q}/2$ and $Q^2 = \mathbf{q}^2$, one has simply

$$\frac{G_4(t, \mathbf{q}/2; -t, -\mathbf{q}/2)}{G(2t, \mathbf{q}/2)} \rightarrow F_\pi(Q^2). \qquad (6.168)$$

Notice that the l.h.s. of Eq. (6.168) should not depend on t, for t large enough; for the pion, this is achieved already for $t \sim 0.6$ fm.

We do not describe similar work in SIA on the nucleon form factor [416], which is currently subject to a renewed experimental interest at JLAB. Let me just make a qualitative remark that the N $\rightarrow$ Δ formfactor decreases with Q more rapidly, indicating that Δ has a large size and perhaps a different wave function. This can be related to the missing well-bound scalar diquark in the Δ channel.

Another potential source of valuable information about hadronic structure are weak decays. As a recent example, emphasizing the role of scalar diquarks in baryon weak decays, see M. Cristoforetti *et al.*, hep-ph10402180. In this work the enhanced weak decays are quantitatively explained, using again the instanton liquid model.

.

CHAPTER 7

High Energy Hadronic Collisions

7.1. Introduction

7.1.1. *Reggions and the Pomeron*

The history of experimental and theoretical studies of high energy collisions is a long story, which I would not be able to discussed here in full. The reader may consult, e.g. a rather recent book on the subject [419].

The main object of the scattering theory is the S matrix, which is a unitary $(SS^+ = 1)$ transformation from the initial to the final states. The so called T matrix is more often used: its matrix element between states a, b is defined by

$$S_{ab} = \delta_{ab} + i(2\pi)^4 \delta^4 \left(\sum p_a - \sum p_b \right) T_{ab} \tag{7.1}$$

and unitarity takes the form of the so called Cutkosky rule

$$2 \operatorname{Im} T_{ab} = (2\pi)^4 \delta^4 \left(\sum p_a - \sum p_b \right) \sum_c T_{ac} T_{cb}^+ = F \sigma_{\text{tot}}. \tag{7.2}$$

Here F is the flux factor, for large energies it is just $2s$, and σ_{tot} is the total cross section.

Before QCD was discovered, there was no dynamical basis for theoretical development and people tried to put to the maximal use general properties of the scattering amplitudes, such as *analyticity* and the *crossing symmetry* of the S matrix. The latter implies that if an analytic continuation from the physical region $s > 0, t, u < 0$ to, say, the region $t > 0, s < 0$ is made, one should get the correct amplitude for the "crossing" reaction $a + \bar{c} \to \bar{b} + d$ as well.

Standard partial wave expansion is done in the s-channel, with the scattering angle θ related to invariants by $\cos \theta = 1 + 2t/s$. A similar expansion in the t-channel, now with $\cos \theta = 1 + 2s/t$, can also be made:

$$T_{ab}(s, t) = \sum_{l=0}^{\infty} (2l + 1) a_l(t) P_l(1 + 2s/t). \tag{7.3}$$

The next standard (Sommerfeld–Watson) trick changes the sum into an integral, by inserting in the denominator the function $\sin l\pi$ which provides poles at all integers l and by taking the integral over a contour C surrounding the real positive axis. Small complication appears because of the $(-)^l$ factor in $a_l(t)$, preventing closing the contour in the imaginary direction. It is however easily overcome by the introduction of two functions, odd and even ones, defined with the explicit signature factor η

$$T_{ab}(s,t) = 1/2i \int_C dl \frac{(2l+1)}{\sin l\pi} \sum_{\eta=\pm 1} \frac{(\eta + e^{il\pi})}{2} a^\eta(l,t) P_l(1+2s/t). \qquad (7.4)$$

Regge poles are poles in the l plane, and at high energies $s \gg -t$ one should look for the highest power of s possible. It means the cross section is dominated by the Regge pole with the highest trajectory $\alpha(t)$. This contribution is usually written as "Reggion exchange"

$$T_{ab}(s,t) \approx \frac{(\eta + e^{i\alpha(t)\pi})}{2\sin(\pi\alpha(t))} \gamma_{ac}(t)\gamma_{bd}(t) \frac{s^{\alpha(t)}}{\Gamma(\alpha(t))}, \qquad (7.5)$$

where the last factor is just an asymptotics of the P_l, and a *factorized* form of a_l in terms of two coupling constants $\gamma(t)$ is assumed.

We have already discussed Regge trajectories for hadrons in Section 2.1.5, and one can see both the Chew–Frautschi linear trajectories and the non-linear Pomeron in Fig. 2.2. Multiple experiments have shown that the analytic continuation of the same linear trajectories from physical masses $t > 0$ to the scattering domain $t < 0$ does in fact reproduce quite accurately the scattering amplitudes. Say for the isospin exchange one needs the ρ trajectory with the intercept seen in Fig. 2.2, namely $\alpha(t = 0) \approx 1/2$; and indeed such cross sections all decrease as $s^{\alpha(0)-1} \sim s^{-1/2}$.

The *Pomeron*, named after Pomeranchuck, is a hypothetical Regge pole with vacuum quantum numbers. In the 1950s and 1960s it was thought that cross sections are constant at large s, so that its intercept $\alpha_P(t = 0)$ was thought to be unity. When energies higher than $s \sim 1000\,\mathrm{GeV}^2$ became available, it was found that it slowly grows. It is however true that $\bar{p}p$ and pp cross sections seem to become equal at high enough s with a universal power growth

$$\sigma_{pp} \approx \sigma_{\bar{p}p} \approx 21.7\,\mathrm{mb}\, s^{0.08}. \qquad (7.6)$$

Therefore the Pomeron was thought of as a *"supercritical"* pole slightly above 1. Its trajectory was depicted in Fig. 2.2: in this case it is measured at $t < 0$ and extrapolated to positive ones (probably way too far).

Among few solid theoretical statements about the cross section is the *Froissart bound* which limits the cross section growth

$$\sigma(s) < \mathrm{const}\, \log^2 s, \qquad (7.7)$$

which obviously excludes any power growth. Note however that we are still far from the Froissart bound on the total cross section; but in principle we should keep in

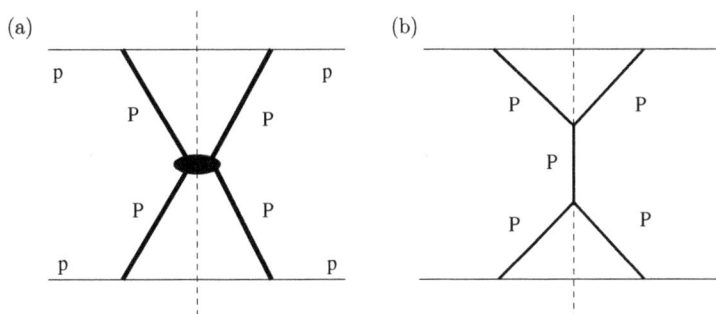

Fig. 7.1. Examples of the double-Pomeron processes: (a) with the four-Pomeron vertex, and (b) two 3-Pomeron ones.

mind that eventually this growth should be tamed by the unitarity corrections at very high energies.

What it means in terms of Pomerons is that they are expected to become interacting at very high s. The effective theory of interacting Reggions and Pomerons was developed in the 1970s, mostly by Ter Martirosyan, Gribov and collaborators. We would not go into it and just provide here an example (which will be discussed below in this chapter). It is the so called double-Pomeron processes, see Fig. 7.1. The vertical-dashed line in the middle corresponds to the unitarity cut. In Fig. 7.1a the cut goes through the four-Pomeron vertex, which means that experimentally one would observe both protons scattered elastically, with two large rapidity gaps and a cluster of hadrons corresponding to the vertex cut. Figure 7.1b has a cut going through the central Pomeron, which means that secondaries would include two elastically scattered protons also separated by two "rapidity gaps" without secondaries from the "plateau" at mid-rapidity region corresponding to the Pomeron imaginary part. This amplitude is of course proportional to two three-Pomeron vertices. The empirical values of 3-P and 4-P vertices can thus be obtained from a fit to the data: their small values indicate that some small parameter should exist which makes the Pomeron diagrammatic expansion possible. It was not clear till recently why are they small; we will return to this question at the end of the chapter.

7.1.2. *High energy collisions in pQCD and its "phases"*

The very first application of pQCD included studies of "Bjorken scaling violation" or dependence of hadronic structure function on the scale at which they are analyzed, Q^2. To that end pQCD diagrams containing $\alpha_s(Q) \log(Q^2)$ should be resumed following the so called DGLAP evolution [421], which describes $1 \to 2$ "splittings" of partons. A general view of a high energy hadron as a more and more dense cloud of partons at higher energies/smaller x provides a simple explanation of why the cross sections grow with energy.

Another way to describe deep inelastic processes, more convenient for our purposes, is to do so in the lab frame. Here one can describe deep inelastic process

(a)   (b)

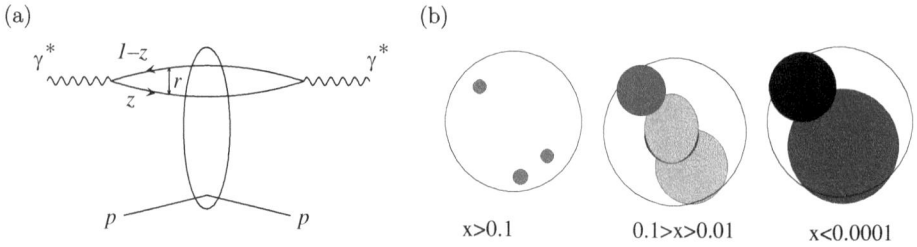

Fig. 7.2. (a) A virtual dipole traversing the nucleon target. (b) A snapshot of the gluon field distribution in a nucleon seen by a dipole with different x, in an impact parameter plane.

as a high energy collision of a virtual photon, evolving into a $\bar{q}q$ dipole, which is frozen in time during its passage through the target, see Fig. 7.2a. If the dipole has small size $r \ll 1$ fm, its interaction with it is proportional to the gluonic field strength in the target. Depending on which energies or x the collisions have, the distribution of the gluonic field in the nucleon has the shape schematically shown in Fig. 7.2b. The experiment can provide both the dependence of the dipole cross section on its size $\sigma(r_\perp)$ and, after some analysis (see e.g. recent work by Kowalski and Teaney [698]) the thickness of the glue in the nucleon at the impact parameter b, called $T(b)$. Both show that the "gluonic spot" one finds inside the nucleon has a surprisingly small size ~ 0.4 fm, much smaller than the nucleon electromagnetic radius.[1] One may wonder how such a small spot may be understood theoretically, and even whether it can be in agreement with pictures shown in Fig. 7.2b.

On top of that, one observes (see discussion in Section 9.7.1) very large fluctuations of the cross section. Although the issue is not yet understood, I have tried to connect it with instanton physics in Ref. [447] and subsequent papers. In particular, due to vacuum diluteness, one probably basically finds that only *one* instanton at any given time is active. (This is indicated by blacker circle in the figure.)

However, $\log Q$ is not the only logarithm which appears in pQCD diagrams; the integral over longitudinal momenta bring in another large log, $\log x$, where x is a small fraction of the hadronic momentum carried by a parton. The leading power of such logs originates from the ladder-like diagrams with reggeizised gluons and "Lipatov effective vertex", converting two virtual gluons into a real[2] one.

---

[1] One important consequence of smallness of the gluonic spot is that in eA and pA collisions even with heavy nuclei, the number of subsequent interactions is not $A^{1/3} \sim 6$ but only about 2 for the heaviest nuclei [698].

[2] Real in the perturbative sense: it is on-shell. Of course, gluon is not the physical final state and will become a set of hadrons due to fragmentation process and confinement. However, all pQCD calculations assume that the probability of this "final state interaction" is just 1, and the cross section is not modified.

Their re-summation has been made by Balitsky, Fadin, Kuraev and Lipatov (BFKL) [422]. A nice pedagogical derivation of this BFKL result in terms of path-ordered exponent of the gauge field can be found in Ref. [424]; further development leading to non-linear evolution with unitarization is also described there.

The result, expressed in terms of Regge theory corresponds to the highest singularity at $t = 0$ located at

$$\alpha_{\text{BFKL}}(0) - 1 = \alpha_{\text{s}} \frac{12 \ln 2}{\pi} + O(\alpha_{\text{s}}^2). \tag{7.8}$$

Such singularity means that the pQCD cross sections should grow as $s^{\alpha_{\text{BFKL}}(0)-1}$. The value of α_{s}, as usual, depends on the underlying scale of momenta: and here is a problem. In the BFKL calculation all virtualities of the momenta are kept above some normalization scale μ, which should at the end be set at some value appropriate to the problem in question. If the μ we started with is high, pQCD is accurate but $\alpha_s(\mu)$ is small and the growth with energy is not so rapid. If one moves μ down, including softer processes, the effective power is larger. In deep inelastic processes we at least know that at the top of the ladder we start with high scale Q^2. However for the total cross section of pp collisions the issue is what is the relevant scale at given x.

The pQCD by itself describes how the initial scale propagates diffusively, both toward the UV and the IR, and the only reason to limit it is that the next pQCD correction gets large. The value of the next-to-leading correction to the BFKL pomeron trajectory has also been calculated by Lipatov *et al.* in Ref. [423]: the resulting correction is very large and pertinently demands that $\alpha_{\text{s}} < 0.05$ or so. This made the situation quite confused: but we know that in some cases higher order corrections have a tendency to compensate each other and in reality the accuracy may be better than judged by those corrections.

The limitations on the pQCD description more often comes from the generic non-perturbative effects. One of those, due to instantons and sphalerons, appears at the so called *semihard scale* $\mu \sim$ few GeV, and we will discuss it below in detail.

However at sufficiently high energies/small x the so called *saturation* phenomenon [425] is expected to happen, which means that pQCD processes can determine their own *saturation scale* $Q_s(x)$. If one assumes that it grows indefinitely with energy, e.g. $Q_s(x) \sim 1/x^\lambda$ with some index $\lambda \sim 0.3$, quite good fits to HERA data can be obtained [430], see Fig. 7.3.

The cutoff scale is given by the magnitude of the virtual field determined from some self-consistency condition. It was further argued that when, at sufficiently small x, the gluon occupation numbers reach the magnitude $\sim O(1/\alpha_s) \gg 1$ the *classical* approach to the YM field becomes possible [426]. If so, any hadrons and/or nuclei approach at small x some universal limit called the Color Glass Condensate (CGC).

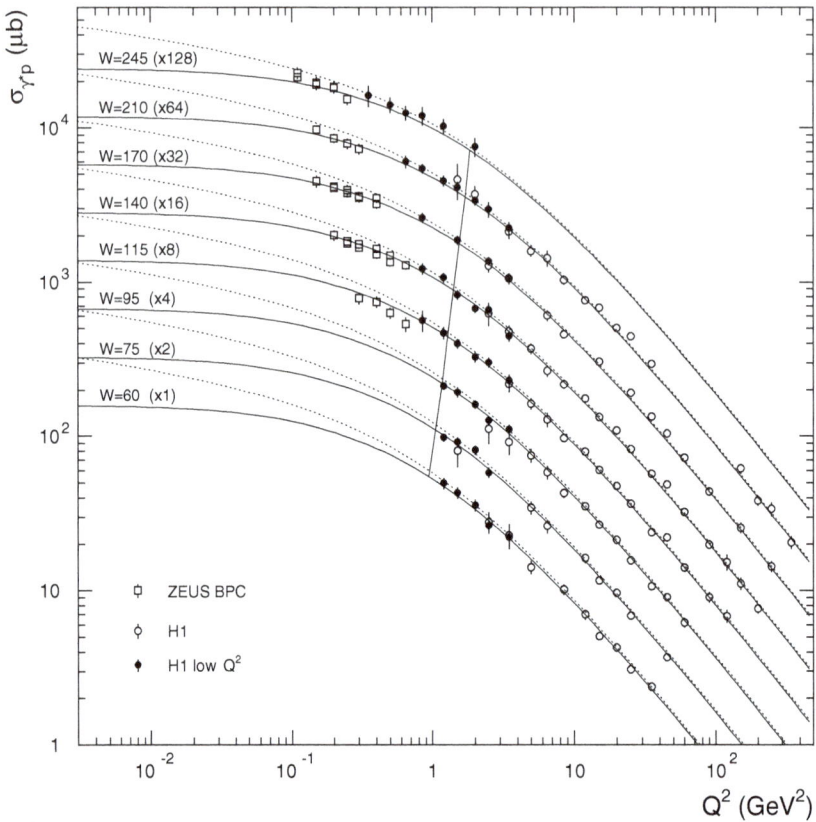

Fig. 7.3. The cross section of virtual photon on nucleon at different energies and Q^2, the data versus the fit by the Golec–Biernat–Wüsthoff model (lines). The solid near-vertical line near $Q^2 \approx 1\,\mathrm{GeV}^2$ is the saturation boundary: its slope is related with the index λ.

Without going into this vast subject, I will try to summarize the current understanding of the problem by using a kind of a "phase diagram",[3] following a recent paper by Kharzeev, Levin and Mclerran [428], shown in Fig. 7.4. The lower right side is the "dilute domain" in which perturbative DGLAP and BFKL approaches work, each in the direction indicated by two respective arrows. Increasing the energy at fixed large scale Q (going up) one finds the "geometrical scaling" region, in which instead of two variables Q^2, x all quantities are a function of their one combination.[4] It separates the dilute and the dense or "liquid" *CGC region* higher up.

---

[3]We put this term in quotation marks since we do not discuss here a macroscopic system which may have true phase transitions. The lines on the figure indicate of course gradual changes from one regime to another.

[4]Note the amusing correspondence between this phase diagram and the hydrodynamical Riemann wave solution with the "explosive edge" and scaling.

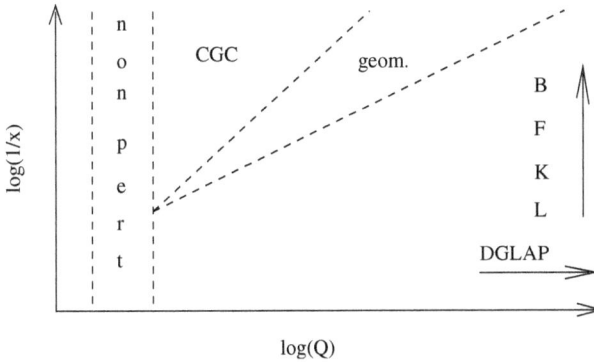

Fig. 7.4. The phase diagram of high energy processes in the coordinates: the scale $\log \log Q$ vs the log of energy $\log(1/x)$.

If the scale Q^2 is reduced and $x > x_{\text{saturation}}$ (the left lower side), one finds that the dilute domain is limited by the generic non-perturbative physics. It happens at the semihard scale of about 1 GeV where we enter the region in which scattering is described by the original "soft Pomeron", discussed at the beginning of this section. And indeed, the growth of the γN cross section is similar in this domain to that of NN scattering.

The open question is whether we have one common Pomeron, in which perturbative and nonperturbative physics smoothly substitute each other as the scale changes, or say two different trajectories [429] with quite different intercepts and slopes, the soft and the hard Pomerons, which intercept each other and define a sharper phase boundary.

Another interesting question is what happens at the upper left of the diagram, where a transition from non-perturbative dynamics to that of the CGC is supposed to take place. We will argue below that in this region CGC is not just a chaotic classical field, but it possesses topological properties. In terms of the filed strength it means that the electric and magnetic field strength are not orthogonal, $\vec{E}\vec{B}$ is substantial, which leads to quark/chirality production due to chiral anomaly. We will show that already in NN collisions there is evidence for production of gluonic clusters possessing such properties, and will argue in Chapter 10 that multiple production of such clusters in heavy ion collisions — especially in the energy range of RHIC — provides spectacular "fireworks" of their multiple and simultaneous explosions.

7.1.3. *Evolving descriptions of soft Pomeron dynamics*

Let me not go into a discussion of the pre-QCD *multiperipheral models* of high energy collisions, based on ladder-type diagrams with various hadrons. The simplest QCD-based model we start with is the so called Law–Nussinov model [420],

which describes the Pomeron by the minimal colorless exchange possible, namely by
the 2-gluon exchange in the t-channel. Its non-perturbative version is developed by
Nachtsmann and Dosch, see e.g. [419]. If a gluon has some effective mass, say about
1 GeV, such an exchange has a range in the transverse plane consistent with the
Pomeron size α'. The intercept of the resulting singularity in the complex momentum scale is $\alpha(0) = 1$, which means the cross section does not grow with energy. The
double-gluon-exchange diagram has the imaginary part of the amplitude, given by
the usual cutting rules: it corresponds to a square of diagrams with a single gluon
exchange. Naively this process corresponds to elastic scattering of the throughgoing
partons. However such a "half-Pomeron" amplitude with a *single* gluon, exchanges
colors of these partons, it would become later a multi-hadron production. Confining
strings have to be produced and their breaking would lead to multiple hadrons with
all possible rapidities. As this is a final state interaction, it has probability $P = 1$
and is thus ignored in the calculation of the cross section.

In order to have a Pomeron with growing cross section one has to re-sum logarithms, which originate from the longitudinal phase space of the produced quanta.
It means that *prompt* production is needed, and the gluon ladder shown in Fig. 7.5a
is such an example. It is the re-summation of such diagrams which have produced
the BFKL result (7.8).

Various incarnations of the old multi-peripheral models with hadronic ladders
neither provided a clear cut explanation of why only particular hadrons should be
used, nor gave really quantitative predictions. They also did not explain why the
pomeron itself can be a small-size object in a transverse plane, as inferred from the
observed t-dependence of pp scattering.

But what if one keeps the two t-channel gluons of the Law–Nussinov model, but
adds prompt non-perturbative production processes rather than gluon radiation?
What would be the most probable states which can be produced promptly, what
can its mechanism be?

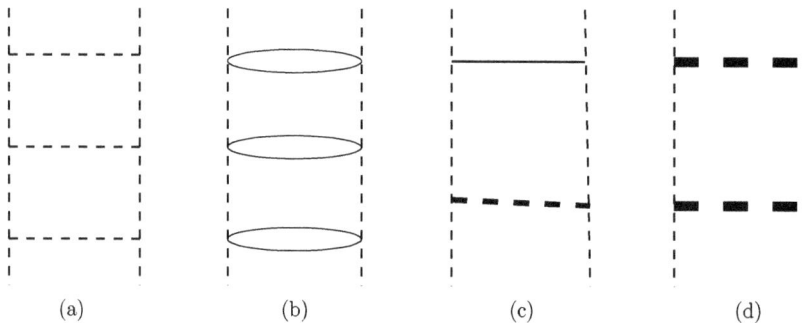

Fig. 7.5. (a) The BFKL ladder with rungs made of gluons; (b) Kharzeev–Levin Pomeron with
rungs made of pion pairs; (c) My model for Pomeron [447] with rungs made of σ mesons and scalar
glueballs. Note that the vertices now are instanton-induced; (d) The "semiclassical Pomeron" with
rungs producing sphaleron-like topological clusters [436, 438].

The development of these ideas took time, and went in several installments. The first step was made by Kharzeev and Levin (KL) [435] who used diagrams shown in Fig. 7.5b with the "rungs" of the ladder made of pion pairs in the scalar isoscalar channel. Their motivation was to use the low energy theorem by Novikov *et al.* [396] which gives the $gg \to \pi\pi$ transition matrix element $\langle 0|(gG_{\mu\nu}^a)^2|\pi\pi\rangle$ near the threshold. It is based on scale anomaly we discussed in Section 1.5.5 and makes use of the fact that the G^2 operator is but a quantum part of the stress tensor, which can be mapped to the stress tensor of soft pion Lagrangian. The resulting answer does not have small α_s, argued KL, and thus it is parametrically larger than perturbative amplitudes like $gg \to gg$. With a schematic scalar spectral density cut off at some duality scale $M_0 \approx 2\,\text{GeV}$, KL have obtained a value for the pomeron intercept close to the phenomenological value 0.08 [418].

My work [447] has modified the KL model, with a sigma meson describing interacting $\pi\pi$ in a scalar channel plus a large scalar glueball contribution into the spectral density, see Fig. 7.5c. The glueball contribution follows from studies of the gluonic correlation functions we discussed in Section 6.5.5. More generally, this paper was important because it introduced a question about the magnitude of the $gg \to$ hadrons amplitudes which can be generated semiclassically, by instantons. It provided further motivation for the model, explaining that hadrons selected for "rungs" of the ladder are not arbitrary but only those which have parametrically strong interaction with instantons, like $\sigma = (0^+ + \pi\pi)$ and the scalar glueball. It also improved the crude schematic model of KL in near-threshold $\pi\pi$ contribution, explaining how the expression for $\alpha(0) - 1$ should get a meaningful chiral limit.

In brief, the spectral density is approximated by two resonances, σ, G_0, and their contribution to the pomeron intercept is

$$\alpha(0) - 1 = \frac{18\pi^2}{b^2} \int \frac{dM^2}{M^6} \left(\rho_{\text{phys}}(M^2) - \rho_{\text{pert}}(M^2) \right), \tag{7.9}$$

where $\rho_{\text{phys}}(M^2), \rho_{\text{pert}}(M^2)$ are physical and perturbative (gg cut) spectral densities, respectively and $b = \frac{11}{3}N_c - \frac{2}{3}N_f$ is the coefficient of the beta function. The integrand in (7.9) is non-zero only for $M < M_0$ because at $M > M_0$ pQCD works and two spectral densities become identical. A remarkably large coupling constant to gg current,[5] which according to [241] is $\langle 0|g^2G_{\mu\nu}^2|G_0\rangle \approx 16. \pm 2\,\text{GeV}^3$. The glueball contribution to the spectral density is then

$$\rho_{G_0} = \left(\frac{b}{(32\pi^2)} \right)^2 |\langle 0|g^2G_{\mu\nu}^2|G_0\rangle|^2 \delta(M^2 - M_{G_0}^2), \tag{7.10}$$

which leads to the contribution to the pomeron intercept $|\alpha(0) - 1|_{G_0} \approx 0.03$, about the same as the contribution of the $\pi\pi$ continuum or σ.

---

[5]We remind the reader that the units in gluodynamics are traditionally defined by setting the string tension to be the same as in QCD.

However the KL model and its improvements were just steps on the road leading to the main focus of this chapter, the *"semiclassical Pomeron"*. Its schematic representation is shown in Fig. 7.5d, in which the wide dashed lines are the production of topological gluonic clusters, of which the scalar glueball is just a small part. It is semiclassical in at least three different ways:

- The two colliding quarks or gluons are substituted by Wilson lines evaluated via instantons.
- The produced objects, the sphaleron-like clusters, can be classically followed into Minkowski evolution.
- The Euclidean part of the path leading from the vacuum wave function to the turning states can also be studied semiclassically.

7.2. Instanton-induced processes at high energies

7.2.1. Toward the "holy grail"

Historically, the semiclassical approach to high-energy reactions originated in a somewhat different context in the early 1990s. A number of very insightful papers [441, 442] have focused on a possibility to calculate the magnitude of the *baryon-number violating tunneling in the electroweak theory*. It waned several years later when it became clear that these fascinating phenomena predicted by the theory cannot be experimentally observed.

A connection between tunneling and baryon number in electroweak theory has been discussed in Section 4.2.4. In a very crude way, one may say that collision of two (virtual) gauge bosons, W, Z in electroweak theory can produce multiple W, Z (about 50) plus 12 fermions, provided a transition to a different classical vacuum takes place.

The instanton-based description started with papers by Ringwald and Espinosa, who noticed that for instanton-based production vertices grows as n with production of n gauge quanta. It is quite unlike the perturbative expressions in which each of those goes with a price of small electroweak coupling α_{ew}. Zakharov re-summed these production processes and showed that in forward scattering the amplitude exponentiates and describes the well known semiclassical dipole interaction between an instanton in the amplitude and an anti-instanton in the conjugated amplitude. He later argued that this interaction cancels *exactly half of the action*, so the probability goes from $\exp(-16\pi^2/g_{ew}^2)$ to $\exp(-8\pi^2/g_{ew}^2)$. However a calculation of the excitation amplitude turned out to be a notoriously difficult problem, named the "holy grail" function, which is not quite solved yet even now. We will not go into discussion of these papers: for their good pedagogical description see a review by Mattis [443].

The first QCD effect of similar origin discussed was the instanton-induced multi-jet production: the search in this direction continues at HERA, see recent work [440].

However, due to the large scale involved and the small-size instantons, this phenomenon is also associated with small cross sections. So, if the predicted signal could be found, it would not be an easy task to prove that it is not due to some perturbative diagrams, not included in the usual jet-modeling event generators.

The situation is different at the so called semi-hard scale $Q \sim 1$–$2\,\mathrm{GeV}$, at which tunneling phenomena are in some cases so large than they simply dominate the perturbative processes. A recent suggestion that the instanton-induced processes may explain why the cross section of high energy hadronic scattering grows with energy made in Refs. [436, 438, 447] will be described in the rest of the chapter. But before we do so, in the pedagogical spirit of this book we turn from gauge theories to the quantum mechanical double-well problem. We show what the quantum-mechanical analog of this process is in this simple context, in which one may calculate everything not only semiclassically but also directly, without any assumptions whatsoever.

7.2.2. *Exciting a quantum system from under the barrier*

The setting for this section can be any problem with a barrier and tunneling, and we use as an example the same double well potential we used so frequently in Chapters 1 and 3, namely

$$V = \lambda(x^2 - f^2)^2. \tag{7.11}$$

One can always set the mass of a particle to unit value $m = 1$. We will use the well parameters $\lambda = 1, f = 2$, for which the "sphaleron mass" of this problem — the maximum of the potential $V(0) = 16$ — is several oscillator quanta, as in the QCD applications to follow. The ground state wave function is well known since the time of Hund [143], it has two maxima corresponding to both wells, at $x \approx \pm f$, with a relatively small probability below the barrier, at $x \sim 0$.

The question we would like to ask, following my paper [431] is what happens if one *rapidly localizes the quantum particle under the barrier*. One may view it pictorially as a narrow beam of particles hitting some spot under the barrier, say near $x = 0$, and exciting the system. More specifically, our question is what are *the final states* generated by such an experiment.

To answer those questions, let us introduce an external periodic perturbation acting on the system

$$\delta V(x,t) = f(x) \exp[-it\omega], \tag{7.12}$$

with $f(x)$ well localized under the barrier, at $x \sim 0$. The specific shape of $f(x)$ does not matter as long as it does not extend to the wells, where the ordinary oscillation quanta (analogs of gluons) can be excited. I have used several of them and will show results for $f = \exp(-4x^2)$ and $f = 1, |x| < 0.2; f = 0, |x| > 0.2$.

The frequency of the excitation can easily be tuned to excite the subsequent nth levels of the system, $\omega_n = E_n - E_0$, with the excitation probability

$$P_n \sim |\langle 0|f(x)|n\rangle|^2, \qquad (7.13)$$

then calculated directly from the numerically known wave functions and energies of the nth states. For even excitation functions $f(x) = f(-x)$ used, only the even levels $n = 2, 4$ etc. can be excited.

The result of the calculation [431] is shown in the Fig. 7.6a. Note the strong peaking of the excitation happens near the "sphaleron mass" $\omega \approx 16$, indicating that the main final state is sitting *at* the barrier top. The reasons for the peak are as follows. For energies much less than $V(0)$ both the ground and the final wave functions are small under the barrier. For energies well above it $\psi_n(x)$ is not small but rapidly oscillating, so that its overlap with $f(x)\psi_0(x)$ is small.

Let me summarize the main lesson of this section in one sentence: if the particle under the barrier is hit, it jumps *at the barrier* with the same coordinate.

7.2.3. *Semiclassical production of sphaleron-like clusters*

Now we return to the gauge theories and will argue that the same phe-nomenon should occur there as well: the instantons, when strongly hit by colliding partons, evolve into a sphaleron-like clusters.

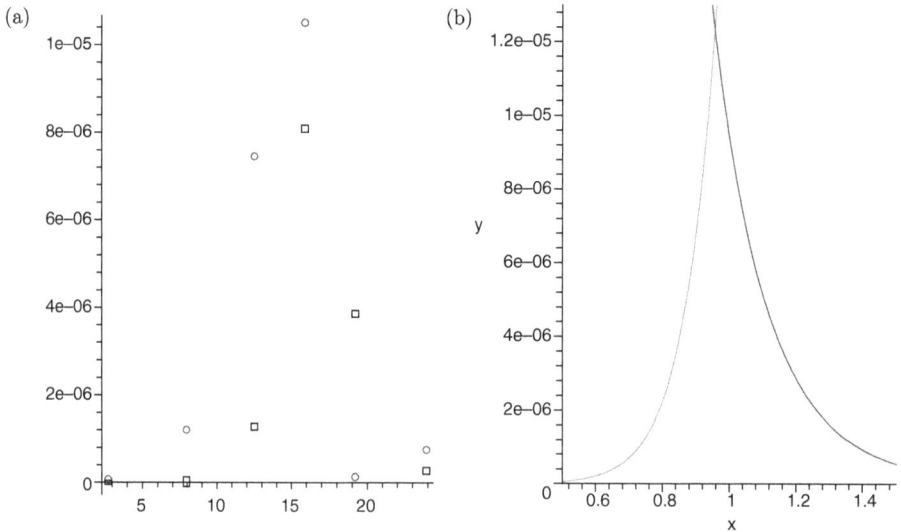

Fig. 7.6. (a) The excitation probability P_n of the double-well system versus the excitation energy. Two sets of points are mentioned in the text for two excitation functions $f(x)$. Note that the peak excitation energy corresponds to the maximum of the potential, $V \approx 16$. (b) The inclusive gluon cross section of the process $gg \to$ sphaleron $\to g + \cdots$ versus the energy, in units of the sphaleron mass $x = Q/M_s$, from Ref. [444].

In very general terms, a high energy collision allows virtual fields — part of the wave function of the target or projectile — to become real. In the QED context such idea is the basis of the so called *Weizsacker–Williams approximation*, describing how the boosted virtual Coulomb field may become real photons. The pQCD Lipatov vertex plays the same role, describing how a virtual gluon may become a real (on-shell) one, with just a slight scattering.

Similar phenomena take place non-perturbatively. The virtual fields in hadrons and in the vacuum itself, if promptly excited by collisions, may produce certain real objects. A sudden excitation of the part of the vacuum wave function *under the barrier* also produces certain real objects[6]: those are states *on* the barrier. Surprisingly, only recently have they been studied in QCD [175, 444]. So, promptly excited glue is *not* just several gluons: it should appear first as gluomagnetic *topological clusters* of well defined structure. The main reason for that is the same as in the double-well example of the previous section: there is no time to change the values of the coordinates. This argument becomes even more convincing if the coordinate in question is related with the non-trivial topology.

Furthermore, a real breakthrough in the realization that the whole process can be described by a continuous semiclassical path, starting in the vacuum *under the barrier*, proceeding to a *turning point* and then to a *real (Minkowskian) evolution*. Such paths are schematically shown in Fig. 7.7.

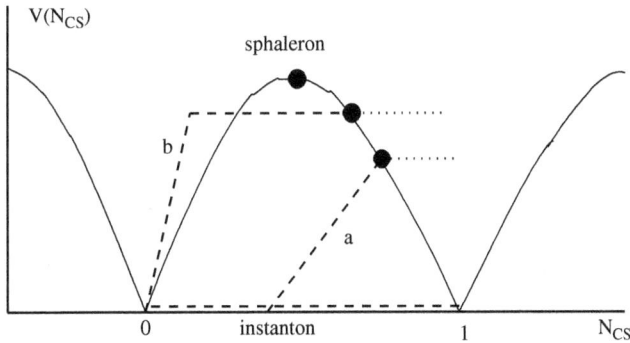

Fig. 7.7. Schematic plot of the energy of Yang–Mills field versus the Chern–Simons number N_{cs}. It is a periodic function, with zeros at integer points. The *instanton* (shown by the lowest horizontal dashed line) is a transition between such points. However if some nonzero energy is deposited into the process during the transition, the virtual paths (the dashed lines) emerge from the barrier, via the *turning points* (black circles). The later real time motion outside the barrier (shown by horizontal dotted lines) conserves energy, as the driving force is switched off. The maximal cross section corresponds to the transition around the top of the barrier, the *sphaleron*.

---

[6] They are real in the same sense as gluons: namely at times shorter than those at which confinement sets in.

Whichever way the system is driven, it emerges from under the barrier via what we will call "*a turning state*", familiar from the WKB semiclassical method in quantum mechanics. The turning states, released into the unitarity cut or Minkowski world, are the states we studied in Section 3.4.4. From there starts the real time motion outside the barrier. Here the action is real and $|e^{iS}| = 1$. That means that whatever happens at this Minkowski stage has probability 1 and cannot affect the total cross section of the process; this part of the path is only needed for understanding of the properties of the final state.

7.2.4. *Explosion of the turning states*

Provided one knows the initial (turning) states, their subsequent fate can be studied by solving the YM equation starting from them. In this section we will describe this evolution, following the paper by Ostrovsky, Carter and myself [175].

The first study of the kind was made a decade ago in electroweak theory [174] for the sphaleron, where it was found that it decays in about 51 W, Z, H quanta. The difference in QCD is that there is no Higgs scalar and its non-zero VEVs, so gluons are massless. This makes the process much more *explosive* because all harmonics with different momenta move together, with the speed of light. In the paper [175], we solved this both *numerically* and *analytically* (based on works by Luescher and Schekhter [455]).

Omitting all the details, let me say that we found that the initial pure magnetic YM field in sphaleron-like configurations happens to be extremely "explosive". Not only do they quickly expand spherically into an expanding fireball, they reach the

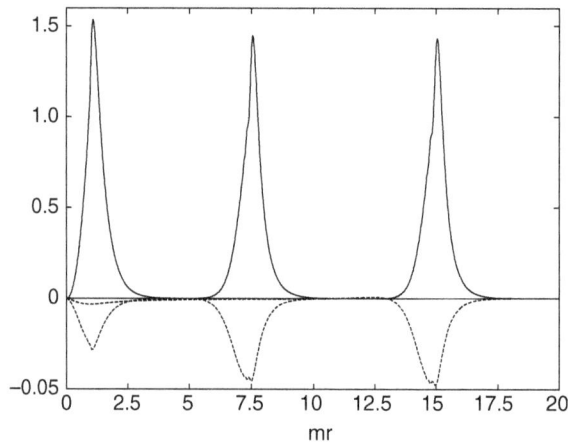

Fig. 7.8. Results of a numerical solution of YM equations, as the energy density $r^2\epsilon(r)$ at times $0, 7.5, 15$ fm and the density of the Chern–Simons number (the dashed lines at negative values). One can see that the profile is stabilized and no longer changes after a short time $\sim\rho$.

speed of light very quickly. Furthermore, all the energy is found to be contained in the thin spherical wave, with nothing (but a pure gauge) field inside it.

The details can be found in the original paper, let me just show the promised spherical shell at several values of the time. Those have the energy density

$$4\pi r^2 e(r,t) = \frac{8\pi}{g^2\rho^2}(1-\kappa^2)^2 \left(\frac{\rho^2}{\rho^2+(r-t)^2}\right)^3.$$ (7.14)

Of course, at large times the field becomes a weak field which can be decomposed into gluons: the Fourier transform of the field provides the energy distribution of the resulting gluons. Numerical studies of the problem has been reported in Ref. [175] as well. Those are naturally more flexible than analytic and allow for a more realistic initial shape of instantons and sphalerons, with exponential (rather than power-like) tails of the fields at large distances.[7]

An alternative derivation of the same explosive solution, using specific conformal mapping from the $d = 4$ spherically symmetric Euclidean solution has been found in the paper by Janik, Zahed and myself [444].

Now, what about quark pair production? Again, this problem has been much debated in the literature on electroweak theory since the 1980s, but it is not yet really solved. In its general form, the issue is to derive an analog of the index theorem for the case when there are outgoing fields: it is supposed to tell us how many fermionic levels have crossed zero (and are actually produced) based on some topological properties of the gauge field alone. Its usual form, involving a change in the Chern–Simons number, is obviously incorrect because the variation of N_{CS} is in general (and specifically for all the time-dependent solutions we speak about) not even an integer. However the number of the Dirac levels which cross zero during the evolution from the original cluster to the final state must be an integer.

Further progress in this direction was achieved by derivation of the explicit solution to the Dirac equation in the background of exploding sphalerons, by Zahed and myself [445]. This solution starts with the static sphaleron, for which the Dirac equation has one zero energy solution. Then one can see from the solved Dirac equation how the quarks in that initial zero mode get accelerated by the electric field, into the finite-energy spectrum of the emitted quarks. The energy spectrum of the outgoing quarks has been even found to be simpler than for gluons, namely

$$\mathbf{n}_R(k) = \frac{4\pi k^2}{2k}|\mathbf{q}_R^\dagger(\vec{k})|^2 = \rho(2k\rho)^2 e^{-2k\rho}.$$ (7.15)

---

[7]The phenomenological reasons for exponential rather than power instanton tail are discussed e.g. in Ref. [456].

The distribution integrates to exactly one produced quark. It is close to the Planck spectrum with the effective temperature $T = 1/2\rho$, which is about 300 MeV for a standard $\rho = 1/3$ fm.[8]

This phenomenon can be viewed as a continuation of the 't Hooft process into real time, a production of the whole set of light quark pairs, $\bar{u}u\bar{d}d\bar{s}s$. So tentatively we estimate the yield of partons per cluster to be about 3 gluons and up to 6 quarks-and-antiquarks.

7.2.5. *Semiclassical evaluation of the cross section*

Unfortunately, there is no equally nice description of the Euclidean part of the excitation path yet. In Ref. [175] we have shown that the so called Yung ansatz describes so-to-say slow excitation paths, which indeed lead from the vacuum to all the minimal energy states at the potential. Unfortunately in the collisions we rather expect some rapid-excitation paths to dominate, which we still hope to find.

The original semiclassical approach with vacuum (undeformed) instantons pioneered by Ringwald and Espinosa and then developed by others [441] is only applicable at low energies of partonic sub-collisions, much below the sphaleron mass.[9] The problem to find a general semi-classical answer has been known as the so called *holy grail* problem. Three methods toward its solution proposed in the literature may eventually provide the complete solution.

(i) *Unitarization* of the multi-gluon amplitude when it becomes strong was first suggested by Zakharov and worked out by Shifman and Maggiore [441]. Basically one can treat a sphaleron as a resonance, and even the resulting expression for the cross sections in Ref. [436] looks similar to Breit–Wigner formula. This is the most worked out approach, but it still cannot guarantee the parametrically accurate numerical values of the cross sections.

(ii) The so called *Landau method with singular instantons* was applied by Diakonov and Petrov [442] (following some earlier works which are cited in their work). They were able to find the cross section in the opposite limit of high energies. It follows from the comparison of the two limits, that the peak is indeed very close to the sphaleron mass, and the cross section is very close to be *first order* in instanton diluteness, not the second order as the initial probability. Unfortunately they were not able to find the solution at intermediate times which would provide the turning points of this approach.

(iii) *Classical solution on the complex time plane* [454] is another possible direction, in which a zig-zag shaped path in complex time includes classical

---

[8]Accidentally, this is close to the initial temperature of quark–gluon plasma in the RHIC energy domain.

[9]Let me remind the reader that it is about 10–15 TeV in electroweak theory and about 3 GeV in QCD.

evolution and tunneling in one common solution. Unfortunately, this interesting idea also has not been fully implemented, even for toy models with only scalar fields considered in this paper.

It would not be possible to describe here any of those approaches in more detail. Recently Janik, Zahed and myself [444] have been able to find the states which are excited in the Diakonov–Petrov high energy approximation: those turned out to be YM sphalerons with slightly rescaled (and energy-dependent) size. In Fig. 7.6b we have already shown the comparison from this paper of the low energy and high-energy approximations for gluon production. Note the strong resemblance to the quantum-mechanical excitation curves shown in the Fig. 7.6a. This confirms the main idea, according to which all that is happening is "jumping on the barrier". The action is about $1/2$ of the tunneling one, also in agreement with it.

7.2.6. *Semiclassical Wilson lines*

On our way toward the "semiclassical Pomeron" we still have one part of quantum diagrams left: the excitation is produced by exchange of two t-channel gluons with the target and the projectile. This is clearly the minimal necessary number, but why not more? Unlike in the pQCD, there is no good answer to this question, as all the non-perturbative objects have $A_\mu \sim 1/g$ and so extra gluon exchanges are not associated with extra powers of α_s. The answer to this question, suggested by Zahed and myself [436] is to use the Wilson lines rotated to Euclidean domain where at least for instantons they are easily evaluated. In this section we will explain how it works.

But we first have to explain the history of this idea. The eikonalized expressions for the scattering amplitude in terms of a correlator of two Wilson-lines (quarks) or Wilson-loops (dipoles) are well known. The first systematic step developed toward a semi-classical but non-perturbative formulation of high-energy scattering in QCD was suggested by Nachtmann [432], who has related the scattering amplitude to expectations of pairs of Wilson lines. Semi-classical expressions with a similar pair of Wilson lines for DIS structure functions were also proposed by Muller [434]: in contrast to their traditional interpretation as partonic densities, they were treated as cross sections for targets penetrated by small dipole-like probes at high energy.

Meggiolaro [433] has argued that one can in principle bring these expressions to Euclidean space–time, by interpreting the angle between the Wilson lines $\theta = iy$ where y as the Minkowski rapidity.

Generically, we will refer to quark–quark scattering as

$$Q_A(p_1) + Q_B(p_2) \to Q_C(k_1) + Q_D(k_2). \tag{7.16}$$

We denote by AB and CD respectively, the incoming and outgoing color and spin of the quarks (polarization for gluons). Using the eikonal approximation and

LSZ reduction, the scattering amplitude $\mathcal{T}$ for quark–quark scattering reads [432]

$$\mathcal{T}_{AB,CD}(s,t) \approx -2is \int d^2b\, e^{iq_\perp \cdot b} \langle (\mathbf{W}_1(b) - 1)_{AC}\, (\mathbf{W}_2(0) - 1)_{BD} \rangle, \qquad (7.17)$$

where as usual $s = (p_1 + p_2)^2$, $t = (p_1 - k_1)^2$, $s + t + u = 4m^2$ and

$$\mathbf{W}_{1,2}(b) = \mathbf{P}_c \exp \left(ig \int_{-\infty}^{+\infty} d\tau\, A(b + v_{1,2}\tau) \cdot v_{1,2} \right). \qquad (7.18)$$

The 2-dimensional integral is over the impact parameter b with $t = -q_\perp^2$, and the averaging is over the gauge configurations using the QCD action. The color bearing amplitude allows for scattering into a singlet or an octet configuration, i.e.

$$\mathcal{T} = \mathcal{T}_1 \mathbf{1} \otimes \mathbf{1} + \mathcal{T}_{N_c^2-1}(\tau^a \otimes \tau^a) \qquad (7.19)$$

following the decomposition $N_c \otimes N_c = 1 \oplus (N_c^2 - 1)$. For gluon–gluon scattering the lines are doubled in color space (adjoint representation) and further gauge-invariant contractions are possible. For quark–quark scattering the singlet exchange in the t-channel is the 0^+ (pomeron) while for quark–antiquark it is the 0^- (odderon) as the two differ by charge conjugation.

A quark with large momentum p travels on a straight line with four-velocity $\dot{x} = v = p/m$ and $v^2 = 1$. In the the eikonal approximation an ordinary quark transmutes to a scalar quark. The argument applies to any charged particle in a background gluon field, with the following amendments: for anti-quarks the four-velocity v is reversed in the Wilson line and for gluons the Wilson lines are in the adjoint representation. Quark–quark scattering can be also extended to quark–antiquark, gluon–gluon or scalar–scalar scattering. For quark–antiquark scattering the elastic amplitude dominates at large $\sqrt{s}$ since the annihilation part is down by $\sqrt{-t/s}$.

It can be described in Minkowski geometry in the CM frame with $p_1/m = (\cosh\gamma/2, \sinh\gamma/2, 0_\perp)$ and $p_2/m = (\cosh\gamma/2, -\sinh\gamma/2, 0_\perp)$ with the rapidity γ defined through $\cosh\gamma/2 = \sqrt{s}/2m$. For $s \gg m^2$ the rapidity gap between the receding scatterers becomes large with $\gamma \approx \log(s/m^2)$. The momentum transfer between the scatterers is $q = p_1 - k_1$, with $q_0 \approx q_3 \approx t/\sqrt{s}$ and $q_\perp^2 = tu/(s - 4m^2) \approx -t$. Hence $q = (0, 0, q_\perp)$ with $q^2 = -q_\perp^2 = t$. Although the partons or dipoles change their velocities after scattering, this change is small for $s \gg -t$. This is the kinematical assumption behind the use of the eikonal approximation.

In Euclidean geometry, the kinematics is fixed by noting that the Lorenz contraction factor translates to

$$\cosh\gamma = \frac{1}{\sqrt{1 - v^2}} = \frac{s}{2m^2} - 1 \to \cos\theta. \qquad (7.20)$$

Scattering at high-energy in Minkowski geometry follows from scattering in Euclidean geometry by analytically continuing $\theta \to -i\gamma$ in the regime

$\gamma \approx \log(s/m^2) \gg 1$. It is sufficient to analyze the scattering for $p_1/m = (1, 0, 0_\perp)$, $p_2/m = (\cos\theta, -\sin\theta, 0_\perp)$, $q = (0, 0, q_\perp)$ and $b = (0, 0, b_\perp)$. The Minkowski scattering amplitude at high-energy can be altogether continued to Euclidean geometry through

$$\mathcal{T}_{AB,CD}(\theta, q) \approx 4m^2 \sin\theta \int d^2 b \, e^{iq_\perp \cdot b} \langle (\mathbf{W}(\theta, b) - 1)_{AC} (\mathbf{W}(0, 0) - 1)_{BD} \rangle, \quad (7.21)$$

where

$$\mathbf{W}(b, \theta) = \mathbf{P}_c \exp\left(ig \int_\theta d\tau \, A(b + v\tau) \cdot v \right) \quad (7.22)$$

with $v = p/m$. The line integral in (7.22) is over a straight-line sloped at an angle θ away from the vertical.

One can test how it works in the lowest order of the QCD perturbation theory, which gives in the Euclidean version the result

$$\mathcal{T}(\theta, b) = \frac{g^2}{2\pi^2} \frac{\pi}{\tan\theta} \log\left(\frac{T}{b}\right), \quad (7.23)$$

where T in the log is the time cutoff showing the infrared sensitivity of the quark–quark scattering amplitude in perturbation theory. The more familar expression for $\mathcal{T}$ follows after integrating over the impact parameter b. The result in Euclidean geometry is

$$\mathcal{T}(\theta, q) = 4m^2 \sin\theta \int d^2 b \, e^{iq \cdot b} \mathcal{T}(\theta, b) = -\cos\theta \frac{g^2}{2} \frac{4m^2}{q^2} \int_0^\infty dx \, J_0(x) \log x, \quad (7.24)$$

which can be translated into Minkowski geometry by analytical continuation through $\theta \to -i\gamma$ with $q^2 = -t$. In both geometries, $\mathcal{T}$ is purely real and divergent as $t \to 0$, leading to a differential cross section of the order of $d\sigma/dt \approx g^4/t^2$ with a corresponding divergent Coulomb cross section $\sigma \approx g^4/(-t_{\min})$. The Euclidean perturbative analysis can be carried out to higher orders as well, in close analogy with analytically continued Feynman diagrams. It is instructive to calculate a dipole–dipole scattering also, which is or course IR finite.

The instantons are special in the sense that the Wilson line in the instanton field can be calculated analytically. The untraced and tilted Wilson line in the one-instanton background reads

$$\mathbf{W}(\theta, b) = \cos\alpha - i\tau \cdot \hat{n} \sin\alpha, \quad (7.25)$$

$$n^a = \mathbf{R}^{ab} \eta^b_{\mu\nu} \dot{x}_\mu (z - b)_\nu = \mathbf{R}^{ab} n^b, \quad (7.26)$$

and $\alpha = \pi\gamma/\sqrt{\gamma^2 + \rho^2}$ with

$$\gamma^2 = n \cdot n = \mathbf{n} \cdot \mathbf{n} = (z_4 \sin\theta - z_3 \cos\theta)^2 + (b - z_\perp)^2.$$

In the purely Euclidean domain one can use these expressions and easily calculate the correlator of two parallel Wilson lines, deriving the instanton-induced static quark potential [214]. Similarly, in Ref. [436] we have derived the instanton-induced static dipole–dipole potential. It is amusing to find out that while in the former case the result happens to be quite small compared to the perturbative one-gluon exchange, it is in fact dominant in the latter case. Simple generalization, introduction of the non-zero angle θ between the lines promotes the calculation into that of the scattering amplitude. The continuity between static potential and the low-energy scattering amplitude $\theta \to 0$ for two very heavy dipoles is then apparent.

The one-instanton contribution to the untraced QQ-scattering amplitude follows from the following correlator

$$\langle \mathbf{W}_{AC}(\theta, b)\mathbf{W}_{BD}(0,0)\rangle$$

$$\approx n_0 \int d^4z \left(\cos\alpha \cos\underline{\alpha}\, \mathbf{1}_{AC}\mathbf{1}_{BD} - \frac{1}{N_c^2 - 1}\hat{\mathbf{n}}\cdot\underline{\hat{\mathbf{n}}}\,\sin\alpha\sin\underline{\alpha}\,(\tau^a)_{AC}\,(\tau^a)_{BD}\right),$$

$$(7.27)$$

where the (under) bar notation means the same as the corresponding un-bar one with $\theta = 0$ and $b = 0$.

Furthermore,

$$\left\langle \frac{1}{N_c}\mathrm{Tr}(\mathbf{W}(\theta, b)\mathbf{W}(0,0))\right\rangle = \frac{2n_0}{N_c}\int d^4z (\cos\alpha\cos\underline{\alpha} - \hat{\mathbf{n}}\cdot\underline{\hat{\mathbf{n}}}\sin\alpha\sin\underline{\alpha}). \quad (7.28)$$

The integrand in (7.28) can be simplified by changing variables $(z_4\sin\theta - z_3\cos\theta) \to z_4$ and dropping the terms that vanish under the z-integration. Hence

$$\left\langle \frac{1}{N_c}\mathrm{Tr}(\mathbf{W}(\theta, b)\mathbf{W}(0,0))\right\rangle$$

$$= \frac{2n_0}{N_c}\int d^4z \left(\frac{1}{\sin\theta}\cos\tilde{\alpha}\cos\underline{\tilde{\alpha}} - \frac{1}{\tan\theta}\sin\tilde{\alpha}\sin\underline{\tilde{\alpha}}\frac{z_\perp^2 - z_\perp\cdot b}{\tilde{\gamma}\underline{\tilde{\gamma}}}\right). \quad (7.29)$$

The tilde parameters follow from the un-tilde ones by setting $\theta = \pi/2$. We note that $\tilde{\gamma} = \gamma = |\vec{z}|$. After analytical continuation, the first term produces the elastic amplitude which decays as $1/s$ with the energy. The second term corresponds to the color-changing amplitude. It is of order s^0 and dominates at high energy. Specifically

$$\left\langle \frac{1}{N_c}\mathrm{Tr}(\mathbf{W}(\theta, b)\mathbf{W}(0,0))\right\rangle$$

$$= \frac{2n_0}{N_c}\left(\frac{1}{\sin\theta}F_{cc}(b/\rho_0) - \frac{1}{\tan\theta}F_{ss}(b/\rho_0)\right). \quad (7.30)$$

Note that the second function (which describes color-inelastic collisions and survives in the high energy limit) changes sign, before decreasing as a power law to zero at large b.

One can directly generalize the calculation of the quark–quark scattering amplitude to that of any number of partons. For that, we assume that they all move with high energy in some reference frame and opposite direction: in Euclidean space those would propagate along two directions, with parton numbers N_1 and N_2 respectively. Any one of them, passing through the instanton field, is rotated in color space by a different angle α_i around a different axis $\vec{n}_i$, depending on the shortest distance between its path and the instanton center. Integration over all possible color orientations of the instanton leads then to global color conservation.

Before discussing specific cases in detail, let us make a general qualitative statement about such processes. We have found in the previous section that (the color-changing) quark–quark instanton-induced scattering has a finite high energy limit. For perturbative n-gluon exchange a factor of α_s^n is paid, while for an instanton mediated scattering the diluteness factor of $n_0\rho_0^4$ is paid (the price to find the instanton at the right place), no matter how many partons participate. Since the instanton vacuum is dilute, the one-gluon mediated process dominates the instanton one. However, the situation dramatically changes for two or more gluon exchanges: the instanton-induced amplitude is about the same for any number of partons, provided that all of them pass at a distance $\approx\rho_0$ from the instanton center.

We will not go into dipole–dipole scattering but note that explicit algebra shows it to be similar to quark–quark scattering. Namely, the (color) elastic dipole–dipole amplitude scales as $1/\sin\theta$ and vanishes at high energy after analytical continuation. However the (color) inelastic part of the amplitude does not. After performing the change of variables described earlier, the θ dependence drops from all the angles α. There is a remaining θ dependence in the four combinations $\mathbf{n}\cdot\underline{\mathbf{n}}$. In general, the θ dependence in the latter is linear in $\sin\theta$ or $\cos\theta$, and one may worry that the last term may involve higher powers of the trigonometric functions, which would yield an unphysical cross section growing as s after analytical continuation. In the cases we have checked that this is not the case. Moreover, the $\cos\theta$ term drops in the integral over z (odd under $z_3 \to -z_3$), making this contribution subleading at high-energy after analytical continuation. As a result, the pertinent octet contribution to the scattering amplitude is proportional to $\cotan\theta$ which is $1/i\tan y = 1/iv$ after analytical continuation, which is correct.

We have found that the instanton contribution at large s but small t behaves in a way similar to one-gluon exchange: only color-inelastic channels survive in the high energy limit. This means that the contribution to the total cross section appears in the amplitude squared, leading naturally to the concept of two-instanton exchange. The latter contribution to each Wilson-line is more involved. To streamline the discussion we will present the analysis of the two instanton contribution to the differential cross-section of quark–quark scattering at high energy. Similar considerations apply to dipole–dipole scattering.

Inserting the substitution (7.27), we obtain

$$\frac{d\sigma}{dt} \approx \left(\frac{4n_0}{N_c}\right)^2 \int db\, db'\, e^{iq\cdot(b-b')} \left(\mathbf{J} + \frac{1}{(N_c^2-1)}\mathbf{K}\right), \tag{7.31}$$

with two instanton-induced form factors

$$\mathbf{J} = \int d^4z\,(\cos\alpha - 1)(\cos\underline{\alpha} - 1) \int d^4z'(\cos\alpha' - 1)(\cos\underline{\alpha}' - 1),$$

$$\mathbf{K} = \int d^4z\,\hat{\mathbf{n}}\cdot\underline{\hat{\mathbf{n}}}\sin\alpha\sin\underline{\alpha} \int d^4z'\hat{\mathbf{n}}'\cdot\underline{\hat{\mathbf{n}}}'\sin\alpha'\sin\underline{\alpha}'. \tag{7.32}$$

The primed variables follow from the unprimed ones through the substitution $z, b \to z', b'$. For large $\sqrt{s}$, $\mathbf{J} \approx (1 - F_{cc})(1 - F'_{cc})/s^2$ [10] and $\mathbf{K} = F_{ss}F'_{ss}$, so that

$$\frac{d\sigma}{dt} \approx \frac{16n_0^2}{N_c^2(N_c^2-1)}\left|\int db\, e^{iq\cdot b}\, F_{ss}\left(\frac{b}{\rho_0}\right)\right|^2. \tag{7.33}$$

In particular, the forward scattering amplitude in the two-instanton approximation is

$$\sigma(t=0) \approx \frac{16n_0^2}{N_c^2(N_c^2-1)}\int_0^\infty dq_\perp^2 \left|\int db\, e^{iq\cdot b}\, F_{ss}\left(\frac{b}{\rho_0}\right)\right|^2, \tag{7.34}$$

which is finite at large $\sqrt{s}$. Hence, for forward scattering partons in the instanton vacuum model, we have

$$\sigma_{qq} \approx \left(n_0\rho_0^4\right)^2 \rho_0^2. \tag{7.35}$$

Clearly, the present analysis generalizes to the dipole–dipole scattering amplitude and also produced a finite scattering cross section,

$$\sigma_{dd} \approx \sigma_{qq}\frac{(d_1 d_2)^2}{\rho_0^4}. \tag{7.36}$$

It is instructive to compare our instanton results to those developed by Dosch and collaborators [419] in the context of the stochastic vacuum model (SVM). In brief, in the SVM model the Wilson-lines are expanded in powers of the field-strength using a non-Abelian form of Stokes theorem in the Gaussian approximation. A typical hadronic cross section in the SVM model is

$$\sigma \approx \langle(gG)^2\rangle^2\, a^{10}\, \mathbf{F}(R_h/a), \tag{7.37}$$

where the first factor is the "gluon condensate", a is a fitted correlation length, $\mathbf{F}$ is some dimensionless function depending on the hadronic radius R_h. Although our assumptions and those of Ref. [419] are very different regarding the character of the vacuum state, it is amusing to note the agreement between (7.36) and (7.37).

---

[10]Up to self-energies.

Indeed, the correlation length a of the SVM model is related (and in fact numerically close) to our instanton radius $\rho_0 \approx 1/3$ fm, while the gluon condensate $\langle (gG)^2 \rangle$ of the SVM model is simply proportional to the instanton density n_0 in the instanton model.

The most significant difference between these two approaches apart from their dynamical content and the way we have carried out the analytical continuation, is the fact that we *do not expand in field strength.* In fact, in the instanton model there is no parameter which would allow to do so for strong instanton fields. This difference is rather important as it is on it that our conclusion regarding multiple color exchanges is based. (In the SVM model with Wick-theorem-like decomposition, those would be just products of single exchanges, as in pQCD.)

7.2.7. *Pomeron from instantons*

This section is actually based on three papers, one by Kovchegov, Kharzeev and Levin [438] and two by Zahed, Nowak and myself [436], although the presentation follows the last one. In it the quasi-elastic (color exchanged) scattering we discussed in the previous subsection is generalized to "truly inelastic" parton–parton scattering amplitudes, with prompt gluon production from the instanton vertex into the unitarity cut.

It turns out that including multi-gluon states in the intermediate state (the cut) we obtain a cross section which is parametrically *much larger* than the quasi-elastic one, with only the color exchange, considered in the preceding section. The main parameter of the instanton physics, much discussed in the corresponding chapter, is *instanton diluteness* of the QCD vacuum, $\kappa_0 = n_0 \rho_0^4 \approx 0.01$ [226]. The quasi-elastic processes considered above have this small parameter in the amplitude and thus κ_0^2 in the cross section. However, as we will show below, the sum of all inelastic diagrams will have a cross section suppressed by only the first power of κ_0. The physical meaning of that is that in the former case there is a tunneling through the barrier while in the latter we instead jump *on* its top, saving half of the action.

Although we do not use the concept of t-channel gluon exchanges (as they are summed into the eikonalized phases), some of its features remarkably survive. Already in the quasi-elastic process only the octet color exchange survives the high-energy limit [436]. The same feature carries over to the inelastic processes we now consider where only the octet color is transmitted from each parton line. Of course they should be in the same SU(2) subgroup to interact with a given instanton. In general, these restrictions on the possible color representations disagree with the exponentiation of multi-gluon production, leading to the semiclassical theory we use.

Another generic comment is that surprisingly we found an unexpected bonus: this theory of the Pomeron does not produce the *Odderon*. In the early model by Low and Nussinov [420] the near-forward high-energy scattering amplitude is ascribed to a perturbative two-gluon exchange in the t-channel, which is C-parity even. Hence the qq and $\bar{q}q$ cross sections are the same to this order, a result that is very well supported by experiment. Indeed, all special experiments devoted to a search for odderon have so far failed.

This is strange in the perturbation theory which allows for higher order scatterings, e.g. SU(3) allows for a colorless combination of 3 gluons. Perturbatively, the Odderon/Pomeron ratio is expected to be suppressed by $O(\alpha_s)$, or much weaker than the data shows. To fix this problem, a number of ideas have been put forward some of which rely on nucleon specifics to cancel the Odderon. If that is the case, the Odderon should still be observable in hadronic reactions other than pp scattering.

In contrast, the instanton-induced processes at high energy do predict the same cross section for $\bar{q}q$ and qq. Indeed, even though our quasi-elastic and inelastic (see below) amplitudes sum up an indefinite number of gluons, switching a quark to antiquark on the external line amounts to flipping the sign of the corresponding $\sin \alpha$ contribution. As there is no interference between these and the $\cos \alpha$ terms at high energy, there is no Odderon in the instanton induced amplitudes. This is easily understood by noting that an instanton is an SU(2) instead of an SU(3) object, for which the fundamental (quark) and the adjoint (antiquark) representations are equivalent.

Let me now outline the derivation of the inelastic amplitude. To address it with instantons, the eikonal approximation has to be relaxed. To achieve that and elucidate further the character of the s-channel kinematics, we first derive a general result for on-shell quark propagation in a localized background field in Minkowski space. We then show how this result can be applied to instanton dynamics to analyze inelastic parton–parton scattering at high energy beyond the eikonal approximation and ladder graphs. For simplicity, all the instanton algebra will be carried out explicitly for $N_c = 2$. An on-shell massless quark propagating through a localized background $A(x)$ with initial and final momenta p_1 and p_2 follows from the LSZ reduction

$$S(p_1, p_2; A(x)) \equiv \langle p_2 | i\vec{\partial}\, \mathbf{S}_F(x)\, i\vec{\partial} | p_1 \rangle, \tag{7.38}$$

with $i\vec{\nabla}\, \mathbf{S}_F = -1$ the background (Feynman) propagator in the instanton field. At large p_+ momentum, the quark propagates on a straight line along the light cone. This limit can be used to organize (7.38) in powers of $1/p_+$. The result is

$$S(p_2, p_1; A(x)) = e^{i(p_2 - p_1)x}\, \bar{u}(p_2) g\!\!\!/A \sum_{n=0}^{\infty} \left(\frac{i}{2p_1 \cdot \nabla}\vec{\nabla} \frac{\not{p}_1 \gamma_0}{2p_{10}}\vec{\nabla} \right)^n$$

$$\times\, \mathbf{W}_-(x_{1+}, x_{1-}, x_\perp) u(p_1), \tag{7.39}$$

where

$$\mathbf{W}_-(x_{1+}, x_{1-}, x_\perp) = \mathbf{P}_c \exp\left(-\frac{ig}{2} \int_{-\infty}^{x_{1+}} dx'_+ A_-(x'_+, x_{1-}, x_\perp) \right). \tag{7.40}$$

The line-integral is carried along the p_1-direction of the original quark line with $x_{1\pm} = (p_0 x_0 \pm \vec{p} \cdot \vec{x})/p_0$. In the limit $p_{10} \to \infty$, only the $n = 0$ term contributes with $x_{1\pm} = x_\pm$ being just the light-cone coordinates, thereby reproducing the eikonal result (7.40). The higher-order terms are corrections to the eikonal result, with the $n = 1$ term accounting for both recoil and spin effects.

The imaginary part of the quark–quark inelastic amplitude follows from unitarity. Schematically,

$$\mathrm{Im}\mathcal{T}_{if} = \mathcal{T}_{in}\sigma_{nn}\mathcal{T}_{nf}^*, \tag{7.41}$$

where σ_{nn} accounts for the phase space of the propagating quarks and emitted intermediate gluons. The total cross section follows then from the optical theorem $\sigma = \mathrm{Im}\,\mathcal{T}/4s$. Using the result (7.39), we have for the total cross section in Minkowski space

$$\sigma = \frac{1}{4s}\mathrm{Im}\sum_{CD}\int d[A]\,d[A']\,e^{i(S[A]-S[A'])}\int \frac{d^3K_1}{(2\pi)^3}\frac{d^3K_2}{(2\pi)^3}\frac{1}{2K_{10}}\frac{1}{2K_{20}}$$

$$\times \left(\sum_{n=0}^{\infty}\frac{1}{n!}\prod_{i=1}^{n}\int\frac{d^3k_i}{(2\pi)^3}\frac{1}{2k_{i0}}\mathcal{A}(k_i)\mathcal{A}'^*(k_i)\right)\frac{1}{VT}\int dx\,dy\,dx'\,dy'$$

$$\times \exp[i(K_1-p_1)x + (K_2-p_2)y - (K_1-p_1)x' - i(K_2-p_2)y']$$

$$\times \mathcal{S}_{AC}(K_1,p_1;A(x))\,\mathcal{S}_{AC}^*(K_1,p_1;A'(x'))$$

$$\times \mathcal{S}_{BD}(K_2,p_2;A(y))\,\mathcal{S}_{BD}^*(K_2,p_2;A'(y')). \tag{7.42}$$

The functional integration is understood over gauge fields (to be saturated by instantons in Euclidean space after proper analytical continuation), with $\mathcal{A}(k)$ ($\mathcal{A}'(k)$) the Fourier transform of the relevant asymptotic of A (A') evaluated on the mass shell. Similar expressions were used for sphaleron-mediated gluon fusion [441]. The difference with the present case is the occurrence of quarks in both the initial, intermediate and final states. The sum in (7.42) exponentiates into the so-called R-term, which acts as an induced interaction between the A and A' configurations in the double functional integral (7.42). We will refer to it as $S(A, A')$.

The gauge fields carried inside the on-shell quark propagators $\mathcal{S}$ involve virtual exchange of background quanta with no contribution to the cut. In contrast, the on-shell gluons $\mathcal{A}(k)$ are real and the sole contributors to the cut. The $n = 0$ term in (7.42) in the large p_+ limit reduces to the quasi-inelastic contribution discussed above. The term of order n involves n-intermediate on-shell gluons plus two on-shell quarks, and contributes to the bulk of the inelastic amplitude.

The general result (7.42) involves no kinematical approximation regarding the in/out quark states. At high-energy, all $1/p_+$ effects in (7.39) can be dropped to leading $\ln s$ accuracy except in the exponent. As a result (7.42) simplifies dramatically,

$$
\sigma \approx \frac{1}{4VT} \operatorname{Im} \sum_{CD} \frac{1}{(2\pi)^6} \int dq_{1+} dq_{1\perp} dq_{2-} dq_{2\perp}
$$

$$
\times \int [dA][dA'] \exp[iS(A) - iS(A') + iS(A, A')]
$$

$$
\times \int dx_- dx_\perp dy_+ dy_\perp \exp\left[\frac{i}{2}q_{1+}x_- - iq_{1\perp}x_\perp + \frac{i}{2}q_{2-}y_+ - iq_{2\perp}y_\perp\right]
$$

$$
\times (\mathbf{W}_-(\infty, x_-, x_\perp) - 1)_{AC} (\mathbf{W}_+(y_+, \infty, y_\perp) - 1)_{BD}
$$

$$
\times \int dx'_- dx'_\perp dy'_+ dy'_\perp \exp\left[\frac{i}{2}q_{1+}x'_- - iq_{1\perp}x'_\perp + \frac{i}{2}q_{2-}y'_+ - iq_{2\perp}y'_\perp\right]
$$

$$
\times (\mathbf{W}_-(\infty, x'_-, x'_\perp) - 1)^*_{AC} (\mathbf{W}_+(y'_+, \infty, y'_\perp) - 1)^*_{BD}. \tag{7.43}
$$

Overall, the scattering amplitude follows from the imaginary part of a retarded four-point correlation function in Minkowski space. This correlation function follows from a doubling of the fields, a situation reminiscent of the so called thermo-field dynamics.

To proceed further, some dynamical approximations are needed. Let us assume that the double-functional integral in (7.43) involves some background field configurations characterized by a set of collective variables (still in Minkowski space), say $I = Z, \mathbf{R}, \rho$, for position, color and orientation, respectively. Let $z = Z - Z'$ be the relative collective position. Simple shifts of integrations, produce

$$
e^{iQz} = \exp\left[\frac{i}{2}q_{1+}z_- + \frac{i}{2}q_{2-}z_+ - i(q_1 + q_2)_\perp z_\perp\right] \tag{7.44}
$$

with no dependence on Z, Z' in the $\mathbf{W}$'s. The integration over the location of the CM $(Z+Z')/2$ produces VT which cancels the $1/VT$ in front, due to overall translational invariance. The integration over the relative coordinate z produces a function of the invariant $Q^2 = q_{1+}q_{2-} - (q_1 + q_2)^2_\perp$ because of Lorentz invariance. With this observation and to leading logarithm accuracy, we may rewrite (7.43) as follows:

$$
\sigma \approx \frac{1}{4} \ln s \operatorname{Im} \sum_{CD} \frac{1}{(2\pi)^6} \int dQ^2 \, dq_{1\perp} dq_{2\perp}
$$

$$
\times \int dz \, d\dot{I} \, d\dot{I}' \exp[iQz + iS(\dot{I}) - iS(\dot{I}') + iS(\dot{I}, \dot{I}', z)]
$$

$$
\times \int dx_- dx_\perp dy_+ dy_\perp \exp[-iq_{1\perp}x_\perp - iq_{2\perp}y_\perp]
$$

$$
\times (\mathbf{W}_-(\infty, x_-, x_\perp) - 1)_{AC} (\mathbf{W}_+(y_+, \infty, y_\perp) - 1)_{BD}
$$

$$
\times \int dx'_- dx'_\perp dy'_+ dy'_\perp \exp[-iq_{1\perp}x'_\perp - iq_{2\perp}y'_\perp]
$$

$$
\times (\mathbf{W}_-(\infty, x'_-, x'_\perp) - 1)^*_{AC} (\mathbf{W}_+(y'_+, \infty, y'_\perp) - 1)^*_{BD}. \tag{7.45}
$$

Note that the omitted exponents $e^{(i/2)q+x-}$ etc. are sub-leading in leading logarithm accuracy. The dotted integrations no longer involve the collective variables Z, Z'. The only dependence on the relative variable z resides in the induced R-term. The appearance of $\ln s$ underlines the fact that the integrand in (7.45) involves only Q^2 which is the transferred mass in the inelastic half of the forward amplitude, and $q_{1,2\perp}$ which are the transferred momenta through the quark form factors.

All kinematical approximations were carried out in Minkowski space, a point stressed in our earlier work [436]. The outcome (7.45) is now ripe for an analysis in Euclidean space. Performing the analytical continuation back to Minkowski space shows that only the combination $(\mathbf{n} \cdot \underline{\mathbf{n}}\, \mathbf{n}' \cdot \underline{\mathbf{n}}' + \mathbf{n} \cdot \underline{\mathbf{n}}'\, \mathbf{n}' \cdot \underline{\mathbf{n}})$ survives. The result is

$$\mathcal{W}(Q, q_{1\perp}, q_{2\perp}) = (8\pi^5)^{1/2} \, \mathbf{K}(q_{1\perp}, q_{2\perp})$$
$$\operatorname{Im} n_0^2 \int_0^\infty dR \left(\frac{R}{Q}\right)^{3/2} \int_0^\pi d\chi \, \sin^6 \chi \exp[QR - \mathbf{S}(R, \cos^2 \chi)],$$
$$(7.46)$$

with the induced kernel including two form factors introduced earlier

$$\mathbf{K}(q_{1\perp}, q_{2\perp}) = |\mathbf{J}(q_{1\perp}) \cdot \mathbf{J}(q_{2\perp}) + \mathbf{J}(q_{1\perp}) \times \mathbf{J}(q_{2\perp})|^2. \qquad (7.47)$$

Apart from the unphysical (perturbative) singularity at small $q_\perp$, the instanton-induced form factor can be parametrized by a simple exponential

$$\mathbf{J}(q_\perp) \approx -i\hat{q}_\perp 50 \, e^{-1.3q_\perp \rho_0}. \qquad (7.48)$$

We note that this is very different from just the Fourier transform of the instanton field used as a form factor in Ref. [438]. Throughout, the tail of the instanton will be subtracted resulting into a renormalization of the perturbative result.

The inelastic contributions to the quark–quark scattering outlined above still have one serious problem: their amplitude grows rapidly with the invariant mass of the produced gluons M^2. Zakharov [441] has argued that for M approaching the sphaleron mass the rise in the cross section has to stop simply because of unitarity constraints. Maggiore and Shifman [441] studied this phenomenon further and indeed found that re-scattering (unitarization, or multi-instanton effects) have to become important. Unitarization can be simply enforced by re-summing a chain of alternating instanton–anti-instanton configurations, leading to a unitarized amplitude confirming Zakharov's observations. Following Maggiore and Shifman's work we have re-summed these contributions and get an expression very similar to the Breit–Wigner expression for the two-body resonance. Specifically, its imaginary part reads

$$\sum_{n=1}^\infty \kappa_0 (\kappa_0 \, iB(Q))^n \qquad (7.49)$$

for alternating insertions of instantons and anti-instantons (chain). Each factor of iB results from the insertion of an extra instanton or anti-instanton on the chain, producing a bond with an extra unstable mode. Hence, the total cross section now reads

$$\sigma \approx \pi \rho_0^2 \ln s \, \frac{16}{15} \frac{1}{(2\pi)^8} \int dq_{1\perp} \, dq_{2\perp} \, \mathbf{K}(q_{1\perp}, q_{2\perp})$$

$$\times \kappa_0 \int_{(q_1+q_2)_\perp^2}^{\infty} dQ^2 \frac{\kappa_0 B(Q)}{1 + \kappa_0^2 B(Q)^2}. \tag{7.50}$$

The integrand in (7.50) rises with $B(Q)$ as expected in the small Q regime and falls off as $1/B(Q)$ due to unitarization. The dominant contribution takes place at the sphaleron invariant mass

$$B(Q_s) \approx 1/\kappa_0, \tag{7.51}$$

for which the total cross section becomes

$$\sigma \approx \pi \rho_0^2 \ln s \, \kappa_0 \frac{16}{15} \frac{1}{(2\pi)^8} \int dq_{1\perp} \, dq_{2\perp} \, \mathbf{K}(q_{1\perp}, q_{2\perp}). \tag{7.52}$$

Note that under the condition (7.51) the Q^2-integration amounts to a number of order 1, a measure of the area under the curve picked at Q_s with maximum $1/2$ and width of order 1. The rise in the partial inelastic cross section due to multi-instanton effects results into an increase of the cross section by one power of the diluteness factor which is about a 100-fold increase.

The energy following from (7.51) implies that half of the original instanton–anti-instanton action of $2 \times (8\pi^2/g^2)$ is compensated by their attraction. In other words, the s-exchange in the inelastic process starting from the vacuum is *half-instanton*. This is exactly the transition from vacuum to a static QCD sphaleron.

Now we are ready to discuss the resulting properties of the *Semi-classical Soft Pomeron*. For its trajectory we obtain the expression

$$\Delta(t) = \kappa_0 \frac{16}{15} \frac{1}{(2\pi)^8} \int dq_{1\perp} \, dq_{2\perp} \, \mathbf{H}(q_{1\perp}, q_{2\perp}; t), \tag{7.53}$$

with the new t-dependent induced kernel

$$\begin{aligned}
\mathbf{H}(q_{1\perp}, q_{2\perp}; t) \equiv \; & (\mathbf{J}(q_{1\perp} - q_\perp/2) \cdot \mathbf{J}(q_{2\perp} - q_\perp/2) \\
& + \mathbf{J}(q_{1\perp} - q_\perp/2) \times \mathbf{J}(q_{2\perp} - q_\perp/2)) \\
& \times (\mathbf{J}(q_{1\perp} + q_\perp/2) \cdot \mathbf{J}(q_{2\perp} + q_\perp/2) \\
& + \mathbf{J}(q_{1\perp} + q_\perp/2) \times \mathbf{J}(q_{2\perp} + q_\perp/2))^*,
\end{aligned} \tag{7.54}$$

with $t = -q_\perp^2$. The form factor $\mathbf{J}$ was defined earlier. The intercept is small because of the low diluteness of the instanton ensemble. But the non-linear trajectory depends on the instanton size ρ. For small t, we have $\Delta(t) \approx \Delta(0) + t\Delta'(0)$.

The slope parameter $\Delta'(0)$ follows from a Taylor expansion of (7.53) *after* integration. Using (7.48) which removes the unphysical singularity at $q_\perp \approx 0$ (related to the perturbative singularity discussed earlier in qq scattering), we can perform the double integrations in (7.53) to obtain $\mathbf{H}(q_\perp^2)$. Modulo the pre-factors in (7.53) this is just the Pomeron trajectory for $t < 0$ (physical region). The upper curve refers to instantons with unmodified vacuum sizes ($\mathbf{a} = 0$), while the lower curve corresponds to slightly smaller instantons ($\mathbf{a} = 0.25$). The trajectory never crosses zero, implying that the cross section grows in the physical region with increasing $\sqrt{-t}$.

We note that that the trajectories fall quickly with increasing $-t$, showing that most of the variation is located at small $-t$. The induced trajectories are sensitive to the instanton size through a rescaling of the induced form factor by the parameter $\mathbf{a}$. Indeed, the slope α' of $\Delta(t)$ (albeit $\mathbf{H}(t)$ around $t \approx 0$ relates to the Pomeron slope) is of order

$$\Delta'(0) = \alpha' = (0.5 - 0.2)\,\text{GeV}^{-2}, \tag{7.55}$$

with 0.5 corresponding to the unmodified instanton-induced form factor ($\mathbf{a} = 0$) and 0.2 corresponding to a scaled down instanton induced form factor ($\mathbf{a} = 0.25$). Qualitatively, the trajectory is not linear and curves at large negative t.

The specific trajectory depends on the new physics issues, of the *instanton shape modification*. In the singular gauge, the instanton gauge field falls off as $A \sim \rho_0^2/x^3$ at large x. By keeping the topological properties at the center, a modification of the "tail" of the gauge field at large distances may be allowed. Moreover, since the semi-classical analysis holds only for strong fields, these tail modifications should be in general expected.

There are many effects which can modify the tail of the instantons through interactions. Indeed an early variational estimate of the instanton vacuum energy using exponentially modified instanton form factors such as $e^{-\mathbf{a}\,x/\rho_0}$ [220] shows a minimum at $\mathbf{a} \approx 0.5$. Other possible reasons for instanton-shape modifications can be due to confinement as discussed in the context of the dual superconductor [63]. Lattice studies of various gluonic correlators also show rapid exponential fall-off. Thinking about the lowest glueballs, with their $2\,\text{GeV}$ mass scaling as a 2-gluon bound state, may imply an even larger sizing down with $\mathbf{a} \approx 1$.

Interestingly enough, the phenomenological issue regarding the instanton shape has never been addressed previously in this book. It is because nearly all previous applications of instantons are insensitive to it. All the applications related to light quarks rely on the fermionic zero mode, which exists and are normalizable independently of the shape of the instanton field. Also, the correlators of the scalar field strength combination $G_{\mu\nu}^2$ are also instanton-shape insensitive since they fall-off rapidly at large x even in perturbation theory, typically as $1/x^8$. This is in sharp contrast to the present case, where the induced form factors fall-off as $1/x_\perp^2$ and their integrated effect in the transverse plane (through $dx_\perp$) diverges logarithmically.

Instanton-shape modifications cause our results to change quantitatively, especially our estimate of the pomeron intercept and slope discussed earlier, showing

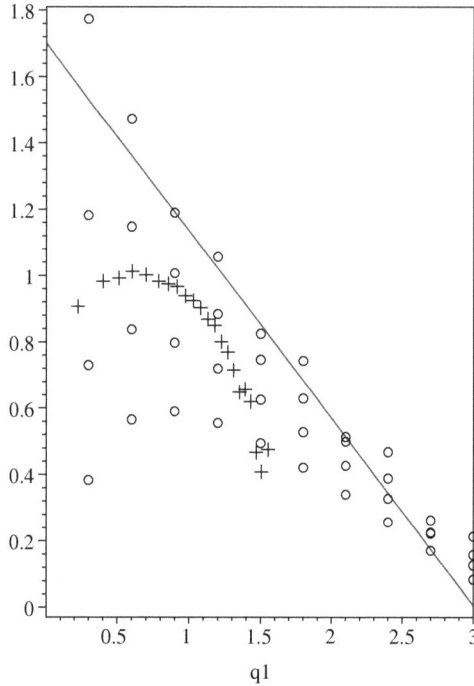

Fig. 7.9. The log of the form factor $|\mathbf{J}(q_\perp)|$ versus $q1 = \rho\sqrt{-t}$ where ρ is the instanton size. The 4 theoretical curves are for the instanton shape parameter $\mathbf{a} = 0, 1/4, 1/2, 3/4$ (top to bottom) [436]. The crosses show the UA8 data points (without error bars) as $(d\sigma/dt)^{1/4}$ with the same units of $q1$ but arbitrary normalization of the magnitude.

some of the limitations in our current analysis. To illustrate this, consider changing the description from regular to singular gauge and inserting the exponential modifications in the instanton tails. Hence, the argument in the definition of the induced instanton form factor $\mathbf{J}$ changes as follows:

$$\frac{\pi|x|}{\sqrt{x^2 + \rho_0^2}} \rightarrow \pi\left[\frac{|x|}{\sqrt{x^2 + \rho_0^2}} - 1\right] e^{-\mathbf{a}|x|/\rho_0}. \tag{7.56}$$

The effect on the form factor $\mathbf{J}$ is illustrated in Fig. 7.9. In the unmodified case with $\mathbf{a} = 0$ (top points) the form factor rises toward small momentum transfer, which is not the case for $\mathbf{a} > 0$. As expected the modifications due to $\mathbf{a}$ are small at large $q_\perp^2$.

7.3.   Pomeron structure and interactions

7.3.1.   *Clustering in inclusive pp collisions*

Already experiments done at late 1970s at CERN ISR have shown puzzling results related with strong correlations in multiparticle production. Common assumption was that secondary particles appear as a result of QCD string fragmentation. Lund-based codes have done a rather good job for e^+e^- annihilation (in which one string is formed) at all levels, while for pp (with two strings) only the inclusive single-body distributions were OK, but not the two-body correlations. Their magnitude indicated that some clusters are produced in pp, which are heavier than those produced in e^+e^- and heavier than the usual resonances, as they decay to more secondaries. Unfortunately, these clusters have not been studied in detail and their spectrum, quantum numbers etc. were not identified. A recent review of these "clans" can be found in Ref. [446].

With the advance of new models of the Pomeron discussed in this chapter one may ask if the mysterious clusters (absent in e^+e^-) can be related with glueballs and sphalerons.

In a specific case of *scalar glueballs*, the papers [435, 447] have included ladders with rungs made of such particles. They agree that its contribution to the Pomeron intercept is about 0.05, which directly translates into prediction of the rapidity density of these particles

$$\frac{dN_{\sigma,0++}}{dy} \approx 0.05. \tag{7.57}$$

In preliminary STAR data from a pp run at RHIC [437] there are indications for a scalar glueball resonance in the K^+K^- mode, roughly consistent with this estimate, while being at least an order of magnitude above the usual resonance production systematics provided by the statistical models. If confirmed by subsequent studies, this could indicate that rungs with scalar glueballs do really exist.

Now, what about the sphaleron production? As we have emphasized already, the latter should have a mass and a size determined by the typical size of the instantons in the QCD vacuum, or $M \sim 3\,\text{GeV}, \rho \sim 1/3\,\text{fm}$.

There are however additional complications. When a topological cluster is produced in hadronic collisions, as a *colored* object it is still connected by the QCD strings to some receding partons, and thus can only appear on top of some debris of the usual string fragmentation. It can in principle be located from correlation measurements, but it is difficult to do. The process to be discussed in the next section seems to solve all those problems.

7.3.2. *Inclusive production of clusters in double-Pomeron processes*

We have already explained in Section 7.1.1 above what the four-Pomeron vertex is and how its unitarity cut describes the diffractive production of one central cluster, separated by large rapidity gaps from two protons, see Fig. 7.1a.

The UA8 experiment at CERN studied the reaction $\bar{p}p \rightarrow \bar{p}Xp$ where X is a set of hadrons at mid-rapidity. There are two sets of data: one in which both nucleons were detected (AND) and one in which only one nucleon was detected

(OR). Since the two triggers are different, the two data sets were measured at different kinematics. UA8 used the following model-dependent parametrization of their measured differential cross section

$$\frac{d^6\sigma_{\mathrm{DPE}}}{d\xi_1 d\xi_2 dt_1 dt_2 d\phi_1 d\phi_2} = F_{\mathcal{P}/p}(t_1,\xi_1) \cdot F_{\mathcal{P}/p}(t_2,\xi_2) \cdot \sigma_{\mathrm{PP}}(M_X) \qquad (7.58)$$

where the variables (ξ_i, t_i, ϕ_i) describe the fraction of the longitudinal momentum, momentum transfer squared and its azimuthal direction for each Pomeron. All the parameters are uniquely given by the measured parameters of the outgoing $p, \bar{p}$. The Pomeron flux factor or structure function is defined as

$$F_{\mathcal{P}/p}(t,\xi) = K |\mathbf{F}_1(t)|^2 \, e^{bt} \xi^{1-2\alpha(t)}, \qquad (7.59)$$

where $|\mathbf{F}_1(t)|^2$ is the so called Donnachie–Landshoff [418] nucleon form factor

$$\mathbf{F}_1(t) = \frac{4m_\mathrm{p}^2 - 2.8t}{4m_\mathrm{p}^2 - t} \frac{1}{(1 - t/0.71)^2}. \qquad (7.60)$$

The parameters were defined from single-pomeron data with $b = 1.08 \pm 0.2\,\mathrm{GeV}^{-2}$, and nonlinear Pomeron trajectory

$$\alpha(t) = 1 + \epsilon + \alpha't + \alpha''t^2 = 1.035 + 0.165t + 0.059t^2. \qquad (7.61)$$

The parameter $K = 0.74/\,\mathrm{GeV}^2$ was not measured and was set from the Donnachie–Landshoff fit. The specific parameterizations were used to set up the acceptances and so on. However, in the UA8 paper to be discussed below, it was pointed out that the difference between the AND and OR data sets may suggest that the above parametrization with factorizable flux factors may be oversimplified.

The uncertainties related to the empirical extrapolation from the covered to the full kinematical range notwithstanding, the UA8 data show a striking and an unexpected shape and magnitude[11] for the Pomeron–Pomeron cross section $\sigma_{\mathrm{PP}}(M_X)$ shown in Fig. 7.10. Only at the central cluster mass $M_X > 10\,\mathrm{GeV}$ was the cross section small ≈ 0.1 mb, and more or less in agreement with standard Pomeron calculus (more specifically with the Pomeron factorization appended by some Reggeon contributions decreasing with M_X). At smaller masses $M_X < 10\,\mathrm{GeV}$ the observed cross section is an order of magnitude larger than what is expected from factorization, Fig. 7.1b. This was neither predicted prior to the experiment, nor explained (to our knowledge) after the experiment. Another crucial finding is that the low-mass clusters decay isotropically (in their rest frame). In addition, the momentum transfer to both protons is rather large, $|t| = 1$–$2\,\mathrm{GeV}^2$, which should ensure that the colliding Pomerons are indeed small-size gluonic objects.

Production of clusters of a few GeV is qualitatively consistent with the idea of sphaleron excitation. We start with a qualitative argument of why the factorization does not work in our approach. This happens because instead of a single universal

---

[11]The extracted cross section is based on the value of K quoted above.

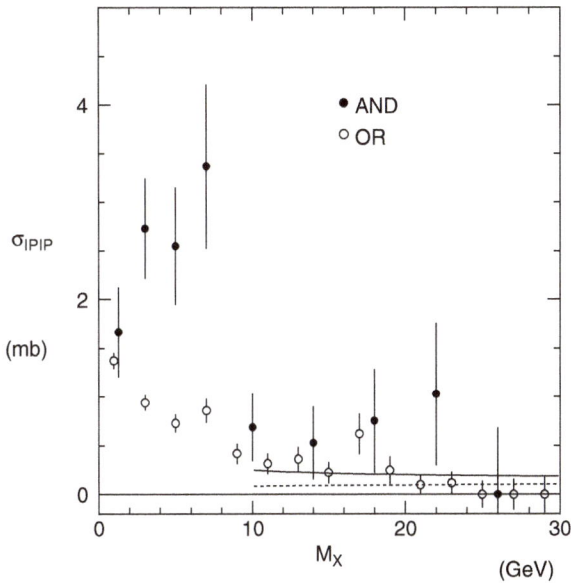

Fig. 7.10. Mass dependence of the Pomeron–Pomeron total cross section σ_{PP}, derived from the AND and OR triggered data, respectively. The dashed curve is the factorization prediction (independent of K) for the Pomeron-exchange component of σ_{PP}. The solid line is the fit to the OR points of a Regge-exchange term, $1/(M_X^2)^{0.32}$, added to this Pomeron-exchange term.

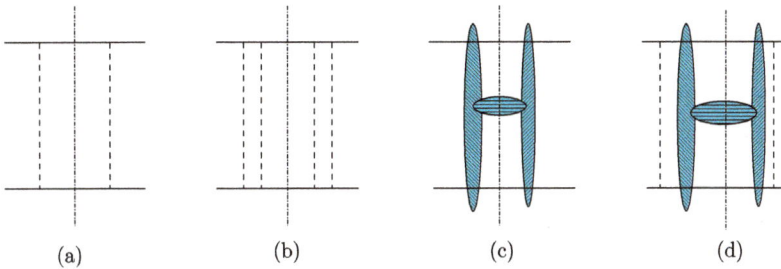

Fig. 7.11. Schematic diagrams for the cross sections of different processes associated with high energy collusion of two quarks, shown by horizontal solid lines. The vertical dash-dotted lines are unitarity cuts, they separate the amplitude from its complex conjugate. (a) Low–Nussinov or single-gluon exchange, leading to inelastic collisions due to color exchange; (b) Low–Nussinov cross section, with no color exchange; (c) instanton-induced inelastic collision with color transfer and prompt cluster production; (d) combined instanton–gluon process leading to double pomeron like events with a cluster.

Pomeron pole we view high energy scattering in the semi-hard regime as a superposition of two different phenomena: color exchanges (which lead to a constant cross section, that does not grow with $\sqrt{s}$) and the topological cluster production. The single Pomeron may be caused by the former component alone, while the double Pomeron may produce a visible cluster which we will try to associate with the latter.

Zahed and myself attempted a quantitative study of this process, see Ref. [448]. Our general setting is a instanton + gluon exchange, as explained in Fig. 7.11. We will not go into details of the calculation and predicted expressions, as they are quite involved. We predicted all qualitative features of the UA8 data correctly, as well as the right magnitude of the cross section.

7.3.3.   *Exclusive production of hadrons in double-Pomeron processes*

It would be very interesting to know if heavy and apparently small-size gluonic clusters seen in the UA8 experiment decay into particular channels with 1, 2 or more hadrons. It is also very important to see if there is a difference between the overall parity odd and even states, as the calculation [448] just mentioned predicts different distributions of the azimuthal angle between two momentum transfers to two protons, namely $\cos^2 \phi$ and $\sin^2 \phi$, respectively.

The issue of azimuthal dependence has an interesting history. The WA102 collaboration at CERN carried a fixed target pp experiment at $\sqrt{s} = 28\,\text{GeV}$, focusing on the double-pomeron exclusive production into few hadron states. This experiment was the first to discover a strong dependence of the cross section on the azimuthal angle between the momenta transferred to two protons, a feature that was not expected from standard Pomeron phenomenology.[12] This result inspired some phenomenological works [450,452,453] pointing out a possible analogy between the Pomeron and vector particles. Close and his collaborators have even suggested to use this azimuthal distribution as a glueball filter, selecting the hadronic states which peak at a small difference in the transverse momentum dP_T of the protons. In particular, the production of scalars and tensors such as $f_0(980)$, $f_0(1500)$, $f_J(1710)$, and $f_2(1900)$ was found to be considerably enhanced at small dP_T, while the production of pseudoscalars such as η, η' was found to be peaked at mutually orthogonal momentum transfers of the protons. In our approach the produced QCD sphalerons can be regarded as precursors of glueballs or η' pseudoscalars strongly coupled to glue.

What can be the hadronic final states into which the outgoing quarks and gluons from the topological clusters should be projected? One may think that if we discuss single hadrons, those should be glueballs, the scalar and pseudoscalar ones[13]: the best modes to look for those are K^+K^- and $K_s K^{\pm}\pi^{\mp}$, respectively. Unfortunately, the pseudoscalar glueball has not yet ever been seen, but one can use η' as

---

[12]By pure chance I happened to be in Cambridge on some unrelated conference in the summer of 1997 and came to one seminar in the DAMPF. It happened to be the F. Close's talk about unexpected azimuthal dependences seen by WA102. I asked a question, whether anyone has ever considered that a Pomeron, like, a photon, might have a polarization vector along the kick direction. Frank answered, "why don't we both work on that." I decided that I at the time knew too little about Pomerons to contribute and declined.

[13]We remind the reader that the tensor glueball does not couple to instanton-like fields.

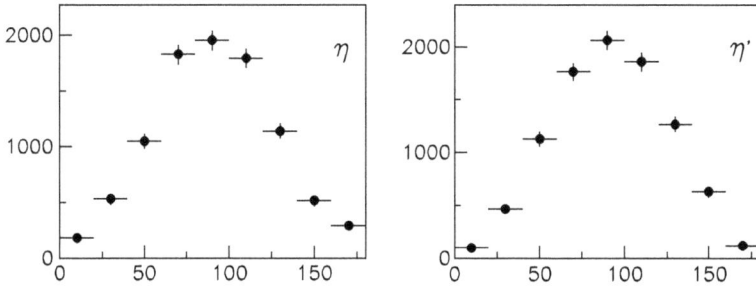

Fig. 7.12. Dependence of the double-Pomeron cross section on the azimuthal angle ϕ for 0^{-+} production in the final states from Ref. [451].

a substitute. And indeed, both states are copiously produced, and the angular distributions are exactly as expected. Pseudoscalarity appears in vector theories as an $\vec{E}\vec{B}$ product, which implies that two kicks which two protons obtain are most probably mutually *orthogonal* in the transverse plane. This is possible because of the fact that the electric and magnetic fields of the instanton are parallel, not orthogonal as in the plane waves.[14] The measured dependence on the transverse momentum $|t|$ can be compared to instanton form factors: and again qualitative agreement in shape and magnitude is seen. The absolute cross section for the O^{++} glueball and η' is also reproduced well.

---

[14] Although in a diagram there is also a gluon, but its weak field is needed for color compensation, otherwise the gluon plays no dynamical role since its field is weak compared to the instanton one.

CHAPTER 8

QCD at Finite Temperatures

8.1. Introduction

8.1.1. *Brief history and the basic scales*

Application of perturbative methods borrowed from quantum field theory to various problems of statistical mechanics were worked out in late 1950s. The main ideas (to my knowledge) and formalism came from Refs. [460–463]. Important examples of graphical re-summation of some series of diagrams worked out by Gell-Mann and Bruckner [465]. All of that quickly found multiple applications in condensed matter and nuclear physics.

The first use of statistical mechanics in the context of the hadronic matter started somewhat earlier. In 1950 Fermi suggested a statistical model of multiparticle production. Soon Landau [583] argued that hydrodynamics should then also be applicable: we will return to those works in the next chapter. In 1956[1] Landau and Belenky [490] suggested to account for strong interaction via the so called *resonance gas*.[2] Then Hagedorn, in the 1960s and beyond, developed this idea into a new philosophy called *the dual bootstrap* approach for resonances, in which resonances were made of other resonances. It had the exponentially rising spectrum of states which led to Hagedorn's famous *upper possible temperature* $T_{\text{Hagedorn}} \approx 160\,\text{MeV}$. As it was understood later, it is not the maximal temperature possible in Nature but just the phase transition above which the *hadronic phase*, with its confinement, QCD strings and chiral symmetry breaking, no longer exist. During the 1970s Hagedorn [491] and others made many practical applications of the statistical model to various hadronic reactions. For example in my work [493] Einstein's expression for the probability of fluctuations $P \sim \exp S$, where S is the entropy, was successfully

---

[1]The time was not accidental; it was soon after the first hadronic resonance, Δ, had been discovered.

[2]The idea itself, that resonances enter the statistical sum in the same way as bound states goes back to classic Beth–Ulenbeck work of 1936 [489], which provided the well-known expression for the *second virial coefficient* one can find in the textbooks.

used to describe all multi-body exclusive channels for reactions like low energy p̄p annihilation. I also argued [492] that the power spectrum rather than Hagedorn's exponential mass spectrum is better approximation to a reality, leading to a very simple EoS of the resonance gas $\epsilon \sim p \sim T^6$ widely used since. A special case of low T — the interacting pion gas — was studied in a realm of chiral Lagrangians, see Refs. [457, 488, 500, 501, 503]. We will discuss this issue below in Section 8.3.3.

From the early 70s onwards were the days when QCD was discovered, and its ideas were immediately used for hot/dense matter. Based on asymptotic freedom, Collins and Perry [467] suggested that the high T/density matter should be close to ideal gas at high T/density. They used a simple argument: in dense matter constituents are almost always close to each other, and therefore, due to the asymptotic freedom, their interaction should be sufficiently weak.

The first perturbative corrections to the ideal gas were evaluated by Baym and Chin [468], Kislinger and Morley [469], Freedman and McLerran [470], myself [471] and Kapusta [472]. First reviews, by myself and Gross, Pisarski and Yaffe [457, 458] followed, and so around 1980 a new field was more or less formed.

The calculation of the polarization tensor done in my paper [471] showed that unlike the virtual gluon loops which anti-screen the charge, the real in-matter gluons behave more reasonably and *screen* the charge: therefore this new phase was called *Quark–Gluon Plasma*, QGP for short. Its excitations are quark and gluon quasiparticles plus collective "plasmon" modes similar to those in the usual QED plasma.

What this means, is that apart from the obvious

- energy-momentum scale $p \sim T$

there is also the so called

- electric or Debye scale $p_E \sim g(T)T$

at which the static electric charge is screened. Significant progress in understanding what happens at this scale was achieved by Braaten and Pisarski [477], who made a complete re-summation of hot thermal loops. We will not discuss its later form, as an effective Lagrangian, but proceed instead to a somewhat more intuitive kinetic formulation due to Blaizot and Iancu [478].

Furthermore, it was found in the same work [471] that static magnetic field is *not* screened: thus infrared divergences and other non-perturbative phenomena survive in the magnetic sector. First re-summation of the so called ring diagrams produced a finite plasmon term [471,472], but higher order diagrams are still infrared divergent. Polyakov [484] argued that there should be a new scale at which magnetic fields are also screened

- magnetic mass scale $p_E \sim g^2(T)T$

Linde [485] was quick to suggest some monopole-related mechanism to that. It was however soon realized that the magnetic sector at high T is basically a problem of $3d$ Yang–Mills theory, which is a nontrivial confining theory, and strictly speaking

Table 8.1. A comparison of the basic scales in high-T gauge theories in (a) the weak coupling with (b) the strong coupling limit of the N $= 4$ SUSY gauge theory. (All constants are ignored except in the last entry.)

Quantity	Small g, N	Large $g^2 N$
Inter-particle $n^{-1/3}T$	~ 1	$\sim 1/N^{2/3}$
Debye screening radius $R_\mathrm{D}T$	$\sim 1/g$	~ 1
Magnetic screening radius $R_\mathrm{M}T$	$\sim 1/g^2$	~ 1
Sound attenuation $\Gamma_s T$	$\sim 1/g^4 \log(1/g)$	$1/3\pi$

we still do not understand it and only know that magnetic mass exists from lattice simulations, both in $3d$ and $4d$.

We will follow the brute force pQCD evaluation of multi-loop diagrams such as was done in Ref. [479]: these works demonstrated that convergence of pQCD series is not reached in the region of interest. Multiple attempts to do a different re-summation have also been done [482, 483] and seemed successful, but the overall situation is not clear since the strong-coupling results [895] are not that different.

Of course, thermodynamical quantities are not the only ones which we would like to know. Many other practically important quantities can be calculated perturbatively in QGP, such as dilepton and photon production rates or viscosity: we will turn to their discussion later in the book.

Recently a new development has appeared on the interface of string theory and QCD, allowing one to calculate the finite-T quantities for $\mathcal{N} = 4$ supersymmetric Yang–Mills theory, known also as conformal field theory (CFT), in the *strong coupling* regime. We will discuss it in Section 12.5. Table 8.1 compares the main QGP length scales (in $1/T$) in both limits.

8.1.2. *From field theory to thermodynamics*

A general introduction into the application of field-theory methods at non-zero temperatures and densities can be found in standard texts, such as Refs. [463, 464], so we will only discuss some simplest examples which help the reader to get familiar with the subject, and then jump to the main results relevant for the gauge theories.

Thermodynamics has many formulations and people use many different potentials, depending which variables are most natural for the problem they consider. Since in hadronic matter many reactions are possible which change the number of particles, it is convenient to consider as such variables the temperature T and a set of chemical potentials μ_i (so that the number of particles in physical states is undetermined). We will use the so called *grand partition function* introduced by Gibbs. So our definition for the statistical sum will be

$$Z = \mathrm{Tr} \, \exp \left(-\frac{\hat{H}}{T} - \sum_i \frac{\mu_i \hat{N}_i}{T} \right), \tag{8.1}$$

where Tr indicates the sum taken over all states of the system, with any energies and particle numbers. $\hat{H}$ is the Hamiltonian, μ_i and operators $\hat{N}_i$ are the chemical potential and the conserved quantities conjugated to them such as the total number of particles of the type i which have such a quantum number. In QCD applications the conserved numbers are the global baryon number N_B and strangeness S (usually zero). Their input values determine the corresponding chemical potentials μ_B, μ_S, as the input energy or entropy determine T.

The grand (or Gibbs) potential Ω is related to Z by taking its log

$$\Omega = -T \log Z. \tag{8.2}$$

Because of the minus sign in this definition the maximum of Z corresponds to the minimum of Ω. Standard thermodynamical relations provide the following alternative expressions for Ω:

$$\Omega(V, T, \mu_i) = E - TS - \sum_i N_i \mu_i = -pV, \tag{8.3}$$

where E is the total energy, S is the entropy, N_i are particle numbers and p is the pressure. Note that V is the only extensive quantity among arguments of $\Omega(V, T, \mu_i)$, and since Ω (like all other potentials) is an extensive quantity as well; as a result its dependence on V can only be just linear (as expressed in the last equality of the above equation). So the criterion used in practice for selecting the true equilibrium configurations is as follows: *the physical state always corresponds to the maximum of the pressure.*

The next important set of relations is provided by the *full differential* of Ω

$$-d\Omega(V, T, \mu_i) = pdV + SdT + \sum_i N_i d\mu_i, \tag{8.4}$$

which is a short-hand notation implying that p, S and N_i are the partial derivatives of Ω over the corresponding variables,

$$p = -\frac{\partial \Omega}{\partial V}\Big|_{T,\mu_i}, \qquad S = -\frac{\partial \Omega}{\partial T}\Big|_{V,\mu_i}, \qquad N_i = -\frac{\partial \Omega}{\partial \mu_i}\Big|_{T,V}. \tag{8.5}$$

As soon as Z is calculated, all thermodynamics follows by trivial differentiations.

Exercise 8.1.   For a simple parameterization of the hadronic resonance gas (8.83) calculate the energy density and entropy, and verify the expression for the sound velocity given in that formula.

8.1.3.   *A quantum particle at finite T*

In Section 1.3 we made an extensive study of quantum motion described by path integrals in Euclidean time. We also showed there that by taking the duration of the

Euclidean time finite $\tau = 0 \cdots \tau_{\text{Matsubara}}$ with the periodic paths, we may describe the same quantum system at the temperature

$$kT = \frac{\hbar}{\tau_{\text{Matsubara}}}, \qquad (8.6)$$

where we explicitly indicated the Boltzmann constant k and the Planck one $\hbar$. (Of course we use conventions in which both are taken to be 1, and they will not appear in all other expressions.)

In this section we will describe some elementary calculations in the context of the quantum anharmonic oscillator, which we hope can provide a good opportunity for a reader using these methods for the first time: people familiar with it can just skip this subsection.

We recall that the density matrix for a quantum particle has been represented by the Euclidean path integral

$$\rho(x, x', \beta) = \int Dx(\tau) \exp(-S[x(\tau)]) \qquad (8.7)$$

over paths stating at the initial coordinate x and ending at coordinate x' and taking Euclidean time $\beta = 1/T$. The weight is given by the Euclidean action

$$S[x(\tau)] = \int_0^\beta d\tau \left(\frac{m}{2} \left(\frac{dx}{d\tau} \right)^2 + V(x) \right). \qquad (8.8)$$

Periodicity of the paths $x(\beta) = x(0)$ appears when one calculates the statistical sum, which in terms of the density matrix is just

$$Z = \text{Tr} \exp(-\beta H) = \int dx \rho(x, x, \beta). \qquad (8.9)$$

A Green function is the simplest correlation function possible, the average of two coordinates $x(\tau)$ (later to become quantum or statistical fields $\phi(x)$) taken at *different* time moments.

$$G(\tau, \tau') = \langle x(\tau) x(\tau') \rangle = \frac{\int Dx(\tau) x(\tau) x(\tau') \exp(-S[x(\tau)])}{\int Dx(\tau) \exp(-S[x(\tau)])}, \qquad (8.10)$$

where the angular brackets denote averaging with the path integral weight.

The good news is that *the only modifications of the Feynman rules which appear at finite T, μ relative to vacuum are inside the Green function.* So, it is worth spending some time to understand what these modifications are and what they mean, using as usual the simplest setting possible.

Let us calculate $G(\tau, \tau')$ for a linear oscillator, for which the potential is $V(x) = m\omega^2 x^2/2$. Since it does not depend on time, everything can only depend on time difference, and so $G(\tau, \tau') = G(\tau - \tau', 0)$, or we can simply put τ' to zero.

The action has potential energy, which is quadratic in x, and kinetic one, which is quadratic in the *derivative* of x. Going to the Fourier representation we express the derivative in terms of x itself, and make the path integral Gaussian. In other words, Fourier series will diagonalize the operator we have in the action.

The Fourier series is

$$x(\tau) = \sum_n c_n \exp(i\omega_n \tau),$$

where $c_{-n} = c_n^*$ because the coordinate is real. The sum runs over discrete *Matsubara frequencies*

$$\omega_n = \frac{2\pi n}{\beta} = 2\pi n T, \tag{8.11}$$

with integer n, from minus to plus infinity. The corresponding completeness condition reads

$$T \sum_n \exp(i\omega_n(\tau - \tau')) = \delta(\tau - \tau'),$$

which at zero T becomes the usual integral over continuous frequency

$$\int \frac{d\omega}{2\pi} \exp(i\omega(\tau - \tau')) = \delta(\tau - \tau').$$

Using it, we re-write the oscillator action as a *single* sum over Matsubara frequencies

$$S = (m\beta/2) \sum_n (\omega_n^2 + \omega^2)|c_n|^2.$$

Putting it into the partition function weight one finds that the amplitude of each nth harmonics fluctuates independently: the path integral is now easy since it is now splits into the product of independent Gaussian ones. It is now easy to find the *average* amplitudes, $\langle c_n \rangle = 0$ and

$$\langle |c_n|^2 \rangle = \frac{\int dc_n |c_n|^2 \exp[-(m\beta/2)(\omega_n^2 + \omega^2)|c_n|^2]}{\int dc_n \exp[-(m\beta/2)(\omega_n^2 + \omega^2)|c_n|^2]} = \frac{1}{m\beta(\omega_n^2 + \omega^2)}.$$

Now it is easy to find the Green function

$$G(\tau) = \sum_n \exp(i\omega_n \tau)|c_n|^2. \tag{8.12}$$

In particularly,

$$G(0) = \langle x^2 \rangle = \frac{1}{m\beta} \sum_n \frac{1}{(2\pi T n)^2 + \omega^2}$$

$$= \frac{1}{2m\omega} \coth(\omega/2T) = \frac{1}{m\omega}\left(\frac{1}{2} + \frac{1}{\exp(\omega/T) - 1}\right). \tag{8.13}$$

The last expression emphasizes that it is just twice the average potential energy,

$$\langle m\omega^2 x^2 \rangle = \hbar\omega\left(\frac{1}{2} + \langle n(T) \rangle\right), \tag{8.14}$$

where the last term is nothing else but the thermally averaged occupation number. This is of course the thermal energy of the oscillator. (Note that here I again indicated $\hbar$ as an exception: we use elsewhere the natural units in which $\hbar = c = k_B = 1$.)

Another instructive way of deriving the Euclidean finite-T Green function is via the differential equation it satisfied. From its definition (and the role of "propagator" in general, to propagate the external forces) it satisfies the equation

$$\left(-\frac{d}{d\tau^2} + \omega^2\right) G(\tau) = \frac{2}{m}\delta(\tau). \tag{8.15}$$

One can see (using the completeness sum mentioned above) that the solution we found indeed satisfies it.

On the other hand, it is instructive to find the solution from the equation itself. The l.h.s. suggests that time dependence should be just a combination of exponentials

$$G(\tau) = A\exp(-\omega\tau) + B\exp(\omega\tau). \tag{8.16}$$

One condition follows from the periodicity condition which I would like to write now as $G(\tau = 0) = G(\tau = \beta)$. The other follows from the fact that the derivative of the function is not periodic, and its jump in derivative between $\tau = 0 + \epsilon$ and $\tau = \beta - \epsilon$ fits the coefficient of the delta-function

$$\left.\frac{dG}{d\tau}\right|_\beta - \left.\frac{dG}{d\tau}\right|_0 = \frac{2}{m}, \tag{8.17}$$

which leads to

$$B = \frac{1}{\omega m}\frac{1}{\exp(\beta\omega) - 1}, \qquad A = \frac{1}{\omega m}\frac{\exp(\beta\omega)}{\exp(\beta\omega) - 1}. \tag{8.18}$$

Two examples of its behavior at small and large T/Ω are shown in Fig. 8.1. One can see from Fig. 8.1a that at low T, $G(\tau)$ approaches its zero energy limit $\sim \exp(-\Omega|\tau|)$ and decays away from the origin. (This would be more evident if one re-plotted this periodic function in an interval $\tau = -\beta/2, \beta/2$.)

In the high-T limit exemplified in Fig. 8.1b one can see that $G(\tau)$ changes slightly, it mostly stays constant (note a very extended scale). In terms of Matsubara frequencies it means that the main contribution is given by the zeroth term $\omega_n = 2\pi T n = 0$. One may also note that if only this zero mode is kept, $\hbar$ does not appear anywhere which signals that we are in the classical limit. And indeed, it is easy to see that all the results in this approximation correspond to the classical Boltzmann oscillator at finite T, e.g. $\langle x^2 \rangle = G(0) = T/\omega^2$ etc.

With the finite-T Green function now at hand, one can go on and calculate any terms of the Feynman perturbation series for the anharmonic oscillator. The rules and the diagrams are exactly the same as described in Chapter 3: the only difference is the periodicity in τ which is already incorporated into the modified $G(\tau)$.

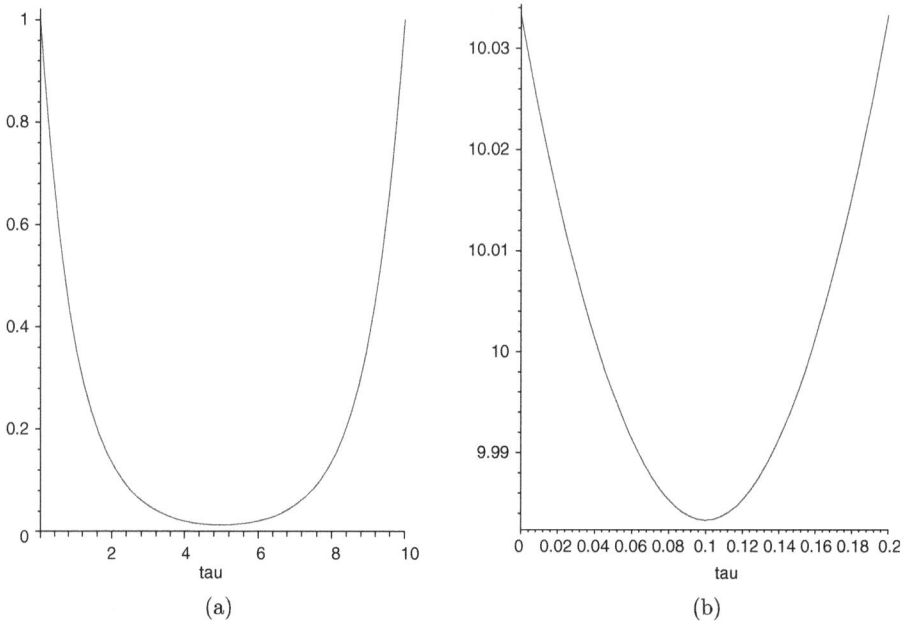

Fig. 8.1. (a) Green function at $m = \Omega = 1$ and temperature $T = 1/\beta = 0.1$ and (b) $T = 1/\beta = 5$. Note that the scales of these two figures are very different.

For example, introducing a quartic coupling λx^4 to the potential one gets the first order correction from a diagram with two closed loops, the first diagram in Fig. 1.6a

$$Z = Z_0(1 - 3\lambda G(0)^2), \tag{8.19}$$

where $G(0)$ is now given by (8.13). Further diagrams involve various powers of the Green function, connecting different time moments: but with very simple — exponential — dependence on time (8.16) there is no problem to do any of them directly in the time representation. In fact it is even slightly easier than at $T = 0$ where we had to think about different analytical forms at positive and negative time difference, while now there is a single expression on a whole interval $\tau = 0 \cdots \beta$.

Nevertheless, we will now return to the frequency representation and describe some methods which will be very useful in the field theory applications to follow. Let us define for any function of τ its Fourier transform and the inverse

$$\tilde{f}(\omega_n) = \int_0^\beta e^{i\omega_n \tau} f(\tau), \qquad f(\tau) = \frac{1}{\beta} \sum_n e^{-i\omega_n \tau} \tilde{f}(\omega_n). \tag{8.20}$$

For example, the transform of the oscillator Green function is

$$\tilde{G} = \frac{1}{m} \frac{1}{\omega_n^2 + \omega^2}. \tag{8.21}$$

The following "completeness" identities are useful:

$$\int_0^\beta d\tau e^{i(\omega_n - \omega_{n'})\tau} = \beta\delta_{nn'}, \qquad \frac{1}{\beta}\sum_n e^{-i\omega_n(\tau-\tau')} = \delta(\tau-\tau'). \qquad (8.22)$$

For example, using the first of them it is easy to see that *in each vertex the sum of all frequencies is zero*, reflecting energy conservation.

The Feynman rules for each diagram is summation over all frequencies unrestricted by such conservation. Let me give as examples expressions which appear for "balloon" and "loop" diagrams for the Green function in Fig. 1.6a

E

$$\text{"balloon"} \sim g^2 \left(\frac{1}{\omega_n^2 + \omega^2}\right)^2 \frac{1}{\omega^2} \sum_{n1} \frac{1}{\omega_{n1}^2 + \omega^2},$$

$$\text{"loop"} \sim g^2 \left(\frac{1}{\omega_n^2 + \omega^2}\right)^2 \sum_{n1} \frac{1}{\omega_{n1}^2 + \omega^2} \frac{1}{\omega_{n-n1}^2 + \omega^2}. \qquad (8.23)$$

The sum in the first case is nothing but $G(0)$ in coordinate representation: we already learned that

$$\sum_n \frac{1}{x^2 + n^2} = \frac{\pi}{x}\coth(\pi x),$$

which translates into average occupation number of quanta as we discussed before. The sum with two factors for $n = 0$ can also be worked out from it by differentiation over x, but the general bubble is not so simple to do.

The general method to calculate sums like that is the Zommerfeld–Watson method based on writing it as a contour integral over a contour C which goes around the real axes, from below to above it. The introduced function has poles at all integers, and its residua give back the sum

$$\sum_n F(n) = \oint_C dz F(z)\frac{1}{e^{2\pi i z} - 1}. \qquad (8.24)$$

After that, one can manipulate the contour. The exponent decays lower part of the complex plane, but not for the upper part. Here one uses the following "magic trick":

$$\frac{1}{e^{2\pi i z} - 1} = -1 - \frac{1}{e^{-2\pi i z} - 1}, \qquad (8.25)$$

which is rewritten as a constant plus the function which decays upward. Moving the contour C up and down to infinities one picks up singularities (in our examples, the poles) of the function F. On top of that, we are left with the integral originating from 1 in the magic trick formula. The answer is then

$$\sum_n F(n) = \sum_{\text{Im}(z_i)<0} \text{Res}(F)_i \frac{1}{e^{2\pi i z_i} - 1}$$

$$+ \sum_{\text{Im}(z_i)>0} \text{Res}(F)_i \frac{1}{e^{-2\pi i z_i} - 1} + \int F(z)\, dz, \qquad (8.26)$$

where the sum runs over all poles of the function. In our applications, these three terms would be identified with absorption of a quantum from a heat bath, emission of a quantum, and interaction with the virtual quantum, respectively. At $T = 0$ there are no sums and the last integral is what is given by the ordinary Feynman rules. For example, the sum we met in the "loop" diagram has two factors, which have together four poles away from the real axis, $n1 = \pm i w/(2\pi T), n - n1 = \pm i w/(2\pi T)$ which will contribute their residua.

Exercise 8.2.   Calculate the expression for the diagrams (8.23) in time and frequency representations. Take the classical limit $T \gg w$ and compare it to what you get from classical statistical mechanics.

Exercise 8.3.   Calculate a one-loop "polarization operator" appearing as a correction to a Green function in second order of the triple coupling. The remaining integral over the real axis is of the type

$$\int dz \frac{1}{(2\pi T z)^2 + w^2} \frac{1}{(2\pi T(n - z))^2 + w^2}.$$

Hint: although no exponential factors are present it can still be done by moving the contour to infinity and picking up the pole residua.

8.1.4.   *Gauge and fermion fields at finite T*

The definition of Z as a sum over all physical states is quite sufficient for, say, atomic systems, but for the gauge theories under consideration it should be more accurately defined. First of all, there are complications connected with the gauge fixing problem. The unphysical degrees of freedom create the unphysical states (like the longitudinal photons or gluons) which may carry some energy. Summing over all of them makes no sense. Furthermore, ghosts are never emitted or absorbed in any on-shell physical process, so it is quite meaningless to imagine them getting into equilibrium with the statistical "heat bath".

There are basically three practical ways to proceed. (i) As discussed already in Chapter 1, on the lattice one simply sums over all fields with finite Matsubara time period. (ii) In a perturbation theory one however has to fix the gauge. Since the Lorenz invariance is broken by heat bath anyway, one can use non-covariant gauges, like $A_0 = 0$ or the Coulomb gauge, which have no ghosts. In this approach the only one thing to worry about is not to forget that the generalized momentum conjugated to the eliminated coordinate (e.g. A_0) should be then put to zero by hand: this generates the so called *Gauss law condition*

$$(D_m E_m - j_0)|\text{state}\rangle = 0 \tag{8.27}$$

on the allowed states. Only those should be in the statistical sum, or are "physical". (iii) The last alternative which many authors still use in practice is to keep all the familiar covariant Feynman rules with ghosts, following from the Faddeev–Popov trick, etc.

The detailed presentation of all these steps would need many pages, but, fortunately, all steps are essentially identical to those made in the derivation of ordinary Feynman rules. Perturbatively, the only change is finiteness of the Euclidean time, taken care of by the Matsubara sums in place of the frequency integrals.

We gained some insight in the method of quantum mechanical paths in the preceding section. In QCD we have not only bosonic gauge fields but also quarks which are fermions. When only the temperature is nonzero the statistical sum can be rewritten as the functional integral satisfying the following periodic/anti-periodic conditions (again, $\beta = 1/T$):

$$A_\mu(\tau + \beta) = A_\mu(\tau), \qquad \psi(\tau + \beta) = -\psi(\tau) \tag{8.28}$$

(the meaning of the fermionic minus will be clarified shortly). Going to frequency representation one finds that the fourth component of the Euclidean momenta consistent with those conditions are

$$\omega_n = 2\pi T n \text{ (bosons)}, \qquad \omega_n = 2\pi T(n + 1/2) \text{ (fermions)}, \tag{8.29}$$

where n is an integer, denoting positive or negative or 0 Matsubara frequency, respectively. All the frequency integrals are changed to the sums according to the rule

$$\int \frac{d\omega}{2\pi i} \cdots \to T \sum_n \cdots . \tag{8.30}$$

As $T \geq 0$ the Matsubara sums disappear, and we return back to the "vacuum" theory.

In some papers the short-hand notation

$$\sumint_P \to \mu^{2\epsilon} T \sum_{p_0} \int \frac{d^{3-2\epsilon}p}{(2\pi)^{3-2\epsilon}} \tag{8.31}$$

is used for bosonic momenta and

$$\sumint_{\{P\}} \to \mu^{2\epsilon} T \sum_{\{p_0\}} \int \frac{d^{3-2\epsilon}p}{(2\pi)^{3-2\epsilon}} \tag{8.32}$$

for fermionic momenta, where

$$\sum_{p_0} \rightarrow \sum_{p_0=2\pi nT} , \qquad \sum_{\{p_0\}} \rightarrow \sum_{p_0=2\pi(n+\frac{1}{2})T} . \tag{8.33}$$

Although we will not use it, we included in this definition the non-zero ϵ for variable dimensions, used in popular dimensional regularization methods.

In order to master this formalism it is convenient to consider some simple examples. The single-body observable $O_{mn}(\omega, k)$ (such as energy) of the transverse gluons in the $A_0 = 0$ gauge can be written as

$$\epsilon = T\sum_n \int \frac{d^3k}{(2\pi)^3} O_{mn}(\omega, k) D_{mm}(\omega_n, k), \tag{8.34}$$

where $D = (\delta_{mn} - k_m k_n/k^2)/(\omega_n^2 + k^2)$ is the usual zeroth order gluonic propagator. Evaluating the sum first, by the method described in the preceding section one gets the famous "bosonic summation" formula

$$\sum_n \frac{1}{x^2 + n^2} = \frac{2\pi}{x}\left(\frac{1}{2} + \frac{1}{\exp(2\pi x) - 1}\right), \tag{8.35}$$

and one finds the result $\boxed{\text{E}}$

$$\epsilon(T) = 2(N_c^2 - 1)\int \frac{d^3k\, k}{(2\pi)^3}\left(\frac{1}{2} + \frac{1}{\exp(k/T) - 1}\right) = 2(N_c^2 - 1)\frac{\pi^2 T^4}{30}. \tag{8.36}$$

The first term — the divergent energy of zero point oscillations of all oscillators in the vacuum — is omitted, and only the second thermal term is kept. Note that up to an obvious statistical weight it is the famous Planck black body formula.

It is also worth repeating the evaluation for zeroth order single-body observables for fermions, again by using the zeroth order propagator $S = (\gamma_\mu p_\mu + m)/(p^2 + m^2)$ in place of the $\bar{\psi}$ and ψ fields. $\boxed{\text{E}}$

In this case the same manipulation with loops gives the following generic *fermionic* identity:

$$\sum_n \frac{1}{x^2 + (2n+1)^2} = \frac{\pi}{2}\left(\frac{1}{2} - \frac{1}{\exp(\pi x) + 1}\right) = \frac{\pi}{2x}\tanh\left(\frac{\pi x}{2}\right). \tag{8.37}$$

The r.h.s. in our context has some rather transparent physical meaning. Note first of all that antisymmetry in Matsubara time leads to Fermi instead of Bose distribution. Note also how the extra factor 2, from 2 poles of the l.h.s. appears due to $\boxed{\text{E}}$ both quarks and antiquarks in the heat bath. Finally note a curious detail: unlike the bosonic case, with its standard combination familiar from the harmonic oscillator $\langle n(\text{quanta}) + 1/2\rangle$, we have $\langle 1/2 - n(\text{quanta})\rangle$ now. Furthermore, at very high temperatures $T \rightarrow \infty$ the exponent $\exp(\omega_k/T) \rightarrow 1$ and $\langle 1/2 - n(\text{quanta})\rangle \rightarrow 0$. It will have an important effect on high-T behavior of the fermionic propagator.

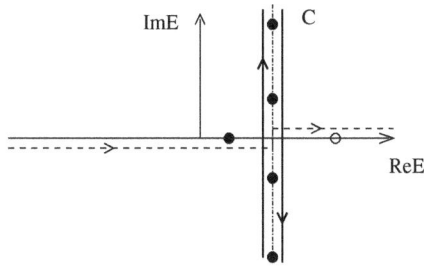

Fig. 8.2. The complex plane of the energy (real is Minkowskian and imaginary Euclidean). The vertical dash-dotted line corresponds to chemical potential μ, small black circles on it are fermionic Matsubara frequencies over which summation is made. The solid line with arrows around it indicate the contour C used in the Zommerfeld–Watson integral. As explained in the text, it includes the integral over the dash-dotted line, which is the only thing which remains at $T = 0$. If all poles are on the real axis, this integral can be transformed to an integral over the contour shown by the dashed line. Note that it treats a pole on the real axis shown by the black circle differently from the one shown by the open circle.

So far we have considered only the temperature, let us now introduce a nonzero chemical potential. In the statistical sum it enters as some constant part of the fourth component of the vector potential, and during lectures on that students often ask why is a "pure gauge" contribution relevant at all, and cannot be gauged away. But there is no mistake here: by its physical definition, the chemical potential indeed means that our system is indeed put into some potential well, so that the energies of the quark states are shifted by μ.

Consider as an example free massless fermions. In vacuum one has the Dirac sea: all states with negative energy occupied, all states with positive energy empty. By introducing nonzero potential we shift the levels, and for some states with $E_{\rm kin} - \mu < 0$ the total energies becomes negative, so quarks "from infinity" may come and occupy them. Now we have the occupied Dirac sea plus the Fermi sphere.

Formally, the chemical potential makes a difference because the contour C for the Zommerfeld–Watson integral is shifted. The situation is shown in Fig. 8.2.

Let us specifically comment on the nonzero μ and zero T cases. The Matsubara sum is now just an integral over the vertical contour shown by the dash-dotted line. It can be rotated to the real axis, provided all poles are on the real axis, and becomes as shown by the dashed line. Note however that it is not the same contour as in vacuum because between zero energy and μ it runs *below* the real axis, not above it as at $\mu = 0$. It means that those states are treated as occupied, like negative energy states. So, the chemical potential determines the asymptotics of the fields considered. In particular, at $T = 0$ it can be simply inserted into the *modified "iϵ"* Feynman prescription for the (Minkowskian) propagators

$$S \sim \frac{1}{(\omega + i\epsilon\,\mathrm{sign}(\omega - \mu))^2 - \mathbf{k}^2 - m^2}. \tag{8.38}$$

As a result, all integrals include the ordinary "vacuum" part plus an extra term corresponding to the closed contour around the $[0, \mu]$ representing the Fermi sphere.

Exercise 8.4. Derive the zero order result for the total energy of the gauge field (8.36). Do the same calculation using the covariant gauge formalism, in which one should sum over all *four* polarizations. Check that the wrong coefficient is then cured by the "ghost" term, which appears in such gauges. Note that this calculation explains the role of ghosts in a most transparent manner.

Exercise 8.5. Derive the zero-time expression for the spatial propagator of a free massless fermion

$$S_T(r, 0) = \frac{i\vec{\gamma} \cdot \vec{r}}{2\pi^2 r^4} z \exp(-z) \frac{(z+1) + (z-1)\exp(-2z)}{(1 + \exp(-2z))^2},$$

where $z = \pi r T$. This result shows that propagation in the space-like direction at finite temperature is exponentially suppressed by the screening mass $m = \pi T$. This is a simple consequence of the fact that the lowest Matsubara frequency for fermions is πT, and that this energy acts like a (chiral) mass term for space-like propagation. For bosons, on the other hand, the lowest Matsubara frequency is zero and the propagator is not screened.

Exercise 8.6. Calculate the zeroth order energy of free quarks and antiquarks at temperature T using the identity (8.37). Check that the total baryonic density is zero at zero μ.

Exercise 8.7. Calculate the energy of quarks in a cold fermion gas at nonzero μ following $T = 0$ and finite T rules.

8.2.   QCD at high temperatures

8.2.1.   *Screening versus anti-screening*

In the mid 1970s (when I approached this problem) the following facts had been well known:

- In QED plasma the charge is screened
- In QED vacuum the charge is screened
- In QCD vacuum the charge at small distances is *anti-screened* — it is of course the celebrated asymptotic freedom we discussed in Chapter 1.

The remaining question was then: *What happens in a QCD heat bath at high enough T at large distances?* The answer came from a calculation [471] and it was:

- In QED plasma the charge is *screened* at large distances.

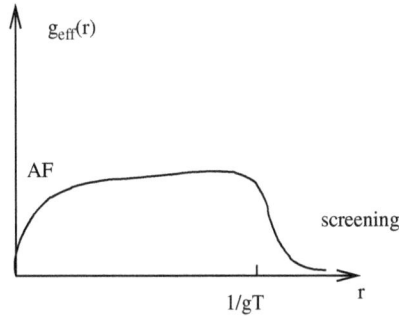

Fig. 8.3. Schematic behavior of the effective electric interactions in QGP, as a charge versus distance.

That is why the Quark Gluon Plasma was called a *Plasma*. A schematic behavior of the effective electric charge is shown in Fig. 8.3: so a combination of screening and anti-screening keeps it small at all distances, provided T is high enough.

The calculations are easier to perform in the "physical" non-covariant gauges than in the covariant ones, as was done first in Ref. [471] using the Coulomb gauge. Not only in this case are there no ghosts, but also one can show that there is no vertex renormalization of the static charge. As a result, all one has to do in this gauge is to calculate the polarization tensor $\Pi_{\mu\nu}(\omega, k)$, in close analogy to what is done in QED and the ordinary plasmas.

As usual, matter polarization has two structures corresponding to the electric (Coulomb) and the magnetic (transverse) fields. The former is described by $\Pi_{00}(\omega, k)$ and the latter by the spatial part of the tensor which has the following structure:

$$\Pi_{mn}(\omega, k) = \left(\delta_{mn} - \frac{k_m k_n}{k^2} \right) \Pi_\perp(\omega, k) + \frac{\omega^2 k_m k_n}{k^2} \Pi_{00}(\omega, k). \qquad (8.39)$$

Exercise 8.8. Prove that it is necessary to have the last term in order to ensure $4d$ transversality or current conservation, $k_\mu \Pi_{\mu\nu} = 0$.

These two functions enter the corresponding two Green functions as follows:

$$D_{00} = \frac{1}{k^2 + \Pi_{00}(\omega, k)}, \qquad D_{mn} = -\frac{\delta_{mn} - k_m k_n/k^2}{\omega^2 - k^2 - \Pi_\perp(\omega, k)}. \qquad (8.40)$$

The one-loop diagrams together with their combinatorial weights were shown in Fig. 1.5, and their logarithmic contributions due to vacuum virtual fields have already been given in (1.27).

Now we evaluate the matter contribution, originating from the same diagrams. We start with the simplest case $T = 0$ and the *non-zero* chemical potential for quarks. Of course at $T = 0$ there are no Matsubara sums, but it is still a good idea to calculate everything in the Euclidean domain first and then analytically continue the result to Minkowski notations. The alternative is to calculate it directly in Minkowski notation using the modified $i\epsilon$ prescriptions. In this case the only additional contribution of matter (on top of the standard vacuum loop) comes from the same quark loop diagram. We recall that the difference between the modified prescriptions for $i\epsilon$ at non-zero μ (8.38) and in vacuum corresponds to on-shell particles inside the Fermi sphere. There is no real difference between QCD and QED at this point, so the plasma-like behavior of the cold quark matter is quite obvious.

The result for the electric polarization tensor in the "soft" limit $\omega, k \ll \mu$ looks as follows

$$\Pi_{00} = -\frac{g^2}{2\pi^2} \left(\sum_f \mu_f^2 \right) \left(1 - \frac{\omega}{2k} \ln \left| \frac{\omega + k}{\omega - k} \right| + \frac{i\pi\omega}{2k} \theta(\omega)\theta(k - \omega) \right), \qquad (8.41)$$

where the sum runs over all quark flavors, if their chemical potentials are all different.

Let us try to understand what this expression means, starting with its real part. Note that the function in square brackets is a function of ω/k so there are various "long-wave" limits depending on this ratio. One is the static limit $\omega = 0, k \to 0$, $\omega/k = 0$ in which

$$D_{00}(\omega, k) \to -\frac{1}{k^2 + \kappa_D^2}, \qquad \kappa_D^2 = \frac{g^2}{2\pi^2} \left(\sum_f \mu_f^2 \right). \qquad (8.42)$$

The Fourier transform of this Coulomb-field propagator is now a screened potential between two charges separated by distance r, $V(r) \sim \exp(-\kappa_D r)/r$, and so the inverse $R_d = 1/\kappa_D$ is the so called *Debye screening radius*.

The opposite limit $k = 0, \omega \to 0, k/\omega = 0$ is called the plasma oscillation limit

$$D_{00}(\omega, k) \to \frac{\omega^2}{k^2(k^2 + \omega_D^2)}, \qquad \omega_D^2 = \frac{g^2}{6\pi^2} \left(\sum_f \mu_f^2 \right) = \frac{\kappa_D^2}{3}, \qquad (8.43)$$

where ω_D is the frequency of the longitudinal long-wavelength plasma oscillations. We do not present here an analogous expression for the magnetic polarization tensor, and only note that in this case the most interesting physical limit is the so called *gluon effective mass* defined by the third "on shell" limit $k = \omega \to 0, k/\omega = 1$,

$$m_g^2 = \Pi_\perp|_{\text{on shell}} = \frac{g^2}{4\pi^2} \left(\sum_f \mu_f^2 \right) = \frac{\kappa_D^2}{2}. \qquad (8.44)$$

The last important point is that the static magnetic field in quark Fermi gas is not screened because in the static limit $\omega = 0, k \to 0, \omega/k = 0$ the magnetic polarization $\Pi_\perp \to 0$ vanishes.

The imaginary part of the tensor technically originates from the log in (8.41), since in the Euclidean domain $w \to iw$ there is no imaginary part. Its physics is the so called Landau damping of waves in a plasma, which may occur if their w/k matches the longitudinal velocity of some particles inside the Fermi sphere. If this happens, the particle may "surf the wave", continuously draining its energy.

Let me repeat once again, that all these results are quite identical to those in QED, for example for the electron plasma in ordinary metals, and they are well known. What was not known in 1970s was the contribution of the "valence" gluons present at high temperatures. However, as was found by the explicit calculations [471] all the results for the polarization tensors $\Pi_{00}, \Pi_{\perp}$ turned out to be nearly identical to those given above, with the same dependence on w/k in the soft limits given by the square brackets in (8.41).

The Debye screening parameter obtains the following contributions from the three diagrams (a–c) of Fig. 1.5 plus that of the fermionic loop

$$\kappa_{\mathrm{D}}^2 = g^2 T^2 \left(\frac{1}{2} + 0 + \frac{1}{2} + \frac{1}{6} N_f \right). \tag{8.45}$$

Note that the most mysterious non-QED diagram (b) with a magnetic–electric loop, which was responsible for the unexpected sign and gave the asymptotic freedom, in this case does not contribute at all! The relation between this parameter and plasma oscillation frequency and effective on-shell gluon mass are the same as above, at $T = 0$. Very importantly, perturbatively there is again *no static magnetic screening* as in QED.

8.2.2. *Thermodynamical potential in the lowest order*

We now turn to explicit evaluation of the perturbative corrections to the thermodynamical quantities. Technically those are given by all "closed loop" diagrams without external legs. Although they are the same as those for the "vacuum energy" in the field theories, in field theory courses those are usually discarded without consideration as strongly divergent and unphysical. To see that it is not at all true, recall the effort we made in Chapter 1 is worthwhile, where we introduced Feynman diagrams in the context of quantum mechanics. Let me just remind here the reader that vacuum loop diagrams have a specific feature, distinguishing them from more familiar diagrams for the Green functions: the corresponding analytic expressions contain *combinatorial factors* making naive "graphical summation" of subdiagrams impossible.

One more thing which looks confusing at the beginning is how one formally obtains the zero order expressions from such diagrams. Of course the functional path integral for non-interacting fields is Gaussian, resulting in determinants of free motion operators. These determinants can be related to *one-loop* diagrams in which the propagator terminates at the same point, where some free operators (like the kinetic energy) is placed. Let me save time by explaining that, since the free results are obvious anyway. The reader may find a good discussion, e.g. in the book [459].

The nontrivial pQCD corrections $\sim g^2$ appear starting with the two loop diagrams. Let us start with the cold quark matter in which there is only one[3] diagram, a quark loop with a gluon exchanged. We do not want to calculate the divergent vacuum part, and apply the modified $i\epsilon$ prescription to each quark propagator, putting both quarks on shell $\omega_k^2 = k^2 + m^2$, with $\omega_k < \mu$ and k inside the Fermi sphere for each flavor. Its physical meaning is a part of forward scattering amplitude due to the "exchange" interaction, in which two quarks have to *exchange their momenta, colors and spins* in order to find the vacant place in the Fermi sphere. The corresponding correction to the thermodynamic potential is as follows [470,471]

$$\delta\Omega = \frac{g^2}{4\pi^4} \sum_f \left\{ \frac{3}{2}\left[\mu_f p_f - m_f^2 \ln\left(\frac{\mu_f + p_f}{m_f}\right)\right]^2 - p_f^4 \right\}, \tag{8.46}$$

where as above the sum runs over the flavors f. The reason I give this expression is that in the nonrelativistic and ultrarelativistic plasma it leads to corrections which look very similar, but upon closer inspection shows that they have opposite signs,

$$\delta\Omega|_{\mu_f \gg m_f} = \frac{g^2}{8\pi^4}\sum_f p_f^4, \qquad \delta\Omega|_{\mu_f \ll m_f} = -\frac{g^2}{4\pi^4}\sum_f p_f^4. \tag{8.47}$$

The sign of the latter non-relativistic expression corresponds to attractive electric interaction. As for the latter, ultrarelativistic case, magnetic interaction are as large as the electric ones, and the sign is no longer obvious. One may however ask here how it can be possible to violate the well-known quantum-mechanical theorem, according to which the ground state energy of any system can only decrease in the second order of the perturbation theory. The answer is that this theorem is not applicable in the relativistic case: the necessity of renormalization — subtraction of the divergent vacuum contribution — nullifies this argument.

A similar calculation of all closed 2-loop diagrams includes all qq, qg, gg scattering processes to order g^2, and in hot quark–gluon plasma this leads to the following result:

$$\delta\Omega = g^2 T^4 \frac{N_c^2 - 1}{144}(N_c + \tfrac{5}{4}N_f), \tag{8.48}$$

which is positive and thus corresponds to attraction.

8.2.3. *Ring diagram re-summation*

In higher orders at high T there appear soft (IR) divergences of the perturbation theory, which are absent in vacuum. The simplest example of that are the "ring"

---

[3]There is no analog of the "dumbbell" diagram of quantum mechanics, with two fermion loops connected by a gluon, since matter can only be colorless and the $\mathrm{tr}\, t^a = 0$ makes it zero.

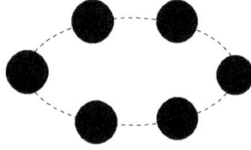

Fig. 8.4. Example of a ring diagram, with $n = 6$ gluon polarization operators in one single gauge field loop.

diagrams exemplified in Fig. 8.4. Their electric contribution to the Matsubara frequency zero is

$$\delta\Omega|_n = \frac{1}{n} \int \frac{d^3k}{(2\pi)^3} \left(\frac{\Pi_{00}}{k^2} \right)^n \tag{8.49}$$

and for $n > 2$ those are clearly divergent at small k at which $\Pi_{00} = \kappa_D^2$.

However, following the example of Gell-Mann and Bruckner [465] in their discussion of the cold QED plasma of electrons, one may sum all such diagrams over n from 2 to infinity first [471, 472] obtaining the expression

$$\delta\Omega^{\text{ring}} = \frac{1}{2} \int \frac{d^3k}{(2\pi)^3} \left[\ln \left(1 + \Pi_{00}/k^2 \right) - \Pi_{00}/k^2 \right], \tag{8.50}$$

which is infrared-finite. The resulting correlation energy or "plasmon term" at high T and μ is equal to

$$\delta\Omega^{\text{plasmon}} = -\frac{N_c^2 - 1}{12\pi} g^3 T^4 \left(\frac{N_c}{3} + \frac{N_f}{6} + \sum_f \frac{\mu_f^2}{6\pi^2 T^2} \right)^{3/2}. \tag{8.51}$$

This nice example of re-summation shows that sometimes divergence of diagrams is only the way Nature tells us that the answer is not necessarily an analytic function of g^2, but g^3 one gets is OK as soon as g is small.

The physical nature of this correlation energy is the interaction of a charge with its Debye cloud, and its magnitude per particle is $\sim g^3 T$, as one can get from dividing this contribution by particle density. The same magnitude has a correlation energy for a static quark in a plasma. Following Debye, one can derive it much more simply by writing the potential as

$$4\pi V(r) = -\frac{4}{3} \frac{g e^{-\kappa_D r}}{r} \approx \frac{4}{3} \frac{g}{r} - \frac{4}{3} g \kappa_D + \cdots, \tag{8.52}$$

and noticing that the second constant potential times g of a point static charge is the correlation energy

$$E_{\text{corr}} = -\frac{4}{3} \alpha_s \kappa_D. \tag{8.53}$$

8.2.4. *IR divergences in general*

Unfortunately, a resummation for ring-type diagrams is not the general case, as the following consideration shows. A bosonic nature of the gluon field together with its Abelian (or "charged") properties leads to IR problems absent, e.g. in the QED plasma. Its presence can in fact be seen already on dimensional grounds. Each loop at the non-zero temperatures contains the factor T and a sum over discrete Matsubara frequencies. Consider for a moment only the sector with zero frequencies (which corresponds to the classical limit or the vanishing Planck constant or $3d$ gauge theory). The factor T enters any loop and in fact makes the coupling dimensional, its power grows with the number of loops, and something should compensate for it in the denominator. It can only be particle momenta, so the growing infrared divergences are expected in higher orders. In fact, the four-loops or 8th order already produce such power divergence and therefore

$$\delta\Omega^{(8)} \sim g^6 T^4 \left(\frac{g^2 T}{q_{\min}} \right), \tag{8.54}$$

where a cutoff is put explicitly. Polyakov [484] has suggested[4] that this disease can only be cured by the development of some magnetic screening length for the gluomagnetic field of the order of $\kappa_{\mathrm{M}} \sim g^2 T$. If so, using it as an IR cutoff in the formula above we would conclude that all diagrams above the 8th order would contain a contribution of the same amount, $\delta\Omega \sim g^6 T^4$. Although small at high T, those cannot be calculated perturbatively.

In lectures I like to compare the situation in high T QCD with that of a star, e.g. the Sun. If we would like to evaluate its total thermal energy, normal plasma physics would be enough. It is however true that the Sun has very complicated magnetic field structure, with spots of opposite polarities and magnetic fluxes protruding from those into space. We do not know how to calculate those: but of course however interesting those may be, their total energy is negligible.

One can separate, by the so called dimensional reduction, soft fields of a *magnetic scale* $g^2 T$ from all harder scales. It has been shown in Ref. [486] how, in principle, an effective theory could be constructed to deal with this particular problem by marrying analytical techniques (to determine the coefficients of the effective theory) and numerical ones (to solve the non-perturbative 3-dimensional effective theory). The resulting effective theory is a 3-dimensional theory of static fields, with Lagrangian:

$$\mathcal{L}_{\mathrm{eff}} = \frac{1}{4}(F_{ij}^a)^2 + \frac{1}{2}(D_i A_0^a)^2 + \frac{1}{2}m_{\mathrm{D}}^2(A_0^a)^2 + \lambda(A_0^a)^4 + \delta\mathcal{L}, \tag{8.55}$$

with $D_i = \partial_i - ig\sqrt{T}A_i$. This strategy has been applied recently to the calculation of the free energy of the quark–gluon plasma a high temperature [487]. This technique of dimensional reduction puts a special weight on the static sector (it singles out

---

[4]An alternative suggestion by Kajantie and Kapusta [475] and others has been abandoned, as far as I know.

the contributions of the zero Matsubara frequency), and a major effort is devoted to the calculation of the coefficients of the effective Lagrangian (which contain the dominant contribution to the thermodynamical functions).

8.2.5. *Are perturbative series useful in practice?*

In the next chapter we will discuss QGP produced experimentally, and the region of temperatures we can hope to reach is never larger than 3–4 times the critical temperature T_c, or $T \lesssim 1\,\text{GeV}$. The corresponding momenta are about $3T$, but $\alpha_s \sim 1/3$ or $g \sim 2$ and one can ask how good perturbative series are in this regime.

Due to efforts of many people which took many years, the calculable terms are now known. Following Ref. [479], we present the results for the free energy $F(T)$ (the same as Ω at zero density or μ):

$$
F = d_A T^4 \frac{\pi^2}{9} \left\{ -\frac{1}{5} \left(1 + \frac{7d_F}{4d_A} \right) + \left(\frac{g}{4\pi} \right)^2 \left(C_A + \frac{5}{2} S_F \right) \right.
$$

$$
- \frac{16}{\sqrt{3}} \left(\frac{g}{4\pi} \right)^3 (C_A + S_F)^{3/2} - 48 \left(\frac{g}{4\pi} \right)^4 C_A (C_A + S_F) \ln \left(\frac{g}{2\pi} \sqrt{\frac{C_A + S_F}{3}} \right)
$$

$$
+ \left(\frac{g}{4\pi} \right)^4 C_A^2 \left[\frac{22}{3} \ln \frac{\bar{\mu}}{4\pi T} + \frac{38}{3} \frac{\zeta'(-3)}{\zeta(-3)} - \frac{148}{3} \frac{\zeta'(-1)}{\zeta(-1)} - 4\gamma_E + \frac{64}{5} \right]
$$

$$
+ \left(\frac{g}{4\pi} \right)^4 C_A S_F \left[\frac{47}{3} \ln \frac{\bar{\mu}}{4\pi T} + \frac{1}{3} \frac{\zeta'(-3)}{\zeta(-3)} - \frac{74}{3} \frac{\zeta'(-1)}{\zeta(-1)} - 8\gamma_E + \frac{1759}{60} + \frac{37}{5} \ln 2 \right]
$$

$$
+ \left(\frac{g}{4\pi} \right)^4 S_F^2 \left[-\frac{20}{3} \ln \frac{\bar{\mu}}{4\pi T} + \frac{8}{3} \frac{\zeta'(-3)}{\zeta(-3)} - \frac{16}{3} \frac{\zeta'(-1)}{\zeta(-1)} - 4\gamma_E - \frac{1}{3} + \frac{88}{5} \ln 2 \right]
$$

$$
+ \left. \left(\frac{g}{4\pi} \right)^4 S_{2F} \left[-\frac{105}{4} + 24 \ln 2 \right] + O(g^5) \right\}.
\tag{8.56}
$$

Two last calculable terms, $O(g^5)$ and $O(g^6 \log g)$, can be found in recent work by Kajantie *et al.*, hep-ph/10211321.

Evaluated numerically for QCD with n_f quark flavors, this is

$$
F = -\frac{8\pi^2 T^4}{45} \left\{ 1 + \frac{21}{32} n_f - 0.09499\, g^2 \left(1 + \frac{5}{12} n_f \right) + 0.12094\, g^3 \left(1 + \frac{1}{6} n_f \right)^{3/2} \right.
$$

$$
+ g^4 \left[0.08662 \left(1 + \frac{1}{6} n_f \right) \ln \left(g \sqrt{1 + \frac{1}{6} n_f} \right) \right.
$$

$$
- 0.01323 \left(1 + \tfrac{5}{12} n_f \right) \left(1 - \frac{2}{33} n_f \right) \ln \frac{\bar{\mu}}{T}
$$

$$
+ 0.01733 - 0.00763\, n_f - 0.00088\, n_f^2 \left. \right] + O(g^5) \right\}.
\tag{8.57}
$$

For QED with n_f massless charged fermions with charges $q_i e$, the free energy is

$$F = -\frac{\pi^2 T^4}{45} \left\{ 1 + \frac{7}{4} n_f - 0.07916\, e^2 \sum q_i^2 + 0.02328\, e^3 \left(\sum q_i^2 \right)^{3/2} \right.$$

$$\left. + e^4 \left[\left(-0.00352 + 0.00134 \ln \frac{\bar{\mu}}{T} \right) \left(\sum q_i^2 \right)^2 + 0.00193 \sum q_i^4 \right] + O(e^5) \right\}. \quad (8.58)$$

With such results at hand, one can investigate whether the perturbative expansion of the QCD free energy is well-behaved for the physically realized values of couplings. Fig. 8.5 shows the result for six-flavor QCD when $\alpha_s(T) = 0.1$ (which corresponds to scales of order a few 100 GeV). The free energy is plotted vs. the choice of renormalization scale $\bar{\mu}$. We have taken

$$\frac{1}{g^2(\bar{\mu})} \approx \frac{1}{g^2(T)} - \beta_0 \ln \frac{\bar{\mu}}{T} + \frac{\beta_1}{\beta_0} \ln \left(1 - \beta_0 g^2(T) \ln \frac{\bar{\mu}}{T} \right), \quad (8.59)$$

where

$$\beta_0 = \frac{1}{(4\pi)^2} \left(-\frac{22}{3} C_A + \frac{8}{3} S_F \right), \quad \beta_1 = \frac{1}{(4\pi)^4} \left(-\frac{68}{3} C_A^2 + \frac{40}{3} C_A S_F + 8 S_{2F} \right). \quad (8.60)$$

If the expansion is well-behaved, the result for F should become more independent of $\bar{\mu}$ as higher-order corrections are included. Unfortunately it obviously does not.

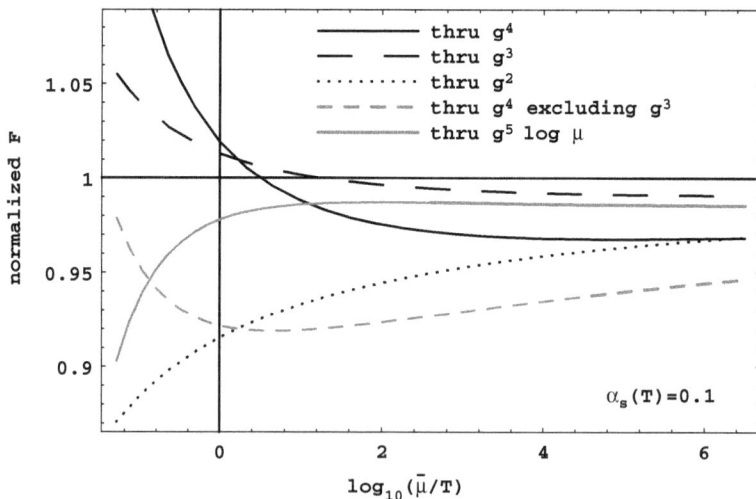

Fig. 8.5. The dependence of the free energy F on the choice of renormalization scale $\bar{\mu}$ for six-flavor QCD with $\alpha_s(T) = 0.1$. The free energy is normalized in units of the ideal gas result $-\left(\frac{1}{45} d_A + \frac{7}{180} d_F \right) \pi^2 T^4$. The thick solid, dashed, and dotted lines are the results for F including terms through g^4, g^3, and g^2 respectively. The light solid curve is the g^4 result plus the $g^5 \ln(\bar{\mu}/T)$ term required by renormalization group invariance. The light dashed curve is the g^4 result minus the g^3 term.

8.2.6. *HTL re-summations and the quasiparticle gas*

The ring diagram re-summation above is an example in which hard momenta $p \sim T$ inside the polarization tensor were separated from the soft scale $p \sim gT$ in the loop. The program initiated and carried out by Braaten and Pisarski [477] was to identify *all* diagrams containing Hard Thermal Loops (HTL) and re-sum them, obtaining some effective Lagrangian (Frenkel and Taylor [477]) to be consistently used at this softer scale.

Instead of going into this rather technical discussion we note that it can be alternatively explained from classical kinetics. The fluctuations at the *soft* scale $k \sim gT \ll T$ can be described by classical fields because the associated occupation numbers $N_k \sim T/E_k \sim 1/g$ are large. As emphasized by Blaizot and Iancu [478], these soft excitations can be described in terms of average fields which obey classical equations of motion. In QED these equations are Maxwell equations:

$$\partial_\mu F^{\mu\nu} = j^\nu_{\text{ind}} + j^\nu_{\text{ext}} \qquad (8.61)$$

with a source term composed of an external perturbation j^ν_{ext}, and an extra contribution j^ν_{ind} referred to as the *induced current*, can be obtained using linear response theory,

$$j^\mu_{\text{ind}}(x) = \int dy\, \Pi_{\mu\nu}(x - y) A^\nu(y). \qquad (8.62)$$

which generalizes the usual dielectric and diamagnetic constants. In leading order in weak coupling, this polarization tensor is given by the one-loop approximation we discussed above, $\Pi \sim g^2 T^2 f(\omega/p)$.

In a non-Abelian theory, linear response is not sufficient: constraints due to gauge symmetry force us to take into account specific non linear effects. The relevant generalization of the Yang–Mills equation reads [478]:

$$D_\nu F^{\nu\mu} = \Pi^{ab}_{\mu\nu} A^\nu_b + \frac{1}{2}\Gamma^{abc}_{\mu\nu\rho} A^\nu_b A^\rho_c + \cdots, \qquad (8.63)$$

where the induced current in the right hand side is non-linear: when expanded in powers of A^μ_a, it generates an infinite series of HTMs not only in self-energy but also in vertices with any number of soft gluonic legs.

Physically the need for re-summation arises from the existence of collective excitations in the system, whose properties are not well captured by perturbation theory. One may hope that a "gas of quasiparticles" would do a better job in reproducing the results. The main spirit of these works is that quasiparticles have effective masses which are functions of g and T which one can evaluate perturbatively first, and only then include their motion (e.g. [481]) and interactions [482].

As an example of such an approach I would present a calculation by Blaizot and Iancu [483], which will be compared to lattice data. For some technical reasons they prefer to focus on entropy $s(T)$ rather than, say $p(T)$, which at least is not defined up to arbitrary constant (the bag term).

The entropy density of their *quasiparticle gas* is defined as the entropy of the ideal gas of quasiparticles:

$$s = \int \frac{d^3k}{(2\pi)^3} \left[(1 + N_k) \log(1 + N_k) - N_k \log N_k \right], \tag{8.64}$$

$$N_k = \frac{1}{e^{E_k/T} - 1}, \quad E_k = \sqrt{M^2(T) + k^2},$$

where the T-dependent effective mass is not yet defined.

The pressure P and the energy density ϵ are then given by the corresponding ideal gas expressions, corrected by a function $B(T)$ adjusted so as to satisfy thermodynamic identities. That is, one sets:

$$P = -T \int \frac{d^3k}{(2\pi)^3} \log \left(1 - e^{-E_k/T} \right) - B(T), \tag{8.65}$$

$$\epsilon = \int \frac{d^3k}{(2\pi)^3} N_k E_k + B(T). \tag{8.66}$$

Such a parametrization obviously fulfills the identity $\epsilon + P = Ts$. The function $B(T)$ is then determined by requiring that $s = dP/dT$.

The effective mass used is the HTL one. The quasiparticle gas can be viewed as an approximation to a more general approach, known as *hard thermal loop perturbation theory* [482] in which one adds the HTL mass term into tree-level Lagrangian and then subtracts it from "perturbation" part.

Blaizot and Iancu noticed that for a special case of the entropy the calculations can be pushed one order further, to make better "Φ-derivable" self-consistent approximations. The stationarity property of the free energy entails important simplifications in the calculation of the entropy. Indeed we have

$$\mathcal{S} = -\frac{d\mathcal{F}}{dT} = -\frac{\partial \mathcal{F}}{\partial T}\bigg|_D \tag{8.67}$$

where the last derivative is at fixed propagator. Due to that expression the temperature dependence of the propagator can be ignored here. Explicitly one gets:

$$\mathcal{S} = -\int \frac{d^4k}{(2\pi)^4} \frac{\partial N(\omega)}{\partial T} \mathrm{Im} \log D^{-1}(\omega, k)$$

$$+ \int \frac{d^4k}{(2\pi)^4} \frac{\partial N(\omega)}{\partial T} \mathrm{Im}\, \Pi(\omega, k) \mathrm{Re}\, D(\omega, k) + \mathcal{S}', \tag{8.68}$$

where $N(\omega) = 1/(e^{\beta\omega} - 1)$, and

$$\mathcal{S}' \equiv -\frac{\partial(T\Phi)}{\partial T}\bigg|_D + \int \frac{d^4k}{(2\pi)^4} \frac{\partial N(\omega)}{\partial T} \mathrm{Re}\, \Pi \, \mathrm{Im}\, D.$$

At two-loop in the skeleton expansion, the entropy takes then the simple form

$$\mathcal{S} = -\int \frac{d^4p}{(2\pi)^4} \frac{\partial N}{\partial T} (\mathrm{Im} \ln D^{-1} - \mathrm{Im}\, \Pi \, \mathrm{Re}\, D). \tag{8.69}$$

The simplifications discussed above have led to important cancellations leaving for the entropy an expression which is effectively a one-loop expression, thus emphasizing the direct relation between the entropy and the quasiparticle spectrum. Residual interactions start contributing at order 3-loop. This expression is also manifestly ultraviolet-finite.

Now, in the regime where the loop momenta are soft, we can use as an approximation for Π the corresponding hard thermal loop (HTL). As an illustration of the quality of the results which were obtained in this direction, let me show the entropy of pure SU(3) gauge theory (normalized to the ideal gas limit). The agreement with the parameterized lattice results[5] at large T ($T \gtrsim 2.5T_c$) is quite good. Note also that in going from one level of approximation to the next (i.e. from $\mathcal{S}_{\mathrm{HTL}}$ to $\mathcal{S}_{\mathrm{NLA}}$), the changes are moderate, in contrast to what happens in ordinary perturbation theory.

On the other hand, a radically different approach — the strong coupling limit of $\mathcal{N} = 4$ SUSY theory [896] — gives the value $s/s_0 = 3/4$, which is also not very far off. So it is difficult to conclude at the moment which of the two limits are more appropriate at $T \sim 2 - 3T_c$. For more discussion of this refer to two recent works by Zahed and myself [hep-ph/0307267] and [hep-th/0308073], which provides alternative explanation to data shown in Fig. 8.6.

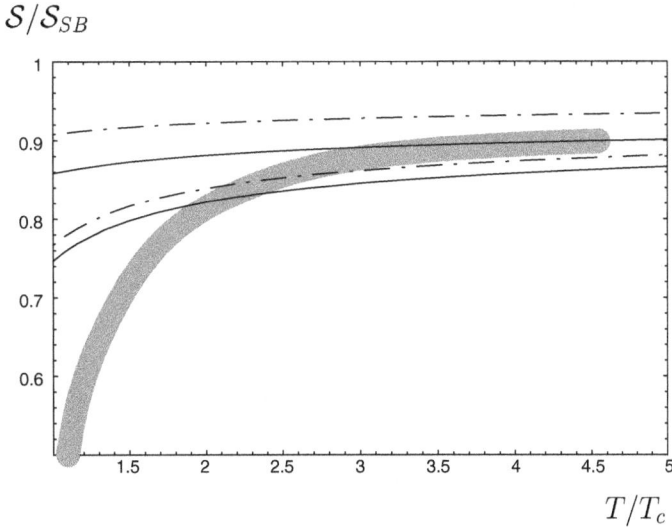

$\mathcal{S}/\mathcal{S}_{SB}$

T/T_c

Fig. 8.6. The entropy of pure SU(3) gauge theory normalized to the ideal gas entropy $\mathcal{S}_{SB}$. Full lines: $\mathcal{S}_{\mathrm{HTL}}$ from Blaizot *et al.* Dashed-dotted lines: $\mathcal{S}_{\mathrm{NLA}}$. For each approximation, the two lines correspond to the choices $\bar{\mu} = \pi T$ and $\bar{\mu} = 4\pi T$ of the running coupling constant $\alpha_s(\bar{\mu})$ in the $\overline{\mathrm{MS}}$ renormalization scheme. The dark gray band represents the lattice results.

[5]We will discuss those in more detail in Section 8.4.4.

8.2.7. *Viscosity of the QGP*

Viscosity is an example of the so called transport properties of matter. For dilute gases and other weakly interacting systems they are all related to mean free paths of constituents and therefore are inversely proportional to rescattering cross sections. Thus weak coupling means large viscosity, while strong coupling leads to small viscosity. (For example, in a supernova explosion, viscosity is dominated by the component with the weakest interaction, the neutrinos.)

Outside of the realm of quasiparticles and mean free paths, there is a general expression for viscosity in terms of correlators known as Kubo formulae,

$$\eta = \frac{1}{20} \lim_{\omega \to 0} \frac{1}{\omega} \int d^4x \, e^{i\omega t} \langle [T_{lm}(t,\mathbf{x}), T_{lm}(0)] \rangle_{\text{eq}}. \tag{8.70}$$

The perturbative theory is well summarized by P. Arnold *et al.* [505]. The high temperature shear viscosity in a gauge theory with a simple gauge group (either Abelian or non-Abelian) has the leading-log form

$$\eta = \kappa \frac{T^3}{g^4 \ln g^{-1}}, \tag{8.71}$$

where g is the gauge coupling. For the case of SU(3) gauge theory the leading-log shear viscosity coefficient κ for various numbers of light $m \ll T$ fermion species are shown in the Table 8.2.

Strong coupling calculation by Policastro, Son and Starinets [898] in the $\mathcal{N} = \triangle$ supersymmetric Yang–Mills theory. The method is too far from the current discussion to go into it: so we only discuss the result for shear viscosity

$$\eta = \frac{\pi}{8} N_c^2 T^3. \tag{8.72}$$

Table 8.2. Leading-log shear viscosity as a function of the number of (fundamental representation) fermion flavors with $m \ll T$, for gauge group SU(3).

n_f	$\eta \times (g^4/T^3) \ln g^{-1}$
0	27.126
1	60.808
2	86.473
3	106.664
4	122.958
5	136.380
6	147.627

In order to compare it properly with the QCD we recall that on top of the $SU(N_c)$ gauge fields this theory has four gluino-like fermions and six scalars.[6] For 3 colors this is 16 gluons, plus 64 fermionic d.o.f. (to be compared to $12N_f$ with anti-fermions in QCD) plus 48 scalars. The thermodynamics of this is discussed extensively in Ref. [895] with the conclusion that in the strong coupling limit $\epsilon = 3p$ reaches 3/4 of its Stephan–Boltzmann value.

The sound attenuation length includes the ratio of viscosity to entropy

$$L_{\text{sound}} = \frac{4}{3}\frac{\eta}{sT}. \tag{8.73}$$

What was found in Ref. [897] is

$$\omega = \frac{q}{\sqrt{3}} - \frac{iq^2}{6\pi T} + O(q^3), \tag{8.74}$$

which is in complete agreement with hydrodynamics, using the value of $\eta/(\epsilon + P) = (4\pi T)^{-1}$ computed previously. This implies that strongly coupled matter has a very small mean free path.

In Section 9.4.6 we will compare it with data on viscosity effects in hydro and will see that the strong coupling value actually agrees much better with the data.

8.3. Hadronic matter

8.3.1. *Pion gas at low T*

In this section we jump from asymptotically high T to the opposite case of "lukewarm" hadronic matter with low T. As pions — the Goldstone modes — are the lowest excitations, this matter is basically a dilute gas of pions. At sufficiently low T we are sure it is indeed a gas and not a liquid, because interaction amplitudes of the Goldstone modes must be proportional to their momenta, and so soft pions hardly interact with each other. Its pressure should then be close to the ideal gas one

$$p_{\text{ideal}} = (N_f^2 - 1)\frac{\pi^2}{90}T^4, \tag{8.75}$$

for N_f massless flavors. For simplicity we ignored the pion mass and the non-zero scattering lengths proportional to them, in a chiral limit.

Furthermore, as pion interaction is governed by effective chiral Lagrangian, one can evaluate the corrections. The first study of the kind was made in the appendix

---

[6]This is needed by supersymmetries, to balance the number of degrees of freedom $(N_c^2 - 1) *$ $(2 - 8 + 6) = 0$, where we use the negative sign for fermions.

of my review [457]. For binary collisions the correction to pressure is

$$\delta p = \int \prod_{i=1,2} \frac{d^3 k_i}{(2\pi)^3 \omega_i} \sum_{I=0,1,2} (2I+1)\, T_I(t=0, s=(p_1+p_2)^2)$$

$$\times \frac{1}{\exp(\omega_i/T)-1}, \tag{8.76}$$

where the sum runs over the total isospin of two pions. The T_I are the corresponding forward scattering amplitudes, which follow from the Weinberg Lagrangian to be

$$T_0 = \frac{4s}{f_\pi^2}, \qquad T_1 = \frac{2(t-u)}{f_\pi^2}, \qquad T_2 = \frac{-2s}{f_\pi^2}. \tag{8.77}$$

We need the forward amplitude only, $t=0, u=-s$, and combining all three into the combination which enters δp I found that the result vanishes. So, there is no correction $\delta p \sim T^6/f_\pi^2$. By going one order further and iterating these amplitudes I obtained the first non-zero corrections of the order $T^8 \log(T)/f_\pi^4$. However, $\log(T)$ is not really a good parameter and terms of the same order without it include also the higher order "Leutwyler terms" in the chiral Lagrangian (2.49) as well as the three-pion scattering: those were included later by Leutwyler and collaborators [32]. Combined, all of those produced a correction like

$$\frac{p(T)}{p_{\text{ideal}}(T)} = 1 + \frac{T^4}{T_0} + \cdots \tag{8.78}$$

with $T_0 \approx 150\,\text{MeV}$ which can be taken as a weak hint for chiral restoration at about this temperature.

One may get a more direct hint by calculating the chiral condensate at small T [488] by including the chiral symmetry breaking effects $O(m_q)$ and differentiating the total free energy over m_q. The result is

$$\langle \bar{q}q \rangle(T) = \langle \bar{q}q \rangle(0) \left(1 - \frac{T^2}{8 f_\pi^2} - \frac{T^4}{384 f_\pi^4} + \cdots \right), \tag{8.79}$$

noting now that the effect $O(T^2/f_\pi^2)$ is present.

8.3.2. *Resonance gas*

As the temperature grows, the elementary excitations other than pions (say vector mesons ρ, ω) are thermally excited. Naively one may think that happens when $T \sim m_\rho \sim 800\,\text{MeV}$ in QCD and when $T \sim m_{\text{scalar glueball}} \sim 1.6\,\text{GeV}$ in gluodynamics without quarks. Not at all! It happens much earlier because the next excitations have large statistical weights.[7]

---

[7]One more example of the kind: the Sun is made of plasma even up to its photosphere, which has $T \approx 6000° \sim 0.5\,\text{eV} \ll \delta E \sim 10\,\text{eV}$.

The theoretical approach widely used for $T = 100-160$ MeV is that of the *resonance gas* approach, suggested very early by Landau and Belenky [490]. They have shown, using the lowest order virial expansion, that resonances[8] seen in scattering phases in fact contribute to thermodynamical parameters exactly as stable particles. The Beth–Ulenbeck formula is a correction to the partition function due to binary interaction between particles, which has the form

$$\delta Z = \frac{1}{\pi} \int dp \sum_l (2l+1) \frac{d\delta_l(p)}{dp} \exp\left(-\frac{p^2}{mT}\right), \qquad (8.80)$$

where the sum is over angular momenta and there is mT rather than $2mT$ in the exponent because relative motion has reduced mass. The reason for that is that the scattering phase shift enters the boundary condition for the wave function at its boundary R as

$$\sin(kR + \delta_l(k)) = 0, \qquad (8.81)$$

and a resonance in which δ_l makes another 2π literally adds one more state to the statistical sum.

One more comment about the Beth–Ulenbeck formula from Ref. [495] can be made here. Note that this sum can be identically rewritten as

$$\sum_l (2l+1) \frac{d\delta_l(p)}{dp} = \frac{d}{dp}[p \operatorname{Re} f(p, \theta = 0)] + \frac{i}{4\pi} \int d\Omega \left(f \frac{\partial f^*}{\partial p} - f^* \frac{\partial f}{\partial p} \right). \qquad (8.82)$$

The first term related to the real part of the forward scattering amplitude is a "collective potential". The second term (which is of course also real) is the one which generates an extra contribution from narrow binary resonances. So, there is no "double counting" when one includes both together.

The idea to use the resonance gas was later used by Hagedorn in his statistical bootstrap studies of the 60s: his main point was that bags made of hadrons are hadrons themselves. Putting this complicated idea aside, let me mention another of Hagedorn's points which was well taken. If the particle mass spectrum is *exponential* $\rho(m) \sim \exp(m/T_{\max})$, it would lead to the uppermost possible temperature of the hadronic gas because its combination with the Boltzmann factor $Z \sim \int dm\, \rho(m) \exp(-m/T)$ would be divergent as $T \to T_{\max}$. We will return to a discussion of this issue in connection with deconfinement in Section 8.4.1 below.

However, I argued in 1972 [492] that if one excludes too inelastic resonances $\Gamma > T$ (as the Beth–Ulenbeck formula demands) the observed resonance mass spectrum fits better with the power of the mass than the exponent. It leads to a rather

---

[8]Those should be narrow enough: $\Gamma \ll T$.

simple EoS, for resonance gas at zero baryon number

$$p = (20 \, \text{GeV}^{-2})T^6, \qquad c_s^2 = \frac{dp}{d\epsilon} = \frac{1}{5}. \tag{8.83}$$

Later much more detailed calculations with actual scattering phases [494] confirmed it.

In applications to be discussed in the next chapter one also has to include the non-zero baryon density, and so our thermodynamics has two variables, T and μ_b.[9] Simple generalization of the resonance gas to the non-zero chemical potential is widely used. Some problems are avoided (e.g. [622]) by the so called excluded volume correction, which effectively reduces the baryonic pressure at high μ. Specifically the thermodynamically consistent way to do it is to include it in the canonical partition function

$$Z^{\text{excl}}(T, \{N_i\}, V) = \sum_i Z(T, N_i, V - V_0 N_i) \theta(V - V_0 N_i),$$

from which

$$P^{\text{excl}}(T, \{\mu_i\}) = \sum_i P_i^{\text{ideal}}(T, \mu_i - V_0 P^{\text{excl}}(T, \{\mu_i\})) = \sum_i P_i^{\text{ideal}}(T, \tilde{\mu}_i),$$

V_0 is the excluded volume, which we assume to be the same for all fermions, while $V_0 = 0$ for bosons. The excluded volume radius is taken to be $r_0 = 0.7 \, \text{fm}$. We do it just for completeness of the phase diagram to be shown below: however we actually do not discuss the corner of the T, μ_b phase diagram, with small T and large μ in this book. Apart from the excluded volume factor in p (which is also log Z), we use standard thermodynamical formulae for the ideal gas of hadrons and stable resonances. Typically one uses all resonances till $m = 2 \, \text{GeV}$.

8.3.3. *Pion liquid*

There is an alternative way to include the interaction in hadronic matter. The idea originates from my paper [500] which in turn is strongly influenced by the famous Landau paper about quasiparticles in liquid He[4], which I should explain next. In it the effect of all the resonances between pions is included in a very simple manner: just the propagation of the pions slows down, as they waltz in pairs (or larger groups) on their way. Thus the true elementary excitations — "quasipions" — which propagate straight have smaller velocity.

When Landau wrote it, the specific heat (and thus all thermodynamics) was just measured at low $T \sim 1$ kelvin for the first time. What was found was that the power dependence at very small T, consistently with the expected gas of phonons, changed by a much more rapid rise at somewhat larger T, approximately as $\exp(-\Delta/T)$ with

---

[9]The chemical potential for strangeness μ_s is a *dependent* variable, with its value always fixed from the total strangeness $S = 0$ condition.

some constant $\Delta \approx 8$ kelvin. That was enough for Landau to guess that there exist another regions of the quasiparticle spectrum — he called them rotons — with Δ being their minimal energy. Only decades later was it confirmed by direct neutron scattering experiments in which one quasiparticle was excited at a time.

Note now that such T dependence as was seen long ago in liquid He^4 is very similar to the low-T behavior of QCD thermodynamical quantities. At low T it is $p \sim T^4$ as expected from the free pions, with a more rapid rise at $T > 100$ MeV, $p \sim T^6$. This rise may be not the new excitations but just a modification of the pion dispersion curve. Since the pion mass is protected by chiral symmetry, my natural guess was that it is the *pion velocity* which is modified, in a presumed dispersion relation of the form

$$\omega_k^2 = u^2(T)k^2 + m_\pi^2. \tag{8.84}$$

With this form the calculation of thermodynamic quantities is very simple: rescaling the momenta in d^3k one can see that the energy density is just

$$\epsilon(T) = \frac{\epsilon_{\text{ideal}}(T)}{u^3(T)} \tag{8.85}$$

where $\epsilon_{\text{ideal}}(T) = 3p_{\text{ideal}}$ is the usual pion gas (8.75). Note now that *decreasing* pion velocity with increasing T leads to increasing $\epsilon(T)$: it happens because one can populate more phase space at a given T. Furthermore, in order to reproduce the fit to the resonance gas (8.83) one needs $u(T) \sim T^{2/3}$.

How can one observe it in principle? For example by studying the shape near the threshold of the process $\pi^+\pi^- \to e^+e^-$ from late stages of heavy ion collisions.

At low T explicit calculations in the chiral perturbation theory were performed by Pisarski and Tytgat [501], who introduced two different pion decay constants

$$\langle 0|A^{0a}|\pi^b(P)\rangle_T = if_\pi^t \delta^{ab} p^0, \qquad \langle 0|A^{ia}|\pi^b(P)\rangle_T = if_\pi^s \delta^{ab} p^i. \tag{8.86}$$

As in the vacuum, the pion mass shell is defined using current conservation,

$$\partial^\mu \langle 0|A^{\mu a}|\pi^b\rangle = 0 \longrightarrow f_\pi^t \omega^2 = f_\pi^s p^2. \tag{8.87}$$

Then, quite trivially, $f_\pi^t \neq f_\pi^s$ implies for the velocity u

$$u^2 = \Re(f_\pi^s/f_\pi^t) < 1. \tag{8.88}$$

and the dispersion law is as above. To order T^2/f_π^2 one can see that $u = 1$, and one needs the next order calculation to see how the pion velocity changes. The result for the f_π is

$$f_\pi^t \sim (1 - t_1 + 3t_2 + it_3)f_\pi \qquad f_\pi^s \sim (1 - t_1 - 5t_2 - it_3)f_\pi, \qquad (8.89)$$

$$t_1 = \frac{T^2}{12 f_\pi^2}, \quad t_2 = \frac{\pi^2 T^4}{45 f_\pi^2 m_\sigma^2}, \quad t_3 = \frac{m_\sigma^4}{32\pi f_\pi^2 \omega^2} \exp\left(-\frac{m_\sigma^2}{4\omega T}\right), \qquad (8.90)$$

so that

$$u^2 \sim 1 - 8t_2. \qquad (8.91)$$

Alternatively, results to that order for pressure are reproduced by the pion gas with a modified velocity such that

$$u \sim 1 - \frac{1}{3} \frac{T^4}{108 f_\pi^4}\left(7 \ln \frac{\Lambda_p}{T} - 1\right). \qquad (8.92)$$

Numerically these effects are small, reaching only of about 10 percent decrease of u and 30 percent of thermodynamic quantities by $T \sim 150$ MeV. One may also look at a paper by Rapp and Wambach [502], in which the free energy in terms of self-consistent skeleton diagrams, with realistic π–π interactions, is calculated.

The last important work I would like to mention in this section is a paper by Son and Stephanov [503], in which a very detailed analogy between pions and another collective excitation — the hydrodynamical sound — is made. One of their important statements is that the pion velocity u is the ratio of two statically measurable quantities, the *temperature-dependent* pion *decay constant* f^2 and the *axial isospin susceptibility* χ_{I5},

$$u^2 = \frac{f^2}{\chi_{I5}}. \qquad (8.93)$$

The axial isospin susceptibility χ_{I5} can be defined as the second derivative of the pressure with respect to the axial isospin chemical potential: so it can be determined from thermodynamics. The numerator, f, vanishes as $T \to T_c$, while the denominator does not. The final conclusion is that *the pion velocity goes to zero at the transition point*. Furthermore, they argued that in the exact chiral limit $m_\pi = 0$ the dispersion relation takes the form $\omega = |\mathbf{p}| - \frac{1}{2}iD'\mathbf{p}^2 + \cdots$, where D' is a temperature-dependent parameter which also diverges at $T \to T_c$.

8.4.   QCD phase transitions at finite T

8.4.1.   *Deconfinement*

The QCD-like theories with a variable number of colors N_c and (light) flavors N_f have a very rich phase structure at large temperatures and densities, which only starts to emerge from the theoretical ideas and numerical simulations.

In the preceding sections we have argued that the highly compressed or very hot hadronic matter is in the so called "quark–gluon plasma" state, being quite different from what we observe under normal conditions. The question is whether it is indeed the phase transition in the strict sense — the discontinuity in the thermodynamical quantities — or whether the transition is in fact continuous (as is the case for ordinary plasma created by the ionization of the atoms). In the spirit of the Landau theory of the phase transitions one usually asks in such situations whether the two phases are indeed qualitatively different, so that some "order parameter" can be introduced. (We remind the reader that such a parameter should be zero in one of the phases.) Therefore, considering phase transitions in QCD one naturally starts with the main qualitative aspects of the QCD vacuum, the confinement and the chiral symmetry breaking.

The first point to note is that in general one may expect two qualitative changes to happen, as the temperature T grows: (i) deconfinement; and (ii) chiral symmetry restoration. The deconfinement transition can be seen in pure gauge theory, as a transition in which the order parameter is e.g. the *string tension*.

The observable can thus be the expectation value of the Wilson loop, which should follow the *area law* at $T < T_c$, i.e. Eq. (2.57), and the *perimeter law* at $T > T_c$

$$\langle W(C) \rangle \sim \exp[-(T + L)\text{const}], \qquad T > T_c. \tag{8.94}$$

One may ask how exactly we can envision the vanishing of the string tension at $T \to T_c$. This was answered in the paper of Polyakov [506], who had considered thermal excitations of the strings. As usual, the level of excitations is a compromise between the energy and the entropy. In order to count the number of states of a string of length L one may consider the discretized space in the form of the cubic lattice with the spacing a. At any site there are $2 * d - 1 = 7$ links to which the string can turn,[10] therefore the number of states grows with L as

$$N(\text{states}) \sim 7^{L/a} \sim e^{L(\log 7/a)}. \tag{8.95}$$

The energy of a string is also linear in L, therefore the statistical sum for the strings looks as

$$Z = \sum_L \exp\left[-\frac{KL}{T} + L(\log 7/a)\right], \tag{8.96}$$

where the sum runs over different string lengths, while the sum over all shapes at fixed L is already included. One can see that since a string has an exponentially growing number configurations, in agreement with Hagedorn's observation its statistical sum diverges at all temperatures above the critical one

$$T_c = Ka/\log 7. \tag{8.97}$$

---

[10] The negative one here is to cancel the possibility that the string turns 360° back, which is ignored.

Although in the argument a was called the lattice spacing, it is clear that effectively it is some minimal radius of curvature a string may dynamically have. We do not really know what that is, probably a string width times some coefficient. Still one may speculate that this a is related to the tension K, $Ka \sim \sqrt{K}$ dimensionally. If the deconfinement critical temperature T_c is a function of the string tension alone, a simple consequence is as follows. Since K is used by definition to set the scale in lattice calculations, it would mean that T_c is numerically the same for all N_c. The known values for $N_c = 2, 3, 4$ agree with this idea. The value of this Hagedorn temperature is $T_c \approx 260\,\mathrm{MeV}$.

What is rather unusual in this nice argument is that at $T \to T_c$ nothing happens with the string microscopically. The strings do not really disappear: they just loose tension because their large entropy causes them to expand against the tension more and more, eventually becoming arbitrarily long and loosing notice of the original charges they are still attached to. One may then ask if a string language can still be used for a description of *deconfined* phases. At least one example of that is known, see Section 12.5 in which we will discuss how the Maldacena duality works at finite T.

Qualitative explanations of why the QCD phase transition takes place are often done in an over-simplified way, for example suggesting that deconfinement happens when hadrons overlap is space, and in a bag-model-type picture one naturally expects a coalescence into a single common bag. Not only does such a picture give all numbers[11] wrong, the physics is wrong as well. Pure gluodynamics is an especially good example to show that. Indeed, one can easily see that at $T_c \approx 260\,\mathrm{MeV}$ the fraction of the volume occupied by glueballs is negligible since even the lightest one has a mass of about 1.7 GeV. The real reason for deconfinement is internal string excitation, as we just discussed.

A different and more informative order parameter is the so called Polyakov line [506]

$$P = \left\langle \mathrm{Tr}\, P \exp\left[\left(ig \int_0^{1/T} d\tau\, A_0 \right) \right] \right\rangle, \tag{8.98}$$

which has zero expectation value at $T < T_c$, while a finite one otherwise. This finite value is proportional to $\exp(-M_Q^*/T)$ where M_Q^* is (T-dependent) effective mass of a static quark in QGP. The reason the Polyakov line is more informative than the Wilson loop [508] is related with the center group $Z(N_c)$ symmetry which pure gauge theory possesses. As the Polyakov line is transformed non-trivially under a phase multiplication for quark fields,[12]

$$P \to \exp(i2\pi k/N_c)P, \tag{8.99}$$

---

[11] For example, the MIT bag model literally predicts QGP formation in heavy ion collisions at unrealistically small (BEVALAC/GSI) energies.

[12] Note that for a baryon, made of N_c quarks there is no transformation at all, as should be the case. It explains why the center group has only such discrete phases.

with integer k, the deconfinement can be viewed as spontaneous breaking of the $Z(N_c)$ symmetry. This argument, made by Svetitski and Yaffe [508], then implies that the deconfinement transitions in $SU(N_c)$ gauge theory should belong to the same universality classes as $Z(N_c)$ spin systems. And indeed, for $N_c = 2$ one finds the second order transition [513] with standard indices of the Ising model [514], while for $N_c = 3$ deconfinement is the 1st order transition [515, 516].

It is also so for the SU(4) gauge theory [517]. It was noticed in Ref. [518] that $Z(N_c)$ groups at larger N_c may be reducible: for example, Z(4) has the Z(2) subgroup. The question is whether these subgroups are broken simultaneously with the whole group, or in separate phase transitions. If the latter possibility is the case, it leads to a very peculiar intermediate phase in which quarks are confined but diquarks are deconfined.

For a more recent discussion of the Polyakov lines and the effective spin description of the pure gauge phase transition see Ref. [509]. Interesting relations of the deconfinement to ideas of percollation and percollation clusters have been studied by Satz and collaborators, see e.g. Ref. [510].

In theories with dynamical quarks like QCD the formulation of confinement at finite T becomes obscure. No qualitative order parameter is available now. The strings can be screened by the dynamical quarks due to quark–antiquark pair production. No "center group transformation" is also possible: the anti-periodic boundary conditions for quark fields do not allow them.

One can however start by introducing quarks with some heavy masses m and decrease it. Obviously, the 1st order deconfinement transition for $N_c = 3$ cannot disappear immediately, but it does so for impressively large $m_c \approx 800\,\mathrm{MeV}$, according to lattice results.

Another approach is to look more closely at the static quark potential at finite distances: we will return to this issue below in Section 8.4.3.

8.4.2. *Chiral symmetry restoration*

In order to explain qualitatively why the chiral symmetry at high T gets restored, we may follow the logic of the NJL model and emphasize the analogy between this phenomenon and the superconductivity. Both in the QCD vacuum and the superconductors the interaction between fermions modify the states near the Fermi surface. In the QCD case the fermion states near the surface of the "Dirac sea" get modified due to $\bar{q}q$ pairing, with a nonzero "quark condensate" and a gap, the nonzero quark effective mass. Furthermore, the usual superconductivity can only exist below some critical temperature, so one may suspect that the breaking of chiral symmetry also has this property. Following the analogy, one may indeed see that the gap equation of the NJL model gets the same T-dependent modifications as that for a superconductor, and that at some critical T_c the nontrivial solution disappears. A generic reason is the following: pairing shifts the states below the Fermi level downward and states above it upward. At $T = 0$ only the former are occupied, so pairing is beneficial. At high enough $T \gg \Delta$, much larger than the gap, both sets of states are populated about equally, and the advantage of a paired state is gone.

Unlike for the deconfinement transition, there is a good order parameter, the quark condensate $\langle \bar{q}q \rangle$. One may think that if it disappears above a certain critical temperature, the chiral symmetry must be restored. This is a correct conjecture in most cases, but not a rigorous one! In principle, chiral symmetry breaking is actually possible even *without* the quark condensate. In Chapter 11 we will discuss an important example of such a phase [836], in QCD at high density.

More direct manifestations of the chiral symmetry breaking phenomenon are associated with the pions. Due to their Goldstone nature, they remain massless. Furthermore, the long-wave pions do not interact with any kind of hadrons and so they easily propagate even in hot/dense matter and dominate the long-distance correlators of axial currents to all $T < T_c$. Thus the pion mass, or rather its coupling f_π to the axial current, is actually a more rigorous order parameter than $\langle \bar{q}q \rangle$. So, one should test whether $f_\pi(T) \to 0$ at the transition point. In QCD-like theories, however, lattice simulations have shown that both vanish at the same T_c.

What is more surprising, the maximal variation in the Polyakov line also happens at this same T_c, so it looks as though deconfinement and chiral restoration coincide. There has been quite a bit of discussion in the literature, e.g. [521, 522], on whether the deconfinement and the chiral symmetry restoration should be one or two different phase transitions. In effect, they are different, as will become clear from the discussion to follow.

My first point is that a very significant difference between numerical values of the transition temperature is found on the lattice. While in pure gauge calculations ($N_f = 0$) the "deconfinement" phase transition occurs at[13] $T_c \approx 260$ MeV, simulations with dynamical quarks and $N_f = 3$ massless flavors show "chiral restoration" already at $T_c \approx 150$ MeV, while for real QCD with physical strange quark mass it is about 170 MeV.

The energetics of the "deconfinement" and "chiral restoration" transitions is entirely different, suggesting different physics. The former has a huge latent heat, a few GeV/fm^3, consistent with the bag model ideology in which all the non-perturbative vacuum energy density (proportional to the "gluon condensate" (2.22) is completely "melted" at $T > T_c$. This in not the case for "chiral restoration" in QCD: as G. Brown and V. Koch [511] have emphasized in their discussion of lattice data, a significant portion of the gluon condensate should actually survive the transition and be there at $T > T_c$. What is this remaining "hard glue" ("epoxy" as Gerry Brown called it)? Why, unlike "soft glue", does it not produce a quark condensate? We will return to those questions in Section 8.5 below.

The expected order of chiral symmetry restoration transition is expected to depend on the number of flavors N_f, not colors. More specifically, Pisarski and

---

[13]We remind the reader that by convention the units are always provided by fixing phenomenological value for the string tension.

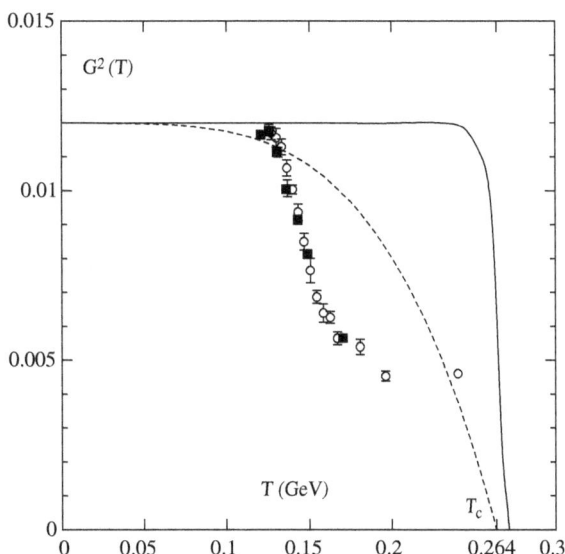

Fig. 8.7. Lattice data from Ref. [512] on the change in the gluonic condensate, or scale anomaly, near the critical temperature. The lines illustrate the behavior of the pure gauge theory, while open (closed) points refer to simulations with the lightest (medium mass) dynamical quarks. Note that only about one half of the condensate melts in the chiral phase transition, while the other half, or *epoxy*, survives.

Wilczek [523] argued that for $N_f = 2$ the restoration should be the second order transition, but become the first order for $N_f = 3$. They also argued that the universality class in the former case should correspond to $O(4)$ symmetric spin models, since this is the symmetry of the "chiral sphere" made of 3 pions and a sigma. More detailed information about our current understanding of the QCD phase diagram of 3-flavor QCD as a function of quark masses one can get from the sketch shown in Fig. 8.8. As one can see from it, on the plane of light-versus-strange quark mass there are two disconnected regions of the 1st order deconfinement (at large masses) and chiral (at small masses).

More generally, we know that chiral breaking may be the case without confinement: the high density case to be discussed in Chapter 11 is an example.

Finally, concluding this section on the qualitative discussion of chiral restoration, we have to address the following question: *What is the fate of the* $U(1)_A$ *chiral symmetry at high T?* One can find a relatively extensive discussion of this issue in my paper [525]. A short answer is that since it is not broken spontaneously but dynamically, in cannot be exactly restored (as suggested, e.g. in Ref. [526]). For example, an instanton of arbitrarily small size violates it, while any finite T can only affect instantons of size $\rho > 1/T$. In particular, in a gauge theory with only one light quark flavor, there will always be non-zero $\langle \bar{q}q \rangle$ no matter how high the T may be.

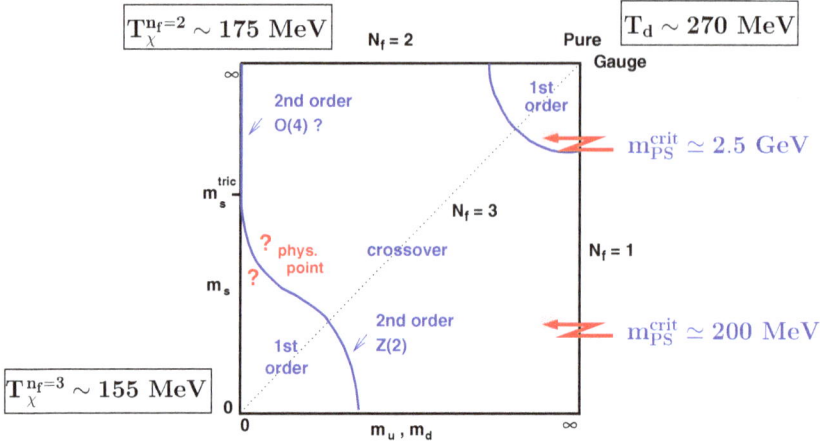

Fig. 8.8. The QCD phase diagram of three-flavor QCD with degenerate (u, d)-quark masses and a strange quark mass m_s.

On a more practical level, it turns out that restructuring of the instanton ensemble in high-T QCD leads to a dramatic reduction of $U(1)_A$ violating effects at $T > T_c$. In particular, it means that differences between the η and η' correlators are reduced.

8.4.3. *Static quark potential at high T*

As a very important observable related to both deconfinement and chiral restoration, we will discuss now the static quark effective potential at finite T, following a recent review by Karsch [542]. This free energy can be extracted from Polyakov loop correlation functions

$$\frac{F_{q\bar{q}}(r, T)}{T} = -\ln\langle L(\vec{x})L^{\dagger}(\vec{y})\rangle + \text{const.}, \quad |\vec{x} - \vec{y}| = r. \qquad (8.100)$$

The potential energy should then be calculated[14] by the usual relation $V = F + TS$ where S is associated entropy given by $\partial F/\partial T$.

In Fig. 8.9 one can see the data for 3 flavor of dynamical fermions. With increasing T the string breaking (which was so difficult to see at $T = 0$) happens at smaller and smaller distances. The deconfinement should manifest itself as a decrease of the string tension K, the slope of the potential in the linear region. The data shown in the figure hardly show any change in K at all.

Instead one finds that the flat part of the free energy reaches lower and lower as T increases. The decrease is a combination of two different and important phenomena. For $T < T_c$ this flat potential means that two static quarks separates into two nearly non-interacting $D, B \ldots$ heavy–light mesons, and $V \approx 2M_D^*$. The fact that it goes

---

[14] The reason the entropy term, or heat, should be subtracted is that motion of two quasiparticles are at high enough frequencies. For usage of these potentials for discussing quasiparticle binding, see Shuryak and Zahed, hep-ph/0403127.

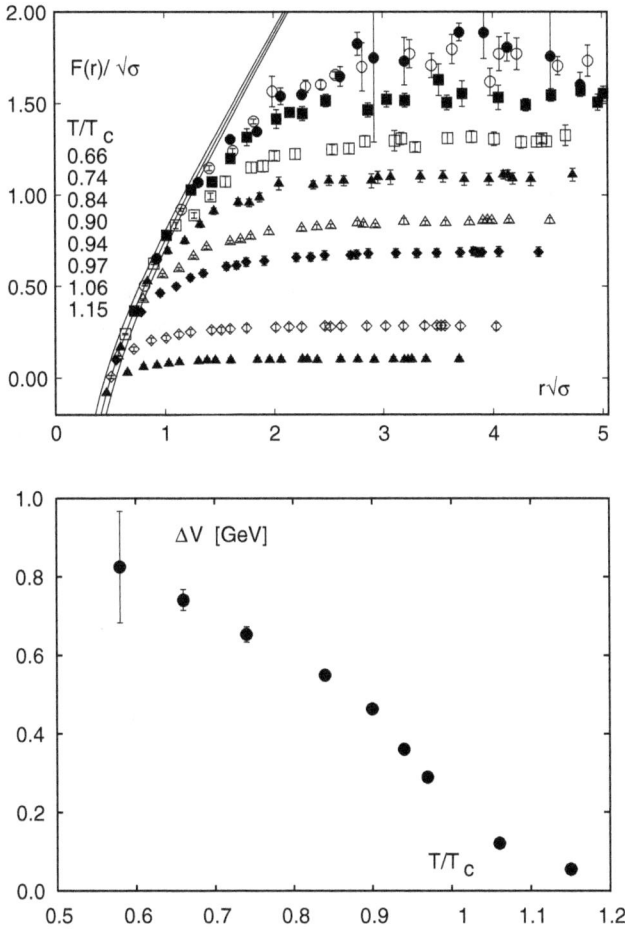

Fig. 8.9. The top panel shows the temperature dependence of the heavy quark free energy for three-flavor QCD with a quark mass $m_q = 0.1$. The band of lines gives the Cornell potential in units of the square root of the string tension, $V(r)/\sqrt{\sigma} = -\alpha/r\sqrt{\sigma} + r\sqrt{\sigma}$ with $\alpha = 0.25 \pm 0.05$. They have been normalized at short distances such that they agree with the zero temperature Cornell potential at $r = 1/4T$. The bottom panel shows the temperature dependence of the depth of the heavy quark free energy.

down is thus a direct manifestation of the *chiral symmetry restoration*, which drives the effective masses of the light quarks down.

At $T = T_c$ it is supposed to melt away completely. In Chapter 6 we discussed the QCD sum rules for heavy–light mesons and concluded that those "hydrogen atoms of the hadronic world" have energy of about 450 MeV in excess of the heavy quark mass. Figure 8.9b shows that the effect of the reduction by the same magnitude but a bit smaller, namely $\Delta V = 2M^*(0) - 2M^*(T_c) \approx 700$ MeV.

At $T > T_c$ the static charge is also screened, at very high T by a Debye cloud of the opposite charge. The perturbative expression for the corresponding "correlation energy" (8.53) we have already derived.

To quantify the screening effect in the presence of light dynamical quarks we define the depth of the potential as the difference between the asymptotic value at large distances and the potential at distance $r\sqrt{\sigma} = 0.5$,

$$\Delta V \equiv \lim_{r \to \infty} V(r) - V(0.5/\sqrt{\sigma}), \qquad (8.101)$$

which is the point at which the Cornell-potential vanishes if one chooses the coupling of the Coulomb term as $\alpha = 0.25$. It approximately corresponds to a distance $r \simeq 0.23$ fm. In Fig. 8.9b from Ref. [542] we show this difference of potential energies as a function of T. Note a break of the dependence near $T = T_c$, which is hinted by these data: indeed if anything, one can be surprised why a transition from the confined light quarks to the Debye cloud is so smooth.

The issue is not academic and is closely related to the rates of the J/ψ and D mesons in heavy ion collisions to be discussed in Chapter 10, in which we proceed from high to low T through the transition. The difference in their binding is crucial for a decision in which direction the reaction $J/\psi \leftrightarrow \bar{c}c$ goes. More generally, the issue is whether quasiparticles are or are not bound.

8.4.4.   *Equation of state in the transition region*

In this section we start discussing the physics of the QCD phase transitions. We start with rather simple estimates, locating the transition temperature.

We just discussed the EoS of the hadronic matter above, and we also would need that for QGP. Although we have seen that pQCD corrections are not reliably known, lattice data tell us that the ideal gas approximation is not that bad at high enough T. Following tradition, we include the effect of *non-perturbative* effects present in the vacuum and absent (or somewhat suppressed) in QGP simply by addition of the bag-type term B to the EoS of the ideal quark–gluon plasma

$$\epsilon_{\text{QGP}} = \frac{\pi^2 T^4}{15} \left(16 + \frac{7}{8} 6 N_f \right) + \frac{3 N_f}{2} \left(T^2 \mu_b^2 + \frac{\mu^4}{2\pi^2} \right) + B,$$

$$p_{\text{QGP}} = \frac{\pi^2 T^4}{45} \left(16 + \frac{7}{8} 6 N_f \right) + \frac{N_f}{2} \left(T^2 \mu_b^2 + \frac{\mu^4}{2\pi^2} \right) - B. \qquad (8.102)$$

At the phase transition point, both the QGP pressure and temperature should be equal to that of the hadronic phase,

$$p_{\text{QGP}}(T_c) = p_{\text{resonance gas}}(T_c) \qquad (8.103)$$

which is the equation for T_c. For the r.h.s. one can use, e.g. the simple parameteri- $\boxed{\text{E}}$ zation of (8.83). Note first of all, that crossing of the two curves is guaranteed even without B. However it is needed phenomenologically: one can actually obtain from this equilibrium condition the value of B if the phase transition point is tuned to get $T_c = 160$ MeV for zero baryon density, as lattice and heavy ion data indicate.

The resulting value is rather large, $B \approx 320$ MeV/fm^3, which is about six times the original constant of the MIT bag model. At the same time it is only about a 1/2–1/3 of what one would get if all gluon condensate or all instantons would be eliminated in QGP. This is something to think about, if one wants to understand what this non-perturbative glue left in the QGP phase is.

Exercise 8.9. Solve the equation for T_c using the parameterization (8.83) for the hadronic gas, and derive the magnitude of the corresponding bag constant and the latent heat $L = \epsilon(T = T_c^+) - \epsilon(T = T_c^-)$ across the transition.

Quantitative data on the thermodynamics in the phase transition region has been a subject of intense lattice studies. It was found that the overall temperature dependence of pressure in QCD with 2 and 3 light quark flavors is very similar to the case of the pure gauge theory. Lattice data from the Bielefeld group [542] are shown in the left hand part of Fig. 8.10. Although after rescaling, the thermodynamic

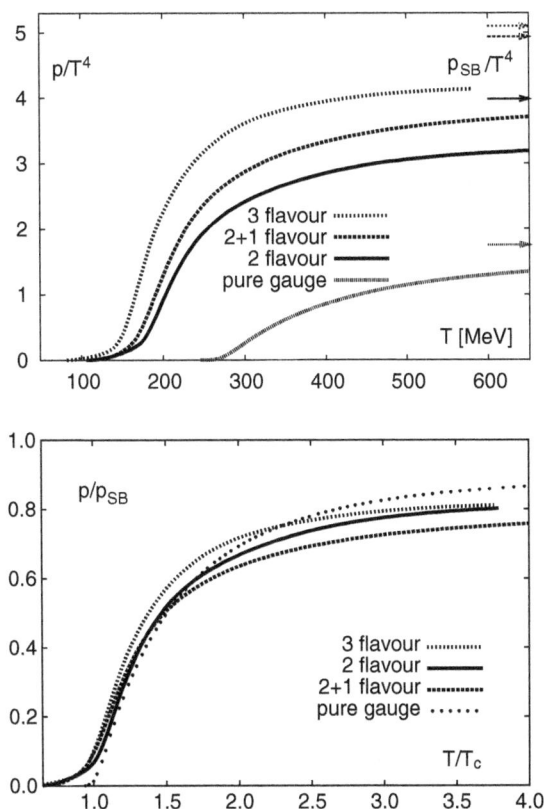

Fig. 8.10. The pressure in QCD with $n_f = 0, 2$ and 3 light quarks as well as two light and a heavier (strange) quark. In the top panel we show p/T^4, and different theories have different curves. They nearly fall into a universal behavior, shown in the bottom panel, if for each theory one uses free case as a normalization. For $n_f \neq 0$ calculations have been performed on a $N_\tau = 4$ lattice using improved gauge and staggered fermion actions. In the case of the SU(3) pure gauge theory the continuum extrapolated result is also shown. Arrows indicate the ideal gas pressure p_{SB}.

observables with the corresponding Stefan–Boltzmann constants look quite alike in units of T/T_c, they still differ in details. In particular, it is apparent from the right hand part of Fig. 8.10 that at T_c the rescaled pressure of QCD with light quarks is significantly larger than in the pure gauge theory. This is also the case for the energy density which is found to be $\epsilon_c/T_c^4 \simeq 6$ in QCD with light quarks while it is only $\epsilon_c/T_c^4 \simeq 1$ in the SU(3) gauge theory. The energy density is discontinuous in the latter case and the latent heat $\Delta\epsilon/T_c^4 = 1.40(9)$ with $\epsilon/T_c^4 \simeq 2$ in the high temperature phase. Much of this factor 6 difference in ϵ_c/T_c^4, however, seems to arise from the difference in T_c between QCD with light quarks and the purely gluonic theory. The critical energy densities turn out to be quite similar. Unfortunately, the current error on T_c, which is about 10%, amplifies in the calculation of the energy density, which makes ϵ_c still badly determined in lattice calculations,

$$\epsilon_c = (0.3 - 1.3)\text{GeV}/\text{fm}^3. \qquad (8.104)$$

Unfortunately the non-zero chemical potential enters with the complex weight into the path integral, and thus it is in general difficult to obtain any lattice results at non-zero density. (We will discuss this case in Chapter 11 below.) However if the chemical potential is relatively small $\pi T \gtrsim \mu$, the re-weighting methods can still be used. By using it, Fodor and Katz [543] were able to derive the location of the critical endpoint, $\mu_{\text{end point}} \approx 700\,\text{MeV}$, which has since been confirmed by a number of other approximations,[15] such as the re-weighting technique on the phase diagram [544]. The direction of the re-weighting lines is shown in Fig. 8.11a. This endpoint will be important in our discussion of the event-by-event fluctuations in Chapter 9.

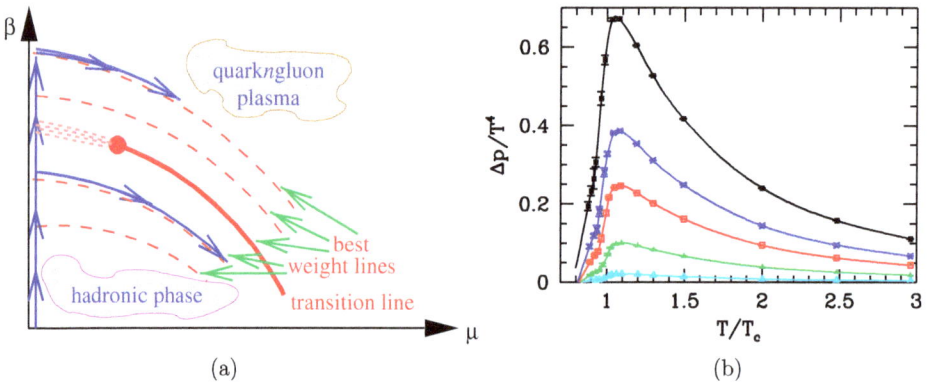

(a)   (b)

Fig. 8.11. (a) Schematic directions of the best approach toward the increasing chemical potential, in the vicinity of the QCD phase transition line. Its first dotted part is the crossover region. The blob represents the critical endpoint, after which the transition is of first order. (b) $\Delta p = p(\mu \neq 0, T) - p(\mu = 0, T)$ normalized by T^4 as a function of T/T_c for $\mu_B = 100, 210, 330, 410\,\text{MeV}$ and $\mu_B = 530\,\text{MeV}$ (from top to bottom).

---

[15]The value of μ_{endpoint} is however substantially reduced in later works, see e.g. Fodor and Katz hep-lat/0402006.

In Fig. 8.11b we show also how nonzero quark density and chemical potential affects the thermodynamical observables, such as the pressure (from Fodor et al. [543]).

In hydrodynamical applications to be discussed in the next Chapter 9 one traditionally uses a somewhat simplified EoS with a variable latent heat in the e and n_B plane. For applications to be discussed in detail in the next chapter the so called cooling trajectory on the T–μ plane should be determined. Furthermore, the relevant EoS is the one along this cooling curve. If expansion of matter[16] is slow enough, the *entropy is conserved*. If so, we should follow the *adiabatic paths*. Furthermore, the two conjugates to the T–μ_b pair of variables — the entropy s and the baryonic density n_b — provide a more natural description. As in cosmology, both the global entropy S and the baryon number N_B are conserved during expansion, and the ratio of their densities stays constant, from the beginning to the end. Therefore, the paths can be marked by this ratio, and its value can be determined at the end of the evolution.

Let me describe as an example the EoS used in the paper by Teaney, Lauret and myself [673]. First, note the following two derivatives which apply along the path where n_B/s is constant

$$\left(\frac{dp}{de}\right)_{n_B/s} \equiv c_s^2, \tag{8.105}$$

$$\left(\frac{ds}{de}\right)_{n_B/s} = \frac{s}{p+e}. \tag{8.106}$$

The first of these is simply the definition of the speed of sound. The second relation is surprising: it does not contain the chemical potential μ_B explicitly. (It follows by noting that

$$\left(\frac{ds}{de}\right)_{n_B/s} = \left(\frac{ds}{de}\right)_{n_B} + \left(\frac{ds}{dn_B}\right)_e \frac{n_B}{s} \left(\frac{ds}{de}\right)_{n_B/s}$$

and solving for $(ds/de)_{n_B/s}$, by using thermodynamic identities.) Given the speed of sound everywhere and the entropy on a single arc in the e, n_B plane, these derivatives may be integrated to determine the entropy, $s(e, n_B)$. From the entropy, all other thermodynamic functions (e.g. T and μ_B) may be determined. Below, only the speed of sound is specified.

The EoS thus consists of three pieces: a hadronic phase, a mixed phase, and a QGP phase. In strong interactions, Baryon number (B), Strangeness (S), and Isospin (I) are conserved and therefore the EoS depends on T and μ_B, μ_S, and μ_I. In the hadronic phase, the thermodynamic quantities — the pressure (p), the energy

---

[16] Note that we do not discuss the compression stage here: it is not slow and therefore entropy is in fact produced here.

density (e), the entropy density (s), and number densities $(n_Q$ where $Q = B, S, I)$ —
are taken as ideal gas mixtures of the lowest $SU(3)$ multiplets of mesons and baryons.
The mix includes the pseudo-scalar meson octet (π, η, K) and singlet (η'), the vector
meson octet (ρ, K^*, ω) and singlet (ϕ), the $\frac{1}{2}^-$ baryon and anti-baryon octets and the
$\frac{3}{2}^-$ baryon and the anti-baryon decuplets. Specifically, p, e, s and n_Q are given by

$$n_Q = \sum_i Q_i n_{id}^{\sigma_i}(T, \mu_B B_i + \mu_S S_i + \mu_I I_i), \qquad (8.107)$$

$$p = \sum_i p_{id}^{\sigma_i}(T, \mu_B B_i + \mu_S S_i + \mu_I I_i), \qquad (8.108)$$

$$e = \sum_i e_{id}^{\sigma_i}(T, \mu_B B_i + \mu_S S_i + \mu_I I_i), \qquad (8.109)$$

$$s = \sum_i s_{id}^{\sigma_i}(T, \mu_B B_i + \mu_S S_i + \mu_I I_i). \qquad (8.110)$$

Here the sum is over the hadrons species, B_i, S_i, I_i are the quantum numbers of the
ith hadron, σ_i is $+$ for bosons but $-$ for fermions, and for example, $p_{id}^+(T, \mu)$ is the
pressure of a simple ideal Bose gas. This hadronic EoS is taken up to a temperature
of $T_c = 165$ MeV or an energy density $e_H \approx 0.45$ GeV/fm^3 (see Fig. 8.12). The
squared speed of sound is approximately $1/5$ in this hadronic gas [492]. Above the
hadronic phase, only the speed of sound squared, c_s^2, is specified. For the mixed
phase the speed of sound was made approximately zero, $c_s^2 = 0.02\,c$. The width of
the mixed phase (see Fig. 8.12) is the Latent Heat (LH), LH$= e_Q - e_H$. LH is taken
as a parameter and is adjusted to form phase diagrams LH8, LH16, . . . with latent

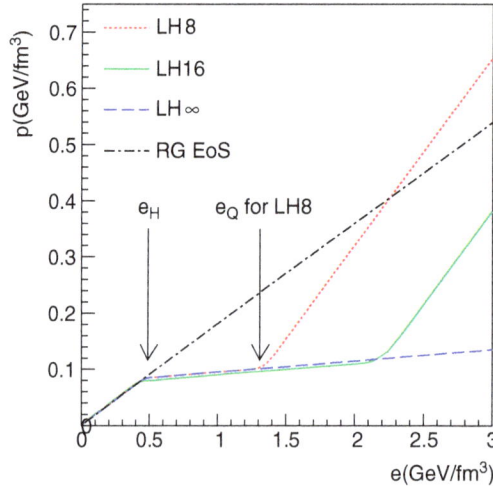

Fig. 8.12. The pressure (p) versus energy density (e) for different EoS. EoS LH8, LH16 and LH∞
become increasingly soft and have latent heats 0.8 GeV/fm^3, 1.6 GeV/fm^3, and ∞. The EoS are
shown along the adiabatic path for SPS initial conditions, $s/n_B = 42$. For RHIC initial conditions,
$s/n_B = 150$, the changes are small.

heats, $0.80\,\text{GeV/fm}^3$, $1.6\,\text{GeV/fm}^3$ .... Above the mixed phase, $e > e_Q$, the degrees of freedom are taken as massless and the speed of sound is accordingly, $c_s = \sqrt{1/3}$. We also consider two limiting cases: a Resonance Gas (RG) EoS and LH∞. For a RG EoS, the speed of sound is constant above e_H. For LH∞, the mixed phase continues forever ($e_Q = \infty$) and there is no ideal plasma phase.

With the speed of sound specified in all phases, Eqs. (8.105) and (8.106) are integrated to find the pressure and entropy along the adiabatic path specified by the initial conditions, with given n_B/s. The full phase diagram for SPS initial conditions is shown in Fig. 8.12.

Although the T–μ plane is rather convenient for the determination of the thermodynamical parameters in both phases, the mixed phase domain is hidden behind the transition line. As is well known, in the mixed phase new thermodynamical variable appear — e.g. the fraction f of the volume occupied by the QGP phase. That is why the adiapatic paths look like a complicated zig-zag on a T, μ plane, see Fig. 9.1a.

8.5. Instantons at finite T

8.5.1. *Finite temperature field theory and the caloron solution*

In the Chapters 4 and 6 we have shown that the instanton liquid model provides a mechanism for chiral symmetry breaking in vacuum and quantitatively describes a large number of hadronic correlation functions. Clearly, it is of interest to generalize the model to finite temperature. Extending these methods is fairly straightforward, basically one replaces all the gauge potentials and fermionic zero modes by their finite temperature (periodic or anti-periodic) counterparts.

Periodic instanton configurations can be constructed by lining up zero temperature instantons along the imaginary time direction with the period β. Using the 't Hooft multi-instanton solution, the explicit expression for the gauge field is given by [547]

$$A_\mu^a = \bar{\eta}_{\mu\nu}^a \Pi(x) \partial_\nu \Pi^{-1}(x), \tag{8.111}$$

$$\Pi(x) = 1 + \frac{\pi\rho^2}{\beta r} \frac{\sinh(2\pi r/\beta)}{\cosh(2\pi r/\beta) - \cos(2\pi\tau/\beta)}. \tag{8.112}$$

Here, ρ denotes the size of the instanton. The solution (8.111) is usually referred to as the *caloron*. Its topological charge $Q = 1$ and the action $S = 8\pi^2/g^2$ are independent of temperature.[17] A caloron with $Q = -1$ can be constructed making the replacement $\bar{\eta}_{\mu\nu}^a \to \eta_{\mu\nu}^a$. Of course, as $T \to 0$ the caloron field reduces to the field of an instanton. In the high temperature limit $T\rho \gg 1$, however, the field looks very different [546]. In this case, the caloron develops a core of size $O(\beta)$ where the fields

---

[17]The topological classification of smooth gauge fields at finite temperature is more complicated than at $T = 0$ [546]. Topological charge is quantized only in the absence of magnetic charges, but the presence of magnetic charges requires QCD to be in the Higgs phase at $T \neq 0$. There is no evidence that this is the case.

are very strong $G_{\mu\nu} \sim O(\beta^2)$. In the intermediate range $O(\beta) < r < O(\rho^2/\beta)$ the caloron looks like a (T-independent) dyon with unit electric and magnetic charges,

$$E_i^a = B_i^a \simeq \hat{r}^a \hat{r}^i / r^2. \tag{8.113}$$

In the far region, $r > O(\rho^2/\beta)$, the caloron resembles a three-dimensional dipole field, $E_i^a = B_i^a \sim O(1/r^3)$. The caloron has one left-handed fermionic zero mode, given by [223]

$$\psi_i^a = \frac{1}{2\sqrt{2}\pi\rho}\sqrt{\Pi(x)}\partial_\mu\left(\frac{\Phi(x)}{\Pi(x)}\right)\left(\frac{1-\gamma_5}{2}\gamma_\mu\right)_{ij}\epsilon_{aj}, \tag{8.114}$$

where $\Phi(x) = (\Pi(x) - 1)\cos(\pi\tau/\beta)/\cosh(\pi r/\beta)$. Note that the zero-mode wave function also shows an exponential decay $\exp(-\pi rT)$ in the spatial direction, despite the fact the "energy" is zero.

8.5.2.   *Instanton density at high temperature*

After introducing the classical instanton field at finite temperature, we can study quantum fluctuations around the caloron solution. Analogous to the 't Hooft calculation at $T = 0$, quantum fluctuations determine the instanton density at finite temperature. At zero temperature, the instanton density is controlled by the instanton action $8\pi^2/g^2$ in the exponent. It was argued in my work [216] that instantons should be strongly suppressed at high temperature, since the gluoelectric fields in the instantons must be Debye screened. Normal $O(1)$ electric fields are therefore screened at distances $1/gT$, while the much stronger non-perturbative fields of the instantons $O(1/g)$ should be screened for sizes $\rho > 1/T$.

An explicit calculation of the quantum fluctuations around the caloron was performed by Pisarski and Yaffe [548]. Their result is

$$d(\rho, T) = d(\rho, T = 0)\exp\left(-\frac{1}{3}(2N_c + N_f)(\pi\rho T)^2 - B(\lambda)\right),$$

$$\tag{8.115}$$

$$B(\lambda) = \left(1 + \frac{N_c}{6} - \frac{N_f}{6}\right)\left(-\log\left(1 + \frac{\lambda^2}{3}\right) + \frac{0.15}{(1 + 0.15\lambda^{-3/2})^8}\right),$$

where $\lambda = \pi\rho T$. At high temperatures the contribution of large instantons with $\rho \gg 1/T$ is exponentially suppressed. As a result, the instanton related contribution to physical quantities like the energy density (or pressure, etc.) is on the order of

$$\epsilon(T) \sim \int_0^{1/T}\frac{d\rho}{\rho^5}(\rho\Lambda)^{(11N_c/3)} \sim T^4(\Lambda/T)^{(11N_c/3)} \tag{8.116}$$

which is small compared to the energy density of an ideal gas $\epsilon(T)_{\mathrm{SB}} \sim T^4$.

It was emphasized in Ref. [556] that although the Pisarski–Yaffe result contains only one dimensionless parameter λ, its applicability is controlled by two separate conditions:

$$\rho \ll 1/\Lambda, \qquad T \gg \Lambda. \tag{8.117}$$

The first condition ensures the validity of the semi-classical approximation, while the second justifies the perturbative treatment of the heat bath. In order to illustrate this point, we would like to discuss the derivation of the semi-classical result (8.115) in somewhat more detail. Our first point is that the finite temperature correction to the instanton density can be split into two parts with different physical interpretation. For this purpose, let us consider the determinant of a scalar field in the fundamental representation.[18] The temperature dependent part δ of the one loop effective action

$$\log \det \left(\frac{-D^2}{-\partial^2} \right) \bigg|_T = \log \det \left(\frac{-D^2}{-\partial^2} \right) \bigg|_{T=0} + \delta \tag{8.118}$$

can be split into two pieces, $\delta = \delta_1 + \delta_2$, where

$$\delta_1 = \mathrm{Tr}_T \log \left(\frac{-D^2(A(\rho))}{-\partial^2} \right) - \mathrm{Tr} \log \left(\frac{-D^2(A(\rho))}{-\partial^2} \right), \tag{8.119}$$

$$\delta_2 = \mathrm{Tr}_T \log \left(\frac{-D^2(A(\rho,T))}{-\partial^2} \right) - \mathrm{Tr}_T \log \left(\frac{-D^2(A(\rho))}{-\partial^2} \right). \tag{8.120}$$

Here Tr_T includes an integration over $R^3 \times [0,\beta]$, $A(\rho,T)$ is the caloron field and $A(\rho)$ is the instanton field. The two terms δ_1 and δ_2 correspond to the two contributions appearing in the exponent in the semi-classical result.

It was shown in Ref. [552] that the physical origin of the first term is the scattering of particles in the heat bath on the instanton field. The forward scattering amplitude $T(p,p)$ of a scalar quark can be calculated using the standard LSZ reduction formula

$$\mathrm{Tr}\, T(p,p) = \int d^4x\, d^4y\, e^{ip\cdot(x-y)} \,\mathrm{Tr}(\partial_x^2 \Delta(x,y)\partial_y^2), \tag{8.121}$$

where $\Delta(x,y)$ is the scalar quark propagator. By rescaling variables in (8.121) as $\xi = px, \eta = py$, subtracting the trace of the free propagator and going to the physical pole $p^2 = 0$ one gets

$$\mathrm{Tr}\, T(p,p) = \int d^4\xi\, d^4\eta\, e^{i\eta\cdot(x-y)} \frac{\rho^2}{2\pi^2(\xi-\eta)^2} \left(\frac{\xi\cdot\eta}{\xi^2\eta^2} - \frac{1}{2\xi^2} - \frac{1}{2\eta^2} \right)$$

$$= -4\pi^2\rho^2. \tag{8.122}$$

---

[18] The quark and gluon (non zero mode) determinants can be reduced to the determinant of a scalar field in the fundamental and adjoint representation [546].

Since the result is just a constant, there is no problem with analytic continuation to Minkowski space. Integrating the result over a thermal distribution, we get

$$\delta_1 = \int \frac{d^3p}{(2\pi)^3} \frac{1}{2p(\exp(p/T) - 1)} \operatorname{Tr} T(p,p) = -\frac{1}{6}(\pi\rho T)^2 = -\frac{1}{6}\lambda^2. \quad (8.123)$$

The constants appearing in the result are easily interpreted, ρ^2 comes from the scattering amplitude, while the temperature dependence enters via the integral over the heat bath. Also note that the scattering amplitude has the same origin (and the same dependence on N_c, N_f) as the Debye mass.

As usual the validity of this perturbative calculation requires that $g(T) \ll 1$, although it is not clear what this means in practice (in QCD, $g \ll 1$ is only satisfied at ridiculously high temperature). We would argue, however, that the accuracy of the calculation sketched above is controlled by the same effects that determine the validity of the perturbative result for the Debye mass. Available lattice data suggest that perturbative screening works well above $T > 3T_c \simeq 500\,\text{MeV}$. Near the critical temperature one expects (and observes on the lattice) that the Debye mass vanishes: screening disappears together with the plasma. Below T_c the instanton density should be determined by the scattering of hadrons on the instanton. We will discuss this question in detail in the next section.

The second term in (8.120) in the finite temperature effective action has a different physical origin. It is determined by the quantum correction to the colored current (this effect was first considered in Ref. [180]), times the T-dependent variation of the instanton field, the difference between the caloron and the instanton. For small λ it gives the correction

$$\delta_2 = -\frac{1}{36}\lambda^2 + O(\lambda^2). \quad (8.124)$$

It has the same sign and parameter dependence as δ_1, but a smaller coefficient. Thus, finite T effects not only lead to appearance of the usual (perturbative) heat bath, but they also modify the strong $O(1/g)$, classical gauge field of the instanton. In Euclidean space, we can think of this effect as arising from the interaction of the instanton with its periodic "mirror images" along the imaginary time direction. Below T_c, color singlet gluon correlators are strongly suppressed since glueballs states are very heavy. We would therefore argue that below T_c, the instanton field is not modified by the boundary conditions, and that there is no instanton suppression due to the change in the classical field below T_c.

With such theoretical prediction at hand, one may now have a look at what lattice simulations can say about the matter. So far, there have been very few lattice simulations of QCD instantons at finite temperature, and most of the results that have been published were obtained in the quenched approximation. As two representative examples I have used Ref. [560] by Chu and Schramm and the more recent work by Gattringer [561]. The main calculated quantity is the topological susceptibility $\langle Q^2 \rangle / V$, which is quite stable in the cooled configurations since $\bar{I}I$ pairs do

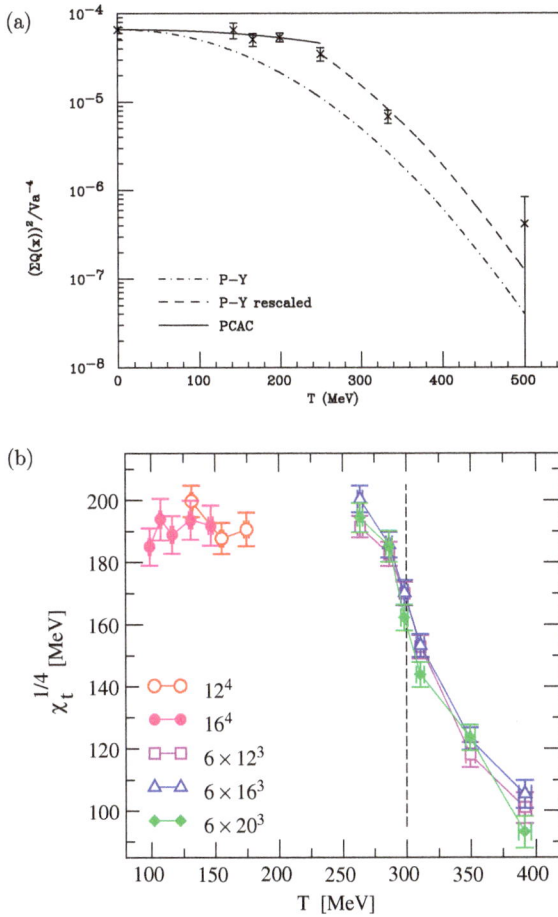

Fig. 8.13. Lattice measurements of the temperature dependence of the topological susceptibility in pure gauge SU(3), (a) from Ref. [560] and (b) Ref. [561]. The dash-dotted line labeled P–Y in (a) corresponds to Pisarski–Yaffe result (8.115), while the dashed curve shows the rescaled formula discussed in the text.

not contribute to global topological charge Q. Figure 8.13 shows their results as a function of the temperature. At $T = 0$, $\chi_{\text{top}} \simeq (180 \, \text{MeV})^4$, in agreement with the phenomenological value. The topological susceptibility is almost temperature-independent below the critical temperature ($T_c \simeq 260 \, \text{MeV}$ in the quenched theory), but drops fast above T_c.

Note that in the quenched theory, the correlations between instantons are not so strong (no molecules), and thus the ensemble is nearly random and the topological susceptibility is expected to be close to the instanton density. Indeed, the temperature dependence of χ_{top} above T_c is consistent with the semi-classical suppression factor (8.115), but with a *shifted* temperature, $T^2 \rightarrow (T^2 - T_c^2)$. The naive Pisarski–Yaffe formula (the dashed line) does not agree with this trend.

Chu and Schramm also extracted the average instanton size from the correlation function of the topological charge density. Below T_c, they find $\rho = 0.33$ fm independent of temperature while at $T = 334\,\text{MeV} > T_c$, they get a smaller $\rho = 0.26$ fm. It is also in qualitative agreement with the Debye-type suppression factor (provided we use the shifted temperature dependence proposed above). Finally, they considered instanton contributions to the pressure and space-like hadronic wave functions. They find that instantons contribute roughly 15% of the pressure at $T = 334\,\text{MeV}$ and 5% at $T = 500\,\text{MeV}$.

8.5.3. *Instantons at low temperature*

The modification of the instanton density at low temperatures was derived by Velkovsky and myself [556]. At temperatures well below the phase transition the heat bath consists of weakly interacting pions. The interaction of the pions in the heat bath with instantons can be determined from the effective 't Hooft Lagrangian. For two flavors, this Lagrangian consists of four fermion operators multiplied by the semi-classical single instanton density. We have argued in the last section that that the semi-classical instanton density is not modified at small temperature. The temperature dependence is then determined by the T dependence of the vacuum expectation value of the four fermion operators. For small T, the situation can be simplified even further, because the wavelength of the pions in the heat bath is large and the four fermion operators can be considered local.

The T dependence of a general expectation value of the form $\langle (\bar{q}Aq)(\bar{q}Bq) \rangle$ can be determined using soft pion techniques. The leading contribution is given by the one-pion matrix element, integrated over a thermal pion distribution. The one-pion matrix element can be calculated by reducing out the pion states and using current algebra commutators. The general result is [554]

$$\langle (\bar{q}Aq)(\bar{q}Bq) \rangle_T = \langle (\bar{q}Aq)(\bar{q}Bq) \rangle_0$$

$$-\frac{T^2}{96f_\pi^2} \langle (\bar{q}\{\Gamma_5^a, \{\Gamma_5^a, A\}\}q)(\bar{q}Bq) \rangle_0$$

$$-\frac{T^2}{96f_\pi^2} \langle (\bar{q}Aq)(\bar{q}\{\Gamma_5^a, \{\Gamma_5^a, B\}\}q) \rangle_0$$

$$-\frac{T^2}{48f_\pi^2} \langle (\bar{q}\{\Gamma_5^a, A\}q)(\bar{q}\{\Gamma_5^a, B\}q) \rangle_0 \qquad (8.125)$$

where A, B are arbitrary flavor-spin-color matrices and $\Gamma_5^a = \tau^a \gamma_5$. Using this result, the final result for the instanton density is given by

$$n(\rho, T) = d(\rho) \left(\frac{4}{3}\pi^2\rho^3 \right)^2 \left[\frac{\langle K_1 \rangle_0}{4} \left(1 - \frac{T^2}{6f_\pi^2} \right) - \frac{\langle K_2 \rangle_0}{12} \left(1 + \frac{T^2}{6f_\pi^2} \right) \right], \qquad (8.126)$$

where we have defined the following two operators:

$$K_1 = \bar{q}_L q_L \bar{q}_L q_L + \frac{3}{32} \bar{q}_L t^i q_L \bar{q}_L t^i q_L - \frac{9}{128} \bar{q}_L \sigma_{\mu\nu} t^i q_L \bar{q}_L \sigma_{\mu\nu} t^i q_L,$$

$$K_2 = \bar{q}_L \tau^a q_L \bar{q}_L \tau^a q_L + \frac{3}{32} \bar{q}_L \tau^a t^i q_L \bar{q}_L \tau^a t^i q_L - \frac{9}{128} \bar{q}_L \sigma_{\mu\nu} \tau^a t^i q_L \bar{q}_L \sigma_{\mu\nu} \tau^a t^i q_L.$$
(8.127)

Although the vacuum expectation values of these two operators are unknown, it is clear that (baring unexpected cancellations) the T-dependence should be rather weak, most likely $n = n_0(1 \pm T^2/6f_\pi^2)$. Indeed, if one estimates the expectation values using the factorization assumption [352], the T-dependence exactly cancels.

Summarizing the last two sections we conclude that the instanton density is expected to be essentially constant below the phase transition, but exponentially suppressed at large temperatures.

8.5.4. *Chiral symmetry restoration and instantons*

In order to study an interacting system of instantons at finite temperature, we need to determine the bosonic and fermionic interaction between instantons at $T \neq 0$. This subject was studied in detail, in Ref. [553], which the reader can consult for definitions.

In the previous sections, we have argued that the instanton density remains roughly constant below the phase transition. If this is correct, then the chiral phase transition has to be caused by a rearrangement rather than the disappearance of instantons. Furthermore, we have shown that the gluonic interaction between instantons remains qualitatively unchanged until rather high temperatures. This means that, most likely, the phase transition is driven by fermionic interactions between instantons [245, 557].

The mechanism for this transition is most easily understood by considering the fermion determinant for one instanton–anti-instanton pair [245]. Using the asymptotic form of the overlap matrix elements specified above, we have

$$\det(\hat{D}) \sim \left| \frac{\sin(\pi T \tau)}{\cosh(\pi T r)} \right|^{2N_f}.$$
(8.128)

This expression is maximal for $r = 0$ and $\tau = 1/2T$, which is the most symmetric orientation of the instanton–anti-instanton pair on the Matsubara torus. Since the fermion determinant controls the probability of any given configuration, we expect polarized molecules to become important at finite temperature. The effect should be largest, when the IA pairs exactly fit onto the torus, i.e. $4\rho \simeq \beta$. Using the zero temperature value $\rho \simeq 0.33$ fm, we get $T \simeq 150$ MeV, close to the expected transition temperature for two flavors. The picture of such molecules and the corresponding Dirac eigenvalue spectra was shown in Fig. 2.6.

In general, the formation of molecules is controlled by the competition between minimum action, which favors correlations, and maximum entropy, which favors randomness. Determining the exact composition of the instanton liquid as well as the transition temperature requires the calculation of the full partition function, including the fermion induced correlations. We will do this using numerical simulations in Section 8.5.5.

Before we come to this we would like to study the physical effects caused by the presence of molecules. Qualitatively, it is clear why the formation of molecules leads to chiral symmetry restoration. If instantons are bound into pairs, then quarks can only jump from one instanton to the corresponding anti-instanton and back. All eigenstates are localized, and no quark condensate is formed. Another way to see this is by noting that the Dirac operator will essentially decompose into 2×2 blocks corresponding to to the instanton–anti-instanton pairs. This means that the eigenvalues will be concentrated around some typical $\pm\lambda$ determined by the average size of the pair, so the density of eigenvalues near $\lambda = 0$ goes to zero.

The existence of molecular configurations on the lattice have been reported in many papers. A decisive proof that they do indeed dominate the lowest Dirac eigenvalues at $T > T_c$ is still not there, unfortunately. The closest discussion of the lowest Dirac states in this case one can find in Ref. [562].

The effect of molecules on the effective interaction between quarks at high temperature was studied in Ref. [245]. The effective interaction between quarks is convenient to calculate by rearranging the exchange terms and into a direct interaction. The resulting Fierz symmetric Lagrangian reads [245]

$$
\mathcal{L}_{\text{mol sym}} = G \left(\frac{2}{N_c^2} \left[(\bar{\psi}\tau^a\psi)^2 - (\bar{\psi}\tau^a\gamma_5\psi)^2 \right] \right.
$$
$$
\left. - \frac{1}{2N_c^2} \left[(\bar{\psi}\tau^a\gamma_\mu\psi)^2 + (\bar{\psi}\tau^a\gamma_\mu\gamma_5\psi)^2 \right] + \frac{2}{N_c^2} (\bar{\psi}\gamma_\mu\gamma_5\psi)^2 \right) + \mathcal{L}_8,
$$

$$(8.129)$$

with the last complicated term containing color-octet $\bar{q}q$ pairs only. The coupling constant is now proportional to the density of molecules

$$
G = \int n(\rho_1, \rho_2) \, d\rho_1 d\rho_2 \frac{1}{8T_{IA}^2} (2\pi\rho_1)^2 (2\pi\rho_2)^2.
$$

$$(8.130)$$

Here, $n(\rho_1, \rho_2)$ is the tunneling probability for the IA pair and T_{IA} is the corresponding overlap matrix element. τ^a is a four-vector with components $(\vec{\tau}, 1)$, and thus this Lagrangian has both isoscalars and isovectors. The effective Lagrangian (8.129) was determined by averaging over all possible molecule orientations. Near the phase transition, molecules are polarized and all vector interactions are modified according to $(\bar{\psi}\gamma_\mu\Gamma\psi)^2 \to 4(\bar{\psi}\gamma_0\Gamma\psi)^2$.

Like the zero temperature effective Lagrangian, the interaction (8.129) is SU(2)$\times$ SU(2) symmetric. However, in addition to that it is also U(1)$_A$ symmetric. This is a consequence of the fact that molecules are topologically neutral. It does not preclude the possibility that there is still a small $O(m^{N_f})$ fraction of random instantons present above T_c leading to explicit $U(1)_A$ breaking. The effective interaction (8.129) is attractive in the pion channel, but because it is chirally and $U(1)_A$ symmetric, the same attraction also operates in the other scalar–pseudoscalar channels σ, δ and η'. Furthermore, unlike the 't Hooft interaction, the effective interaction in the molecular vacuum also produces an attractive interaction in the vector and axial vector channels. If molecules are unpolarized, the corresponding coupling constant is a factor 4 smaller than the scalar coupling. If they are fully polarized, only the longitudinal vector components are affected. In fact, the coupling constant is equal to the scalar coupling. A more detailed study of the quark interaction in the molecular vacuum was performed in Refs. [242, 245], where hadronic correlation functions in the spatial and temporal direction were calculated in the schematic model mentioned above. We will discuss the results in more detail below.

Instanton–anti-instanton molecules also play a decisive role in formation of deeply bound states of $\bar{q}q, qg, gg$ at $T > T_c$: see recent work by Brown, Lee, Rho and myself, hep-ph/0312175.

8.5.5. *Instanton ensemble in the phase transition region*

In this section we study statistical mechanics of the instanton liquid at finite temperature, especially the radical changes which take place at $T \approx T_c$.

Historically it was first done by extending the mean field approximation to finite temperatures. For pure gauge theory, the variational method was extended to finite temperature by Diakonov and Mirlin [555]. The gluonic interaction between instantons changes very little with temperature, and the only difference as compared to the zero temperature case is the appearance of the perturbative suppression factor (8.115) (at $T > T_c$, although in Ref. [555] it was used for all T). Since the interaction is unchanged, so is the form of the single instanton distribution, and all that is happening is that the ensemble gets more dilute at higher T. The situation is somewhat more interesting if one extends the variational method to full QCD [228, 231]. In this case, an additional temperature dependence enters through the T dependence of the average fermionic overlap matrix elements. More importantly, the average determinant depends on the instanton size,

$$\det(\hat{D}) = \prod_I (\rho m_{\det}), \quad m_{\det} = \rho^{3/2} \left[\frac{1}{2} I(T) \int dn_a(\rho)\rho \right]^{1/2}, \quad (8.131)$$

where $I(T)$ is the angle and distance averaged overlap matrix element T_{IA}. The additional ρ dependence modifies the instanton distribution and introduces an additional nonlinearity into the self-consistency equation. As a result, the instanton density at large T depends crucially on the number of flavors. For $N_f = 0, 1$, the density drops smoothly and N/V goes like $1/T^{2a}$ with $a = (b - 4 + 2N_f)/(2 - N_f)$ for large T. For $N_f = 2$, the instanton density becomes zero at some critical temperature T_c, whereas for $N_f > 2$, the density goes to zero discontinuously at the critical temperature. This behavior can be understood from the form of the gap equation for the quark condensate. We have $\langle \bar{q}q \rangle \sim \text{const} \langle \bar{q}q \rangle^{N_f - 1}$, which, for $N_f > 2$ cannot have a solution for arbitrarily small $\langle \bar{q}q \rangle$.

However the main assumption of the mean-field approximation remains that the instanton ensemble remains random (weakly correlated) at all temperatures. Then chiral restoration of course implies that there cannot be any instantons present above T_c, which is obviously impossible. The way out of this puzzle is that the mean field approximation must get broken at $T \sim T_c$, even if it is accurate at $T = 0$. *The chiral phase transition is not caused by the disappearance of instantons but their binding into $\bar{I}I$ molecules* (or larger clusters). Those clusters are precisely the correlations in the ensemble ignored by the mean field method.

In order to include these aspects into a variational calculation, Ilgenfritz and myself introduced the so called *cocktail model* [228, 557], in which the instanton ensemble consists of a random plus a molecular component. The composition of the instanton liquid is determined by minimizing the free energy with respect to the two concentrations. Although this model is analytical and not very complicated, let me skip it here and proceed directly to results of the numerical simulations of the interacting instanton liquid. In those, one does not have to make any assumptions about what the important configurations are (molecules, larger clusters, ...). Let us also discuss only the physically relevant case of two light and one intermediate mass flavor. It is straightforward to extend the simulations to different numbers of flavors and the resulting phase diagram will be discussed in more detail later in Section 12.2.

The finite temperature suppression factor (8.115) in the semi-classical instanton density is used only above the phase transition, not below. The resulting instanton density, free energy and quark condensate are shown in the left panel of Fig. 8.14. In the ratio ansatz the instanton density at zero temperature is given by $N/V = 0.69\Lambda^4$. Taking the density to be $1\,\text{fm}^{-4}$ at $T = 0$ fixes the scale parameter $\Lambda = 222\,\text{MeV}$ and determines the absolute units. The temperature dependence of the instanton density is shown in Fig. 8.14a. It shows a slight increase at small temperatures,[19] starts to drop around $115\,\text{MeV}$ and becomes very small for $T > 175\,\text{MeV}$. The free energy closely follows the behavior of the instanton density. This means that the instanton-induced pressure first increases slightly, but then drops and eventually vanishes at high temperature. This behavior is expected for a system of instantons, but in a complete theory with perturbative effects, the pressure should always increase as a function of the temperature.

The temperature dependence of the quark condensate is shown in Fig. 8.14c. At temperatures below $T = 100\,\text{MeV}$ it is practically temperature-independent. It then starts to drop fast and becomes very small around the critical temperature $T \simeq 140\,\text{MeV}$. Note that at this point the instanton density is $N/V = 0.6\,\text{fm}^{-4}$, slightly more than half the zero temperature value. This means that the phase

---

[19]The zero temperature point is not a true $T = 0$ calculation, but corresponds to a simulation with $T = 0$ matrix elements in a periodic box $V = (2.828\Lambda^{-1})^4$.

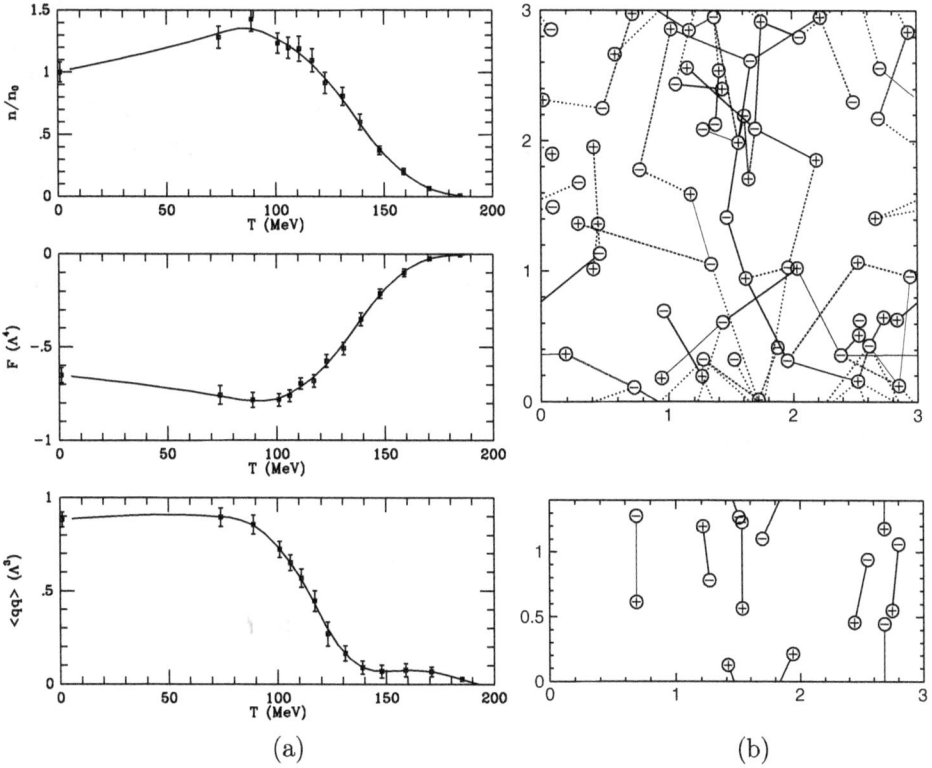

Fig. 8.14. (a) Instanton density, free energy and quark condensate as a function of the temperature in the instanton liquid with two light and one intermediate mass flavor, from Ref. [242]. The instanton density is given in units of the zero temperature value $n_0 = 1\,\text{fm}^{-4}$, while the free energy and the quark condensate are given in units of the Pauli–Vilars scale parameter, $\Lambda = 222\,\text{MeV}$. (b) Typical instanton ensembles for $T = 75$ and $158\,\text{MeV}$, from Ref. [242]. The plots show projections of a four-dimensional $(3.0\Lambda^{-1})^3 \times T^{-1}$ box into the 3–4 plane. The positions of instantons and anti-instantons are indicated by $+$ and $-$ symbols. Dashed, solid and thick solid lines correspond to fermionic overlap matrix elements $T_{IA}\Lambda > 0.40, 0.56, 0.64$, respectively.

transition is indeed caused by a transition within the instanton liquid, not by the disappearance of instantons. This point is illustrated in the right panel of Fig. 8.14, which shows projections of the instanton liquid at temperatures $T = 74, 158\,\text{MeV}$, below and above the chiral phase transition. The figures are projections of a four-dimensional cube $V = (3.00\Lambda^{-1})^3 \cdot 1/T$ into the 3–4 plane. The positions of instantons and anti-instantons are denoted by $+/-$ symbols. The lines connecting them indicate the strength of the fermionic overlap ("hopping") matrix elements. Below the phase transition, there is no clear pattern. Instantons are unpaired, part of molecules or larger clusters. As the phase transition progresses, one clearly observes the formation of polarized instanton–anti-instanton molecules.

Completing this section, let us add a few general remarks concerning the thermodynamics of the instanton liquid. In this section, we have only determined the free energy associated with instantons. Our calculation includes the contributions from the zero mode determinant, which describes the excitation of low energy collective modes (the pions). However, it does not include non-zero mode contributions from quarks or gluons. In particular, it does not include the Stefan–Boltzmann contribution at large temperature, or perturbative $O(\alpha_s)$ corrections to it. In a more realistic description of the thermodynamics of the chiral phase transition, these contributions should certainly be included (like in the schematic model discussed in the last section). Here, it was our intention to study the physics contained in the partition function of the instanton liquid, without making further assumptions or inputs. Remarkably, the instanton liquid undergoes a phase transition in the expected temperature regime even if no additional ingredients are added.

8.5.6. *Critical behavior in the instanton liquid*

We have already mentioned that for two massless flavors QCD is expected to have a second order phase transition with $O(4)$ critical indices. If the strange mass is sufficiently heavy, the phase transition in real QCD should still be governed by the same effective theory. Before we study this question in the instanton liquid, we want to review some of the universality predictions. The mass dependence of the condensate at the phase transition is governed by the critical index δ

$$\langle \bar{q}q \rangle|_{T=T_{\max}} \sim m^{1/\delta}, \tag{8.132}$$

where $T_{\max}$ denotes the pseudo-critical temperature, corresponding to the position of the peak in the scalar susceptibility. The $O(4)$ Heisenberg magnet has $1/\delta = 0.21$ while mean field scaling would give $1/\delta = 1/3$. Using the Banks–Casher relation

$$\langle \bar{q}q(T) \rangle = -\int d\lambda \rho(\lambda, m, T) \frac{2m}{\lambda^2 + m^2}, \tag{8.133}$$

one can try to convert this relation into a prediction for the spectral density of the Dirac operator. The problem is that there are two sources for the mass dependence of the condensate, the explicit mass term in the integral (the "valence mass") and the implicit mass dependence of the spectral density (the "sea mass"). Combining these two effects, one can only conclude that

$$\rho(\lambda, T = T_{\max}) \sim \lambda^{1/\delta - \kappa} m^{\kappa}, \tag{8.134}$$

where κ characterizes the explicit mass dependence of the spectral density. Nevertheless, it has been argued [230] that the influence of the sea mass might be small and that information on the critical behavior can be obtained from the spectral density near zero at the phase transition. The spectrum does not look like a fractional exponent, but consists of two parts, a smooth part behaving like $\rho(\lambda) \sim \lambda^2$ and a spike near zero. We already mentioned that this spike is important in connection with U(1)$_A$ breaking. In Ref. [559], we also studied the distribution of eigenvalues for $N_f = 2$ and somewhat smaller quark masses. In this case the spectrum near T_c is consistent with a fractional exponent, $\rho(\lambda) \sim \lambda^{0.3}$.

Additional information about critical phenomena is provided by mesonic susceptibilities. The susceptibilities are defined as the integral of the corresponding mesonic correlation function

$$\chi_\Gamma = \int d^4x \langle \bar{q}(x)\Gamma q(x)\bar{q}(0)\Gamma q(0)\rangle, \tag{8.135}$$

where Γ is a spin–isospin matrix with the appropriate quantum numbers. They characterize the response of the system to slowly varying external perturbations. For example, the scalar–isoscalar susceptibility can also be defined as the second derivative of $\log Z$ with respect to the quark mass. Near a second order phase transition, the susceptibilities associated with order parameter fluctuations are expected to diverge in the chiral limit in a universal manner. In particular, we have the predictions by Wilczek [524]

$$\chi_\sigma|_{T=T_{\max}} \sim m^{1/\delta-1}, \qquad \chi_\sigma/\chi_\pi|_{T=T_c} = 1/\delta. \tag{8.136}$$

Above the phase transition, chiral symmetry is restored and one has $\chi_\sigma = \chi_\pi$ as the quark mass is taken to zero. Below the phase transition, $\chi_\sigma/\chi_\pi \to 0$ as $m \to 0$, since χ_π has a $1/m$ singularity in the chiral limit, while χ_σ only has a logarithmic singularity.

The susceptibilities associated with flavored mesons like the pion and δ meson are directly related to the spectrum of the Dirac operator. Inserting the general decomposition of the quark propagator in terms of eigenfunctions into the definition (8.135), one finds

$$\chi_\pi = 2 \int d\lambda\, \rho(\lambda, m, T)\frac{1}{\lambda^2 + m^2}, \tag{8.137}$$

$$\chi_\delta = 2 \int d\lambda\, \rho(\lambda, m, T)\frac{\lambda^2 - m^2}{(\lambda^2 + m^2)^2}. \tag{8.138}$$

Comparing (8.137) with the Banks–Casher relation (8.133) one notes that $\chi_\pi = -\langle \bar{q}q\rangle/m$, a relation that can also be obtained by saturating the pion correlator with one-pion intermediate states and using PCAC. The susceptibility in the scalar–isoscalar (σ meson) channel receives an additional contribution from disconnected diagrams, which cannot be expressed in terms of the spectral density

$$\chi_\sigma = \chi_\delta + 2V(\langle(\bar{q}q)^2\rangle - \langle\bar{q}q\rangle^2). \tag{8.139}$$

The second term measures the fluctuations of the quark condensate in a finite volume V (as the volume is taken to infinity).

If the $U(1)_A$ symmetry is restored during the transition, one expects (for $N_f = 2$) two additional degrees of freedom to become light at T_c, the η' and δ. The η' susceptibility is connected with fluctuations of the number of zero modes of the Dirac operator and cannot be measured directly in a system with total topological charge zero.[20] The δ susceptibility, however, can easily be studied in our simulations. If $U(1)_A$ symmetry is restored, we expect χ_δ to peak at the transition and diverge as the quark mass is taken to zero. Furthermore, above the transition one would have $\chi_\pi = \chi_\delta$ as $m \to 0$. This also requires that the disconnected part of χ_σ has to vanish for $T > T_c$ as $m \to 0$.

The mesonic susceptibilities clearly show a peak in the sigma susceptibility, indicating the chiral phase transition. The delta susceptibility does not show any pronounced enhancement. This is also seen from the lower part of the figure, where we show the disconnected part of the scalar susceptibility and the difference $\chi_\pi - \chi_\delta$. Clearly, the peak in χ_σ comes completely from the disconnected part. Furthermore, $\chi_\pi - \chi_\delta$ becomes very small above the phase transition, but shows no tendency to go to zero in the range of temperatures studied. In Ref. [559], we also studied the susceptibilities at somewhat smaller masses. In this case, there is a broad enhancement in χ_δ, but there are also some signs that finite size effects become important. We also extended our calculations to larger masses $m_u = m_d = (0.10 - 0.20)\Lambda$. The dependence of χ_σ on m is well described by a single exponent. Fitting the dependence of the peak height on the light quark mass, we find $\chi_{dis,max} \sim m^{-0.84}$, quite consistent with the ($N_f = 2$) universality prediction $\chi_{\sigma,max} \sim m^{-0.79}$.

8.6. Hadronic correlation functions at finite temperature

Clearly, the behavior of hadronic correlation functions is of great interest in connection with possible modifications of hadrons in hot and dense matter. If the Operator Product Expansion (OPE) can be used as a guide, for most correlation functions the dominant power corrections are determined by the quark condensate. So, one may expect a tendency for resonance masses and continuum thresholds to decrease, as chiral symmetry is restored. Our discussion above of static potentials and the heavy–light hadron masses is a good example that this indeed happens.

At finite temperature, Lorenz invariance is broken, and the correlation functions in the spatial and temporal direction are no longer the same. Since the former is not limited, the correlators have been studied in spatial direction first: their exponents are known as *screening masses*. While these states are of theoretical interest and have been studied in a number of lattice calculations, they do not correspond to poles of the spectral function in energy. In order to look for real bound states, one has to study temporal correlation functions. However, at finite temperature the periodic boundary conditions restrict the useful range of temporal correlators to the interval $\tau < 1/(2T)$ (about 0.6 fm at $T = T_c$), so that there is no direct procedure to extract information about the ground state. The underlying physical reason is clear: at finite temperature excitations are always present. In addition to that, mesonic

---

[20]Instead, one can study the topological susceptibility by considering fluctuations of the topological charge in sub-volumes of a large system with total charge zero [243].

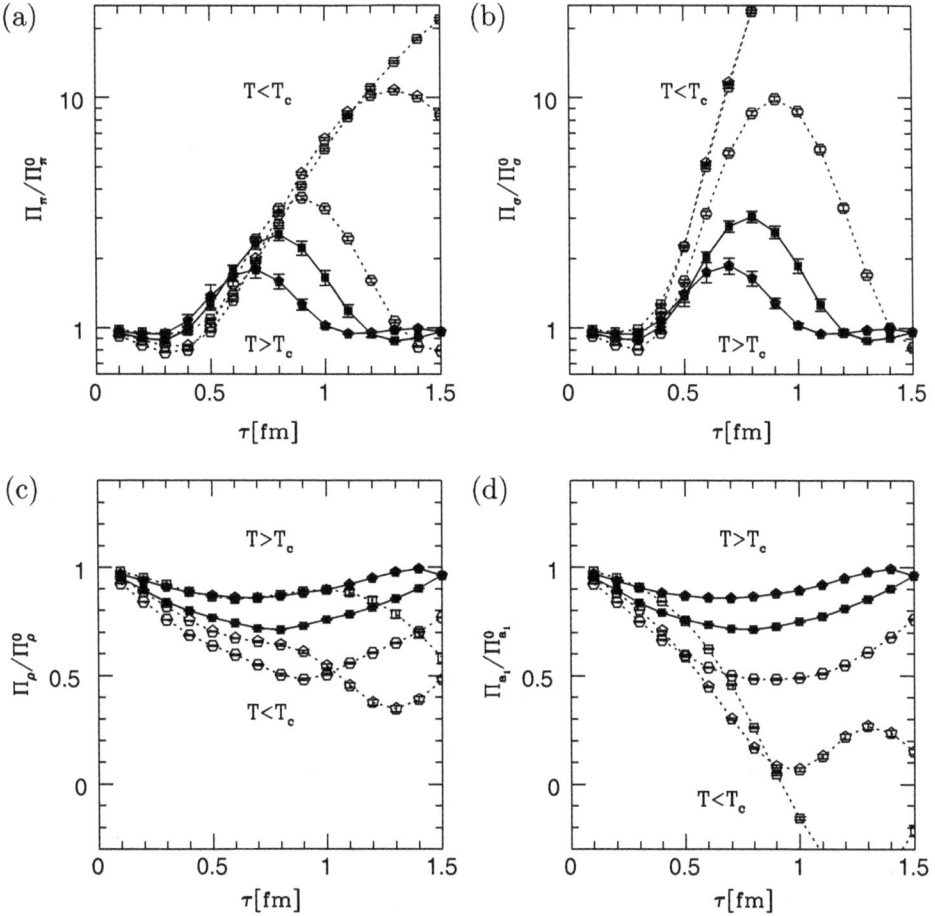

Fig. 8.15. Temporal correlation functions at $T \neq 0$, normalized to free thermal correlators. (a) The pseudo-scalar (pion) correlator, (b) the isoscalar scalar σ, (c) the isovector vector (ρ), and (d) the axial vector (a_1). Correlators in the chirally symmetric phase ($T \geq T_c$) are shown as solid points, below T_c as open points. The open triangles, squares and hexagons correspond to $T = 0.43$, 0.60 and $0.86 \, T_c$, while the closed triangles and squares show the data at $T = 1.00 \, T_c$ and $1.13 \, T_c$.

and baryonic correlation functions have to obey periodic or anti-periodic boundary conditions, respectively, in the temporal direction.

8.6.1. *Screening masses*

Correlation functions in the spatial direction can be studied at arbitrarily large distance, even at finite temperature. This means that contrary to the temporal correlators, the corresponding spectrum can be determined with good accuracy. The first papers such as that by DeTar and Kogut [563] have found that for the ρ meson the mass and the wave function is not changed significantly, as one crosses from hadronic phase to QGP.

However it has been pointed out by Koch, Brown, Jackson and myself [564] that this observation can be easily accounted for theoretically. At high T the Euclidean space is limited in the fourth direction by the Matsubara time, while other three are unrestricted. Let us rename the axes, calling the direction z in which the correlator is measured the new time, while calling the former time the new z. What we now have is a zero-T theory in a $1d$ box, with its size decreasing as $1/T$ as T grows.

One may think that at high T we simply will get a $(2+1)$-dimensional theory, but it is true only subject to an important additional observation. At finite temperature, the fermions should obey the *anti-periodic* boundary conditions in the temporal direction, which requires the lowest Matsubara frequency for fermions to be πT. This energy acts like a mass term for propagation in the spatial direction, so in a new $(2+1)$-dimensional theory quarks effectively are massive. At asymptotically large temperatures, quarks are very heavy (only propagate in the lowest Matsubara mode) and the spectrum of the screening masses is then determined by a quarkonium-like theory of heavy (chiral) mass πT, interacting via the 2-dimensional Coulomb law and the non-vanishing space-like string tension. Dimensional reduction at large T predicts almost degenerate multiplets of mesons and baryons with screening masses close to $2\pi T$ and $3\pi T$. The splittings of mesons and baryons with different spin can be understood in terms of the non-relativistically reduced spin–spin interaction. The resulting pattern of screening states is in qualitative agreement with lattice results even at moderate temperatures $T \simeq 1.5T_c$.

The most notable exception is the pion, whose screening mass is significantly below $2\pi T$. This deviation shows that non-perturbative effects do not disappear in the QGP phase.

Exercise 8.10.   Calculate spectra and spin splitting of high-T screening masses, using the $2+1$ effective theory and a purely perturbative (one-gluon-exchange) interaction.

Naturally we would like to study whether the spectrum of screening states can also be understood in the instanton liquid model. One should note that dimensional reduction does not naturally occur in the instanton model. Instantons have fermionic zero modes at arbitrarily large temperature, and the perturbative Coulomb and spin–spin interactions mentioned above are not included in our model.

The screening masses are determined from the exponential decay of the space-like correlators, $\Pi(r) \overset{r\to\infty}{\longrightarrow} \exp(-m_{\text{scr}}r)$. Our results are summarized in Fig. 8.16. First of all, the screening masses clearly show the restoration of chiral symmetry as $T \to T_c$: chiral partners like the π and σ or the ρ and a_1 become degenerate. Furthermore, the mesonic screening masses are close to $2\pi T$ above T_c, while the baryonic ones are fairly close to $3\pi T$, as expected. Most of the screening masses are shifted slightly upward as compared to the most naive prediction. Considering the vector channels ρ, a_1, Δ, this shift corresponds to a residual "chiral quark mass" on the order of 120–140 MeV.

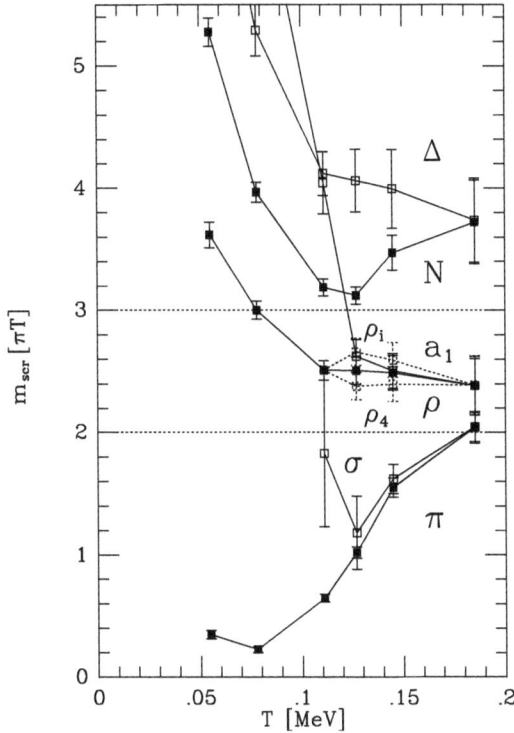

Fig. 8.16. Spectrum of space-like screening masses in the instanton liquid as a function of the temperature, from Ref. [242]. The masses are given in units of the lowest fermionic Matsubara frequency πT. The dotted lines correspond to the screening masses $m = 2\pi T$ and $3\pi T$ for mesons and baryons in the limit when quarks are non-interacting.

The most striking observation is the strong deviation from this pattern seen in the scalar channels π and σ, with screening masses significantly below $2\pi T$ near the chiral phase transition. This effect persists to fairly large temperature. We also find that the nucleon–Delta splitting does not disappear at the phase transition, but decreases smoothly. Note that in the limit of a dilute system of fully polarized molecules, ρ_4 is expected to be degenerate with the pion [245]. The presence of unpaired instantons, interactions among the molecules and deviations from complete polarization cause the pion to be significantly lighter than the longitudinal rho.

8.6.2. *Temporal correlation functions*

Analogous to our procedure at $T = 0$, we normalize all correlation functions to the corresponding noninteracting correlators, calculated from the free $T \neq 0$ propagator. The corresponding resonance parameters are extracted from a simple pole plus continuum parameterization. In the case of a scalar meson, it reads

$$\Pi_\Gamma(\tau) = \lambda_\Gamma^2 D(T, m_\Gamma, \tau) + \text{Tr}[\Gamma S^V(T, m, \tau)\Gamma S^V(T, m, -\tau)], \qquad (8.140)$$

where $D(T, m, \tau)$ is the finite temperature massive boson propagator and $S^V(T, m, \tau)$ is the Dirac vector part of the finite T massive fermion propagator. The second term corresponds to the contribution of massive constituent quarks at finite temperature. As $m \to 0$, the continuum threshold moves down to zero energy.

Finite temperature mesonic temporal correlation functions are shown in Figs. 8.15. In Figs. 8.15(a,b) we show the pion and sigma correlators for various temperatures. Here and in the following figures, open points correspond to temperatures below T_c, while the solid points show results near and above T_c. The π and σ correlators are larger than the perturbative one at all temperatures, implying that the interaction is attractive even above T_c. The peak height decreases strongly with T, but by a too large part; this is a simple consequence of the fact that the length of the temporal direction shrinks. Below T_c, the disconnected part of the sigma correlation functions tends to the square of the quark condensate at large distance. Above T_c, chiral symmetry is restored and the σ and π correlation functions become equal.

Vector and axial vector correlation functions are shown in Figs. 8.15(c,d). At low T the two are very different while above T_c they become indistinguishable, again in accordance with chiral symmetry restoration. In the vector channel, the changes in the correlation function indicate the "melting" of the resonance contribution. At low T (e.g. $T = 0.43T_c$, shown by the open triangles) one can see a peak in the correlation function at $x \simeq 1$ fm, indicating the presence of a bound state well separated from the two-quark (or, more realistically, two-pion continuum) continuum. However, this signal disappears at $T \sim 100$ MeV, implying that the ρ meson coupling to local current becomes small. This is consistent with the idea that the resonance "swells" (or completely dissolves) in hot and dense matter. A similar phenomenon is also observed in the Delta baryon channel (not shown).

The dominant effect at small temperature is mixing between the vector and axial-vector channels [617]. This means, in particular, that there is a pion contribution to the vector correlator at finite T. This contribution is most easily observed in the longitudinal vector channel $\Pi_{44}^V(\tau)$ (note that in Fig. 8.15 we show the trace $\Pi_{\mu\mu}^V(\tau)$ of the vector correlator). We find a sizable enhancement in this channel at $T = 77$ MeV, but the effect disappears at larger temperatures $T > 110$ MeV.

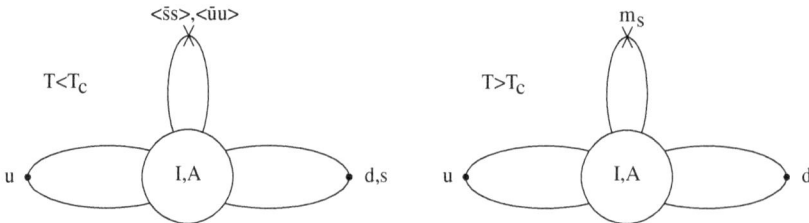

Fig. 8.17. Leading contributions to flavor mixing in the η–η' system below (left) and above (right) the chiral phase transition.

For $T > 110\,\mathrm{MeV}$ both the ρ and Δ correlation functions are well described by a continuum contribution only, modeled by the propagation of two or three massive quarks. The T dependence of this effective "chiral" mass was studied in more detail in Ref. [559], with the conclusions that it does not disappear at T_c, but decreases smoothly at high temperature.

Temporal correlation functions have been calculated on the lattice. Their understanding was significantly improved with the help of the so called Maximal Entropy Method (MEM) which provides evaluation of the relevant spectral densities. It has been found, by Asakawa and Hatsuda [hep-lat/0309001] and Petretczky *et al.* [hep-ph/0305189], that the lowest charmonium states, η_c and J/ψ, survive as a bound state till $T \approx 2T_c$. Zahed and myself, hep-ph/0307267, explained it by sufficiently strong coupling constant. They also introduced the so called "zero binding lines" for other binary states, colorless or colored, of $\bar{q}q, qg, gg$, and argued that those go deep into the QGP domain, up to $T \approx 4T_c$ for scalar glueball. If so, there are hundreds of bound states in "strongly coupled QGP", a quite different picture from an asymptotic gas of quasiparticles.

Summarizing this section we find that most chiral even correlation functions (in the restricted range in which they are accessible at finite temperature) are remarkably stable as a function of temperature, despite the fact that the vacuum fields and the quark condensate change significantly. Phenomenologically this means that the melting of resonances is compensated by the continuum threshold moving down in energy. Roughly speaking, we find three different types of behavior in chiral even correlators. In "attractive" channels, like the π and σ meson or the nucleon, resonance contributions seem to survive the phase transition. In channels that do not have strong interactions at $T = 0$ (like the ρ meson and the Δ), the resonances disappear quickly, and the correlators can be described in terms of free quark propagation with a certain effective chiral mass.

8.6.3.   U(1)$_\mathrm{A}$ *breaking at high T*

In this section we would like to concentrate on correlation functions related to the fate of U(1)$_\mathrm{A}$ symmetry at finite temperature. Here we consider the δ and η_{ns} channels, the U(1)$_\mathrm{A}$ partners of π and σ. The temperature dependence of these correlation functions is shown in Figs. 8.18(a,b). We remind the reader that the amount of repulsion in the δ and η_{ns} was too strong in the random model (dashed lines), but the behavior is improved in the interacting model. In contrast to the cases considered above in which correlation functions at finite temperature were reduced in magnitude, now one observes the opposite trend. As the temperature increases, the amount of repulsion is clearly reduced. This effect is more pronounced in the δ than in the η_{ns} channel, partly because the $T = 0$ ensemble is over-correlated and produces an η_{ns} mass that is already too light.

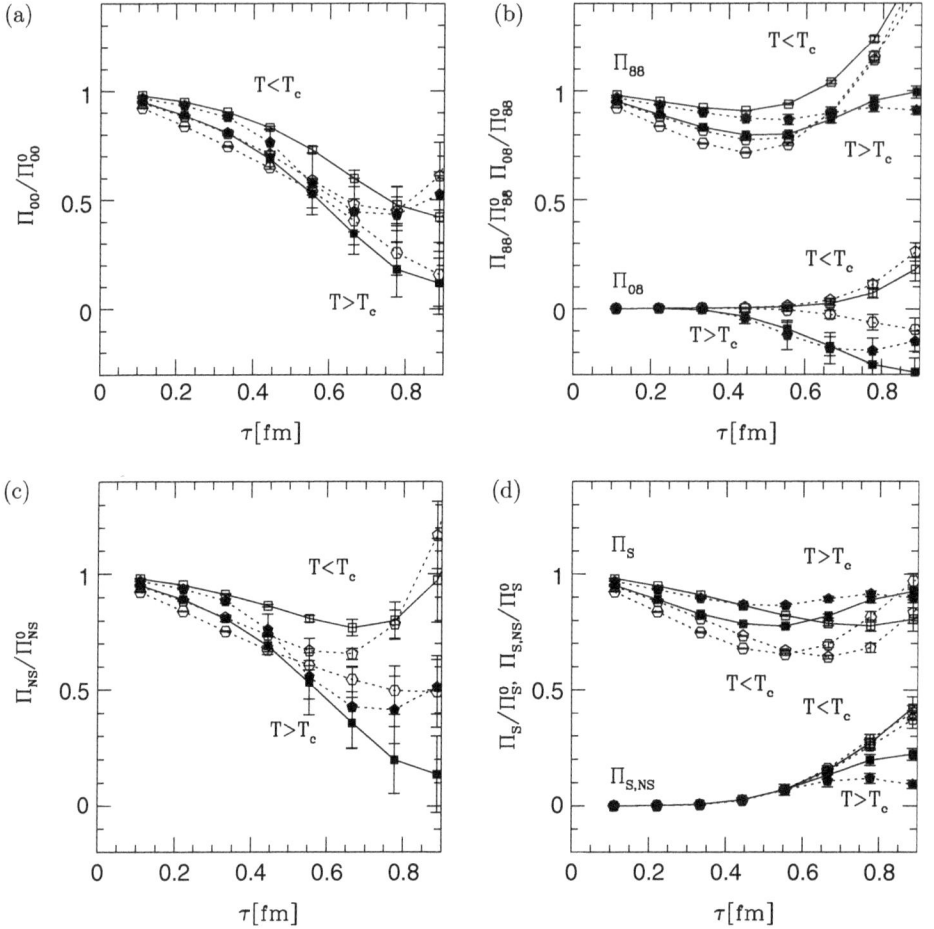

Fig. 8.18. Eta meson correlation functions at $T \neq 0$, normalized to free thermal correlators. Figure (a) shows the flavor singlet, (b) the octet and off-diagonal singlet-octet, (c) the non-strange and (d) the strange and off-diagonal strange-non-strange eta correlation functions. The correlators are labeled as in Fig. 8.15.

The question of $U(1)_A$ restoration at $T = T_c$ (or above) can be studied by comparing the η and δ correlators with their $U(1)_A$ partners, the π and σ. A serious complication for this comparison is the strong dependence of the η and δ correlators on the current quark mass. In Fig. 8.18d we show the δ correlator at fixed temperature $T = 0.86T_c$ for a number of different valence quark masses $m = 5.2, 8.6, 13, 26\,\text{MeV}$. For a very small quark mass $m = 5.2\,\text{MeV}$ (open triangles), there is a significant enhancement in the δ channel, which is indeed equal to that in the pion channel. This would imply chiral and $U(1)_A$ restoration. However, this is not a physical result, but a finite size effect. If the quark masses in a system with finite volume become too small, chiral (and $U(1)_A$) symmetry breaking is lost.

In order to be more quantitative one would like to determine the precise masses and coupling constants of the η_{ns} and δ. Again, given the restricted information available and the uncertainty concerning the form of the spectrum, it is hard to obtain reliable results for these parameters. In practice, we need to make some assumptions. Here, we first fix the non-resonant continuum, assuming it to be given by free propagation of quarks with the effective masses determined above. In the temperature region of interest, $T \geq 100\,\text{MeV}$, the pion and delta masses are roughly consistent with zero $m_\pi \simeq m_\delta \simeq 0 \pm 200\,\text{MeV}$. In the following we therefore fix the masses to be zero and study the temperature dependence of the coupling constants $\lambda_{\pi,\delta}$. In Fig. 8.18c we compare the coupling constants for the pion (stars) and Delta. At $T = 130\,\text{MeV}$ we show the value of λ_δ for the different quark masses discussed above. For the lowest mass $m = 5.2\,\text{MeV}$, the $U(1)_A$ symmetry is restored even below T_c, but this is definitely a finite size effect. For a somewhat larger mass $m = 8.6\,\text{MeV}$, $U(1)_A$ symmetry appears to be restored at or just above T_c. At even larger current quark masses, no evidence for $U(1)_A$ restoration is seen. To settle this important question will require significantly larger simulations.

Possible experimental signatures of (partial) $U(1)_A$ chiral symmetry restoration were discussed in Ref. [525] and, more recently, in Ref. [619]. Basically, the idea is that if the η_{ns} becomes degenerate with the π at $T \simeq T_c$, one would expect a significant (low p_T) enhancement of η and η' production in relativistic heavy ion collisions. There are experimental hints: e.g. the WA80 collaboration at CERN found a (modest) enhancement of the η/π^0 ratio in central relative to peripheral (or pp) collisions [722]. Furthermore, the enhancement of the low-mass dilepton production observed by the CERES [723] and Helios [724] collaborations could be explained in terms of the Dalitz decays of (more abundant) η' mesons [619]. To study these suggestions more quantitatively would not only require a better determination of the η_{ns} mass, but also an understanding of η–η' mixing and η' absorption as a function of temperature.

CHAPTER 9

Excited Hadronic Matter in Heavy Ion Collisions

9.1. Introduction

9.1.1. *Toward the macroscopic limit*

After we have considered properties of highly excited hadronic matter theoretically in the previous chapter, in this one we are going to review those which have been already observed in real-time experiments. But before we come to specifics, we have to answer a more general question: *Is hadronic matter really produced in heavy ion collisions?* Or, to put it more quantitatively, how do we know that our heavy ions are heavy enough, and the high energy is sufficiently high? What does it mean, to produce new form of matter? What is really new and different from what is going on during "elementary" pp collisions, at comparable energies?

By using heavy ion collisions with many more secondaries produced per event (several thousands, at RHIC), we study a *qualitatively different dynamical regime*, characterized by the opposite relation between the microscopic scale l (e.g. mean free path) as compared to the macro scale L (the system size). The excited system produced in heavy ion collisions is indeed large enough to be treated as a macroscopic body, since $L \gg l$, while for that created in elementary pp reactions the opposite is the case. As a result, the former are collective "bangs" while the latter are rather *collision-less fireworks* of debris created in a collision. (For example, deep-inelastic lepton–nucleon scattering results in a knocked out single quark, evolving into a jet of hadrons as it flies away from the rest of the struck nucleon and decay in free space.)

How do we know that this is indeed the case? One possible approach is to compare the observed behavior of the produced fireball to predictions which follow from this assumption. If the system is macroscopic then its description via *thermo-dynamics* of its static properties (like matter composition) and *hydrodynamics* for space–time dynamic evolution should work. The other approach would be to understand the microscopic dynamics, evaluate all the necessary microscopic quantities and then compare it to the macroscopic scales of the problem, the system's size or the fireball lifetime.

As the reader will see from the material below, at this time we have tried the former approach and have seen that it basically works well, but are still far from being able to carry out the latter one. We only have indirect phenomenological hints on what actually the microscopic scales of QGP l really are, from corrections to ideal macroscopic approach by effects like viscosity.

One argument in favor of the macroscopic approach is that the observed particle ratios are remarkably close to the equilibrium ones in a resonance gas of hadrons at the critical temperature, $T \approx 165$–$170\,\mathrm{MeV}$ [622–625]. This suggests that the system is in a state close to thermal equilibrium, as it crosses the phase transition boundary from QGP to the hadronic phase and the particle composition is determined.

The convincing power of this argument is however undermined by the observed fact that nearly the same thermal description reproduces particle ratios in proton–proton and e^+e^- collisions as well (strangeness is different). Obviously in those cases the system size is not large enough to reach the thermal equilibrium in the usual sense. The success of thermodynamics seems to reflect some specific dynamics of the hadronization we do not yet understand[1] rather than the macroscopic nature of the system.

Therefore, it was important to test whether the elementary proton–proton and e^+e^- collisions *do or do not* exhibit also collective hydrodynamic behavior. An analysis of the early hadronic spectra obtained at CERN ISR from pp collisions performed by Zhirov and myself [603] concluded that little or no sign of the transverse expansion, predicted by hydrodynamics, is seen. Thus, the apparent lack of collectivity shows that multi-body excited systems produced in elementary collisions are *not* macroscopic.

For heavy ion collisions, on the other hand, we do have an impressive set of evidence of a macroscopic behavior. The available evidence includes a number of "*flows*", various forms of a collective expansion. Those are observed at all energies, and most recent data from RHIC have shown it even in a most spectacular way. The accuracy of its hydrodynamical description is getting much better at the highest energies.

In Pb–Pb (SPS) or Au–Au (RHIC) collisions the matter emerges from the collision with a collective transverse velocity of up to $0.7\,c$. This radial flow is firmly established from a combined analysis of particle spectra, HBT correlations, a deuteron coalescence and other observables. Furthermore, in non-central collisions, the flow develops an ellipticity quantified by v_2, the asymmetry of the angular distribution

$$v_2 = \langle \cos(2\,\phi) \rangle, \tag{9.1}$$

---

[1] It is often emphasized that thermodynamics is "nothing else but the phase space" and thus is somehow trivial. I strongly disagree: any reaction has a phase space, and the thermal description means that the matrix element is very specific, it does not depend (or depends in a very complicated way averaged to a constant) on momenta but it does depend on, say, particle multiplicity. After all, the observed T is a dimensional quantity which must have a certain dynamical origin.

where ϕ is measured around the beam axis with respect to the impact parameter direction $\vec{b}$. Radial and elliptic flow data are measured as a function of transverse momenta, particle type, impact parameter, and collision energy. This wealth of momentum correlations severely constrains viable models of the heavy ion collision; we will discuss those in detail.

Another way to demonstrate the radically different behavior of elementary pp and heavy ion collisions is revealed by the different character of *event-by-event fluctuations* to be discussed at the end of this chapter. Heavy ion collisions show only rather small fluctuations, $\delta X / X \sim 1/\sqrt{N}$ (with N being the observed multiplicity), of the order of a few percent. Those are crudely consistent with the thermodynamic or statistical fluctuations in the resonance gas. Moreover, as first shown by the CERN NA49 [701] experiment, nearly all quantities have event-by-event distributions in quite a nice narrow Gaussian, valid for several orders of magnitude without any visible "tails". So, *all heavy ion events are about the same*, provided some global quantities like centrality of the collision are fixed. They are drastically different from the elementary pp collisions, which exhibit unexpectedly strong fluctuations (to be discussed in Section 9.7.1).

Let me now briefly characterize the main microscopic models developed over the years to explain the available heavy-ion data. The microscopic model widely used before the RHIC era was mostly based on the Lund-type string model, which is quantified with the RQMD, UrQMD and other (more jet-related) event generators. In those models in the early stages a set of nearly independent strings is produced by through-going partons, which then independently break into hadrons. Only then do hadrons re-scatter and interact, producing some collectivity. However the RHIC data ruled out those models, as longitudinally stretched strings generate too small a transverse pressure at early times, and completely miss the early explosion observed at RHIC.

The other set of models includes various descendants of the original parton model and pQCD-based ideas. About the simplest of those is the HIJING event generator [768] which combines the parton model (in a leading or next-to-leading order, with gluon bremsstrahlung) and string fragmentation approaches with nuclear geometry. There are also other versions of "parton cascades". Previous analysis of pp data has led to optimistic ideas that the parton model can be extrapolated down to a rather small scale, e.g. $\sim 1\,\mathrm{GeV}$. If this would be true for Au–Au collisions at RHIC, there would be relatively small mini-jet multiplicity of $dN^g/dy \sim 200$ for central collisions, with essentially *no* collective effects. This strongly contradicts all the RHIC data.

So we did learn that whatever the interaction may be at such a scale as a few GeV, it is not just pQCD. Therefore practitioners of the "parton cascade models" strongly enhanced the cross sections of parton rescattering, relative to pQCD, so as to describe data. However so far no dynamical explanation of any possible dynamics which might be responsible for that phenomenon is known. Perhaps the strong coupling limit (to be discussed in Section 12.5) can offer an explanation in the future.

What we will, however, show below, is that these modified parton cascades are no different from hydrodynamics. In fact, with their enhanced cross sections just mentioned, they do predict small viscous corrections to it.

9.1.2. *"Little Bang" versus Big Bang*

Before we go to specifics of heavy ion physics, let me make a more general introduction or perhaps a diversion, emphasizing multiple amusing analogies between the heavy ion physics and the Big Bang cosmology.

First of all, the same hot/dense hadronic matter, which heavy ion physics is trying to produce in the laboratory, was present at an appropriate time (microseconds) after the Bing Bang. More specifically, in both cases the matter goes through the QCD phase transition, from the *quark–gluon plasma* we discussed in the preceding chapter to the hadronic phase. On the phase diagram shown in Fig. 9.1a the Big Bang path proceeds along the T axis downward, since the baryonic density in it is tiny, $n_B/T^3 < 10^{-9}$. For heavy ion collisions the baryonic charge is small but still important, and in the sequel we will discuss (see Fig. 9.1) the location and shape of the paths on the phase diagram corresponding to various heavy ion experiments.

The fireball created in the heavy ion collision of course explodes, as its high pressure cannot be contained; this is the *Little Bang* we referred to in the title of this section. The *Big Bang* is a cosmological explosion, which proceeds against the pull of gravity. So, the Little Bang is a laboratory simulation of its particular stage, and thus the first obvious similarities between the "Little Bangs" and the cosmological "Big Bang" is that both are violent explosions.

Expansion of the created hadronic fireball approximately follows the same Hubble law as its bigger relative, $v(r) = Hr$ with $H(t)$ being some time-dependent parameter, although the Little Bang has rather anisotropic (tensorial) H. However, by the end of the expansion the anisotropy is nearly absent and local expansion at freezeout at RHIC is nearly Hubble-like.

The *final* velocities of collective motion in the Little Bang are measured in the spectra of secondaries: the transverse velocities now are believed to be reasonably well known, and are not small, reaching about $0.7\,c$ at RHIC. For the Big Bang there is no end and it proceeds till today; the current value of the Hubble constant, $H(t_{\rm now})$ is, after significant controversies for years, believed to be reasonably well measured.

However already the next important question one would obviously ask, "*how exactly are such expansion rates achieved*", remains a matter of hot debate in both cases. The observed velocity of matter expansion at freezeout of the collisions or in Big Bang today is determined by the earlier acceleration, determined by the Equation of State (EoS) of the matter.

For the Big Bang the EoS includes gravity of all matter forms, including the still mysterious *dark matter*, as well as the (really shocking) *dark energy* (or the *cosmological constant*). Recent experimental observations of supernovae in very distant galaxies have concluded that the Bing Bang is *accelerating* at this time due to it, but it was decelerating earlier.

Amusingly, experts in the Little Bang have to deal with the cosmological constant as well. Of course, gravity is not important here, but going from QGP to hadronic matter one has to think about a definition of the zero energy density and pressure, or the so called bag constant between the two phases. As we will see, the cosmological term exists in this case, but it has the opposite sign and tries to *decelerate* the expansion of QGP. The EoS is thus effectively very soft near the QCD phase transition, and the magnitude of the flow observed at AGS/SPS energies is not that large. With RHIC we had the first chance to go well beyond into the QGP domain, with a harder EoS $p \approx \epsilon/3$ and a more robust expansion.

The hydro models with the expected EoS have predicted the observed spectra very well (we will be discussing them later in detail in this chapter). However interferometric studies of the absolute size of the radiation region (see Section 9.5) have shown that they are somewhat smaller than expected. It means that the exact acceleration history of the Little Bang has to be studied still better.

In a Little Bangs the observed hadrons (like microwave cosmic photons) are seen at the moment of their last interaction, or as we call it technically, at their *freeze-out stage*. In order to look deeper, one uses rare hadron particles with smaller cross sections, such as Ω^- hyperons, which decouple earlier, or even *penetrating probes* [457,709] leptons or photons which penetrate through the whole system. We return to their discussion in the next chapter.

The next comparison I would like to make here deals with the issue of fluctuations. Very impressive measurements of the microwave background anisotropy made in the last decade have taught us a lot about cosmological parameters. First the dipole component was found — motion of the solar system relative to the microwave heat bath, and then very small ($\delta T/T \sim 10^{-5}$) chaotic fluctuations of T originated from plasma oscillations at the photon freeze-out. It has been possible lately to measure some interesting structures in fluctuations of the cosmic microwave background, with angular momenta $l \sim 200$. The theoretical predictions for these have been available since they are related to the primordial plasma-to-gas transition at temperature $T \sim 1/3$ eV. Primordial fluctuations of sufficiently long wavelength are attenuated by the gravitational instability, until they reach the stabilization moment $\tau_{\text{stabilization}}$ when the instability is changed to a regime called *Sakharov acoustic oscillations*. Hydrodynamics tells us that fluctuations disperse during this period with the (current) speed of sound, reaching the so called *sound horizon* scale

$$r_{\text{s.h.}} = \int_{\tau_{\text{stabilization}}}^{\tau_{\text{observation}}} c_s(t) \, dt. \tag{9.2}$$

This scale is the physical size of the spots observed, and corresponds to the recently observed peak at $l \approx 200$.

In heavy ion collisions similar studies are in their infancy. There exists of course the *global ellipticity* of the event, related to elliptic flow and non-zero impact parameters: we will discuss it below in detail in Section 9.4.5. We do not yet see reliable signals for either mean values of higher harmonics, or their fluctuations. Nevertheless, since in the mixed phase the transition of the high density QGP phase into low density hadronic gas should also be characterized by some instabilities, the sound horizon is also a useful concept to define the observed spectrum of the fluctuations.[2] Although we do not yet have any observations of higher harmonics and their fluctuations, I think there is a chance to see eventually "frozen plasma oscillations" in this case as well. After all, RHIC produced millions of events, while the Big Bang remains the only one! At the end of the day, the amount of information obtained about the Little Bang and the Big Bang are comparable.[3]

9.1.3. *Experimental centers, present and future*

The field of high energy heavy collisions is often looked at as an intersection of high energy and traditional nuclear physics. However we will show below that it has its own goals and methods that are quite different from both of those: in brief, it is a production and study of new forms/phases of matter at the extremes of reachable temperature and density.

It originated in such well established centers of nuclear physics as LBL, Berkeley and GSI, Darmstadt, Germany in the early 1980s. Then its main centers became two laboratories: Brookhaven National Laboratory (BNL) in Upton NY, and the European Center for Nuclear Research (CERN) in Geneva, Switzerland. Both are home to many celebrated experiments in high energy physics, but with heavy ion physics playing an increasingly important role these days.

The heavy ion program at BNL started with a series of experiments at its veteran accelerator AGS ($E = 2\text{--}11\,\text{GeV}^*A$), while CERN has used ion beams at another veteran, SPS ($E = 168\,\text{GeV}^*A$, or $\sqrt{s_{NN}} = 17\,\text{GeV}$). The first dedicated facility, Relativistic Heavy Ion Collider (RHIC) had its first run in the summer of 2000. It is a collider, with $100 + 100\,\text{GeV}^*A$ in the heavy ion mode and up to $250 + 250$ in the pp mode. Most of the data to be discussed below are from the first two runs of RHIC.

The community of heavy ion physicists is now working at a dedicated heavy ion experiment, ALICE, to be placed at the world's largest collider LHC at CERN, to start operating around 2008.

In February 2003 the German government approved a new (1 bn $ scale) project at GSI in Darmstadt. It includes a heavy ion facility to address the extremes of baryon density and around the so called "endpoint" of the QCD phase transition.

---

[2]The paper on that subject on which I am working is in progress at this point.

[3]In the final analysis, both are determined by the number of pixels in the detectors and computer storage, which is limited by common technology.

In short summary, the field has a firmly established experimental base at least for another 1–2 decades or so.

9.1.4. *Mapping the phase diagram*

In the previous chapter we looked at the phase diagram as a theoretical concept and we will further discuss it as such in Chapter 11, see Fig. 11.1; now is the time to look at it from more practical angle. Two views of it, on the plane temperature T-chemical potential for baryon charge μ_b are shown in Fig. 9.1.

The first of them is the theoretical view of how the cooling paths should look on it. If the produced system is macroscopically large and its evolution is sufficiently slow on the scale of microscopic reactions, its paths on the phase diagram follow the so called *adiabatic* lines, on which *entropy* is kept constant. Since the baryon number is also conserved, the ratio of both densities n_B/s is constant along the line. Examples of such lines are shown in Fig. 9.1a. Note the characteristic zigzag shape [607], appearing because lines with the same n_B/s value enter and exit the phase transition line at two different points.[4] The EoS used in hydrodynamical models should be calculated along such adiabatic paths.

How good is the adiabatic approximation in practice? Such studies have been based on hadronic cascades like URQMD or partonic cascades, which have shown that the non-equilibrium corrections to the total entropy differ by only about a few percent from the ideal conservation, so the approximation is very good. Basically all models agree that most of the entropy is produced rather early, and then its value is approximately unaffected by rescattering.

With the small and Gaussian event-by-event (EBE) fluctuations we have already mentioned, and rather accurate parameterizations of the multiple particle yield

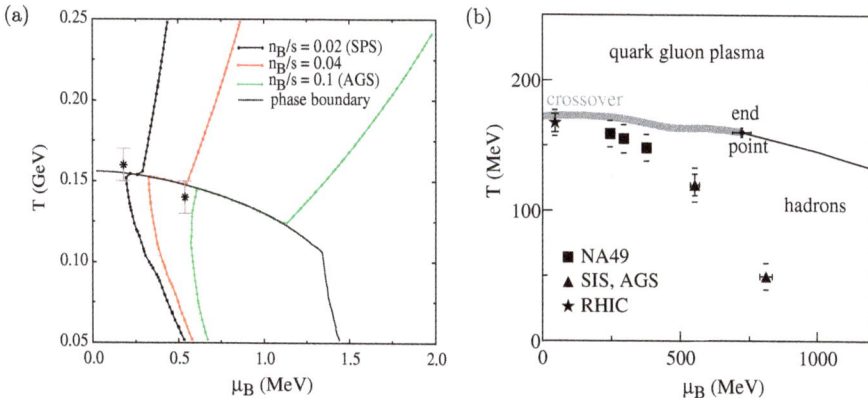

Fig. 9.1. (a) Example of the adiabatic cooling path. (b) Compilation of experimental data on chemical freezeout parameters from different experiments, together with the phase transition line and the expected tricritical point.

[4]One may then naively worry that they got mixed up inside the line; this is not so, the same ratio n_B/s distinguishes curves from each other in the mixed phase, although all of those collapse to the same line on this particular plane.

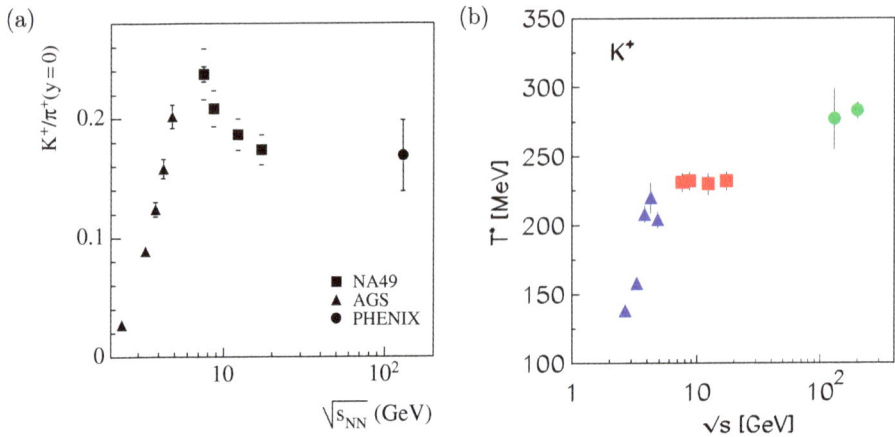

Fig. 9.2. (a) The K^+/π^+ ratio at mid-rapidity and (b) the effective slope of the K^+ at mid-rapidity versus the collision energy.

ratios, we are quite confident in mapping of the resulting data into the phase diagram.[5] The experimentally extracted line of "chemical freezeout points" for different collision energies is shown in Fig. 9.1b, together with lattice-based estimates about the crossover line and the tricritical point. One can see that all points are safely on the hadronic side of the transition, although quite close to it.

Comparing those two plots the reader should get the message that (i) various experiments probe quite different regions of the phase diagram, (ii) we do know well at least one point for each line; and (iii) that many of those lines do indeed cross the QCD phase transition line.

The plot suggests that the chemical freezeout line crosses the QGP boundary at about a collision energy of 20 GeV (in the lab). One may further ask if there is some interesting change of the observed behavior reported near this energy domain. And indeed, the NA49 collaboration [652] did find some. As an example we show here the behavior of K^+ production, in Fig. 9.2 which seems to exhibit such singular behavior. Of course, any sceptic would immediately say that a break between data points of different experiments may or may not be a real effect. Indeed, more studies of this are needed.

9.2. Relativistic hydrodynamics

9.2.1. *Equations of the ideal hydrodynamics*

If in high energy collisions of hadrons and/or nuclei a macroscopically large excited system is produced, its expansion and decay can be described by relativistic

---

[5]There is still no shortage of people completely hostile to temperature and density variables, thinking that any description of heavy ion collisions short of mentioning each secondary by name and giving its momentum must be wrong.

hydrodynamics. Its first application for such a purpose was made in the classical work by L.D. Landau [583] soon followed in 1953 the pioneer statistical model by Fermi in 1951 [582]. In between there was an important paper by Pomeranchuck [584] who pointed out that even if Fermi was right and the conditions he predicted did happen, one would be only able to see the final stage of the collisions since strong interaction in the system would persist. Pomeranchuck proposed some universal freezeout temperature $T_f \sim m_\pi$.

Let me add at the outset, that often hydrodynamics is considered as some consequence of kinetic equations. In fact, for the former approach the stronger the interaction in the system, the better. The latter approach, on the contrary, was never formulated but for weakly interacting systems.

The conceptual basis of the hydrodynamics is very simple: it is just a set of local conservation laws for the stress tensor $(T^{\mu\nu})$ and for the conserved currents (J_i^μ),

$$\partial_\mu T^{\mu\nu} = 0, \qquad \partial_\mu J_i^\mu = 0. \tag{9.3}$$

The main assumption is that $T^{\mu\nu}$ and J_i^μ are related to the bulk properties of the fluid by the relations

$$T^{\mu\nu} = (\epsilon + p)u^\mu u^\nu - pg^{\mu\nu}, \qquad J_i^\mu = n_i u^\mu. \tag{9.4}$$

Here ϵ is the energy density, p is the pressure, n_i is the number density of the corresponding current, and $u^\mu = \gamma(1, \mathbf{v})$ is the proper velocity of the fluid. In strong interactions, the conserved currents are isospin (J_I^μ), strangeness (J_S^μ), and baryon number (J_B^μ). For the hydrodynamic evolution, isospin symmetry is assumed and the net strangeness is set to zero; therefore only the baryon current J_B is considered below.

Let me add a simple heuristic argument why the first term in the stress tensor has $(\epsilon + p)$ and not any other combination. The point is that ϵ and p themselves are defined up to a constant $\pm B$ (which depending on the context we call the bag constant, or cosmological term or dark energy). The combination $(\epsilon + p)$ does not have it, and it is also proportional to the entropy which is defined uniquely.[6]

In order to close up this set of equations, one needs also the equation of state (EoS) $p(\epsilon)$. One should also be aware of two thermodynamical differentials

$$d\epsilon = Tds, \qquad dp = sdT, \tag{9.5}$$

and the definition of the sound velocity

$$c_s^2 = \frac{\partial p}{\partial \epsilon} = \frac{s}{T}\frac{\partial T}{\partial s}, \tag{9.6}$$

---

[6]How does this argument comply with the last term in (9.4)? Well it has B but without the velocity, so in hydrodynamic equations there is only a pressure gradient and this B term disappears as well.

and that $\epsilon + p = Ts$. Using these equations and the thermodynamical relations in the form

$$\frac{\partial_\mu \epsilon}{\epsilon + p} = \frac{\partial_\mu s}{s}, \tag{9.7}$$

one may show that these equations imply another non-trivial conservation law, namely, the conservation of the entropy

$$\partial_\mu (s u_\mu) = 0. \tag{9.8}$$

Therefore in the idealized adiabatic flow all the entropy is produced only in the discontinuities such as shock waves.

Before we change coordinates, let me remind the reader a few general formulae for that, following Ref. [629]. In an arbitrary coordinate system the equations of motion can be written as

$$T^{mn}{}_{;m} = 0, \qquad j^m{}_{;m} = 0, \tag{9.9}$$

where the semicolon indicates a covariant derivative. For tensors of rank 1 and 2 it reads explicitly

$$j^i{}_{;p} = j^i{}_{,p} + \Gamma^i_{pk} j^k, \tag{9.10}$$

$$T^{ik}{}_{;p} = T^{ik}{}_{,p} + \Gamma^i_{pm} T^{mk} + \Gamma^k_{pm} T^{im}, \tag{9.11}$$

where the comma denotes a simple partial derivative and the Christoffel symbols Γ^s_{ij} are given by derivatives of the metric tensor $g^{ab}(x)$:

$$\Gamma^s_{ij} = (1/2)g^{ks}(g_{ik,j} + g_{jk,i} - g_{ij,k}). \tag{9.12}$$

In high energy collisions, we should first worry about the longitudinal motion, so we discuss the $1d$ case first. As an example, let us do the following transformation from Cartesian to light cone coordinates:

$$x^\mu = (t, x, y, z) \rightarrow \bar{x}^m = (\tau, x, y, \eta),$$

$$t = \tau \cosh \eta \qquad \tau = \sqrt{t^2 - z^2}, \tag{9.13}$$

$$z = \tau \sinh \eta \qquad \eta = (1/2) \ln \frac{t+z}{t-z}. \tag{9.14}$$

In the new coordinate system the velocity field (after inserting $v_z = z/t$) is given by

$$\bar{u}^m = \bar{\gamma}(1, \bar{v}_x, \bar{v}_y, 0), \tag{9.15}$$

with $\bar{v}_i \equiv v_i \cosh \eta$, $i = x, y$, and $\bar{\gamma} \equiv 1/\sqrt{1 - \bar{v}_x^2 - \bar{v}_y^2}$.

Now we turn to the metric of the new system. We have

$$ds^2 = g_{\mu\nu}dx^\mu dx^\nu = dt^2 - dx^2 - dy^2 - dz^2$$
$$= d\tau^2 - dx^2 - dy^2 - \tau^2 d\eta^2 \tag{9.16}$$

and therefore

$$g_{mn} = \begin{pmatrix} 1 & 0 & 0 & 0 \\ 0 & -1 & 0 & 0 \\ 0 & 0 & -1 & 0 \\ 0 & 0 & 0 & -\tau^2 \end{pmatrix}. \tag{9.17}$$

The only non-vanishing Christoffel symbols are

$$\Gamma^\eta_{\eta\tau} = \Gamma^\eta_{\tau\eta} = \frac{1}{\tau}, \qquad \Gamma^\tau_{\eta\eta} = \tau. \tag{9.18}$$

9.2.2. *Dissipative terms*

These equations are valid if the microscopic length (such as mean free paths of constituents l) is small compared to the size of the system considered[7] L. Inclusion of dissipative effects, to the first order in l/L, is possible. Of course, the modified stress tensor and the currents no longer conserve the entropy, which is monotonically increased during the evolution.

Let me first comment that the relativistic generalization of the Navier–Stokes equations have created some polemics in literature. The key issue is what one should call the "rest frame of the matter". In the standard Landau and Lifshits textbook [580] the definition is based on the "natural assumption" of the zero *energy* flow in the rest frame, or $T_{0m} = 0, m = 1, 2, 3$. However, some authors adopt a definition due to Eckart [581], according to which it is the reference frame in which a *conserved charge* (say, the baryonic charge in our case) is at rest $N_m = 0$. Generally speaking, there is no reason for these two definitions to coincide in the relativistic flow, and this ambiguity has led to some misunderstanding. In what follows we adopt the former point of view (being much more suitable for high energy applications in which the role of baryonic charge is in general very small).

Dissipative corrections to the stress tensor and the current can be written as

$$\delta T_{\mu\nu} = \eta \left(\nabla_\mu u_\nu + \nabla_\nu u_\mu - \frac{2}{3}\Delta_{\mu\nu}\nabla_\rho u_\rho \right) + \xi(\Delta_{\mu\nu}\nabla_\rho u_\rho), \tag{9.19}$$

$$\delta J_\mu = k \left(\frac{\eta T}{\epsilon + p} \right)^2 \nabla_\mu(\mu_B/T), \tag{9.20}$$

---

[7]Note that this is the same condition as for other macroscopic approaches, e.g. thermodynamics; however historically the hydrodynamical models met much more resistance than, say statistical models.

where the three coefficients η, ξ, k are called the shear and the bulk viscosities and the heat conductivity, respectively. In this equation the following projection operator onto the matter rest frame was used:

$$\nabla_\mu \equiv \Delta_{\mu\nu} \partial_\nu, \qquad \Delta_{\mu\nu} \equiv g_{\mu\nu} - u_\mu u_\nu. \tag{9.21}$$

Two general comments may be made in connection with these definitions. First, for matter with vacuum quantum numbers, characterized by the temperature only, one cannot even introduce the notion of heat conductivity. Second, bulk viscosity effects are very elusive, in particular, they vanish both in the ultrarelativistic and nonrelativistic gas. So, the main non-equilibrium effect is connected with *shear* viscosity. Since, as we will show below, the late-time flow approaches a radially symmetric Hubble flow without shear, the viscosity effect has a curious self-quenching property.

It is further useful to relate the magnitude of the viscosity coefficient η to a more physical observable. As such one can use the sound attenuation length. If a sound wave has frequency ω and wave vector $\mathbf{q}$, its dispersion law (the pole position) is

$$\omega = c_s q - \frac{1}{2} i \mathbf{q}^2 \Gamma_s, \qquad \Gamma_s \equiv \frac{4}{3} \frac{\eta}{\epsilon + p}. \tag{9.22}$$

The magnitude of the viscosity coefficient or Γ_s in the quark–gluon plasma will be discussed in Section 9.4.6, and it can be used in order to evaluate the amount of entropy created by this term.

9.2.3. The Bjorken solution

The $(1+1)d$ equations for a boost-invariant solution can be written in the following way:

$$\frac{\partial}{\partial t}(s \cosh y) + \frac{\partial}{\partial z}(s \sinh y) = 0, \tag{9.23}$$

$$\frac{\partial}{\partial t}(T \sinh y) + \frac{\partial}{\partial z}(T \cosh y) = 0. \tag{9.24}$$

Exercise 9.1. Derive those equations using the example for the curvature correction in Section 9.2.1.

The so called Bjorken [588] solution[8] is obtained if the velocity is given by the velocity $u_\mu = (t, 0, 0, z)/\tau$ where $\tau^2 = t^2 - z^2$ is the proper time. In this $1d$-Hubble

---

[8] We call it following established tradition, although the existence of such simple solution was first noticed by Landau and it was included in his classic paper as some intermediate step. The space-time picture connected with such scaling regime was discussed in Refs. [586, 587] before Bjorken, and some estimates for the energy above which the transition to the scaling regime were expected to happen were also discussed in my paper [711] as well.

regime there is no longitudinal acceleration at all: all volume elements are expanded linearly with time and move along straight lines from the collision point. The spatial $y = \tanh^{-1}(z/t)$ and the energy–momentum rapidity $y = \tanh^{-1} v$ are just equal to each other. Exactly as in the Big Bang, for each "observer" (the volume element) the picture is just the same, with the pressure from the left compensated by that from the right. The history is also the same for all volume elements, if it is expressed in its own proper time τ.

Thus the entropy conservation becomes the following (ordinary) differential equation in proper time τ:

$$\frac{ds(\tau)}{d\tau} + \frac{s}{\tau} = 0, \tag{9.25}$$

which has the obvious solution

$$\boxed{E}$$

$$s = \frac{\text{const}}{\tau}. \tag{9.26}$$

Furthermore, as a volume of each all of matter grows as τ, the entropy in each volume element is conserved.[9]

Exercise 9.2. Check that it also agrees with the second equation (9.24).

Let us compare three simple cases: (i) hadronic matter, (ii) quark–gluon plasma and (iii) the mixed phase (existing if there are first order transitions in the system). In the first case we adopt the equation of state suggested in Ref. [492] $c^2 = \partial p/\partial \epsilon = \text{const}(\tau) \approx 0.2$. If so, the decrease of the energy density with time is given by

$$\epsilon(\tau) = \epsilon(0) \left(\frac{\tau_0}{\tau}\right)^{1+c^2}. \tag{9.27}$$

In the QGP case the same law holds, but with $c^2 = 1/3$.

In the mixed phase the pressure remains constant $p = p_c$, therefore

$$\epsilon(\tau) = (\epsilon(0) + p_c) \left(\frac{\tau_{mix}}{\tau}\right) - p_c. \tag{9.28}$$

So far all dissipative phenomena were ignored. Including first dissipative terms into our equations one has

$$\frac{1}{\epsilon + p}\frac{d\epsilon}{d\tau} = \frac{1}{s}\frac{ds}{d\tau} = -\frac{1}{\tau}\left(1 - \frac{(4/3)\eta + \xi}{(\epsilon + p)\tau}\right). \tag{9.29}$$

Note that ignoring ξ one finds in the r.h.s. exactly the combination which also appears in the sound attenuation, so the correction to the ideal case is $(1 - \Gamma_s/\tau)$. Thus the length Γ_s directly tells us the magnitude of the dissipative corrections. At

---

[9]Here actually Bjorken made a mistake: he assumed that it is the energy density, not the entropy, which is conserved for each matter element. The reason it is incorrect is that non-zero work is done by pressure on matter during the expansion.

time $\tau \sim \Gamma_s$ one has to abandon the hydrodynamics altogether, as the dissipative corrections cannot be ignored.

Since the correction is negative, it reduces the rate of the entropy decrease with time. Another way to say that, is that the total positive sign shows that some amount of entropy is generated by the dissipative term. Danielewicz and Gyulassy [589] have studied this equation with various coefficients.

9.2.4. *Further simplifications and solutions*

The other simple geometry one can think of is of course the $3d$ spherical expansion. In this case the equations read

$$\frac{\partial}{\partial t}(s \cosh y) + \frac{1}{r^2}\frac{\partial}{\partial r}(sr^2 \sinh y) = 0, \tag{9.30}$$

$$\frac{\partial}{\partial t}(T \sinh y) + \frac{\partial}{\partial r}(T \cosh y) = 0. \tag{9.31}$$

Numerical solutions for this case were studied in Refs. [590–592], with approximately spherical clusters created in low energy $\bar{p}p$ or e^+e^- annihilation in mind.[10] They also have a scaling radial solution

$$v_r = \frac{r}{t}, \qquad s \sim t^{-3}, \tag{9.32}$$

which is however not reached in the cases studied numerically prior to freezeout.

The next approximation (which is being widely used) is to combine the Bjorken longitudinal boost-invariant solution with the arbitrary transverse expansion. It is convenient (as was done already by Landau) to change longitudinal variables from t, z to more appropriate variables $\tau = \sqrt{t^2 - z^2}$ and $\eta = \frac{1}{2}\log[(t+z)/(t-z)]$, which are respectively referred to as the *proper time* and the *spatial rapidity*. Boost invariance then means that the solution for any value of η may be found by boosting the solution at $\eta = 0$ to a frame moving with velocity $v = \tanh(\eta)$ in the negative z-direction. With this assumption, the equations of motion become two-dimensional [665] and are given at $\eta = 0$ by

$$\begin{aligned}
\partial_\tau(\tau T^{00}) + \partial_x(\tau T^{0x}) + \partial_y(\tau T^{0y}) &= -p, \\
\partial_\tau(\tau T^{0x}) + \partial_x(\tau T^{xx}) + \partial_y(\tau T^{xy}) &= 0, \\
\partial_\tau(\tau T^{0y}) + \partial_x(\tau T^{xy}) + \partial_y(\tau T^{yy}) &= 0, \\
\partial_\tau(\tau J_B^0) + \partial_x(\tau J_i^B) + \partial_y(\tau J_B^y) &= 0.
\end{aligned} \tag{9.33}$$

Integrating over the transverse plane, one finds that net baryon number per unit spatial rapidity, $\int dx\, dy\,(\tau J_B^0)$, and the transverse momentum per unit rapidity, $\int dx\, dy\,(\tau T^{0x})$ as well as $\int dx\, dy\,(\tau T^{0y})$, are conserved. The energy per unit rapidity,

---

[10]Note that in the final multi-particle state occurring in heavy quark annihilation, the decay of psions and upsilons, are also approximately spherical. It may be there is some collective radial flow there as well; I am not however aware of any work in this direction.

$\int dx\, dy\, (\tau T^{00})$, decreases due to the work done per unit time [634] by the pressure in the longitudinal direction, $\int dx\, dy\, p$,

$$\partial_\tau(\tau S^0) + \partial_x(\tau S^x) + \partial_y(\tau S^y) = 0. \tag{9.34}$$

Integrating over the transverse plane, we find that

$$\int dx\, dy\, \tau\, s\gamma = \frac{dS_{\text{tot}}}{d\eta} \tag{9.35}$$

is a constant of the motion.

The simplest case is that of *axially symmetric* expansion. Its nice form due to [593] closely resembling the $1d$ case appears if one also uses the *transverse rapidity* variable $\alpha = \tanh^{-1} v_r$

$$\frac{\partial}{\partial t}(rts\cosh\alpha) + \frac{\partial}{\partial r}(rts\sinh\alpha) = 0, \tag{9.36}$$

$$\frac{\partial}{\partial t}(T\sinh\alpha) + \frac{\partial}{\partial r}(T\cosh\alpha) = 0. \tag{9.37}$$

One can also find in this paper [593] and also in another work [596] a nice discussion of the simplest numerical solutions of this equation, as well as of the $1d$ and $3d$ spherical ones. We will return to real applications below. Some new analytic solutions resembling Hubble expansion in the universe have been discussed recently; we will discuss those in Section 9.3.

9.2.5. *Singularities: shocks, rarefaction waves, and the "explosive edge"*

As is well known, hydrodynamics admits solutions with discontinuities. Shocks are solutions with matter compression, while deflagration/detonation are examples of rarefaction discontinuities. There are certain conditions which such solutions should obey: those are continuity of energy and momentum flows, as well as all conserved charges through the discontinuity. There is however one important *non*-conserved quantity, the entropy, which must be *produced* in the discontinuity. The relativistic version of these conditions, known as Gugonio/Taub adiabats, can be found in textbooks such as Ref. [580]. We discussed those in some details in the first edition of the book, but now the application of those looks much less probable and I would not go into those.

Let me just mention that shocks at the matter formation stage have been discussed already by Belenky and Landau [490]: but this treatment does not seem to be valid in QCD since asymptotic freedom precludes too strong interaction at short time. The rarefaction waves were extensively discussed by van Hove and others [604–606] in the 1980s and then discussed again in the 1990s [607, 608], when it was proposed that a long-lived fireball could be formed near the so called "softest point" of the EoS. If so, a very long lifetime of QGP could have been seen, as low pressure led to zero expansion and decay by deflagration propagating inward. However in all realistic applications it seems that pressure is always large enough to generate a matter flow velocity larger than that of the deflagration front, making it unimportant.

Unusual phenomena at large p_t observed at RHIC are clearly related with the surface of the excited system produced in the collision. This is clear already from the significant quenching of spectra, effectively excluding the central region, but also from a large azimuthal asymmetry, and absence of the backward jets in events with high-p_t trigger. Furthermore, the data on $v_2(p_t)$ show that hydro-generated behavior at $p_t < 2\,\mathrm{GeV}$ joins very smoothly to a new (so far unexplained) regime at $p_t > 2\,\mathrm{GeV}$. Also, it was observed that v_2 for baryons is larger than for mesons: natural for the hydrodynamic regime, it seems to be true at higher $p_t = 2\text{–}6\,\mathrm{GeV}$ as well. All this hints to a possibility that the basic physics in the hydrodynamic domain and that at higher p_t is somewhat related.

There is however a generic problem, preventing us from combining hydrodynamic and jet physics into any composite model: their quite different *time scales*. Indeed, jets move with the speed of light, and cannot be emitted too far from the surface: so escaping jet fragments can only interact with matter for a short time, $t_{\mathrm{quenching}} \sim 1\,\mathrm{fm}/c$. Hydrodynamic flow, on the other hand, needs longer time to be developed, about $t_{\mathrm{hydro}} \sim 10\,\mathrm{fm}/c$ for radial and about half of that for elliptic flow.

The way out of this dilemma discussed in this section is to focus on the specific collective phenomena which may develop during the short time $t_{\mathrm{quenching}}$ at the edge of the fireball. We will do so first in the hydrodynamical context, where analytic answers can be obtained. This however can only be used as qualitative indications since the size of the system (or the particle number) is not large enough to justify this approach quantitatively. In the second subsection we suggest that multiple explosion of clusters may also "jump-start" collective phenomena at the surface.

Let us start with the generic problem, of an explosion of a system which has a sharp edge. The main point to make here at the outset is that phenomena near the edge basically drive explosions in $1d$, which are much more robust that those in $2d$ and $3d$.

Do we have an edge in heavy ion collisions? Indeed, if one ignores the "skin" of nuclei and treats them as drops of nuclear matter with a sharp spherical surface, the almond-shaped overlap region of two nuclei colliding with the impact parameter b will have a sharp edge at which the energy (also particle and entropy) density vanishes as $\epsilon(x) \sim \sqrt{x - x_{\mathrm{edge}}}$. From the hydrodynamical equations it then follows that the acceleration at this point is formally infinite[11]

$$\frac{\partial v}{\partial t} \approx \frac{1}{\epsilon + p}\frac{\partial p}{\partial x} \sim \frac{1}{x - x_{\mathrm{edge}}}. \tag{9.38}$$

---

[11]If the edge is regulated by the usual skin with a Fermi-type distribution, the acceleration is finite and constant at large x, but inversely proportional to small skin width.

It is well known how the problem is resolved, at least mathematically, in the framework of hydrodynamical equations. The singularities in the *initial conditions* are extensively discussed in textbooks such as Ref. [580]: let me recall the main idea only. Although hydrodynamics admits discontinuities (such as shocks), they always come with certain conditions on them, as we mentioned, which the initial conditions in general do not obey. Therefore, a generic situation is that a singular point in the initial conditions opens up into a whole region of special solutions, separated by two discontinuities from other regions.

A singularity of the type we discuss opens up into a region filled with the so called Riemann wave in which rapidity and thermodynamical quantities can be viewed as functions of each other. The region of this solution is schematically shown in Fig. 9.3, it is separated from vacuum by a light cone $t = x - x_{\text{edge}}$ at which matter content vanishes, and from the non-trivial hydrodynamic solution on the left by a deflagration front. (The latter curves are at later time because the developed hydrodynamic rapidity reaches values larger than that of the deflagration.)

The Riemann wave solution itself in relativistic flow is discussed in Ref. [580], problem 1 in Section 134. It can be summarized by two equations:

$$y = \int \frac{c_s \, d\epsilon}{p + \epsilon}, \tag{9.39}$$

$$x - x_{\text{edge}} = t * \tanh(y - y_s), \tag{9.40}$$

where y, y_s is matter and sound rapidity, $v = \tanh(y), c_s = \tanh(y_s)$. ϵ is energy density, p is pressure and $c_s^2 = dp/d\epsilon$ is squared sound velocity. If p, ϵ is a function of one variable (e.g. temperature) the meaning of the integral in the former equation is clear: if there are more conserved quantities and chemical potentials, it should be taken along the adiabatic path in the phase diagram the systems take while cooling down. In particular, for QGP with the simple black-body-like EoS $p = \epsilon/3 \sim T^4, c_s^2 = 1/3$ this first equation establishes a direct relation between

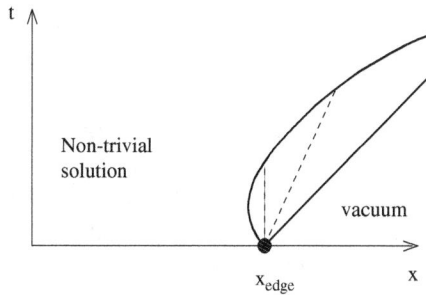

Fig. 9.3. The region in which the Riemann wave solution develops, out of the "edge" singularity of the initial conditions.

parton density $\sim T^3$ and rapidity

$$\frac{n}{n_0} = \exp(-y/c_s), \tag{9.41}$$

each of which holds at the whole line originating from the edge. The light cone corresponds to $y \to \infty, n \to 0$, n_0 is the density at rapidity zero, etc. Of course, the exponent in rapidity is a power law in momenta/energies, with a rather unusual index. For QGP EoS it gives quite unusual power $1/p^{\sqrt{3}}$, with a somewhat larger power 2–3 if one uses the more appropriate EoS of the resonance gas. However in all cases the spectrum generated by the Riemann wave is much harder than the observed particle spectrum, which has a larger effective power of about 10–12.

Let us now discuss the limitations on hydrodynamics stemming from the fact that the particle multiplicity in our problem is not like 10^{23} as in air/water but rather limited. At central Au–Au collisions at RHIC at rapidity $y \sim 0$ one actually finds[12]

$$\frac{dN_{\text{hadrons}}}{d\eta\, d\phi} \sim \frac{dN_{\text{partons}}}{d\eta\, d\phi} \sim 150, \tag{9.42}$$

where η, ϕ are pseudorapidity and azimuthal angle. This means that when we go about two orders of magnitude down the spectrum we reach only about one particle per unit solid angle, and should stop using hydrodynamics. More elaborate estimates using viscosity corrections [678] lead to the same conclusion.

This simple idea put a cutoff to hydrodynamics to $p_t \sim 1.7$–$2\,\text{GeV}$, which is indeed where deviations from hydrodynamic predictions are observed. It also indicates to us that the above discussion of the Riemann wave can only be used as a qualitative indication, and its comparison to the data is not really justified.

We concluded at the end of the previous section, that hydrodynamics can provide only qualitative ideas about matter evolution at the edge of the system, and thus some microscopic approach is needed.

This is of course very much limited by our poor knowledge of the matter conditions at the early time. However the generic idea — that particles at the edge of the system experience collisions with a stream of particles from one side only and are therefore accelerated more rapidly than any others — holds in one form or another for any models under consideration.

In a perturbative mini-jet scenario with a modest rescattering one would simply find a fraction of outward moving mini-jets which is never re-scattered, but fragment into secondary hadrons independently. Naturally, those would spatially separate from those which do re-scatter, and create a thin collisionless "atmosphere" receding from the fireball with (transverse) rapidity $y_t \sim 1$.

---

[12]We mean all kinds of hadrons together, charged and neutral. The partons are not directly observed, of course, but believed to be about the same to justify the approximately adiabatic expansion which works well in hydrodynamics applications and is confirmed by cascades.

In a CGC picture one should study the behavior of the classical Yang–Mills equations which start with the initial conditions having a sharp edge. Although it is not yet done, by analogy to the discussion of the hydrodynamic equation in the preceding section, one may expect the edge also to act as a focal plane of an explosive expansion into vacuum. Moreover, in terms of the effective pressure the non-linear term in the YM equation is even stronger than in the EoS of QGP, so more dramatic phenomena can be expected.

Finally, we come to the scenario with multiple production of topological clusters, which go into expanding shells of glue. Some part of those shells going from the edge toward the empty space cannot be stopped.

So, if in all pictures of the initial stage one finds partons receding from the system edge outward, it is obviously interesting to know what happens with them? The answer to this question depends on their density. In pp collisions, with a couple of mini-jets or a single cluster produced, confining flux tubes would slow down their propagation and drain their energy into fragmentation of flux tubes (hadronization a la Lund model).

We would argue that the situation should be different if the outward moving partons are sufficiently numerous to neutralize each other in color and thus fly away *without strings attached*. In other words, we suggest that parton fragmentation in AA collisions is different from the independent one observed in e^+e^- and pp collisions.

We will further suggest that even fragmentation of high-p_t jets should be affected, as they fly through the co-moving matter originating in the explosive edge, see Fig. 9.3. A number of detailed models of such kind are under numerical investigation now, with results to be reported. Hopefully these provide some insight into the puzzles of large v_2 and high baryon content.

9.3. Chemical and thermal freezeouts

9.3.1. *Why two freezeouts?*

Looking at the fireball by a detector one has the same problem as an astronomer has, looking at a star (such as the sun). Only the photosphere can be seen, having $T \sim 6000\,\mathrm{K}$, but not the much hotter interior. It took years to detect elusive solar neutrinos to test directly what the "standard solar model" predicts for the properties at the Sun's interior. However, people studying the Little Bang are luckier than those who study solar physics: instead of just two species of detectable radiation (photons and neutrinos) there is a whole zoo of particle species available to them. Since they have quite different scattering cross sections, they can provide (in principle) a set of pictures at different depths of the "photosphere".

The general freezeout condition can be formulated in several different ways. According to the first way, the microscopic hydrodynamic scale — the mean free

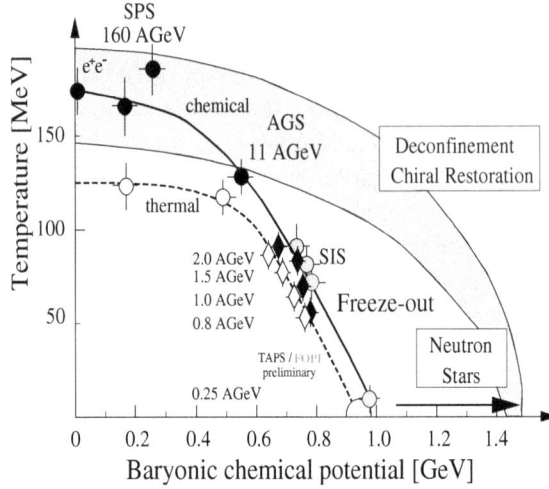

Fig. 9.4. Solid (dashed) lines indicate chemical (thermal) freezeout. (Adopted from U. Heinz.)

path l — becomes comparable to the macroscopic ones, with respect to either the system size L, or its inverse expansion rate $d \log(s)/d\tau$, or the approximation to it defined in terms of the 4-velocity as

$$\nu_{\text{expansion}} = \partial_\mu u_\mu. \tag{9.43}$$

According to the second way, the escape probability for a particle is comparable to that of its rescattering. They are both needed because the freezeout surface has in general both space-like and time-like parts. We will return to exact definitions and examples of this surface below.

What is extremely important to emphasize is that for low energy hadronic reactions *the rates for inelastic reactions are much smaller than for the elastic.*

One of the reasons for that is that pions interact with derivatives due to their Goldstone nature, so it is difficult to re-scatter or produce a soft pion. Thus, in a low T pion gas the elastic and inelastic reactions have even different powers of T.

As a result, one has to identify *two separate freezeout temperatures*, (i) the *"chemical"* (fixing the composition of secondaries) and (ii) the *"thermal"* or kinetic one, at which the last collisions happen and the particle momenta are finally fixed to their observed values. Two temperatures approximately[13] associated with them are denoted as T_{ch}, T_{th}.

Phenomenologically, the case for two separate freezeouts was argued especially in Ref. [607], emphasizing the fact that at AGS/SPS collision energies studied at the time, the main contribution to the radial flow came from the collision stage in between $T_{\text{ch}} < T < T_{\text{th}}$. Although fits to the observed spectra have produced

---

[13]The freezeout conditions are not coincidental with isotherms, as will be clear below.

a wide range of acceptable values for the freezeout T and radial flow velocity v, we argued that quite low $T_{th} = 120\text{--}100\,\mathrm{MeV}$ should be the case. Later this was directly confirmed (see, e.g. Ref. [690]) by the HBT data to be discussed later in Section 9.5. In other words, AGS/SPS collisions produce basically a chemically frozen but kinetically equilibrated "resonance" gas.

Furthermore, the naive freezeout prescription $T = \mathrm{const.}$ fails in a number of respects, in the late hadronic stages chemical freezeout must be correctly accounted for. Second, different particle types freeze out at different times and with different collective velocities. With a universal freezeout temperature, the transverse flow of the strange particles Λ, Ξ, Ω, has never been reproduced [633].

To model the hadronic stage, Bass and Dumitru [631] replaced the hadronic phase of the hydrodynamics with a hadronic transport model, UrQMD. In this approach, the switch from hydrodynamics to cascade is made at some (arbitrary) switching temperature, $T_{\mathrm{switch}} < T_c$. The same approach also adopted in the work by Teaney, Lauret and myself [673] from which most plots shown below are taken.

9.3.2. Chemical freezeout

The particle composition is a stronghold of thermodynamics, and even that in pp and e^+e^- can also be well explained by the statistical model. There is one significant difference though: in pp and e^+e^- the *strangeness* is significantly suppressed, unlike in AA collisions it takes exactly the equilibrium value. The QGP-based scenario of the evolution explains it naturally, while in the string scenario one needs "color ropes" or other exotic devices to explain why the strings fragment differently in AA.

Significant experimental efforts has been made to locate the transition between the two extremes, the pp-type and heavy-ion-type regimes of strangeness production. Excellent data from WA97, NA49 for Λ, Ξ, Ω and their antiparticles have shown *weak dependence on centrality*. Lighter ion data obtained previously also confirm the impression that strangeness production is *constant* throughout the whole SPS domain, with transition being lower, at AGS energies. To me it does indicate that strangeness "unsuppression" is clearly related with the approach of the hadronic phase boundary and production of the mixed phase with at least some QGP.

Well, adding ψ' which melts in between and decomposing J/ψ into χ and proper J/ψ, one would get a whole sequence of "melting" phenomena happening as matter becomes hotter/denser. We will see at RHIC how members of the Υ family do the same later on.

Which of them is the QGP signal then? Well, this still depends on details which we still have to work out.

9.3.3. Between chemical and kinetic freezeouts

By definition, in this part of space–time all reactions changing the particle composition are assumed to be *frozen*, while the elastic rescattering is still going on.

This implies that although a common matter temperature T still exists, we can no longer put all chemical potentials to zero (other than associated with conserved quantities).

In fact now *any* particle number N_i is a conserved quantity, and so at this stage of the evolution one has to introduce the chemical potentials for *all particle species* conjugated to their number and determined by the pre-fixed values of N_i at $T = T_{ch}$. As the fireball expands and T decreases, all those chemical potentials grow in magnitude and compensate the difference between real and equilibrium (at current T) particle numbers.

For pions this was described in detail in Ref. [614], see also more recent works [607, 615]. The thermodynamical relation written in the form

$$\frac{\epsilon + p}{nT} - \frac{s}{n} = \frac{\mu}{T} \tag{9.44}$$

is especially useful. For slow (adiabatic) expansion the s/n ratio is not changing, while in the first term on the l.h.s. the chemical potential *nearly* cancels (provided the Boltzmann approximation is used). So, one can read the T-dependence of μ directly from the r.h.s. The notorious exceptional case worth mentioning is that for massless particles, the whole l.h.s. is just a constant. Therefore $\mu/T = $ const. and so if $\mu = 0$ at the beginning it remains so for any T.[14]

Accounting for the non-zero pion mass and Bose statistics one finds $\mu_\pi(T)$, see Ref. [614]. For example, if one assumes that $\mu_\pi(T_c = 160\,\text{MeV}) = 0$, one finds that by a thermal freezeout (which happens for Pb–Pb collisions at CERN at $T = 110$–$120\,\text{MeV}$) the pion chemical potential $\mu_\pi = 60$–$80\,\text{MeV}$.

In order to demonstrate that such an effect really occurs in experiment, we will give two examples. The first example comes from the paper of Hung and myself [607] can use the difference between Boltzmann and Bose distributions at small momenta, sensitive to chemical potential. We show in Fig. 9.5 (from Ref. [607]) the ratio of p_t spectra for Pb–Pb collisions (in which we expect thermal freezeout at $T = 100$–$120\,\text{MeV}$, and thus the formation of a significant pion chemical potential) to our reference point, central S–S collisions (for which the effect should be much smaller). The data sets are both for *positive* pions from the NA44 experiment, in the same experimental settings (and thus systematic errors should somewhat cancel). One finds, that there is a significant enhancement of this ratio at small p_t, which agrees with the formation of the non-zero pion chemical potential. Moreover, as one can see, the magnitude of the effect is in approximate agreement with our estimates.

The second example comes from the direct measurements of the phase space density of pions at the freezeout, by HBT correlations, see Fig. 9.30.

The secondaries other then pions can be to a good accuracy treated as a non-relativistic Boltzmann gas, and so one can easily derive the following relation

---

[14]This is what happens in the case of background radiation in the expanding universe: although the photons do not collide after the Big Bang, they still have the Planck spectrum, with $\mu = 0$.

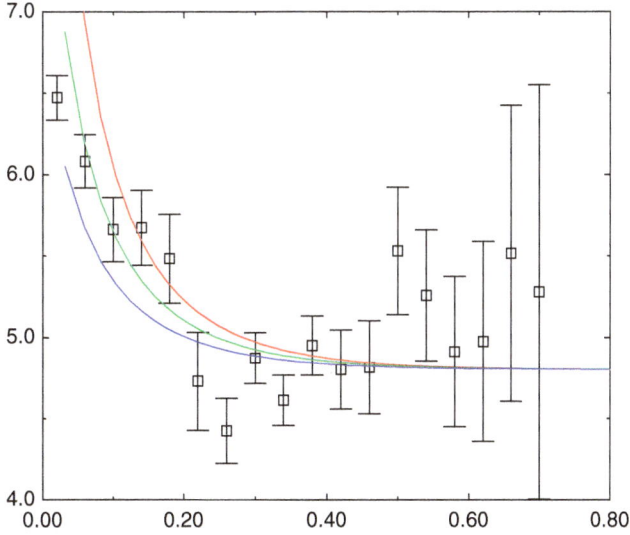

Fig. 9.5. The ratio of $\pi^+ p_t$ spectra for Pb–Pb to S–S collisions. Points are experimental data from the NA44 experiment, three curves correspond to pion chemical potential $\mu_\pi = 60$, 80 and 100 MeV (from bottom up).

between the chemical potentials at chemical and thermal freezeout:

$$\mu_{th} = \mu_{ch} \frac{T_{th}}{T_{ch}} + m \left(1 - \frac{T_{th}}{T_{ch}} \right). \tag{9.45}$$

Note that for very large systems $T_{th} \to 0$, the chemical potential $\mu_{th} \to m$ as it should, and one can then proceed to normal non-relativistic notation.

Before going any further let me note that the notion of chemical freezeout (as any idealization) is not absolute and has exceptional cases. Those are particle-number-changing reactions which have such a large rate that they proceed in a hadronic phase, after $T = T_{ch}$. One such case to be discussed below in this section is annihilation of antibaryons. The other case is short-lived strong resonances leading to reactions such as $\pi\pi \to \rho \to \pi\pi$; $\pi K^+ \to K^{+*} \to \pi K^+$; $\pi K^- \to K^{-*} \to \pi K^-$; $\pi N \to \Delta \to \pi N$; $\pi\Upsilon \to \Sigma^* \to \pi\Upsilon$; $\pi\Xi \to \Xi^* \to \pi\Xi$ where Υ denotes the hyperons Λ, Σ.

The general rule how to proceed in such cases is to separate two independent combinations, affected and unaffected by the reaction. For example, the reactions $\pi\pi \to \rho \to \pi\pi$ do not change the "orthogonal" combination $N_\rho - 2N_\pi$. As a result, it is frozen at T_{ch} and one can introduce an (increasing with time) chemical potential for it keeping its value. However in the "longitudinal" combination $\mu = 0$ and the reaction is unfrozen.

We will now follow Teaney [615] and show which changes in the EoS and matter evolution chemical are caused by the freezeout. If only these reactions and elastic collisions are included, then the system has 16 conserved currents: baryon number (J_B^μ),

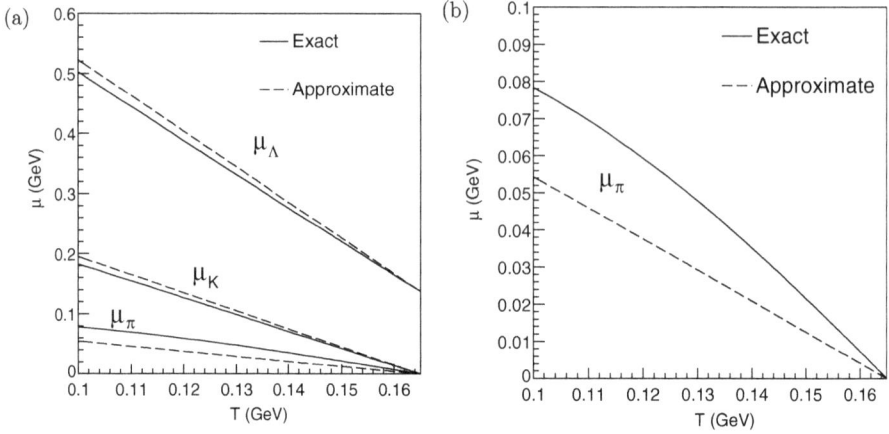

Fig. 9.6. Chemical potentials as a function of temperature at the SPS ($s/n_B = 42$) for (a) π, K and Λ and (b) π only.

strangeness (J_S^μ), isospin (J_I^μ) and 13 other conserved numbers, $J_{H_i}^\mu$, where H_i runs over the hadron species, $\pi, \bar{N}, \Upsilon, \bar{\Upsilon}, \Xi, \bar{\Xi}, K^-, \eta, \eta', \omega, \phi, \Omega, \bar{\Omega}$. (The nucleon number and K^+ number are not included in the list of hadrons, but their conservation follows from the conserved hadron currents already specified and from baryon and strangeness conservation.) All those conserved currents satisfy the continuity equation, all of which should be solved together with hydrodynamics. Fortunately, particle and entropy conservation give similar equations as $n_i/s = \text{const}(x, t)$.

We can derive an approximate formula for the chemical potentials as a function of T, for a collection of non-relativistic ideal gases. For non-degenerate and non-relativistic ideal gases the partial pressure p_i, and partial energy density e_i of the ith species are given by, $p_i = n_i T, e_i = n_i m_i + (\text{const}) n_i T$. Since

$$Ts = \sum_i \epsilon_i + p_i - \mu_i n_i,$$

we have

$$1 = \sum_i \frac{n_i}{s} \left(\frac{m_i + \mu_i}{T} + \text{const.} \right).$$

The n_i/s is constant below $T < T_c$. To keep all the terms in parentheses constant below T_c, we require

$$\mu_i = m_i \left(1 - \frac{T}{T_c} \right) + \frac{T \mu_c}{T_c},$$

where μ_c is the chemical potential at the critical temperature. The approximation is shown by the dashed lines and works well for all particles except pions. Being the most important, pions are shown separately in Fig. 9.6b. Thus, a pion chemical potential of nearly 80 MeV may be acquired.

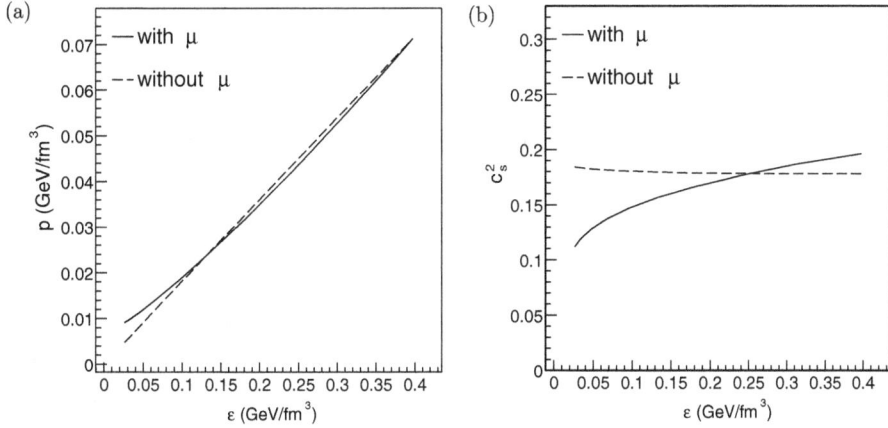

Fig. 9.7. The (a) pressure, (b) sound speed squared as functions of the energy density with and without chemical freezeout at the SPS ($s/n_B = 42$). The analogous curves at RHIC are only slightly different.

What remains to be determined is the EoS for the chemically frozen hadronic gas, which should then be compared to the equilibrium one as well as to the effective EoS of hadronic cascades (RQMD in practice).

The energy density as a function of temperature, with and without the chemical potentials, is shown in Fig. 9.8a. The principal observation is clear: without chemical freezeout the energy density drops very rapidly as a function of temperature, since the total number of particles drops.

Furthermore, for the hydrodynamic solution the most important relationship is not $\epsilon(T)$ but $p(\epsilon)$ — the EoS. Figure 9.7 shows the pressure and speed of sound along the adiabatic path.

Unlike the energy density versus temperature, the pressure is scarcely modified by the chemical potentials. From the point of view of dynamics, this means that the hydrodynamic solutions, with and without chemical freezeout, are nearly the same when expressed as a function of energy density. The flow velocities on a freezeout surface of constant energy density are independent of whether or not chemical potentials are included. However, the temperature on that freezeout surface depends dramatically on the chemical freezeout.

Finally, let us make a comparison with the hadronic cascade as a simpler alternative: Refs. [631, 673] with RQMD after burner. The difference with hydrodynamics for chemically frozen EoS is really small. Let me show just one figure which shows the mean emission time of pions, as the switching temperature to cascade is lowered. As with the spectra (not shown), the mean emission time after cascading is insensitive to the switching temperature. (Note that what this calculation also shows, is that the viscosity effects present in the cascade but absent in the ideal hydrodynamics should not be large: otherwise no agreement between spectra and freezeout time would be observed.

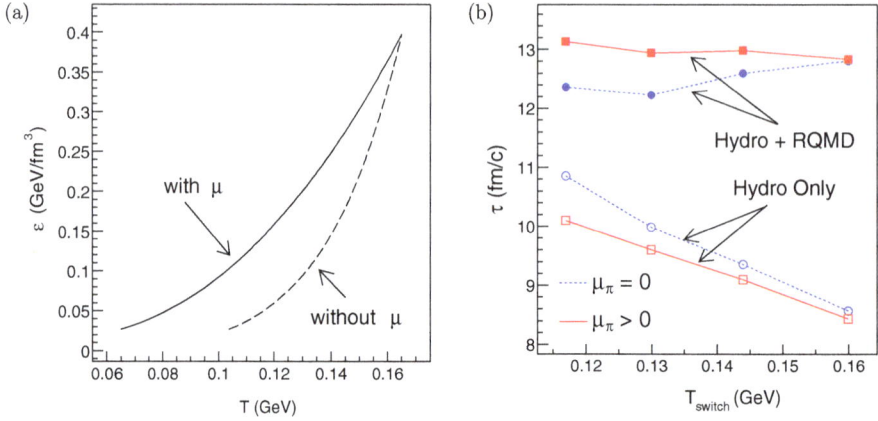

Fig. 9.8. (a) The energy density as a function of temperature with and without chemical freezeout for the SPS ($s/n_B = 42$). (b) The mean emission time $\langle\tau\rangle$ at the SPS (PbPb, $\sqrt{s} = 17\,\mathrm{GeV}A$, $b = 6\,\mathrm{fm}$, $s/n_B = 42$) as a function of T_{switch} with chemical freezeout $\mu_\pi > 0$, and without chemical freezeout, $\mu_\pi = 0$.

9.3.4. *Thermal freezeout*

Now we switch to a discussion of the *elastic* reactions in the resonance gas, in order to determine their rates and the shapes of the surfaces at which thermal freezeout takes place.

The major processes are the low energy elastic $\pi\pi, \pi K$ and πN scattering. Those have especially large cross sections due to the existence of the low energy resonances, ρ, K^*, Δ. The estimates of the $\pi\pi$ collision rate using the chiral Lagrangian was made in my work [620] and, in more detail, by Goity *et al.* [621]. The simple result obtained,

$$\nu_{\pi\pi} = \frac{T^5}{12F_\pi^4}, \tag{9.46}$$

displays a very strong T-dependence, which can be obtained by dimensional considerations alone. This feature remains true when one includes the resonances [614]; basically in the interval we are dealing with ($T = 120\text{–}150\,\mathrm{MeV}$) the pion–pion scattering rate increases by the factor 2.[15] These rates are increased further by the inclusion of the non-zero value of the pion chemical potential discussed in the preceding subsection.

Strong T-dependence leads to the following qualitative feature of freezeout: relatively modest changes in the freezeout temperature correspond to quite significant changes in the duration of the hadronic collision-dominated stage.

Let us now provide more quantitative information about the rates (see details in Refs. [494, 607]) tuned for the AGS/SPS region. For RHIC see, e.g. Ref. [689].

---

[15]Furthermore, the inclusion of resonances changes the dependence on the pion momentum p: in contrast to the chiral result the rate becomes basically flat for $p < 700\,\mathrm{MeV}$ which we need, and decreases for larger p (now, in contrast to the lowest order chiral result which predicts the unphysical rise with p).

The scattering rate for $\pi\pi$ can be obtained from the following expression, including resonances with the parameterization of the chemical potential term (the exponent):

$$\nu_{\pi\pi} = 40\,\mathrm{MeV}\left(\frac{120\,\mathrm{MeV}}{T}\right)^{3.1}\exp\left[190\,\mathrm{MeV}\left(\frac{1}{T}-\frac{1}{T_c}\right)\right] \tag{9.47}$$

(which is not valid at T below 80–100 MeV, where it would predict Bose condensation of pions).

The πN cross section is very large, reaching about 200 mb at the Δ resonance peak. The naive radius of the interaction $R = \sqrt{(\sigma/\pi)} \approx 2.6\,\mathrm{fm}$ is so large that one may question simple cascades and think about collective effects ("pi-sobars"). Absolute scattering rates depend on the density of nucleons at the decoupling stage. At low collision energies such as AGS the (isospin averaged) rate is of the order of $1/\tau_{\pi N} \approx 100\,\mathrm{MeV}$, which is larger than $1/\tau_{\pi\pi}$. Since the nucleon-to-pion ratio is about one, the rates are very close also. At SPS energies the situation is quite different: the nucleon/pion ratio is about 1/5. It makes the πN scattering less important for pions, but nucleons have very large collision rate and thus should freezeout very late.

Kaon and other strange secondaries have smaller collision rates. We have already mentioned a special case of ϕ with the scattering and absorption cross sections in a few mb range. Clearly one can completely ignore their re-scattering in the hadronic phase: we assume therefore that their thermal freezeout (as well as chemical one) coincides with the end of the mixed phase.

The general formula for the averaged collision rate of particle a resulting from binary collision with particle b is given by

$$\Gamma_{ab}^{a}(T) = \left(\int\frac{d^3\mathbf{p}_a}{(2\pi)^3}\frac{1}{e^{E_a/T}\pm 1}\right)^{-1}\int\frac{d^3\mathbf{p}_a}{(2\pi)^3}\int\frac{d^3\mathbf{p}_b}{(2\pi)^3}\frac{1}{e^{E_a/T}\pm 1}\frac{g_b}{e^{E_b/T}\pm 1}$$
$$\times\,\sigma_{ab}[(p_a+p_b)^2]\left|\frac{\mathbf{p}_a}{E_a}-\frac{\mathbf{p}_b}{E_b}\right|, \tag{9.48}$$

where E_a is the energy of a (minus any chemical potential for E_a within the thermal exponent), and similarly for b. g_b is the multiplicity of b and the signs in the denominator of the thermal weights are chosen based on whether a, b is a fermion or boson.

A natural definition of the kinetic freezeout condition: a secondary (e.g. a pion) emitted at this time has *equal chances* to be either re-scattered or escape:

$$P_{\mathrm{escape}} = P_{\mathrm{rescattering}} = \frac{1}{2}. \tag{9.49}$$

We will combine it with the escape probability

$$P_{\mathrm{escape}} = \exp\left(-\int_{\tau_{\mathrm{th}}}^{\infty}dt'\sum_j n_j(t')\langle\sigma_{ij}v_{ij}\rangle\right) = \frac{1}{2}, \tag{9.50}$$

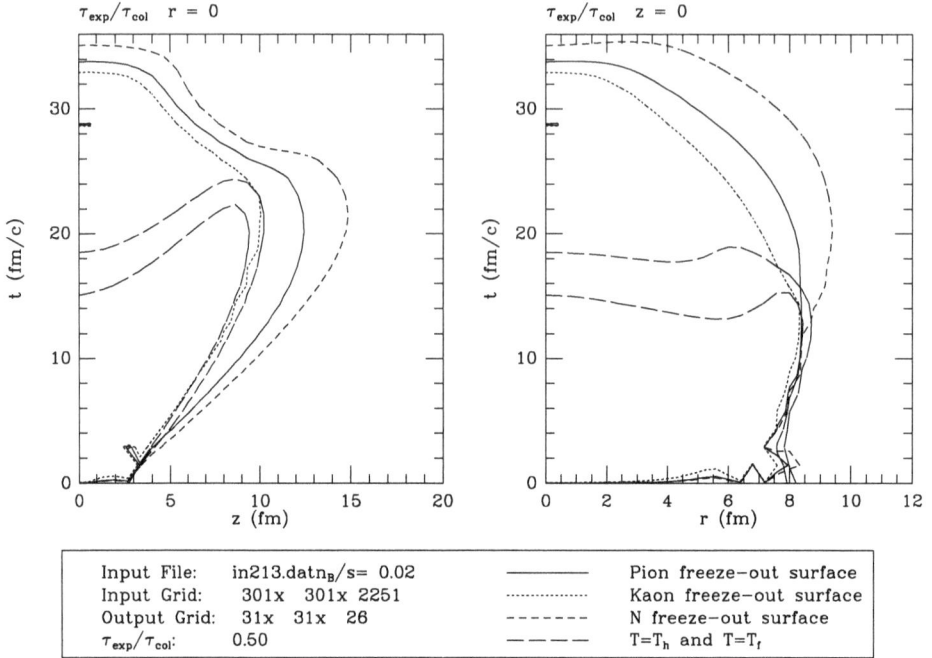

Fig. 9.9. Freezeout surfaces for 160A GeV Pb+Pb.

where $n_i(t)$ is the time-dependent density of the relevant secondary in matter, and the $\langle \sigma_{ij} v_{ij} \rangle$ include the thermally averaged cross section and velocity for scattering on type-j particle.

Let us now, following the paper by Hung and myself [607], show what the resulting thermal freezeout ($3d$) surface looks like. Some representative cases are shown in Fig. 9.9 as a section by time t-longitudinal coordinate z plane (at transverse coordinate $r = 0$) and the t–r plane ($z = 0$). We have already commented on the dependence on the particle kind above. Note also a significant difference between heavy and light ions. As expected, one finds that the larger the system, the lower the T_{th} on this surface.

Furthermore, the shape of the freezeout surface is very different from simple isotherms. It means that there is a significant variation of this temperature over the surface itself. The main conclusion of the paper by Hung and myself [607] is thus somewhat like a paradox: in order to find the *coolest* pion gas, one should look at *the very center of central collisions of heaviest nuclei at highest available energy!* (If the reader needs the experimental confirmation of this statement, have a look e.g. at Fig. 9.30 and related discussion.)

Summarizing this section we conclude that the improved thermal freezeout condition leads to a huge difference compared with simple isotherms. Different particles freeze out at different times. Also even for the same fixed value of the $\tau_{\mathrm{exp}}/\tau_{\mathrm{coll}}$ one gets very different conditions for different A, or for large and small y (the central region cools further). These observations provide natural explanations to strong A- and y-dependence of the radial flow flow at AGS/SPS energies.

9.3.5. *After freezeout*

The reader may be puzzled why we have to discuss the stage after freezeout, when there is nothing interesting but the free streaming of all produced secondaries. However, the particle density is never zero and for some observables (e.g. deuteron formation) we have to know how exactly this density depends on space–time variables.

In particular, the observation of resonances (such as ρ, f_2 in $\pi^+\pi^-$ mode, etc.) is done via the peak in invariant mass distribution. This in turn demands that both outgoing decay pions (or other secondaries) do not have a *single* rescattering.

The same problem appears in the production of nuclear fragments (such as d, He^3, He^4 or their antiparticles), which are easily destroyed on the way out by collisions.

Let me start with quite a interesting solution of the hydrodynamic equations, as well as even the collisionless Boltzmann equation, recently found by T. Csorgo and collaborators [594]. In this work a non-relativistic spherically symmetric hydrodynamics is considered, with EoS of a simple Maxwellian gas

$$\epsilon = \frac{3}{2}p = \frac{3}{2}Tn. \tag{9.51}$$

The idea is to take a Gaussian ansatz for the density and decreasing T

$$n(r,t) \sim \exp\left(-\frac{r^2}{2R(t)}\right), \qquad T = T_0\frac{R_0}{R(t)}, \qquad v_r = r\frac{dR(t)/dt}{R(t)}, \tag{9.52}$$

with time-dependent radius and velocity, which is consistent with all hydrodynamic equations provided $R(t)$ satisfies some second-order equation of motion, the solution of which is a quadratic polynomial in t. As a result, the effective Hubble constant $R(t)/dt/R(t) \to 1/t$ at large time, while $T \to 1/t^2$. So, on one hand one can view this solution as a normal collisional hydrodynamic solution, in which all of the thermal energy is eventually given to the hydrodynamic flow.

What however is special for this solution is that the final spectrum of particles is exactly the same as the initial thermal one, and in fact the integral over space gives the same particle spectrum at any time moment. (This is of course to be expected because it retains its Gaussian form and thus its width is held constant by the energy conservation.) So what this solution describes is basically a free streaming of particles.

The authors were indeed able to show it explicitly, by noticing that the phase space distribution for their solution,

$$f(r,p,t) \sim \exp\left[-\frac{r^2}{2R^2(t)} - \frac{(p - mv(r,t))^2}{2mT(t)}\right], \tag{9.53}$$

actually satisfies the *collisionless* Boltzmann equation, which is quite remarkable.

A related discussion of an elliptic solution to the hydrodynamic equation (but unfortunately not the Boltzmann equation in this case) with three different radii in three directions, is discussed in Refs. [595, 609]. These authors have found simple solutions for self-similar elliptic flows, in which v_x, v_y, v_z have Hubble-like behavior with different Hubble constants. They can also go through the equations provided the initial conditions and the EoS (which is that for non-relativistic Maxwellian gas) are tuned accordingly, so that the ellipticity of the system does not change throughout the expansion.

9.3.6. *Freezeout of resonances and nuclear fragments*

Let me start with few simple pedagogical points about the observability of resonances and fragments, or "composites" as we may call them collectively, for brevity.

We will discuss below the rate equations which can be solved and determine the number of composites $N(t)$ at time t. The "observability condition" of a resonance can be written as

$$
\nu_{\text{visible}}(t) = \Gamma N(t) \exp\left(-\int_t^\infty \nu(t')\,dt'\right),
\tag{9.54}
$$

where the l.h.s. is the production rate of "visible" resonances, Γ is the resonance decay width and the exponent is the so called optical depth factor containing integrated $\nu(t)$, the combined scattering rate for all decay products. The $N(t)$ decreases with time due to expansion and cooling, while the exponent changes from 0 at early time to 1 at late time: so the product naturally has a maximum at the time t_{m} such that

$$
\frac{1}{N(t_{\text{m}})}\frac{dN(t_{\text{m}})}{dt} + \nu(t_{\text{m}}) = 0.
\tag{9.55}
$$

This condition means that for observable resonances the freezeout condition is different from that for stable particles. It reads: *the rate of their number change is equal to the absorption rate of all the decay products.* For example, for ρ and σ we should not know their scattering rates but just that of two pions. For short-lived ρ and σ the first factor is close to the overall expansion rate of matter at late time, which follows from Hubble-like late-time regime $d\log N(t_{\text{m}})/dt \approx 3/t$, and the second is the same. For that reason we conclude that "visible" ρ and σ are produced at the same time.

The formation rate for a fragment made of A nucleons is made by some coalescence, and such rate is obviously proportional to nucleon density to that power, $\sim (n_N(t))^A$. After it is produced, however, it still has very small probability to survive. Assuming that the destruction rate for a fragment $\nu_A \approx A\nu_N$, where ν_N is a scattering rate for one nucleon, one finds that for A-fragment the time distribution is approximately the Ath power of the same universal function

$$n_{\text{fragments}}(t) \sim \left[n_N(t) \exp \left(- \int_t^\infty \nu_N(t')\, dt' \right) \right]^A . \tag{9.56}$$

So, *the maximum of production of any visible fragment happens at the same time for all A.* Furthermore, the width of the distribution over production time decreases as A grows, as $1/\sqrt{A}$.

The equations themselves describing the dynamics of resonances are well known, generically they contain the sink (the decay) and the source terms

$$\frac{\partial n(t, \vec{r})}{\partial t} = -\Gamma n(t, \vec{r}) + S(n_i) \tag{9.57}$$

(where for an expanding source the time should be understood as the proper time in the rest frame of all volume elements). In many papers in the literature (e.g. Ref. [618]) the source term is ignored citing "instantaneous hadronization", but (especially for resonances we consider as an example, ρ, σ) it is not true: in fact the primary generation of resonances die out long before the "observable" ones are born.

We use an approximate power fit of the source time dependence $\int d^3r\, S = \Gamma N_0 \, (t_0/t)^P$. Its power can be related to fireball expansion. In the volume $V(t) \sim 1/n(t) \sim t^a$ the integrated source is proportional to $V(t)[n_\pi(t)]^N$ where N is the multiplicity in resonance decay ($N = 2$ for σ, ρ). The pion number is "chemically frozen", $N_\pi = V(t)n_\pi(t) = \text{const}(t)$ from which we conclude that the source term power is $P = (N - 1)a$ (for σ, ρ and other binary resonances $P = a$). An example of such an equation solved is shown in Fig. 9.10.

We then evaluate the *optical factors* for pions and nucleons, using realistic rescattering rates, with chemically frozen composition, using papers by Hung and myself [607] and by Tomasik and Wiedemann [689]. An example for the final time distributions for visible ρ and d is shown in Fig. 9.11. Note that both distributions have maxima as we discussed above, and that the "visible" d are indeed produced very late. This is our main point: we are speaking about a very dilute matter, after the freezeout of all the basic ingredients of the fireball.

9.3.7. *Resonance modification*

It has been argued over the years that in matter the resonances should be modified, with shifted mass, increased width and even significantly changed shape. With the very late stages of relevant RHIC collisions, we can now access very dilute matter in which those effects must be calculable in the lowest order of the density, providing a benchmark test to all such discussions. In such a case hadron modification is expressed in terms of their *forward scattering amplitude* $M_{ij}(t = 0, s)$. Note that the scattering amplitude is complex, so this approach gives both the real and imaginary part of the dispersion law modification, also known as the optical potential.

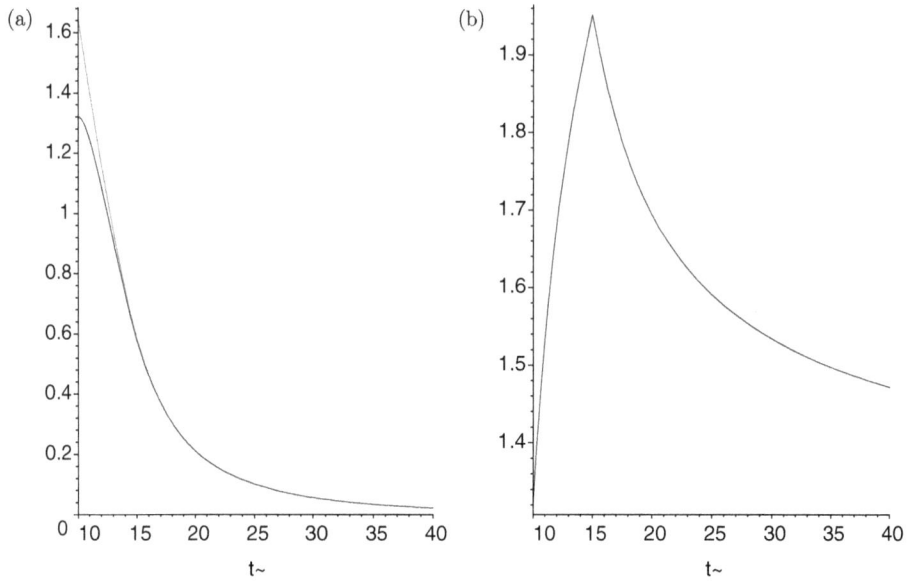

Fig. 9.10. (a) The time dependence of the ρ-meson density, starting from chemical equilibrium at $t = 5$ and 10 fm. Plotting its ratio to the pion density in (b) one observes the transition from hydro to free streaming regimes and that the ρ density decreases power-like rather than exponentially, because of the source term.

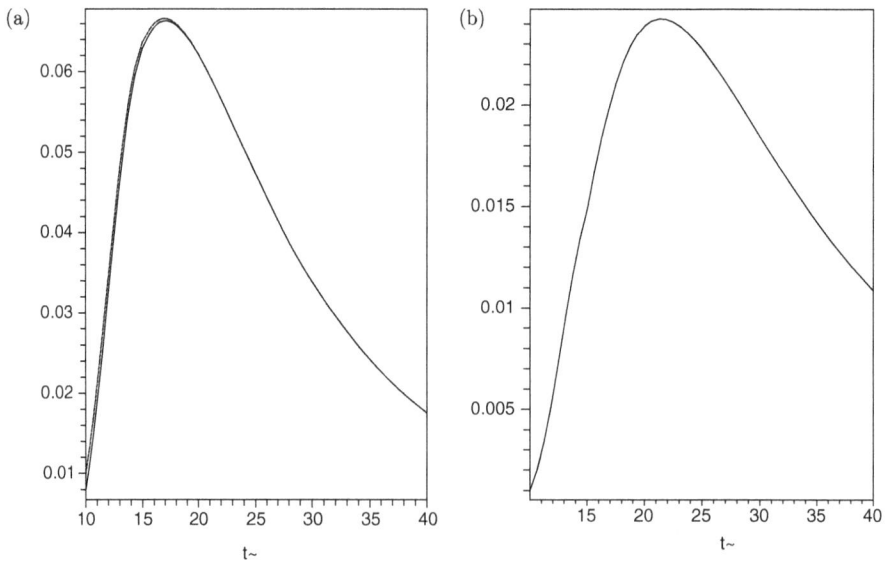

Fig. 9.11. The time dependence of the visible ρ-meson (a) and deuteron (b) production, for the $r = 0$ point in central AuAu collisions at RHIC.

Table 9.1. A set of resonances considered for
ρ-meson mass shift, from [611].

Name/Mass	Width	Branching	Mass shift
$a_1(1260)$	400	0.6	−19
$a_2(1320)$	104	0.7	−15
$K_1(1270)$	90	0.4	+1.6
$K_2(1430)$	100	0.087	−0.4
$N^*(1520)$	120	0.2	10

There are two major theoretical approaches to the issue discussed in literature, to be called an *s-channel* and a *t-channel* one. The former approach assumes that the scattering amplitude is dominated by *s*-channel resonances which are known to decay into the $i + j$ channel. For most mesons such as π, ω, ρ, K in a gas made of pions such a calculation has been made, e.g. in Ref. [495] related with such $\pi\rho$ resonances as a_1 or $N\rho$ resonance $N^*(1520)$ [721] for ρ at SPS. Note that the signs of the effects are opposite in those examples, as seen from Table 9.1, from the paper by Brown and myself [611].

The majority of the particles in the matter are Goldstone bosons π, K, η which do not interact at small momenta. However the usual attraction between other particles is there. Using a simple sigma-model-like expression for the mass shift one gets [611] $\delta m_\rho^N \approx -28\,\text{MeV}$ due to all $\bar{B} + B$. An additional shift $\delta m_\rho^{\rho,\omega,K^*} \approx -10\,\text{MeV}$ comes from scalar exchanges between ρ and all other vector mesons ρ, ω, K^*. The main difference between the two mechanisms of the mass shift discussed above is that the *t*-channel attraction is not associated with the broadening, while the *s*-channel resonances increase the width by about 50 MeV. On the other hand, there is a "kinematic" effect working in the opposite direction. The negative mass shift discussed above automatically reduces the width, both because of the reduced phase space and also due to the power of p in the P-wave matrix element. The magnitude of this effect for the predicted mass shift is

$$\delta\Gamma_\rho = 3\frac{\delta m_\rho}{m_\rho}\Gamma_\rho \approx -50\,\text{MeV}. \tag{9.58}$$

So, inside the accuracy these two effects cancel each other.

The invariant mass distributions in *pp* and mid-central AuAu of the $\pi^+\pi^-$ system, with a transverse momentum cut $0.2 < p_t < 0.9\,\text{GeV}$ have been measured by STAR [612]. I should skip here a discussion of the shift in *pp* (see Ref. [611]), and I only comment that in AuAu the ρ peak is found to be shifted by an additional $\sim -40\,\text{MeV}$ in mass, with the same width. This agrees well with the estimates above.

The same approach should of course be applied to many other resonances. For narrower resonances, like K^*, we expect smaller shifts, while for broader resonances like σ we predict a complete change of shape. At small freezeout $T \approx 100\,\text{MeV}$ sigma was predicted to be deformed into a much narrower structure at mass of about 400 MeV, see figures in Ref. [611].[16] Another confirmation of a very late freezeout and low T, is that the σ/ρ ratio is found to be strongly growing toward the most central collisions.

Let me briefly comment on the issue of nuclear fragment coalescence, such as $p + n \rightarrow d$, discussed in many papers over the years when and authors struggled with the question how to calculate its rate. In particular, it is clear that when the level crosses zero the wave function at the origin vanishes, and so the production should do so too. And if the production rate is small compared to two other relevant rates, ν_{abs} and $d \log N/dt$, there is never thermal equilibrium and one should *not* use the statistical models for d and other fragments.

Significant progress has been made in a recent paper by Ioffe *et al.* [613] who have pointed out how to use consistently the in-matter widths of all particles and obtain the production rate. The recipe is to write all propagators keeping instead of the Feynman $i\epsilon$ the real in-matter widths of all particles involved. For example, in vacuum the $p+n \rightarrow d$ reaction is obviously forbidden by energy conservation, while including in-matter widths of p, n, d (which is of the order of hundred MeV) one finds that to get 2.2 MeV of deuteron binding energy is no problem. Ioffe *et al.*'s rate does satisfy the condition mentioned above, that at zero binding energy the rate vanishes because of a vanishingly small wave function at the origin. Hopefully this rate can be soon compared with the growing amount of RHIC data on fragments.

9.4. Hydrodynamic description of heavy ion data

9.4.1. *A long road to unveiling the transverse flow*

Landau has aimed his hydrodynamical theory mainly at a description of the longitudinal momenta. However, with the advent of QCD, Fermi's idea of the initial condition in the form of a Lorentz contracted and *stopped* disk of equilibrated matter, is in obvious contradiction with the asymptotic freedom[17] at small times. Nevertheless, as was noted by a number of people (including myself) in the early 1970s [598–601], the specific shape and width of the rapidity distribution predicted by Landau provided a surprisingly good fit to these data. The multiplicity prediction by Fermi $\langle n \rangle \sim s^{1/4}$ worked also. Does it really mean that we have hydrodynamics (with complete "stopping") in pp collisions, or is it a mere coincidence?

To get a better understanding of what is going on one may try to look for other effects predicted by hydrodynamics. The transverse expansion and its effect on the transverse momenta of secondaries was studied by Milekhin [602] (who by the way has already used the first lamp computer available to him in the late 1950s!).

---

[16]Exactly such a peak has been recently seen by STAR but the official presentation has to wait till larger data sample.

[17]In fact Landau had been motivated by the opposite behavior of effective charge in QED and scalar theories at small distances, which was then considered the only possible behavior of any QFT.

His main result for EoS $p = \epsilon/3$ was that the mean transverse momentum grows very slowly,

$$\langle p_t \rangle \sim s^{1/12}. \tag{9.59}$$

The first attempts to connect the experimental information on the collective transverse flow with matter properties was made independently at about the same time by Siemens and Rasmussen [626] for low energy (BEVALAC) and by Zhirov and myself [603] for high energy pp collisions at CERN ISR. The idea was exactly the same: the collective velocity of explosion can be measured by comparing spectra of light and heavy particles. For light pions the thermal spectrum is exponential in p_t, and being boosted by flow it remains the exponential one with a "blue shifted" temperature $T^* = T\gamma_t$ with the gamma factor corresponding to the transverse flow. For heavy particles (such as baryons) the effect is quite different. In particular, an infinitely heavy particle has zero thermal velocity, and, if captured by the flow, it just has the momentum mv_t, where v_t is the transverse flow velocity. So very heavy particles should follow the distribution of the flow velocity: real nucleons are in between the two limits.

The findings of these two papers were however completely different. BEVALAC spectra discussed by Siemens and Rasmussen [626], for heavy ions at $E \sim 1 \, \text{GeV} \cdot A$, indeed have shown the expected difference for the pion and proton. They were well fitted with two parameters, the freezeout temperature $T_f \sim 30 \, \text{MeV}$ and the velocity of what they have called the "blast wave" $v_t \approx 0.3$. (A long discussion of whether hydrodynamics is applicable to derive it followed: we should not go into that now.)

In our paper [603] the conclusion was negative: π, K, N spectra showed a very good m_t-scaling

$$\frac{dN}{dp_t^2} \sim \exp\left(-\frac{m_t}{T}\right), \qquad m_t^2 = p_t^2 + m^2, \tag{9.60}$$

with no sign of any corrections due to transverse collective flow. Presumably it simply means that the system created in pp collisions is too small for the macroscopic theory to become applicable. So we thought we had to give up all hopes of a hydrodynamical description of the pp collisions, and had to wait for heavy ion collisions at similar energies.[18] It did indeed happen as planned, only two decades later, at SPS and at RHIC.

(The puzzle remained why the pp data do admit such a good description by the thermodynamical model. We thus suggested that a hydrodynamic expansion is compensated by the bag pressure effect, which may stabilize the system into a metastable drop of QGP. We still do not know if this is the case.)

---

[18] Incidentally, let me recall that around 1980–1982 it looked quite possible that CERN ISR would be used for that purpose. Unfortunately the CERN leadership of the time decided otherwise, and even destroyed that steady and reliable collider "to store LEP magnets in the building". The extra irony of that announcement was the fact that LEP magnets were very low field ones and were made mostly of concrete, with a small admixture of iron, so there could be no problem with their storage.

Let us now jump many years ahead and explain the modern version of the same argument. In Fig. 9.13 we show a nice compilation of the results from the NA49 experiment at SPS [652], compared to blast wave parameterization with two basic parameters — the freezeout temperature and the mean flow velocity. One can see that modification of spectra of all secondaries are nicely explained by the model. As the mass of the particle increases, its distribution is less influenced by random thermal motion and more by the distribution over hydrodynamic velocity: therefore spectra shift from thermal exponential shape to those with flat or even increasing ones at small p_t. The parameter values, shown in the figure, show smooth variation with the collision energy: T_f decreases and velocity increases.

If the reader is not yet convinced that such a thing as radial flow does really exist, let me suggest one simpler but quite convincing argument. Suppose the flow interpretation is wrong and that the different m_t slopes in AA and pp collisions observed are due to something else (e.g. to "initial state" parton rescattering as many authors suggested). Let us then calculate the slope of the m_t spectra for *deuterons*. One can see in a back-of-the-envelope calculation that if the transverse momenta of the two nucleons had Gaussian shape and were uncorrelated, there should be *the same slope for deuterons* when two nucleons coalesce together.

| E |

Exercise 9.3.   Derive the last statement about the slope of the deuteron.

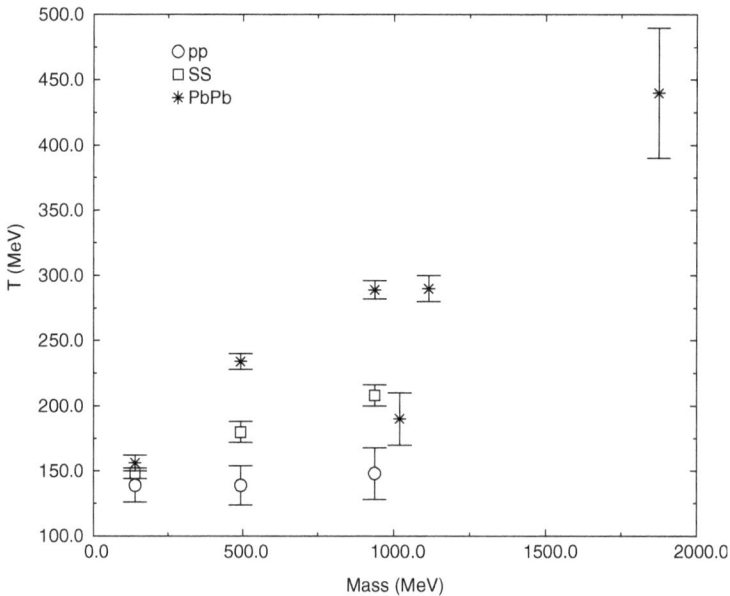

Fig. 9.12. Inverse slope parameters at SPS calculated from the m_t spectra for pp collisions (ISR) (open circles), SS and PbPb collisions at SPS (squares and stars, respectively versus the particle mass.)

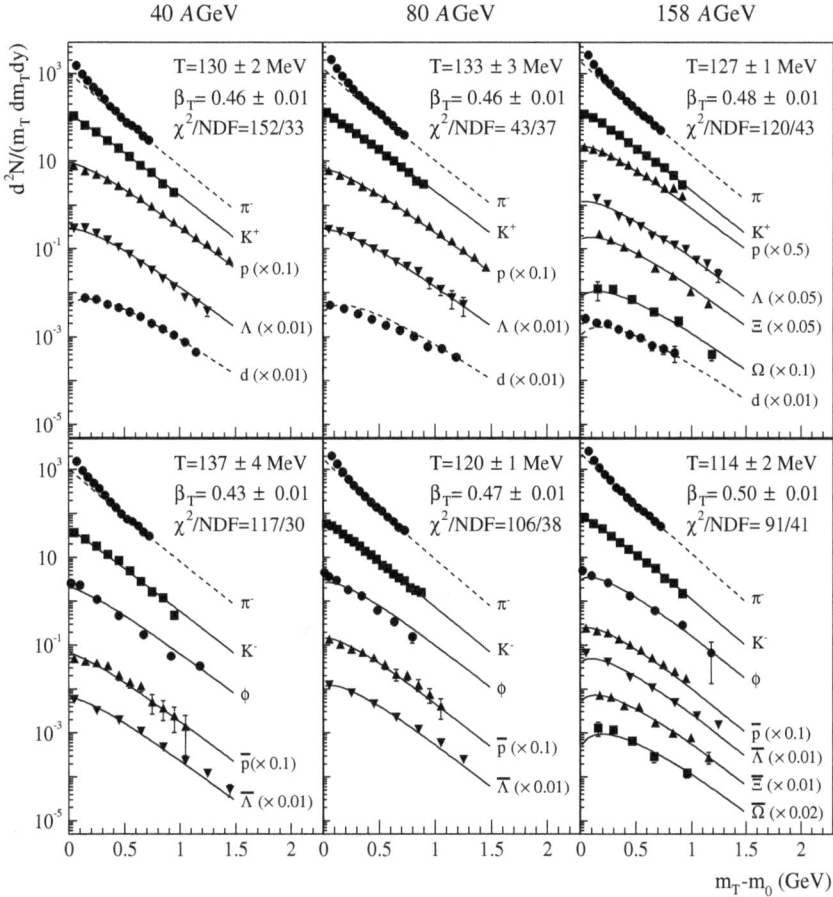

Fig. 9.13. Transverse mass spectra in central 40 (left), 80 (middle) and 158 GeV (right) Pb+Pb collisions at SPS by NA49 collaboration. The lines show the result of the transverse flow fit with parameters shown in the top of the figures.

This simple conclusion is contrary to observations: as Fig. 9.12 shows, the d slope is about twice as large. Only some existing correlation between the momenta of the two nucleons can explain that, and its magnitude is exactly as reproduced by the flow.

Let me now briefly explain the reason for the strong A-dependence of the slopes. While the pp data show perfect thermal-looking spectra without the slightest trace of radial flow, for SS collisions the slopes start growing with the mass of the secondary particle, and for PbPb the effect is about twice as large. So, *the larger the nuclei, the stronger the flow is.*

One possible interpretation of this fact is that the S nucleus is too small and there is no hydrodynamics in SS collisions, as in pp. This interpretation was widely accepted, since hydrodynamics practitioners at the time mostly predicted the same flow for all A, as a simple consequence of scale invariance of the hydrodynamics equations. The loophole in the argument however was in the universal *A-independent* freezeout temperature used in those works which is incorrect.

It has been shown in our work [607] that hydrodynamics describes these data very well. One simply has to use the correct freezeout conditions, and then it follows that the hydrodynamic evolution is cut off by freezeout earlier in SS than in PbPb collisions, consequently the collective effects are smaller. For the same reason the systematics of the *rapidity* dependence of the radial flow worked as well.

9.4.2. *Qualitative effects of the QCD phase transition*

For EoS with a phase transition there are three phases and three corresponding stages in the acceleration history. (i) an explosive QGP phase $(e > e_Q)$, in which the matter accelerates rapidly, (ii) a soft mixed phase $(e_H < e < e_Q)$, in which the matter streams freely with constant velocity and (iii) a hadronic phase $(e < e_H)$, in which the hadronic pressure produces additional acceleration.

The effective EoS calculated along the appropriate adiabatic paths [607] from resonance gas plus QGP (from lattice) is shown in Fig. 9.14a in the form p/ϵ versus ϵ. The effect of the bag pressure on QGP is seen in this figure as a strong dip of this ratio, toward the so called **softest point**, the minimum of p/ϵ. Since the gradient of p is the driving force and ϵ is the mass to be moved, the acceleration of matter is proportional to this ratio. Its small value means that the mixed phase is much softer than both the relativistic pion gas and the high-T QGP on both sides. What this picture shows is that one can expect more robust hydrodynamic flow when the

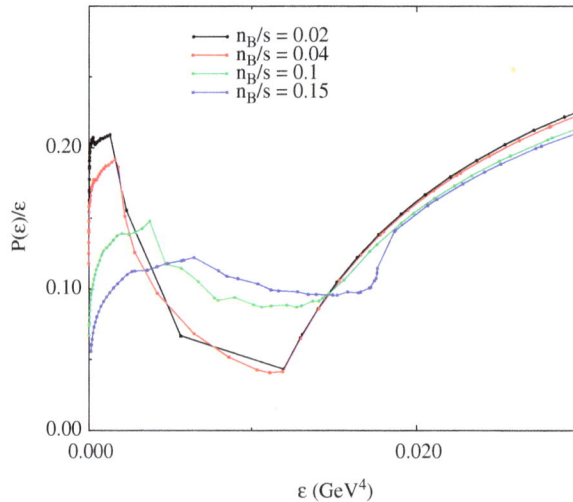

Fig. 9.14. The dependence of the pressure-to-energy-density p/ϵ ratio on the energy density along the adiabatic paths with different baryon density to entropy ratio.

QGP produced is significantly more dense than at $T = T_c$, the "soft QGP". In a nutshell, that is why high energies of RHIC are needed so much.

One version of a similar argument, suggested by van Hove [604], was that the $\langle p_t \rangle(s)$ dependence should first rise, then level off, then rise again, as the initial conditions scan the hadronic, mixed and QGP phases, respectively. Unfortunately, more detailed studies have not found this behavior. First of all, the pion $\langle p_t \rangle$ is rather insensitive to flow. But even for hydrodynamic velocity $\langle v_t \rangle(s)$ itself it is not quite so, mostly because the final velocity is still the integral of acceleration over time, and softer matter EoS automatically leads to longer time.

Compared to these strong trends, the observed dependence of the flow on the *collision energy* appears to be weak. Unlike for BEVALAC/SIS energies, in which v_t steadily grows, for the AGS (10–15 GeV/A) and SPS (160–200 GeV/A) domain the radial flow velocity is (within uncertainties) about the same. It must be a mere coincidence, since both the meson/baryon ratio, the EoS and even the general picture of space–time development of the collisions are radically different.

At a more practical level, the fits to spectra with "hydro-motivated formulae" have generally produced vastly varying (although correlated) values for the transverse flow velocity v_t and the thermal freeze-out temperature T_f. In addition, there is strong A- and rapidity dependence of these parameters which needed to be explained.

It was for the most part worked out in kinetic calculations [607]. Let me just mention the *motto* of this work, "the larger the system, the further it cools", explaining in effect both strong A and y dependence of the flow. Very large $v_t \approx 0.6$ and very low $T_f \approx 120$ MeV in PbPb was predicted.[19] Later analysis of the HBT radii, combined with spectra, had confirmed such selection, see e.g. Ref. [690].

Let us now proceed and introduce terminology for other forms of collective flow. So far we discussed (i) axially symmetric *radial* and (ii) *longitudinal* flow that exist even for central collisions. For non-zero impact parameter experiments we have also seen clear signals for at least two non-zero harmonics in the angle ϕ, known as (iii) *dipole* and (iv) *elliptic* flow.

That the shape of the "initial almond" for non-central collisions leads to flow enhancement toward the impact parameter, see Ref. [665]. Now most works use the particle *number* v_i harmonics defined as

$$\frac{dN}{d\phi} = \frac{v_0}{2\pi} + \frac{v_2}{\pi}\cos(2\phi) + \frac{v_4}{\pi}\cos(4\phi) + \cdots, \qquad (9.61)$$

rather than "energy flow moments", with momenta in the weight. In some works v_2 is additionally normalized to the spatial asymmetry of the initial state (the

---

[19]Only very elementary kinetics of low energy pion and nucleon rescattering is actually involved here, so any event generator like RQMD which included it also generates it. Just nobody put it in the right words...

"almond") $s_2 = \langle y^2 - x^2 \rangle / \langle y^2 + x^2 \rangle$ at the same b, in order to cancel out this trivial factor and make the effect roughly b-independent.

The reason why elliptic flow is even more important than radial one was pointed out in Ref. [667]: as it is developed *earlier* than the radial one, it may shed some light on whether we do or do not have QGP at SPS.

9.4.3. *Solutions to hydrodynamic equations*

It is clear that the collision of two nuclei at some arbitrary impact parameter is a rather complicated process, even if we have reduced it to the solution of hydrodynamic equations. Thus, it is reasonable to use any simplification possible, and most studies in the literature for high energy collisions (SPS/RHIC) are done for *boost independent* or $(2+1)d$ equations (9.33). Much less effort has been made in order to understand the complete $(3+1)d$ hydrodynamic equations (such as Ref. [645]), partly because of very uncertain initial conditions in the longitudinal direction. In applications hydrodynamic equations are usually solved numerically, e.g. with a Godunov method as was the case in Ref. [635].

At lower energies (GSI/AGS) boost-invariance cannot be used, and hydrodynamic works have used another $(2+1)d$ simplification, that of axial symmetry, and thus gave up the elliptic flow phenomenon (which is very small, of the order of a couple of percent only, at those energies.)

We will follow in this section mostly the paper [673] which systematically studied how the hydrodynamic evolution depends on the different EoS used, although it also used plots from other groups. The space–time picture of matter evolution at the SPS and RHIC is summarized in Fig. 9.15. The so called switching isotherm, e_H (shown in the middle), is where the hadronization happens. In the hydro+cascade approach, the particles are injected into a hadronic cascade (RQMD) with the velocity distribution of this isotherm.

The QGP phase dictates the duration and transverse size of the collision. At RHIC, the QGP pressure drives the matter outwards, rapidly increasing the radius, which in turn shortens the overall lifetime. Therefore, approximately doubling the total multiplicity from the SPS to RHIC increases the total lifetime only slightly. All the additional multiplicity is absorbed by a larger longitudinal and slightly larger transverse radius.

By contrast, for very soft EoS with large latent heat, the stiff QGP phase is absent and the mixed phase is dominant up to higher energy densities. The strong transverse acceleration is replaced with a slow evaporative process — this picture was known as the "burning log". It is excluded by RHIC data.

Summarizing: the QGP drives a transverse expansion while the "dark energy" of the bag constant works against it. The last effect is very important in the "mixed phase" region of the energy density, and therefore only with sufficiently high collision energy of RHIC the transverse expansion is sufficiently robust.

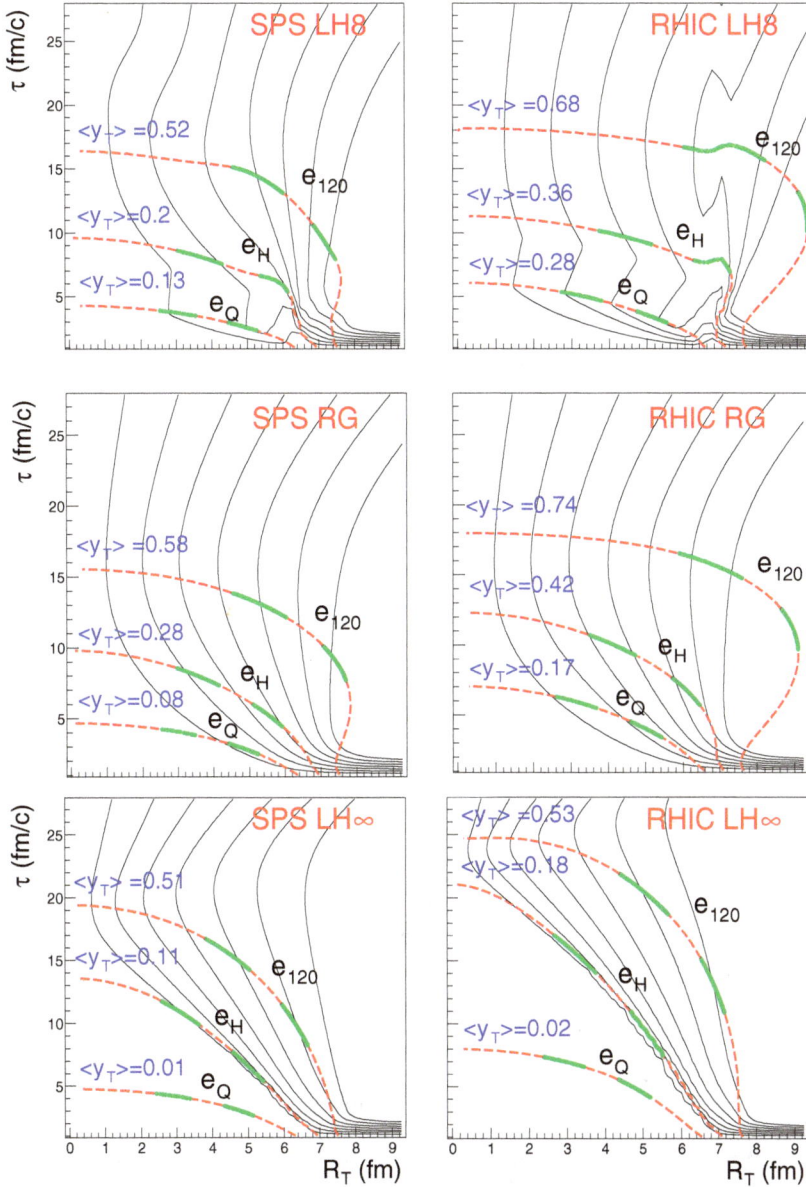

Fig. 9.15. The left and right sides show the hydrodynamic solution at the SPS and RHIC for different EoS. The thin lines show contours of constant transverse fluid rapidity ($v_T = \tanh(y_T)$) with values $0.1, 0.2, \ldots, 0.7$. The thick lines show contours of constant energy density. e_{120} denotes the energy density where $T = 120$ MeV. e_H and e_Q denote the energy density (for LH8) where the matter shifts from hadronic to mixed and mixed to a QGP, respectively. The shift to RQMD is made at e_H. $\langle y_T \rangle$ denotes the mean transverse rapidity weighted with the total entropy flowing through the energy density contours. Walking along these contours, the line is broken into segments by dashed and then solid lines. 20% of the total entropy passing through the entire arc passes through each segment.

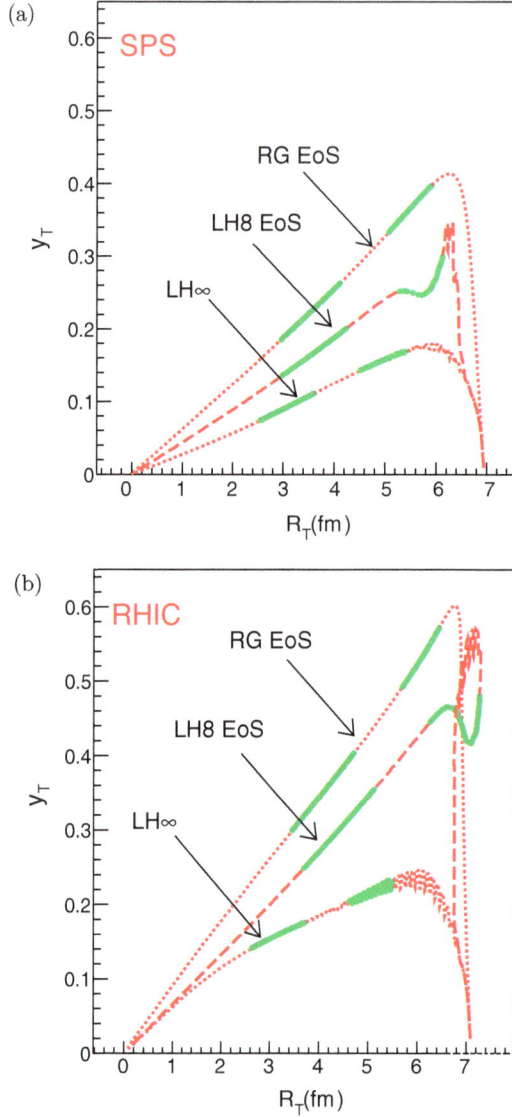

Fig. 9.16. Walking along the e_H contours in Fig. 9.15 (where the pure hadronic stage starts and the switch to hadronic cascade can be made), the transverse rapidity is traced as a function of radius at (a) the SPS and (b) RHIC for the three EoS with different latent heat. The dashed and solid segments of the curve have the same meaning as in the previous figure.

To quantify the velocity distributions obtained by the end of the QGP/mixed phases, we show in Fig. 9.16 the transverse fluid rapidity versus the radius along the switching isotherm e_H at the SPS and RHIC. For LH8 and RG EoS, the transverse rapidity shows a linear Hubble-like rise with radius. For an EoS with a phase transition to the QGP (LH8), the acceleration is initially large but subsequently stalls in the mixed phase.

Nevertheless, it should be noted that if freezeout is taken as some constant $T_f \approx 120\,\text{MeV}$, as was done in many works, then the differences between the flow velocities of the EoS are smeared out by the hadron phase, as can be seen by examining the mean flow velocities on the e_{120} curves in Fig. 9.15.

9.4.4. *Radial flow at SPS and RHIC*

Let us now show how the hydrodynamical calculations described above relate to experimental data. In Fig. 9.17 we show NA49 data on negative hadrons ($\pi^- K^- \bar{p}$) and net proton spectra. The lines correspond to the calculated net proton spectrum for the resonance gas EoS and for EoS LH4–LH16. The experimental and theoretical spectra are absolutely normalized. The two numbers parameterizing the initial conditions of total entropy and baryon number are adjusted to match the height of these spectra. The model curves have been multiplied by a factor of 0.93 to account for the fact that the data is 5% central. Once the height of the spectrum is tuned, the shape of the spectrum is determined by the course of the hydrodynamic evolution, or more generally, by pressure gradients and the duration of the collision. Therefore, it is significant that hydrodynamics generates a flow $v_T \approx 0.5$, which is needed to explain the spectra.

For EoS with large latent heats (e.g. LH16), the p_T spectrum is too soft. This is because the hydrodynamic system spends a long time in the mixed phase in which pressure gradients do not generate collective motion.

How can one see which fraction of the flow velocity was generated in the QGP phase and which during hadronic one? The slope systematics of multi-strange particles like Ω provides the necessary information, since their hadronic cross section is smaller and they freeze out earlier [633]. The slope parameter, T_{slope}, defined by

$$\frac{1}{M_T}\frac{dN}{dM_T} = C\exp(-\frac{M_T}{T_{\text{slope}}}), \tag{9.62}$$

where $M_T = \sqrt{m^2 + p_T^2}$, is shown versus the particle mass in Fig. 9.18.

Note that the dip at Ω is really observed at SPS.

Now we are ready to discuss qualitative changes at RHIC. It was argued above that LH8 provides the best description of the radial flow at SPS: the same EoS worked best for RHIC. It is far from trivial, since RHIC, the initial energy density, is well above the phase transition, and the large early pressure is expected to drive collective motion. In Fig. 9.19, the nucleon M_T spectrum for the SPS and RHIC are shown with and without the hadronic rescattering in RQMD. Two features are immediately observed: 1. The $\langle M_T \rangle$ increases as the collision energy is increased from the SPS to RHIC. 2. The spectra without hadronic rescattering are reasonably well described by a single exponential (i.e. they look linear on the log plot shown).

(a)

(b)

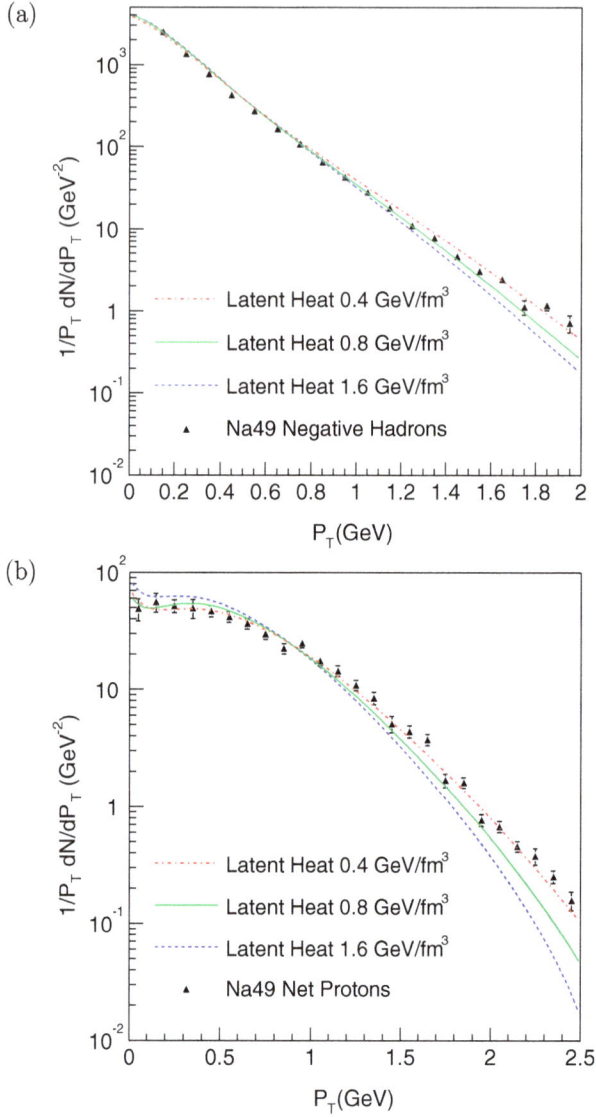

Fig. 9.17. A comparison to NA49 [660] (a) negative hadron and (b) net proton spectra for different EoS.

To summarize the bulk energy transport in the model we show in Fig. 9.18: in Fig. 9.18a the predicted slopes, as a function of particle mass, for different EoS. One can see that the actual SPS data (the closed points) agree best with LH8 predictions.

How the effect grows with the collision energy, or rather with the entropy produced or the rapidity density $\frac{dN}{dy}$, is shown in Fig. 9.18b.

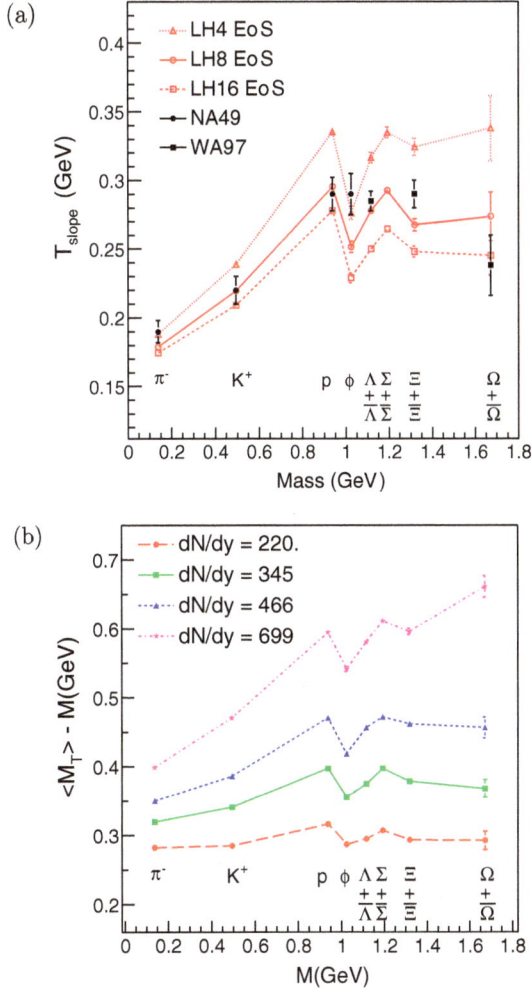

Fig. 9.18. (a) A compilation of slope parameters from the SPS experiments and a comparison to the model for different EoS. The points are from NA49 and WA97 publications. (b) For PbPb collisions at $b = 6$ fm, $\langle M_T \rangle - M$ for different particle species as the total multiplicity is increased from the SPS to RHIC and beyond.

Figure 9.18b demonstrates how the increase in the mean M_T influences the spectra of different particles by plotting $\langle M_T \rangle$ versus mass [633]. At the SPS the flow velocity at the end of the mixed phase is relatively small — $v_T \approx 0.4$. The slopes before the RQMD phase show a linear rise characteristic of the hydrodynamic system, $T_{\text{slope}} = T_{\text{th}} + m\langle v^2 \rangle$. When the flow velocity is small, hadronic rescattering changes the linear mass dependence significantly, giving the characteristic shape observed at the SPS. As the flow velocity increases from the SPS to RHIC and beyond, the linear rise with mass becomes increasingly steep and hadronic rescattering, while still contributing to 20% of the $\langle M_T \rangle$ for nucleons, does not change the

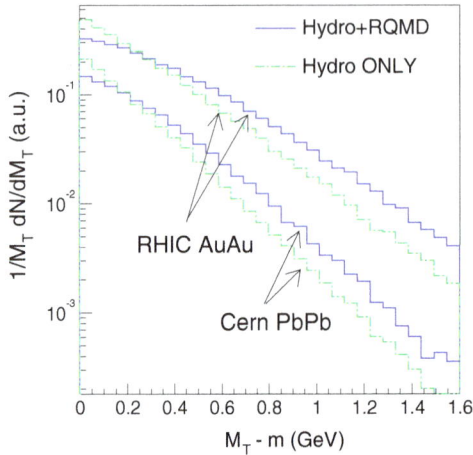

Fig. 9.19. The nucleon M_T spectrum at the SPS and RHIC with and without the RQMD after-burner.

overall mass dependence. The qualitative shape of the mass dependence of $\langle M_T \rangle$ therefore gives a good measure of the flow velocity at the end of the mixed phase.

Now we compare model predictions to the first RHIC spectra, keeping in mind the two major predictions of hydrodynamics. First, $\langle M_T \rangle$ should increase significantly. Since the EoS LH8 was found to give the best agreement to SPS flow data, LH8 should give the best agreement at RHIC. Out of all the EoS studied, the flow velocity increases the most for LH8. Second, the spectra should show the about linear flow profile, sensitive to the particle mass and flow velocity. At RHIC therefore, LH8 predicts a significant change in slope from low M_T to high M_T.

Figures 9.20 and 9.21 show the absolutely normalized model spectra for three different EoS compared to the data for π^-, K^- and $\bar{p}$. The data indicate a strong macroscopic transverse response, as expected of an EoS with a speed of sound $c_s^2 \approx 1/3$.

It is worthwhile to plot the π^- and $\bar{p}$ spectra on the same plot. The spectra almost cross for $p_T \approx 2.3\,\mathrm{GeV}$. It was recently pointed out that the measured $\pi^-/\bar{p}$ ratio is several times above the expected ratio from jet fragmentation. The ratio is readily explained in a simple hydrodynamic/thermal model with additional hadronic scattering. Above $M_T > 2.0\,\mathrm{GeV}$ without rescattering, the slope parameters of pions and nucleons approach the universal value $T_{\mathrm{slope}} \approx 250\,\mathrm{MeV}$. This slope is given by the "blue-shifted" spectrum with the parameters for $T_{\mathrm{th}} = 160\,\mathrm{MeV}$ and $v_T = 0.45\,c$. Accounting for hadronic rescattering, the nucleon slope at large p_T approaches $T_{\mathrm{slope}} \approx 300\,\mathrm{MeV}$ and is better described by $T_{\mathrm{th}} = 160\,\mathrm{MeV}$ and $v_T = 0.55\,c$. Hadronic rescattering therefore increases the nucleon flow velocity slightly, from $v_T \approx 0.45\,c$ to $v_T \approx 0.55\,c$. We then adjust μ_B/T to match the experimental $\bar{p}/p$ ratio [656], $\bar{p}/p = \exp(-2\mu_B/T) = 0.65$. Then with all the parameters specified, we draw spectra for $\bar{p}$ and π^- in Fig. 9.22. A source expanding with a collective velocity

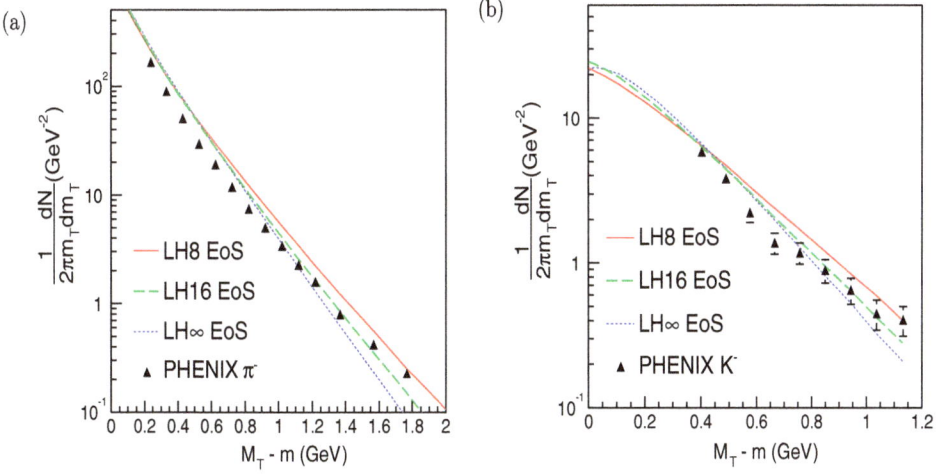

Fig. 9.20. A comparison to PHENIX spectra [664]: (a) compares π^- spectra. (b) compares K^- spectra. Both the model and the experimental spectra are absolutely normalized.

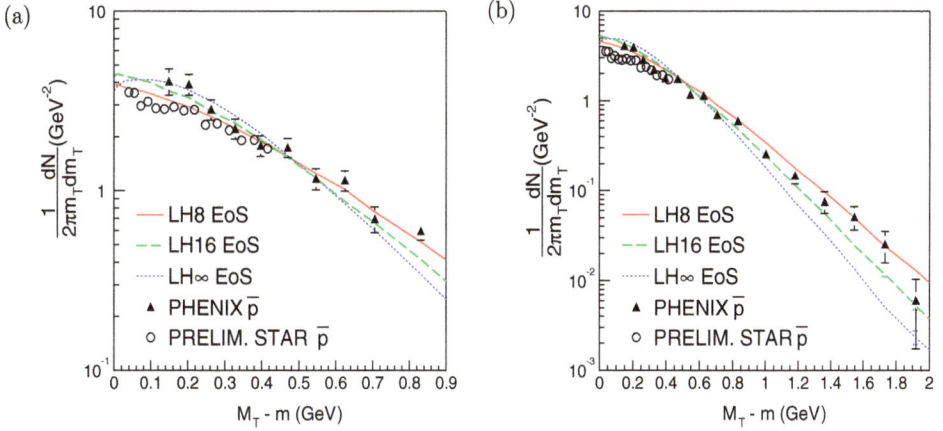

Fig. 9.21. A comparison to PHENIX [664] and STAR [663] $\bar{p}$ spectra for (a) low and (b) high M_T, respectively. Both the model and the experimental spectra are absolutely normalized.

of $v_T \approx 0.5\,c$ and hadronizing according to a thermal prescription at a temperature of $T \approx 160\,\mathrm{MeV}$, generates the observed $\pi^-/\bar{p}$ ratio once pion–nucleon scattering is taken into account.

9.4.5. *Elliptic flow*

Let us recall why in non-central collisions the particles emerge with an elliptic flow. The spectator matter flies down the beam pipe and the excited nuclear matter is formed in the transverse plane with an almond shaped distribution. Subsequently,

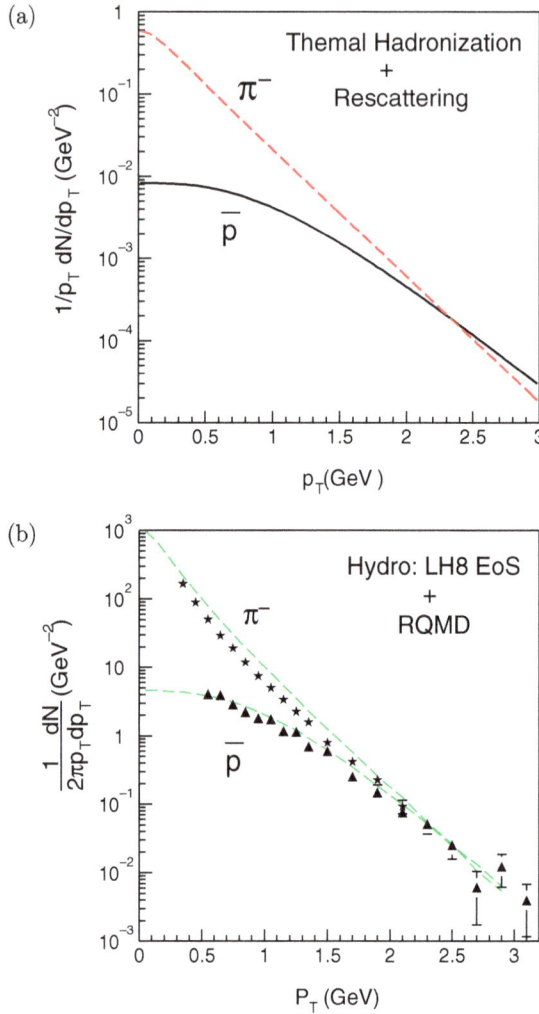

Fig. 9.22. A comparison of π^- and $\bar{p}$ spectra. (a) shows a simple thermal model with parameters discussed in the text. The spectra are relatively normalized. (b) shows an absolutely normalized comparison of the complete model spectra and PHENIX spectra [664].

if pressure develops in the system, the pressure gradients are larger in the impact parameter direction (the x-direction) than in the y-direction. Then, the excited matter expands preferentially in the x-direction. The magnitude of this elliptic response is quantified experimentally by expanding the distributions in a Fourier series

$$\frac{dN}{p_T\,dp_T\,dy\,d\phi} = \frac{dN}{2\pi\,p_T\,dp_T dy} \times (1 + 2\,v_2(p_T, y)\,\cos(2\phi) + \cdots),$$ (9.63)

where ϕ is measured around the z-axis relative to the impact parameter, which points in the x direction.

The *elliptic* flow, $v_2(p_T, y) \equiv \langle \cos(2\phi) \rangle_{p_T, y}$, gives a measure of the dynamic response of the excited nuclear matter to the initial anisotropy.

In non-central collisions the matter distribution in the transverse plane is initially almond shaped. After the collision, the pressure gradient is larger along the impact parameter direction (the x-direction). The pressure gradient then drives the matter preferentially in the x-direction and converts the initial spatial anisotropy into a final momentum anisotropy. The momentum asymmetry characterizes the fireball response.

The initial spatial anisotropy is quantified by the parameter ϵ introduced as

$$\epsilon \equiv \frac{\langle y^2 - x^2 \rangle}{\langle y^2 + x^2 \rangle}. \tag{9.64}$$

The hydrodynamic response is linear in ϵ [665] and therefore v_2 is sometimes divided by ϵ to compare different impact parameters and nuclei [669, 674]. As the system expands, the eccentricity ϵ decreases. Since ϵ is the driving force behind the elliptic flow, the elliptic development finishes before the radial development. Therefore, elliptic flow is generated by the early pressure, although this statement must be qualified (see below). The spatial anisotropy that remains after the collision is quantified by s_2

$$s_2 \equiv \frac{\langle x^2 - y^2 \rangle}{\langle x^2 + y^2 \rangle}. \tag{9.65}$$

Figure 9.23b shows the integrated elliptic flow of pions as a function of the total multiplicity for different EoS. Figure 9.23b shows the relative contribution of RQMD to the integrated elliptic flow. Note that elliptic flow increases for all EoS and dramatically so for LH8. Assume momentarily that elliptic flow for Hydro+RQMD stops developing at a temperature of $T \approx T_c \approx 165 \, \text{MeV}$. (Note however that the radial flow develops well below this temperature.) The dramatic increase of elliptic flow in Fig. 9.23a can be understood as the dynamic response of the QGP pressure.

At the SPS, the anisotropy of the stress tensor increases rapidly at first and then stalls. The final stress tensor anisotropy is small at the end of the mixed phase. At RHIC, elliptic flow develops more rapidly and stalls only when the anisotropy is large. Thus, provided the elliptic flow stops developing at T_c, the elliptic flow increases dramatically as the QGP pressure appears. For a RG EoS at the SPS, there is no mixed phase and no stall and consequently the RG elliptic flow is significantly larger than LH8 and the data. However with RHIC collision energies, LH8 begins to behave as an ideal QGP. Consequently, at RHIC the RG elliptic flow is only 20% larger than that of LH8 and of the data.

To understand when elliptic flow stops developing, it is important to track the spatial geometry of the underlying source. When the spatial anisotropy, s_2,

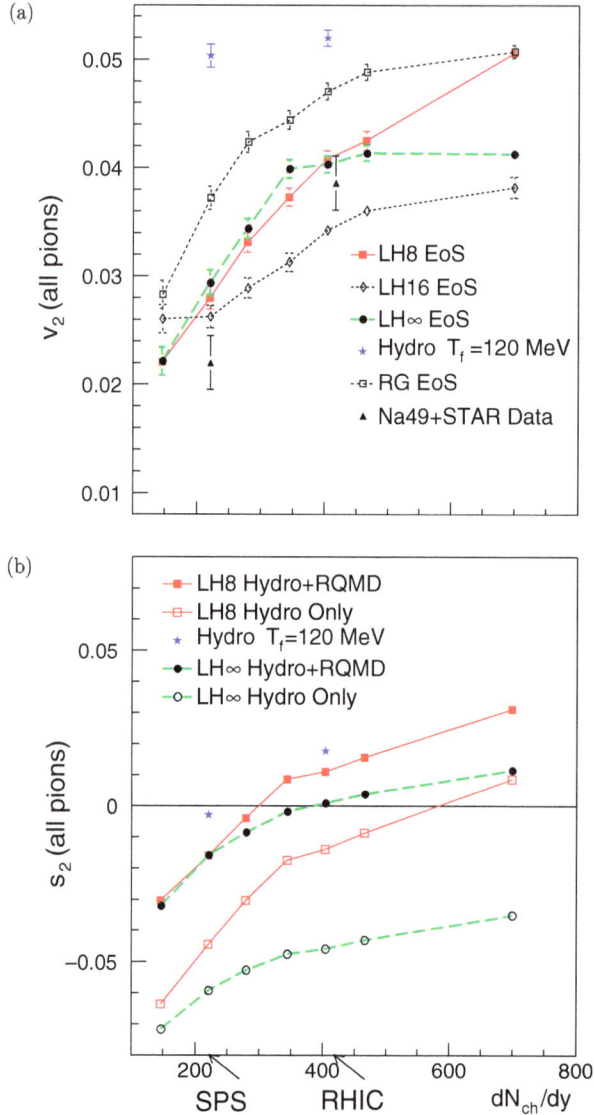

Fig. 9.23. Panels (a) and (b) show related quantities as a function of the total multiplicity in a PbPb collision at $b = 6$ fm. (a) Integrated elliptic flow v_2 of pions for different EoS and freezeout conditions; (b) spatial anisotropy s_2 with and without hadronic rescattering.

is negative, the pressure drives the elliptic flow. However, as s_2 approaches zero, the pressure gradients drive the radial motion rather than the elliptic motion. The elliptic development then stops. The next Fig. 9.24 is taken from Ref. [638].

Figure 9.23b shows the spatial anisotropy, s_2, for pions as a function of multiplicity from the SPS to RHIC. Compare the LH8 curves (the stars, the solid squares, and the open squares) seen in Fig. 9.23b. The open squares (LH8 Hydro Only)

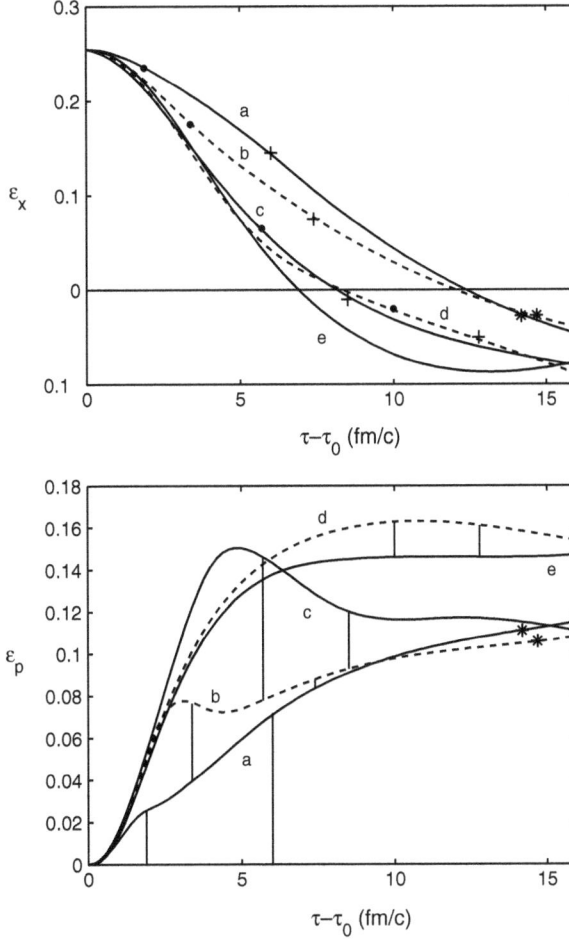

Fig. 9.24. Time evolution of the spatial ellipticity ϵ_x, the momentum anisotropy ϵ_p, and the radial flow $\langle v_\perp \rangle$. The labels a, b, c and d denote systems with initial energy densities of 9, 25, 175 and 25000 GeV/fm^3, respectively, expanding under the influence of the EoS Q. Curves e show the limiting behavior for EoS I as $e_0 \to \infty$ (see text). In the bottom panel the two vertical lines below each of the curves a–d limit the time interval during which the fireball center is in the mixed phase. In the top panel the dots (crosses) indicate the time at which the center of the reaction zone passes from the QGP to the mixed phase (from the mixed to the HG phase). For curves a and b the stars indicate the freezeout point; for curves c–e freezeout happens outside the diagram.

show the spatial anisotropy, s_2, at the end of the mixed phase, the closed squares (LH8 Hydro+RQMD) show s_2 after the cascade, and the stars show s_2 when the hydrodynamic evolution is continued to $T_f = 120$ MeV. After the mixed phase (LH8 Hydro Only), the matter retains some of its initial almond shape. Continuing the hydrodynamics destroys the initial almond shape completely and increases v_2 by a factor of ≈ 2–3 (see Fig. 9.23a). Cascading also changes the almond shape but increases v_2 by only a factor of ≈ 1.5. In either case, s_2 crosses zero between the

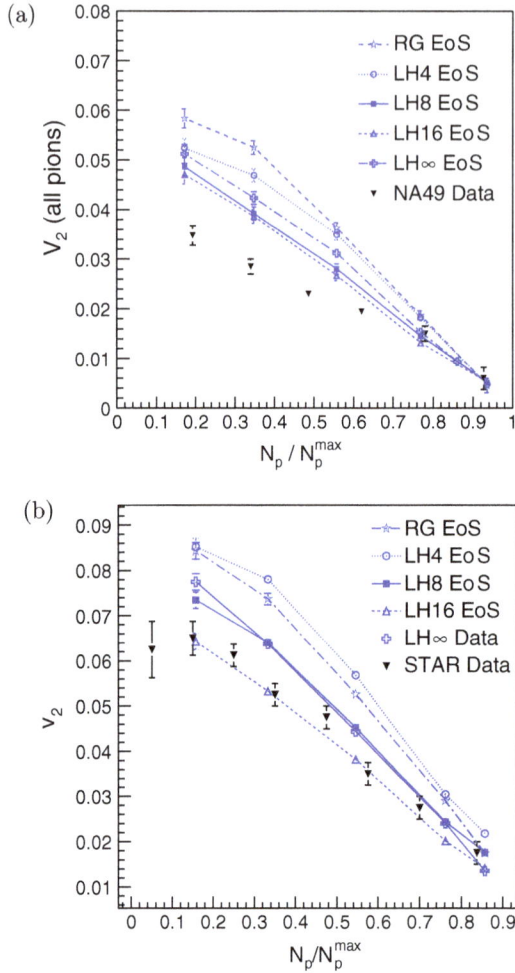

Fig. 9.25. (a) v_2 for pions at the SPS as a function of participants (relative to the maximum) compared to NA49 data [675]. (b) v_2 for charged particles at RHIC as a function of participants (relative to the maximum) compared to STAR data [676].

SPS and RHIC and therefore elliptic development in the hadronic stage ceases to be significant between the SPS and RHIC.

The centrality dependence of the elliptic flow is shown in Fig. 9.25 as integrated pion elliptic flow at the SPS and RHIC depends on the number of participants N_p.

The data restrict the underlying EoS. At the SPS, the data favor a soft EoS — LH $\geq 0.8\,\mathrm{GeV/fm}^3$. LH4 and RG EoS generate too much elliptic flow. For LH8-LH16, the model is about 20% above the data. However, the model to data comparison is not completely fair — the data points are integrated over rapidity, while the model points are only strictly valid for mid-rapidity. This probably accounts for the residual model/data discrepancy. As the latent heat is increased beyond LH32 to LH∞, the elliptic flow begins to rise. The origin of this elliptic flow was described in Ref. [673] and results from the slow evaporation of particles in an asymmetrical fashion over a long time. This elliptic flow is generated without radial flow [646] and the p_T dependence of v_2 for nucleons (see below) is modified accordingly [646]. At RHIC (Fig. 9.25b), the comparison is fair and the data again favor a relatively soft EoS, LH8-LH16. Thus the elliptic flow data at the SPS and RHIC are consistent with a single underlying EoS.

Note that the ordering of the EoS in Fig. 9.25 differs at the SPS and RHIC. At the SPS, LH8 and LH16 generate approximately the same elliptic flow. At RHIC, the hard QGP phase lives substantially longer with LH8 than with LH16 and therefore generates more elliptic flow. Additionally at RHIC, LH4 generates more elliptic flow than a RG EoS. Thus, the elliptic flow indicates that at high energy densities LH4 (with $c_s^2 \approx 1/3$) has a larger speed of sound than a RG EoS (with $c_s^2 \approx 1/5$). At asymptotically, high energy densities all EoS in the LH(x) family approach the massless ideal gas limit.

Having discussed qualitative changes from the SPS to RHIC, we explore the p_T dependence of elliptic flow. Experimental measurements are performed over a range of impact parameters. To find $v_2(p_T)$ in a specific impact parameter range, $b_{\min} < b < b_{\max}$, the following integrals must be performed,

$$v_2(p_T, y)_{b_{\min}}^{b_{\max}} \equiv \frac{\int_{b_{\min}}^{b_{\max}} v_2(p_T, y; b) \frac{dN}{dy \, dp_T}(b) \, 2\pi b \, db}{\int \frac{dN}{dy \, dp_T}(b) \, 2\pi b \, db}. \tag{9.66}$$

Again, we drop the y, $b_{\min}$ and $b_{\max}$ labels below when it is not confusing. $v_2(p_T)^{\min-\text{bias}}$ denotes the elliptic flow integrated over all events, or $v_2(p_T)_0^\infty$.

Figures 9.26a and b show $v_2(p_T)^{\min-\text{bias}}$ for negative hadrons (mostly pions) and nucleons at RHIC. Look first at the negative hadrons (a): Although LH8 and LH16 both show a strong linear rise, the slope is smaller for LH16. For LH∞, $v_2(p_T)$ is curved and bends over. For small p_T, LH∞ is above LH8, but by $p_T \approx 2.0 \, \text{GeV}$, LH∞ is substantially below LH8. The data show a strong linear rise and agree remarkably well with the slope of LH8. $v_2(p_T)$ slightly favors LH8 over LH16. The kaon $v_2(p_T)$ curve has the same shape and magnitude as the h^- spectrum.

For nucleons, the v_2 spectral shape is different and is initially curved upward. For nucleons, LH8 and LH16 are concave up, indicating a strong radial expansion. By contrast, LH∞ shows a linear rise in $v_2(p_T)$, indicating a weak transverse expansion. As discussed above, soft EoS like LH∞ slowly evaporates particles and generates elliptic flow only at small p_T: thus it is excluded by data. (It is also excluded by the radial flow.)

9.4.6. *Limits to ideal hydrodynamics*

We have already discussed dissipative corrections to hydrodynamical equations in Section 9.2.2, and now is the time to see how it works in practice. Let me recall that

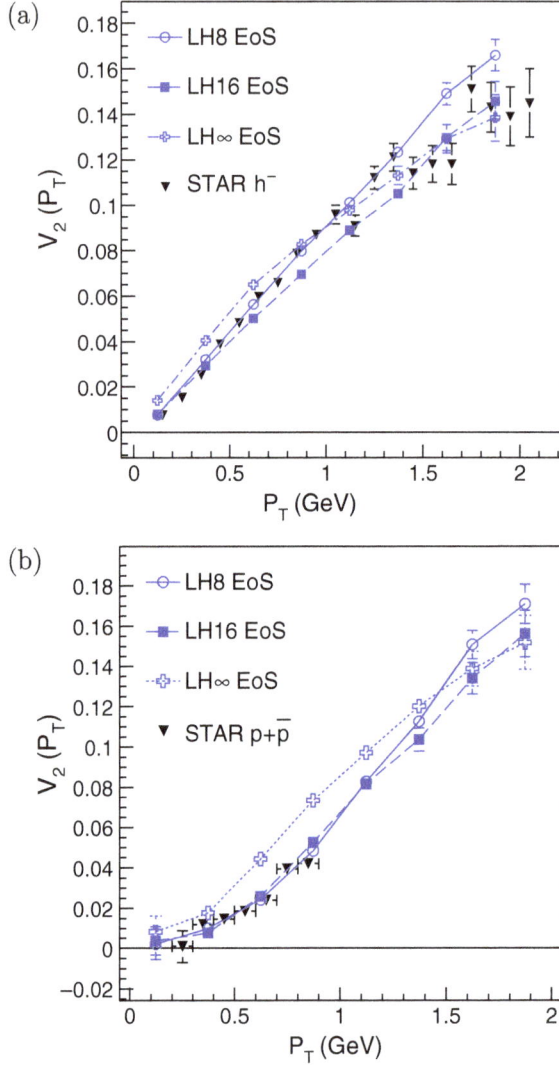

Fig. 9.26. $v_2(p_T)^{\mathrm{min-bias}}$ (see Eq. (9.66)) for three different EoS compared to data. Panels (a), (b) are for negative hadrons and $p + \bar{p}$, respectively.

the viscous corrections are given by the ratio of the sound attenuation parameter $\Gamma_s = \frac{4}{3}\eta/(\epsilon+p)$ to the time or space values under consideration. As we have discussed in the Section 8.2.7, theoretically the weak coupling (perturbative) calculations predict large mean free paths for quasiparticles corresponding to large viscosity (8.71), while strong coupling results (for $\mathcal{N} = 4$ supersymmetric Yang–Mills theory so far) give a surprisingly small value $\Gamma_s = 1/(3\pi T)$.

Thus far only one paper [678] tried to extract the viscosity coefficient from the data, by looking at deviations from the ideal hydrodynamic predictions. Before we look at the results of this work, let me first try to explain qualitatively what viscous effects on flow are.

Ideal hydrodynamics assumes that all the dynamics is local, and energy and currents of all conserved quantities are conserved locally. Recognizing the finiteness of the mean free path l (or other correlation lengths) one however finds that a particle may jump between the matter cells and transfer energy and density etc. non-locally. The effect of that is non-zero only if there is a difference between the cells: thus extra gradients of velocity and densities appear.

In order to understand qualitatively what the viscous term would do it is convenient to use the simplest analytic example, Bjorken boost independent flow, and calculate what is added to the ideal isotropic pressure terms in the stress tensor. The result is very different in the longitudinal z and transverse directions,

$$p_z \to p - \frac{4}{3}\frac{\eta}{\tau}, \qquad p_\perp \to p + \frac{2}{3}\frac{\eta}{\tau}. \tag{9.67}$$

Viscosity works against the longitudinal pressure, this is understandable: but why does it help the transverse one? The reason is that the particle distribution over the momenta is deformed, in the longitudinal direction it has less particles with large p_z than the thermal one because those "jump" to other cells. But then the distribution over $p_\perp$ should be somewhat wider, to keep the mean chaotic energy the same. (Note that the sum of all three pressures is unchanged.)

Teaney has shown that in fact the viscous correction to the particle distribution is completely determined,

$$f = f_0 \left(1 + \frac{C}{T^3}p^\alpha p^\beta \langle \nabla_\alpha u_\beta \rangle \right), \tag{9.68}$$

where the gas is assumed to be a Boltzmann gas (otherwise $f_0(1 \pm f_0)$ is needed for Bose/Fermi ones). The coefficient for a Boltzmann gas is just the viscosity to entropy ratio[20] $C = \eta/s$.

Substituting the distorted local particle distribution into the Cooper–Frye formula for spectra, one finds corrections, which for large enough $p_\perp$ become

$$\frac{\delta\,dN}{dN_o} \approx \frac{\Gamma_s}{4\tau}\left(\frac{p_T}{T}\right)^2. \tag{9.69}$$

Deviations from hydrodynamic spectra of such shape are indeed seen in the data, and they reach about a factor 2 at $p_\perp \approx 1.7\,\text{GeV}$ at RHIC. This implies that at the time of the freezeout $\Gamma_s/\tau \approx 1/5$. It agrees perfectly well with the mean free paths of secondary hadrons at this stage about 1.5 fm or so.

Now that we see that the procedure is reasonable, we can go on and try to evaluate the effect of viscosity at earlier times. To do so one can evaluate the viscosity correction to the elliptic flow parameter v_2, since we saw above that its value is

---

[20] For a Bose gas this expression is corrected but only by an insignificant factor 1.04.

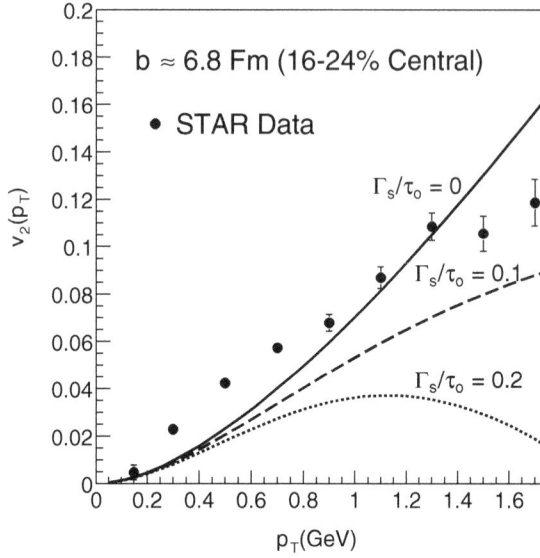

Fig. 9.27. Elliptic flow v_2 as a function of p_T for different values of Γ_s/τ_0. The data points are four particle cumulants data from the STAR collaboration [648]. Only statistical errors are shown.

determined only by sufficiently early times when the fireball is spatially strongly anisotropic. For RHIC energies such a time — about $3\,\mathrm{fm}/c$ — corresponds to the time when matter was in the QGP phase. The results for different Γ_s/τ are shown in Fig. 9.27. The experimental data (not shown in this figure) deviate from the ideal hydrodynamic curve at $p_\perp \approx 1.6\,\mathrm{GeV}$ which indicates that $\Gamma_s/\tau \sim 0.05$ or so. Substituting here the relevant time $\tau \sim 3\,\mathrm{fm}/c$ we get $\Gamma_s \sim 0.15\,\mathrm{fm}$. The strong coupling result for typical $T \sim 200\,\mathrm{MeV}$ at the time gives $\Gamma_s \sim 0.1\,\mathrm{fm}$, while the weak coupling one would predict much a larger value for $\Gamma_s \sim 2\,\mathrm{fm}$ or so. This implies, that QGP is the most ideal fluid ever observed.

Why can the viscosity be so low? Zahed and myself [hep-ph//0307267] related the issue of bound states at $T < T_c$ and small viscosity. We argued that near the "zero binding lines" the scattering length of quasiparticles becomes very large, which should drastically reduce the mean free path.

This idea had remarkable confirmation in a very different setting, with trapped Li atoms. In this case the magnitude of the scattering length can be tuned to very large values by applying a magnetic field which creates the so called Feshbach resonances between 2 atoms, when their bound states is close to zero binding. Remarkably, it was found that such regime indeed leads to a hydrodynamical behavior. The way it was demonstrated is precisely the "elliptic flow" pattern, when atoms are released from a deformed trap, see K.M. O'Hara *et al.*, *Science* **298**, 2179, 2002; T. Bourdel *et al.*, *Phys. Rev. Lett.* **91** (2003) 020402.

9.5. Interferometry of identical secondaries or HBT method

9.5.1. *Main idea of the method*

At the end of the chapter devoted to studies of highly excited hadronic matter using various features of the data as *"barometers"*, *"speedometers"* and *"thermometers"* we come to an *"interferometric microscope"* which in principle provides information about the space–time picture of the fireball at its freezeout.

The main idea of the two-quanta interferometry was first suggested in radioastronomy by R. Hanbury-Brown and R.Q. Twiss [679] in 1954, and it is often referred to as the HBT method. Its first application was to measure radii of distant stars, or to measure distances of close binaries which ordinary optical telescopes could not resolve, see e.g. Ref. [680]. It is used since for multiple purposes.[21]

The first appearance in hadronic physics was a decade later, when Goldhaber, Goldhaber, Lee and Pais [682] ascribed a specific correlation between pions produced in $\bar{p}p$ annihilation to their Bose–Einstein statistics.[22] To use it as a research tool aimed at measuring sizes of fireballs was first advocated by Kopylov and Podgoretsky [683]. A theoretical description in terms of a correlator of sources was initiated by my paper [684] and developed later in Refs. [685–687]. Its main concept is the existence of completely random sources, with phases averaged over the "heat bath" — all the unobserved particles in a many-body reaction. A reprint of all relevant papers is available as a book [691], for a recent detailed review see e.g. Heinz and Jacak [690].

Although known for decades, it was only occasionally used for analysis of the hadronic collisions because it needs very high statistics. As it is more readily available for heavy ion collisions, it naturally gets much more attention in this case.

Let us now describe the main idea. Consider two point sources, S1 and S2, emitting an infinitely long train of the monochromatic radiation (see Fig. 9.28) and two detectors, D1 and D2, measuring the *intensity* (not amplitude!) of radiation at two distant points, which are then correlated with each other. What interferes with one another is actually not the wave themselves but two "diagrams" shown in this figure: for identical secondaries one cannot tell which source emits each quantum. The total amplitude of the coincidence experiment can be written as a sum of two terms

$$M \sim e^{ik(R_{11'}+R_{22'})} + e^{ik(R_{12'}+R_{21'})}. \tag{9.70}$$

---

[21]There should be plenty of applications since the simplest version of HBT correlations is even used as demonstrations in graduate labs. Let me still mention my own little suggestion [681] for beam technology. The idea was to use the HBT correlations of two X-rays from the synchrotron radiation, to measure very small (sub-micron) beam size of modern electron storage rings. I know it has been indeed tried at Argonne National Lab. and has worked well, but I am not sure it is routinely used.

[22]It seems apparent from the paper that they did not know about the two-photon HBT correlations mentioned above.

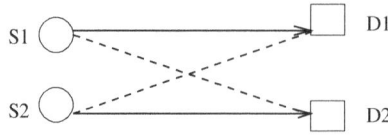

Fig. 9.28. Two quanta are produced by sources S1, S2 and detected by two detectors D1, D2. The process can go via two different possible paths, indicated by the solid and dashed lines, which interfere in the quantum-mechanical amplitude.

The probability is a modulus of the amplitude squared, which is equal to

$$|M|^2 \sim 1 + \cos[k(R_{11'} + R_{22'} - R_{12'} - R_{21'})]. \tag{9.71}$$

When the distance from the sources to the detectors is much larger than their separation, one may rewrite the argument of the cosine as $kL\theta$ where L is the "baseline", the distance between the detectors and θ is the angular size of the source. Substituting here the radioastronomical wavelength $1/k \sim 1\,\mathrm{m}$ and the distance between the observatories, $L \sim 10^4\,\mathrm{km}$, one finds an angular resolution of the order of $\theta \sim 10^{-7}$. In principle optical wavelengths can also be used but then L is also much shorter as it is impossible to correlate signals with sufficient time resolution.

In a very similar way the two-body spectrum of, say, a pair of identical π^- pions one cannot say where in the fireball each pion comes from. If R is the source size, the condition for HBT correlation is

$$|\vec{k}_1 - \vec{k}_2|R < 1. \tag{9.72}$$

Another aspect of the problem deals with the time structure of the signal we so far ignored. It is clear that even if the momenta of both quanta are the same and the condition given above is satisfied, their interference is impossible if these two quanta do not overlap in time. Using the language of four-momenta, one finds an additional term in the argument of the cosine, with times and frequencies substituting for distances and wave vectors. The interference is absent if the two frequencies measured are too different, namely

$$|\omega_1 - \omega_2|\tau < 1, \tag{9.73}$$

where τ is the time between the emission moments of two quanta, about the lifetime of the source. Basically these two "second order coherence" conditions determine where the HBT effect is seen in 2-body spectra.

Experimentally, the correlation was first identified by comparison to non-identical (e.g. $\pi^+\pi^-$) pion pairs, and now mostly by comparing the true two-body spectrum to that extracted from a "mixed event" sample, in which two pions come from different events. The latter is better since the former method brings in corrections for (quite different) interaction (e.g. attractive rather than repulsive Coulomb or σ resonance) of two non-identical pions.

It should be emphasized from the start, that the two-quanta interference effects to be considered below are very different from more familiar interference effects at the one-quantum level. In particular, the "thermal" sources of radiation considered below are not assumed to give a "coherent" radiation in the ordinary meaning, but show the effect in the coincidence-type measurements with two (or more) intensity-sensitive detectors.[23]

9.5.2. *Correlator and random source model*

These ideas can be put into a more definite form in the context of "random source" model [684]. It starts with the introduction of some (scalar) currents or sources emitting pions at some $4d$ point $j(x)$. Let us refer to the positions of the two sources as x_1, x_2, which can also be expressed via the mean position $X = (x_1+x_2)/2$ and the separation $R = x_1 - x_2$. One can imagine a multi-pion production amplitude as (a set of) Feynman diagrams, in which positions of each emission vertex is integrated over d^4x. Let us now select only positions of our two detected identical pions x_1, x_2 and integrate over *all others*. It would give the *correlator* of two sources

$$\langle j^*(x_1)j(x_2)\rangle = I(X, R). \tag{9.74}$$

Here the angular bracket refers to all unused (maybe even unobserved) particles as a kind of a "heat bath". If one introduces the Fourier transform of the correlator

$$\tilde{I}(P, Q) = \int d^4X \, d^4R \, e^{iXP+iQR} I(X, R), \tag{9.75}$$

one can see that $\tilde{I}(P, Q = 0)$ is just the average single-body spectrum. One often also uses the mixed representation, also known as the Wigner or *source* function

$$S(P, R) = \int d^4X \, e^{iXP} I(X, R), \tag{9.76}$$

which tells us where the particles with momentum P come from.

The main result of Ref. [684] is that writing down the amplitude as the sum of two diagrams indicated in Fig. 9.28, making a square and averaging over random sources, one finds that the two-body spectrum contains the product of two such intensities plus/minus the interference effect if particles are bosons/fermions

$$\frac{d^6N}{d\mathbf{p}_1 d\mathbf{p}_2} \sim \tilde{I}(p_1, 0)\tilde{I}(p_2, 0) \pm |\tilde{I}((p_1 + p_2)/2, p_1 - p_2)|^2. \tag{9.77}$$

Exercise 9.4. Derive the formula above, assuming that the sources producing the pions have random phases and only terms in which they do not appear can survive.

Note that if the four-momenta of both particles are identical the second term is the same as the first: so at this point the effect is expected to be twice as large for

---

[23]The "coherent" sources such as lasers, on the contrary, do not produced HBT correlation effects, which needs extermination of the cross terms due to chaoticity of phases, see below.

bosons and zero for fermions. The latter is easy to understand: the Pauli exclusion principle would demand so.[24] The former factor 2 is a factor 2^2 in amplitude squared times $1/2$ from symmetrization of the phase space.

The shape of the measured correlation function is related by the Fourier transform with the source space–time distribution, the information we are looking for. So, naively, just by measuring the correlator we would be able to do a Fourier transform back and recover the source. It does not however work quite like that because the measured spectra are 6-dimensional and for the source we need an 8-dimensional transform. Two are missing because the observed particles are on-shell.

Lacking direct inversion, one can do what people always do in such cases, parameterizing the answer and extracting parameters from comparison to data. The simplest parametrization in which a Fourier transform is easy to carry out is a Gaussian, which is mostly used in practice.

$$S(R, P) = N(P)S(\bar{R}(P), P) \exp\left(-\sum_{\mu\nu} B_{\mu\nu}(P)\delta x_\mu \delta x_\nu\right), \tag{9.78}$$

where the overbar means the effective source center for a given mean momentum of the pair P, and $\delta x_\mu(P) = R_\mu - \bar{R}_\mu$ is the deviation. Now denoting the spatial $\int d^4 R\, S \cdots$ average over the source by angular brackets, one finds that the quadratic form of the coordinates is the inverse of B,

$$\langle \delta x_\mu \delta x_\nu \rangle = B_{\mu\nu}^{-1}(R). \tag{9.79}$$

Note also, that performing the Fourier transform of the source function and plugging it back into the main expression for the correlator one finds a convenient expression

$$C(Q, P) = 1 + \exp\left(-\sum_{\mu\nu} q_\mu q_\nu \langle \delta x_\mu \delta x_\nu \rangle(P)\right). \tag{9.80}$$

Note that all the information about the center position $\bar{x}$ is lost in the correlator; one can learn from it only about the relative distances and emission duration. Furthermore, standard Cartesian parameterization for the tensor is based on three coordinates called *long* (longitudinal or z direction is along the beam), *out* or y is along the vector $\vec{P}_\perp$ and *side* or x is orthogonal to both. For azimuthally symmetric central collisions the non-zero components of the quadratic form are the following four radii (all functions of $\vec{P}$)

$$R_s^2 = \langle \delta y^2 \rangle, \qquad R_o^2 = \langle (\delta x - \beta_\perp \delta t)^2 \rangle$$
$$R_o^2 = \langle (\delta z - \beta_l \delta t)^2 \rangle, \qquad R_{ol}^2 = \langle (\delta z - \beta_l \delta t)(\delta x - \beta_\perp \delta t) \rangle, \tag{9.81}$$

where $\vec{\beta} = \vec{P}/P_0$ is mean velocity of the pair.

---

[24] Note however that for the usually observed *unpolarized* fermions the effect is a dip only down to $1/2$ of the original level, not zero.

These four parameters are usually reported by experimental groups reporting their fitted correlators. For non-central collisions the source is elliptic, thus there are more non-zero components which depend on absolute orientation of out and side directions with respect to the impact parameter plane.

9.5.3. *Issue of "coherency" and long-lived resonances*

The "random source" model assumes that the phases of the sources are random. For stars, the currents in the excited atoms which emit two observed photons are at different points at the star photosphere, huge distances apart from each other and obviously they are completely uncorrelated.

In high energy heavy ion collisions the number of emitted pions is now counted in thousands per event, and also the size of the fireball produced is large compared to the microscopic correlation length in the excited matter. It seems quite clear that two observed pions are likely to come from different places in the fireball and their mutual incoherence should hold in this case as well.

On the other hand, if both pions would be produced from the same point-like source, both quanta would be produced in the same partial wave with the same (zero) angular momentum. If so, the production amplitude would be a product of one-body amplitudes, the same would be true of the probabilities, and there would be no place for any additional correlation effect.

In hadronic collisions one may question the random phase assumption: there are only a few pions (as in the original Goldhaber *et al.* data, with multiplicity about 4–6) and the fireball is only 1 fm in size. And yet, in spite of a lot of searches (see Ref. [691]) the experiment shows so far no sign of the pion coherence effect, and recent data for 3-pion interferometry has confirmed that.

It may sound confusing since in the parameterizations used in practice one does not use (9.80) directly but one usually includes the so called "coherency factor" λ in front of the correlator

$$C(q, K) = 1 + \lambda \exp\left(-\sum_{\mu\nu} q_\mu q_\nu \langle \delta x_\mu \delta x_\nu \rangle (K)\right).$$

However in fact its meaning is different and related with the fraction of pions originating from the decay of the *long-lived resonances* $\omega, K_s, \ldots$ from which some fraction f_{llr} of the observed pions originates. The interference is observed only if both are "prompt" pions, so $\lambda = (1 - f_{\text{llr}})^2$. Of course, if the resolution of the detector was ideal one would see the second component in the correlator $\delta C_{\text{llr}} \sim (1 - \lambda)$ (narrow peak) so that if two pions are completely identical one would still have $C(q = 0, K) = 2$, as Bose statistics demands. However, even if we were able to resolve the contribution of the long-lived resonances, we would not be including (well known) lifetimes into the "lifetime of the source".

Our second point is that there exists the "final state interaction" in a more direct sense, the observed interaction between the two particles. In particular, there is the Coulomb interaction and the low energy strong interaction. The Coulomb forces are the long-range ones. Particles with close momenta travel for a long time is close proximity, therefore even small force may produce large effect. This indeed happens if the Coulomb parameter is no longer small,

$$\gamma_c \equiv m_\pi e^2/q \sim \alpha_{EM}/\delta v \gtrsim 1, \qquad (9.82)$$

and the correlation term should be corrected by the well known "Gamow factor"[25]

$$G = \frac{2\pi\gamma_c}{\exp(2\pi\gamma_c) - 1}. \qquad (9.83)$$

It is not a problem to take into account strong $\pi^-\pi^-$ interaction because at small relative momentum q it is negligible due to the Goldstone nature of the pions. For the protons the scattering length is not small and one does account for it in practice.

Finally, another type of final state interaction is due to the known interaction of the observed pair with other particles. In particular, for the heavy ion collisions one may correct for the electric field outside the nuclei.

9.5.4.   *Expanding sources and regions of homogeneity*

One of the first dynamical issues discussed in the context of HBT interferometry was the dependence of the system dimensions on the multiplicity. The first evidence for such an effect came from the compilation of the data, that was made already in my first work [684], with much better data to come later.

Theoretically, the expectation that more pions should freeze out at larger volume sounds quite trivial. It was predicted that radii would be so large at RHIC that it would be nearly impossible to measure them. And yet this is *not* the case; the heavy ion data have shown that with growing collision energy and growing multiplicity the growth of the HBT radii has actually stopped. Their values measured at RHIC, shown in Fig. 9.29, are not very different from the size of the colliding nuclei. Although it is still very true that the emitting source is very large, especially in the longitudinal directions (pions are emitted with rapidities from about -5 to 5), all the radii are very close to each other as if the source was spherically symmetric.

In order to explain why it is happening we should first introduce the idea of *the region of homogeneity* introduced by Makhlin and Sinyukov [688]. If the source from which pions originate is expanding, pions with different momenta come from its different parts. The maximum would come from a spatial point in which the mean *velocity* of the two pions $\vec{\beta}$ is exactly equal to the collective velocity $\vec{v}_{coll}$. The points from a region around can only contribute if the magnitude of random thermal

---

[25]Note that this factor has poles, corresponding to the bound states of the pionic atoms.

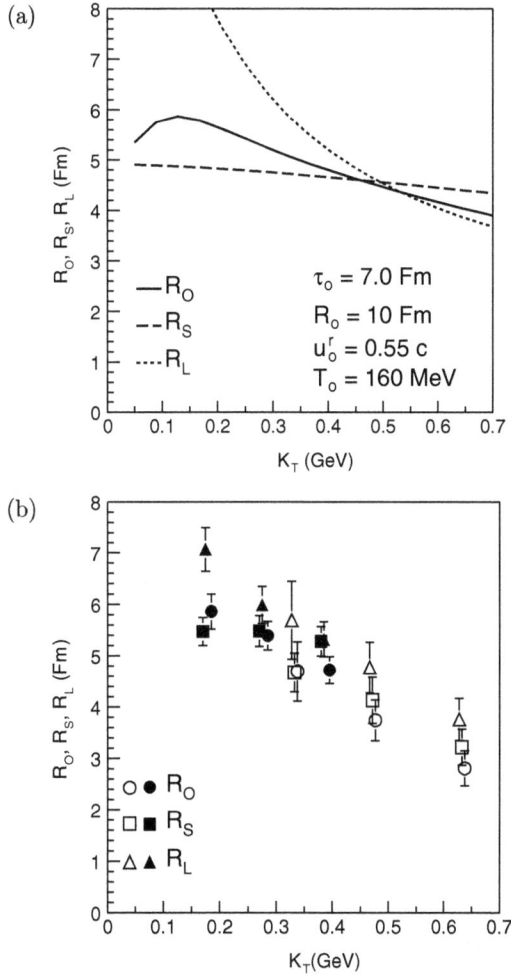

Fig. 9.29. (a) Ideal blast wave fit to the experimental HBT radii R_O, R_S, and R_L shown in (b) as a function of transverse momentum K_T. The solid symbols are from the STAR collaboration and the open symbols are from the PHENIX collaboration. For clarity, the experimental points have been slightly shifted horizontally.

motion is strong enough, so that the gradient of the flow field can be compensated by it,

$$(\delta x_\mu \partial_\mu) v_{\text{coll}} < v_{\text{therm}}. \qquad (9.84)$$

For example, if we speak about the longitudinal radius, the r.h.s. of this condition tells us that $R_l \sim 1/\sqrt{(M_\perp T)}$ where $M_\perp^2 = p_t^2 + m^2$ is the transverse mass. We have just predicted that the pions with larger p_t show smaller longitudinal radius, which indeed has been observed, see Fig. 9.29b. Another consequence of this argument,

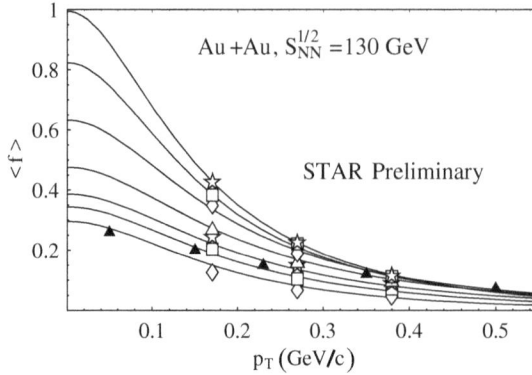

Fig. 9.30. Average π^- phase space density versus p_T for seven centralities from STAR and from NA49 for top central events (solid triangles) with BE distribution model fits (solid curves) (see text).

also confirmed experimentally, is that the R_l HBT radii for $\pi^\pm\pi^\pm$ and $K^\pm K^\pm$ identical pairs with *the same* $M_\perp$ should be the same.

The next step toward the understanding of RHIC HBT radii is the realization of the fact, that although the total flow pattern is explosive and very anisotropic (in general, not even azimuthally symmetric), the local flow near the freezeout is close to isotropic Hubble-like flow. Nearly isotropic HBT radii are a very good manifestation of that, as they actually show us the size of the homogeneity region or the derivatives of the flow velocity.

Unfortunately, at the time of this writing, the quantitative theory of HBT radii is still incomplete. The naive hydrodynamical calculations tend to over-predict the radii, and an understanding of the viscosity effects discussed in Section 9.4.6 above and possibly other corrections have not yet been achieved. Another issue to be completed soon is the HBT correlators for non-azimuthally symmetric sources, to be correlated with the event parameter plane and elliptic flow.

The magnitude of the correlation itself provides another important piece of information, the *phase-space density* (PSD) at freezeout. A compilation of STAR data [696] and also some NA49 data (solid triangles) is shown in Fig. 9.30. One can see from these data that a rather high phase-space density is reached for central collisions. This directly supports the idea that the non-zero pion chemical potential grows for more central collisions. The reason for that is a larger separation between the chemical and thermal freezeouts.

9.6. Correlation of non-identical hadrons

The issue of correlations between secondaries is in general a rather wide subject, since those are created by all kinds of physical effects — resonances, clusters and jets, to name a few — all of which use different methods and variables. Therefore I would just briefly describe two specific examples, which are both related to the issue of correct production timing for different kinds of secondaries.

9.6.1. *Ordering the production time for all hadronic species*

Suppose we would like to study a correlation of a pair of different species, e.g. $i = \pi$, $j = K$. The correlation can be caused by string interaction, including a resonance decaying into an ij pair, or simply by a Coulomb interaction: the magnitude and origin of the correlation should not be important.

On general grounds, one may first argue that it is expected to be maximal when both particles have the same *velocities* $\vec{v}_i = \vec{v}_j$, since in this case they have obviously larger chance to travel together for a long time and interact.

The second important argument by Lednický *et al.* [695] deals with the timing of the correlation. Assuming for simplicity that the directions of both particles are the same we would ask how the correlation depends on the magnitude and sign of the velocity difference $\delta v_{ij} = v_i - v_j$. If particle i is produced later than j at the same spatial position, it should have a larger velocity to catch it and interact. Alternatively, particle i can be emitted at the same time as j, but at a different distance from the collision point. In both cases, one expects an asymmetric peak of the correlation function $C_{ij}(\delta v_{ij})$ near $\delta v_{ij} = 0$.

This method has been used by the STAR collaboration [696] which studied correlations between $\pi, K; \pi, p$ and K, p pairs. The asymmetry is indeed observed, and fitting it with a "blast wave model" they concluded that nucleons are indeed on average emitted at larger distances than pions, $r_\pi - r_N \approx -7.1$ fm. This of course agrees well with our discussion of the hadronic freezeout, as nucleons have a larger cross section of rescattering than pions and are thus released later.

9.6.2. *Balance functions*

The idea of a balance function has been put forward by Bass, Danielowicz and Pratt [693]. Consider a conserved quantity — charge, strangeness or baryon number — which are conserved locally. Suppose the pairs of opposite charges are produced locally at some (proper) production time τ_{charge}. Later random diffusion amplified by the gradients of the collective motion may take the charges apart. So, by the measuring correlation of a particle with its "twin" one can understand how much time has passed between their creation and freezeout. Alternatively, one may think of it as a measure of the diffusion constant.

Of course, an event has hundreds or even thousands of particles, and not all of them are observed. (For example strangeness may be carried by neutral K^0, baryon number by neutrons, etc.) Therefore, one cannot obviously recover the "twin" individually. And nevertheless, as shown in Ref. [693], one can do it statistically. The

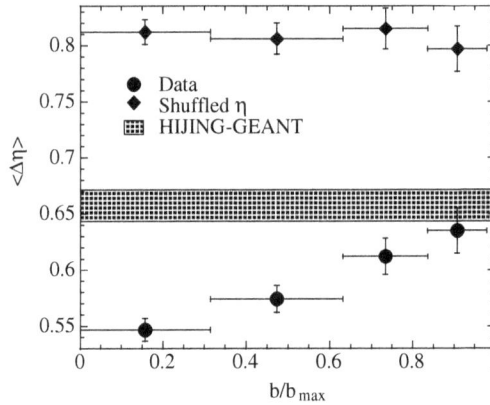

Fig. 9.31. The width of the balance function for charged particles, $\langle \Delta\eta \rangle$, as a function of normalized impact parameter (b/b_{max}). Error bars shown are statistical. The width of the balance function from HIJING events (basically a superposition of the pp events) is shown as a band whose height reflects the statistical uncertainty. Also shown are the widths from the shuffled pseudorapidity events, in which correlations are killed but the overall p_t distribution is preserved.

result is achieved by the so called *balance function* defined as

$$B(p_2|p_1) \equiv \frac{1}{2}\{\rho(b, p_2|a, p_1) - \rho(b, p_2|b, p_1) + \rho(a, p_2|b, p_1) - \rho(a, p_2|a, p_1)\}, \quad (9.85)$$

where $\rho(b, p_2|a, p_1)$ is the conditional probability of observing a particle of type b in bin p_2 given the existence of a particle of type a in bin p_1. The label a might refer to all negative kaons with b referring to all positive kaons, or a might refer to all hadrons with a strange quark while b might refer to all hadrons with an anti-strange quark. The conditional probability $\rho(b, p_2|a, p_1)$ is generated by first counting the number $N(b, p_2|a, p_1)$ of pairs that satisfy both criteria and dividing by the number $N(a, p_1)$ of particles of type a that satisfy the first criterion. $\rho(b, p_2, a, p_1) = N(b, p_2|a, p_1)/N(a, p_1)$.

The function is normalized to 1, and has usually a Gaussian-like shape, so the only parameter is its width. The experimental data from STAR [694] for this width are shown in Fig. 9.31. They clearly show that for more central collisions (smaller impact parameter b) the balance function in rapidity becomes narrower.

One possible interpretation of this suggested in Ref. [693] is that hadronization happens rather late, so that "twins" are not separated by the time of the final freezeout. If so, one may think that QGP should be mostly gluonic.

However another possibility is that charges were created early, but QGP was so impenetrable ("molasses") that diffusion in it was negligible. This fits well with very low viscosity of QGP.

It is at the moment difficult to see whether those are the only possible interpretations, to say nothing about which one is correct. Clearly more thought and modeling should be put into this issue before we will know the answer.

9.7. Event-by-event fluctuations

This field is still in its infancy. We start it with very brief summary of what we know about fluctuations of hadronic cross sections, to be contrasted with first heavy ion data from SPS and RHIC. We end up discussing two interesting theoretical ideas about event-by-event (EBE) fluctuations (although none of them has as yet been shown to work): (i) possible manifestations of the second order "endpoint" as the source of increased fluctuations, and (ii) reduced charge fluctuations due to incomplete equilibration in hadronic matter.

9.7.1. *Fluctuations of hadronic cross sections are surprisingly large*

We have argued in Chapter 7 that high energy hadronic cross sections are basically measuring a frozen snapshot of a glue distribution in the nucleon, see Fig. 7.2 and the discussion surrounding it. The cross section one usually discusses (as we did in Chapter 7) is the *average* one, while now we will focus on the deviation from this average. In principle, as we discussed in Chapter 2, many different views on the nucleon have been suggested. In particular, one can view the nucleon as: (i) multi-parton cloud, assuming that all of them are basically uncorrelated in the transverse plane; (ii) three "constituent quarks"; (iii) quark+diquark picture; (iv) a single associated tunneling event. Because fluctuations should scale about as $1/\sqrt{N}$ with the number of independent degrees of freedom N, these pictures predict very different fluctuations, from small for (i) to large for (iii, iv).

Not going into its long history, let me start from QCD-related ideas and, as one of the motivations, mention the idea of *color transparency*. It assumes that with a certain probability hadrons may have small size and thus can be viewed as small color dipoles. If so, the cross section is small, namely proportional to the squared matrix element of this dipole $\sigma \sim r_\perp^2$.

We cannot discuss here the details of how the event-by-event fluctuations of hadronic cross section were determined: see Ref. [697] for explanations and references. In brief, the idea is to trace inelastic rescattering and shadowing in hadron–nuclei collisions. They allow to determine two key parameters,

$$\omega_\sigma \equiv \frac{\langle \sigma^2 \rangle - \langle \sigma \rangle^2}{\langle \sigma \rangle^2}, \qquad \kappa_\sigma \equiv \frac{\langle \sigma^3 \rangle - \langle \sigma \rangle \langle \sigma^2 \rangle}{\langle \sigma \rangle^3}, \tag{9.86}$$

from which the underlying distribution $P(\sigma)$ can be approximately determined.

What was found in this analysis is that hadronic fluctuations are very large, for example at $s \sim 400\,\text{GeV}^2$ $\omega_\sigma^N \approx 0.25, \kappa_\sigma^N \approx 0.5$. To put it differently, although the average inelastic NN cross section is 35 mb, the distribution $P(\sigma)$ is basically flat from 0 to 70 mb. Equally strong fluctuations were also found for the pion–nucleon cross section.

Independent observations, pointing towards the same general direction, was a surprising discovery at HERA of the significant $\sim 1/10$ fraction of high-Q diffractive events.

What all these results imply is that picture (i), of dozens of uncorrelated partons in the transverse plane, is completely ruled out. Similarly large fluctuations for π, N indicate possibilities (iii, iv). The rather small size of the "black spot" of the glue distribution in a nucleon, r.m.s. $r_\perp \approx 0.4\,\text{fm}$ [698] provides further input into the emerging picture.

9.7.2. *All heavy ion collisions are (about) the same!*

Central collisions of two heavy nuclei (e.g. AuAu) have a large number (~ 400) of participating nucleons. Therefore, a simple statistical argument tells us that all fluctuations in the incoming nucleons should be reduced by a factor of about $\sqrt{400} \sim 20$ and e.g. the dispersion becomes of the order of 0.01.

On the other hand, a system undergoes QCD phase transitions, and one may expect some collective instabilities, formation of hadronic clusters of various masses, formation of "disoriented chiral condensate" and the like, all leading to increased fluctuations. A search for "unusual" events (e.g. with very large multiplicity or larger mean $\langle p_t \rangle$ of secondaries) had attracted significant attention more than a decade ago, when the heavy ion program at SPS and RHIC was suggested.

What was found instead is that *all events are about equal.* First data from NA49 [701] have shown that EBE fluctuations in such observables as $\langle p_t \rangle$ are not only relatively small, but they also stay Gaussian for 3–4 orders of magnitude, without noticeable non-Gaussian tails. More recent data of such type, from STAR [696], are shown in Fig. 9.32. One can see that it is indeed Gaussian, but its width is not the same as for purely statistical fluctuations but 14% larger. The widths of these Gaussians can be measured very accurately. Can we learn from them something useful?

Long before these data appeared it had been suggested to apply classical theory of thermodynamical fluctuations to their description. For example, for the description of the probabilities of exclusive channels for low energy $\bar{p}p$ reactions I have used in Ref. [493] the famous Boltzmann formula used backwards by Einstein, $P \sim e^S$ back in 1972. In the mid-1990s attempts were made to use this expression for small fluctuations, in particular Stodolsky [699] and myself [700] proposed to measure the specific heat of a hadronic matter at freeze-out by looking at the event-by-event temperature fluctuations. Unfortunately, as explained in Ref. [702], these arguments

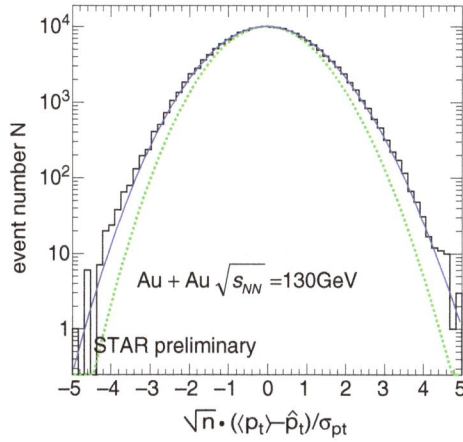

Fig. 9.32. Mean p_T distribution for $\sqrt{s_{NN}} = 130\,\text{GeV}$ Au+Au central collisions with respect to $\overline{p_T}$ in units of $\sigma_{\hat{p}_T}/\sqrt{N}$ compared to the Gamma distribution reference expected in the absence of non-statistical fluctuations (dotted curve) and a Gamma distribution calculated with r.m.s. width increased by 14% (solid curve).

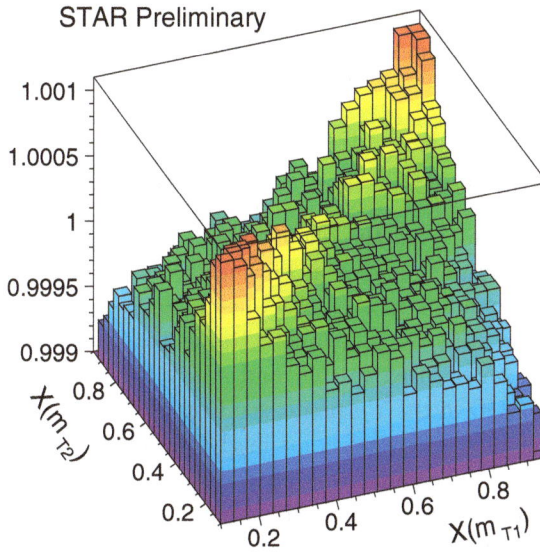

Fig. 9.33. Perspective view of the two-dimensional $X(m_{T1}) \otimes X(m_{T2})$ charge independent correlations for $\sqrt{s_{NN}} = 130\,\text{GeV}$ Au+Au collision data.

appeared to be too naive, and it is not so simple to invent a "thermometer" capable of measuring the temperature fluctuations. The thermodynamics of the "resonance gas" however does describe the EBE fluctuations of $\langle p_t \rangle$ and multiplicity.

In general, EBE fluctuations are quite different for *intensive* and *extensive* observables. An example of the former is $\langle p_t \rangle$: deviations from pure statistics (mixed events) are in such variables surprisingly small. An example of the latter is the fluctuations in particle multiplicity: in this case the Gaussian width is twice as large as for random emission. As shown from Ref. [702], about half of the effect comes from resonances at freezeout. (The rest comes from fluctuations in the initial conditions related to what was discussed in the preceding section.) In order to understand what happens here, imagine for a moment that all pions are only produced via σ, ρ or other $\pi\pi$ resonances, so that they always appear in pairs. By measuring the dispersion of the multiplicity distribution[26] one can see it, detecting resonances without measuring a single momentum.

Apart from resonances, there can be EBE fluctuations of the radial flow. One particular reason for that can be clustering, as suggested in my work [431]. The result of flow fluctuations should be a change in the slope of the whole distribution. I argued that in a 2-body distribution the correlation should thus be a "saddle-shaped" one, with negative correlation for one p_t small and the other large, and positive otherwise. Remarkably, this is indeed how the differential correlator observed by STAR looks, see Fig. 9.33.

We will not go into specifics, as the RHIC data on EVE fluctuations still have to be studied in a systematic way and published. They will certainly increase our knowledge of matter composition at the freezeout. I think one should pay attention to non-trivial correlations between fluctuations, such as between momenta and multiplicities $\langle \Delta p_t \Delta N \rangle$ [702] or fluctuations in particle composition [703].

9.7.3. *Critical opalescence near the tricritical point*

Can we learn something from the magnitude of these small fluctuations and their dependence on the parameters of the collision? Do they contain any more information than the corresponding moments of one-particle inclusive distribution?

One idea put forward by Stephanov, Rajagopal and myself [702] is that the EBE fluctuations can be used to find and study the critical end-point E on the phase diagram of QCD in the $T\mu$ plane, see Fig. 11.1. The point E can be thought of as a descendant of a tricritical point in the phase diagram for two-flavor QCD with *massless* quarks. The signatures proposed in Ref. [702] are based on the fact that such a point is a genuine thermodynamic singularity at which susceptibilities diverge and the order parameter fluctuates on long wavelengths. The resulting signatures all share one common property: they are *non-monotonic* as a function of an experimentally varied parameter such as the collision energy, centrality, rapidity or ion size. Once experimentalists vary a control parameter which causes the freeze-out point in the (T, μ) plane to move toward, through, and then past the vicinity of the endpoint E, they should see all the signatures we describe first strengthen, reach a

---

[26]Triggering on centrality should be independent of the multiplicity as such, for example based on the number of "spectator nucleons".

maximum, and then decrease, as a non-monotonic function of the control parameter. It is important to have a control parameter whose variation changes the μ at which the system crosses the transition region and freezes out. The collision energy is an obvious choice, since it is known experimentally that varying the collision energy has a large effect on μ at freezeout. Other possibilities should also be explored.[27]

We assume throughout that freezeout occurs from an equilibrated gas of pions, nucleons and resonances. If freezeout occurs "to the left" (lower μ; higher collision energy) of the critical end point E, it occurs after the matter has traversed the crossover region in the phase diagram. If it occurs "to the right" of E, it occurs after the matter has traversed the first order phase transition. Another example of non-monotonic signatures in a different but analogous context is the rise and fall in the number of large fragments as a function of total observed multiplicity in multifragmentation experiments [704] in low energy nuclear collisions. These experiments allow us to confirm the existence and study the properties of another critical point — the endpoint of the first-order nuclear liquid–gas transition (boiling of the nuclear matter liquid to yield a gas of nucleons) [704, 705]. This point occurs at a temperature of order 10 MeV, much lower than the one we are studying [702]. The analogy which is perhaps most familiar is with the phenomenon of critical opalescence observed in most liquids, including water. As the fluid cools down under conditions such that it passes near the endpoint of the boiling transition, it goes from transparent to opalescent to transparent as the endpoint is approached and then passed. This non-monotonic phenomenon is due to the scattering of light on critical long wavelength density fluctuations, and thus signals the universal physics unique to the vicinity of the critical point.

The sigma field is the order parameter for the transition and near the critical point it therefore develops large critical long wavelength fluctuations. These fluctuations are responsible for singularities in thermodynamic quantities. It would be strange if the properties of the pions remained regular in the thermodynamic limit in the presence of the non-analytic behavior of the sigma field. We will see that the fluctuations of the mean transverse momentum of the pions do in fact diverge at the critical point.

Let us consider small fluctuations of the field σ around the minimum of $\Omega(\sigma)$. We can then expand the effective potential $\Omega(\sigma)$ around $\sigma = 0$. The first terms will be

$$\Omega(\sigma) = \int d^3x \left\{ \frac{m_\sigma^2 \sigma^2}{2} + G\sigma : \pi^2 : +\mathcal{O}(\sigma^3) \right\}, \qquad (9.87)$$

---

[27] If the system crosses the transition region near E, but only freezes out at a much lower temperature, the event-by-event fluctuations will not reflect the thermodynamics near E. In this case, one can push freezeout to earlier times and thus closer to E by using smaller ions.

where we have temporarily omitted terms independent of σ (such as $m_\pi^2 \pi^2 / 2$).[28] The second term is the interaction between sigmas and pions. The coupling G has the dimensions of mass, and its magnitude near the critical point will be estimated below. The notation :: signifies tadpole subtraction: $: \pi^2 := \pi^2 - \langle \pi^2 \rangle$, which makes sure that the minimum of σ is not shifted (see below). Thus we have:

$$dP(\sigma) = d\sigma \exp \left\{ -\frac{m_\sigma^2 \sigma^2}{2T} - \frac{G}{T} \sigma : \pi^2 : \right\}. \qquad (9.88)$$

Now, the field π also fluctuates. Let us determine the corresponding (joint) probability distribution. In the previous section we used the probability distribution for the occupation numbers, and we begin by translating the fluctuations of the field π into fluctuations of the occupation numbers. We write, doing the usual Fourier transform: $\pi^2 = \sum_p |\pi_p|^2$. One can relate the Fourier components π_p to the occupation numbers n_p. It is clear that $n_p \sim |\pi_p|^2$. The coefficients can be determined, for example, by using

$$Z = \int \mathcal{D}\pi \exp \left[-\int_0^{1/T} dt \int dV \left(\frac{1}{2}(\partial_\mu \pi)^2 + \frac{1}{2} m_\pi^2 \pi^2 \right) \right]$$
$$= Z_{T=0} \prod_p \left(1 - e^{-\omega_p / T} \right). \qquad (9.89)$$

Differentiating $\ln Z$ with respect to m_π^2 we find

$$\langle \pi^2 \rangle = \sum_p \frac{1}{\omega_p} \langle n_p \rangle, \qquad (9.90)$$

up to the temperature independent vacuum contribution (equal to $\sum_p \omega_p / 2$) from $\ln Z_{T=0}$. So we have $\pi^2 = \sum_p (1/\omega_p) n_p$. Note now that $\langle \pi^2 \rangle = \sum_p \langle n_p \rangle / \omega_p \neq 0$. So, unless we subtract $\langle \pi^2 \rangle$ the minimum of σ will be shifted from the origin (this subtraction will also take care of the vacuum fluctuations). We have finally:

$$\pi^2 - \langle \pi^2 \rangle = \sum_p \frac{\Delta n_p}{\omega_p}. \qquad (9.91)$$

Now, putting everything together, we find the joint probability distribution for the sigma field and for the pion occupation numbers:

$$dP(\sigma, n_p) = d\sigma \left(\prod_p dn_p \right) \exp \left\{ -\sum_p \frac{1}{2v_p^2} (\Delta n_p)^2 - \frac{G\sigma}{T} \sum_p \frac{\Delta n_p}{\omega_p} - \frac{m_\sigma^2}{2T} \sigma^2 \right\}. \qquad (9.92)$$

---

[28] Clearly, the fluctuations of σ are not small. We shall proceed with the assumption that the higher-order terms in $\Omega(\sigma)$ yield sub-leading contributions to the singular effect we seek. Also note that we consider only the zero momentum mode of the field σ. This can be justified in a diagrammatic approach, which can also handle non-zero momentum modes of σ.

This is a very important formula which will allow us to calculate the fluctuations. The measure $dP(\sigma, n_p)$ is Gaussian, which is very helpful. Completing the squares, we find

$$dP(\sigma, n_p) = d\sigma \left(\prod_p dn_p \right) \exp \left\{ -\sum_p \frac{1}{2v_p^2} \left(\Delta n_p + \frac{G\sigma}{T} \frac{v_p^2}{\omega_p} \right)^2 \right\}$$

$$\times \exp \left\{ - \left(\frac{m_\sigma^2}{2T} - \frac{G^2}{T^2} \sum_p \frac{v_p^2}{2\omega_p^2} \right) \sigma^2 \right\}. \tag{9.93}$$

Finally, we can read off the following expectation values from the probability distribution (9.93):

$$\left\langle \left(\Delta n_p + \frac{G\sigma}{T} \frac{v_p^2}{\omega_p} \right) \left(\Delta n_k + \frac{G\sigma}{T} \frac{v_k^2}{\omega_k} \right) \right\rangle = v_p^2 \delta_{pk}, \tag{9.94}$$

$$\langle \sigma^2 \rangle = \frac{T}{m_\sigma^2}; \qquad \langle \sigma \Delta n_p \rangle = -\langle \sigma^2 \rangle \frac{G}{T} \frac{v_p^2}{\omega_p}. \tag{9.95}$$

This gives

$$\langle \Delta n_p \Delta n_k \rangle = v_p^2 \delta_{pk} + \frac{1}{m_\sigma^2} \frac{G^2}{T} \frac{v_p^2 v_k^2}{\omega_p \omega_k}. \tag{9.96}$$

We see that the coupling of the pions to the sigma field leads to a *singular* contribution to the correlator of the pion fluctuations as we approach the critical point at which $m_\sigma = 0$. The first term on the right hand side describes the variance of the inclusive distribution and the Bose enhancement effect. The singular term is due to the exchange of the sigma in the process of forward pion–pion scattering. This results in a characteristic $1/m_\sigma^2$ singularity.

Using the experimental coupling constant and including a finite size and lifetime of the fireball, it was concluded that all observables including the EBE fluctuations of the mean transverse momentum would be affected by the tricritical point sufficiently strongly to be detected. Hopefully this region of the phase diagram will be reachable by the newly approved GSI heavy ion facility near Darmstadt.

9.7.4. *Can QGP charge fluctuations survive the hadronic phase?*

Let me first confess that I was strongly hesitant as to whether this discussion should or should not be in this book. The answer to the question put into the above title is basically *no*, as far as we can tell from the data, and so strictly speaking the idea to be discussed is all but dead. However, negative information can also be very instructive, which I take as a justification for including this section.

The idea itself has been proposed in two simultaneous papers, by Asakawa, Heinz and Muller [706] and Jeon and Koch [707]. They argued that the primordial QGP-based fluctuations may be observed in fluctuations of certain conserved charges, rather than the equilibrium value at freezeout. More specifically, they argued that early-time fluctuations can possibly be seen via the *long-range* fluctuations of *conserved* quantities, like the electric (or the baryon) charge. Their idea is based on the well known phenomenon of *kinetic slowdown* of fluctuation relaxation, provided *sufficiently long-range harmonics* of those are considered. If the relaxation time happens to be shorter than the lifetime of the hadronic stage of the collisions, the authors argue, the values of such fluctuations should deviate from their equilibrium (resonance gas) values toward their earlier, primordial value, typical of QGP, which happen to be smaller by a factor 2 to 3.

Although this idea should work for parametrically long-range effects, whether a number of necessary conditions can indeed be fulfilled in realistic heavy ion collisions depends on the answers to the following questions: (i) How fast is the relaxation due to final-state re-scattering in hadronic gas? (ii) How large should the detector acceptance really be, in order to counter the relaxation? (iii) Is there a window of parameters, forbidding resonance gas relaxation but allowing it at the QGP stage? (iv) If not, which fluctuations should follow from non-equilibrium parton kinetics, at very early pre-QGP stages of the collisions?

Relaxation of long-range fluctuations has been studied by various methods, ranging from direct cascade simulations to analytic ones. In particular, a diffusion equation in rapidity space was derived and studied in a paper by Stephanov and myself [708]. The conclusion was that it is not likely that QGP magnitude of fluctuations can survive, since diffusion is rather robust and can still erase the QGP signal. Experimental data on charge fluctuations observed by STAR (still to be published) have reached values close to predictions for equilibrium resonance gas, and very far from equilibrium values for QGP.

In summary, one learns from this story that hadronic phase is long enough to relax even the long-range fluctuations of conserved quantities. All models of "instantaneous hadronization" from QGP are simply wrong.

CHAPTER 10

Early Diagnostics of Hadronic Matter

10.1. Penetrating probes: dileptons and photons

10.1.1. *Basic rates and space–time profile*

One of the first suggestions in studying the earlier stages is to look for "penetrating probes" [457], the particles which do not suffer final state interaction such as *dileptons*[1] or *photons*. It was argued in this work that the spectra of dileptons produced in the QGP phase should be different from those originating from hadronic matter. In particular, those should show no peaks corresponding to familiar vector mesons, $\rho, \omega, \phi, J/\psi, \ldots$, but be just smooth decreasing functions of the dilepton invariant mass $M^2 = (p_{l+} + p_{l-})^2$. Furthermore, the dileptons produced in the hadronic phase are expected to show all these resonances, but with their parameters — masses and widths — being *modified*. It seems quite possible now that some bound states survive well into QGP as well.

The theoretical evaluation of the dilepton production is usually made in two steps: the first is the determination of the *production rate* in equilibrium matter, per time per volume at a given T. The second step is the space–time integration over the four-volume of the excited matter during its evolution, including temperature variation. The latter is usually done using the hydrodynamic description of the explosion following high energy heavy ion collisions, discussed in detail in the preceding chapter.

The simplest (I would call the 0th approximation) models for the dilepton production rate are based on two well-tested processes:

- $\pi\pi$ annihilation in hadronic phase (small T)
- $\bar{q}q$ annihilation[2] in QGP (large T).

These two basic processes can be both accounted for by the "standard rate" formula:

$$\frac{dR}{d^4q} = \frac{\alpha^2}{48\pi^4} F \frac{1}{\exp\left(\frac{q_0}{T}\right) - 1},\tag{10.1}$$

---

[1]A dilepton is a lepton pair, e^+e^- or $\mu^+\mu^-$, originating from a virtual photon.

[2]This process is similar to the well-known Drell–Yan process, only with the weight coming from the thermal ensemble rather than from the nucleon structure functions.

where the rate R is counted per 3-volume per unit time, and q is the four-momentum of the virtual photon ($q^2 = M^2_{e^+e^-} = M^2$). The remaining factor F for $\pi\pi$ annihilation is the usual pion form factor, e.g. written in standard vector-dominance form. For the QGP case the factor F is just a constant, up to small corrections, containing the sum over electric charges squared of all quarks,

$$
F = \begin{cases} F_{\mathrm{H}} \stackrel{\mathrm{def}}{=} \dfrac{m^4_\rho}{[(m^2_\rho - M^2)^2 + m^2_\rho \Gamma^2_\rho]}, & \text{(Hadronic)}, \\[4mm] F_{\mathrm{Q}} \stackrel{\mathrm{def}}{=} 12 \sum_q e^2_q \left(1 + \dfrac{2m^2_q}{M^2}\right)\left(1 - \dfrac{4m^2_q}{M^2}\right)^{1/2}, & \text{(QGP)}. \end{cases} \qquad (10.2)
$$

At low density one can relate the modification of mesons (e.g. of ρ) to the $\pi\rho$ and Nρ forward scattering amplitudes, or momentum-dependent optical "potentials", which predict a relatively modest shift of m_ρ downward and some broadening as we discussed in Chapter 9. Further developments along such lines lead to what I would call the *1st approximation models*. An example is e.g. the Li–Ko–Brown model [726] in which all hadronic masses, including those of the vector mesons, are shifted proportional to Walecka-type mean field. Those models were implemented as codes, self-consistently describing the motion of particles in the space–time dependent mean field. What is not so easy to achieve is to have a complete energy conservation and "curved" trajectories of particles.

The "*2nd approximation*" *models* try to describe what happens in between these two limits, especially in hadronic matter at $T \approx T_c$ loosely called "the mixed phase". It is also based on the notion of "meson modification" in matter which is no longer considered small. We have discussed in Section 8.4.3 how the "constituent quark mass" in heavy-light D, B... mesons decreases toward T_c due to chiral symmetry restoration; one may assume that other masses do about the same.

When matter is no longer dilute and modifications are no longer small, one should do some re-summations and try to minimize the free energy consistently. This leads to a set of coupled gap equations. As a good example of such kind of work let me mention the Rapp–Wambach approach [721], in which very strong broadening of the ρ meson was predicted based on properties of ρ–N interaction and N^* resonances. One can also understand this broadening as due to mixing between ρ with excitations of the lowest baryon resonances, such as[3] $N^*(1520)N^{-1}$.

Let us now see how these models compare to each other. In Fig. 10.1, from Rapp [725] one can see the two elementary rates plus state-of-the-art manybody calculation. The pion annihilation curve peaks at the (unmodified) ρ peak, while the $\bar{q}q$ annihilation curve just monotonically decreases with M. Interestingly enough, the "realistic" curve which includes in-matter effects in the *hadronic phase* (not QGP)

---

[3]This notation means the baryonic resonance *minus* a nucleon, meaning that the nucleon was there in the initial state.

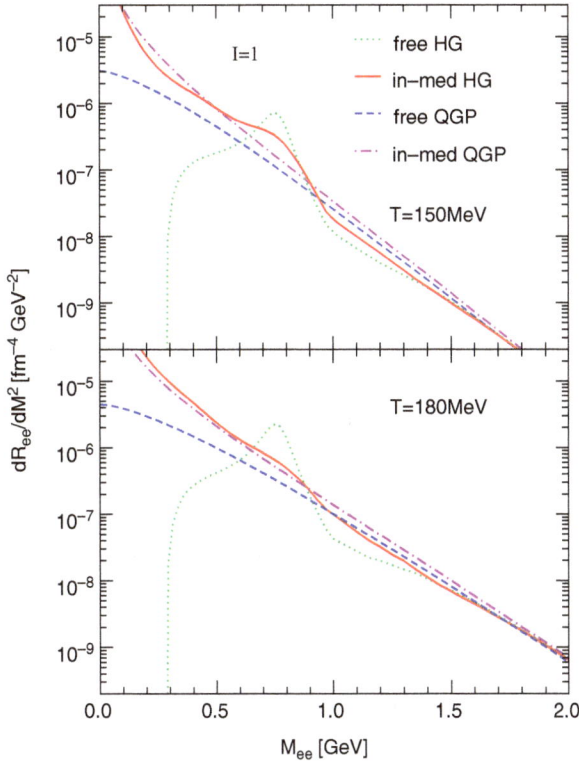

Fig. 10.1. Equilibrium dilepton production rates, from hadronic (a) and QGP (b) matter.

is very close[4] to the $\bar{q}q$ annihilation curve, and it shows very little trace of the ρ peak. Presumably this fact is a manifestation of "hadron–parton" duality which we discussed in Section 6.2.2 in connection with vector correlation functions. One may expect that as T rises, all masses decrease while all widths grow, and therefore the amplitude of the duality-violating oscillations we discussed in Section 6.2.5 is also decreasing.

Whatever the reasons may be, one finds that such theoretical evaluations of the rates that the dilepton production, both in QGP and dense hadronic matter close to T_c, lead to about the same as the 0th approximation "partonic" rate. If so, it provides a "standard candle" to calibrate the temperature profile of the exploding matter.

Let me now briefly comment on the space–time evolution issue. Already in Ref. [709] a universal *temperature profile function* has been introduced, defined by

$$\Phi(T) = \int d^4x\, \delta(T(x) - T), \tag{10.3}$$

---

[4]Except for the low-mass region in which the Rapp–Wambach curve grows as $1/M$ at small M. This happens due to bremsstrahlung and also to important processes like $a_1 \to \pi\gamma$ [729].

where the integral is done over the four-volume of expanding excited matter, between $3d$ surfaces of its production and freezeout. The space–time dependence of the temperature $T(x)$ is determined e.g. from hydrodynamics or kinetic models. This function is universal in the sense that it can be convoluted with thermal rates of various processes, each of which depends on T only.

$$\text{Yield} = \int dT \, \Phi(T) \, \text{Rate}(T). \qquad (10.4)$$

Two features of this function are worth mentioning. First, it contains a significant contribution of the mixed phase close to the QCD critical temperature, which looks like $\sim\delta(T - T_c)$ because the system spends a finite (and sometimes a large) fraction of space–time in it. Apart from that, the profile function $\Phi(T)$ is rapidly falling with T, as a large power of T. All rates rapidly grow with T due to the Boltzmann factor $\sim\exp(-M/T)$. As a result, the T-integrals in the yields usually have well-defined saddle points, which means that a particle of a given mass $M(\text{or } p_t)$ corresponds with a well defined temperature $T = \text{const } M$. The constant in this relation is usually small numerically, $\sim 1/10$, due to the rapid expansion of the system: that is why dileptons of mass $\sim 2\,\text{GeV}$ are most useful for accessing the QGP phase with $T \sim 200\,\text{MeV}$, etc.

10.1.2. *Dilepton data versus expectations*

Let me start with a general remark about the history of the dilepton experiments in general. This kind of experiments is generally much more difficult as compared to measurements of hadronic observables. In addition to the large background which needs to be rejected, they are related with smaller cross sections and are significantly limited by statistics and thus put severe demands on the luminocity. Historically, the dilepton experiments have a sad tendency to come "too late". At Berkeley BEVALAC the DLS spectrometer was built to study dileptons, but its detector efficiencies were finalized years after the last run. A striking dilepton signal was found, but my guess is its reality will probably never be understood, although this mass/energy region will be soon tested by the HADES experiment at SIS (GSI). The Brookhaven AGS program had no dilepton experiments at all. The CERN SPS program has three experiments mentioned below, and those have produced exciting data. The RHIC luminosity is still not yet high enough to allow for quality dilepton measurements, and so far only some J/ψ results have been obtained. The dileptons are supposed to be the main part of the PHENIX detector program: let us hope it will have the luminosity to do it in time.

Experimental studies of dilepton production in heavy ion collisions have been performed at CERN SPS by three experiments, (i) CERES (NA45), which studied the low mass ($M = (0\text{--}1.0)\,\text{GeV}$) e^+e^- pairs, (ii) HELIOS-3, which studied medium mass $\mu^+\mu^-$, $M = (1\text{--}2)\,\text{GeV}$, and (iii) NA38/50 concentrated on high mass $\mu^+\mu^-$. All three see quite significant enhancement over "standard sources", ranging from factor 5 at CERES (in some kinematic region) to about 3 at NA38/50[5] at $M = 2$–$3\,\text{GeV}$. These numbers are maximal, corresponding to the most central heavy ion collisions, like Pb–Au.

I have no place to explain how experimentalists have done their homework, they have measured the dilepton production in pp (or p–Be) and found the results completely consistent with a "cocktail" of known effects, such as π, η and resonance decays. The reader should either just take my word for that, or look up the original papers.

In Fig. 10.2 one can see the (space–time integrated) rates compared to data from the CERES experiment at CERN. The "cocktail" of known sources is clearly unable to reproduce the data, especially at dilepton masses $M \sim 0.5\,\text{GeV}$. The models using the reduced mass of vector mesons and their increased width describe the enhancement rather well. They also describe the measured distribution over $q_\perp$ of the dilepton (not shown).

These data clearly tell us that *the spectral density of in-matter excitations is indeed qualitatively different from the in-vacuum one*. The ρ peak seems to be gone, and the spectral density looks close to the smooth "partonic" quark continuum we expect to see in QGP. The central point is that the observed effective "ρ melting" indicates an approach toward chiral symmetry restoration and QGP.

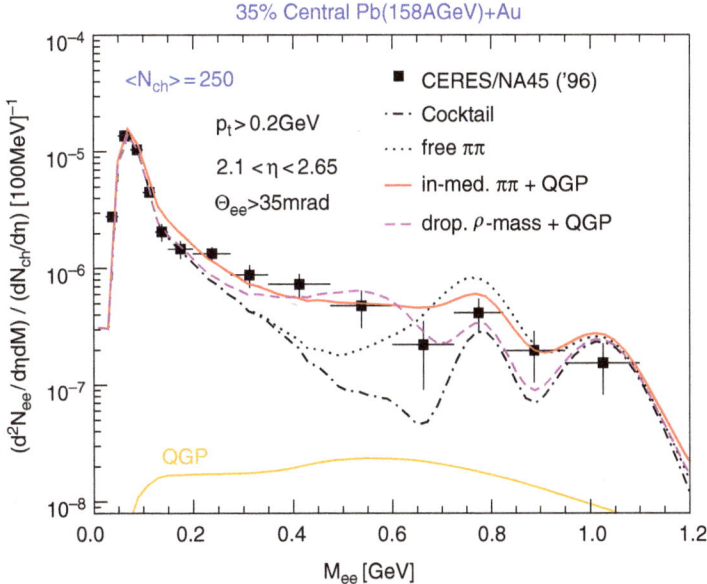

Fig. 10.2. Comparison of the CERES full energy (96 data) for mass spectrum of the observed dileptons with several theoretical calculations.

---

[5]In the "very high" mass region, $M > 3\,\text{GeV}$, the dilepton production is well described by the simple partonic Drell–Yan process.

This was quite puzzling, since at the SPS only a small fraction of the space–time volume contributing to dilepton production is expected to originate from QGP, while most of it should still be from the mixed or hadronic phase. Apparently the observed spectral density looks partonic already at $T < T_c$.

The "duality" argument emphasized in the preceding section is the best explanation we have for this, so let me reiterate it. In a very crude way, one can characterize it by a change in the so called *quark–hadron duality* parameter, E_{cont}. By definition, for the dilepton masses $M > E_{\text{cont}}$ the spectral density is reasonably well described by the zeroth order parton model. From e^+e^- collisions and τ decays we know that *in vacuum*, at $T = 0$, its value is $E_{\text{cont}} \approx 1.5\,\text{GeV}$. Below this value there is only the isolated ρ, ω peaks, while all states above it can be well approximated by a "partonic" continuum. As analysis of the CERES data shows the corresponding "in matter" spectral density corresponds to much smaller value, maybe as low as $E_{\text{cont}} \approx 0.5\,\text{GeV}$. If this is indeed true, it is the strongest drop in E_{cont} possible since this parameter cannot really reach zero even in QGP.[6]

One more thing to note here is that a very interesting *mixing phenomenon*, between vector and axial currents and spectral densities [616] can be studied via dileptons. The a_1 resonance can decay into a dilepton plus a pion: this was not included in the standard hadronic cocktail. The chiral restoration demands that vector spectral density should become identical to that of the *axial* current. Indeed, the only difference between them is related with chirality-flipping terms.

General properties of the correlators following from OPE can be written as a set of Weinberg-type sum rules [727], relating certain moments of the difference between vector and axial spectral densities to a particular chirally-odd order parameters like $\langle \bar{q}q \rangle$ and f_π. As all of them should vanish at $T = T_c$, these sum rules demand that both spectral densities be the same at $T > T_c$.

The so called intermediate mass (1.5–3 GeV) dileptons, IMD, is also a region where a significant enhancement of the dilepton production has been observed, by the NA38/50 experiment. One possible interpretation (put forward by the experimentalists themselves) is the enhancement of the open charm production. In order to explain the data it should however be quite large, up to factor 3 for the most central collisions. It has been however shown by Rapp and myself [738] (see also Ref. [739]) that there is no need for this anomalous charm production because the thermal radiation of dileptons can naturally describe it, but for the invariant mass and the transverse momentum distributions. What is important to note, is that the same thermal model works well for CERES (low mass region) and NA50 (intermediate mass region). In these calculations, about 25–30% of the dilepton yield comes from the QGP phase, while the rest is emitted from hadronic matter. So the QGP component for intermediate mass dileptons at SPS is sizable, although not dominant.

---

[6]Because of the bremsstrahlung at small M, and also due to the fact discussed in detail in Chapter 8, namely that the quasi-free massless quarks are "dressed" by a plasma and obtained their effective masses $\sim gT$.

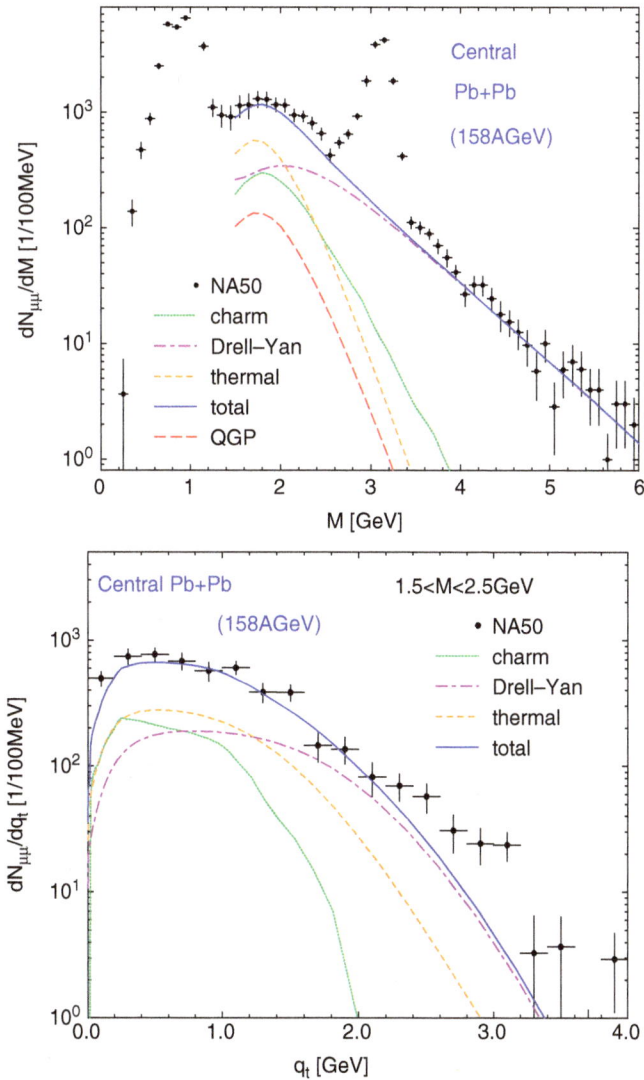

Fig. 10.3. IMR dilepton spectra in comparison to NA50 data from central Pb(158 AGeV)+Pb collisions with $N_{part} \simeq 380$ (top panel: invariant mass spectra, bottom panel: transverse momentum spectra). The short-dashed and dashed-dotted curves are the calculated thermal and Drell–Yan yields, respectively, using the approximate acceptance as given in the text. The long-dashed curve in the top panel is the part of the thermal contribution originating from the plasma phase. The dotted curves represent the open charm contribution as obtained by the NA50 collaboration in a PYTHIA simulation without anomalous enhancement. The full lines are the sum of thermal, open charm and Drell–Yan dileptons.

(a)   Au+Au @ $\sqrt{s_{NN}}$ = 130 GeV : 0–10 % central

(b)   Central Au+Au $S^{1/2}$ = 130AGeV

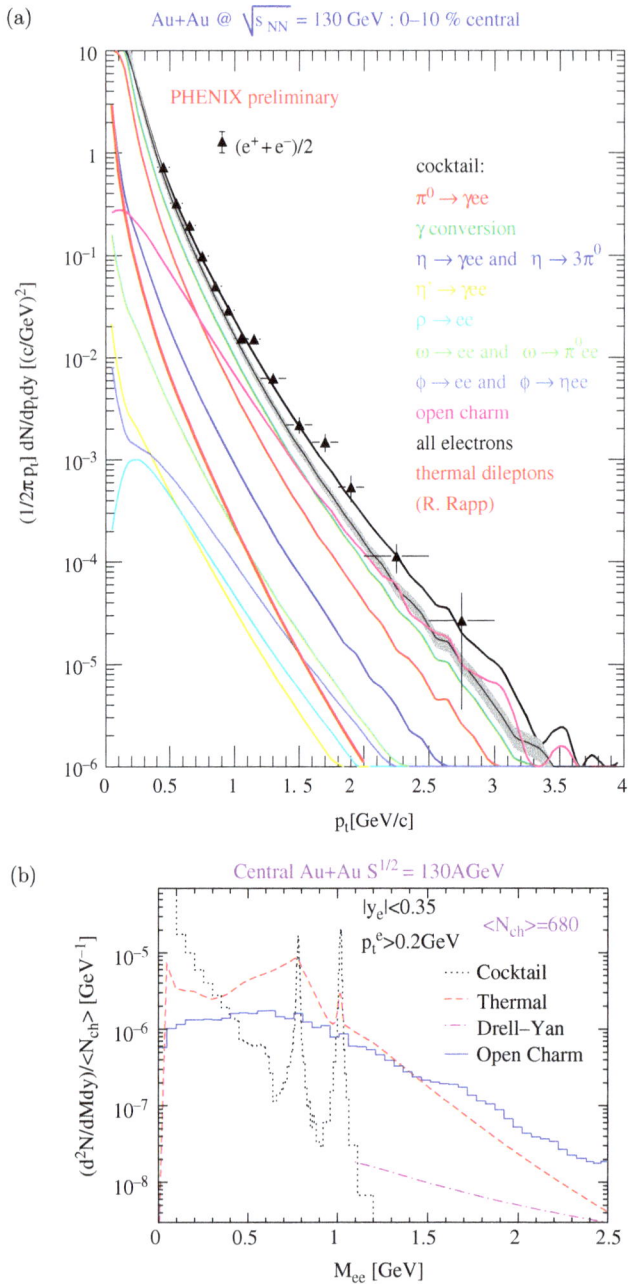

Fig. 10.4. (a) The single-electron p_t spectra from PHENIX compared to expected sources, the thermal dileptons and charm decay contributions. (b) The corresponding spectrum of the dilepton mass, based on the same ingredients as the singles in (a).

The NA50 experiment now evolved into to NA60 which is supposed to look for charm production directly, and thus settle the issue. However PHENIX at RHIC, at much higher energy, observes single electron production [728] consistent with charm production, being in agreement with the parton model, *without* factor 3 enhancement.

The parameterization of the space–time picture used by Rapp and myself was further simplified by Kempfer *et al.* [739], who replaced functions by their average $T(t, \vec{x}) \to \langle T \rangle \equiv T_{\rm eff}$, $u(t, \vec{x}) \to \langle u \rangle$ and $\int dt \, d^3x \to \int dt \, V(t) \to N_{\rm eff}$. They have used the same parameterization to describe all electromagnetic data together, as shown in Fig. 10.5.

Let me now briefly describe the dilepton predictions for RHIC, following Rapp [725]. Although charm production cross section at RHIC energies is rather

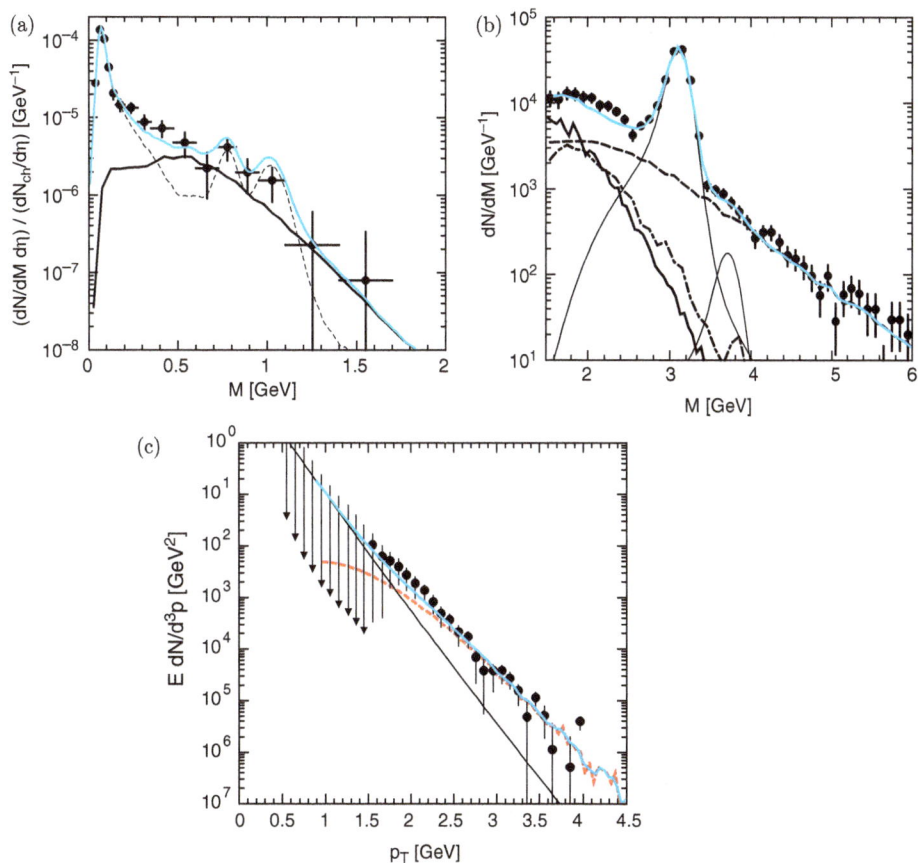

Fig. 10.5. Comparison of the thermal model by with dilepton data (top left panel for the CERES data [736]; top right panel for the NA50 data [737], thin lines: parameterizations of the J/ψ and ψ') and the photon data (bottom panel for WA98 data [734]; hadron decay contributions are subtracted). The thermal contribution is characterized by the unique set of parameters $T_{\rm eff} = 170\,{\rm MeV}$ and $N_{\rm eff} = 3.3 \times 10^4\,{\rm fm}^4$. Sum of all contributions: gray curves.

uncertain; the first indirect evidence that it complies well with pp-based extrapolations from lower energies is provided by single-electron spectra measured by PHENIX at $\sqrt{s} = 130\,A\,\mathrm{GeV}$. The charm decays nicely account for the excess over observed sources at $p_t^e \geq 1.5\,\mathrm{GeV}$. At the same time, the thermal contribution to the single-$e^\pm$ spectra turns out to be negligible.

This situation changes in the pair spectra: the open-charm contribution is relatively suppressed to thermal radiation (especially at lower masses) by an additional semileptonic branching ratio, their harder spectral shape and the typically large rapidity gaps between e^+ and e^- (within the rather narrow PHENIX acceptance). One can conclude from the last figure that thermal dilepton production remains an important contribution, hopefully to be observed at RHIC soon. Whether one can get information about quasi-particle masses and bound states in QGP remains to be seen.

In conclusion of this section, let me express my views that "3rd generation models" are still missing; i.e. those to be based on more direct lattice correlators and more fundamental understanding of the modified basic interactions between quarks, both at T just below and above T_c. Those should be able not only to predict changes of masses, but also of condensates themselves and other parameters of the spectral density. I think the most important issue is actually the modification of the "duality scale", the "threshold" E_0, above which parton results are dual to hadronic calculations.

10.1.3. *Direct photon production*

We discussed above how the dileptons can be produced by both a partonic prompt Drell–Yan process and in-matter thermal annihilation. The production of direct (i.e. not from hadronic decays) photons can also be made from a number of direct and thermal sources. The partonic production of photons is well known and studied in pp collisions: we would not describe those in detail.

The rate of photon production in matter, per unit time and volume, of a real photon of momentum p is

$$\frac{E\,dN}{d\vec{p}\,dt\,d\vec{x}} = -\frac{1}{(2\pi)^3}\,n_B(E)\,\mathrm{Im}\,\Pi_\mu^{R\mu}(E,\vec{p}), \qquad (10.5)$$

where $\Pi_\mu^{R\mu}(E,\vec{p})$ is the retarded photon polarization tensor. The pre-factor $n_B(E)$ provides the expected exponential damping $\exp(-E/T)$ when $E \gg T$.

In the original paper, suggesting thermal photons from QGP [709], only two lowest order reactions were considered: (i) the "Compton" scattering $gq \to \gamma q$ and (ii) annihilation $\bar{q}q \to \gamma g$. The corresponding rates for hard photons are

$$\mathrm{Im}\,\Pi^R(E,\vec{p}) \sim e^2 g^2 T^2 \left(\ln\left(\frac{ET}{m_q^2}\right) + C \right). \qquad (10.6)$$

Note that there is an IR logarithm here: its exact form up to a constant C was calculated in Ref. [719]. However it is clear that in general the photon production, even at large photon energy $E \gg T$, is more difficult to evaluate because of possible bremsstrahlung-like phenomena. In particular, a process of next order in g such as $qq \to qq\gamma$ can be as large as the leading one of the momentum transfer which is soft $q \sim gT$. So, a consistent re-summation of hard thermal loops is needed. For a recent review of its status, see Ref. [733]. In brief, the next order diagrams with bremsstrahlung lead to rates of the order

$$\text{Im } \Pi^{\text{R}}(E, \vec{p})\big|_{\text{brems}} \sim e^2 g^2 T^2, \tag{10.7}$$

$$\text{Im } \Pi^{\text{R}}(E, \vec{p})\big|_{\text{annil}} \sim e^2 g^2 T E. \tag{10.8}$$

The $q\bar{q}$ annihilation with scattering (a "3 to 2" process crossed from bremsstrahlung) actually grows with the energy E and dominates when $E/T \gg 1$. Phenomenological applications of these results have been carried out and the two-loop processes have been included in hydrodynamic evolution codes to predict the rate of real photon production. It is found that the two-loop processes (especially the annihilation with scattering) lead to a numerical increase of the rate [740–742,745]. The enhancement mechanism operative at two-loop could also be at work at the multi-loop level especially in ladder diagrams. It was indeed found that the higher order ladder diagrams contribute to the same order as the two-loop ones [743], but these higher order effects have been combined in a consistent way [744] and reduction of the two-loop photon production rate is found. From the phenomenological point of view the reduction is less that 30% in the p_T range of interest.[7]

The issue of photon production from hadronic gas has also been rather extensively discussed in the literature. Kapusta *et al.* [719] included all diagrams with ρ, π mesons and found that it is about the same as the lowest-order QGP. Xiong, Brown and myself [729] have argued that one process $(\pi\rho \to)a_1 \to \gamma\pi$ contributes about a factor 3 more to the rate. As a result, one again sees an approximate "duality", as for dileptons; namely that *the photon production rates in hadronic and QGP phases at close T seem to be very similar*. This fact makes it more difficult to claim "QGP radiation" as a signal, but makes theoretical predictions less uncertain.

Finally, let me present as an example a comparison between the WA80/98 data and a direct + thermal photons calculation by Kempfer *et al.* [739] shown in Fig. 10.5c. (Figures 10.5a and b are for low mass and intermediate mass dileptons, which are also described using the same space–time evolution parameters.)

One can see that direct photon production from partonic processes is in good agreement with data. However, in spite of strong efforts, large background from π^0 decays only allowed the upper limits in the region crucial for thermal production. Whether or not the theory is describing data well enough, is hard to tell.

---

[7]All these calculations are done for a plasma in chemical equilibrium which is not a realistic approximation.

There is still one way to beat the π^0 decay background: it is the two-photon interferometry [746]. Indeed, photons created from the decay cannot interfere, as they appear far from each other, only photons directly produced in the fireball can. At QM02 there was a poster [747] from WA80/98 claiming that the two-photon data do indeed show the interference peak. Unfortunately, by the time this text was to print, the collaboration had not yet made this result official so it still has a preliminary status, see hep-ph/0403274. Potentially, the two-photon measurements can answer the question about soft photon production from hadronic matter.

10.2.   Quarkonia in heavy ion collisions

10.2.1.   *Charmonium suppression, the mechanisms*

In a very schematic way, the idea is as follows: if the observed charmonium originates at early hard processes, and then goes through all stages of evolution, by comparing the yield in AA with pp (where matter is absent) we can learn something about properties of the matter.

Several mechanisms of this suppression were proposed over the years: (i) gluonic "photo-effect" in QGP, by myself [709, 753], exciting bound states into continuum; (ii) the bound states simply do not exist in QGP because of the Debye screening, by Matsui and Satz [752]; (iii) hadronic co-movers kill them; (iv) non-monotonic variation of the QGP lifetime [755]; (v) charm quarks can percolate in the transverse plane if the produced strings overlap into a large cluster [756]; (vi) new version of photoeffect, a quasi-free gluon scattering [757].

To add to this complicated issue, the usual assumption that all charmonia originate from hard reactions has also been questioned. At RHIC one definitely expects that a significant fraction of charmonia is produced via a coalescence of charm quarks or D-mesons, see original references and a discussion e.g. in Ref. [757].

Let me start by explaining the main idea of Matsui and Satz [752]. They have argued that because of the Debye screening, the modified potential between two charm quarks with increasing T becomes more and more shallow, and is loosing the bound states, one by one. The lowest ones, $J/\psi, \eta_c$ are obviously the last ones to melt, while for example ψ' turned out to be unbound for any $T > T_c$.

I would suggest to modify their argument, which can be explained by a simple pedagogical analogy. Consider for example a molecule of the ordinary salt, NaCl, which is well bound due to electric attraction of positive Na and negative Cl charges of the effective ions. However if this molecule is put into water it dissolves into two separate ions, because attraction of ions to water is larger than to each other.

We expect to see the same with the charmonium in QGP at very high T. The correlation energy for a static quark in QGP 8.53 we derived previously is $E_c \sim g^3 T$ and grows with T. If the usual binding energy of the charmonium states B_i changes, it goes down. So for all bound states i of heavy quarks, at some T_i one would get an equality

$$2|E_{\rm c}(T_i)| = |B_i(T_i)|, \tag{10.9}$$

and at $T > T_i$ the ith state should be dissolved into two ions with separate Debye clouds.

Non-relativistic studies in the framework of the Schrödinger equation with a screened potential predicts that all quarkonia states dissolve at $T_{\rm c}$ with the ground states, $J/\psi, \eta_{\rm c}$ dissolving at $T > T_\psi \approx 1.1 T_{\rm c}$. However recent lattice results on quarkonia states at finite T seem to move this dissolution temperature T_ψ higher. In particular, in two papers [758] the calculation of the charmonium correlation functions have shown that $J/\psi, \eta_{\rm c}$ remain very well-defined resonances at least at $T = 1.5 T_{\rm c}$, and put the dessolution probably at $T_\psi \sim 2 T_{\rm c}$ or higher. If so, the Matsui-Satz explanation for J/ψ suppression at SPS and RHIC is quite questionable.

In summary, the heavy quarkonia provide a sequence of states which can be used as diagnostic tools, from ψ' to χ then to the J/ψ proper, displaying a spectacular sequence of "melting" phenomena happening at different densities. Presumably with the luminosity upgrade at RHIC we will be also able to see how members of the Υ family do the same, but at higher densities.

10.2.2. *Charmonium suppression, the data*

Experimentally one can vary the probe (bottonium versus charmonium) or on the binding energy (compare $\psi', \chi, J/\psi$). One can also change the matter properties, by changing centrality, collision energy or atomic number.

Let me comment on the A dependence first. Experiments done at CERN have found quite contrasting A dependence of the dilepton enhancement (the ρ melting/ shifting discussed at the beginning of this chapter) with the J/ψ suppression. The former, studied in the CERES/NA49 experiment display comparable effects for all A, such as the S and Pb. On the other hand the J/ψ suppression studied by NA38/50, was found to be drastically A-dependent. All light ions including S behave similarly to the pp collision, with "non-anomalous" nuclear absorption only as in pA collisions.

Only the central PbPb data seem to have large "anomalous" effect. A dramatic centrality dependence observed by NA38/50 is shown in Fig. 10.6 from Ref. [759], which shows that the "anomalous" (different from extrapolations based on pA and SA collisions) J/ψ suppression is only seen for *central* ($b < 8\,{\rm fm}$) PbPb collisions. Figure 10.6b shows a comparison to a conventional cascade-like model including some cross section of J/ψ scattering. Figure 10.6c shows a comparison with the "instantaneous" dissolution model for χ and then J/ψ in two "jumps", due to Satz and collaborators. It appears that the data are explained better by the latter model, although skeptics question the reality of the "jumps" in the data sample.

Of course, we are waiting for RHIC experiments to clarify the issue. At the time of this writing, however, RHIC charmonium data have too poor statistics to make

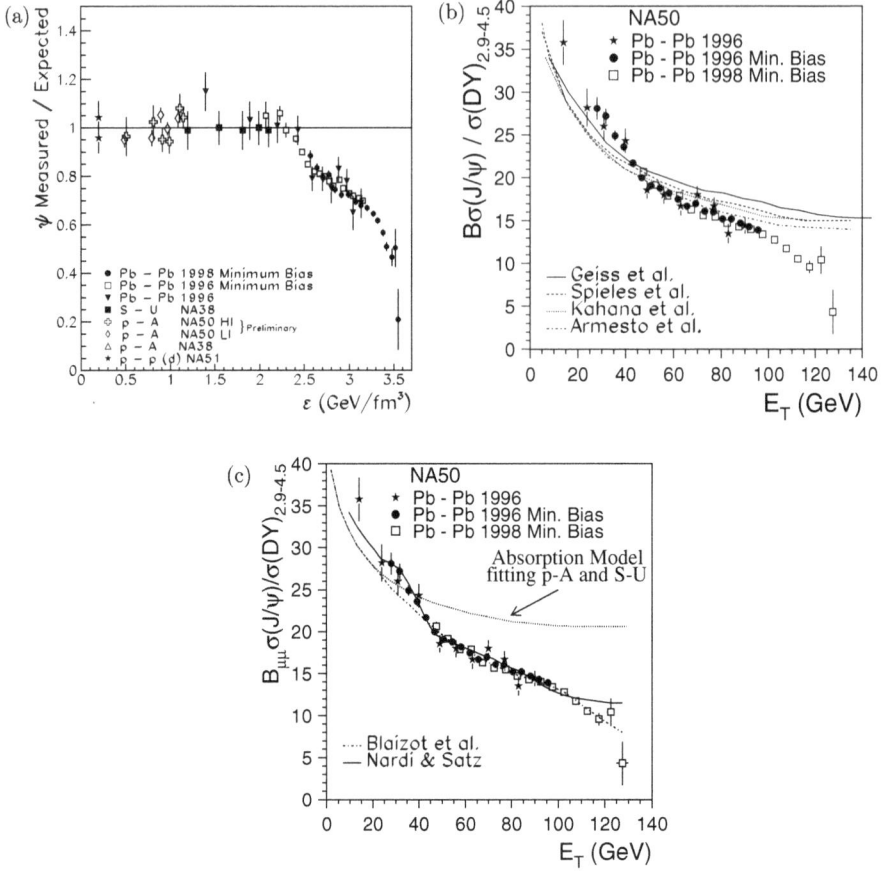

Fig. 10.6. (a) J/ψ production normalized to the expected yield (absorption fit from p-p to S-U) as a function of the energy density ϵ. The J/ψ/DY data with stepwise pattern compared with conventional models (b) and the one assuming rapid deconfinement (c).

a meaningful comparison: let me just say that they do not show either very strong suppression or enhancement.

What can be done to differentiate experimentally between these ideas? The old idea is to study suppression dependence on p_t. Unfortunately in changing p_t we also change the kinematics: e.g. the destruction by gluons or hadrons goes better if p_t grows. Maybe a better idea is to use the azimuthal dependence of the suppression. Instantaneous suppression should show *no* asymmetry, but if it takes a few fm/c the anisotropy should show up. For example, if ψ can form from two charm quarks floating with QGP, the usual elliptic flow will be seen. The problem is that the initial "almond" at $b < 8$ fm is not very anisotropic, and for larger b there is no anomalous suppression.

Additional hints may be provided by the ψ' suppression. The data show that the ψ'/ψ ratio first fell and then stabilized at some small value. According to an observation made by Sorge, Zahed and myself [760] this final ratio is consistent with the thermal value $\psi'/\psi = \exp(-(M_{\psi'} - M_{J/\psi})/T_c)$ for the usual chemical freezeout. In this paper we also argued that all ψ' are most probably first suppressed and then re-created from J/ψ.

10.2.3. ϕ-related puzzles

The $\bar{s}s$ vector meson ϕ is a little brother of the $\bar{c}c$ vector meson J/ψ. However, we do know (both from $\mu^+\mu^-$ data from NA50 and KK data of the NA49 experiments at CERN SPS) that its production in AA is *enhanced* as compared to what one would get from a set of NN collisions, in contrast to J/ψ. This is not because it is somehow unaffected by the destructive effects killing the J/ψ, but because strangeness is close to equilibrium at chemical freezeout and ϕ readily regenerates from the readily available sea of strange quarks floating in matter.

It was shown by Lissauer and myself [496] that because m_ϕ nearly coincides with $2m_K$, modification of kaons even on a few percent level by their attraction to nucleons or pions should strongly (by factor 2 or more) affect the probability for the ϕ decay. This implies that some fraction of ϕ should decay inside the fireball and be modified at least in width. Theoretical estimates suggest it should be a significant fraction, while the experiments do not see any modification of the ϕ mass or width in the KK channel. This is consistently observed at AGS, SPS and RHIC. This puzzle can probably be explained by absorption of the decay products, K^+K^-, or their scattering in a collective potential arising from $K\pi$, KN interaction.

Lately, the SPS experiments added two more pieces to the puzzle. The m_t slopes of the ϕ spectra measured by NA49 in the KK channel are large, close to those of the $p, \bar{p}$ (particles of similar mass). It suggests that the ϕ meson fully participates in the radial flow. It is possible only if ϕ coalescence from kaons happens at a very late time. The rapidity distribution of ϕ is close to the product if K^+ and K^- ones,[8] which also hints at the same conclusion.

The NA50 dilepton experiment reveals however quite a different m_t slope for ϕ, much smaller value, of $T_{\text{eff}} \approx 220\,\text{MeV}$. It is measured in a difference acceptance (larger p_t) than NA49, and this is unusual. It seems to indicate that the ϕ seen in the leptonic channel has different features from those seen in KK: in the case of leptons we should not worry about rescattering. So, maybe NA50 has also found the other ϕs, which decay inside the fireball *before* flow develops.

We are expecting more details on that from the PHENIX experiment at RHIC, which will be able to observe ϕ in the same kinematical domain in both dilepton and KK modes.

---

[8]Those are quite different at SPS because large baryon density has strong preference for hyperons relative to anti-hyperons.

10.3.   Evolving views on the initial stage

10.3.1.   *Perturbative processes and minijets*

At relatively low collision energies, such as AGS with up to 11 GeV*A of lab momentum, the produced matter is always in the hadronic state. One may expect that multiple production of hadrons occurs via ordinary cascade of hadronic reactions. The usual models describing well this region are based on the color exchanges + string fragmentation picture. Popular event generators for this region, RQMD and UrQMD, are based on these ideas.

Matter produced in heavy ion collisions at the higher energies of CERN SPS fall into the critical region $T \sim T_c \approx 160$ MeV, at lower SPS energies, and slightly venture into the QGP phase at its higher end. At RHIC the initial energy density is of order of magnitude higher than critical values, and the initial temperature $T_i \sim 350$ MeV, while it is estimated to be $T_i \sim 1$ GeV at LHC.

Strings clearly cannot exist in such hot deconfined matter, as e.g. lattice results for equilibrium matter of such density have shown. So, one thought about the early time evolution of matter in terms of perturbative processes with quarks and gluons.

The typical energies of quanta being about $3T$, one might think that for $3T \sim 1$ GeV one might get a quasi-perturbative QGP, well within the domain of small coupling constants and pQCD processes. This is how the problem was addressed in the late 70s [457, 711], using the lowest order (Born approximation) cross sections for processes like gg, gq scattering.

Hard parton processes accompanied by bremsstrahlung-like multiplication of quarks and gluons have been studied in detail in the high energy physics context, with really high p_t jets. The extrapolation of the same approach to lower p_t of few GeV is known as the "mini-jet" models. An example of the kind is the popular event generator HIJING by Gyulassy and Wang [768]. Rescattering of partons up to equilibration has been carried out by "parton cascade" models, such as the PCM by Geiger and Muller [730].

Among the main qualitative predictions of all such approaches was a statement that *the gluons dominate the partonic system created early in the collisions*. First of all, there are more gluons at small x, and also their production cross sections are larger than for $\bar{q}q$ pairs. Rough estimates made in my work [761] lead to the so called "hot glue scenario" in which the *kinetic* equilibrium is reached promptly, at about 0.3 fm/c after the collision, with the temperature $T \sim 500$ MeV (here and below we mean RHIC conditions, 200 GeV/nucleon Au + Au central collisions). Furthermore, the way a kinetically equilibrated plasma moves towards the *chemical* equilibrium was studied perturbatively. It was done through gluon number changing precesses $gg \leftrightarrow ng$ in Ref. [762], concluding that it happens in time of about 2 fm/c, which is still smaller than the QGP lifetime $\tau_{QGP} \sim 5$ fm/c at RHIC.

Another study performed by Xiong and myself [762] was based on (seductively magical) Parke–Taylor formulae for the tree-level cross sections of the n-gluon processes.[9] The matrix element averaged over indices is, for one chirality amplitude, as simple as

$$|M_n^{PT}|^2 = g_s^{2n-4} \frac{N_c^{n-2}}{N_c^2 - 1} \sum_{i>j} s_{ij}^4 \sum_P \frac{1}{s_{12} s_{23} \cdots s_{n1}}. \tag{10.10}$$

In the above expression the binary invariants are defined as usual $s_{ij} = (p_i + p_j)^2$, the summation P is over the $(n-1)!/2$ non-cyclic permutations of a string of numbers $(1 \cdots n)$. Clearly the answer is much simpler than one may think, looking at huge number of possible diagrams.

We defined the resolution of the jets to be such that *all* binary invariants for any pair of gluons in the process be large enough, $(p_i + p_j)^2 > s_0$. In the limit $s \gg s_0$, the cross sections form a geometric series in the leading log approximation, $\sigma_n \approx \sigma_4 (\alpha_s c_n \log(s/s_0))^{n-4}$. This means that the last gluon can be emitted from any of the $n-1$ previous gluons. The secondary gluon distribution has several interesting features [762]. The rapidity distribution piles up in the middle when n increases. For $gg \rightarrow ggg$, it is almost flat, quite different from the QED case where forward and backward emission is expected. The p_t distribution for larger n actually looks exponential and Boltzmann-like in the mid-p_t region. These two facts suggest that something like a sphaleron is created already in re-summation of perturbative multi-gluon processes. The slope of the spectrum is quite weakly dependent on the cutoff. It follows a simple rule "T"$= 0.01 E_0$. So for RHIC energy collision, the initial p_t spectrum of the gluons will have a slope $2\,\text{GeV}$. This is not very different from some numerical results from the transverse lattice and the classical Color Glue Condensate approach [427], to be discussed in Section 10.3.2 below.

However attempting to use these expressions for LHC we were disappointed because large n processes give larger and larger gluon production. For example, for $n = 4, 5, 6$, and 7 at the regular cut-off $p_0 = 2\,\text{GeV}$ the numbers are 0.35, 1.32, 3.35, 15.11, respectively. Only at much larger cutoff $p_0 = 10\,\text{GeV}$ did the series become manageable, but it is difficult to expect such large scale to develop by itself. Most probably, the lowest order minijet processes (considered e.g. in HIJING, with $n = 6$) is nothing but the beginning of a divergent series. Whether inclusion of loop diagrams helps or not remains unknown, but I think a new approach such as saturation and the Color Glass Condesate should be used in order to understand it. Hopefully, LHC data are not so far away....

10.3.2. *Classical fields in heavy ion collisions*

Another widely discussed idea is that the initial scale of the process can actually be provided by the wave function of the heavy nuclei itself. We have already discussed in Section 7.1.2 that people argued that all hadronic processes at sufficiently high

---

[9]Since in- or out-going momenta are treated similarly, n refers to the sum of all legs. For example, $gg \rightarrow ggg$ is a $n = 5$ process, and for it the Parke–Taylor expression given below is actually exact.

energy may possibly be described via classical $A_\mu \sim 1/g$ color fields, or the Color Glass Condensate (CGC).

A reasoning, suggested by McLerran and Venugopalan [426] goes as follows. Consider a small element $\delta x\, \delta y$ of the transverse plane, inside the disk-shaped region of boosted heavy nuclei. Because structure functions grow as $x = p/E_N$ (parton momentum fraction of the energy per nucleon) goes to zero, there is a growing $(2d)$ density of color charges n. Eventually their number becomes large $N = n\, \delta x\, \delta y \gg 1$, in which case the total color charge $Q \sim \sqrt{N}$ is large as well.[10] And large charges create classical fields. To complete the idea, one should evaluated the size of statistically-independent surface elements: $\delta x, \delta y \sim 1/Q_s$ where Q_s is the so called saturation scale [425] to be derived from some self-consistency condition. In summary, *virtual* classical fields are actually a part of nuclear wave functions, which becomes apparent only when those are shed off (radiated) and get real in heavy ion collisions.

After the collision happens, in this scenario the initial excited glue is still not a set of incoherent gluons but a coherent classical field with large occupation numbers $n \sim O(1/g^2)$. Therefore its evolution can be studied by solving classical Yang–Mills equations, which should be terminated when the occupation numbers decrease to $O(1)$ magnitude and the classical field approach would no longer be possible. It was found that CGC is a rather explosive substance and very quickly the initial random field transforms into transverse propagating fields which freeze into some set of gluons.

Such studies have been performed numerically on the transverse lattice. The SU(3) version of such expanding CGC were carried out by Krasnitz, Nara and Venugopalan [427]. Let me mention a few numbers from this work, for orientation. At RHIC energies the saturation scale Q_s was found to be 1.3 GeV. The initial classical CGC field was found to abelianize in a time τ_{CGC} of order $Q_s\tau_{CGC} \sim 3$. In this regime, the gluon energy density was found to be $\epsilon/Q_s^4 \approx 0.17/g^2$ with an approximately thermal momentum distribution. The transverse energy per quantum was found to be $1.66\, Q_s$, resulting in an effective temperature of 1 GeV. This is of course only apparent as the underlying evolution is classical and originates from a coherent state. The field strength F (the r.m.s. combination of electric and magnetic fields) is about

$$gF \sim 0.58\, Q_s^2 \sim 1\, \text{GeV}^2. \qquad (10.11)$$

What happens next with such gluons? They are not yet ready to become a QGP, as is seen from the relatively high "T" and small gluon number.

A scenario of the equilibration based on pQCD has been recently proposed by Baier *et al.* [763], which was called a "bottom-up" equilibration. The basic point of this elegant paper is that one can start from the CGC and the hard partons (promptly generated in the initial collision at some saturation scale), then re-scatter

---

[10]The square root assumes statistically independent addition or random colors, ensured by the fact that all partons originated from different nucleons.

them on each other and calculate the radiation rate for the next-generation softer gluons, including the Landau–Pomeranchuck–Migdal effect. Eventually these softer gluons absorb all the energy of the hard ones, and get thermalized by rescattering. The length of the time of several stages of this process and all the parameters of matter are given in terms of the original scale of hard momenta, Q, times the powers of α_s at this scale. (The numerical pre-factors are not known.) For example, the time till equilibration is $\tau_{eq} \sim Q/\alpha_s^{13/5}$.

If the initial scale is, say 2 GeV, so that $\alpha_s \sim 1/3$ is small and pQCD applicable, it predicts the equilibration time of about 3 fm. It may look short compared to the duration of the heavy ion collisions at RHIC, ~ 10 fm/c but not compared to "initial time" $\tau_i \sim 0.6$ fm/c at which one has to start hydrodynamics in order to explain elliptic flow and other collective phenomena, discussed in the preceding chapter. Therefore, one has to conclude that *all the pQCD-based evolution scenarios failed to describe the observations at RHIC.*

10.3.3. *Non-perturbative equilibration and topological clusters*

A different source of classical color fields also comes from *virtual classical fields getting real* during the collision. As we have discussed in Chapters 3 and 4, classical virtual fields are in general a significant part of the vacuum structure, and they have a non-trivial topology that can be described semiclassically by *instantons*. We have also discussed in Chapter 7 that in the collisions, which can provide sufficient excitation energy, such virtual classical fields can be excited, from deep under the topological barrier into real physical gluonic clusters at or above the barrier. We have shown there that the objects are born onto Minkowski space–time as purely gluo-magnetic clusters of a particular structure, the relatives of the *sphalerons* of the electroweak theory, with a mass of about 3 GeV. As these states are classically unstable, they explode into a thin shell of a coherent color field.

In Chapter 7 we considered pp or ep collisions, in which at most one such object is produced. Indeed, since the QCD vacuum has relatively small instantons, with the average instanton size, $\rho \sim 1/3$ fm small compared to their separation, it has a small "instanton diluteness parameter" $\kappa = n\rho^4 \sim 0.01$. Because of that an excitation of this kind is a rare event and a simultaneous production of more than one is unlikely.

The situation is different in heavy ion collisions, in which the collision energy is deposited into a volume of hundreds of fm³. It seems natural that nearly all instantons which happen to be in the appropriate space–time volume are excited into such clusters. As a result, a set of sphaleron-like clusters produced by their instantaneous transformation, is dilute as well, and a time of the order of 1 fm is available for their independent evolution, till the shells collide and get destroyed.

One may ask whether this scenario is really different from the CGC one we discussed in a previous section, since classical expanding fields appear in both of them. The answer is that the color fields are qualitatively different in both cases, because sphaleron-like clusters have topology and this brings in prompt fermion production.

Massive production of light $\bar{q}q$ pairs can be viewed as "prompt evaporation" of the vacuum chiral condensate. In total, those of $\langle\bar{q}q\rangle$ for u, d, s quarks together make about 5 quark–anti-quark *pairs* per fm^{-3}, to be compared to about 0.5 valence quarks per fm^{-3} in nuclear matter. In sudden collisions all condensates should disappear, thus all these vacuum quarks should become real as well. A central AuAu collision at RHIC energies produced QGP in a volume at its maximum of about $V_{\text{QGP}} \approx 1000 \,\text{fm}^3$. Considering now the total energy needed to kill instantons (non-adiabatically) is of the order of 1 TeV, the total number of quark pairs from eliminated quark condensates is about 5000. This is comparable to the total *transverse* energy and to the total hadron multiplicity observed: thus the phenomenon clearly is non-negligible or may be even dominant.

More details about this topological scenario and possible other features following from it one can find in my paper [431].

10.3.4.   *Dilemma of weakly versus strongly coupled QGP*

The main conceptual problem in all of those approaches is however that *the boundary of the classical/perturbative QCD is rather uncertain*. In practice people have first used some *ad hoc* cutoffs, which are quite crucial for all the results obtained. In particular, in both HIJING and PCM models just mentioned the chosen cutoffs have been around 1–5–2 GeV for pp collisions at RHIC energy, based on some model-dependent fits to *pp* data.

In Chapter 7 we have discussed high energy collisions and emphasized that the boundary of pQCD should only be determined by some non-perturbative phenomena related to chiral symmetry breaking or confinement.

However the situation is different for the hot/dense matter under discussion now. One can imagine that inside the pQCD ideology some *dynamically generated cutoff* be developed, which would regulate all these IR divergences and provide physical results. For example, the hard thermal loops and the Debye mass (which we have discussed in Chapter 8) are perturbatively defined, and play a role of IR cutoffs in equilibrium. Since these objects can be defined in terms of the forward scattering amplitudes, they can be easily generalized to non-equilibrium matter as well.

The estimates we discussed above show that this possible regulator scale, or the saturation scale Q_s *in the wave function of the colliding nuclei* is about 1 GeV for RHIC and 3 GeV for LHC. What is very important (and shown in Fig. 10.7 of my paper [431]) is that in expanding and cooling QGP the scale falls further, to "thermal masses" of quarks and gluons of the order of only $M \sim 0.3$–$0.4\,\text{GeV}$, before rebounding back to the in-vacuum non-perturbative scale.

These values may look ridiculously small for pQCD practitioners: but their intuition is based on in-vacuum physics which is completely inadequate in QGP. (Recall

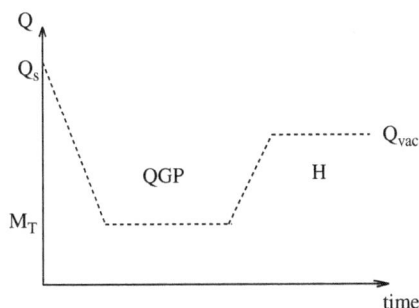

Fig. 10.7. Schematic time evolution of the basic dynamical scale during the heavy ion collisions. The lowest scale $\sim(2\text{–}3)T_c$ is reached in the QGP phase just above the transition temperature.

e.g. a dramatic drop in the duality parameter E_{cont} seen in the dilepton spectrum we discussed at the beginning of this chapter.)

The big question is: *can one use pQCD and Feynman diagrams in QGP at such a small scale?* Logically speaking, the dilemma is as follows:

- One possibility is that the pQCD is still applicable in QGP, and some clever resummation will eventually let us get rid of all frightening higher order corrections. One argument in its favor used over the years is that although the transition temperature to QGP is by itself quite low, only 160 MeV, the energy density and pressure just above it are roughly equal to those of the ideal gas of quarks and gluons. As we have shown in Chapter 8, lattice data undershoot the ideal gas pressure by only about 15% which is roughly of the scale of the $O(g^2)$ correction. So it may be that all higher order effects just cancel somehow.
- The second alternative is that the coupling constant runs into sufficiently large values at the pertinent scale, and a radically different picture and tools (such as the strong coupling limit related with AdS/CFT correspondence, see Chapter 12) would be needed. The pressure argument may still be proven misleading, as the strong coupling limit from the AdS/CFT correspondence gives 3/4 of the pressure also. This alternative would imply that in QGP there are strong non-perturbative effects, although qualitatively different from those in vacuum and unrelated with the chiral symmetry breaking and confinement.

As we mentioned in other chapters, other observables — like viscosity — may turn out to be crucial in deciding what kind of QGP is produced at the early stages of the heavy ion collisions, the weakly or strongly coupled one.

10.4. Jet quenching

Hard collisions typically produce a pair of jets, propagating back to back from the collision point. The main idea [766] is that a "quenching" of these jets by matter may provide a kind of "tomography" of the excited system, created in the early stages of high energy heavy ion collisions. Even very hard jets radiate and lose some energy during their passage through the system, and even relatively small losses chain the observed spectrum appreciably. If observed, all of that provides information about very early stages of the conditions.

In early papers on the subject [766], rather modest jet re-scattering in Quark–Gluon Plasma (QGP) has been considered. Accounting for radiation [767] has significantly increased expectations of the magnitude of the effect, while accounting for the Landau–Pomeranchuck–Migdal (LPM) effect in Ref. [769] had somewhat decreased them. We will not discuss this complicated subject, but just mention that the expected quenching factor from such a mechanism is expected to be $Q = 0.5$–0.7 for jets with $p_t = 10$–$20\,\mathrm{GeV}$.

10.4.1. *Jet quenching in experiment*

Experimentally, a relatively modest jet quenching has been first observed in deep inelastic scattering, for a forward jet going through cold nuclear matter (for recent discussion see Ref. [770]). The heavy ion data from CERN SPS have shown basically no or very little jet quenching, while already the very first RHIC data have produced a set of unexpected phenomena.

Let me first enumerate those: (i) The quenching is very strong, reaching nearly an order of magnitude; (ii) at $p_t > 2\,\mathrm{GeV}$ protons and anti-protons are as copious as pions; (iii) the azimuthal asymmetry remains very strong even at large p_t; (iv) baryons and mesons have different amounts of azimuthal asymmetry.

At the time of this writing, there is no model which can explain all of these phenomena. There are several papers in the market proposed to this effect, but it is too early to tell whether any of them would survive further scrutiny. As a result, I will only show some data, then point out several important points made, and then proceed to a discussion of various regimes of radiation in matter.

The so called *quenching factor* $Q(p_t)$ is ideally defined as the observed number of jets divided by the *expected* number calculated in the parton model *without* accounting for final[11] state interaction. This implies that hard QCD probes are under good theoretical control, and that one can evaluate the initial jet production itself with confidence.

Experimentally jet reconstruction in a heavy ion environment is very difficult to implement, thus all results reported by now on jet quenching refer to the observed/expected ratio of the yields of *single hadrons*. Furthermore, at this point large p_t means $p_t = 3$–$10\,\mathrm{GeV}$ and which part of those actually comes from jets remains unknown.

---

[11]Effects due to *initial* state interaction should be included. It means that parton distribution functions should be nuclear ones, from lepton–nuclei experiments. Parton rescattering in nuclei, leading to the so called Cronin effect, should also be included in the expected yield.

The so called nuclear modification factor R_{AA} quantifies the effect of $A + A$ compared to $p + p$ collisions on particular particle yields for point-like processes. It is defined as the ratio of the particle yield in an $A + A$ collision to the yield in a $p + p$ collision scaled by the mean number of binary (nucleon + nucleon) collisions N_{coll} in the $A + A$ event sample. Because hard processes are generally believed to scale with N_{coll} ("binary scaling"), this ratio is expected to be one at high p_T in the absence of any nuclear effects,

$$R_{AA}(p_T) = \frac{\text{(Yield per } A + A \text{ collision)}}{\langle N_{coll} \rangle \text{(Yield per } p + p \text{ collision)}}. \tag{10.12}$$

Shown in Fig. 10.8 is R_{AA} for identified π^0 as a function of p_T for central Pb+Pb collisions at $\sqrt{s_{NN}} = 17\,\text{GeV}$ from WA80/98 [734] and central Au + Au collisions at $\sqrt{s_{NN}} = 130\,\text{GeV}$ from PHENIX Run I [765]. Already in this comparison, a qualitative change between the SPS and RHIC is quite striking. It looks as if matter produced at SPS is, for some reason, mostly transparent to jets, and the only thing observed is the enhancement known as the "Cronin effect" due to initial state scatterings. At RHIC the matter suddenly becomes quite black. Figure 10.9 shows R_{AA} for central and peripheral Au+Au collisions at RHIC full energy $\sqrt{s_{NN}} = 200$ GeV and based on a more extensive data set [765]. One can see that π^0 is suppressed by a factor 5 or so, and the suppression continues to the highest measured momenta.

The other part of the puzzle is the unexpected particle composition observed at RHIC at large p_t. PHENIX can measure the p/π ratio up to almost $4\,\text{GeV}/c$

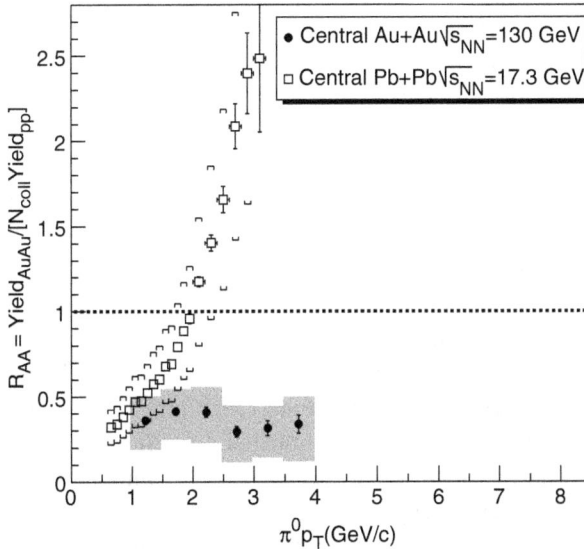

Fig. 10.8. $R_{AA}(\pi^0)$ for central Pb+Pb collisions at $\sqrt{s_{NN}} = 17\,\text{GeV}$ and central Au+Au collisions at $\sqrt{s_{NN}} = 130\,\text{GeV}$. The error bars are the statistical $\oplus$ p_T-dependent systematic errors. The brackets/boxes are the errors on the normalization of this ratio.

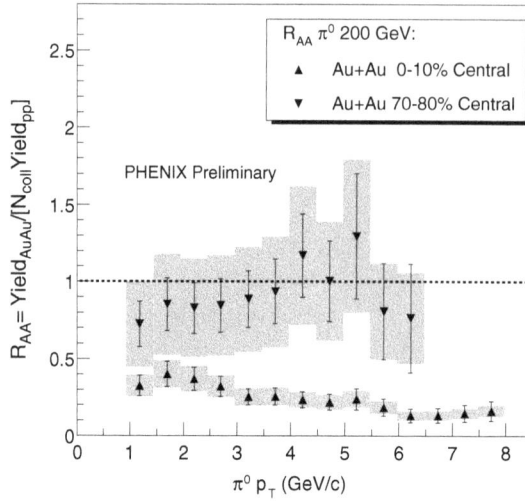

Fig. 10.9. $R_{AA}(\pi^0)$ for central and peripheral Au+Au collisions at $\sqrt{s_{NN}} = 200\,\text{GeV}$. The error bars are the statistical $\oplus$ p_T-dependent systematic errors. The shaded boxes are the errors on the normalization of the ratio.

in p_T, at which point the protons can no longer be distinguished from other particles via their time of flight. Since neutral pions are identified to much higher p_T, we can look at the ratio of π^0 to non-identified charged hadrons $(h^+ + h^-)/2$ for $p_T > 4\,\text{GeV}/c$. This ratio is shown in Fig. 10.10a for minimum bias events. The surprising feature that this ratio does not increase for transverse momenta greater than $4\,\text{GeV}/c$, but remains nearly constant at a value around 0.5, indicates that approximately half of the charged hadrons at high p_T are protons and/or kaons, assuming $\pi^0 = (\pi^+ + \pi^-)/2$. This result is rather different from other hard processes as measured in pp or e^+e^- collisions. The data shown in Fig. 10.10b exclude a large kaon contribution and show that indeed the proton and antiproton production is comparable to the pion one.

First direct evidence for jets has been found by the STAR and PHENIX collaborations, e.g. Ref. [764]. It was observed well inside a relatively narrow cone typical for jets. The data are taken as follows: there is a trigger particle with the largest p_t (6–8 GeV in the figure below), and then a distribution over the azimuthal angle between this particle and the others (with $p_t > 2\,\text{GeV}$ below) is plotted. In Fig. 10.11 from STAR we show such azimuthal distributions in Au+Au at the most central bin. A peak near zero angle comes from a companion particle, originating from the same jet as the trigger. The histogram corresponds to a fit to pp data, appended by a flow (too small in the central bin to be shown). The pp data show clear back-to-back correlations, at $\phi \approx 180°$, which originate from the second jet, balancing the global p_t. One can see that in AuAu data (points) such a correlation is absent. The obvious explanation to that is that if the first jet is emitted from

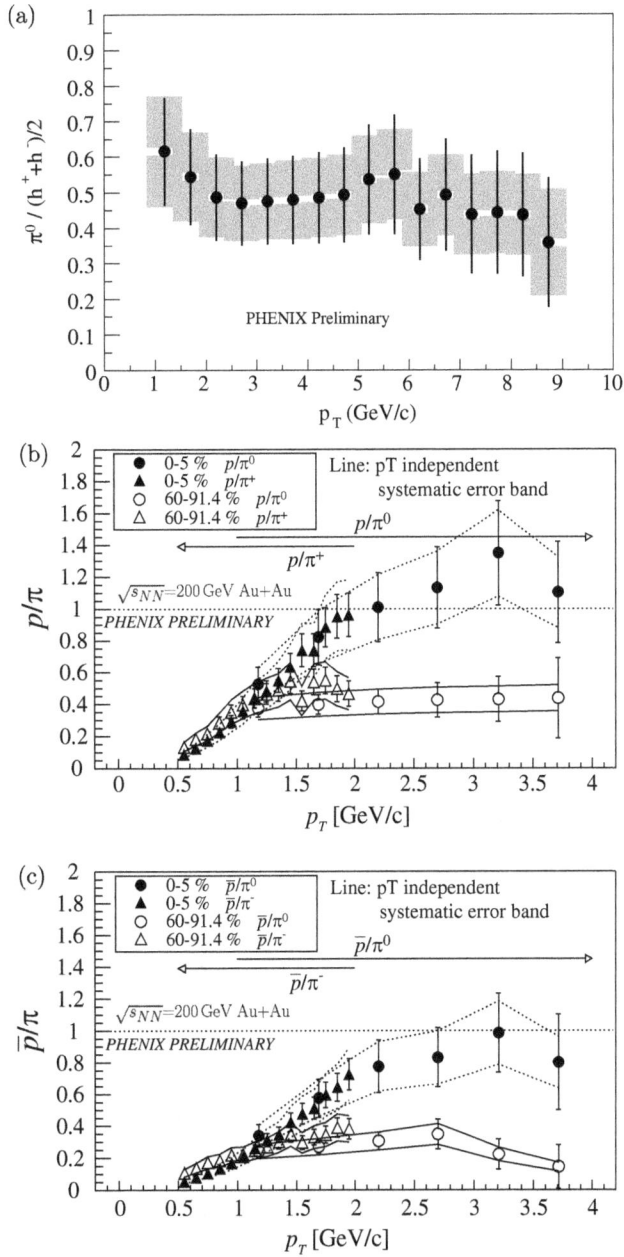

Fig. 10.10. (a) Ratio of π^0 to $(h^+ + h^-)/2$ as a function of p_T for minimum bias Au+Au collisions. (b) p/π and (c) $\bar{p}/\pi$ ratios versus p_t for most central and most peripheral bins.

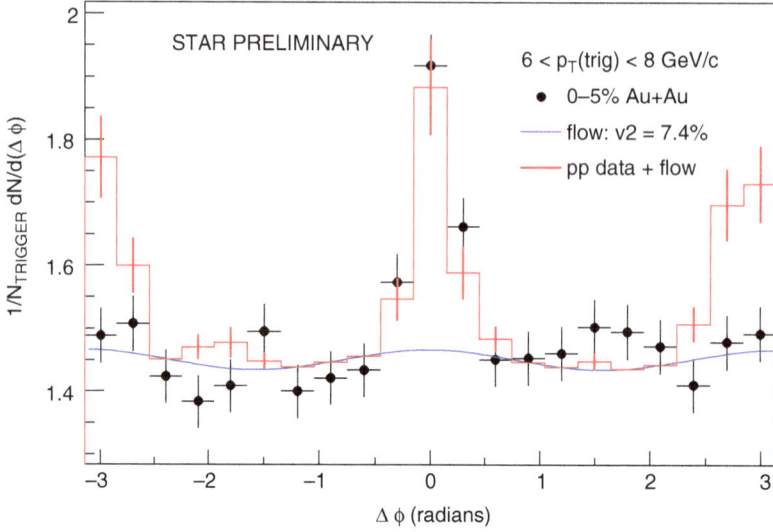

Fig. 10.11. Azimuthal distributions ($0 < |\Delta\eta| < 1.4$) for central $Au + Au$ collisions (solid circles) compared to the expected azimuthal distribution (histogram). The trigger particle threshold is $6 < p_T^{\mathrm{trig}} < 8\,\mathrm{GeV}/c$ and the associated particle threshold is $2\,\mathrm{GeV}/c < p_T < p_T^{\mathrm{trig}}$. Also shown is the expected elliptic flow contribution (solid curve).

the surface of the almond outward (to avoid quenching), the second one must be emitted into dense matter and therefore must be killed.

To quantify deviations one can form a ratio of the $Au + Au$ correlation excess beyond that expected from elliptic flow and the $p + p$ correlation excess,

$$I_{\mathrm{AA}}(\Delta\phi_1, \Delta\phi_2) = \frac{\int_{\Delta\phi_1}^{\Delta\phi_2} d(\Delta\phi)[C_2^{\mathrm{AuAu}} - B(1 + 2v_2^t v_2^a \cos(2\Delta\phi))]}{\int_{\Delta\phi_1}^{\Delta\phi_2} d(\Delta\phi)C_2^{pp}}. \tag{10.13}$$

This ratio is plotted as a function of the number of participating nucleons for the small-angle and back-to-back azimuthal regions in Fig. 10.12. The left panel shows the ratios for $3 < p_T^{\mathrm{trig}} < 4\,\mathrm{GeV}/c$, the middle panel shows the ratios for $4 < p_T^{\mathrm{trig}} < 6\,\mathrm{GeV}/c$, and the right panel shows the ratios for $6 < p_T^{\mathrm{trig}} < 8\,\mathrm{GeV}/c$. The horizontal bars show the systematic error on the ratio due the $+5/20\%$ systematic uncertainty on the v_2 measurement from the reaction plane method. For all three trigger particle thresholds, $I_{\mathrm{AA}}(0, 0.75)$ is smaller than unity for the most peripheral collisions, and increases toward central collisions. This is quite puzzling and clearly indicates that not only quenching (absorption) is at work here, but also some mechanisms (coalescence, pickup) which enhance the same-side jets. (This point will be important in the next section.)

In contrast, $I_{\mathrm{AA}}(2.24, \pi)$ (the 2nd jet) decreases with increasing N_{part}, and approaches near zero for the most central collisions, for all three trigger particle

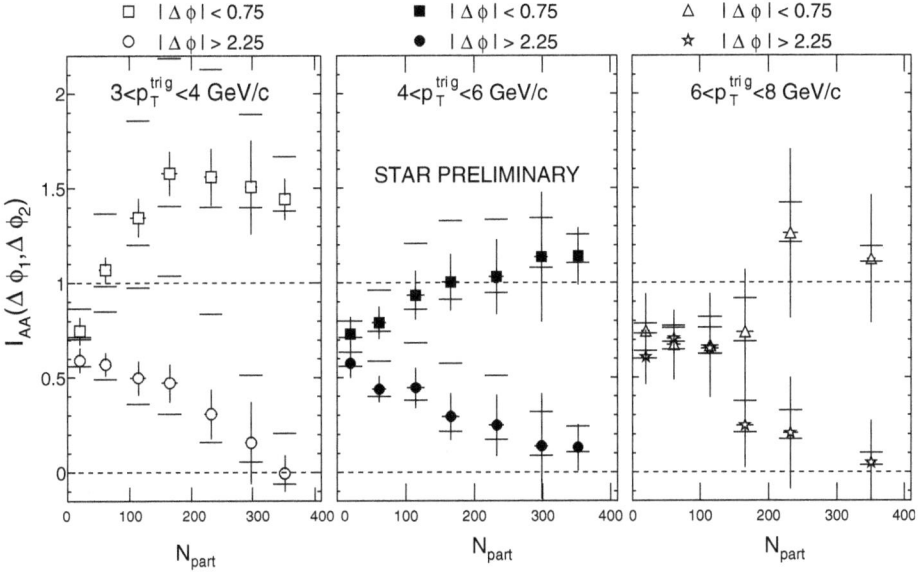

Fig. 10.12. I_{AA} versus the number of participants for small-angle ($|\Delta\phi| < 0.75$) and back-to-back ($|\Delta\phi| > 2.24$) azimuthal regions.

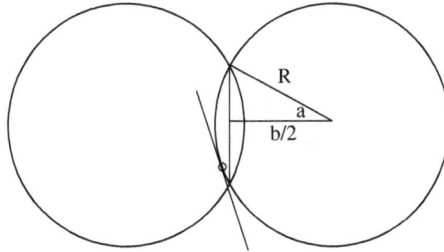

Fig. 10.13. In the case of non-central collisions with the non-zero impact parameter b the excited matter is produced inside the almond on the intersection of two nuclei. In the "black almond" model [771] only the surface radiates: e.g. the point shown by a small circle can only radiate in the half plane to the left of the tangent line. The definition of the angle a used in formula (10.15) is also shown.

thresholds. This further shows the *"blackness of the almond"*, which is remarkably the case even at the largest p_t observed.

10.4.2. *Azimuthal asymmetry at large p_t*

It has been pointed out in my paper [771] that the very high degree of the azimuthal asymmetry observed at $p_t = 2$–6 GeV (e.g. Ref. [764]) cannot be reproduced by *any* amount of jet quenching, no matter how strong.

Consider first a simplified picture of a nucleus, as a homogeneous sphere with a sharp edge. The almond-like overlap region created in the collision at non-zero impact parameter b has an edge as well. With increasing quenching of jets, we observed those which are emitted further from the center of the system and closer to the almond's edge. So, in any purely absorptive model there exists an upper limit,

$$v_2(b) < v_2^*(b),   (10.14)$$

corresponding to the infinite absorption limit. If only jets originating from the surface are seen, "the almond is completely black", the limit is given by the expression

$$v_2^* = \frac{\int_{-a}^{a} d\beta \int_{-\pi/2+\beta}^{\pi/2+\beta} d\phi \cos(2\phi)}{\int_{-a}^{a} d\beta \int_{-\pi/2+\beta}^{\pi/2+\beta} d\phi} = \frac{\sin(2a)}{6a},   (10.15)$$

where $\cos(a) = b/2R$, the angle β is the angle of an emission point on the surface and ϕ is the angle of emission. At central collisions $v_2^*(0) = 0$, but it is equal to $1/3$ at the most peripheral collisions.

The fact that preliminary STAR data were above this limit is reflected in the title of that paper. However, soon afterwards the analysis of four-body correlations [648] has shown that the original data were contaminated by some two-body effects which had nothing to do with multi-body ellipticity and v_2. The corrected data agree with the "black almond" limit (10.15). Unfortunately, when a realistic shape of the nuclei is used instead of a homogeneous sphere with a sharp edge, the theoretical predictions for $v_2^*(b)$ get smaller than the data once again, due to a "halo" around them which radiates jets more or less isotropically and reduces v_2. An example of the distribution of the originating point of all jets which had *survived quenching* is shown in Fig. 10.14b, from Ref. [772]. We will not go into the details of the model here and just note that the amount of quenching is tuned to reproduce the data. Figure 10.14a shows how the distribution changes if on top of the quenching one includes a quark (or diquark) pickup from matter, during the fragmentation process. We can observe that this reduces the contribution of the halo and increases the emission from the almond surface, boosting v_2 back toward the "black almond" geometric prediction (10.15).

We conclude this section, emphasizing that highly excited matter produced in heavy ion collisions is remarkably "black", even for jets/hadrons with momenta as large as $10\,\mathrm{GeV}$. Many puzzling facts have been revealed by RHIC experiments, which we do not yet understand. They indicate that on top of strong quenching, by a factor ~ 10 or more, there are other phenomena (coalescence, pickup of quarks or even diquarks from matter, or something else) which changes jet formation and enhances the forward correlations seen in Fig. 10.12.

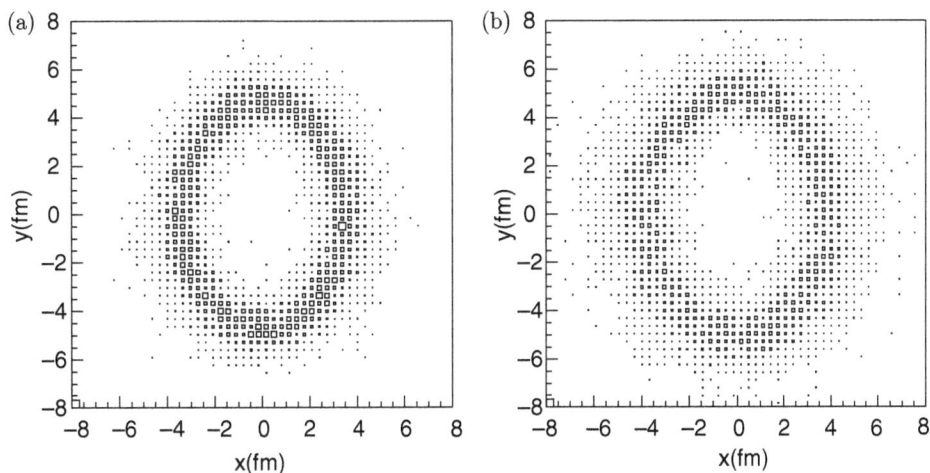

Fig. 10.14. Transverse plane distribution of the jet originating points, for AuAu collisions at impact parameter $b = 6$ fm (along the x axis). The figure (b) is after quenching (absorption) and (a) after the q/qq pick up is included in the fragmentation function.

10.4.3. *Radiation in matter*

The issue of a fast charge propagating in matter, suffering rescattering and radiative energy losses is well known and developed in QED. The important issue is that the bremsstrahlung-type radiation takes quite significant time to be separated from the charge which makes it. The so called radiation length grows with energy. If it exceeds the distance between subsequent re-scatterings, one no longer can consider those as independent radiation events and has to consider the interference of all rescattering + radiation. Such interference leads to cancellations, and basically the softest radiation only follows from the period of time after the last rescattering. An accurate account of this effect is rather complex: it was done in works by Landau, Pomeranchuck, and Migdal in the 1950s and thus the effect is known as the LPM effect. It was actually tested for ordinary matter under a high energy electron beam in SLAC in the 1990s, and the theory was found to be in excellent shape. (Unfortunately, none of the authors lived to see this happen.)

The LPM phenomenon exists in QCD as well, but it is described by a different formula because gluons are now charged and radiate more than quarks, even if the original jet is a quark. We will not describe here this complicated theory, which the reader can find in the original work by Baier, Dokshitzer, Peigne and Schiff [769] and vast subsequent literature. Let me just mention that in QCD, unlike QED, gluons produce more gluons which radiate etc., and as a result the losses grow quadratically with distance L traveled in the medium.

Let me just mention here that one technical but important point, made in the last paper in Ref. [769], is that when quenching is strong it cannot be evaluated using the *mean* energy loss. Let us recall it. The quenching factor is the ratio of the produced to observed spectra

$$Q(p_t) = \frac{\int d\epsilon\, D(\epsilon)\, dN^{\text{hard}}/dp_t^2(p_t + \epsilon)}{dN^{\text{observed}}/dp_t^2(p_t)}, \tag{10.16}$$

where ϵ is the energy lost in the medium and $D(\epsilon)$ is its normalized distribution. For small ϵ one can expand it to first order, obtaining a correction proportional to the mean energy loss $\langle \epsilon \rangle = \int d\epsilon\, D(\epsilon)\epsilon$. However, because spectrum is so steep, it can only be true when ϵ/p_t is not larger than few percent.

Therefore, to illustrate the needed magnitude of the effect, let us suggest a different simple approximation. Using power parameterization of the spectrum

$$\frac{dN}{dp_t^2(p_t)} \sim \frac{1}{p_t^n} \tag{10.17}$$

for both observed and "hard" distributions, one gets

$$Q(p_t) = \int d\epsilon\, D(\epsilon) \left(\frac{1}{1+\epsilon} \right)^n \tag{10.18}$$

instead.

With the simplest delta-function-like distribution peaked at some fractional loss,

$$D(\epsilon) = \delta(\epsilon - \kappa p_t) \tag{10.19}$$

one gets $Q = 1/(1+\kappa)^n$. With $n \approx 12$ in the few-GeV domain at RHIC energies, a jet quenching by one order of magnitude would correspond to $\kappa \sim 1/4$. This means that a mean loss of about 15–20 percent of the produced jet momentum would be sufficient. However, this conclusion is oversimplified. As one can see from (10.18) the quenching factor is dominated by small losses $\epsilon < 1/n \sim 1/12$. It means that what matters is the probability to escape with as small losses as possible.

This conclusion changes the relative role of early (synchrotron-like) versus late (multiple bremsstrahlung with LPM) effects. While the latter can be very large for a specific geometry (the LPM energy loss [769] is $\Delta E \sim L^2$ where L is the path in matter), the integral (10.18) would be dominated by surface emission with small $L \sim 1$ fm or so. Early effects emphasized in this paper, even if producing less average losses, are expected to affect the probability $D(\epsilon)$ at small ϵ, preventing easy escape of some jets. Detailed numerical simulations (which are well beyond the limits of this work) are needed to understand their relative role.

At the moment it is not clear whether the traditional approach, considering matter to be gas of multiple uncorrelated quarks or gluons which act as scattering centers, will be sufficient to describe the very strong jet quenching observed at RHIC. Naturally one wonders which effects can possibly enhance the radiation: this is the main motivation for considering the synchrotron-like radiation discussed by Zahed and myself [777], as we will discuss next.

10.4.4. *Synchrotron-like QCD radiation*

By "synchrotron-like" QCD radiation in general we mean one emitted by a charge moving in some external field, which is strong enough so that this motion cannot be treated perturbatively. Classically it means that we have to solve for trajectories in the field, while in the quantum field theory setting we have to know the propagators in the background fields. We also restrict this term to radiation from within the field region: the radiation from free motion on incoming and outgoing lines, if present, we will call a bremsstrahlung-like radiation. For ultrarelativistic charges we will treat in this paper, the latter is restricted to two narrow ($\theta \sim 1/\gamma$) cones around the initial and the outgoing directions of the charge, while synchrotron-like radiation goes elsewhere.

Synchrotron radiation in QED occurs e.g. when a high energy charged particle is flying through a magnet, see Fig. 10.15a. The classical trajectory in the field includes an arc in between the initial and outgoing directions, and radiation is emitted in all directions *tangent* to it. In the geometric optics limit photons move along straight lines, since in QED they are not affected by the magnetic field.

Another interesting case of classical synchrotron-like radiation (considered in Ref. [773] and vast subsequent literature) is that emitted by a relativistic charge similarly rotating in a circle, but in a *gravitational* field, e.g. close to a horizon of a black hole, see Fig. 10.15b. In this case *both* the charge and the photons are similarly deflected by the field, especially in the ultrarelativistic limit. This leads to a significant reduction of radiation properties: total yield (compared to synchrotron radiation for the same rotation radius) reduces by γ^2, and the radiation length in each particular direction is actually the whole circle, not a sector with $\delta\phi \sim 1/\gamma$ as in a magnet.

We will focus on a QCD charge (parton, quark or gluon) traversing a region of strong field, a "QCD magnet" of a kind, see Fig. 10.15c. In general motion of *both* the initial charge and the radiated gluons is affected by it. If the geometric optic limit can be used, this motion is described by some trajectories. In the simplest case we would discuss most, that is a quasi-Abelian magnetic field,

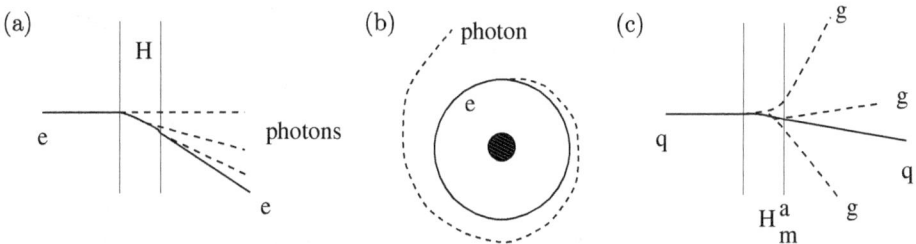

Fig. 10.15. Schematic representation of synchrotron-like radiation in three cases: (a) in the usual magnet; (b) by a charge rotating ultra-relativistically in a gravity field (e.g. around a black hole); (c) the layer of gluo-magnetic gauge field.

those are just circles. Usually the momenta of the radiated gluons are small compared to that of the original charge, so they make circles with smaller (curvature) radii. Furthermore, the radiated gluons generally have different color charges, and thus they would be kicked in different directions in a sufficiently complex color field. One can summarize those qualitative observations by saying that the QCD magnet acts as a kind of Newtonian prism *squared*, in which gluons are analyzed both in their *momenta* and *color*. It is also possible that some part of the radiation would remain trapped in the field and never go to infinity, so the usual method of calculating radiation at large sphere is not in general applicable in this setting.

Other instructive pictures to think about classically are *instantaneous* pictures of the radiation field. For ultrarelativistic cyclotron radiation in QED the non-zero field strength is concentrated mostly at a particular spiral-like curve known as *evolute of the circle*, which is defined by the simple condition that both electron and photon move with the speed of light and thus the distances RE (radiation time–electron now) along the circle and RP (radiation time–photon now) are equal. Details of the field distribution can be found in Ref. [775]. Its analog for QCD synchrotron-like radiation can also be easily derived. Considering the simplest case when the original jet has such a high energy that the curvature of its path can be neglected, one may notice that the radiated gluons rotate around the field in some plane (depending on gluon and field relative colors): therefore it never reaches the distance sphere we usually use in the QED case. The instantaneous picture of the field strength thus has the shape of a *cycloid*, the same at any time moment but sliding (together with the charge) with the speed of light.

The number of soft prompt gluons per frequency ω expected from bremsstrahlung and synchrotron radiation can be estimated classically. The Lienard expression for the dipole radiation intensity,

$$P = -\frac{2e^2}{3m^2}\left(\frac{dp_\mu}{ds}\right)^2, \tag{10.20}$$

was derived in 1898, but is correct even for the ultrarelativistic case. The acceleration appearing in it is proportional to the field strength F and the particle energy E, so classical energy losses depend quadratically on both of them, $P \sim e^4 F^2 E^2/m^4$.

This expression however is not applicable for too large energies: the applicability condition on it (see textbooks e.g. Jackson or Landau–Lifshitz) is

$$e^3(F/m^2)(E/m) \ll 1, \tag{10.21}$$

where we have identified three dimensionless factors: the coupling constant, the field in m^2 units and the relativistic gamma factor. In cases when one uses classical QED such as synchrotron radiation in accelerators, this condition of course holds, in spite of the fact that γ can be as large as 10^5 (LEP): the reason is small coupling and especially very small fields H/m_e^2.

This would not be the case in the applications we have in mind, and so we have to look at the differential spectrum of the radiated frequency w. It would be also instructive to include bremsstrahlung in the comparison, which produced photon spectra of the magnitude

$$\frac{dN_B}{dw} \sim \frac{\alpha}{\pi} \frac{1}{w}. \tag{10.22}$$

Synchrotron radiation has a different spectrum: total number of quanta per circle is

$$\frac{dN_S}{dw} \sim \frac{\alpha}{\pi} \frac{(w/w_0)^{1/3}}{w}, \tag{10.23}$$

where $w_0 = gH/E$ is the cyclotron frequency with E the energy of the radiating charge. At low frequencies bremsstrahlung of course dominates, but not at larger ones. Integrating (10.23) up to some maximal frequency yields

$$N_S \sim \frac{\alpha}{\pi} (w_{\max}/w_0)^{1/3}, \tag{10.24}$$

or the energy loss $\Delta E/E \sim \alpha(w_{\max}^{4/3}/w_0)^{1/3}$. The upper limit of this expression for synchrotron emission can be bounded by the classical "characteristic frequency" $w_c = w_0 * 3\gamma^3$: substituting it in (10.24) we find that one circle of radiation emits $N_S \sim \alpha\gamma$ photons, while the total energy loss recovers what one can get directly from the Lienard expression mentioned above: it again grows quadratically with energy, or $\Delta E/E \sim E$.

Furthermore, from the angular distribution and phase one can see that in the ultrarelativistic case the $1/\gamma$ sectors of the circle radiate photons independently and in different directions. One can further see that condition (10.21) can be viewed as a statement that the energy loss *per segment* should be small compared to the charge energy E.

This condition is entirely classical: but now we are ready to discuss *quantum* effects. One of the obvious limitations is that not only the total radiated energy, but *each* emitted photon should satisfy $\hbar w \ll E$. Although in modern accelerators like LEP $\gamma \sim 10^5$ and this enhancement in the radiation frequency over the rotation one reaches 15 orders of magnitude, the energy of the radiated photons is still small compared to that of the rotating charge, so that its back reaction can be ignored.

This would not be the case in the applications we are thinking of, and so let us look at what happens if we cut off the classical spectrum at this quantum limit $w_{\max} = E$ instead, assuming $E < w_c$. The number of prompt gluons stemming from bremsstrahlung is then logarithmically divergent in IR, $N_B \sim \alpha \log(E/w_{\min})$ while the energy loss is simply $\Delta E_B/E \sim \alpha$. These results agree with what one can get from the standard quantum treatment, such as Feynman diagrams.

For cyclotron radiation a cut at $\omega_{max} \sim E$ leads to

$$N_S \sim \alpha(E^2/eH)(1/3), \qquad (10.25)$$

and the energy loss per fixed[12] width of the field region Δz (rather than a circle)

$$\Delta E_S/E \sim e^2(eH)^{(2/3)}\Delta z/E^{1/3}, \qquad (10.26)$$

which is *decreasing* with energy, although raised to a low power.

Although these simple estimates appear to disfavor the synchrotron mechanism over the bremsstrahlung at large energies E, however subsequent bremsstrahlung at small angle interfere with each other and are subject to Landau–Pomeranchuck–Migdal (LPM) suppression, while the quanta emitted in synchrotron-like radiation are mostly lost beyond recovery. We will return to more detailed estimates of these effects below.

The problem of quantum synchrotron radiation in QED was addressed in a fundamental way by Schwinger [776] using the mass operator formalism. In this section, we extend this approach to the quantum synchrotron radiation in QCD. Two essential differences between the QED and QCD problems: (i) the non-Abelian nature of the charge in QCD; (ii) the emitted radiation also undergoes magnetic deflection. A quantum calculation is required in strong chromomagnetic fields owing to potentially large recoil corrections, essential for large jet quenching. The power radiated will be sought through the mass operator as

$$-\frac{1}{E}\,\mathrm{Im}\,\mathbf{M}_{aa} = \int \frac{d\omega}{\omega}\,\mathbf{P}_{aa}(w) \qquad (10.27)$$

after pertinent kinematical identifications.

For simplicity, in the work by Zahed and myself [777] which we follow here, the case of constant chromomagnetic field is discussed,

$$G^a_{\mu\nu}(x) = \delta^{a8}G_{\mu\nu}, \qquad (10.28)$$

where the Abelian field strength corresponds to a constant magnetic field in the three-direction, $G_{12} = -G_{21} = H$. The background gauge field associated to (10.28) is

$$A^a_\mu(x) = \delta^{a8}A_\mu(x) = \delta^{a8}\delta_{\mu2}Hx_1. \qquad (10.29)$$

With our choice of the chromomagnetic background along the 8th color direction, the quarks and gluons can be diagonalized. The diagonal quarks in the fundamental

---

[12]It cannot be small compared to the radiation length: $\Delta z > m/eH$.

representation carry color $(a = 1, 2, 3)$,

$$e_a = g\,(T^8)_{aa} = \frac{g\sqrt{3}}{6}\,(1, 1, -2), \tag{10.30}$$

and the diagonal gluons in the adjoint representation carry color

$$g_A = (-1)^A\,\frac{g\sqrt{3}}{2}, \tag{10.31}$$

for $A = 4, 5, 6, 7$ and $g_A = 0$ for $A = 1, 2, 3, 8$.

These two cases, as will be shown below, lead to qualitatively different radiation. The second case is basically QED-like.

Quantum synchrotron radiation will be sought for quarks and gluons interacting to **all** orders in H but to leading order in $\alpha = g^2/4\pi$ between the quantized fields, we now present briefly the spin-1/2 case and discuss extensively the spin-0 case. In the semiclassical limit, both spins radiate at the same rate.

Following Schwinger [776], to lowest order in perturbation theory the quark mass operator in the chromomagnetic field reads

$$\mathbf{M}_{aa} = ig^2\,(T^A)_{ab}\,(T^A)_{ba} \int \frac{d^4k}{(2\pi)^4} \int_{0-i0}^{\infty-i0} \frac{ds}{\cos(g_A H s)}$$
$$\times\,\exp[-is(k^2 - k_\perp^2\,(\tan(\,g_A H s)/(g_A H s) - 1))]\,(e^{2g_A s G})_{\mu\nu}\,\Phi(g_A)$$
$$\times\,\gamma^\mu\,(\gamma\cdot(\Pi_b - k) - m)^{-1}\gamma^\nu. \tag{10.32}$$

We have defined the quark four-momentum operator as

$$\Pi_{b\mu}(x) = i\partial_\mu - e_b A_\mu(x) \tag{10.33}$$

and the Bohm–Aharanov line

$$\Phi(g_A; x, y) = \exp\left(i\frac{g_A H}{2}\,(x_1 + y_1)(x_2 - y_2)\right). \tag{10.34}$$

The Bohm–Aharanov phase enforces gauge-invariance in the mass operator, but does not contribute to the radiation. Indeed, for an initial color-a quark emitting a color-b quark plus a color-A gluon,

$$\Phi(e_a; x, y) = \Phi(e_b; x, y) \times \Phi(g_A; x, y), \tag{10.35}$$

showing that the Bohm–Aharanov line in (10.32) on the gluon, can be redistributed to compensate for the analogous ones on the quarks. This procedure will be assumed throughout, and thereby the gluon Φ contribution reshuffled.

Since G is an antisymmetric matrix, its eigenvalues $\pm iH$ are complex. The color precession factor is

$$e^{2g_A sG} \rightarrow e^{\pm is\,(2g_A H)}. \tag{10.36}$$

The occurrence of the cyclotron poles in the gluon propagator (10.32) implies that the s-integration is infinitesimally shifted below the real axis in the complex s-plane. The prescription follows the causal prescription for the free propagator,

$$\frac{1}{k^2 + i0} = i \int_{0-i0}^{\infty-i0} ds\, e^{-isk^2}. \tag{10.37}$$

There is subtlety due to the positive sign in (10.36) for $k = 0$, which is the analogue of the tachyonic mode of a spin-1 coupled to a constant chromomagnetic field in the first quantized approach. This mode is at the origin of the well-known Savvidi instability in QCD. What it says, is that chromomagnetic fields are in general unstable against quantum fluctuations, here gluon emission. Since we are interested in energy losses through time-evolving sphalerons, the instability time caused by this tachyon $\tau_T \approx 1/\sqrt{g_A H}$ is much larger than the synchrotron emission time $\tau_S \approx 1/E$, as well as the sphaleron gluon emission time $\tau_E \approx \rho$, i.e.

$$\tau_S \ll \tau_E \ll \tau_T, \tag{10.38}$$

and should be ignored.

The technique developed by Schwinger [776] can now be applied to (10.32) to derive the power radiated in a QCD synchrotron process whereby an energetic quark radiates through a chromomagnetic field. To avoid the unnecessary algebra triggered by the spin content of the quark, we present the results for the spin-0 case instead. For a scalar quark in the fundamental representation, the analogue of (10.32) is

$$\begin{aligned}
\mathbf{M}_{aa} = ig^2 (T^A)_{ab}\, (T^A)_{ba} \int \frac{d^4k}{(2\pi)^4} \int_{0-i0}^{\infty-i0} \frac{ds}{\cos(g_A Hs)} \\
\times \exp[-is(k^2 - k_\perp^2\,(\tan(g_A Hs)/(g_A Hs) - 1))]\,(e^{2g_A\,sG})_{\mu\nu}\,\Phi(g_A) \\
\times (\vec{\Pi}_a - \Pi_b)^\mu\,((\Pi_b - k)^2 - m^2)^{-1}(\vec{\Pi}_a - \Pi_b)^\nu,
\end{aligned} \tag{10.39}$$

modulo counter-terms. The arrows on the Π indicate the direction of the derivative. On mass-shell we expect $\Pi_a \sim \Pi_b + k$, this will hold in the classical limit.

For spin-0, the power radiated follows from (10.27). Following Schwinger [776] we obtain the chromomagnetic synchrotron emission by a scalar quark in the classical limit in the form

$$\mathbf{P}_{aa}(\omega) = -\frac{\alpha}{\pi} (T^A)_{ab} (T^A)_{ba}$$

$$\times \omega \operatorname{Im} \int_{0-i0}^{\infty-i0} \frac{d\tau}{\tau} \frac{e^{-i(E\omega_A)^2 \tau^3/(24\omega)}}{\cos(E\omega_A\tau/(2\omega))}$$

$$\times \left(\frac{m^2}{E^2} + \frac{1}{2} \omega_b^2 \tau^2 \right) \exp\left[-i\omega \left(\frac{m^2\tau}{2E^2} + \frac{\omega_b^2\tau^3}{24} \right) \right] \tag{10.40}$$

where the $H = 0$ subtraction in (10.40) is subsumed. The quark cyclotron and gluon rescaled frequencies are $\omega_a = e_a H/E$ and $\omega_A = g_A H/E$ respectively.

In carrying out (10.40) the emitted gluon recoil effect on the jet was ignored. We have checked that the gluon recoil effect amounts to the shift

$$\frac{1}{\omega} \rightarrow \left(\frac{1}{\omega} - \frac{1}{E} \right) \tag{10.41}$$

in the energy $\mathbf{P}_{aa}/\omega$ thereby generalizing Schwinger's first quantum correction in QED to the QCD case. The size of this substitution on the energy loss of the gluon jet will be discussed below.

The first contribution in the integrand of (10.40) reflects on the gluon cyclotron effect in the chromomagnetic field. The phase stems from the transverse contribution follow from the gluon propagator as is evident in (10.39). The denominator shows poles for

$$\tau_n = \frac{\pi\omega}{E\omega_A}(2n+1), \tag{10.42}$$

which are the gluon cyclotron orbits (classically the gluon spin and the tachyon problem drop). The synchrotron emission time is short $\tau \approx 1/E$ but so is τ_n. The second contribution in (10.40) stems from the quark synchrotron contribution as in QED.

Rewriting $1/\cos A$ as the geometrical sum of e^{iA}, we may bring (10.40) in the form

$$\omega^{-1} \mathbf{P}_{aa}(\omega) = \frac{\alpha}{\pi} (T^A)_{ab} (T^A)_{ba} \frac{2m^2}{\sqrt{3}\, E^2}$$

$$\times \sum_{n=0}^{\infty} e^{-i\pi n(1-i0)} \left(\int_{\xi_n}^{\infty} dt\, \mathbf{K}_{5/3}(t) + 2\kappa_n\, \mathbf{K}_{2/3}(\xi_n) \right), \tag{10.43}$$

$$\left(\frac{\xi_n}{\xi} \right)^2 = \frac{(1 + (2n+1)(2\beta)/(3\xi^{2/3}))^3}{(1+\lambda^2)},$$

$$\kappa_n + 1 = \frac{1 + (2n+1)(2\beta)/(3\xi^{2/3})}{(1+\lambda^2)}, \tag{10.44}$$

$$\frac{\beta}{\lambda} = \left(\frac{3\omega_b}{2\omega} \right)^{1/3}, \qquad \lambda = \frac{E\omega_A}{\omega\omega_b}, \qquad \xi = \frac{2\omega}{3\omega_b} \left(\frac{m}{E} \right)^3. \tag{10.45}$$

Again the quantum corrections follow from (10.43) through the substitution (10.42) on the RHS. The **K** are modified Bessel functions. The sum over n in (10.43) sums over cyclotron orbits of width $-i0$ except for the lowest orbit which is zero. It is reminiscent of the sum over "Landau levels" in the Schrödinger formulation.

The preceding results are easily analyzed for large and small frequencies ω. Since the Abelian analysis with $g_A = 0$ is known from QED, we focus on the non-Abelian part with $g_A \neq 0$. We will show that the non-Abelian contribution to the synchrotron radiation is strongly suppressed at small ω due to strong "incoherence" effects produced by the deflected radiation. At large ω the non-Abelian contribution is equal to the Abelian contribution. Specifically, the $g_A \neq 0$ contribution for small ω reads

$$
\mathbf{P}_{aa}(\omega) \approx \frac{\alpha}{\pi} (T^A)_{ab} (T^A)_{ba} \left(\frac{4\sqrt{\pi}}{9\sqrt{\omega}} \right) (E\omega_A)^{3/4} \left(\frac{m}{E} \right)^2
$$

$$
\times \sum_{n=0}^{\infty} (-1)^n (2n+1)^{9/4} \exp\left(-\frac{2}{3}(2n+1)^{3/2} \sqrt{E\omega_A}/\omega \right), \qquad (10.46)
$$

which is characterized by an essential singularity at $\omega = 0$. The $g_A \neq 0$ contribution for large ω reads

$$
\mathbf{P}_{aa}(\omega) \approx \frac{\alpha}{\pi}(T^A)_{ab}(T^A)_{ba} \left(\frac{2\sqrt{\pi}}{9} \right) \left(\frac{m\omega}{E\omega_b} \right)^{5/2} \omega_b \exp[-(m/E)^3 (2\omega/3\omega_b)].
$$

$$
(10.47)
$$

At small ω the non-Abelian radiation is of order $e^{-\#/\omega}/\sqrt{\omega}$ which is much smaller than the Abelian radiation of order $\omega^{1/3}$. The reason is that the emitted non-Abelian gluon brings about its own phase which strongly adds to the phase incoherence at small ω as is clearly seen in (10.32) and (10.39). At large ω both the Abelian and non-Abelian radiations are comparable and of order $\omega^{5/2} e^{-2\omega/(3\omega_b\gamma^3)}$. Indeed, we note that (10.47) is independent of the gluon charge g_A. For ultrarelativistic jets, the radiation frequencies are in the range $0 \leq \omega \leq E$. Thus $\omega/\omega_b \leq \gamma$ which is way below the maximum of γ^3. In light of the present observations we conclude that most of the jet radiation is emitted through the "Abelian" part of the gluon charge and in the small frequency range way below the synchrotron maximum since $E/\omega_b \ll \gamma^3$. The "non-Abelian" part only enters as a correction.

The "Abelian" part of the jet radiation follows from the summation over $A = 1, 2, 3, 8$ for which $g_A = 0$, thus $e_a = e_b$. In this case (10.43) simplifies to

$$
\omega^{-1}\mathbf{P}_{aa}(\omega) = \frac{\alpha}{\pi} (T^A)_{ab} (T^A)_{ba} \frac{m^2}{\sqrt{3}\,E^2} \int_{\xi}^{\infty} dt\,\mathbf{K}_{5/3}(t), \qquad (10.48)
$$

since $\xi_n = \xi$ and $\kappa_n = 0$. In this case, the total power emitted follows by integrating (10.48) over the gluon frequency, including the recoil effects (10.43),

$$
\mathbf{P}_{aa} = \frac{\alpha}{\pi} (T^A)_{aa} (T^A)_{aa} \frac{m^2}{\sqrt{3}} \int_0^E \frac{d\omega}{E} \frac{\omega}{E} \int_{\xi}^{\infty} dt\,\mathbf{K}_{5/3}(t), \qquad (10.49)
$$

with the quantum corrected (in denominator) ξ

$$\xi = \frac{2}{3}\left(\frac{m}{E}\right)^3 \frac{\omega/\omega_a}{1 - \omega/E}. \tag{10.50}$$

In Fig. 10.16 we compare two spectra for some particular selection of parameters: the ratio of the integrals of these two spectra is 0.48.

In an exploding sphaleron of size ρ the power radiated has to be corrected by the finiteness of the exploding shell, which is reflected in the fact that the jet trajectory has a finite arc length $R_a \Delta\phi$ with $R_a = 1/w_a$ the curvature radius. Specifically,

$$\frac{d\mathcal{E}_a}{dx} = \frac{\Delta\phi}{2\pi}\mathbf{P}_{aa} = \frac{\rho}{2\pi R_a}\mathbf{P}_{aa}. \tag{10.51}$$

The total energy lost in going through the sphaleron is

$$\Delta\,\mathcal{E}_a = \frac{\rho^2}{2\pi R}\mathbf{P}_{aa}. \tag{10.52}$$

The abelian-like contribution is

$$\frac{\Delta\mathcal{E}_a}{E} = \frac{\alpha}{\pi}\sum_{A=3,8}(T^A)^2_{aa}\,\frac{\mathbf{C}}{2\pi}\,\frac{(e_a\,H\,\rho^2)^{5/3}}{(E\rho)^{4/3}}. \tag{10.53}$$

The non-Abelian contribution is a smaller correction to the Abelian result (10.53).

The details can be found in the original paper, we briefly summarize here the radiation predictions for two popular models of pre-QGP matter, the Color Glass Condensate (CGC) and sphaleron model.

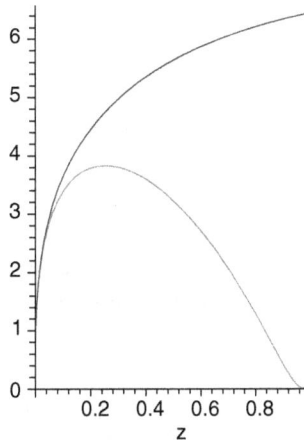

Fig. 10.16. The integral in (10.49) versus radiated energy fraction $z = \omega/E$ for classical (upper) and quantum corrected (lower) cases. The example shown corresponds to $eaH/m^2 = 1$ and $\gamma = E/m = 5$.

The CGC parameters have been mentioned in Section 10.3.2 above.[13] With those numbers, we find that the relative energy loss of a quark by synchrotron radiation in a time τ_{CGC} is

$$\frac{\Delta E_{\text{CGC}}}{E} \approx 0.3 \left(\frac{H}{1\,\text{GeV}^2} \right)^{2/3} \left(\frac{\Delta \tau_{\text{CGC}}}{0.5\,\text{fm}} \right) \left(\frac{1\,\text{GeV}}{E} \right)^{1/3}. \tag{10.54}$$

The gluon loss is about twice the quark loss.

For the sphaleron model, we recall that at large time $t \gg \rho$ the corresponding gauge field is purely transverse, with equal chromoelectric and chromomagnetic fields. The prompt sphaleron configuration released in parton–parton scattering carries initially a very strong chromomagnetic field,

$$\sqrt{H} \sim \left(\frac{2M_S}{\rho^3} \right)^{1/4} \sim 1\,\text{GeV}. \tag{10.55}$$

In the early phase of the prompt process in heavy ion collisions, the escaping sphalerons form a dilute gas. So unlike the CGC they cannot affect most of the jets initially for times $t \sim \rho$. They do affect them as they expand into exploding shells. The net synchrotron radiation loss involves also the transverse density of sphalerons per unit rapidity n_S and their typical collision volume $\sigma(t)\,dt$. Specifically,

$$\Delta E = \int \mathbf{P}(t) n_S \sigma(t)\,dt \sim \int dt\, t^{-2/3+2}, \tag{10.56}$$

where we used the cross section $\sigma \sim t^2$ and the radiation loss $\mathbf{P} \sim H^{2/3} \sim t^{-2/3}$. The result formally diverges for large times. However, the above reasoning is only valid as long as the single shell expansion remains coherent. As we now show, the originally dilute gas of shells quickly evolves into a foam-like structure for times $t \sim 2$–3ρ providing a natural cutoff in the time integral.

Recent estimates of the number of clusters produced in pp and AA collisions as a function of the collision energy and centrality, are still rather uncertain. The theoretical calculations of the cross section such as in Ref. [444] are carried to only exponential accuracy, while phenomenological studies such as in Ref. [456] have only resulted in an crude estimate. Assuming the whole growth with $\sqrt{s}$ of the pp cross section to be due to the release of 1 sphaleron, and ignoring nuclear shadowing, about $\mathbf{S} = 400$ sphalerons are produced in central AuAu collision at $\sqrt{s} = 136\,\text{GeV}$ at RHIC. Since the total rapidity interval is $\Delta y = 4$, this amounts to about 100 sphalerons per unit rapidity. The transverse density of sphalerons per

---

[13] Of course the field is not homogeneous, with a correlation length of the order of $1/Q_s$, while we only evaluated the radiation loss for a constant field. We have ignored that for the estimates.

unit rapidity in central AuAu collisions is

$$n_S = \frac{\mathbf{S}}{\pi R^2 \Delta y} \sim 1\,\text{fm}^{-2}, \tag{10.57}$$

which results into a foam-like structure for the exploding sphalerons for times $t \sim 2\rho \sim 0.7\,\text{fm}$. This time increases as the cubic root of the transverse density for smaller densities. For a jet piercing a sphaleron wall with an integrated kick is about $0.5\,\text{GeV}$ for a quark with $e_a = g\sqrt{3}/6$, and about $1\,\text{GeV}$ for a gluon.

Substituting the expression for the field strength of the exploding shell into the synchrotron radiation loss, one gets $\int F^{2/3} dt = F_{\text{max}}^{2/3} \pi \rho$. For two overlapping shells we add the fields in quadrature and estimate that the maximal field is given by $F_{\text{max}} \approx 0.2\,\text{GeV}^2$ in the foam phase at time $t_{\text{foam}} \approx 2\rho$. Substituting all this into the synchrotron energy loss formulae, we finally obtain for the quark loss

$$\frac{\Delta E}{E} \sim 0.21 \left(\frac{H}{0.2\,\text{GeV}^2}\right)^{2/3} \left(\frac{1\,\text{GeV}}{E}\right)^{1/3}. \tag{10.58}$$

The gluon loss scales with the pertinent color Casimir and is about twice as large. Taking into account the strongly falling p_t spectra, one finds that these losses are indeed sufficient to produce about one order of magnitude of quenching, which is in the right ballpark.

CHAPTER 11

QCD at High Density

11.1. From nuclear to quark matter

Figure 11.1 shows the phase diagram of hot/dense matter. In Chapter 8 we discussed its l.h.s., related with high T and small or moderate chemical potential μ, up to about the endpoint of the first order phase transition marked by E on the figure. In this section we extend our discussion deeper into the region of dense matter, to the right in Fig. 11.1, where a dramatic development took place in the last few years. We indicated two Color Superconducting phases, two-flavor-like (called CSC2 or 2SC) or three-flavor-like (called CSC3 or CFL) as the best established ones. However, as the reader will see below, many more phases have been proposed which are not on this figure.

11.1.1. *Nuclear matter*

We are *not* going to discuss finite nuclei in this book, but provide some brief introduction to the theory/phenomenology of nuclear matter. Two special cases are discussed most frequently: (i) a symmetric p–n matter with QED switched off; and (ii) pure n matter as approximation to neutron stars: we will comment on both below.

Before we do so, we make some more general comments. In general, nuclear matter is a rather dilute Fermi liquid: a volume per nucleon in it is about $6\,\text{fm}^3$, which is significantly larger than the volume of a nucleon. As any liquid, it has a liquid–gas phase transition [793] which is the first order line in the lower part of Fig. 11.1. Its endpoint is marked by M which is supposed to remind us of the so called *multifragmentation*, associated with nuclear reactions passing in the vicinity of this point on the phase diagram and related with the critical opalescence phenomenon observed at this *second*-order transition point.

The understanding of the very basis of nuclear physics — the nuclear forces — is a long-standing theoretical problem. The NN interaction consists of attractive long-range forces and a repulsive core. While the former are understood due to

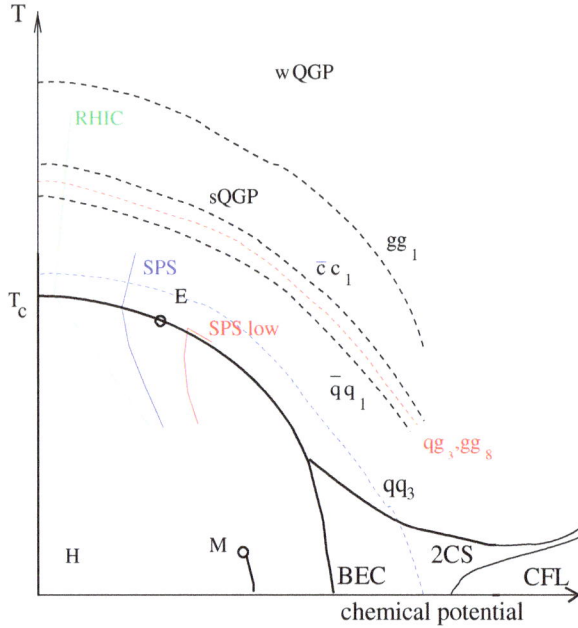

Fig. 11.1. Schematic QCD phase diagram, on a plane the baryonic chemical potential μ_B versus the temperature T, from Shuryak and Zahed, hep-ph/0403127. E and M are endpoints, the tricritical one and of the nuclear liquid-gas transition. Positions of several "zero binding lines" for quasiparticle binary states are also shown, with subscript indicated the total color representations (e.g. gg_1 and qq_3 are two-gluon color singlet and diquark color triplet). sQGP and wQGP are a strongly and weakly coupled quark-gluon plasma, the former having bound states. The qq_3 zero binding line indicate qualitative change from Bose–Einstein Condensate of diquarks (BEC) to a color superconductor, with 2CS and CFL indicating 2-flavor and 3-flavor-type color superconductors. Approximate positions of RHIC and SPS adiabatic paths are also indicated.

light meson exchanges (see, e.g. a well known book [778]), I think the latter is still not yet completely understood. (Our discussion in Chapter 2 of whether there can be a deeply bound $\Lambda\Lambda = H$ dibaryon or not is a piece of this puzzle.) Successful phenomenology follows e.g. from a simple Walecka model [779] where the core is described by the effective ω exchanges.

The core is also more difficult to take into account, as one cannot obviously do it perturbatively. In nuclear literature one may find multiple *potentials* — different parameterizations of the NN scattering data — with complicated names with a display of geography of the nuclear physics centers of the world. However recently the use of the *renormalization group* has allowed one to integrate out all sufficiently hard momenta and to show that all these parameterizations actually converge into the *unique* low-energy effective interaction $V_{k<k_0}$. While the Seattle group [782] tries to run the cutoff k_0 very low and use vacuum scattering directly, the Stony Brook one [781] keeps it intermediate with $k_0 \sim 200$–400 MeV where there is a rather weak dependence on k_0.

Following the Landau–Fermi liquid theory, the phenomenology of nuclear matter is based on a set of certain parameters defined at the Fermi surface. In modern terminology, this theory is also derived from the renormalization group equation, as a limit when the parameter δk (defining a shell around the Fermi momentum in which effective theory is defined) goes to zero. The equation defines how the effective interaction of both particles and holes *inside the shell* varies, as δk is decreased by a small amount. For an introduction see, e.g. good pedagogical reviews by Shankar and Polchinski [780].

This technology substitutes for much more complicated direct re-summation of explicit classes of diagrams, as was done starting from classic papers by Bethe, Bruckner and others [783] in the 1950s, see reviews [778, 784, 785]. The theoretical methods, refined over the years, led to smaller and smaller binding energy per baryon. For quite a long time it was believed that the Bethe–Brueckner–Brandow (BBB) theory is good enough even in its lowest order $n = 2$ (n corresponds to the number of the hole lines in the diagrams used). With computer power rising in the last decades, very powerful *variational* and *Green function Monte Carlo* methods were developed, leading to even lower binding for the same nuclear forces, up to 8–10 MeV/nucleon. However, phenomenologically this binding energy is too small. This "under-binding" problem thus is not a technical problem and can only be solved with 3-body forces. More information on the role of the admixture of isobar and other excited N^* states is also needed to conclude the calculation.

For the necessary estimates we will use the standard phenomenological parametrization of the equation of state near the minimum which looks as follows

$$\epsilon(n) = m_N n + \frac{3}{2}Tn + \frac{1}{18}Kn_0\left(1 - \frac{n}{n_0}\right)^2 \tag{11.1}$$

where ϵ, n are the energy and the baryon density, T is the temperature and $K \approx 200$ MeV is the so called "nuclear compression modulus", and $n_0 = 0.16\,\text{fm}^{-3}$ is the equilibrium density of the nuclear matter.

In the framework of this book, nuclear matter (as well as the dilute pion–resonance gas at the latest stages of heavy ion collisions) can be viewed rather as an example of a *weakly perturbed QCD vacuum*. Indeed, even the total contribution of the nucleon masses $m_N n_0 = 0.15\,\text{GeV/fm}^3$ is several times smaller than the non-perturbative "bag energy" of the QCD vacuum we discussed in Chapters 2 and 4.

More specifically (similarly to the discussion in Section 8.4.3 on the low-T matter) one may again ask how all the vacuum phenomenology — VEVs or condensates of various operators, correlation functions, particle masses and widths — are modified in the nuclear matter.

As a very important example, consider the modification of the quark condensate in nuclear matter [418]

$$\frac{\langle \bar{q}q \rangle^*}{\langle \bar{q}q \rangle} = 1 - \frac{\Sigma_{\pi N} n}{f_\pi^2 m_\pi^2} + \cdots \approx 1 - 0.36\frac{n}{n_0}, \tag{11.2}$$

where $\Sigma_{\pi N}$ is the parameter of πN scattering known as the nucleon sigma term. The resulting depletion of the condensate is surprisingly strong for such a dilute matter: about $1/3$ of the condensate is gone. So the nucleons are quite an effective "vacuum cleaner": more than pions are in the low-T matter.

How does such a modification affect other observables such as hadronic masses? Naively, a view of a sigma as a "Higgs boson of strong interaction" would suggest that all hadronic masses are proportional to its VEV, as are all masses in the Standard Model. If so, all the masses would also be reduced by $1/3$. A somewhat less radical suggestion, known as the *Brown–Rho scaling* [786] prediction is

$$m^* = m \left(1 - 0.36 \frac{n}{n_0} + \cdots \right)^{1/2}, \qquad (11.3)$$

where an asterisk indicates effective in-matter mass. It leads to good agreement with experimental information on the effective nucleon mass, and also on the downward shifted ρ-meson mass. We would not discuss the details of this phenomenology but just note that this formula, being extrapolated to higher density, suggests that the chiral symmetry is probably restored somewhere at $n \sim 3n_0$.

Cold nuclear matter is also subject to pairing instability, like electrons in super-conductors. In symmetric p–n matter the main pairing is in the p–n channel, while in ordinary heavy nuclei (as well as in neutron stars to be discussed shortly) we have a very different density of protons and neutrons. Therefore the Fermi spheres for p and n have different radii and thus they pair separately, forming a neutron superfluid and a proton superconductor. The typical gaps are in the $1\,\mathrm{MeV}$ range.

Before leaving the subject of nuclear matter, let me in passing remind the reader of two observations about the nuclear matter which (in my opinion) have not attracted sufficient attention to be really explained. One is the so called "cumu-lative effect" studied extensively in Dubna [794]: any particle at any energy gener-ates universal backward and large angle spectra of nuclear fragmentation, which is exponential and extends smoothly well beyond the limits for a collision with a single nucleon, 2 nucleons, etc. Another is the so called "EMC effect" found at CERN [795] at large and medium[1] $x > 0.1$. It shows how nuclear structure functions are modi-fied as compared to that of a single nucleon. Both phenomena clearly demonstrate that at, say, the probability level of few percent, one cannot consider nuclei to be just a set of independent nucleons. Nowadays these studies are pushed further by current experiments in the Jefferson Lab, which were able to measure the structure functions on dibaryon clusters in nuclei, for above 1 and in the range $x = 1$–1.4. Some of these data may be already telling us something important about the nature of the nuclear core.

---

[1]Do not confuse it with the "saturation" phenomenon at small $x \lesssim 10^{-3}$.

11.1.2. *Other phases of nuclear matter?*

On the way toward high densities a nontrivial collective phenomenon may happen, resulting in new phases of nuclear matter, prior to deconfinement and transition to quark matter. In this section we briefly discuss some suggestions proposed over the years.

The earliest of them was Lee–Wick's *"abnormal nuclear matter"* [788] in which chiral restoration takes place and baryons become massless. Less radical was a suggestion by A. B. Migdal [787] of the so called *"pion condensation"*. Its main idea is that due to attractive interaction of the pions with the nucleons, the mass operator Σ in the pion propagator

$$D_\pi^{-1}(k,\omega) = \omega^2 - \vec{k}^2 - m_\pi^2 - \Sigma(k,\omega) \qquad (11.4)$$

is negative and proportional to the nucleon density. He argued that at some critical density n_c the r.h.s. may reach zero, at $\omega = 0$ and non-zero critical k_c. At $n > n_c$ the instability would develop, leading to formation of the "pion condensate". Because pions are Goldstone particles and have to interact weakly at small momenta, the instability may only happen at non-zero momentum. This means that this phase is actually a kind of a "crystal" with period $2\pi/k_c$. The interest in such crystals was revived in the 1980s due to the development of the chiral soliton models of the baryons, made of classical pion field (see Section 2.2.3 on Skyrmions). The periodic solutions of the Skyrme Lagrangian have been also considered, see, e.g. Ref. [790].

Attractive interaction of negative kaons may create *kaon condensation* at zero momentum, see the original paper by Kaplan and Nelson [791] and vast subsequent literature, especially on stars where negatively charged condensates can substitute for electrons, thus additionally enhancing the attractive interaction.

11.1.3. *From nuclear to quark matter*

We start with a simple argument, suggesting why the nuclear matter should turn to quark matter at high density. The nuclear matter remains non-relativistic basically until the Fermi momentum p_F reaches m_N, so

$$\epsilon_F \sim p_F^2/2m_N \sim n^{2/3}, \qquad (11.5)$$

while the quark matter is always relativistic,

$$\epsilon_F = p_F \sim n^{1/3}, \qquad (11.6)$$

and therefore at large enough n the latter has less energy per baryon and wins.[2]

In Fig. 11.2(a, b) we schematically explain the behavior of thermodynamic quantities in the discussed phase transition. In Fig. 11.2a we show the energy per baryon versus the density at which the nuclear matter (solid curve marked N) has a minimum at $n/n_0 = 1$, while the curve for quark matter (solid marked $q1$) has a

---

[2]Note also that although quarks go in N_c colors, their baryon number is $1/N_c$ and thus this factor basically drops out of the consideration. Whatever is the N_c, quark density as a function of baryon number density, $p_f(n)$ remains the same.

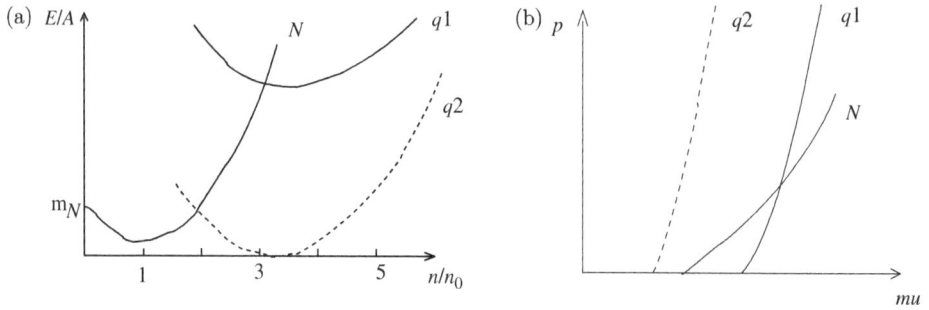

Fig. 11.2. Schematic comparison of the EoS for nuclear and quark matter. In (a) it is plotted as the energy per baryon E/A versus the density in units of nuclear matter density, in (b) as pressure versus the chemical potential.

minimum at higher and higher density. (The hypothetical case with a stable quark matter — shown by the dashed curve marked $q2$ — has a lower minimum: we will return to it below.) Since matter at zero energy is in its ground state, minimizing the energy per baryon, at high enough density there would be a transition from the curve N to $q1$. It will however *not* happen where the curves cross, but when N reaches E/A corresponding to a minimum of $q1$. The corresponding jump in density is a typical first order phase transition.

Minimization of the energy per baryon is correct only at $T = 0$. The more general (although admittedly less intuitive) way to compare any two phases is to see which has the *lowest* grand canonical (Gibbs) potential $\Omega = -pV$ or the *largest pressure at the same chemical potential* μ_B. Such a plot is schematically shown in Fig. 11.2. Zero pressure corresponds to minima in Fig. 11.2a, and now the condition on the critical μ_c is indeed crossing of the corresponding curves, since there are no jumps in μ or p. The condition

$$p_{\text{nuclear/matter}}(\mu_c, T) = p_{\text{quark/matter}}(\mu_c, T) \tag{11.7}$$

is needed for the mechanical, thermal and chemical coexistence of the two phases.

The simplest and most often used model for quark matter is that of ideal relativistic Fermi gas, complemented by the so called bag constant. For N_f massless quark flavors its pressure is

$$p = N_f \frac{\mu^4}{4\pi^2} - B, \tag{11.8}$$

where μ per quark is the same as Fermi momentum. At which μ such quark matter is "self-contained", at the minimum of E/A or at $p = 0.2$, it is obviously at

$$\mu_0 = (B/4\pi^2 N_f)^{1/4}. \tag{11.9}$$

If this is lower than the chemical potential per quark in nuclear matter, $(m_N - E_{\text{binding}})/3$, the curves would never cross and quark matter would be the only truly stable phase, with the nuclear matter being just metastable. The window for this to happen corresponds to $56\,\mathrm{MeV}/\mathrm{fm}^3 < B < 90\,\mathrm{MeV}/\mathrm{fm}^3$, with the lower limit coming from the condition that non-strange quark matter is still above the nuclear one and thus nuclear physics as we know it is still there.

Naively it may seem obvious that this last scenario cannot be true. However Witten [798] has considered such a possibility, which led to much discussion in the literature which continues to this day. This issue deserves to be explained in some detail.

We do know experimentally (e.g. from low energy heavy ion collisions) and theoretically that nuclear matter is rather stiff, has a high pressure or E/A at densities above n_0, see (11.1). However, in this case the time scale involved is that of strong interaction, and weak transitions to strange quarks cannot occur. This corresponds to $N_f = 2$ in (11.9), while the absolute minimum we discuss corresponds obviously to *strange quark matter* with $N_f = 3$.

Why are nuclei stable for billions of years? Witten's answer to this question is that transition of ordinary matter into strange matter becomes possible only at the high orders in weak interactions, and thus ordinary nuclei may be stable enough.

What is then the *smallest* cluster of strange quark matter which can be energetically favorable? This question was addressed by many papers, with Farhi and Jaffe [800] among the first. They worked in the MIT bag model framework and concluded that the smallest A for which the bag constant which admits strange quark matter as the true ground state, is rather large $A_{\min} = 50$–100. Therefore the transition to strange matter may happen only starting from the intermediate nuclei, by the simultaneous transition of $A_{\min}$ quarks $d \to s$. Obviously, such a process of say 50th order in weak interaction is so suppressed that it has no chance of being observed!

The loophole is related with $A = 2$, the "dihyperon H" mentioned in Chapter 2. In the MIT bag model this "alpha particle of quark models", with $(uds)^2$ quarks in the lowest shell, tends to be bound $m_H < 2m_\Lambda$ and with some parameters it can even be *absolutely* stable $m_H < 2m_n$ [99].

This idea has created multiple searches for the H, so far all with negative results, see Ref. [100]. In 1985 a strange anomaly was observed where cosmic muons seem to be correlated in direction with a particular gamma-ray source Cygnus X-3, and it was proposed by Baym *et al.* [101] that those can be due to the H traveling from it to Earth. However it was soon shown by Khriplovich and myself [102] that the second-order weak process would not let it happen. Also the process $nn \to H$ will destroy many isotopes in about a year's time, while in fact they survive billions of years. (It was nevertheless still argued recently by Farrar and collaborators [105] that some small overlap integrals can block the decays by many orders of magnitude, so that very deeply bound Hs are still possible.)

Another direction of research had ignored the issue of strange matter production and simply looked at the experimental limits for its existence. A good summary of those efforts can be found in Ref. [799]. Considering large possible clusters, we should remember that nuclear matter is bound by strong, but unbound by strong plus electromagnetic interactions. For strange quark clusters the charge density is essentially smaller because the symmetric u, d, s matter has zero charge, and only the non-zero strange quark mass leads to some reduction in the s population and thus some positive charge density. Therefore, the strange matter clusters with $Z < 100$ should create the ordinary atoms, but with very heavy nuclei. Since their size is larger than for normal nuclei, the Coulomb barrier is lower (which has been used for experimental search). Still, with the Coulomb barrier of about 10 MeV, the hypothetical strange matter cannot "eat" the ordinary matter under normal conditions.[3] The situation is different for the neutrons, which, if they fall on such cluster, may be "eaten up" with the released energy. Therefore, such matter, if it exists, can be enormously useful for mankind: it can produce energy in the usual nuclear power reactor by absorbing neutrons; see some discussion of that in my lecture [801].

The issue of whether strange quark matter is or is not the true ground state has of course a dramatic effect in the case of compact stars. If it is, all compact stars should be made of it. We return to this issue in the next section.

Summarizing this discussion, let me put Witten's idea in somewhat different words. We know that the free neutrons are unstable, but binding due to strong interaction inside the nuclei stabilizes them. Witten's idea is that a similar phenomenon may happen for strange quarks in dense quark matter. However, while the proton–neutron mass difference is of the order of magnitude of 1 MeV only, the strange quark mass is two orders of magnitude larger. It can in principle be compensated by the binding energy caused by the strong interactions, but perhaps only in bulk matter in compact stars.

From the dynamical point of view, the issue is related with the effective value for the bag constant B. By definition, it represents the *difference* between the non-perturbative effects in vacuum and in the dense matter. Our favorable dynamical model — the instanton liquid — tends to give values for B significantly larger than the original MIT bag constant and that is needed to have a stable strange quark matter.

11.1.4.   *Chiral waves and chiral crystals*

In the previous section quark matter was treated as an ideal gas: but of course interaction, especially near the Fermi surface, may cause many different effects. The most famous of them, winning in the case of weak coupling, is the *particle–particle pairing* which leads to the Bardeen–Cooper–Schrieffer (BCS) instability and *color superconductivity*, as we discuss below in detail. In the strong coupling, however, or if superconductivity is suppressed by mismatched Fermi momenta, other forms

---

[3]Witten wrote, "strange matter is no more dangerous than oxygen". I think he is mistaken here and for biological objects it is dangerous. For example, heavy water with deuterium is not radioactive, but just because the unconventional atomic mass creates mistakes in the highly specialized chemistry of the living substance, it is absolutely deadly.

of pairing may take place. One candidate we will consider here is the *particle–hole* pairing, following the paper by Rapp, Zahed and myself [856]. It is not shown in Fig. 11.1, but is a possible substitution of BEC of diquarks. The fact that we consider quark matter means that we assume deconfinement. On the other hand the pairing channel is the same scalar channel which leads to the quark condensate in vacuum and breaks chiral symmetry. So this phase, like the QCD vacuum, breaks chiral symmetry; just the condensate is modulated in space.

Similar phenomena are known in atomic and condensed matter physics. In $1d$ it is known as the Peierls instability. A specific form of it, the so called *spin-density waves* is relevant for specific electronic materials, especially in the $2d$ case, as originally proposed by Overhauser [857]: for a review see Ref. [858].

The importance of this particular phase for dense quark matter stems from the fact that it leads to a solid crystalline structure. Obviously for the dynamics of compact stars it is crucial to determine whether the quark matter is all in a liquid phase, or solid shells separating liquid nuclear and quark matter may exist.

Particle–hole pairing is characterized by an order parameter of the form

$$\langle \bar{\psi}(x)\psi(y) \rangle = \exp(i\vec{Q} \cdot (\vec{x} + \vec{y}))\Sigma(x - y), \tag{11.10}$$

where $\vec{Q}$ is an arbitrary vector. This state describes a chiral density wave. It was first suggested in [854] as the ground state of QCD at large chemical potential and large N_c. This suggestion was based on the fact that particle–particle pairing, and superconductivity, is suppressed for large N_c whereas particle–hole pairing is not. Particle–hole pairing, on the other hand, uses only a small part of the Fermi surface and does not take place in weak coupling. In the case of the one-gluon exchange interaction these issues were studied in Ref. [855]. The main conclusion is that, in weak coupling, the chiral density wave instability requires very large $N_c \gg 3$.

At moderate densities, and using realistic interactions, this is not necessarily the case. In particular, we know that at zero density the particle–antiparticle interaction is stronger, by a factor $N_c - 2$, than the particle–particle interaction. In a Nambu–Jona–Lasinio type description this interaction exceeds the critical value required for chiral symmetry breaking to take place. For this reason we have recently studied the competition between the particle–particle and particle–hole instabilities using non-perturbative, instanton generated, forces [856]. This work concluded that for quark matter at low density $\mu_q \sim 400\,\text{MeV}$ the chiral density wave state is practically degenerate with the BCS solution. Given the uncertainties that affect the calculation this implies that both states have to be considered as realistic possibilities for the behavior of quark matter near the phase transition. These results are not only relevant to flavor symmetric quark matter at moderate densities, but also in the important case when there is a substantial difference between the chemical potentials for up and down quarks which disfavors ud-pairing, but does not inhibit uu^{-1} and dd^{-1} particle–hole pairing. Unfortunately the crystalline phase in this regime has not yet been discussed.

Let me briefly sketch the standard MFA formalism one can use to address pairing in the particle–hole channel at finite total pair momentum Q. Assuming that the full Green function in the presence of a single density wave takes the form

$$
\hat{G}_{Ovh} = \begin{pmatrix} \langle c_{k\uparrow} c^{\dagger}_{k\uparrow} \rangle & \langle c_{k\uparrow} c^{\dagger}_{k+Q\downarrow} \rangle \\ \langle c_{k+Q\downarrow} c^{\dagger}_{k\uparrow} \rangle & \langle c_{k+Q\downarrow} c^{\dagger}_{k+Q\downarrow} \rangle \end{pmatrix}
$$

$$
\equiv \begin{pmatrix} G(k_0, \vec{k}, \vec{Q}, \sigma) & \bar{S}(k_0, \vec{k}, \vec{Q}, \sigma) \\ S(k_0, \vec{k}, \vec{Q}, \sigma) & G(k_0, \vec{k}+\vec{Q}, \vec{Q}, \sigma) \end{pmatrix}
$$

$$
= \left[\hat{G}_0^{-1} - \hat{\sigma} \right]^{-1}, \tag{11.11}
$$

one can see here the *anomalous* part of the Green function,

$$
S(k_0, \vec{k}, \vec{Q}, \sigma) = \frac{-\sigma}{(k_0 - \epsilon_k + i\delta_{\epsilon_k})(k_0 - \epsilon_{k+Q} + i\delta_{\epsilon_{k+Q}}) - \sigma^2}, \tag{11.12}
$$

and the anomalous self energy σ, to be determined by a self-consistency, or gap, equation

$$
\sigma = (-i)\alpha_{ph} \int \frac{d^4p}{(2\pi)^4} \, S(p_0, \vec{p}, \vec{Q}, \sigma; \mu_q) \, . \tag{11.13}
$$

Notice that the energy contour integration receives non-vanishing contributions only if

$$
\epsilon_p \epsilon_{p+Q} - \sigma^2 < 0, \tag{11.14}
$$

which ensures that the two poles in p_0 are in different (upper/lower) half-planes. This means that one particle (above the Fermi surface) and one hole (below the Fermi surface) participate in the interaction.

The formation of a condensate carrying non-zero total momentum $\vec{Q}$ is associated with nontrivial spatial structures. In the simplest case of particle–hole pairs with total momentum Q this is a density wave of wave length $\lambda = 2\pi/Q$. In three dimensions, however, we can have several density waves characterized by different momenta $\vec{Q}$. In this case, the resulting spatial structure is a crystal. In general, the p–h pairing gap can be written as

$$
\sigma(\vec{r}) = \sum_{j} \sum_{n=-\infty}^{+\infty} \sigma_{j,n} e^{in\vec{Q}_j \cdot \vec{r}}, \tag{11.15}
$$

where the $\vec{Q}_j$ correspond to the (finite) number of fundamental waves, and the summation over $|n| > 1$ accounts for higher harmonics in the Fourier series. The matrix propagator formalism allows for the treatment of more than one density wave through a straightforward expansion of the basis states according to

$$
\hat{G} = \begin{pmatrix}
\langle c_{k\uparrow} c_{k\uparrow}^\dagger \rangle & \langle c_{k\uparrow} c_{k+Q_x\downarrow}^\dagger \rangle & \langle c_{k\uparrow} c_{k+Q_y\downarrow}^\dagger \rangle & \cdots \\
\langle c_{k+Q_x\downarrow} c_{k\uparrow}^\dagger \rangle & \langle c_{k+Q_x\downarrow} c_{k+Q_x\downarrow}^\dagger \rangle & \langle c_{k+Q_x\downarrow} c_{k+Q_y\downarrow}^\dagger \rangle & \cdots \\
\langle c_{k+Q_y\downarrow} c_{k\uparrow}^\dagger \rangle & \langle c_{k+Q_y\downarrow} c_{k+Q_x\downarrow}^\dagger \rangle & \langle c_{k+Q_y\downarrow} c_{k+Q_y\downarrow}^\dagger \rangle & \cdots \\
\vdots & \vdots & \vdots & \ddots
\end{pmatrix}. \tag{11.16}
$$

The possibility of simultaneous BCS pairing can be incorporated by extending the Gorkov propagator to include both particle–hole and particle–particle components. In the following we will consider up to $n_w = 6$ waves in three orthogonal directions with $Q_x = Q_y = Q_z$ and $n = \pm 1$, characterizing a cubic crystal.

Note that in the propagators G_0 we do not include the contribution of antiparticles. This should be a reasonable approximation in the quark matter phase at sufficiently large μ_q, when the standard particle–antiparticle chiral condensate has disappeared. At the same time, since our analysis is based on non-perturbative forces, the range of applicability is limited from above. Taken together, we estimate the range of validity for our calculations to be roughly given by $0.4\,\mathrm{GeV} \gtrsim \mu_q \gtrsim 0.6\,\mathrm{GeV}$. This coincides with the regime where, for the physical current strange quark mass of $m_s \simeq 0.14\,\mathrm{GeV}$, the two-flavor superconductor might prevail over the color-flavor locked (CFL) state so that our restriction to $N_f = 2$ is supported.

Solutions of the gap equations correspond to extrema (minima) in the energy density with respect to the gap σ. However, solutions may exist for several values of the wave vector Q. To determine the minimum in this quantity, one has to take into account the explicit form of the free energy density. In the mean-field approximation,

$$
V_3\,\Omega(\mu_q, Q, \sigma) = \int d^3x \left(\frac{\sigma^2(x)}{2\lambda} + \langle q^\dagger (i\alpha \cdot \nabla - 2\sigma(x)q) \rangle \right), \tag{11.17}
$$

where V_3 is the three-volume. The first contribution removes the double counting from the fermionic contribution in the mean-field treatment.

We have studied the coupled gap equations numerically. We do not find any solutions with simultaneous particle–particle and particle–hole condensates. This reduces the problem to the question whether the BCS or the density wave state is thermodynamically favored. The BCS solution $\Delta = 0.225\,\mathrm{GeV}$ is unique and has a free energy of $\Omega_{\mathrm{BCS}}(\mu_q = 0.4\,\mathrm{GeV}) = 2.3 \times 10^{-3}\,\mathrm{GeV}^4$. Here, we have neglected an irrelevant overall constant that does not affect the comparison with the density wave state.

The situation is more complicated in the case of particle–hole pairing. Let us start with the "canonical" case where the momentum of the chiral density wave is fixed at twice the Fermi momentum, $Q = 2p_\mathrm{F}$. In Fig. 11.3 the resulting minimized free energy is displayed as a function of the number of included waves. The density wave solutions are not far above the BCS ground state, with a slight energy gain for an increased number of waves.

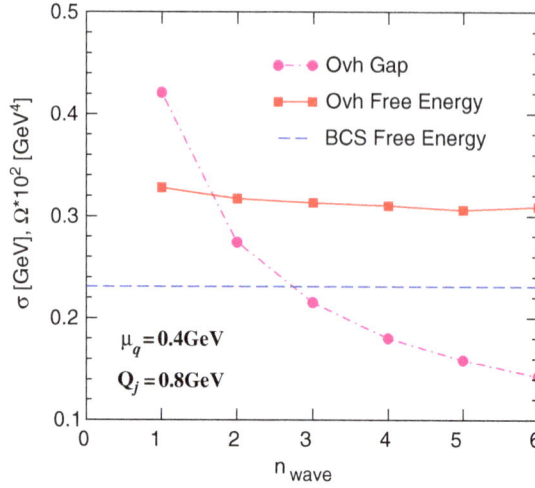

Fig. 11.3. Dependence of the free energy (upper full line) and p–h pairing gap (dashed-dotted) on the number of waves ("patches") with fixed magnitude of the three-momentum $|\vec{Q}_j| = 0.8\,\text{GeV}$. The full line shows the value of the BCS ground state free energy. The results correspond to an instanton calculation with $\mu_\text{q} = 0.4\,\text{GeV}$ and $N/V = 1\,\text{fm}^{-4}$.

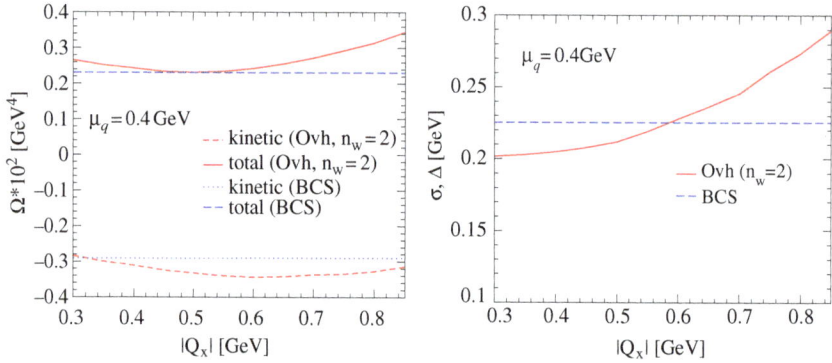

Fig. 11.4. Left panel: wave-vector dependence of the density wave free energy for one standing wave (full line: $\Omega_\text{tot}^\text{Ovh}$, short-dashed line: $\Omega_\text{kin}^\text{Ovh}$) in comparison to the BCS solution (long-dashed line: $\Omega_\text{tot}^\text{BCS}$, dotted line: $\Omega_\text{kin}^\text{BCS}$) at $\mu_\text{q} = 0.4\,\text{GeV}$. Right panel: wave-vector dependence of the density wave pairing gap (full line) compared to the BCS gap (long-dashed line).

However, one can further economize the energy of the chiral density wave state by exploiting the freedom associated with the wave vector Q (or, equivalently, the periodicity of the lattice). For $Q > 2p_F$ the free energy rapidly increases. On the other hand, for $Q < 2p_F$ more favorable configurations are found. To correctly assess them one has to include the waves in pairs $|k \pm Q_j|$ of standing waves ($n_w = 2, 4, 6, \ldots$) to ensure that the occupied states in the Fermi sea are saturated within the first Brillouin Zone. The lowest lying state we could find at $\mu_\text{q} = 0.4\,\text{GeV}$ occurs for one standing wave with $Q_{\min} \simeq 0.5\,\text{GeV}$ and $\sigma \simeq 0.21\,\text{GeV}$ with a free energy $\Omega \simeq 2.3 \times 10^{-3}\,\text{GeV}^4$, practically degenerate with the BCS solution. This density wave has a wavelength $\lambda \simeq 2.5\,\text{fm}$. The minimum in the wave vector is in fact rather shallow, as seen from the explicit momentum dependence of the free energy displayed in Fig. 11.4.

11.2. Compact stars

11.2.1. *Brief introduction*

Already in the 1930s, soon after the discovery of the neutron, Landau [803] and others predicted that there could be compact stars with density exceeding the nuclear matter density. The view held for a long time was that this kind of nuclear matter in the universe should be the highly compressed *neutron* matter, so they were called "neutron stars". The reason protons and electrons are forced to transfer to neutrons by the inverse beta decay is because otherwise relativistic electrons would have very large electron Fermi energy $\epsilon_F = p_F$ rather than much smaller energies for nucleons $\epsilon_F = p_F^2/2m_N$. With a small correction, the electro-neutrality condition is satisfied by a transition to a nearly pure neutron matter. The simplest version for the equation of state of the neutron matter is the ideal Fermi gas model, in which the energy per particle E/A is equal to

$$E/A = m_N + \frac{3p_F^2}{10m_N},\tag{11.18}$$

where the neutron Fermi momentum is related to the baryonic density as usual by $p_F = (3\pi^2 n)^{1/3}$.

Experimentally compact stars are identified with (i) pulsars and/or (ii) compact X-ray sources (e.g. Her X-1 and Vela X-1), which are in close binary orbits with an ordinary star. The first pulsar was discovered in 1982 [804] and by now about 1000 of them are known. Their most remarkable aspect is the high stability of the period, identified as a rotational one. In some cases it is extremely short — the fastest so far observed pulsars have rotational periods of $P = 1.6\,\text{ms}$, or 620 rotations per second. This alone means that the star is very compact, about 10 km in radius or less, or otherwise it would be simply destroyed by the centrifugal force.

Another remarkable property of compact stars is that their masses, measured accurately provided they are members of binary systems, so far fall into a band around 1.4 solar mass, see Fig. 11.5 reproduced from Ref. [805].

Compact stars with observable gamma rays have another important observable: the red shifted annihilation lies in the range 300–511 keV. These are gravitationally red-shifted 511 keV $e^\pm$ pair annihilation lines. A good pedagogical discussion of the related physics one can find in Ref. [802], the bottom line is that the highest

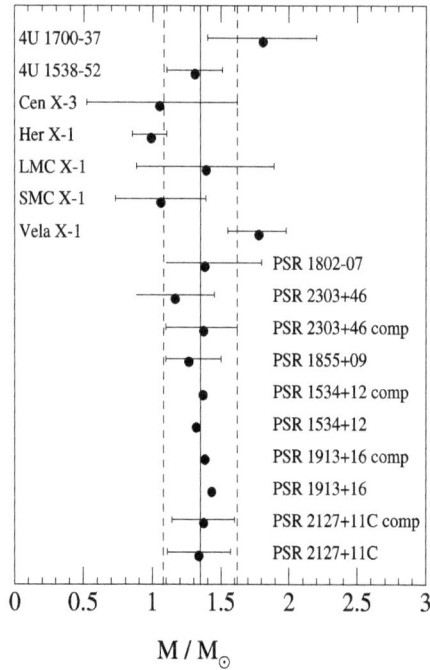

Fig. 11.5. Observed neutron star masses in units of a solar mass $M_\odot$.

concentration of those is in the narrow range $0.25 \leq z \leq 0.35$, which basically tells us the gravitational potential at the star surface.

X-ray compact sources can be compact stars or black holes. Recently a non-pulsing compact star was found via X-ray emission, which happens to be so close that it has been seen in visible light by the Hubble space telescope, and the absolute distance to it has been measured by a classic parallax method [806].

However, in spite of a growing amount of experimental data and multiple theoretical works, we still do not definitely know what they are made of. More precisely, the question is what are the fractions of the nuclear matter, or the mixed phase, or the "quark matter".

The theory of pulsars can be divided into the *statics*, describing them in equilibrium, and *dynamics*, following their life from birth, in a supernova explosion to a cooling curve ranging up to millions of years, during which period the stars can still be seen.

Starting with the former problem, we remind the reader that pulsars are so compact that effects of general relativity are noticeable. Thus, instead of the non-relativistic equilibrium condition $\mu + m\phi_{\text{grav}} = \text{const}$, where ϕ_{grav} is the gravitational potential, one should use their relativistic generalization instead,

$$\mu\sqrt{g_{00}} = \text{const} \qquad (11.19)$$

where g_{00} is the time–time component of the metric. Together with the Einstein equations for the metric this leads to the so called Tallman–Oppenheimer–Volkoff (TOV) equation[4]

$$\frac{dp}{dr} = -\frac{G_N \epsilon(r) m(r)[1 + p(r)/\epsilon(r)][1 + 4\pi r^3 p(r)/m(r)]}{r^2[1 - 2G_N m(r)/r]}, \qquad (11.20)$$

where the mass inside radius r is defined as $m(r) = 4\pi \int_0^r \epsilon(x) x^2 dx$, ϵ, p are the energy density and the pressure and G_N is Newton's gravitational constant.

If the equation of state $p(\epsilon)$ is known, the TOV equation can be integrated from the star center to the surface R where the pressure vanishes, $p(R) = 0$. Then the total mass $M(R)$ of the star is obtained, as a function of the initial central density n_c. It turns out that this mass, expressed as a function of n_c does not grow monotonically, and for some EoS can even show oscillatory behavior with more than one maximum. Stable stars can only exist if the slope of this dependence $M(n_c)$ is positive. (So, for example if there are two maxima in $M(n_c)$, there can be two families of "twin" stars with the same masses.)

There exists always a maximal possible mass $M_{\max}$ of a stable star for each EoS. So the experimental data shown in Fig. 11.5 provide strong constraints on the equation of state. Roughly speaking, the matter should not be too compressible, in order to withstand the forces of gravity and support the observed values of the mass. Of course, the same value of the maximal mass may correspond to quite a different EoS. For that reason, another observable, a better observable being the radius of the star or the red-shift, is badly needed to constrain it significantly. We will not go into the rather vast literature in which various particular EoS have been analyzed.

Coming to the dynamics of the pulsars we note that stars with masses very close to the stability limit have no chance to be created, unless they are members of a close binary with significant mass transfer from the companion star via the accretion process. The reason is that a transition from ordinary (atomic) matter to the nuclear one is connected with such a huge jump in density that it proceeds explosively. A massive protostar looses stability and implodes. Matter falls to the center of the star with a certain velocity and always overshoots the equilibrium configuration, turning for some time to an over-compressed state, which, in turn, may be collapsed by gravity into a black hole. (The recoil from this maximal compression stage produces the outgoing shock and eventually the spectacular supernovae we observe.) This phenomenon was studied for a number of equations of state by P. D. Morley and M. B. Kislinger [814], who concluded that it reduces the maximal allowed mass $M_{\max}$ for the given equation of state by 10–20%.

Another transitory dynamical phenomenon, which explains why the most recent supernova 1987a most probably left no compact star, is additional pressure added to

---

[4]Question: how exactly does this equation reduce to the ordinary hydrostatic equation in the non-relativistic case?

the system while neutrinos are trapped in the system, see a review by Prakash [807]. When they finally leave, about 10 sec later, the pressure falls below critical, and gravitational collapse takes over.

11.2.2.   *Phases of matter in compact stars*

Let me warn the reader that in this section we will discuss the issue in two rounds, first the "naive" one and then the "refined" one. The naive starts with the natural assumption that matter is electrically neutralized (by the necessary amount of electrons) so that the following condition should be satisfied:

$$\frac{2}{3}n_u - \frac{1}{3}n_d - \frac{1}{3}n_s - n_e = 0. \tag{11.21}$$

Another necessary condition is that the non-strange matter is in equilibrium according to weak transitions between quarks. Neutrinos are assumed to leave the system and are therefore ignored. Thus, the chemical potentials for various quarks and electrons are related by

$$\mu_u + \mu_e = \mu_d = \mu_s. \tag{11.22}$$

These two conditions, plus the electro-neutrality leave only one free parameter out of the four μ.

So the following simple picture arises: the EoS has one parameter, such as μ_B and the pressure rises monotonically. The star is made of the dense quark matter below the nuclear matter, and at the boundary the pressures and the chemical potentials are equal. There is no place for a "mixed phase": there is simply a density jump at the surface of the quark matter, like that at the surface of the ocean.

This simple picture has been used by dozens of papers (including mine, I should confess), before N. Glendenning [815] found a serious loophole in the argument. His point is that in the *mixed* phase one should not actually insist that each (quark and nuclear) component is separately electrically neutral: only their mixture should be neutral. In fact, quark matter wants to be somewhat positively charged, due to s-quark suppression by its larger mass, making the hadronic phase a bit negative. Minimizing the resulting Coulomb energy one finds the scale and structure of the mixed phase. The result is that the mixed phase itself goes through 5 phases: (bulk q, drops of N), (bulk q, rods of N), (sheets of q and N), (bulk N, rods of q),(bulk N, drops of q). Details can be found in multiple works, let me give one example from Ref. [802]. A bag constant of $B^{1/4} = 180$ MeV complements some "realistic" nuclear EoS, this leads to the energy per baryon of strange matter at 1100 MeV, well above the energy per nucleon in ^{56}Fe ($\approx$ 930 MeV). Most interestingly, the transition to quark matter sets in already at a density $\epsilon = 2.3\,\epsilon_0$ ($\epsilon_0 \approx m_N n_0$ is the energy density of nuclear matter) and it ends, i.e. the pure quark phase begins, at $\epsilon \approx 15\,\epsilon_0$, which is larger than the central density encountered in the maximum-mass star model constructed from this EoS. So, in this scenario almost the entire star is in the mixed phase. Of course, this is just an example. We do not really know such important parameters as the surface tension between the two phases, etc.

Let us now turn to the question of whether all compact stars can be made of strange quark matter entirely, as would be the case if it is the true ground state of matter. Unfortunately the bulk properties of models of neutron and strange stars of masses that are typical for neutron stars, $1.1 \lesssim M/M_\odot \lesssim 1.8$, are relatively similar and therefore do not allow the distinction between the two possible pictures. The situation changes however as regards the possibility of fast rotation of strange stars. This has its origin in the completely different mass–radius relations of neutron and strange stars. As a consequence of this the entire family of strange stars can rotate rapidly — i.e. considerably below one millisecond — not just those near the limit of gravitational collapse to a black hole as is the case for neutron stars. So for the moment one cannot exclude or confirm it considering the global parameters of the star.

For the purposes of this book it is more instructive to show how the cooling history can be used, as it should be quite different for neutron stars and strange

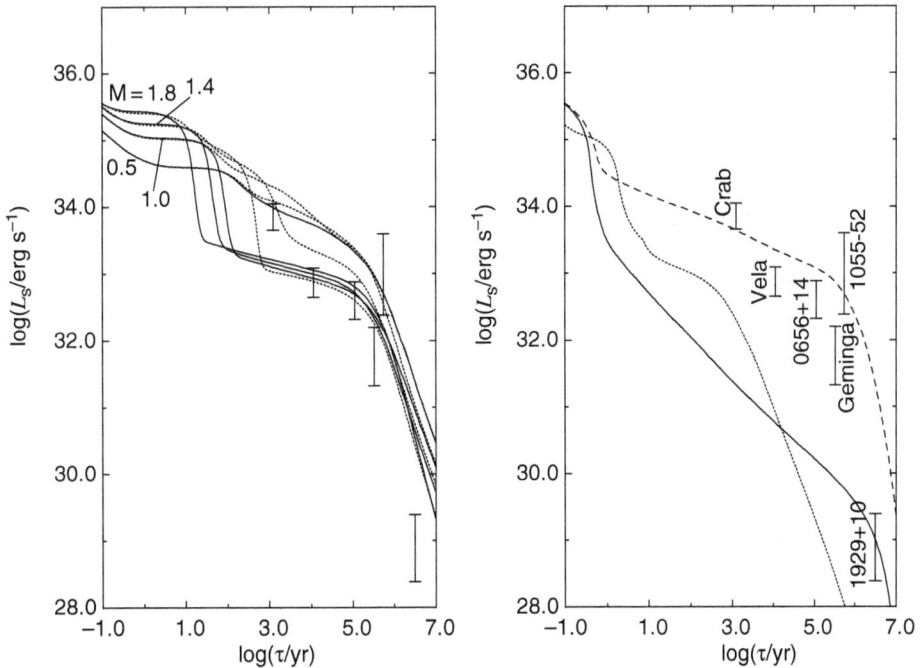

Fig. 11.6. Left panel: Cooling of neutron stars with pion (solid curves) or kaon condensates (dotted curve). Right panel: Cooling of $M = 1.8$ solar mass strange stars with crust. The cooling curves of lighter strange stars, e.g. $M = 1$ solar mass differ only insignificantly from those shown here. Three different assumptions about a possible superfluid behavior of strange quark matter are made: no superfluidity (solid), superfluidity of all three flavors (dotted), and superfluidity of up and down flavors only (dashed). The vertical bars denote luminosities of observed pulsars.

stars. In the next figure from Ref. [802] one finds some data on the star temperature versus its age, compared to theoretical predictions, for neutron stars (left) and strange stars (right). The rapid variations seen in the curves are due to the onset of different forms of color superconductivity; needless to say, the data set is still too small to conclude anything. The physical reason for that is that superconducting gaps effectively eliminate excitations near the Fermi surface. With more data one should be able to select the right theory, and even now some of those seem to be excluded already.

One more interesting issue is the possible existence of *twins*, two stars with the same mass but different internal structure, see Glendenning and Kettner [816]. A possibility of "catastrophic rearrangement" between those has been discussed by Mishustin *et al.* [817], which is one more potential signal.

11.3.   Color superconductivity in very dense quark matter

11.3.1.   *Brief introduction to superconductivity*

In this brief introduction to the superconductivity phenomenon we will explain two main facts: (i) that the so called Cooper pair with zero momentum is an exceptional channel; and (ii) that the interaction in it is analogous to the $1d$ case, when any attraction no matter how weak creates a bound state.

The kinematics of scattering of two quasiparticles $|p| > p_F$, or quasi-holes $|p| < p_F$, or a particle–hole with initial momenta $\vec{p}_1, \vec{p}_2$ is displayed in Fig. 11.7. Two spheres shown in thick solid lines have the radius of Fermi momentum p_F, and a couple of concentric nearby spheres shown in thin lines indicate the range occupied by all thermal excitations, particles and holes respectively. If the total momentum $\vec{p}_1 + \vec{p}_2$ is non-zero, as in the left figure (a), all possible phase space for the final momentum $\vec{p}_3$ consists of two small rhombic intersections of those spherical

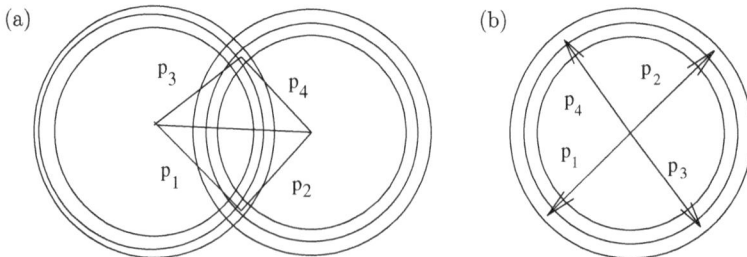

Fig. 11.7. The left side of the figure shows the kinematics of scattering $\vec{p}_1 + \vec{p}_2 = \vec{p}_3 + \vec{p}_4$. When all four momenta are forced to be close to the surface of the Fermi sphere (between two thin circles around the thick circle), the final momentum has to be inside small diamond-shape overlap regions. The exception however is the case when the total momentum is zero: this case is shown in the right figure. In this case the whole region near the Fermi surface is open for final momenta.

shells (in $2d$) or a thin circle obtained by rotation of those intersections around the direction of the total momentum in $3d$. As $T \to 0$ the available volume is $\sim T^2$.

The situation is different if the total momentum is zero, $\vec{p}_1 + \vec{p}_2 = 0$. As shown in Fig. 11.7b, in this case the whole spherical region near the Fermi sphere is available, and as $T \to 0$ the volume goes to zero as $\sim T^1$. In conditions in which ordinary superconductors are used, the temperature-to-Fermi-energy ratio which determines the width of these shells is very small, $T/\epsilon_F \sim 10^{-4}$, and so this difference is very important.

The second point we would like to make is that in the latter situation the direction of $\vec{p}_1$ is irrelevant (as soon as $\vec{p}_2 = -\vec{p}_1$) and in low energy s-wave scattering the area of the Fermi sphere just enters as a common factor. It does not really matter whether we consider the $2d$ or $3d$ problem: only the $1d$ component of the momentum, the *radial* one, really matters and enters the equations. Therefore, all the equations for superconductivity (such as the famous BCS gap equation) are basically the same for any dimensions.

If so, one should now recall a fact we all learned in quantum mechanics courses: in $1d$, unlike higher dimensions, any attractive potential creates at least one bound state. In the field theory language this means that one should calculate a particle-particle scattering "fish" diagram and find that it is logarithmically divergent [463]. A more modern way of doing the same thing is to use the renormalization group approach [780] which sums up the logs. In a book discussing QCD the latter approach is much more appropriate, since it nicely emphasizes the analogy between the asymptotic freedom — the running QCD coupling — with the *running BCS coupling in a superconductor*. We will very closely follow here the original paper by Son [829], who used this method to derive the gap equation for asymptotically high densities.

Let me start by reminding the reader the RG setting at finite density. We would like to formulate an effective theory which contains only the fermionic degrees of freedom located *in a thin shell surrounding the Fermi surface*, $|\epsilon_p| < \delta$: all other components of the pertinent fields have to be integrated out. As noted above, the only relevant interaction between the fermions is the scattering of pairs with opposite momenta. Let us introduce the scattering amplitude from a pair with momenta $(\mathbf{p}, -\mathbf{p})$ to another pair with momenta $(\mathbf{k}, -\mathbf{k})$,

$$f(\theta) \equiv f(\mathbf{p}, \mathbf{k}) = T(\mathbf{p}, -\mathbf{p} \to \mathbf{k}, -\mathbf{k}).$$

To avoid complications with statistics, we will assume that the two particles are of different flavors (or other quantum number). As observed by Landau, near the Fermi surface the scattering amplitude depends only on the angle θ between $\mathbf{p}$ and $\mathbf{k}$. A positive f corresponds to a repulsive interaction, and a negative f means attraction.

In the spirit of the renormalization group, let us now integrate out all states with $e^{-1}\delta < |\epsilon_p| < \delta$. According to quantum mechanics, the scattering through virtual states in this region gives a correction to the scattering amplitude. To account for these virtual processes, we need to correct the scattering amplitude:

$$f(\mathbf{p}, \mathbf{k}) \to f(\mathbf{p}, \mathbf{k}) - \sum_i \frac{T(\mathbf{p}, -\mathbf{p} \to i)T(i \to \mathbf{k}, -\mathbf{k})}{E_i - 2\epsilon_{\mathbf{p}}}, \tag{11.23}$$

where the sum is over all intermediate states i that should be integrated out. We assume that the initial and final particles are almost exactly located at the Fermi surface, so $\epsilon_{\mathbf{p}} = \epsilon_{\mathbf{k}} = 0$.

The scattering through an intermediate state can be of two types:

1. The pair $(\mathbf{p}, -\mathbf{p})$ can scatter to an intermediate pair $(\mathbf{p}', -\mathbf{p}')$, which then goes to $(\mathbf{k}, -\mathbf{k})$. In this case, the intermediate state i is that with two particle excitations with momenta $\pm\mathbf{p}'$. The Pauli principle requires that $\mathbf{p}'$ is located above the Fermi surface. This state has $E_i = 2\epsilon_{\mathbf{p}'}$ and $T(\mathbf{p}, -\mathbf{p} \to i) = f(\mathbf{p}, \mathbf{p}')$, $T(i \to \mathbf{k}, -\mathbf{k}) = f(\mathbf{p}', \mathbf{k})$.
2. Alternatively, first a pair of particles inside the Fermi sea with momenta $(\mathbf{p}', -\mathbf{p}')$ can scatter to make the final pair $(\mathbf{k}, -\mathbf{k})$, and then the initial pair $(\mathbf{p}, -\mathbf{p})$ scatters to fill the holes vacated by the pair $(\mathbf{p}', -\mathbf{p}')$ in the Fermi sphere. In this case, the intermediate state i consists of six elementary excitations: four particles with momenta $\pm\mathbf{p}$ and $\pm\mathbf{k}$, and two holes with momenta $\pm\mathbf{p}'$ located below the Fermi surface, $p' < \mu$. In this case, $E_i = -2\epsilon_{\mathbf{p}'}$, $T(\mathbf{p}, -\mathbf{p} \to i) = f(\mathbf{p}', \mathbf{k})$, $T(i \to \mathbf{k}, -\mathbf{k}) = f(\mathbf{p}, \mathbf{p}')$.

Now that Eq. (11.23) becomes

$$f(\mathbf{p}, \mathbf{k}) \to f(\mathbf{p}, \mathbf{k}) - \int_{\mathbf{p}'} \frac{f(\mathbf{p}, \mathbf{p}')f(\mathbf{p}', \mathbf{k})}{2|\epsilon_{\mathbf{p}'}|}, \tag{11.24}$$

where the integration is over all $\mathbf{p}'$ satisfying $e^{-1}\delta < |p - \mu| < \delta$. The integral over $|\mathbf{p}'|$ can be taken, and Eq. (11.24) reads,

$$f(\mathbf{p}, \mathbf{k}) \to f(\mathbf{p}, \mathbf{k}) - \frac{\mu^2}{2\pi^2} \int \frac{d\hat{\mathbf{p}}'}{4\pi} f(\mathbf{p}, \mathbf{p}')f(\mathbf{p}', \mathbf{k}),$$

where the integration is over the directions of $\mathbf{p}'$.

We have thus derived the RG equation,

$$\frac{d}{dt}f(\mathbf{p}, \mathbf{k}) = -\frac{\mu^2}{2\pi^2} \int \frac{d\hat{\mathbf{p}}'}{4\pi} f(\mathbf{p}, \mathbf{p}')f(\mathbf{p}', \mathbf{k}), \tag{11.25}$$

where $t = -\ln \delta$ goes to $+\infty$ at the Fermi surface. Such RG evolution of the scattering amplitude toward the Fermi surface is very similar to the RG equation for the QCD coupling growing in the IR.

As in Landau–Fermi liquid theory, it is convenient to expand the scattering amplitude in partial waves,

$$f(\theta) = \sum_{l=0}^{\infty} (2l+1) f_l P_l(\cos\theta),$$

or, inversely, $f_l = \frac{1}{2} \int_0^\pi d\theta \sin\theta \, P_l(\cos\theta) f(\theta)$. The partial-wave amplitudes f_l RG-evolve independently,

$$\frac{df_l}{dt} = -\frac{\mu^2}{2\pi^2} f_l^2, \tag{11.26}$$

leading to the solution

$$f_l(t) = \frac{f_l(0)}{1 + (\mu^2/2\pi^2) f_l(0) t}.$$

Note that if at $t = 0$ all $f_l > 0$, the interaction is repulsive in all channels, then the four-fermion interaction vanishes at the Fermi surface. However, if one of $f_l(0)$ is negative, it will develop a singularity (analogous to the Landau pole in QED RG) at $t = -2\pi^2/\mu^2 f(0)$, which is known as the Cooper pairing. The pole is reached first by the channel having the largest negative $f_l(0)$.

This singularity indicates the BCS instability of the Fermi surface with respect to any attractive interaction. The BCS gap is proportional to the energy scale at which the Landau pole is reached, which is

$$\Delta \sim \exp\left(-\frac{2\pi^2}{\mu^2 f(0)}\right). \tag{11.27}$$

11.3.2. *BCS pairing and Gorkov abnormal Green functions*

In order to study competing instabilities we use the standard Nambu–Gorkov formalism, in which the propagator is written as a matrix in the space of all possible

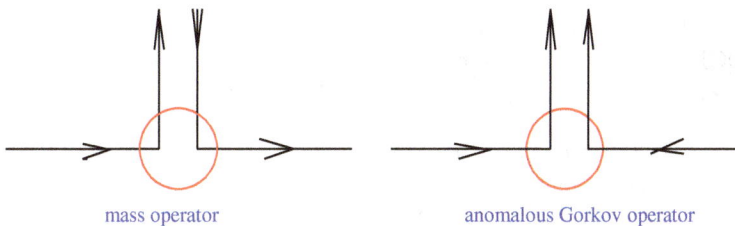

mass operator   anomalous Gorkov operator

Fig. 11.8. Two mass operators; the usual and the anomalous Gorkov one.

pair condensates. The BCS channel is described by the 2×2 matrix

$$
\hat{G}_{\text{BCS}} = \begin{pmatrix} \langle c_{k\uparrow} \, c_{k\uparrow}^\dagger \rangle & \langle c_{k\uparrow} \, c_{-k\downarrow} \rangle \\ \langle c_{-k\downarrow}^\dagger \, c_{k\uparrow}^\dagger \rangle & \langle c_{-k\downarrow}^\dagger \, c_{-k\downarrow} \rangle \end{pmatrix}
$$

$$
\equiv \begin{pmatrix} G(k_0, \vec{k}, \Delta) & \bar{F}(k_0, \vec{k}, \Delta) \\ F(k_0, \vec{k}, \Delta) & \bar{G}(k_0, -\vec{k}, \Delta) \end{pmatrix} . \tag{11.28}
$$

The propagator has the form

$$
\hat{G}_{\text{BCS}} = \frac{1}{G_0^{-1} \bar{G}_0^{-1} - \Delta \bar{\Delta}} \begin{pmatrix} \bar{G}_0^{-1} & -\Delta \\ -\bar{\Delta} & G_0^{-1} \end{pmatrix} , \tag{11.29}
$$

where

$$
G_0 = \frac{1}{k_0 - \epsilon_k + i\delta_{\epsilon_k}} , \qquad \bar{G}_0 = \frac{1}{k_0 + \epsilon_k + i\delta_{\epsilon_k}} \tag{11.30}
$$

are the free particle propagator and its conjugate at finite chemical potential. Here, $\epsilon_k = \omega_k - \mu_q$ and $\delta_{\epsilon_k} = \text{sgn}(\epsilon_k)\delta$ determines the pole position. From this equation we can read off the diagonal and off-diagonal components of the Gorkov propagator. The off-diagonal, anomalous, propagator is

$$
F(k_0, \vec{k}, \Delta) = \frac{-\Delta}{(k_0 - \epsilon_k + i\delta_{\epsilon_k})(k_0 + \epsilon_k + i\delta_{\epsilon_k}) - \Delta^2} . \tag{11.31}
$$

The anomalous self energy Δ is determined by the gap equation

$$
\Delta = (-i)\alpha_{pp} \int \frac{d^4 p}{(2\pi)^4} F(p_0, \vec{p}, \Delta) . \tag{11.32}
$$

Here α_{pp} is the effective coupling in the particle–particle channel.

11.3.3.   *Three mechanisms of quark pairing*

So, the superconductivity needs an attractive interaction between fermions at the Fermi sphere. To find one in a metal was a non-trivial task, since naive expectations were that electrons obviously always repel each other. Eventually it was understood that the Coulomb forces can be screened at large enough distances, where finally attractive phonon exchanges will take over.

In QCD we do not have such a problem. In fact, there are even three strong contenders for this role, which we discuss subsequently:

(i) the electric (Coulomb) interaction;
(ii) magnetic interaction; and
(iii) instanton-induced 't Hooft interaction in the qq channel.

The *Coulomb mechanism* is the most obvious one, and indeed that was the mechanism discussed in the earliest works on the subject [818–821] and also in

Ref. [836]. Unlike two electrons, two quarks can be in two color representations of $SU(3)$, the symmetric 6 and antisymmetric $\underline{3}$, and in the latter they have a *negative* scalar product of their relative color vectors.[5] It means their charges are in fact opposite, and thus they attract each other, as electron and positron do.

The problem with this mechanism is that electric forces are strongly screened by matter, at Debye scale $g\mu$ as we have seen in Section 8.2. Thus, although for superconductivity any attraction will do, they can only produce quite a weak one only.

The second issue to be discussed historically was the *instanton mechanism*, first in the context of the simplest two-flavor[6] problem [827, 828]. These two papers, submitted to the preprint archive on the same day, have created an avalanche of hundreds of papers on the subject, which continues to this day.

In order to explain why it is so let me first mention that forces induced by instantons are so much stronger, than their presence increased the expected gaps (and T_c) from the 1 MeV scale to that of $\sim$100 MeV (and $T_c \sim$ 50 MeV). Many people (including the referee of our paper) could not believe it, but in hindsight it should hardly be surprising, since the *same* interaction in the $\bar{q}q$ channel is responsible for chiral symmetry breaking, producing a gap (the constituent quark mass) as large as 350–400 MeV.

Furthermore, in the *two-color* QCD, there is the so called Pauli–Gursey symmetry which relates these two condensates. So at high density the chiral condensate $\langle \bar{q}q \rangle$ simply rotates into the superconductor one $\langle qq \rangle$, while the gap remains the same.

The symmetries of the 2SC phase are similar to the electroweak part of the Standard Model, with the condensed scalar isoscalar ud diquark operating as Higgs. The colored condensate breaks the color group, making 5 out of 8 gluons massive. The three-flavor-like phase, CSC3, is brand new: it was proposed in Ref. [836] based on the one-gluon exchange interaction, but in fact it is favored by instantons as well [837]. Its unusual features include *color-flavor locking* and *coexistence* of both types of condensates, $\langle qq \rangle$ and $\langle \bar{q}q \rangle$. It combines features of the Higgs phase (8 massive gluons) and of the usual hadronic phase (8 massless "pions").

Finally, (historically) the last mechanism of pairing in dense quark matter is the *magnetic one* pointed out by Son [829]. Indeed, at asymptotically high density the electric part of the one-gluon exchange is Debye screened. The instantons in this case are Debye screened as well, as they have (virtual) electric fields.

The pairing due to magnetic one-gluon exchange forces is interesting by itself, as a quite peculiar example. Note that in this case one has to take care of *time delay*

---

[5]Note that it is the same representation as for a pair of quarks inside a baryon, which classically can be pictured as three color vectors adding to zero, with an angle 120° between each pair. Thus in a heavy quark baryon, like the bbb ones, one would have a nearly-Coulomb three-body problem.

[6]The reader who came that far in this book surely by now knows that an 't Hooft interaction between two quarks of the same flavor does not exist.

effects of the interaction. As we will see below, the result is the indefinitely growing gaps at large $\mu > 10\,\text{GeV}$, as $\Delta \sim \mu \exp(-3\pi^2/\sqrt{2}g(\mu))$.

11.3.4. *Magnetic pairing in asymptotically dense matter*

In this section we follow closely the original paper by Son [829], who very nicely explained the modifications which appear in the case of one-gluon exchange in dense matter in the RG equation.

The first step is to note that the gluon propagator in the magnetic sector can be approximated as

$$D(q_0, q) = \frac{1}{q^2 + \frac{1}{2}\pi m_D^2 |q_0|/q} \tag{11.33}$$

if $q_0 \ll q \ll \mu$. The term $\frac{1}{2}\pi m_D^2 |q_0|/q$ comes from the imaginary part of the ordinary loop, known as Landau damping. However, in the static limit $q_0 = 0$, the magnetic field is not screened. If $q_0 \neq 0$, the field is said to be "dynamically screened"[7] on the scale $q \sim m_D^{2/3} q_0^{1/3}$.

Now let us return to the RG formalism and apply it to the magnetic interaction mediated by one-gluon exchange. We will see immediately that we have serious trouble with the very soft gluons. Indeed, on the Fermi surface, the tree-level small-angle ($\theta \ll 1$) scattering amplitude, due to one-gluon exchange, is

$$f_{\text{tree}}(\theta) = -\frac{2g^2}{3}\left(\frac{1}{\mu^2\theta^2 + m_D^2} + \frac{1}{\mu^2\theta^2}\right). \tag{11.34}$$

The two contributions in the r.h.s. come from the electric and the magnetic interaction, respectively (again, the factor $2/3$ comes from considering only the $\bar{\mathbf{3}}$ channel). All partial amplitudes diverge logarithmically. For example,

$$f_0 = \frac{1}{2}\int_{q_{\min}/\mu}^{\pi} d\theta \sin\theta\, P_l(\cos\theta)\, f(\theta) \approx -\frac{g^2}{3}\ln\frac{\mu}{q_{\min}}, \tag{11.35}$$

where $q_{\min}$ is the smallest allowed momentum exchange that one has to put in by hand to make f_0 finite.

Still we apply the RG procedure, starting the RG evolution at $t = 0$ with $\delta \sim m_D$. The evolution stops when δ is of the order of the gap, so typically $\delta \ll m_D$. At the tree level, the fermions interact via one-gluon exchange, characterized by the momentum of the gluon $(q_0, \mathbf{q})$. Since all fermions have energy less than δ, the energy of the gluon q_0 is naturally of order or less than δ, while the momentum exchange q can be anywhere between 0 and 2μ.

Let us divide the four-fermion interaction that arises from the one-gluon exchange into "instantaneous" and "non-instantaneous" parts. The instantaneous

---

[7]The statements above are the same as in Chapter 8 for QGP. The only difference at high density and zero T is that now there is no nonperturbative "magnetic mass".

interaction is mediated by the gluons that have momenta $q > q_\delta \equiv m_D^{2/3}\delta^{1/3}$. The Landau damping for these gluons is negligible, $m_D^2|q_0|/q \lesssim q^2$. The gluon propagator, which is now simply q^{-2}, does not depend on q_0, which means that the four-fermion interaction they mediate can be considered as instantaneous. This part of the interaction is of the familiar type and will be treated in the conventional way. In particular, one can characterize this part by the partial-wave amplitudes f_l. For $q \lesssim q_\delta$, the Landau damping can no longer be neglected. This part of the interaction has a considerable temporal retardation and should be treated separately.

Now let us reduce δ by a factor of $1/e$ by integrating out fermion degrees of freedom with energy between $e^{-1}\delta$ and δ. During this process the following will occur:

1. The partial-wave amplitudes f_l obtain the conventional renormalization, as written in Eq. (11.26).
2. One could ask if the non-instantaneous coupling is renormalized during this integration. To answer this question, one should compute the correction to the non-instantaneous interaction that comes from integrating out the fermion degrees of freedom. In the Appendix of Ref. [829] it is shown that the non-instantaneous part of the interaction does not get renormalized.
3. Most importantly, and what makes our RG distinctive, *part of the non-instantaneous interaction becomes instantaneous.* Specifically, the gluon exchange with q lying in the interval $(e^{-1/3}q_\delta, q_\delta)$, which was formerly treated as non-instantaneous, now becomes part of the instantaneous interaction and contributes to f_l. Simply speaking, our criterion of what to consider as instantaneous has become more inclusive, since we are now looking at a smaller energy scale, corresponding to a larger time scale.

How much of the non-instantaneous part of the interaction transfers to the instantaneous part during one step of the RG? According to Eq. (11.34) and the non-renormalization of the non-instantaneous interaction, the increment in $f(\theta)$ has the form,

$$\Delta f(\theta) = -\frac{2g^2}{3}\frac{1}{\mu^2\theta^2}, \quad \text{for} \quad e^{-1/3}\frac{\delta}{\mu} < \theta < \frac{\delta}{\mu}, \tag{11.36}$$

and vanishes outside this window of θ. For definiteness, let us concentrate our attention to the s-wave amplitude f_0. This amplitude obtains a constant additive contribution from the soft sector at each RG step,

$$\Delta f_0 = \frac{1}{2}\int_{e^{-1/3}\delta\mu^{-1}}^{\delta\mu^{-1}} d\theta \sin\theta \, \Delta f(\theta) = -\frac{g^2}{9\mu^2}.$$

Therefore, the RG group equation for f_0 now becomes

$$\frac{d}{dt}f_0 = -\frac{g^2}{9\mu^2} - \frac{\mu^2}{2\pi^2}f_0^2. \tag{11.37}$$

The second term in the r.h.s. is the familiar term that gives rise to the BCS effect for short-range interactions. What is new is the first term, which takes into account the fact that softer and softer gluon exchanges contribute to f_l. The non-instantaneous part of the interaction can be considered as an infinite pool, which continuously replenishes the instantaneous part during the RG evolution. Clearly, this should speed up the approach to the Landau pole.

To secure a solution we also need to specify an initial condition on f_0. Recall that $t = 0$ corresponds to $\delta \sim m_D$; from Eq. (11.34) one finds, to the leading logarithm,

$$f_0(0) = -\frac{2g^2}{3\mu^2} \ln \frac{1}{g}. \tag{11.38}$$

The solution to Eq. (11.37) with the initial condition (11.38) is

$$f_0(t) = -\frac{\sqrt{2}\pi g}{3\mu^2} \tan\left[\frac{g}{3\sqrt{2}\pi}\left(t + 6\ln\frac{1}{g}\right)\right].$$

The coupling f_0 hits the Landau pole when the argument of the tangent is equal to $\pi/2$. This happens when

$$t = \frac{3\pi^2}{\sqrt{2}g} - 6\ln\frac{1}{g}.$$

The Fermi liquid description, thus, breaks down at the energy scale

$$\Delta \sim m_D e^{-t} \sim \mu g^{-5} \exp\left(-\frac{3\pi^2}{\sqrt{2}g}\right), \tag{11.39}$$

which will be interpreted as the scale of the gap. Notice that the gap is proportional to $e^{-c/g}$, which is parametrically larger than the naive estimate e^{-c/g^2} at small g. The reason for this enhancement is obviously the singularity of the magnetic interaction.

We will not describe here the explicit equations for the gap, known as the *Eliashberg equation*, borrowed from the physics of electron–phonon systems. It does show that the gap has the rather unusual behavior

$$\Delta \sim \mu \exp\left(-\frac{3\pi^2}{\sqrt{2}g\mu}\right). \tag{11.40}$$

Although the coupling at large chemical potential $g\mu$ is weak, the suppression due to the exponent is weaker than the power in front, so the resulting gap (in absolute units) *grows* with μ. A "realistic model" consistent with Son's regime for the gap at large μ is shown in Fig. 11.11b. This regime probably only starts for $\mu \gtrsim 10\,\text{GeV}$ or so, and thus is rather academic by itself.

11.3.5. *Instanton-induced color superconductivity*

For quark matter inside stars, with $\mu \lesssim 0.5\,\text{GeV}$, the main quark pairing interaction is presumably instanton-induced, same as for chiral symmetry breaking in the QCD vacuum.

My interest was initiated by finding [239] that in the instanton liquid model even without *any* quark matter, the *ud scalar diquarks* are very deeply bound, by an amount comparable to the constituent quark mass. So, the phenomenological manifestations [112] of such diquarks have in fact deep dynamical roots: they follow from the same basic dynamics as the "superconductivity" of the QCD vacuum, the chiral (χ-)symmetry breaking. These spin-isospin-zero diquarks are related to pions, and should be quite a robust element of the nucleon (octet baryons) structure.[8]

Another convincing argument in favor of the existence of *deeply bound scalar diquarks* comes from the bi-color ($N_c = 2$) theory we discussed in Section 1.7.2. In it the scalar diquark is degenerate with pions. By continuity from $N_c = 2$ to 3, a trace of it should exist in real QCD.[9]

The corresponding sigma model describing this χ-symmetry breaking was worked out in Ref. [828]: for further development see Ref. [846]. As argued in Ref. [828], in this theory the critical value of the transition to color superconductivity is simply $\mu = m_\pi/2$, or zero in the chiral limit. The diquark condensate is just the rotated $\langle \bar{q}q \rangle$ one, and the gap is the constituent quark mass. Recent lattice works on two-color QCD at high density display it in great detail, building confidence for other cases.

Explicit calculations with the instanton-induced forces for 2-flavor $N_f = 2$, $N_c = 3$ QCD have been made in two simultaneous[10] papers [827, 828]. Indeed, very robust Cooper pairs and gaps $\Delta \sim 100\,\text{MeV}$ were found. From then on, the field is booming, with about 500 papers instead of about 5 before that.

Instantons lead to the following amusing *triality*: there are three attractive channels which compete, namely:

- the instanton-induced attraction in $\bar{q}q$ channel leading to χ-symmetry breaking;
- the instanton-induced attraction in qq which leads to color superconductivity;
- the *light-quark-induced* attraction of $\bar{I}I$, which leads to pairing of instantons into "molecules" and a Quark–Gluon Plasma (QGP) phase without *any* condensates.

---

[8] As opposed to Δ (decuplet) baryons.

[9] The instanton-induced interaction strength in the diquark channel is $1/(N_c - 1)$ of that for the $\bar{q}\gamma_5 q$ one. It is the same at $N_c = 2$, zero for large N_c, and is exactly in between for $N_c = 3$.

[10] Submitted to hep-ph on the same day. It happened so not by chance but by agreement between us: we agreed not to discuss anything till the first publication day, and fully collaborate after it. It worked nicely, and I can recommend this arrangement.

Of course, the "molecules" can be there in any phase, for example they will be important for the CFL phase to be discussed below.

Technically, the instanton calculus at finite density is similar to that at finite temperature, and instead of discussing it in detail we delegate the reader to our comprehensive work [837]. Its main element is the instanton zero mode at non-zero μ, and all that is determined after that is basically the same. The full-scale simulations are not feasible because the Euclidean determinant with chemical potential is not real. As in the case of the lattice, there are only exceptional cases, like the two-color one, when the non-zero-μ determinant is real and can be used as a weight.

11.3.6. *Two-flavor QCD: 2SC phase*

The instanton-induced forces are very different depending on how many light flavors there are in the theory under consideration. For two flavors, the 't Hooft vertex is a quasi-local four-fermion operator, similar to what was used in the original BCS model for ordinary superconductors. For two flavors there is only one pair, ud, with spin and isospin $S = I = 0$. The diquark has color, which selects a direction in color space. This phase, known as 2SC phase, gaps four Fermi spheres out of 6 available (3 colors times 2 flavors) and does not touch the other two.

The first studies of instanton-induced CSC were made for this theory [827,828]. The gap around 100 MeV was found, and the main issue is the *competition* between the usual χ-symmetry breaking (pairing in the $\bar{q}q$ channel) with CSC (pairing in the qq channel). In the corresponding high density phase we called 2SC, the chiral symmetry is simply restored.

In all these works one more possible phase (intermediate between vacuum and 2SC) — *Fermi gas of constituent quarks*, with both $M, \Delta \neq 0$ — was unstable. However in the last more refined calculation [837] it obtains a small window. Its features are amusingly close to those of nuclear matter: but it is not, of course: to get nucleons one should go outside the mean field. The first attempt to do so in Ref. [837] was for another cluster — the $\bar{I}I$ molecules. At $T = 0$ it is however only a 10% correction to the previous results, but is dominant as T grows.

11.3.7. *Three-flavor QCD: CFL phase*

A much more symmetric *Color-Flavor-Locked* (CFL) phase has been discussed for the three flavor QCD $N_f = N_c = 3$ in the second round of papers by the same groups. Alford, Rajagopal and Wilczek [836] gave up instanton-based interaction in favor of the one-gluon-exchange one, to keep four-fermion operators. They found a very nice phase in which color and flavor indices are locked, so that out of three SU(3) groups (the color plus L- and R-handed flavor ones) one residual SU(3) remains. All nine Fermi spheres are gapped, in a symmetric way. A quite tricky feature of their approach was the fact that in their model there is no quark condensates, $\langle \bar{q}q \rangle = 0$, but the chiral symmetry is nevertheless broken.

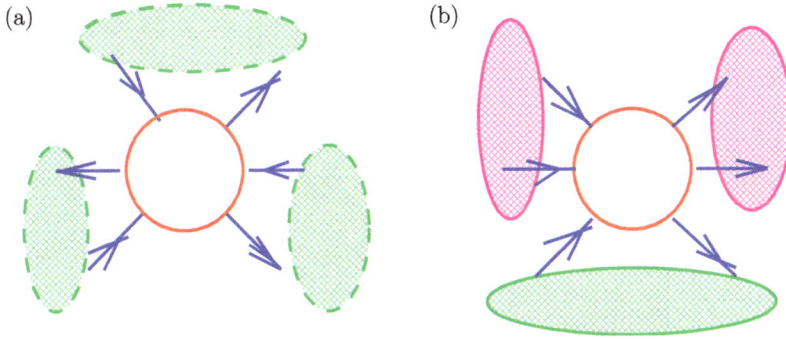

Fig. 11.9. Two contributions to the potential energy for the three-flavor theory, including the cube of the $\langle \bar{q}q \rangle$ and the combination of both types of condensates $|\langle qq \rangle|^2 \langle \bar{q}q \rangle$.

In a second large paper our group [837] has solved the 3-flavor instanton-induced model. It turned out that out of all possible phases the most symmetric CFL wins in this case as well. Furthermore, the chiral-symmetry violating condensates $\langle \bar{q}q \rangle$ are non-zero in this case. It is easy to understand why this happens: the "*potential energy*" in such an approximation is the interaction Lagrangian convoluted with all possible condensates. Specifically, the instanton-induced vertex for $N_f = 3$ leads to two types of diagrams shown in Fig. 11.9, with (a) $\langle \bar{q}q \rangle^3$ and (b) $\langle qq \rangle^2 \langle \bar{q}q \rangle$.

Then one minimizes the potential over all condensates and gets *gap equations*: the algebra may be involved because masses/condensates are *color-flavor matrices*. The diquark condensate has the structure

$$\langle q_i^a C q_j^b \rangle = \bar{\Delta}_1 \delta_{ia}\delta_{bj} + \bar{\Delta}_2 \delta_{ib}\delta_{ja}, \tag{11.41}$$

where the ij are color and ab flavor indices. It is very symmetric, reducing $SU(3)_c SU(3)_f \rightarrow SU(3)_{\text{diagonal}}$, as mentioned above.

Gaps δ_i and masses σ_i (proportional to $\langle \bar{q}q \rangle$), following from an instanton-based calculation [837], are shown as a function of μ in Fig. 11.10 reproduced from Ref. [837].

Physics issues under discussion for the CFL phase include a fascinating idea of possible *hadron–quark continuity*. As proposed by Schafer and Wilczek [838], the CFL phase not only has the same symmetries as hadronic matter (e.g. broken χ-symmetry), but also very similar excitations. Eight gluons become eight *massive* vector mesons, $3*3$ quarks become $8+1$ "baryons". The eight massless pions remain massless in the chiral limit. Furthermore, photon and gluons are combined into a *massless* γ_{inside}. The calculation of masses and coupling constants for all of them is now in progress. Can these phases be distinguished, and should there be *any* phase transition in the $N_f = 3$ theory, separating it from nuclear matter? There is no need for it, at least from the symmetry point of view.

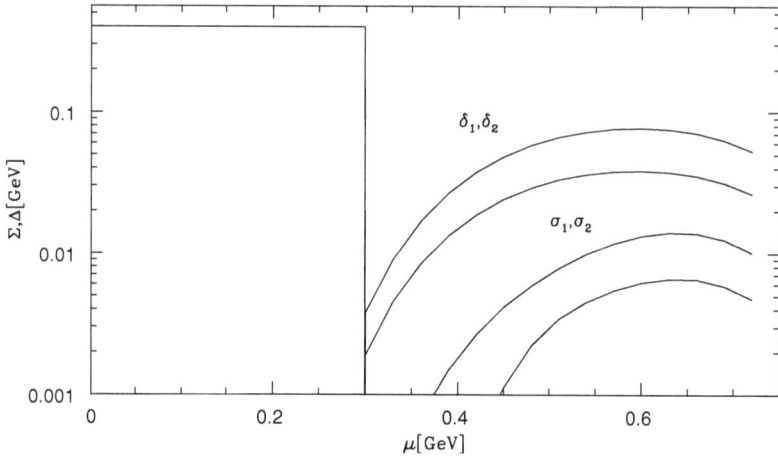

Fig. 11.10. The $\langle qq\rangle$-induced gaps marked by δ and the $\langle \bar{q}q\rangle$-induced gaps marked by σ, as a function of the global (baryon number) chemical potential per quark. There are two gaps of each kind because nine Fermi spheres classify into an octet and singlet, with different gaps. Below the critical value $\mu < \mu_c = 0.3\,\mathrm{GeV}$ only the usual chiral symmetry breaking takes place, and the gap $\sigma \approx 400\,\mathrm{MeV}$ is just the constituent quark mass.

11.3.8. *Excitations of color superconductor*

All variants of CSC have an "internal photon" (a combination of the photon and gluon) for which the condensate is not charged and which therefore is not expelled from matter. This fact initiated the discussion of the following question: is it a superconductor, after all, if light can penetrate through it?

I would still think that the answers to it should be "yes". For example, if one puts a piece of color superconductor into a magnet, it should still levitate: although γ_{inside} is massless, the magnet uses the $\gamma_{\mathrm{outside}}$ field and a part of it is expelled.[11]

A lot of efforts have been addressed to an effective theory of the lowest excitations. For the most symmetric CFL phase, which has an SU(3) multiplet of light pions, one can write down an effective chiral theory of the CFL phase [845]. We will describe it briefly, following Schafer [844].

For excitation energies smaller than the gap the only relevant degrees of freedom are the Goldstone modes associated with the breaking of chiral symmetry and baryon number. The interaction of the Goldstone modes is described by the effective Lagrangian [845]

$$
\begin{aligned}
\mathcal{L}_{\mathrm{eff}} = \frac{1}{4} f_\pi^2 \, \mathrm{Tr}[\nabla_0 \Sigma \nabla_0 \Sigma^\dagger - v_\pi^2 \partial_i \Sigma \partial_i \Sigma^\dagger] + [C\,\mathrm{Tr}(M\Sigma^\dagger) + \mathrm{h.c.}] \\
+ [A_1 \,\mathrm{Tr}(M\Sigma^\dagger)\,\mathrm{Tr}(M\Sigma^\dagger) + A_2\,\mathrm{Tr}(M\Sigma^\dagger M\Sigma^\dagger) \\
+ A_3\,\mathrm{Tr}(M\Sigma^\dagger)\,\mathrm{Tr}(M^\dagger\Sigma) + \mathrm{h.c.}] + \cdots .
\end{aligned}
\tag{11.42}
$$

---

[11]The same would happen with a small piece of Weinberg/Salam vacuum, if one could make magnet with the "original" (or "out-of-this world") field.

Here $\Sigma = \exp(i\phi^a \lambda^a / f_\pi)$ is the chiral field, f_π is the pion decay constant and M is the mass matrix. The field ϕ^a describes pion, kaon, and eta collective modes in the CFL phase. In the CFL phase the flavor and color orientation of the left-handed condensate $X_i^a = \epsilon_{ijk} \epsilon^{abc} \langle (\psi_L)_j^b C (\psi_L)_k^c \rangle$ are locked, $X_i^a \sim \delta_i^a$. The same is true for the right-handed condensate $Y_i^a = \epsilon_{ijk} \epsilon^{abc} \langle (\psi_R)_j^b C (\psi_R)_k^c \rangle$. Low energy excitations of the CFL phase correspond to small fluctuations of X and Y around their equilibrium values. Because color gauge invariance is broken, colored excitations acquire a mass via the Higgs mechanism. The true low energy modes are color neutral fluctuations of X relative to Y, parameterized by $\Sigma = XY^\dagger$. For example, a low energy mode with the quantum numbers of the K^0 is given by $K^0 \sim \epsilon^{abc} \epsilon^{ade} (\bar{u}_R^b C \bar{s}_R^c)(d_L^d C u_L^e)$.

At very high baryon density the effective coupling is weak and the coefficients f_π^2, C, A_i can be determined in perturbative QCD. The pion decay constant is given by (see Fig. 1b of Ref. [847])

$$f_\pi^2 = \frac{21 - 8 \log 2}{18} \left(\frac{p_F^2}{2\pi^2} \right).$$
(11.43)

The coefficient C in (11.42) is related to instantons: at large baryon density $C \sim (\Lambda_{\rm QCD}/p_F)^8$ and the linear mass term is not important. The coefficients A_i of the quadratic mass terms are given by [847]

$$A_1 = -A_2 = \frac{3\Delta^2}{4\pi^2}, \qquad A_3 = 0.$$
(11.44)

Finally, the covariant derivative $\nabla_\mu \Sigma$ was determined in Ref. [849]. The temporal component $\nabla_0 \Sigma$ contains the quark mass matrix,

$$\nabla_0 \Sigma = \partial_0 \Sigma + i \left(\frac{MM^\dagger}{2p_F} \right) \Sigma - i\Sigma \left(\frac{M^\dagger M}{2p_F} \right).$$
(11.45)

We note that Eq. (11.45) is completely fixed by the symmetries of the theory. In addition to that, using the power counting proposed in Ref. [849] we find that the low energy constants A_i are of natural magnitude. This suggests that even though Eqs. (11.43)–(11.45) were obtained using weak coupling methods, the results are more general.

Another interesting work of such series is that by Son, Stephanov and Zhitnitsky [852], which discussed instantons in the limit of very high density for $N_f = 2$. As we discussed above, the $\langle qq \rangle$ condensate is created at very high density by magnetic gluon exchanges, and thus the instantons are relieved from the duty of doing it. As a result, the instanton ensemble is simply a dilute Coulomb gas, with the "photon" being a nearly massless η' exchanged between positive charges (instantons) and negative (anti-instantons). The effective mass and interaction of η' can thus be calculated.

11.3.9. *Quark matter with charge neutrality and realistic m_s*

It was proposed by T. Schafer [861] that the stress created by m_s may be relieved by *adding*[12] strangeness to the system, via the *kaon condensation*, and making the corresponding Fermi momenta come closer. The issue was further studied by

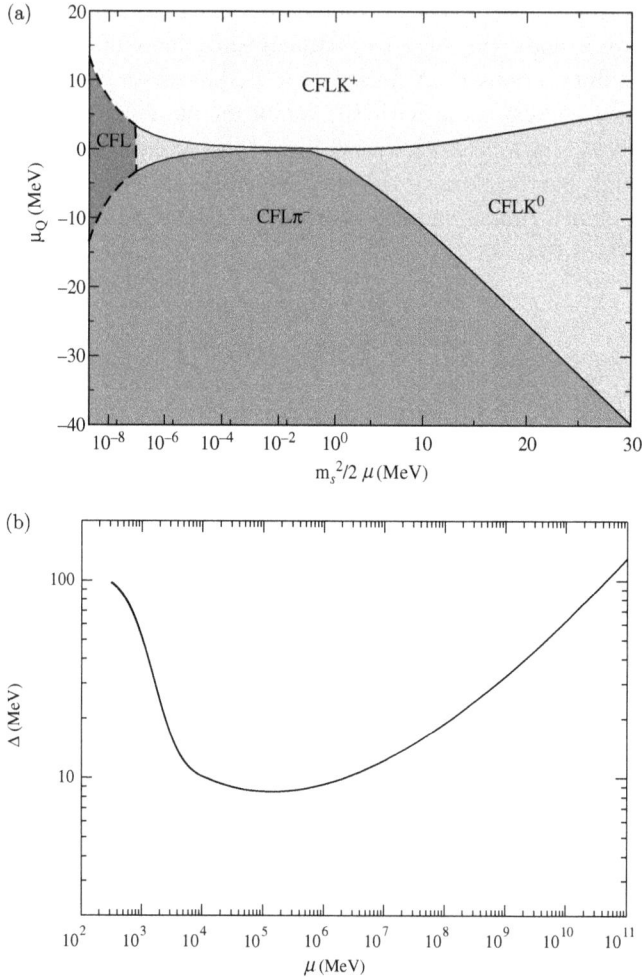

Fig. 11.11. (a) Meson condensed phases in the neighborhood of the symmetric CFL state are shown in the $(m_s^2/2\mu)$–μ_Q plane, where m_s is the strange quark mass (set to 150 MeV), μ is the quark number chemical potential, and μ_Q is the chemical potential for positive electric charge, at five times nuclear density $\mu \sim 400$ MeV and $m_s^2/2\mu \sim 25$ MeV. Solid and dashed lines indicate first- and second-order transitions respectively. (b) A model for superconducting gap Δ as a function of the quark chemical potential μ. The curve agrees with perturbative calculations for $\mu > 10$ GeV.

---

[12]By historical conventions, the strangeness of the s quark is negative, and so literally one might have said here "removing strangeness". Hopefully the reader is not confused by that.

Kaplan and Reddy [862] from whom we borrowed the phase diagram shown in Fig. 11.11a. For reference, their model for an "unstressed" superconducting gap is shown in Fig. 11.11b.

One more option possible is to pair say *us* quarks with different Fermi spheres into Cooper pairs with a non-zero total momentum. This of course is a crystalline phase, originally suggested for ordinary superconductors by Larkin, Ovchinnikov, Fulde and Ferrell [859] and known as *the LOFF phase*. Its possible manifestation for CFL color superconductivity has been worked out by M. Alford, J. Bowers and K. Rajagopal [842]. This phase is found to separate the CFL phase from QGP at the high-T side of it, when the gap becomes weak due to thermal fluctuations.

Finally, if the disparity of the two p_F is too large but the pairing tendency is strong enough, the last resort is the so called *interior gap superconductivity* proposed by Liu and Wilczek [860]. The pairing interactions carve out a gap within the interior of a large Fermi ball, while the exterior surface remains gapless. This defines a system which contains both a superfluid and a normal Fermi liquid simultaneously, with both gapped and gapless quasiparticle excitations.

11.3.10. *Open questions*

The reader may be at this point overwhelmed by many ideas and phases proposed: but the issue of how many species of fermions with non-identical Fermi spheres interact is quite a novel and complicated many-body problem, which so far avoided experimental studies. Hopefully the situation may change in a few years, when trapped fermions can be cooled enough so that not only the celebrated Bose–Einstein condensation, but also superconductivity-related phases would be open to a systematic study.

The ongoing studies of what may happen inside stars, where on top of non-zero strange quark mass one should uphold the charge neutrality of matter, which enforces a reduction of the u quark density relative to d. Probably we have not had the final word on the exact ground state under such conditions yet.

Let me finish this section with a few "homework" questions we have not discussed at all. What is the role of confinement in all these phase transitions? What is the form of the nuclear matter for different quark masses, anyway? Do we have other phases in between, like the diquark–quark phase or (analog of) K condensation, or different crystal-like phases? Is there indeed a (remnant of) the tricritical point which we can find experimentally? And, how can we do finite density calculations on the lattice?

CHAPTER 12

A Wider Picture

We conclude this book with a broader view on "QCD-related" gauge theories, defined as gauge theories with the SU(N_c) color groups with N_f flavors of quarks, with the possible addition of N adjoint fermions (gluinos) and corresponding scalars (in order to include supersymmetric theories). For simplicity, all fermion and scalar mass terms are assumed to be absent, so we focus only on theories with a single dimensional parameter Λ. The exception, i.e. $\mathcal{N} = 4$ supersymmetric theory considered in Section 12.5, is a conformal theory which does not have even Λ and is thus a conformal theory, with the same physics at all scales.

12.1. Hadronic world in alternative or changing universe

The understanding of how exactly the fundamental parameters of the Standard Model enter any observable is certainly one of the most important aims of hadronic/nuclear physics. Two old deep questions drive this discussion:

(i) Can there be "alternative universes" with different sets of parameters, and what are the boundaries of the world we know in the parameter space?

(ii) How to observe cosmological variations of weak and strong scales, if those exist? What are the present limits which can be established from all the known facts?

A discussion of both issues has been significantly revived recently. One can of course be curious about the size of the phase space of the Standard Model parameters which allow for a universe like ours, allowing in particular for such highly tuned phenomena as life based on elements C, H, O in appropriate proportions.

What are the "phase boundaries" separating it from other universes, and what can be their properties?

Too little work has been done so far to answer these fascinating questions, and we will not delve in substance: the interested reader can consult e.g. a recent paper [863] and references therein. Let me only mention one very specific claim [864]: that the space of all parameters has a *fixed point*, at which the production rate of black holes is maximized. The basis for that is (a quite speculative) idea of possible "evolution of universes" via black hole formation, which may drive the parameters toward this fixed point. Whether the idea is correct or not, one can at least try to calculate where the maximum is, if it exists.

The issue of cosmological time variation of major constants of physics has been recently revived by astronomical data which seem to suggest a variation of electromagnetic α at the 10^{-5} level for the time scale 10 bn years, see Ref. [865]. The statistical significance of the effect at the moment obviously excludes any random fluctuations, so the effect definitely exists. Whether it may or may not have a conventional explanation is not yet clear: more experimental work is clearly needed to reach any conclusions.

Nevertheless, it is quite timely to have another look at existing limits on time variation of all the fundamental constants. In particular, since the electromagnetic and weak forces are mixed together in the Standard Model, one may expect a similar modification of the weak couplings, the weak scale in general and quark masses in particular. In fact, one can measure only the variation of dimensionless parameters. Therefore, we obtain limits on the variation of m_s/Λ_{QCD} where m_s is the strange quark mass and Λ_{QCD} is the QCD scale defined as a position of the Landau pole in the logarithm of the running coupling constant. It is convenient to put $\Lambda_{\text{QCD}} = \text{const}$.

The masses of three heavy quarks — c, b, t — are clearly *too large* to be important in hadronic and nuclear physics. The masses of two light ones — u, d — are important, in particular via the "pion cloud" contribution to nucleon properties. These and related issues we have already discussed in Section 5.3.3, so let us just add a few remarks about specific quantities relevant for cosmological applications. Basically there are two such parameters: (i) *nuclear magnetic moments* (which can be measured from optical spectra from distant galaxies) and (ii) *the deuteron binding* which is crucial for Big Bang Nucleosynthesis.[1]

The overall conclusion is however that m_u and m_d are *too small* to be really important, and so one has to focus on the dependence on the *strange quark mass m_s*. This important point, made in particular in a recent paper by Flambaum and myself [868], we will now explain using the deuteron binding issue as the most relevant example.

It is well known that the formation of a deuteron is the first step toward Big Bang Nucleosynthesis of other elements. From observable isotopes created at that stage, from d to Li^7, plus well known kinetics of the reactions involved one can deduce

---

[1] As discussed in detail in Refs. [866, 867], due to competing attraction and repulsion, even the sign of the derivative of the deuteron binding energy over the light quark masses is not known for sure.

that the deviation of the binding from its present day value is limited by [866]

$$\left|\frac{\delta Q_d}{Q_d}\right| < 0.1. \tag{12.1}$$

The main idea of Ref. [868] is that the attractive and repulsive parts of the nuclear forces (represented by sigma and omega exchanges in a popular Walecka model, see review in Ref. [779]) have a different dependence on m_s. Mixing of flavors in the vector channel is very small, while it is much stronger for scalars. Because nuclear forces are small differences of two large numbers, the effect is significantly increased. The estimates of Ref. [868] lead to a large derivative,

$$\frac{\delta Q_d}{Q_d} \approx -48\frac{\delta m_\sigma}{m_\sigma} \approx -26\frac{\delta m_s}{m_s}, \tag{12.2}$$

which leads to rather strong limit on the m_s variation,

$$\left|\frac{\delta(m_s/\Lambda_{\rm QCD})}{(m_s/\Lambda_{\rm QCD})}\right| < 0.006, \tag{12.3}$$

at the time of the Big Bang.

It turns out that the measured isotope ratios from the natural nuclear reactor OKLO give a very strong limit [868] on the same ratio, which must have been less than 10^{-10} about 2 bn years ago.

12.2.   Increasing the number of quark flavors: the first window to conformal world

In this section we do not actually consider a limit of very large N_f: we do not wish to leave the domain of the asymptotic freedom and thus keep negative the combination $(11/3)N_c - (2/3)N_f < 0$ which enters the first coefficient of the beta function. (The solid line labeled $b = 0$ in Fig. 12.2 shows this boundary in QCD, together with the corresponding relation $b = 3N_c - N_f = 0$ in SUSY QCD.) Above this line, the coupling constant decreases at large distances, and the theory is IR-free.

We would focus instead on what happens just below this line. It has been pointed out in Refs. [869,870] that there should be an infrared fixed point here because the sign of the the second coefficient of the beta function $b' = 34N_c^2/3 - 13N_cN_f/3 + N_f/N_c$ is *negative* in this region. As a result, the beta function is zero at the fixed point value

$$g_*^2/(16\pi^2) = -b/b'. \tag{12.4}$$

This value thus determines the limiting value of the charge at large distance, as we already discussed in Section 1.4.1.

The presence of an IR fixed point is familiar from second order phase transitions and implies that the theory is conformal, so that all the correlation functions show a power law decay at large distance. There is no mass gap and the long distance behavior is characterized by the set of critical exponents.

Where is the lower boundary of this conformal domain? What are the properties of the chiral restoration and deconfinement phase transition at this boundary? How different are they from what we expect at more familiar theories with small N_f, since it is the first time we have not a second order transition point but a boundary to complete conformal phase?

Surprisingly little attention has been paid to those questions. A perturbative study based on the $1/N_f$ expansion [871] suggests that the IR fixed point may persist all the way down to $N_f = 6$ (for $N_c = 3$). However, the critical coupling becomes larger and non-perturbative phenomena may become important. For example, it has been argued in Ref. [874] that if the coupling constant reaches a critical value the quark–anti-quark interaction is sufficiently strong to break chiral symmetry. In their calculation, this happens for $N_f^c \simeq 4N_c$ (see the dashed line in Fig. 12.2a).

It was then realized that instanton effects can also be important [875]. If the critical coupling is small, even large instantons have a large action $S = 8\pi^2/g_*^2 \gg 1$ and the semi-classical approximation is valid. As usual, we expect random instantons to contribute to chiral symmetry breaking. According to estimates made in Ref. [875], the role of instantons is comparable to perturbative effects in the vicinity of $N_f = 4N_c$. Chiral symmetry breaking is dominated by large instantons with size $\rho \sim |\langle \bar{q}q \rangle|^{-1/3} > \Lambda^{-1}$, while the perturbative regime $\rho < \Lambda^{-1}$ contributes very little. For even larger instantons $\rho \gg |\langle \bar{q}q \rangle|^{-1/3}$ fermions acquire a mass due to chiral symmetry breaking and effectively decouple from gluons. This means that for large distances the charge evolves as in pure gauge theory, and the IR fixed point is only an approximate feature, useful for analyzing the theory above the decoupling scale.

What happens with instantons in this region? As noticed in Ref. [872], some non-perturbative ultraviolet divergences appear here. The first point is that in the chiral limit ($m = 0$) and without the quark condensate, only instanton–anti-instanton molecules exist. The second point is that counting powers of some global size ρ (e.g. the geometric mean of $\rho_I, \rho_{\bar{I}}$) and integrating over all other variables one sees that the density of the molecules is

$$dn_{\text{molecules}} \sim \frac{d\rho}{\rho^5}(\Lambda\rho)^{2b}. \tag{12.5}$$

It becomes UV divergent at $b = 2$ (the short-dashed line in Fig. 12.2a). Fortunately, this divergence appears only in the *vacuum energy*, not in any other physical observables (so probably just lattice practitioners should worry about it).

However, increasing further the number of quarks and decreasing b to zero, one will find that not only the *density* of the molecules is strongly divergent, but their contribution to the charge renormalization becomes divergent as well. This is easily seen, e.g. from the dipole interaction with the external field [214], which to second order contributes the term

$$\delta S \sim (G_{\mu\nu}^a)^2 \int d\rho \, \rho^4 \, dn_{\text{molecules}}. \tag{12.6}$$

Now, as it becomes *logarithmic* at $b \to 0$, it should be considered as a non-perturbative contribution to b itself.[2]

On the other hand, one can simply study models such as the "instanton liquid". In it the absence of the condensate and quasi-zero modes can only mean that the "liquid" is now broken into finite pieces. The simplest of them are pairs, or the instanton–anti-instanton molecules. This is precisely what instanton simulations have found [559]. The results of the simulations with $N_f = 5$ have already shown that instantons cannot break chiral symmetry, even at $m = 0$. Qualitatively, the reason for this behavior is clear. Increasing the number of flavors lowers the transition temperature because the determinant is raised to a higher power, so fermion-induced correlations become stronger. For $N_f = 5$ we find that the transition temperature drops to zero and the instanton liquid has a chirally symmetric ground state, provided the dynamical quark mass is less than some critical value. Studying the instanton ensemble in more detail shows that in this case, all instantons are bound into molecules.

Unfortunately, little is known from lattice simulations about QCD with different numbers of flavors. There are data by the Columbia group for $N_f = 4$ which has shown that chiral symmetry breaking effects were found to be drastically smaller as compared to $N_f = 0, 2$. In particular, the mass splittings between chiral partners such as $\pi - \sigma$, $\rho - a_1$, $N(\frac{1}{2}^+) - N(\frac{1}{2}^-)$, extrapolated to $m = 0$ were found to be 4–5 times as small. This agrees well with what was found in the interacting instanton model; but since it was a very limited simulation with old (non-chiral) fermions I am not sure how much one should trust it.

Direct simulations for multi-flavor QCD were reported in Ref. [876]. By reducing a time step in their hybrid Monte Carlo, these authors were able to simulate up to $N_f = 240$. As expected, for $N_c = 3$ the theory was found to be trivial for $N_f > 16$. By tracing the sign of the beta function in the weak and strong coupling domains, these authors were able to see the existence of the infrared fixed point, which was found to be the case at $N_f = 7$–16. Clearly those are just first exploratory attempts, and more work in this direction is needed.

12.3. $\mathcal{N} = 1$ supersymmetric theories

In this book we will not describe supersymmetry (SUSY) on a technical level: for that one should consult multiple books and reviews. Whether a part of Nature or not, it is a powerful theoretical concept, a kind of theoretical laboratory, allowing

---

[2]Does it modify the Banks–Zaks conclusions discussed above? No, because near the $b = 0$ line this contribution still is $O(\exp(-16\pi^2/g_*^2))$ and g_* is small.

for tremendous simplifications and sometimes exact solutions. The aim of this and subsequent sections is to draw parallels with physics ideas which were discussed above in the QCD context, since many of them got significant support by exact results derived using supersymmetry.

Briefly, by extending the usual coordinate space to *superspace* with a number of fermionic (Grassmanian) coordinates, one extends a symmetry over coordinate transformations to a large symmetry. Like generators of ordinary rotations (which mix different components of, say, vector physical quantities), the generators of supersymmetry mix together bosonic and fermionic fields. The representation of supersymmetry takes place in specific supermultiplets. The number of supersymmetries is counted by a special integer index $\mathcal{N} = 1, 2, 4$ the only three possibilities in $4d$.

In this section we discuss the first case, and $\mathcal{N} = 1$ means that each field (e.g. gluons) would have one superpartner (gluino), with the same color and other charges but spin $1/2$. The simplest case, called SUSY gluodynamics, has no other fields and its Lagrangian is

$$\mathcal{L} = -\frac{1}{4g^2} F^a_{\mu\nu} F^a_{\mu\nu} + \frac{i}{2g^2} \lambda^a (\slashed{D})_{ab} \lambda^b, \tag{12.7}$$

where the gluino field λ^a is a Majorana fermion in the adjoint representation of SU(2). More complicated theories can be constructed by adding additional matter fields and scalars. N extended supersymmetry has N gluino fields as well as additional scalars. Clearly, supersymmetric gluodynamics has θ vacua and instanton solutions. The only difference as compared to QCD is that the fermions carry adjoint color, so there are twice as many fermion zero modes. If the model contains scalar fields that acquire a vacuum expectation value, instantons are approximate solutions and the size integration is automatically cut off by the scalar VEV. These theories usually resemble the electroweak theory more than they do QCD.

This theory is confining and similar to QCD in many respects: the only significant difference is the quite different pattern of chiral symmetry breaking.

Since there is no flavor structure, there is no $SU(N_f)$ continuous chiral symmetry, pions and all of that. However, in a discrete subgroup of $U(1)_A$ chiral symmetry, its center Z_{N_c}, remains unbroken and therefore there exist N_c distinct vacua which have different phases of the gluino condensate. As a consequence, the simplest topological objects in this theory are the *domain walls* separating pairs of such vacua.

Much attention during the last two decades has been paid to SUSY theories with additional "matter" in the form of fundamental quarks, in N_f flavors. (Obviously complemented by scalar fields with the same charges, called *squarks*.) The most surprising feature was that there is a large variety of phases possible, and that a switch happens abruptly, as the relation between the number of colors and flavors changes. The methods developed using mainly the instanton-based ideas have been put into real "industrial" use by N. Seiberg and collaborators who had clarified the whole picture. He also discovered the so called Seiberg dualities, relating SUSY gauge theories with different N_f, N_c to each other.

12.3.1. *Instantons and exact beta function*

At first glance, instanton amplitudes seem to violate supersymmetry: the number of zero modes for gauge fields and fermions does not match, while scalars have no zero modes at all. However, one can rewrite the tunneling amplitude in manifestly supersymmetric form [877]. We will not do this here, but stick to the standard notation. The remarkable observation is that the determination of the tunneling amplitude in SUSY gauge theory is actually simpler than in QCD. Furthermore, with some additional input, one can determine the complete perturbative beta function from the tunneling amplitude.

The tunneling amplitude is given by

$$n(\rho) \sim \exp\left(-\frac{2\pi}{\alpha(M)}\right) M_0^{n_g - n_f/2} \left(\frac{2\pi}{\alpha(M)}\right)^{n_g/2} d^4x \frac{d\rho}{\rho^5} \rho^k \prod_f d^2\xi_f, \quad (12.8)$$

where all factors can be understood from the 't Hooft calculation discussed in Section 3.5.3. There are $n_g = 4N_c$ bosonic zero modes that have to be removed from the determinant and give one power of the regulator mass M each. Similarly, each of the n_f fermionic zero modes gives a factor $M^{1/2}$. Introducing collective coordinates for the bosonic zero modes gives a Jacobian $\sqrt{S_0}$ for every zero mode. Finally, $d^2\xi$ is the integral over the fermionic collective coordinates and ρ^k is the power of ρ needed to give the correct dimension. Supersymmetry now ensures that all non-zero mode contributions exactly cancel. More precisely, the subset of SUSY transformations which does not rotate the instanton field itself, mixes fermionic and bosonic non-zero modes but annihilates zero modes. This is why all non-zero modes cancel but zero modes can be unmatched. Note that as a result of this cancellation, the power of M in the tunneling amplitude is an integer.

Renormalizability demands that the tunneling amplitude be independent of the regulator mass. This means that the explicit M dependence of the tunneling amplitude and the M dependence of the bare coupling have to cancel. As in QCD, this allows us to determine the one-loop coefficient of the beta function $b = (4 - N)N_c - N_f$. Again note that b is an integer, a result that would appear very mysterious if we did not know about instanton zero modes.

In supersymmetric theories one can even go one step further and determine the beta function to all loops [877, 878]. For that purpose let us write down the renormalized instanton measure,

$$n(\rho) \sim \exp\left(-\frac{2\pi}{\alpha(M)}\right) M_0^{n_g - n_f/2} \left(\frac{2\pi}{\alpha(M)}\right)^{n_g/2}$$

$$\times Z_g^{n_g} \left(\prod_f Z_f^{-1/2}\right) d^4x \frac{d\rho}{\rho^5} \rho^k \prod_f d^2\xi_f, \quad (12.9)$$

where we have introduced the field renormalization factors $Z_{g,f}$ for the bosonic/fermionic fields. Again, non-renormalization theorems ensure that the tunneling amplitude is not renormalized at higher orders (the cancellation between the non-zero mode determinants persists beyond one loop). For gluons the field renormalization (by definition) is the same as the charge renormalization $Z_g = \alpha_R/\alpha_0$. Furthermore, supersymmetry implies that the field renormalization is the same for gluinos and gluons. This means that the only new quantity in (12.9) is the anomalous dimension of the quark fields, $\gamma_\psi = d \log Z_f/d \log M$.

Again, renormalizability demands that the amplitude be independent of M. This condition gives the NSVZ beta function [877] which, in the case $N = 1$, reads

$$\beta(g) = -\frac{g^3}{16\pi^2} \frac{3N_c - N_f + N_f \gamma_\psi(g)}{1 - N_c g^2/8\pi^2}. \tag{12.10}$$

The anomalous dimension of the quarks has to be calculated perturbatively. To leading order, it is given by

$$\gamma_\psi(g) = -\frac{g^2}{8\pi^2} \frac{N_c^2 - 1}{N_c} + O(g^4). \tag{12.11}$$

As far as I know, the result (12.10) was checked by explicit calculations up to three loops.[3]

In theories without quarks, the NSVZ result determines the beta function completely. For N-extended supersymmetric gluodynamics, we have

$$\beta(g) = -\frac{g^3}{16\pi^2} \frac{N_c(N - 4)}{1 + (N - 2)N_c g^2/(8\pi^2)}. \tag{12.12}$$

One immediately recognizes two interesting special cases. For $N = 4$, the beta function vanishes and the theory is conformal. In the case $N = 2$, the denominator vanishes and the one loop result for the beta function is exact. The next interesting theory is $N = 1$ SUSY QCD, where we add N_f matter fields (quarks ψ and squarks ϕ) in the fundamental representation. Let us first look at the NSVZ beta function. For $N = 1$, the beta function blows up at $g_*^2 = 8\pi^2/N_c$, so the renormalization group trajectory cannot be extended beyond this point.[4] The beta function vanishes at $g_*^2/(8\pi^2) = [N_c(3N_c - N_f)]/[N_f(N_c^2 - 1)]$, where we have used the one-loop anomalous dimension. This is reliable if g_* is small, which we can ensure by choosing $N_c \to \infty$ and N_f in the conformal window $3N_c/2 < N_f < 3N_c$. Seiberg showed that the conformal point exists for all N_f in the conformal window (even if N_c is not large) and clarified the structure of the theory at the conformal point [881].

---

[3]Note that the beta function is scheme dependent beyond two loops, so in order to make a comparison with high order perturbative calculations, one has to translate from the Pauli–Vilars scheme to a more standard perturbative scheme, e.g. $\overline{MS}$.

[4]Kogan and Shifman suggested that at this point the standard phase meets the renormalization group trajectory of a different (non-asymptotically free) phase of the theory [754].

Let us now examine the vacuum structure of SUSY QCD for $N_c = 2$ and $N_f = 1$. In this case we have one Majorana fermion (the gluino), one Dirac fermion (the quark) and one scalar squark (or Higgs) field. The quark and squark fields do not have to be massless. We will denote their mass by m, while v is the vacuum expectation value of the scalar field. In the semi-classical regime $v \gg \Lambda$, the tunneling amplitude (12.9) is $O(\Lambda^5)$. The 't Hooft effective Lagrangian is of the form $\lambda^4\psi^2$, containing four quark and two gluino zero modes. As usual, instantons give an expectation value to the 't Hooft operator. However, due to the presence of Yukawa couplings and a Higgs VEV, they can also provide expectation values for operators with less than six fermion fields. In particular, combining the quarks with a gluino and a squark tadpole we can construct a two-gluino operator. Instantons therefore lead to gluino condensation [879]. Furthermore, using two more Yukawa couplings one can couple the gluinos to external quark fields. Therefore, if the quark mass is non-zero we get a finite density of individual instantons $O(m\Lambda^5/v^2)$, see Fig. 12.1.

As in ordinary QCD, there are also instanton–anti-instanton molecules. The contribution of molecules to the vacuum energy can be calculated either directly or indirectly, using the single instanton result and arguments based on supersymmetry. This provides a nice check on the direct calculation, because in this case we have a rigorous definition of the contribution from molecules. Calculating the graph shown in Fig. 12.1, one finds [882]

$$\epsilon_{\text{vac}}^{\text{IA}} = \frac{32\Lambda^{10}}{g^8 v^6}, \tag{12.13}$$

which agrees with the result originally derived in Ref. [879] using different methods. The result implies that the Higgs expectation value is driven to infinity and $N_f = 1$ SUSY QCD does not have a stable ground state. The vacuum can be stabilized by adding a mass term. For non-zero quark mass m the vacuum energy is given by

$$\epsilon_{\text{vac}} = 2m^2 v^2 - \frac{16m\Lambda^5}{g^4 v^2} + \frac{32\Lambda^{10}}{g^8 v^6} = 2\left(mv - \frac{4\Lambda^5}{g^4 v^3}\right)^2, \tag{12.14}$$

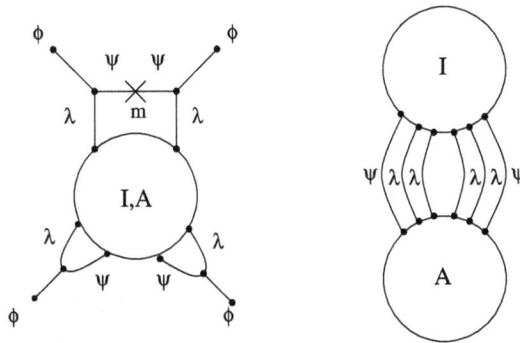

Fig. 12.1. Instanton contributions to the effective potential in supersymmetric QCD.

where the first term is the classical Higgs mass term, the second is the one-instanton contribution and the third is due to molecules. In this case, the theory has a stable ground state at $v^2 = \pm 2\Lambda^{5/2}/(g^2 m^{1/2})$. Since SUSY is unbroken, the vacuum energy is exactly zero. Let us note that in the semi-classical regime $m \ll \Lambda \ll v$ everything is under control: instantons are small $\rho \sim v^{-1}$, individual instantons are rare, molecules form a dilute gas, $nR^4 \sim (v/\Lambda)^{10}$, and the instanton and anti-instanton inside a molecule are well separated, $R_{IA}/\rho \sim 1/\sqrt{g}$. Supersymmetry implies that the result remains correct even if we leave the semi-classical regime. This means, in particular, that all higher order instanton corrections ($O(\Lambda^{15})$ etc.) have to cancel exactly. Checking this explicitly might provide a very non-trivial check of the instanton calculus.

12.3.2. N_c-N_f phase diagram: $\mathcal{N} = 1$ SUSY versus QCD

Significant progress has been made in determining the structure of the ground state of $N = 1$ supersymmetric QCD for arbitrary N_c and N_f [883]. Seiberg showed that the situation discussed above is generic for $N_f < N_c$: A stable ground state can only exist for non-zero quark mass. For $N_f = N_c - 1$ the $m \neq 0$ contribution to the potential is an instanton effect. In the case $N_f = N_c$, the theory has chiral symmetry breaking (in the chiral limit $m \to 0$) and confinement.[5] For $N_f = N_c + 1$, there is confinement but no chiral symmetry breaking. Unlike QCD, the 't Hooft anomaly matching conditions can be satisfied with massless fermions. These fermions can be viewed as elementary fields in the dual (or "magnetic") formulation of the theory. For $N_c + 1 < N_f < 3N_c/2$ the magnetic formulation of the theory is IR-free, while for $3N_c/2 < N_f < 3N_c$ the theory has an infrared fixed point. Finally, for $N_f > 3N_c$, asymptotic freedom is lost and the electric theory is IR-free.

The phase diagram of ordinary and SUSY QCD in the N_c-N_f plane is shown in Fig. 12.2. For simplicity, we have plotted N_c and N_f as if they were continuous variables.[6] We should emphasize that while the location of the phase boundaries can be rigorously established in the case of SUSY QCD, the phase diagram of ordinary QCD is just a guess, guided by some of the results mentioned below.

One general comment is that if one keeps N_f fixed and looks at the limit of up to large N_c, nothing significant happens, and the whole richness of the phase diagram is missed.

Let us first comment on the solid lines, corresponding to the vanishing first coefficient of the beta function, $b = (11/3)N_c - (2/3)N_f = 0$ in QCD

---

[5]There is no instanton contribution to the superpotential, but instantons provide a constraint on the allowed vacua in the quantum theory. To our knowledge, the microscopic mechanism for chiral symmetry breaking in this theory has not been clarified.

[6]In a sense, at least the number of flavors is a continuous variable: one can gradually remove a massless fermion by increasing its mass.

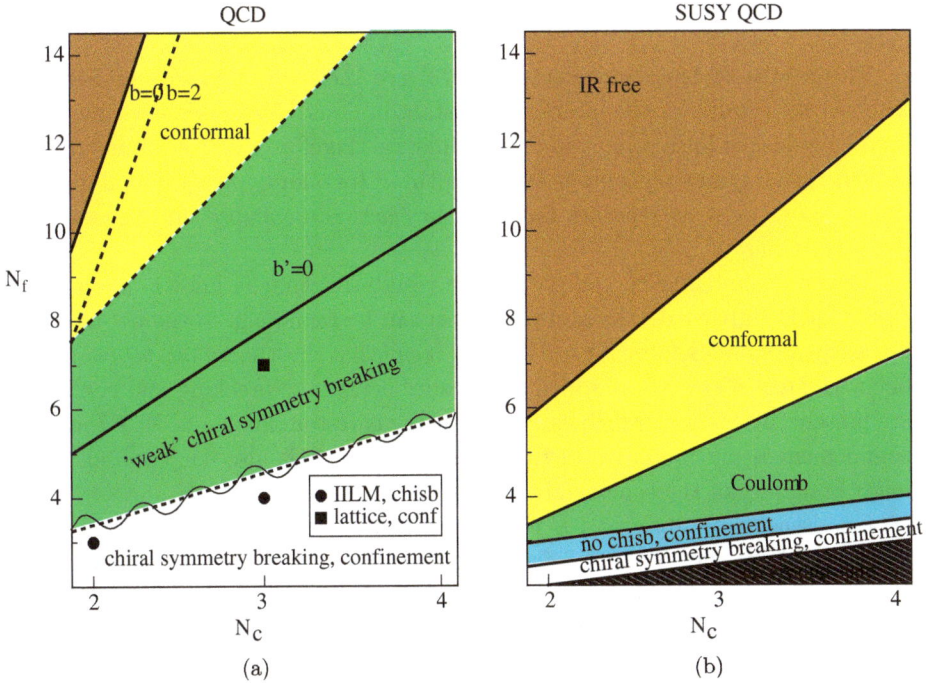

Fig. 12.2. Schematic phase diagram of QCD (a) and supersymmetric QCD (b) as a function of the number of colors N_c and the number of flavors N_f.

and $b = 3N_c - N_f = 0$ in SUSY QCD. Since we are only interested in asymptotically free theories, we consider only regions below these lines.

As we have discussed in Section 1.2.2, just below these solid lines all theories are conformal, they have the *infrared* (IR) *fixed point* [869, 870]. Here the coupling constant is small not only at small distances, but also at large ones. There are *no* bound states of quarks and gluons, the correlators are all power-like at large x.

In Section 1.2.2 we have already discussed the non-perturbative beta functions extracted from lattice studies. These data provide two quite significant hints. First, the sign of the deviation is *negative*, in spite of the fact that for the cases considered ($N_f = 0, 2$) the second beta function coefficient is $b' > 0$. Second, although the data points do not reach zero, the drop is very significant, and it grows when N_f changes from 0 to 2. It is quite possible that the drop is related with the zero of the *non-perturbative* beta function to appear larger N_f.

Let us now return to our map, Fig. 12.2a, and discuss which other phases of QCD may exist. We do know that the real world (which is not far from the point $N_c = 3, N_f = 2.5$) is a confining theory with broken chiral symmetry. These two properties should disappear before the conformal domain: thus, in principle, one may have (i) the chiral symmetry restoration line, $\langle \bar{q}q \rangle = 0$; as well as (ii) the deconfinement line. Furthermore, one more phase (iii) can contain both massless quarks and gluons as excitations *together* with some *bound states* with a non-zero mass.[7]

The results of first lattice studies [873] are shown by triangles in Fig. 12.2a, whereas the results for the interacting instanton liquid [559] are shown by squares. Two long-dashed lines connecting the points are therefore our best boundary of the broken chiral symmetry phase. Whether three (or more) phase boundaries mentioned above exist as separate lines, or they actually collapse into the single line remains unknown.

Now we would like to compare those results to what is known about $\mathcal{N} = 1$ SUSY QCD (further details and references can be found e.g. in the detailed paper by Intriligator and Seiberg [884]). We only recall that this theory has no ground state below the line $N_f = N_c$, at which both chiral symmetry breaking and confinement are present. The line at which chiral symmetry is restored is $N_f = N_c + 1$, although confinement is presumably still there. In Ref. [884] it is also conjectured that the *lower* boundary of the conformal domain is the line $N_f = (3/2)N_c$.

One issue rarely discussed is the role of instantons in these theories. As N_f is increased above the values 2–3 (relevant to real QCD) the two basic components of the instanton ensemble, random (individual) instantons and strongly correlated instanton–anti-instanton pairs (molecules) are affected in very different ways. Isolated instantons can only exist if the quark condensate is non-zero, and the instanton density contains the factor $(|\langle \bar{q}q \rangle | \rho^3)^{N_f}$ which comes from the fermion determinant. As a result, small-size instantons are strongly suppressed as N_f is increased. This suppression factor does not affect instantons with a size larger than $\rho \sim |\langle \bar{q}q \rangle|^{-1/3}$. This means that as N_f is increased, random instantons are pushed to larger sizes. Since in this regime the semi-classical approximation becomes unreliable, we do not know how to calculate the rate of random instantons at large N_f.

For strongly correlated pairs (molecules) the trend is exactly the opposite. The density of pairs is essentially independent of the quark condensate and only determined by the interaction of the two instantons. From purely dimensional considerations one expects the density of molecules to be $dn_m \sim d\rho \Lambda^{2b} \rho^{2b-5}$, which means that the typical size becomes smaller as N_f is increased [872]. If $N_f > 11N_c/2$, we have $b < 2$ and the density of pairs is ultraviolet-divergent (see the dashed line in Fig. 12.2a). This phenomenon is similar to the UV divergence in the $O(3)$ non-linear σ model. Both are examples of UV divergences of non-perturbative nature. Most likely, they do not have significant effects on the physics of the theory. Since the typical instanton size is small, one would expect that the contribution of molecules at large N_f can be reliably calculated. However, since the binding inside the pair increases with N_f, the separation of perturbative and non-perturbative fluctuations becomes more and more difficult.

---

[7]This phase should obviously be outside the conformal domain, because masses produce a scale and exponential correlators.

12.4. $\mathcal{N} = 2$ supersymmetric theories

This set of SUSY theories are often called Seiberg–Witten theories, as they were partially solved by them a decade ago [886]. The main impact of their seminal work is related with a possibility of dual descriptions, in which "fundamental" fields like gauge bosons can be substituted with ease by "composites" like monopoles or dyons. They have also demonstrated that confinement based on dual superconductivity is actually the case, and that monopoles do indeed condense when they are light enough.

However, as all of that is well described in the literature, I will not go into these topics. Instead I concentrate on a few other results which I think are also quite important, but did not get sufficient press. My attitude is that SUSY theories are just ordinary field theories with a very specific matter content and certain relations between different coupling constants. Eventually, we hope to understand non-Abelian gauge theories for all possible matter sectors. In particular, we want to know how the structure of the theory changes as one goes from QCD to its supersymmetric generalizations.

One of the most basic issues we would like to understand is the relation between perturbative and truly non-perturbative dynamics. How much can one learn from calculating and re-summing perturbative diagrams? When a completely new effect unrelated to any diagram comes into play?

SUSY results can shed a lot of light on questions like this since it provides powerful cancellations among diagrams. The $\mathcal{N} = 2$ theory has multiple degenerate vacua with zero energy, as all perturbative (and non-perturbative) corrections to the vacuum energy cancel out. In most cases, the classical vacua are characterized by scalar field VEVs. If the scalar field VEV is large, one can perform reliable semi-classical calculations. Decreasing the scalar VEV, one moves toward strong coupling and the dynamics of the theory is non-trivial. Nevertheless, supersymmetry restricts the functional dependence of the effective potential on the scalar VEV (and other parameters, like masses or coupling constants), so that instanton calculations can often be continued into the strong coupling domain.

Furthermore, the beta function is exactly known and it has only a single-loop log, with all others absent.

The $N = 2$ supersymmetric gauge theory contains two Majorana fermions λ_α^a, ψ_α^a and a complex scalar ϕ^a, all in the adjoint representation of SU(N_c). In the case of $N = 2$ SUSY QCD, we add N_f multiplets $(q_\alpha, \tilde{q}_\alpha, Q, \tilde{Q})$ of quarks $q_\alpha, \tilde{q}_\alpha$ and squarks $Q, \tilde{Q}$ in the fundamental representation. In general, gauge invariance is broken and the Higgs field develops an expectation value $\langle \phi \rangle = a\tau^3/2$. The vacua of the theory can be labeled by a (gauge invariant) complex number $u = (1/2)\langle \text{tr}\,\phi^2 \rangle$. If the Higgs VEV a is large ($a \gg \Lambda$), the semi-classical description is valid and $u = (1/2)a^2$. In this case, instantons are small $\rho \sim a^{-1}$ and the instanton ensemble is dilute $n\rho^4 \sim \Lambda^4/a^4 \ll 1$.

In the semi-classical regime, the effective Lagrangian is given by

$$\mathcal{L}_{\text{eff}} = \frac{1}{4\pi} \text{Im} \left[-\mathcal{F}''(\phi) \left(\frac{1}{2}(F_{\mu\nu}^{sd})^2 + (\partial_\mu \phi)(\partial^\mu \phi^\dagger) + i\psi \partial\!\!\!/ \, \bar\psi + i\lambda \partial\!\!\!/ \, \bar\lambda \right) \right.$$
$$\left. + \frac{1}{\sqrt{2}} \mathcal{F}'''(\phi) \lambda \sigma^{\mu\nu} \psi F_{\mu\nu}^{sd} + \frac{1}{4} \mathcal{F}''''(\phi) \psi^2 \lambda^2 \right] + \cdots, \qquad (12.15)$$

where $F_{\mu\nu}^{sd} = F_{\mu\nu} + i\tilde{F}_{\mu\nu}$ contains both the field strength tensor and its dual. Note that the effective low energy Lagrangian only contains the light fields. In the semi-classical regime, this is the U(1) part of the gauge field (the "photon") and its superpartners. Using arguments based on the electric–magnetic duality, Seiberg and Witten determined the exact prepotential $\mathcal{F}(\phi)$. From the effective Lagrangian, we can immediately read off the effective charge at the scale a

$$\mathcal{F}''(a) = \tau(a) = \frac{4\pi i}{g^2(a)} + \frac{\theta}{2\pi}, \qquad (12.16)$$

which combines the coupling constant g and the θ angle. Also, the Witten–Seiberg solution determines the anomalous magnetic moment $\mathcal{F}'''$ and the four-fermion vertex $\mathcal{F}''''$. In general, the structure of the prepotential is given by

$$\mathcal{F}(\phi) = \frac{i(4 - N_f)}{8\pi} \phi^2 \log\left(\frac{\phi^2}{\Lambda^2}\right) - \frac{i}{\pi} \sum_{k=1}^{\infty} \mathcal{F}_k \, \phi^2 \left(\frac{\Lambda}{\phi}\right)^{(4-N_f)k}. \qquad (12.17)$$

The first term is just the perturbative result with the one-loop beta function coefficient. As noted in Section 12.3.1, there are no corrections from higher loops. Instead, there is an infinite series of power corrections. The coefficient $\mathcal{F}_k$ is proportional to $\Lambda^{(4-N_f)k}$, which is exactly what one would expect for a k-instanton contribution.

For $k = 1$, this was first checked in Ref. [886] in the case of SU(2) and in Ref. [887] in the more general case of SU(N_c). The basic idea is to calculate the coefficient of the 't Hooft interaction $\sim \lambda^2 \psi^2$. The gluino λ and the Higgsino ψ together have 8 fermion zero modes. Pairing zero modes using Yukawa couplings of the type $(\lambda \psi)\phi$ and the non-vanishing Higgs VEV we can see that instantons induce a 4-fermion operator. In an impressive tour de force the calculation of the coefficient of this operator was extended to the two-instanton level[8] [888,889]. For $N_f = 0$, the result is

$$S_{4f} = \int dx (\psi^2 \lambda^2) \left[\frac{15\Lambda^4}{8\pi^2 a^6} + \frac{9! \Lambda^8}{3 \times 2^{12} \pi^2 a^{10}} + \cdots \right], \qquad (12.18)$$

which agrees with the Witten–Seiberg solution. This is also true for $N_f \neq 0$, except in the case $N_f = 4$, where a discrepancy appears. This is the special case where the coefficient of the perturbative beta function vanishes. Seiberg and Witten assume that the non-perturbative $\tau(a)$ is the same as the corresponding bare coupling in the Lagrangian. But the explicit two-instanton calculation shows that even in this

---

[8] Supersymmetry implies that there is no instanton–anti-instanton contribution to the prepotential.

theory the charge is renormalized by instantons [888]. In principle, these calculations can be extended order by order in the instanton density. The result provides
a very non-trivial check on the instanton calculus. For example, in order to obtain
the correct two instanton contribution one has two use the most general (ADHM)
two-instanton solution, not just a linear superposition of two instantons.

Instantons also give a contribution to the expectation value of ϕ^2. Pairing
off the remaining zero modes, the semi-classical relation $u = a^2/2$ receives a
correction [886],

$$u = \frac{a^2}{2} + \frac{\Lambda^4}{a^2} + O\left(\frac{\Lambda^8}{a^6}\right). \tag{12.19}$$

More interesting are instanton corrections to the effective charge τ. The solution of
Seiberg and Witten can be written in terms of an elliptic integral of the first kind

$$\tau(u) = i\frac{K(\sqrt{(1-k^2)})}{K(k)}, \qquad k^2 = \frac{u - \sqrt{u^2 - 4\Lambda^4}}{u + \sqrt{u^2 - 4\Lambda^4}}. \tag{12.20}$$

Again, this result can be written as the one-loop perturbative contribution plus an
infinite series of k-instanton terms. Up to the two-instanton level, we have (for a
more detailed discussion of the non-perturbative beta function, see Ref. [890])

$$\tau(a) = \frac{2i}{\pi}\left(\log\left(\frac{2a^2}{\Lambda^2}\right) - \frac{3\Lambda^4}{a^4} - \frac{3 \times 5 \times 7\Lambda^8}{8a^8} + \cdots\right). \tag{12.21}$$

It is interesting to note that instanton corrections tend to reduce the inverse charge
$\tau(a)$. This is consistent with what was found in QCD by considering how small-size
instantons renormalize the charge of a larger instanton [214]. However, the result
is opposite to the trend discussed in Section 1.2.2 (based on the instanton size
distribution and lattice beta function) which suggests that in QCD the coupling
runs more slowly than suggested by perturbation theory.

If the Higgs VEV is reduced, the instanton corrections in (12.20) start to grow
and compensate for the perturbative logarithm. At this point the expansion (12.21)
becomes unreliable, but the exact solution of Seiberg and Witten is still applicable.
In the semi-classical regime, the spectrum of the theory contains monopoles and
dyons with masses proportional to τ. As the Higgs VEV is reduced, these particles
can become massless. In this case, the expansion of the effective Lagrangian in
terms of the original (electrically charged) fields breaks down, but the theory can
be described in terms of their (magnetically charged) dual partners.

The effective charge itself is not directly observable, unlike the correlators or
a potential between static charges. Nevertheless, all QCD calculations deal with it
in some way or another, and so it is impossible to avoid it in a theory discussion.
So let us ask what constitutes the first non-perturbative correction to the effective
charge. It is defined here as the coefficient in an effective action for some external
weak background field $G_{\mu\nu}^a$, in which *both* perturbative loop corrections and the
non-perturbative effects are to be included.

A very simple non-perturbative effect was suggested long ago by Callan, Dashen, and Gross (CDG) [217]: instantons are simply dipoles and therefore they are polarized by an external field and produce a dielectric constant, exactly as atoms do. The external field is supposed to be normalized at some normalization scale μ, and CDG proposed including all instantons with size $\rho < \rho_{\max} = 1/\mu$. The effective charge is then defined as:

$$
\frac{8\pi^2}{g_{\mathrm{eff}}^2(\mu)} = b \ln\left(\frac{\mu}{\Lambda_{\mathrm{pert}}}\right) - \frac{4\pi^2}{N_c^2 - 1} \int_0^{\rho_{\max}} dn(\rho)\rho^4 \left(\frac{8\pi^2}{g_{\mathrm{eff}}^2(\rho)}\right)^2, \qquad (12.22)
$$

where $b = 11N_c/3 - 2N_f/3$ is the usual one-loop coefficient of the beta function, and $dn(\rho)$ is the distribution of instantons (and anti-instantons) over size.

CDG could not calculate it then, as the instanton size distribution was unknown. Plugging in lattice data for SU(2) and SU(3) gluodynamics we get [891] the curves shown in Fig. 12.3. The curve for QCD with fermions is not from lattice; we do not have it yet, so it corresponds to IILM. Note a rapid departure from the perturbative logarithmic running coupling constant.

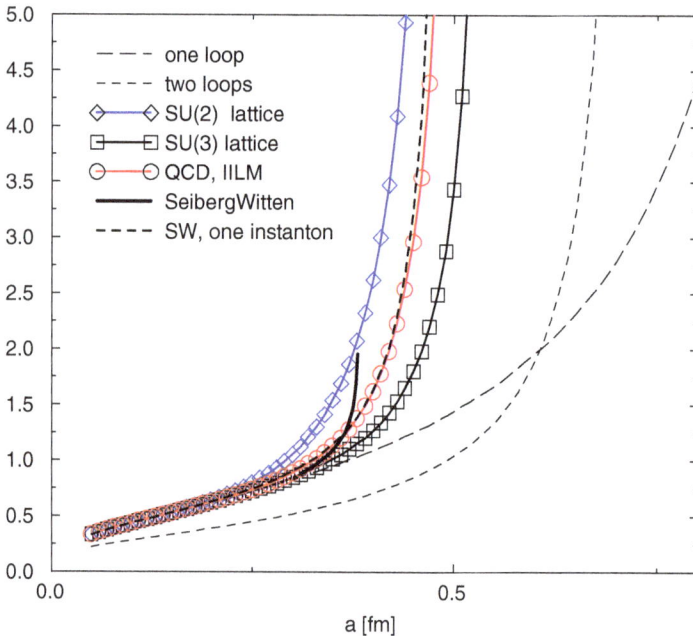

Fig. 12.3. The effective charge $bg_{\mathrm{eff}}^2(\mu)/8\pi^2$ (b is the coefficient of the one-loop beta function) versus normalization scale μ (in units of its magnitude at which the one-loop charge blows up). The thick solid line corresponds to the exact solution [885] for the $N = 2$ SYM, the thick dashed line shows the one-instanton correction. Lines with symbols (as indicated in the figure) stand for $N = 0$ QCD-like theories, SU(2) and SU(3) pure gauge ones and QCD itself. Thin long-dashed and short-dashed lines are one- and two-loop results.

Is this indeed what is happening? In order to answer such questions above I have used either lattice data or hadronic phenomenology. Let me now use another tool from the theory tool-box, namely a *solvable model*. So we [891] are going to compare the charge in QCD to that for the $\mathcal{N} = 2$ supersymmetric theory, for which the famous exact effective Lagrangian was found by Seiberg and Witten [885],

$$\frac{8\pi}{g^2(u)} = \frac{K(\sqrt{1-k^2})}{K(k)}, \tag{12.23}$$

where K is the elliptic integral and the argument

$$k^2 = \frac{(u - \sqrt{(u^2 - 4\Lambda^4)})}{(u + \sqrt{u^2 - 4\Lambda^4})} \tag{12.24}$$

is a function of the gauge invariant vacuum expectation of the squared scalar field

$$u = \frac{1}{2}\langle\phi^2\rangle = \frac{a^2}{2} + \frac{\Lambda^4}{a^2} + \cdots \tag{12.25}$$

and a is just its VEV. For large a there is a weak coupling expansion which includes instanton effects[9]

$$\frac{8\pi}{g^2(u)} = \frac{2}{\pi}\left(\log\left(\frac{2a^2}{\Lambda^2}\right) - \frac{3\Lambda^4}{a^4} + \cdots\right). \tag{12.26}$$

The exact coupling blows up at $u = 2\Lambda^2$, which means that the factor between the exact strong interaction scale and the perturbative one is in this theory $\Lambda_\infty = 2^{3/2}\Lambda$. Actually this is the ratio of the scale $\sqrt{u}$ to the scale a at which the perturbatively evolved coupling (one-loop) blows up. If one were to account for the next term in the expansion of u, the ratio of scales would be reduced to $\sqrt{2}\sqrt{2 + \sqrt{2}}$. The fact that instanton effects can be important at such a high scale was anticipated in Ref. [214] and is presumably due to the significance of the pre-factor in instanton calculations.

The behavior is shown in Fig. 12.3, where we have included both a curve which shows the full coupling (thick solid line), as well as a curve which illustrates only the one-instanton correction (thick dashed one). The units on both axes are chosen so that the naive one-loop running looks the same in both theories. And — surprise, surprise — the instanton-induced corrections seem to be nearly the same as well.

I think this comparison teaches us a few lessons. The main lesson: if the instanton-induced corrections to the charge itself become large, the perturbative expansion in g has to be abandoned. I do not know what exactly one can make out of the (nearly ideal) numerical matching of these two theories, QCD and $\mathcal{N} = 2$ SUSY gluodynamics. Note however the very rapid change of the coupling induced by instantons. Although expressed differently in formulae, the curves are extremely similar. Also of interest is that the full multi-instanton sum makes the rise in the

---

[9]We remind the reader that the first terms in this expansion have been explicitly verified in instanton calculations [886].

coupling even more radical than with only the one-instanton correction incorporated. It is also interesting to observe that at the scale where the true coupling blows up, the perturbatively evolved coupling is still not very large. Individually, the perturbative log and instanton corrections are well defined at this region, however they cancel each other in the inverse charge. This is encouraging from the point of view of developing a consistent expansion for the instanton corrections. The rapid rise in the coupling is also encouraging in that it ensures that perturbation theory is valid almost to the point where it blows up. For a consistent picture of QCD, in which perturbation theory still appears to be applicable at the c-quark scale while the theory is non-perturbative at 1 GeV, such a dramatic effect is essential.

12.5. $\mathcal{N} = 4$ supersymmetric theories and AdS/CFT correspondence

12.5.1.   Conformal field theory

As the number of supersymmetries is increased to 4, the maximal number for $d = 4$, one finds a theory with qualitatively new properties. Its minimal version includes gauge fields ("gluons"), four types of Majorana fermions ("gluinos") and six scalars.[10] All of them are in the same — adjoint — representation of some gauge group, e.g. SU(N_c).

Calculating the beta function for these theories (using instantons, as discussed in the preceding section, or directly from diagrams) one finds that for this theory it just *vanishes*, $\beta(g) = 0$. It means that the charge does not run at all, and the value of the coupling constant g put in the Lagrangian remains valid, at all scales. So, there is no dimensional Λ parameter in this theory. It has conformal invariance instead: physics at all scales is the same.

As we discussed in Chapter 1, the situation when the beta function has a zero is quite common for ordinary second order phase transitions. It has also been discussed in this chapter above, in connection to the QCD with large N_f. A theory with identically zero beta function is just a stronger version of the same behavior.

As at the 2nd order phase transition points, the two-point correlators are just some powers of the distance, with the powers — also called the anomalous dimensions — depending on quantum numbers of the operator in question. The table of such powers is thus as good a substitute for the particle mass table of the usual gauge theories. Of course there is no confinement or chiral symmetry breaking: so if the $\mathcal{N} = 4$ theory reminds us of something QCD-related it should be its QGP phase.

If the coupling is weak, the anomalous dimensions can be calculated perturbatively, from diagrams. A similar procedure has been followed by K. Wilson for phase

---

[10]Which one can easily guess counting the bosonic versus the fermionic degrees of freedom: 2 polarizations for a gluon, 2*4 for gluinos, 6 bosons are needed to balance.

transitions, in $4 - \epsilon$ dimensions at small ϵ: except that one may not worry about indices describing an approach to the fixed point.

The main interest in the $\mathcal{N} = 4$ theory concerns the possibility to study its strong coupling limit. From a pragmatic point of view, it provides a very interesting new window to understand QGP at not-so-high T, when strong coupling is more appropriate. From a theoretical point of view it gives much more, as we will see in the next section.

12.5.2. *A window to the string world: strong coupling*

A description of string theories obviously goes way beyond this book (as well as the ability of the author). Let me only provide a qualitative description of the setting used, needed to give the reader an idea how the results to be discussed were actually obtained.

Historically, string theory originated from QCD "confining flux tubes" and Regge phenomenology in the 1970s. About a decade ago it claimed to be a wonderful candidate for the fundamental "theory of everything", uniting all gauge theories with gravity. Future will show if this is true or not: our approach will be pragmatic. Such a remarkable construction created by very talented people is at the very least a mathematical tool which may help us to learn things we care about in QCD.

Self-consistent string theories live in multiple ($d = 10, 11, 26, \ldots$) dimensions, but they allow for soliton-like objects of smaller dimensions called "branes".[11] If those have $d = 3 + 1$ dimension, they generate $4d$ gauge theories in them. The open strings which end on the brabe can be seen as color/flavor charges. By construction, those theories are supersymmetric. Stacking N_c branes together one may get $U(N_c)$ symmetry.

A specific setting due to Maldacena [892] is a conjecture that there is exact duality between the $\mathcal{N} = 4$ theory and a specific version of the $10d$ string theory in a geometrical background consisting of $5d$ anti-de Sitter (AdS_5) space time and the $5d$ sphere S_5. The metrics of the former depends only on the 5th coordinate we will call U. The physical meaning of this coordinate is the (dimensional) scale, so that moving in the 5th dimension U is like changing the scale in the renormalization group, from IR to UV limits. Many checks has been made, convincing many that this duality is indeed the case.

Further modification of the background metrics [894] with a Schwartzschild (black hole) has allowed us to consider the $\mathcal{N} = 4$ theory at finite temperature T. The metrics used in all calculations to be referred to below are thus

$$ds^2 = \alpha' \left[\frac{1}{\sqrt{G}} \left(-H dt^2 + d\mathbf{x}_\parallel^2 \right) + \sqrt{G} \left(\frac{1}{H} dU^2 + U^2 d\Omega_5^2 \right) \right], \qquad (12.27)$$

---

[11]The name itself stems from membranes.

where α' is the string tension, $t, \mathbf{x}_\parallel$ is the $4d$ coordinate of our space–time, Ω_5 is the $5d$ solid angle of the $5d$ sphere, $G \equiv g_{\text{eff}}^2/\mathrm{U}^4$,

$$H \equiv 1 - \frac{\mathrm{U}_0^4}{\mathrm{U}^4}, \quad \mathrm{U}_0^4 = \frac{2^7 \pi^4}{3} g_{\text{eff}}^4 \frac{\mu}{N^2}. \tag{12.28}$$

Here $g_{\text{eff}}^2 = g^2 N_c$ is the 't Hooft coupling. The parameter μ is interpreted as the free energy density on the near extremal D3-brane, hence, $\mu \approx (4\pi^2/45)N^2 T^4$. So in the metrics without the black hole, $T = 0$ corresponds to $H = 1$.

This statement is rather ironical, for the following reason. String theory has been long thought of as the ultimate "land of confinement", with linear potential guaranteed by construction. What was and still is missing is its connection to gauge theories, although many authors, especially those using the large N_c limit, tried hard to establish it. Now Maldacena duality has done exactly that, but to the $\mathcal{N} = 4$ theory which is conformal and not confining. In particular, in it the interaction between two color charges at distance L is Coulomb-like,

$$V(L) = \frac{f(g, N_c)}{L}, \tag{12.29}$$

with the function f being the usual $O(g^2)$ in weak coupling. News is the *strong coupling* limit, when the 't Hooft coupling $g^2 N_c$ is large. In this limit the Maldacena duality leads to a new astonishing form of the Coulomb law in the strong coupling limit [893]

$$V(L)|_{g^2 N_c \gg 1} = -\frac{4\pi^2}{\Gamma(1/4)^4} \frac{\sqrt{g^2 N_c}}{L}. \tag{12.30}$$

Let me show pictorially one of the settings used for actual calculation [896]. In Fig. 12.4a we show string configuration for inter-quark separation $L < R_{\text{Debye}}$. The string still connects two charges, as in confining theories. The reason one gets a Coulomb-like result [893] mentioned above is because the string is not connecting two points linearly: the gravity induced by AdS metrics pulls it away to the 5th dimension, and away from the brane (our $4d$ world) the scale is different and the tension of the string is much reduced.

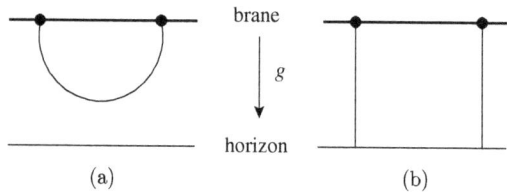

Fig. 12.4. Two types of solutions describing the potential between two static charges (large dots) in the ordinary $4d$ space (on the D3 brane). The string originating from them can either connect them (a) or not (b). In any case the string is deflected by background metrics (the gravity force indicated by the arrow marked g) downward, along the 5th coordinate. After string touches the black hole horizon, a Debye screening of the interaction takes place.

This approach has led to the intriguing results on the finite-T properties of the $\mathcal{N} = 4$ theory in the strong coupling limit: the thermodynamics [895], Debye screening [896] and viscosity [897] (which we have already discussed in Section 8.2.7) and the Wilson loops [899].

Let me first comment on the Debye screening calculation. Furthermore, at sufficiently large L the string touches the Schwartzschild horizon and breaks into two disconnected pieces shown in Fig. 12.4b. That is why the strong-coupling Debye screening potential has an edge, and beyond some distance $r > R_D \sim 1/T$ the potential between static charges is zero.

It is very instructive to compare weak and strong coupling results, which we do in Fig. 12.5. The pressure, shown in Fig. 12.5a, presumably show a monotonic decrease from 1 to $3/4$, but both weak and strong coupling results seem to be quite wrong for $\lambda = g^2 N_c = 1$–5. On the other hand a transition from the weak to strong coupling for the log of the circular Wilson loop is rather smooth and happens at $\lambda = 10$–15 (which is incidentally the range we have for QGP right above T_c). The exact result [899] is

$$\langle W(\text{circle}) \rangle = \frac{2}{\sqrt{\lambda}} I_1(\sqrt{\lambda}). \tag{12.31}$$

Many more results can be obtained this way, including all real-time correlators at any quantum numbers, any space–time separations t, x and the temperature T. Hopefully this will lead to some physical picture of strongly coupled QGP, still

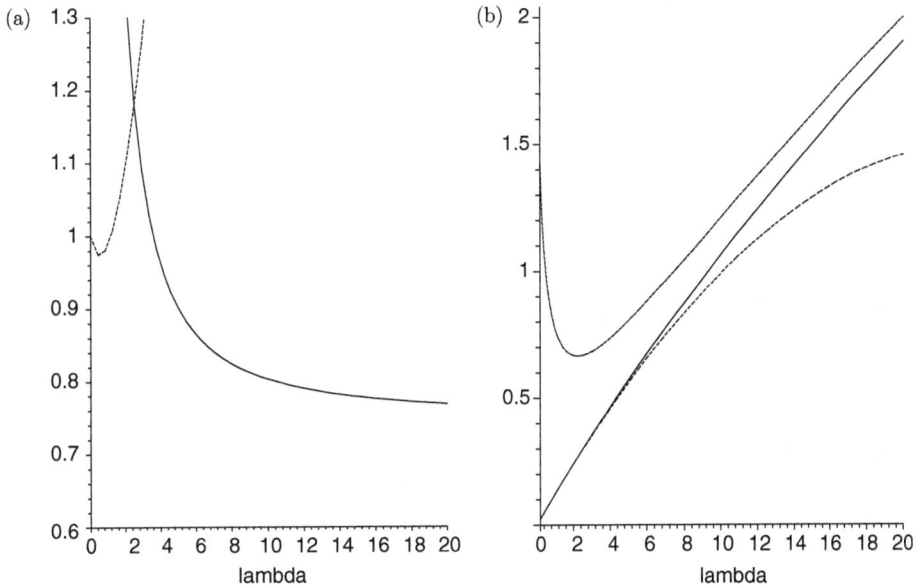

Fig. 12.5. Comparison of the weak and strong coupling results for the $\mathcal{N} = 4$ theory. (a) Ratio of the pressure to the Stephan–Boltzmann value; (b) the circular Wilson loop, for which also the exact result is given.

missing at the moment. Perhaps a step toward it is a work by Zahed and myself, [hep-th/0308073], in which it is argued that a strongly coupled QGP is a gas of strongly bound binary states, with angular momentum growing with a coupling.

Completing this section let me mention that there are other theories with dual string description, and some of them are much more QCD-like, with confinement etc., see e.g. Ref. [898].

12.5.3.   *AdS/CFT duality at weak coupling and instantons*

Although a discussion of this limit got much less attention, I think it has led to equally amazing results that were reviewed in Ref. [888]. Not only have they provided an additional support to Maldacena duality, it hinted at the possible nature of the "master field" as well as the field theoretical origin of the mysterious AdS metrics and the 5th "scale" coordinate associated with it.

First of all, in the $\mathcal{N} = 4$ conformal theory small charge remains small at all scales, and so instanton effects are exponentially small $\sim \exp(-8\pi^2/g^2)$ in this regime. There is a clear parametric distinction between perturbative, one-instanton, two-instanton, etc. effects. Furthermore, contributions of the instantons to any Green function are calculable without any divergences. As usual, the simplest cases to calculate are Green functions which correspond to the 't Hooft interaction and have a number of legs equal to the number of fermionic zero modes. For $N_c = 2$ the gluino has four zero modes, and $\mathcal{N} = 4$ makes it 16 per instanton.

The major distinction between QCD-like fundamental quarks and adjoint gluinos is that in the latter case the number of zero modes grows with N_c. It allows instantons of completely orthogonal color SU(2) subgroups to exchange fermions: something which QCD instantons cannot do. As found by Dorey *et al.*, this leads to a dramatic effect in the large N_c limit, in which k-instanton amplitudes are dominated by remarkable "multi-instanton clusters" in which all instantons are at *the same point and have the same size.*[12]

Any Green function looks like propagators going from various space–time points to the point in AdS_5, which happens to be nothing else but the familiar collective coordinates $d^4z \, d\rho/\rho^5$. Remarkably, even the 5d sphere S_5 appears, from the "di-fermion" condensates induced by zero modes.

The result of their calculation is too technical to be presented here: suffice it to say that it is in truly amazing correspondence with the Maldacena conjecture, and provides a clue on how $AdS_5 \times S_5$ metrics can appear purely in gauge theory context, as a measure of some collective coordinates. I think what it means is that the (long-predicted) "master field" in the large number of colors is not a constant field but a multi-instanton cluster.

---

[12]They fit there without a problem due to the large number of colors.

APPENDIX A

Notations

A.1. Some abbreviations used

QED	Quantum Electrodynamics
QCD	Quantum Chromodynamics, pQCD is its perturbative version
N_c and N_f	are numbers of quark colors and flavors
UV and **IR**	are ultraviolet and infrared, meaning the limits of large and small momenta
RG	Renormalization Group
NJL	Nambu–Jona–Lasinio model
MFA and **RPA**	are the Mean Field and Random Phase Approximations
RILM and **IILM**	are Random Instanton Liquid Model and Interacting Instanton Liquid Model
QGP	Quark–Gluon PLasma
VEV	Vacuum Expectation Value
OPE	Operator Product Expansion
EoS	Equation of State
HTL	Hard Therma Loops
DIS	Deep Inelastic Scattering
2SC=CSC2	Color Superconducting phase with 2 flavors
CFL=CSC3	Color-Flavor Locked phase, or Color Superconducting phase with 3 flavors
SUSY	Supersymmetric
AdS	Anti-de-Sitter space
CFT	Conformal Field Theory

A.2. Units

We use standard "natural units" of high energy/nuclear physics in which the speed of light and the Planck constant are set to $\hbar = c = 1$. Thus length and time has the same dimension, the inverse of momentum and energy. The transition between units occurs by a convenient substitution of 1 according to

$$1 = 0.19732 \, \text{fm} \, \text{GeV},$$

and then cancellation of femto-meters (10^{-15} m, also known as fermis) or GeV (10^9 eV or Giga-electron-volts) as needed.

The discussion of the temperature uses the Boltzmann constant $k_{\mathrm{B}} = 1$, so it is measured in GeV as energy.

A.3.   Space–time and other indices, standard matrices

We follow the standard physics convention according to which an index appearing twice on one side of the equation is a dummy variable with the summation implied, e.g. $a_m b_m \equiv \sum_m a_m b_m$.

We use Latin letters $a, b, \ldots$ to count color generators, 1–8 or 1–$N_c^2 - 1$, and $i, j, \ldots$ to count colors, 1–3 or 1–N_c. We use letters l, m, n also to count spatial vectors 1–3.

Greek letters are generally used for space–time. The standard Minkowski metric $g_{\mu\nu}^{\mathrm{M}} = \mathrm{diag}(1, -1, -1, -1)$ is implied in sums,

$$a_\mu b_\mu \equiv \sum_{\mu\nu} g_{\mu\nu} a_\mu b_\nu.$$

Transition to Euclidean time is done with

$$x_0^{\mathrm{M}} = -i x_4^{\mathrm{E}}, \qquad x_m^{\mathrm{M}} = x_m^{\mathrm{E}}$$

The Euclidean metric is just $g_{\mu\nu}^{\mathrm{E}} = \delta_{\mu\nu} = \mathrm{diag}(1, 1, 1, 1)$.

In the Pauli matrices τ_{ij}^m, all indices 1–3, are twice the generators of the SU(2) rotations. They satisfy the basic relation

$$\tau^a \tau^b = \delta^{ab} + i\epsilon^{abc} \tau^c.$$

Color SU(3) generators are half of the Gell-Mann matrices $T^a = t^a/2$, $a = 1$–8. We also use a notation λ^a for the same set of matrices, and also use those for SU(3) flavor. Their product can be written in a form similar to that for Pauli matrices

$$t^a t^b = \tfrac{2}{3}\delta^{ab} + t^c(d^{abc} + if^{abc}),$$

where d, f are some standard numerical tensors of the SU(3) group.

A.4.   Properties of η symbols

We define 4-vector matrices

$$\tau_\mu^\pm = (\vec{\tau}, \mp i), \quad \tau^a \tau^b = \delta^{ab} + i\epsilon^{abc}\tau^c, \tag{A.1}$$

$$\tau_\mu^+ \tau_\nu^- = \delta_{\mu\nu} + i\eta_{a\mu\nu}\tau^a, \tag{A.2}$$

$$\tau_\mu^- \tau_\nu^+ = \delta_{\mu\nu} + i\bar{\eta}_{a\mu\nu}\tau^a, \tag{A.3}$$

with the η-symbols given by

$$\eta_{a\mu\nu} = \epsilon_{a\mu\nu} + \delta_{a\mu}\delta_{\nu 4} - \delta_{a\nu}\delta_{\mu 4}, \tag{A.4}$$

$$\bar{\eta}_{a\mu\nu} = \epsilon_{a\mu\nu} - \delta_{a\mu}\delta_{\nu 4} + \delta_{a\nu}\delta_{\mu 4}. \tag{A.5}$$

The η-symbols are (anti) self-dual in the vector indices

$$\eta_{a\mu\nu} = \tfrac{1}{2}\epsilon_{\mu\nu\alpha\beta}\eta_{a\alpha\beta}, \qquad \bar{\eta}_{a\mu\nu} = -\tfrac{1}{2}\epsilon_{\mu\nu\alpha\beta}\bar{\eta}_{a\alpha\beta}, \qquad \eta_{a\mu\nu} = -\eta_{a\nu\mu}. \tag{A.6}$$

We have the following useful relations for contractions involving η symbols:

$$\eta_{a\mu\nu}\eta_{b\mu\nu} = 4\delta_{ab}, \tag{A.7}$$

$$\eta_{a\mu\nu}\eta_{a\mu\rho} = 3\delta_{\nu\rho}, \tag{A.8}$$

$$\eta_{a\mu\nu}\eta_{a\mu\nu} = 12, \tag{A.9}$$

$$\eta_{a\mu\nu}\eta_{a\rho\lambda} = \delta_{\mu\rho}\delta_{\nu\lambda} - \delta_{\mu\lambda}\delta_{\nu\rho} + \epsilon_{\mu\nu\rho\lambda}, \tag{A.10}$$

$$\eta_{a\mu\nu}\eta_{b\mu\rho} = \delta_{ab}\delta_{\nu\rho} + \epsilon_{abc}\eta_{c\nu\rho}, \tag{A.11}$$

$$\eta_{a\mu\nu}\bar{\eta}_{b\mu\nu} = 0. \tag{A.12}$$

The same relations hold for $\bar{\eta}_{a\mu\nu}$, except for

$$\bar{\eta}_{a\mu\nu}\bar{\eta}_{a\rho\lambda} = \delta_{\mu\rho}\delta_{\nu\lambda} - \delta_{\mu\lambda}\delta_{\nu\rho} - \epsilon_{\mu\nu\rho\lambda}. \tag{A.13}$$

Some additional relations are

$$\epsilon_{abc}\eta_{b\mu\nu}\eta_{c\rho\lambda} = \delta_{\mu\rho}\eta_{a\nu\lambda} - \delta_{\mu\lambda}\eta_{a\nu\rho} + \delta_{\nu\lambda}\eta_{a\mu\rho} - \delta_{\nu\rho}\eta_{a\mu\lambda}, \tag{A.14}$$

$$\epsilon_{\lambda\mu\nu\sigma}\eta_{a\rho\sigma} = \delta_{\rho\lambda}\eta_{a\mu\nu} + \delta_{\rho\nu}\eta_{a\lambda\mu} + \delta_{\rho\mu}\eta_{a\nu\lambda}. \tag{A.15}$$

A.5. Gauge fields

The QED/QCD gauge part of the Lagrangians is

$$S = -\frac{1}{4}\int d^4x \, (G^a_{\mu\nu})^2,$$

where the QCD field is

$$G^a_{\mu\nu} = \partial_\mu A^a_\nu - \partial_\nu A^a_\mu + g f^{abc} A^b_\mu A^c_\nu,$$

with f^{abc} the structure constant of the SU(N_c) Lee algebra (for SU(2) it is ϵ_{abc}). An alternative form of the gauge fields is a matrix notation, in which the generators are included together with the fields $A_\mu \equiv A^a_\mu T^a$: then the second term is the commutator.

We use mostly the so called perturbative definition in which the coupling constant is explicitly written in non-linear terms while the kinetic terms are free of it. The non-perturbative definition, used in lattice and instanton studies, is obtained by an inclusion of g into $\tilde{A} = gA$ and $\tilde{G} = gG$ so that g no longer appears in front of the nonlinear terms, but is placed instead in front of the action $S = (-1/4g^2) \int d^x \tilde{G}^2$.

The transition to Euclidean time is done by

$$A_0^{\mathrm{M}} = i A_4^{\mathrm{E}}, \qquad A_m^{\mathrm{M}} = -A_m^{\mathrm{E}}.$$

Note the minus sign of spatial components, which is different from what happens with the coordinates themselves. This is done in order not to modify covariant derivatives, so that both E and M read as

$$iD_\mu = i\partial_\mu + \tfrac{1}{2} g A_\mu^a t^a.$$

A.6. Quark fields

We denote quarks fields as ψ or q, usually omitting but implying their spinor index $\alpha = 1\text{–}4$, color $i = 1\text{–}N_c$ and flavor $f = 1\text{–}N_f$. The QED/QCD Lagrangian is

$$S = \int d^4 x \, \bar{q}(i\gamma_\mu D_\mu - m)q.$$

The transition to Euclidean time is done by

$$q^{\mathrm{E}} = q^{\mathrm{M}}, \qquad \bar{q}^{\mathrm{M}} = -i\bar{q}^{\mathrm{E}},$$

and the gamma-matrices change by

$$\gamma_4^{\mathrm{E}} = \gamma_0^{\mathrm{M}}, \qquad \gamma_m^{\mathrm{E}} = -i\gamma_m^{\mathrm{M}}.$$

We recall that the anti-commutators are

$$\{\gamma_\mu^{\mathrm{M}}, \gamma_\nu^{\mathrm{M}}\} = 2g_{\mu\nu}, \qquad \{\gamma_\mu^{\mathrm{E}}, \gamma_\nu^{\mathrm{E}}\} = 2\delta_{\mu\nu}.$$

Another often used notation is a slash or a hat, indicating a convolution of a 4-vector with the gamma matrices. For example, using this notation, the relations just described can be written as

$$\hat{a}\hat{b} + \hat{b}\hat{a} = (ab),$$

where a_μ, b_μ are any 4-vectors.

Finally, the Euclidean fermionic action looks like

$$S^{\mathrm{E}} = -iS^{\mathrm{M}} = \int d^4 x \, \bar{q}(-i\gamma_\mu D_\mu - im)q.$$

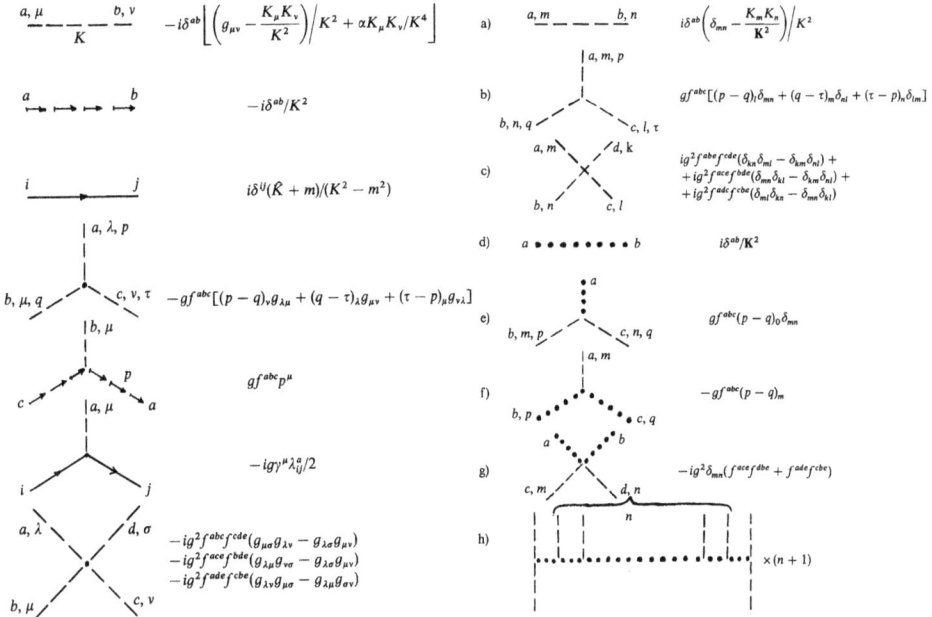

Fig. A.1. The Feynman rules of pQCD, for covariant (left) and Coulomb (right) gauges. The solid lines are quarks. Left: the dashed lines are gluons and arrows indicate Faddeev–Popov ghosts. Right: the dashed and dotted lines indicate the transverse (magnetic) and the Coulomb (electric) gluons.

A.7. QCD Feynman rules

Although we do not discuss perturbative applications of QCD, except at finite temperatures, we found it useful to give the set of Feynman rules in graphical form. Of course, there are rules not included in this figure, such as a minus sign for each fermion and ghost loop.

APPENDIX B

Basic Instanton Formulae

B.1. Instanton gauge potential

We use the following conventions for Euclidean gauge fields: The gauge potential is $A_\mu = A_\mu^a \lambda^a/2$, where the SU($N$) generators are normalized according to $\text{tr}[\lambda^a, \lambda^b] = 2\delta^{ab}$. The covariant derivative is given by $D_\mu = \partial_\mu + A_\mu$ and the field strength tensor is

$$F_{\mu\nu} = [D_\mu, D_\nu] = \partial_\mu A_\nu - \partial_\nu A_\mu + [A_\mu, A_\nu]. \tag{B.1}$$

In our conventions, the coupling constant is absorbed into the gauge fields. Standard perturbative notation corresponds to the replacement $A_\mu \to g A_\mu$. The single instanton solution in regular gauge is given by

$$A_\mu^a = \frac{2\eta_{a\mu\nu}x_\nu}{x^2 + \rho^2}, \tag{B.2}$$

and the corresponding field strength is

$$G_{\mu\nu}^a = \frac{4\eta_{a\mu\nu}\rho^2}{(x^2 + \rho^2)^2}, \tag{B.3}$$

$$(G_{\mu\nu}^a)^2 = \frac{192\rho^4}{(x^2 + \rho^2)^4}. \tag{B.4}$$

The gauge potential and field strength in singular gauge are

$$A_\mu^a = \frac{2\bar{\eta}_{a\mu\nu}x_\nu\rho^2}{x^2(x^2 + \rho^2)}, \tag{B.5}$$

$$G_{\mu\nu}^a = -\frac{4\rho^2}{(x^2 + \rho^2)^2}\left(\bar{\eta}_{a\mu\nu} - 2\bar{\eta}_{a\mu\alpha}\frac{x_\alpha x_\nu}{x^2} - 2\bar{\eta}_{a\alpha\nu}\frac{x_\mu x_\alpha}{x^2}\right). \tag{B.6}$$

Finally, an n-instanton solution in singular gauge is given by

$$A_\mu^a = \bar{\eta}_{a\mu\nu}\partial_\nu \ln \Pi(x), \tag{B.7}$$

$$\Pi(x) = 1 + \sum_{i=1}^{n}\frac{\rho_i^2}{(x - z_i)^2}. \tag{B.8}$$

581

Note that all instantons have the same color orientation. For a construction that gives the most general n-instanton solution, see Ref. [203].

B.2.   Fermion zero modes and overlap integrals

In singular gauge, the zero mode wave function $i\slashed{D}\phi_0 = 0$ is given by

$$\phi_{a\nu} = \frac{1}{2\sqrt{2}\pi\rho}\sqrt{\Pi}\left[\slashed{\partial}\left(\frac{\Phi}{\Pi}\right)\right]_{\nu\mu} U_{ab}\epsilon_{\nu b}, \tag{B.9}$$

where $\Phi = \Pi - 1$. For the single instanton solution, we get

$$\phi_{a\nu}(x) = \frac{\rho}{\pi}\frac{1}{(x^2+\rho^2)^{3/2}}\left(\frac{1-\gamma_5}{2}\right)\frac{\slashed{x}}{\sqrt{x^2}}U_{ab}\epsilon_{\nu b}. \tag{B.10}$$

The instanton–instanton zero mode density matrices are

$$\phi_I(x)_{i\alpha}\phi_J^\dagger(y)_{j\beta} = \tfrac{1}{8}\varphi_I(x)\varphi_J(y)$$
$$\times\left(\slashed{x}\gamma_\mu\gamma_\nu\slashed{y}\frac{1-\gamma_5}{2}\right)_{ij}\otimes\left(U_I\tau_\mu^-\tau_\nu^+U_J\right)_{\alpha\beta}, \tag{B.11}$$

$$\phi_I(x)_{i\alpha}\phi_A^\dagger(y)_{j\beta} = -\tfrac{1}{2}i\varphi_I(x)\varphi_A(y)$$
$$\times\left(\slashed{x}\gamma_\mu\slashed{y}\frac{1-\gamma_5}{2}\right)_{ij}\otimes\left(U_I\tau_\mu^-U_A^\dagger\right)_{\alpha\beta}, \tag{B.12}$$

$$\phi_A(x)_{i\alpha}\phi_I^\dagger(y)_{j\beta} = \tfrac{1}{2}i\varphi_A(x)\varphi_I(y)$$
$$\times\left(\slashed{x}\gamma_\mu\slashed{y}\frac{1+\gamma_5}{2}\right)_{ij}\otimes\left(U_A\tau_\mu^+U_I^\dagger\right)_{\alpha\beta}, \tag{B.13}$$

$$\varphi(x) = \frac{\rho}{\pi}\frac{1}{\sqrt{x^2}(x^2+\rho^2)^{3/2}}. \tag{B.14}$$

The overlap matrix element is given by

$$T_{AI} = \int d^4x\,\phi_A^\dagger(x-z_A)i\slashed{D}\phi_I(x-z_I)$$
$$= r_\mu\,\text{Tr}(U_I\tau_\mu^- U_A^\dagger)\frac{1}{2\pi^2 r}\frac{d}{dr}M(r), \tag{B.15}$$

$$M(r) = \frac{1}{r}\int_0^\infty dp\,p^2|\varphi(p)|^2 J_1(pr). \tag{B.16}$$

The Fourier transform of zero mode profile is given by

$$\varphi(p) = \pi\rho^2\frac{d}{dx}\left(I_0(x)K_0(x) - I_1(x)K_1(x)\right)\Big|_{x=p\rho/2}. \tag{B.17}$$

B.3. Group integration

In order to perform averages over the color group, we need the following integrands over the invariant SU(3) measure:

$$\int dU\, U_{ij}U^{\dagger}_{kl} = \frac{1}{N_{\rm c}}\delta_{jk}\delta_{li},$$

$$\int dU\, U_{ij}U^{\dagger}_{kl}U_{mn}U^{\dagger}_{op} = \frac{1}{N_{\rm c}^2}\delta_{jk}\delta_{li}\delta_{no}\delta_{mp}$$

$$+ \frac{1}{4(N_{\rm c}^2-1)}(\lambda^a)_{kj}(\lambda^b)_{il}(\lambda^a)_{on}(\lambda^b)_{mp}.$$

A different form can be obtained by rearrangement using the SU(N) Fierz transformation

$$(\lambda^a)_{ij}(\lambda^a)_{kl} = -\frac{2}{N_{\rm c}}\delta_{ij}\delta_{kl} + 2\delta_{jk}\delta_{il}. \tag{B.18}$$

A Sample Program for
Numerical Simulation of the
Euclidean Quantum Paths

```
        SUBROUTINE UPDATE(I)
c     updates the i-th coordinate
        COMMON/A/A,delta,N,nit
        COMMON/coord/X(5000)

c checking for the ends, imposing periodic boundary condition
          if(i.eq.1) then
            xleft=x(N)
          else
            xleft=x(i-1)
          endif
          if(i.eq.n) then
            xright=x(1)
          else
            xright=x(i+1)
          endif

c     remember the old value
                  xold=X(I)

C     the main cycle over tries
        ntries=10
          DO 5 L=1,ntries
c colling random number and make displacement of x into x1
        Xnew=DELTA*(2*Rang()-1)+X(I)

c calculating change of the kinetic energy, then adding potential
       Dkin=(Xold-Xnew)*(Xright+Xleft-Xnew-Xold)/A/2.
       Ds=dkin+V(xnew)-V(xold)
```

```
c !!!!! ␣␣␣the␣main␣Metropolis␣block␣!!!!!!!
c␣␣␣if␣action␣is␣gained,␣it␣is␣accepted
␣␣␣␣␣␣␣␣␣IF(DS.LT.0)GO␣TO␣3
c␣␣␣␣␣␣␣if␣action␣is␣increased␣it␣may␣still␣be␣accepted
␣␣␣␣␣␣␣␣␣␣r=␣RANg()
␣␣␣␣␣␣␣␣␣␣␣T=EXP(-DS)
␣␣␣␣␣␣␣␣␣␣␣␣IF(T.GT.r)GO␣TO␣3
c␣␣␣␣␣␣␣␣if␣not␣accepted,␣next␣try
␣␣␣␣5␣CONTINUE
c␣␣␣␣␣␣␣␣␣␣␣non-acceptance:␣nothing␣to␣do
␣␣␣␣␣␣RETURN
c␣␣␣␣␣␣␣␣␣␣␣acceptance:
␣␣␣␣3␣␣␣␣␣␣X(I)=Xnew
␣␣␣␣␣␣RETURN
␣␣␣␣␣␣END
```

Bibliography

1. M. Shifman, ed., At the Frontier of Particle Physics. Handbook of QCD (Ioffefest), WSPC, 2001.

Local gauge invariance
2. H. Weyl, *Ann. Physik* 59 (1919) 101.
3. V. Fock, *Zeit. Physik* 39 (1927) 226.
4. F. London, *Zeit. Physik* 42 (1927) 375.
5. Y. Aharonov and D. Bohm, *Phys. Rev.* 115 (1959) 485.

Non-abelian gauge fields and color
6. C. N. Yang and R. Mills, *Phys. Rev.* 96 (1954) 191.
7. O. V. Greenberg, *Phys. Rev. Lett.* 13 (1964) 598.
8. M. Y. Han and Y. Nambu, *Phys. Rev.* 139 (1965) 1038.

First applications of the gauge theory to strong interactions
9. Y. Nambu, in: Preludes in Theoretical Physics, ed. A. de Shalit, North Holland, Amsterdam, 1966.
10. H. Fritzsch and M. Gell-Mann. 16th Int. Conf. High Energy Physics, Batavia, Vol. 2 (1972) p. 135;
    H. Fritzsch, M. Gell-Mann and H. Leutwyler, *Phys. Lett.* 74B (1973) 365.
11. S. Weinberg, *Phys. Rev. Lett.* 31 (1973) 494.

Quantization: noncovariant gauges free of ghosts
12. J. Schwinger, *Phys. Rev.* 125 (1962) 397, 1043; 127 (1962) 324.
13. N. H. Christ and T. D. Lee, *Phys. Rev.* D22 (1980) 939.

Quantization: covariant gauges
14. R. P. Feynman. *Acta Phys. Polonica* 24 (1963) 697.
15. L. D. Faddeev and V. N. Popov, New difficulties in gauge fixing in the nonperturbative context, *Phys. Lett.* 25B (1967) 29.
16. V. N. Gribov, *Nucl. Phys.* B139 (1978) 1.

Dimensional regularization, lambda (ms), renormalizability
17. G. 't Hooft and M. Veltman, *Nucl. Phys.* B44 (1972) 189;
    G. 't Hooft, *Nucl. Phys.* B61 (1973) 455.

Asymptotic freedom: the levels in magnetic field
18. V. S. Vanyashin and M. V. Terentiev, *ZhETP* 48 (1965) 565. (*Sov. Phys.-JETP* 21 (1965) 375).

19. F. Wilczek, Asymptotic freedom [hep-th/9609099].
20. D. Gross, Vol. 1, pp. 89–120, in Ref. [1].

Asymptotic freedom: electric charge in the Coulomb gauge

21. I. B. Kriplovich, *Yadernaya Fizika* 10 (1969) 409. (*Sov. J. Nucl. Phys.* 10 (1970) 235).
22. J. Frenkel and J. C. Teylor, *Nucl. Phys.* B109 (1976) 185.

Asymptotic freedom: covariant gauges

23. D. Gross and F. Wilczek, *Phys. Rev. Lett.* 26 (1973) 1343.
24. H. D. Politzer, *Phys. Rev. Lett.* 26 (1973) 1346.

Two-loops

25. W. Caswell, *Phys. Rev. Lett.* 33 (1974) 244.
26. A. M. Belavin and A. A. Migdal, *Pisma ZhEtF (JETP Lett.)* 19 (1974) 317.
27. D. R. T. Jones, *Nucl. Phys.* 75B (1974) 531.

Chiral symmetry and quark masses. PCAC

28. M. Gell-Mann, M. Levi, *Nuovo Cimento* 16 (1960) 705.
29. J. Goldstone, *Nuovo Cimento* 19 (1961) 154.
30. M. Gell-Mann, R. J. Oakes and B. Renner, *Phys. Rev.* 175 (1968) 2195.
31. S. Weinberg, *Physica* A96 (1979) 327.
32. J. Gasser and H. Leutwyler, *Phys. Lett.* B125 (1983) 321; *Phys. Rep.* 87 (1982) 79; *Ann. Phys.* (N.Y.) 158 (1984) 142.
33. B. W. Lee, Chiral Dynamics (Gordon and Breach, New York, 1972).
34. H. Leutwyler, in Ref. [1].
35. J. Wess and B. Zumino, *Phys. Lett.* B37 (1971) 95; E. Witten, *Nucl. Phys.* B223 (1983) 422.
36. S. Weinberg, *Phys. Rev.* D11 (1975) 3583.

Chiral symmetry breaking and NJL model

37. Y. Nambu, *Phys. Rev. Lett.* 4 (1960) 380; *Phys. Rev.* 117 (1960) 648. Y. Nambu and G. Jona-Lasinio, *Phys. Rev.* 122 (1961) 345.
38. V. G. Vaks and A. I. Larkin. *ZhETF* 40 (1961) 282.
39. S. P. Klevansky, *Rev. Mod. Phys.* 64 (1992) 649–708.

Chiral symmetry breaking by supercritical one-gluon exchange

40. P. I. Fomin and V. A. Miransky, *Phys. Lett.* B79 (1976) 166. V. A. Miransky, V. P. Gusynin and Yu. A. Sitenko, *Phys. Lett.* B100 (1981) 157.
41. T. Hatsuda and T. Kunihiro, *Phys. Rep.* 247 (1994) 221.

The anomalies

42. J. Steinberger, *Phys. Rev.* 76 (1949) 1180.
43. J. Schwinger, *Phys. Rev.* 82 (1951) 664.
44. S. L. Adler, *Phys. Rev.* 177 (1969) 2426.
45. J. S. Bell and R. Jackiw, *Nuovo Cimento* A60 (1969) 47.
46. G. 't Hooft, in: Recent Developments in Gauge Theories, eds. G. 't Hooft *et al.*, Plenum Press, New York, 1980.
47. V. N. Gribov, preprint, KFKI-1981-66, Budapest (1981), unpublished.
48. J. Ambjorn, J. Greensite and C. Petersson, *Nucl. Phys.* B221 (1983) 381.
49. K. Fujikawa, *Phys. Rev. Lett.* 42 (1979) 1195; 44 (1980) 1733.

Scale invariance and the trace anomaly

50. P. Carruthers, *Phys. Rep.* 1 (1971) 1.
51. R. Crewter, *Phys. Rev. Lett.* 28 (1972) 1421.

52. M. Chanowitz and J. Ellis, *Phys. Lett.* 40B (1972) 397.
53. J. Collins, A. Dunkan and S. Joglekar, *Phys. Rev.* D16 (1977) 438.

Heavy quark symmetry and $1/M$ expansion
54. E. V. Shuryak, *Phys. Lett.* B93 (1980) 134.
55. E. V. Shuryak, *Nucl. Phys.* B198 (1982) 83.
56. N. Isgur and M. Wise, *Phys. Lett.* B232 (1989) 113.
57. I. Bigi, N. Uraltsev and A. Vainshtein, *Phys. Lett.* B293 (1992) 430.
58. M. Neubert, *Phys. Rep.* 245 (1994) 259.
59. M. A. Shifman, in: ITEP Lectures on Particle Physics and Fields Theory, series on: Lecture Notes in Physics, World Scientific, Singapore, Vol. 62.

Confinement
60. Y. Nambu, *Phys. Rev.* D10 4262 (1974);
    S. Mandelstam, *Phys. Rep.* 23C (1976) 145;
    G. 't Hooft, in: Proc. European Physics Society 1975, ed. A. Zichichi (Editrice Compositori, Bologna, 1976) p. 1225.
61. A. A. Abrikosov, *Sov. Phys. JETP* 32, 1442 (1957);
    H. B. Nielsen and P. Olesen, *Nucl. Phys.* B61 (1973) 45.
62. M. Baker, J. S. Ball and F. Zachariasen, *Phys. Lett.* 152B (1985) 351; *Phys. Rev. D* 44 (1991) 3328; 51 (1995) 1968; 56 (1997) 4400;
    S. Maedan and T. Suzuki, *Progr. Theor. Phys.* 81 (1989) 229.
63. E. V. Shuryak, Probing the Boundary of the Non-Perturbative QCD by Small Size Instantons [hep-ph/9909458].
64. I. I. Balitsky, *Nucl. Phys.* B254 (1985) 166.
65. G. S. Bali, The Mechanism of Quark Confinement [hep-ph/9809351].
66. S. Sasaki, M. Fukushima, A. Tanaka, H. Suganuma, H. Toki, O. Miyamura, D. Diakonov, in: *Como 1996, Quark Confinement and the Hadron Spectrum II* 305 [hep-lat/9609043].
67. J. Ambjorn and P. Oleson, *Nucl. Phys.* B170 (1980) 265.
68. J. Greensite, The Confinement Problem in Lattice Gauge Theory [hep-lat/0301023].
69. P. Cea and L. Cosmai, *Phys. Rev.* D43, 620 (1991);
    H. D. Trottier and R. M. Woloshyn, *Phys. Rev. Lett.* 70, 2053 (1993).
70. A. Hanany, M. J. Strassler and A. Zaffaroni, *Nucl. Phys.* B513 (1998) 87.
    M. J. Strassler, *Nucl. Phys. Proc. Suppl.* 73 (1999) 120.
71. J. Ambjorn, P. Olesen and C. Peterson, *Nucl. Phys.* B240 (1984) 189, 533; B244 (1984) 262; *Phys. Lett.* B142 (1984) 410.
    S. Deldar, *Phys. Rev.* D62 (2000) 034509; *JHEP* 0101 (2001) 013.
    G. Bali, *Phys. Rev.* D62 (2000) 114503.
72. B. Lucini and M. Teper, *JHEP* 0106 (2001) 050.
73. L. Del Debbio, H. Panagopoulos, P. Rossi and E. Vicari, K string tensions in SU(n) gauge theories [hep-th/0106185; hep-th/0111090].
74. G. W. Carter and E. V. Shuryak, *Phys. Lett.* B524 (2002) 297 [hep-ph/0101061].
75. M. J. Strassler, Lectures on confinement, in: Proc. Erice 02 School on Subnuclear Physics, ed. A. Zichichi.
76. F. Lenz, J. W. Negele and M. Thies, Confinement from merons, in progress.

Large N_c: general rules
77. G. 't Hooft, *Nucl. Phys.* B72 (1974) 461; Planar Diagram Field Theories, in: Progress in Gauge Field Theory, NATO Advanced Study Institute, eds. G. 't Hooft *et al.* (Plenum, New York, 1984) pp. 271–335.
78. S. Coleman, "1/N", Erice Lectures, 1979.

79. A. V. Manohar, Hadrons in the $1/N$ expansion, in Ref. [1].
80. W. Bardeen and V. I. Zakharov, *Phys. Lett.* 91B (1980) 111.
81. E. Witten, *Nucl. Phys.* B149 (1979) 285.
82. E. Witten, *Phys. Rev. Lett.* 81 (1998) 2862 [hep-th/9807109].
83. M. J. Teper, *Z. Phys.* C5 (1980) 233.
84. H. Neuberger, *Phys. Lett.* B94 (1980) 199.
85. E. V. Shuryak, *Phys. Rev.* D52 (1995) 5370 [hep-ph/9503467].

The $N_c = 2$: a very special theory
86. W. Pauli, *Nuovo Cimento* 6 (1957) 205;
    F. Gursey, *Nuovo Cimento* 7 (1958) 411.
87. M. Peskin, *Nucl. Phys.* B175 (1980) 197.
88. A. Smilga, J. J. M. Verbaarschot, *Phys. Rev.* D51 (1995) 829–837 [hep-th/9404031].

The quark models of hadrons
89. M. Gell-Mann, *Phys. Lett.* 8 (1964) 214.
90. G. Zweig, Cern preprint 8419/th-412, 1964.
91. Review of Particle Properties, *Phys. Rev.* D66 (2002) 010001.

"Constituent" versus "current" quarks
92. M. Gell-Mann, in: Proc 11 Universitätswochen für. Kernphysik, Schladmind 1972 (Schpringer, New York) p. 733.
93. H. J. Mellosh, *Phys. Rev.* D9 (1974) 1095.

Nonrelativistic quark model and spin-spin interaction
94. A. De Rujula, H. Georgy and S. L. Glashow, *Phys. Rev* D12 (1975) 147.
95. N. Isgur and G. Karl, *Phys. Rev.* 18D (1978) 4187.
96. N. I. Kochelev, *Yadernaya Fizika* 41 (1985) 456;
    A. E. Dorokhov, Y. A. Zubov and N. I. Kochelev, *Sov. J. Part. Nucl.* 23 (1992) 522.
97. L. V. Laperashvili and R. G. Betman, *Yadernaya Fizika* 41 (1985) 463.
98. E. V. Shuryak and J. L. Rosner, *Phys. Lett.* B218 (1989) 72.

The dibaryons and the H story
99. R. L. Jaffe, *Phys. Rev. Lett.* 38 (1977) 195.
100. T. Sakai, K. Shimizu and K. Yazaki, *Prog. Theor. Phys. Suppl.* 137 (2000) 121–145 [th/9912063].
101. G. Baym, E. W. Kolb, L. D. McLerran, T. P. Walker and R. L. Jaffe, *Phys. Lett.* B160 (1985) 181.
102. I. B. Khriplovich and E. V. Shuryak, *Sov. J. Nucl. Phys.* 43 (1986) 858 [*Yad. Fiz.* 43 (1986) 1336].
103. M. Oka and S. Takeuchi, *Phys. Rev. Lett.* 63 (1989) 1780; *Nucl. Phys.* A547 (1992) 283C.
104. N. I. Kochelev, *JETP Lett.* 70 (1999) 491–494 [hep-ph/9905333].
105. G. R. Farrar and G. Zaharijas, Transitions of two baryons to the H dibaryon in nuclei [hep-ph/0303047].
106. I. Wetzorke and F. Karsch, The H dibaryon on the lattice [hep-lat/0208029].

$\bar{q}^2 q^2$ mesons and parity doubling
107. M. G. Alford and R. L. Jaffe, *Nucl. Phys.* B578 (2000) 367 [hep-lat/0001023].
108. B. Aubert *et al.*, BABAR Collab. [hep-ex/0304021]; CLEO Collab. (D. Besson *et al.*) [hep-ex/0305017].
109. L. Y. Glozman, Chiral symmetry restoration in hadron spectra [hep-ph/0210216].
110. M. Ishida, The present status on σ and κ meson properties [hep-ph/0212383].

111. L. Y. Glozman and D. O. Riska, *Phys. Rep.* 268 (1996) 263 [hep-ph/9505422].
112. M. Anselmino *et al.*, *Rev. Mod. Phys.* 65 (1993) 1199.

The MIT bag model
113. A. Chodos, R. L. Jaffe, K. Johnson, C. B. Thorn and V. Weisskkopf, *Phys. Rev.* D9 (1974) 3471;
    T. DeGrand, R. L. Jaffe, K. Johnson and J. Kiskis, *Phys. Rev.* D12 (1975) 2060.
114. P. Hazenfratz and J. Kuti, *Phys. Rep.* 40C (1978) 73.
115. R. L. Jaffe, *Phys. Rev.* D15 (1977) 267.

Skyrmions and the chiral bags
116. T. H. R. Skyrme, *Nucl. Phys.* 31 (1962) 556.
117. M. Nowak, M. Rho and I. Zahed, Chiral Nuclear Dynamics (World Scientific, Singapore).
118. D. Finkelstein and J. Rubinstein, *J. Math. Phys.* 9 (1968) 1762.
119. R. A. Leese and N. S. Manton, *Nucl. Phys.* A572 (1994) 575.
120. E. Witten, *Nucl. Phys.* B160 (1979) 57, B223 (1983) 422; B223 (1983) 433;
    G. Adkins, G. Nappy and E. Witten, *Nucl. Phys.* B228 (1983) 552.
121. I. Zahed, U.-G. Meissner and U. F. Kaufluss, *Nucl. Phys.* A426 (1984) 525.
122. J. D. Breit and C. R. Nappi, *Phys. Rev. Lett.* 53 (1984) 889.
123. M. P. Mattis and M. Karliner, *Phys. Rev.* D31 (1985) 2833.
124. I. Zahed and G. E. Brown, *Phys. Rep.* 142 (1986) 1.
125. A. Chodos and C. B. Thorn, *Phys. Rev.* 12D (1975) 2733.
126. G. E. Brown and M. Rho, *Phys. Lett.* 82B (1979) 177.
127. V. Vento *et al.*, *Nucl. Phys.* A345 (1982) 355.
128. S. A. Chin, *Nucl. Phys.* A382 (1982) 355.
129. C. De Tar, *Phys. Rev.* D24 (1981) 752.
130. M. Rho, A. S. Goldhaber and G. E. Brown, *Phys. Rev. Lett.* 51 (1983) 747.
131. D. Diakonov, V. Y. Petrov and P. V. Pobylitsa, *Nucl. Phys.* B306 (1988) 809;
    D. Diakonov, V. Yu. Petrov, in: Ref. [1], Vol. 1* 359–415, also [hep-ph/0009006].

Heavy quarkonia
132. T. Appelquist and H. D. Politzer, *Phys. Rev. Lett.* 34 (1975) 43.
133. M. B. Voloshin, *Nucl. Phys.* B154 (1979) 365.
134. H. Leutwyler, *Phys. Lett.* 98B (1981) 447.
135. V. A. Novikov, L. B. Okun, M. A. Shifman, A. I. Vainshtein, M. B. Voloshin and V. I. Zakharov, *Phys. Rep.* 41 (1978) 1.
136. T. Appelquist, R. M. Barnett and K. D. Lane, *Ann. Rev. Nucl. Part. Science* 28 (1978) 387.

Stochastic vacuum with quasi-constant field
137. G. K. Savvidi, *Phys. Lett.* 71B (1977) 133;
    S. G. Matinyan and G. K. Savvidi, *Nucl. Phys.* B134 (1978) 539.
138. N. K. Nielsen and P. Olesen, *Nucl. Phys.* B144 (1978) 376.
139. M. A. Shifman,*Nucl. Phys.* B173 (1980) 13.
140. H. G. Dosch and Y. A. Simonov, *Phys. Lett.* B205 (1988) 399;
    H. G. Dosch, V. I. Shevchenko and Y. A. Simonov, *Phys. Rep.* 372 (2002) 319 [hep-ph/0007223].

Monopole vacuum
141. S. Mandelstam. *Phys. Rep.* 23C (1976) 245;
    N. P. Nair and C. Rosenzweig, *Phys. Lett.* 131B (1983) 434; 135 (1984) 450.

Tunneling and instantons in quantum mechanics

142. E. Merzbacher, *Physics Today* 55 (2002) 44.
143. F. Hund, *Z. Phys.* 40 (1927) 742.
144. G. Gamow, *Z. Phys.* 51 (1928) 204.
145. R. W. Gurney and E. U. Condon, *Nature* 122 (1928) 439.
146. R. Friedberg, T. D. Lee, W. Q. Zhao and A. Cimenser, *Ann. Phys.* 294 (2001) 67.

Tunneling in the double-well potential and instantons

147. A. M. Polyakov, *Nucl. Phys.* B120 (1977) 429 (one-loop).
148. E. Gildener and A. Patrasciou, *Phys. Rev.* D16 (1977) 423 (two-loop PP density).
149. A. A. Aleinikov and E. V. Shuryak, *Yadernaya Fizika* 46 (1987) 122.
150. S. Olejnik, *Phys. Lett.*, B221 (1989) 372.
151. C. F Wohler and E. V. Shuryak, *Phys. Lett.* B333 (1994) 467.
152. J. Zinn-Justin, *J. Math. Phys.* 22 (1981) 511.

Path integrals in quantum and statistical mechanics

153. R. P. Feynman and H. R. Hibbs, Quantum Mechanics and Path Integrals. (McGraw-Hill, New York, 1965) Chapter 10.
154. R. P. Feynman, Statistical Mechanics (W. A. Benjamin, Massachusetts, 1972).

Numerical simulations of the path integrals in quantum mechanics

155. M. Creutz and B. Freedman, *Ann. Phys.* 132 (1981) 427.
156. E. V. Shuryak and O. V. Zhirov, *Nucl. Phys.* B242 (1984) 394.
157. E. V. Shuryak, Tunneling in the double well potential, *Nucl. Phys.* B302 (1988) 621.

High order effects and tunneling in the double-well problem

158. S. V. Faleev and P.G. Silvestrov, *Phys. Lett.* 197 (1995) 372.
159. E. B. Bogomolny, *Phys. Lett.* B91 (1980) 431.

The double-well problem coupled to fermions

160. P. Salomonson and J. W. Van Holton, *Nucl. Phys.* 196 (1981) 509.
161. F. Cooper, A. Khare and U. Sukhatme, *Phys. Rep.* 251 (1995) 267.
162. E. Witten, *Nucl. Phys.* B188 (1981) 513.

Perturbative series in quantum mechanics

163. C. M. Bender and T. T. Wu, *Phys. Rev.* 184 (1969) 1231; D7 (1973) 1620.
164. F. Dyson, *Proc. Cambridge Phil. Soc.* 48, 625.
165. A. I. Vainshtein, Decaying systems and divergence of the perturbation theory, Preprint of IYaF (1964) reprinted in: Proc. Continuum Advances in QCD (2002). see also discussion in: G. V. Dunne [hep-th/0207046].
166. L. N. Lipatov, *Sov. Phys. JETP* 45 (1977) 216–223 (*Zh. Eksp. Teor. Fiz.* 72 (1977) 411–427).
167. E. Brezin, G. Parisi and J. Zinn-Justin, *Phys. Rev.* D16 (1977) 408.
168. G. C. LeGuillou and J. Zinn-Justin, eds., Large Order Behaviour of the Perturbation Theory (North-Holland, Amsterdam, 1990).

Instantons in gauge theories: texts/reviews

169. R. Rajamaran, Solitons and Instantons (North-Holland, Amsterdam, 1982).
170. S. Coleman, The uses of instantons, Erice lecture (1977).
171. V. A. Novikov, M. A. Shifman, A. I. Vainshtein and V. I. Zakharov, *Uspekhi Fiz. Nauk (Sov. Phys.-uspekhi)* 136 (1982) 553;
also in: M. Shifman, ITEP Lectures on Particle Physics and Fields Theory, World Scientific, Lecture Notes in Physics (World Scientific, Singapore) Vol. 62.
172. T. Schafer and E. V. Shuryak, *Rev. Mod. Phys.* 70 (1998) 323 [hep-ph/9610451].

The topological potential and sphalerons

173. N. Manton, *Phys. Rev.* D28 (1983) 2019;
     F. R. Klinkhamer and N. Manton, *Phys. Rev.* D30 (1984) 2212.
174. J. Zadrozny, *Phys. Lett. B* 284 (1992) 88;
     M. Hellmund and J. Kripfganz, *Nucl. Phys.* B373 (1992) 749.
175. D. M. Ostrovsky, G. W. Carter and E. V. Shuryak, *Phys. Rev.* D66 (2002) 036004
     [hep-ph/0204224].

Discovery of instantons in Yang–Mills theory

176. A. M. Polyakov, *Phys. Lett.* 59B (1975) 82;
     A. A. Belavin, A. M. Polyakov, A. A. Schwartz and Yu. S. Tyupkin. *Phys. Lett.* 59B
     (1975) 85.

One loop quantum effects for Yang–Mills instantons

177. G. 't Hooft, *Phys. Rev.* 14D (1976) 3432; (e) 18D (1978) 2199.
178. C. W. Bernard, N. H. Christ, A. H. Guth and E. J. Weinberg, *Phys. Rev.* D16 (1977)
     2697.
179. C. Bernard, *Phys. Rev.* D19 (1979) 3013.
180. L. S. Brown and D. B. Creamer, *Phys. Rev.* D18 (1978) 3695.

The topological terms in action and theta vacua

181. R. Jackiw and C. Rebbi, *Phys. Rev. Lett.* 37 (1976) 172.
182. C. G. Callan, R. F. Dashen and D. J. Gross.
183. N. Snyderman and S. Gupta, *Phys. Rev.* D24 (1981) 542.
184. G. Schierholz, preprint, DESY (1994) [hep-lat/9409019]; *Phys. Lett.* 63B (1976) 334.

The topological susceptibility

185. W. Bardeen, B. Lee and R. Shrock. *Phys. Rev.* D14 (1976) 985.
186. E. Brezin and J. Zinn-Justin, *Phys. Rev.* B14 (1976) 3110.
187. D'Adda, M. Lusher and P. Di Vecchia, *Nucl. Phys.* B146 (1978) 63.
188. E. Witten, *Nucl. Phys.* B149 (1979) 285.
189. G. Veneziano, U (1) without instantons, *Nucl. Phys.* B159 (1979) 213.

Axions

190. R. D. Peccei and H. R. Quinn, *Phys. Rev. Lett.* 38 (1977) 1440; *Phys. Rev.* D16
     (1977) 1791.
191. S. Weinberg, *Phys. Rev. Lett.* 40 (1978) 223.
192. F. Wilczek, *Phys. Rev. Lett.* 40 (1978) 279.

Experimental limits on the value of theta

193. V. Baluni, *Phys. Rev.* D19 (1979) 2227.
194. P. Di Veccia, Acta *Physica Austriaca Suppl.* 22 (1980) 341.
195. R. J. Crewter *et al.*, *Phys. Lett.* 88B (1980) 123.

Invisible axions and cosmology

196. A. R. Zhitnitsky, *Sov. J. Nucl. Phys.* 31 (1980) 260 (*Yad. Fiz.* 31:4).
197. J. Preskill, M. B. Wize and F. Wilczek, *Phys. Lett.* 120B (1983) 127.
198. L. F. Abott and P. Sikivie, *Phys. Lett.* 120B (1983) 133.
199. M. Dine and W. Fischer, *Phys. Lett.* 120 (1983) 137.

Multiinstantons

200. E. Corrigan and D. Firlie, *Phys. Lett.* 67b (1977) 69.
201. R. Jackiw, C. Nohl and C. Rebbi, *Phys. Rev.* D15 (1977) 1642.

202. E. Witten, *Phys. Rev. Lett.* 38 (1977) 121.
203. M. F. Atiah, N. I. Hitchin, V. G. Drinfeld and Yu. I. Manin, *Phys. Lett.* 65A (1978) 185.
204. E. Corrigan, P. Goddard and S. Templeton, *Nucl. Phys.* B151 (1979) 93.
205. M. Garcia Perez, T. G. Kovacs and P. van Baal, *Phys. Lett.* B472 (2000) 295.

Propagators in the instanton field
206. L. S. Brown, R. D. Carlitz, D. B. Creamer and C. Lee, *Phys. Rev.* D17 (1978) 1583.
207. P. V. Pobylitsa, *Phys. Lett.* B226 (1989) 387.

Instanton interactions
208. D. Foerster, *Phys. Lett.* 66B (1977) 279.
209. E. V. Shuryak, *Phys. Lett.* 153B (1985) 162.
210. E. V. Shuryak, Numerical experiments with the instanton liquid, in: Proc. First Conference on Numerical Experiments in Quantum Field Theories (Alma-Ata, 1985).
211. I. I. Balitsky and A. V. Yung, *Phys. Lett.* 168B (1986) 113.
212. A. V. Yung, *Nucl. Phys.* B297 (1988) 47.
213. J. J. M. Verbaarschot, *Nucl. Phys.* B362 (1991) 33.

The first physical applications of instantons in QCD
214. C. G. Callan, R. Dashen and D. J. Gross, *Phys. Rev.* D17 (1978) 2717; D19 (1979) 1826.
215. R. D. Carlitz and D. B. Creamer. *Ann. Phys.* (NY), 118 (1979) 429.
216. E. V. Shuryak, *Phys. Lett.* 79B (1978) 135.
217. C. G. Callan, R. Dashen and D. J. Gross, *Phys. Rev.* D20 (1979) 3279.
218. M. A. Shifman, A. I. Vainshtein and V. I. Zakharov, *Nucl. Phys.* B163 (1980) 43; B165 (1981) 45.

The hard-core model and the variational MFA
219. E.-M. Ilgenfritz and M. Mueller-Preussker, *Nucl. Phys.* B184 (1981) 443.
220. D. I. Diakonov and V. Yu. Petrov, *Nucl. Phys.* B245 (1984) 259.

Light quarks and instantons
221. M. Atiyah and I. Singer, *Ann. Math.* 87 (1968) 484.
222. A. S. Schwartz, *Phys. Lett.* 67B (1977) 172.
223. B. Grossman, *Phys. Lett.*, A61 (1977) 86.

Chiral symmetry breaking in the mean field approximation (MFA)
224. D. G. Caldi, *Phys. Rev. Lett.* 39 (1977) 121.
225. R. D. Carlitz and D. B. Creamer, *Ann. Phys.* (NY) 118 (1979) 429.
226. E. V. Shuryak, *Nucl. Phys.* B203 (1982) 93.
227. E. V. Shuryak, *Nucl. Phys.* B214 (1983) 237.
228. E.-M. Ilgenfritz and E. V. Shuryak, *Nucl. Phys.* B319 (1989) 511.
229. D. I. Diakonov and V. Yu. Petrov, A Theory of Light Quarks in the Instanton Vacuum, preprint LINP 1053, Leningrad (1985);
230. S. Chandrasekharan, *Nucl. Phys. Proc. Suppl.* B42 (1995) 475.
231. M. A. Nowak, J. J. M. Verbaarschot and I. Zahed, *Phys. Lett.* 217B (1989) 157.
232. M. Musakhanov, *Nucl. Phys.* A699 (2002) 340.
233. P. Faccioli and E. V. Shuryak, *Phys. Rev.* D64 (2001) 114020 [hep-ph/0106019].

Numerical simulations of the instanton liquid
234. E. V. Shuryak, *Nucl. Phys.* B302 (1988) 559, 574, 599.
235. E. V. Shuryak and J. J. Verbaarschot, *Nucl. Phys.* B341 (1990) 1.

236. J. J. Verbaarschot, *Nucl. Phys.* B427 (1994) 534 [hep-lat/9402006].

237. E. V. Shuryak and J. J. M. Verbaarschot, *Nucl. Phys.* B410 (1993) 37.

238. E. V. Shuryak and J. J. M. Verbaarschot, *Nucl. Phys.* B410 (1994) 55.

239. T. Schäfer, E. V. Shuryak and J. J. M. Verbaarschot, *Nucl. Phys.* B412 (1994) 143.

240. T. Schäfer and E. V. Shuryak, *Phys. Rev.* D50 (1994) 478.

241. T. Schäfer and E. V. Shuryak, *Phys. Rev. Lett.* 75 (1995) 1707.

242. T. Schafer and E. V. Shuryak, *Phys. Rev.* D53 (1996) 6522 [hep-ph/9509337].

243. E. V. Shuryak and J. J. M. Verbaarschot, *Phys. Rev.* D52 (1995) 295.

244. T. Schafer, E. V. Shuryak, *Phys. Rev. Lett.* 86 (2001) 3973–3976 [hep-ph/0010116].

245. T. Schäfer, E. V. Shuryak and J. J. M. Verbaarschot, *Phys. Rev.* D51 (1995) 1267.

Instantons in the large N_c limit

246. T. Schafer, *Phys. Rev.* D66 (2002) 076009 [hep-ph/0206062].

Instanton-induced decay modes of charmonium

247. J. D. Bjorken, Intersections 2000: What's new in hadron physics [hep-ph/0008048].

248. V. Zetocha and T. Schafer, *Phys. Rev.* D67 (2003) 114003 [hep-ph/0212125].

Lattice Gauge Theories, Reviews

249. J. B. Kogut, *Rev. Mod. Phys.* 51 (1979) 659; 55 (1983) 775.

250. M. Creutz, *Phys. Rep.* 95 (1983) 201; Lattice gauge theory: a retrospective, *Nucl. Phys. Proc. Suppl.* 94 (2001) 219 [hep-lat/0010047].

Pionear lattice works

251. K. G. Wilson, *Phys. Rev.* D10 (1974) 2445.

252. F. J. Wegner, *Math. Phys.* 12 (1971) 2259.

253. K. G. Wilson, Quark and String on a Lattice — New Phenomena in Subnuclear Physics, ed. A. Zichichi (Plenum Press, New York, 1977) p. 13.

254. N. S. Manton, *Phys. Lett.* 96b (1980) 238.

255. J. Villain, *J. de Physic* 36 (1975) 581.

Hamiltonian formulation

256. J. B. Kogut and L. Susskind, *Phys. Rev.* 11D (1975) 395.

Quantization in a small volume

257. P. van Baal, QCD in a finite volume [hep-ph/0008206].

Renormalization group and action improvement

258. A. A. Migdal, *Soviet Phys.-JETP* 42 (1976) 413, 743.

259. L. P. Kadanoff, *Ann. Phys.* (NY), 100 (1976) 359;
Rev. Mod. Phys. 49 (1977) 267.

260. G. Martinelli and G. Parisi, *Nucl. Phys.* B180/FS2/ (1980) 201.

261. K. G. Wilson, Cargese Lectures (1979), ed. G. 't Hooft ed. (Plenum press, New York 1980).

262. O. K. Symanzik, *Nucl. Phys.* B226 (1983) 187, 205.

263. P. Weisz, *Nucl. Phys.* B221 (1983) 1.

264. G. Curti *et al.*, *Phys. Lett.* 130B (1983) 205.
Strong coupling expansion

265. K. G. Wilson, *Phys. Rev.* D10 (1974) 2445.

266. J. B. Kogut, R. B. Pearson and J. Shigemitsu, *Phys. Rev. Lett.* 43 (1979) 484; *Phys. Lett.* 98b (1981) 63.

Numerical experiments in statistical and quantum mechanics
The "Metropolis" algorithm

267. N. Metropolis, A. Rosenbluth, M. Rozenbluth, A. Teller and E. Teller, *J. Chem. Phys.* 21 (1953) 1087.

Review of the applications to statistical mechanics

268. K. Binder, in: Phase Transition and Critical Phenomena, eds C. Domb and S. Green (Academic Press, New York, 1976), Vol. 5B.

First Monte-Carlo experiments with gauge theories

269. M. Creutz, L. Jacobs and C. Rebbi, *Phys. Rev.* D20 (1979) 1915.
270. M. Creutz, *Phys. Rev. Lett.* 43 (1979) 553; 45 (1980) 313; *Phys. Rev.* D21 (1980) 2308.
271. D. Petcher and D. Weingarten, *Phys. Rev.* D22 (1980) 2465.
272. G. Bhanot and C. Rebbi, *Phys. Rev.* D24 (1980) 3319.

Low Dirac eigenvalues and random matrix theory

273. E. P. Wigner, *Ann. Math.* 53 (1951) 36.
274. C. E. Porter, Statistical Theory of Spectra: Fluctuations (Academic Press, New York, 1965).
275. H. Leutwyler and A. V. Smilga, *Phys. Rev.* D46 (1992) 5607.
276. E. V. Shuryak and J. J. M. Verbaarschot, *Nucl. Phys.* A560 (1993) 306.
277. J. J. Verbaarschot, *Phys. Rev. Lett.* 72 (1994) 2531 [hep-th/9401059].
278. J. J. M. Verbaarschot and T. Wettig, Random matrix theory and chiral symmetry in QCD [hep-ph/0003017].
279. M. E. Berbenni-Bitsch, S. Meyer, A. Schäfer, J. J. M. Verbaarschot and T. Wettig, *Phys. Rev. Lett.* 80 (1998) 1146.

Fermions on the lattice

280. F. A. Berezin, *Commun. Math. Phys.* 40 (1975) 153–174.
281. K. G. Wilson, Quark and string on a lattice — New Phenomena in Subnuclear Physics, ed. A. Zichichi (Plenum press, New York, 1997) p. 13.
282. T. Banks, J. Kogut and L. Susskind, *Phys. Rev.* D13 (1976) 1043.
283. N. H. Nielsen and M. Ninomiya, *Phys. Lett.* B105 (1981) 219.
284. P. H. Ginsparg and K. G. Wilson, *Phys. Rev.* D25 (1982) 2649.
285. H. Neuberger, *Phys. Lett.* B417 (1998) 141.
286. D. Kaplan, *Phys. Lett.* B288 (1992) 342.
287. P. Hasenfratz and F. Niedermayer, *Nucl. Phys.* B414 (1994) 785.
288. P. Hasenfratz [hep-lat/9709110].
289. M. Luscher, *Phys. Lett.* B428 (1998) 342 [hep-lat/9802011].
290. C. Gattringer, *Phys. Rev. Lett.* 88 (2002) 221601 [hep-lat/0202002].
291. T. DeGrand, A. Hasenfratz, P. Hasenfratz and F. Niedermayer, *Nucl. Phys.* B454 (1995) 587; *Nucl. Phys.* B454 (1995) 615; *Phys. Lett.* B365 (1996) 233.
292. M. Teper, Large $N(C)$ physics from the lattice, [hep-ph/0203203];
     N. Cundy, M. Teper and U. Wenger, Topology and chiral symmetry breaking in SU(N_c) gauge theories [hep-lat/0206029].

Glueballs in pure gauge theories

First papers on mass gap in SU(2) theory

293. B. Berg, *Phys. Lett.* 97B (1980) 401.
294. G. Bhanot and C. Rebbi, *Nucl. Phys.* B180 [FS2] (1981) 469.
295. J. Engels, F. Karsch, H. Satz and I. Montvay, *Phys. Lett.* 102B (1981) 332.

296. R. C. Brower, M. Creutz and M. Nauenberg, *Nucl. Phys.* B210 [FS6] (1982) 133.
297. B. Berg, A. Billoir and C. Rebbi, *Ann. Phys.* (NY) 142 (1982) 185; 146 (1983) 470.

First papers on the mass gap in the SU(3) theory
298. H. Hamber and G. Parisi, *Phys. Rev. Lett.* 47 (1981) 1792.
299. B. Berg and A. Billoir, *Phys. Lett.* 113B (1982) 65; 114B (1982) 324; *Nucl. Phys.* B221 (1983) 109.

Recent glueball results
300. D. Weingarten, *Nucl. Phys. Proc. Suppl.* 34 (1994) 29.
301. J. Sexton, A. Vaccarino and D. Weingarten, preprint [hep-lat/9510022].
302. G. S. Bali, K. Schilling, A. Hulsebos, A. C. Irving, C. Michael and P. W. Stephenson, *Phys. Lett.* B309 (1993) 378.
303. F. de Forcrand and K.-F. Liu, *Phys. Rev. Lett.* 69 (1992) 245.
304. C. Morningstar and M. J. Peardon, *Nucl. Phys. Proc. Suppl.* 83 (2000) 887 [hep-lat/9911003].

Quenched approximation
305. H. Hamber and G. Parisi. *Phys. Rev. Lett.* 47 (1981) 1792; *Phys. Rev.* D27 (1983) 208.
306. E. Marinari, G. Parisi and C. Rebbi, *Phys. Rev. Lett* 47 (1981) 1795.
307. S. R. Sharpe, *Phys. Rev.* D46 (1992) 3146 [hep-lat/9205020].
308. C. Bernard, S. Hashimoto, D. B. Leinweber, P. Lepage, E. Pallante, S. R. Sharpe and H. Wittig, Panel discussion on chiral extrapolation of physical observables [hep-lat/0209086].

Unquenched lattice; the virtual quarks are included
309. M. Luscher, On a relation between finite size effects and elastic scattering processes, in: Progress in gauge field theory, eds. 't Hooft *et al.* (Plenum Press, New York, 1984).
310. C. Allton *et al.*, UKQCD collaboration [hep-lat/0203035].
311. S. Ono, *Phys. Rev.* D17 (1978) 888.
312. F. X. Lee *et al.*, *Nucl. Phys. B (Proc. Suppl.)* 119 (2003) 296 (Lattice 2002 proceedings).
313. P. Dreher *et al.* [SESAM Collaborations], Continuum extrapolation of moments of nucleon quark distributions in full QCD [hep-lat/0211021].

Topology on the lattice
314. W. Bietenholz, R. Brower, S. Chandrasekharan and U.-J. Wiese, [hep-lat/9704015].
315. B. Berg and M. Luscher, *Nucl. Phys.* B190 [FS3] (1981) 412;
   M. Lüscher, *Commun. Math. Phys.* 85 (1982) 39.
316. P. di Vecchia, K. Fabricius, G. C. Rossi and G. Veneziano, *Nucl. Phys.* B192 (1981) 392.

"Cooling" of the quantum configurations toward the classical ones
317. Y. Iwasaki and T. Yoshie, *Phys. Lett.* 131B (1983) 159; 131B (1984) 73; 143B (1984) 449.
318. E.-M. Ilgenfritz *et al.*, *Nucl. Phys.* B268 (1986) 693.

The lowest Dirac eigenmodes
319. I. Horváth, N. Isgur, J. McCune and H. B. Thacker, *Phys. Rev.* D 65 (2002) 014502.
320. S. J. Dong *et al.*, *Nucl. Phys. Proc. Suppl.* 106 (2002) 563 [hep-lat/0110037].
321. T. Blum *et al.* [hep-lat/0105006].
322. T. DeGrand and A. Hasenfratz, *Phys. Rev.* D65 (2002) 014503 [hep-lat/0103002].
323. H. Neff *et al.* [hep-lat/0106016].

The correlation functions: review

324. E. V. Shuryak, *Rev. Mod. Phys.* 65 (1993) 1.

Exact inequalities

325. E. T. Tomboulis, *Phys. Rev. Lett.* 50 (1983) 88.
326. E. Witten, *Phys. Rev. Lett.* 51 (1983) 2351.
327. E. Vafa and E. Witten, *NP* B234 (1984) 173; *Commun. Math. Phys.* 95 (1984) 257.
328. D. Weingarten, *Phys. Rev. Lett.* 51 (1983) 1830.
329. S. Nussinov and M. A. Lampert, *Phys. Rep.* 362 (2002) 193 [hep-ph/9911532].
330. P. Faccioli and T. A. DeGrand [hep-ph/0304219].

The quark-hadron duality

331. E. C. Poggio, H. R. Quinn and S. Weinberg, *Phys. Rev.* D13 (1976) 1958.
332. M. A. Shifman, in: Continuous Advances in QCD, ed. A. Smilga (World Scientific, Singapore, 1994), p. 249 [hep-ph/9405246]. B. Chibisov, R. D. Dikeman, M. Shifman and N. Uraltsev, *Int. J. Mod. Phys.* A12 (1997) 2075 [hep-ph/9605465].
333. R. F. Lebed and N. G. Uraltsev, *Phys. Rev.* D62 (2000) 094011 [hep-ph/0006346].
334. M. S. Dubovikov and A. V. Smilga, *Yad. Fiz.* 37 (1983) 984; *Sov. J. Nucl. Phys.* 37 (1983) 585.

Operator product expansion

335. K. G. Wilson, *Phys. Rev.* 179 (1969) 1499.
336. V. A. Novikov, M. A. Shifman, A. I. Vainstein and V. I. Zakharov, *Nucl. Phys.* B249 (1985) 445.

OPE in deep inelastic scattering: the leading and the next twists

337. H. D. Politzer, *Phys. Rep.* 14C (1974) 130.
338. A. De Rujula, H. Georgi and H. D. Politzer, *Ann. Phys.* 103 (1977) 315.
339. E. V. Shuryak and A. I. Vainstein, *Phys. Lett.* 105B (1981) 65; *Nucl. Phys.* B199 (1982) 451; B201 (1982) 141.
340. R. L. Jaffe and M. Soldate, *Phys. Rev.* D26 (1982) 49;
     R. L. Jaffe, *Phys. Lett.* 116B (1982) 437; *Nucl. Phys.* B229 (1983) 205;
     M. Soldate, *Nucl. Phys.* B223 (1983) 61.
341. R. K. Ellis, W. Furmanski and R. Petronzio, *Nucl. Phys.* B207 (1982).

OPE and hard exclusive reactions

342. V. A. Matveev, R. M. Muradyan and A. N. Tavkhelidze. Lett. Nuovo Cimento 7 (1973) 719.
343. S. J. Brodsky and G. R. Farrar, *Phys. Rev. Lett.* 31 (1973) 1153; *Phys. Rev.* D11 (1975) 1309.
344. G. R. Farrar and D. R. Jackson, *Phys. Rev. Lett.* 35 (1975) 1416.
345. P. V. Landshoff, *Phys. Rev.* D10 (1974) 1027.
346. A. I. Vainstein and V. I. Zakharov, *Phys. Lett.* 72B (1978) 368.
347. V. L. Chernyak and A. R. Zhitnitsky, *Pisma ZHETF (JETP Lett.)* 25 (1977) 510; *Phys. Rep.* 112 (1984) 173.
348. G. R. Farrar and D. R. Jackson, *Phys. Rev. Lett.* 43 (1979) 246.

The QCD sum rules: early works

349. T. Appelquist and H. Georgy, *Phys. Rev.* D8 (1973) 4000.
350. A. Zee, *Phys. Rev.* D8 (1973) 4038.
351. V. A. Novikov *et al.*, *Phys. Rev. Lett.* 38 (1977) 626; *Phys. Lett.* 67B (1977) 409.
352. M. A. Shifman, A. I. Vainshtein and V. I. Zakharov, *Nucl. Phys.* B147 (1979) 385.

Non-relativistic and relativistic sum rules at fixed euclidean time

353. M. B. Voloshin, *Yad. Fis.* 29 (1979) 1368.
354. E. V. Shuryak, *Phys. Lett.* 136B (1984) 269.

The OPE-based QCD sum rules

355. V. A. Novikov, M. A. Shifman, A. I. Vainstein and V. I. Zakharov, *Uspechi fis. Nauk* (Soviet physics-Uspechi) 136 (1982) 553; *Fortschr. Phys.* 32 (1984) 585.
356. E. V. Shuryak, *Phys. Rep.* 115C (1984) 151.
357. L. J. Reinders, H. Rubinstein and S. Yazaki, *Phys. Rep.* 127 (1985) 1.
358. S. Narison, *Z. Phys.* C26 (1984) 209.
359. E. Bagan and S. Steele, *Phys. Lett.* B243 (1990) 413.
360. S. Narison, 1995, preprint [hep-ph/9512348].

The fixed-point gauge and OPE

361. V. A. Fock, *Sov. Phys.* 12 (1937) 404.
362. J. Schwinger, Particles, sources and fields, Vol. 1 (Addison-Wesley, 1970).
363. M. S. Dubovikov and A. V. Smilga, *Nucl. Phys.* B185 (1981) 109; A. V. Smilga, *Yadernaya Fizika* 35 (1982) 437.
364. E. V. Shuryak and A. I. Vainstein, *Nucl. Phys.* B201 (1982) 141.

Mesons made of light quarks

365. S. I. Eidelman, L. M. Kurdadze and A. I. Vainstein, *Phys. Lett.* 82B (1979) 278.
366. G. Launer, S. Narison and R. Tarrach, Nonperturbative QCD vacuum from data on e + e- into (I = 1) hadrons. TH 3712-CERN, GENEVA 1983.
367. L. J. Reinders, H. R. Rubinstein and S. Yazaki, *Nucl. Phys.* B196 (1982) 125.
368. S. Narisson and E. De Rafael, *Phys. Lett.* 103B (1981) 57.
369. S. Narisson, N. Paver, E. De rafael and D.'releani. *Nucl. Phys.* B212 (1983) 365; S. Narison,*Phys. Lett.* 104B (1981) 485.
370. D. J. Broadhurst and S. C. Generalis, *Phys. Lett.* 165B (1985) 175; A. G. Grosin and Yu. F. Pinelis *Phys. Lett.* 166B (1986) 429.
371. I. I. Balitsky, D. I. Diakonov and A. V. Yung, *Phys. Lett.* 112B (1982) 71.
372. A. R. Zhitnitsky and I. R. Zhitnitsky, *Yadernaya Fizika* 37 (1983) 6.
373. J. Covaets, F. De Viron, D. Gusbin and J. Weyers, *Phys. Lett.* 28B (1983) 262; *Nucl. Phys.* B248 (1984) 1.

OPE and sum rules for baryons

374. B. L. Ioffe, *Nucl. Phys.* B188 (1981) 317; E B191 (1981) 591; V. M. Belyaev and B. L. Ioffe. ZHETF (Soviet Phys.-JETP) 83 (1982) 876.
375. Y. Chung, H. G. Dosch, M. Kremer and D. Schale, *Nucl. Phys.* B197 (1982) 55.
376. A. V. Smilga, *Yadernaya Fizika* 35 (1982) 473.

Currents with derivatives, hadronic wave functions

377. V. L. Chernyak and A. R. Zhitnitsky, *Nucl. Phys.* B201 (1982) 492.
378. V. L. Chernyak, A. R. Zhitnitsky and I. R. Zhitnitsky, *Nucl. Phys.* B204 (1982) 477.
379. V. L. Chernyak and I. R. Zhitnitsky, *Nucl. Phys.* B246 (1984) 52.
380. M. A. Shifman, Pisma ZHETF (JETP Letters) 30 (1979) 546.
381. V. A. Novikov *et al.*, *Phys. Reports* 41C (1978) 1.
382. Yu. A. Khodjamirian, *Phys. Lett.* 90B (1980) 460.
383. B. L. Ioffe and A. V. Smilga, *Nucl. Phys.* 232B (1984) 109.
384. I. Balitsky and A. Yung, *Phys. Lett.* 129B (1983) 328.

Correlators and sum rules: solvable examples, e.g. constant field

385. J. Schwinger, *Phys. Rev.* 82 (1951) 664.
386. L. S. Brown and W. I. Weisberger, *Nucl. Phys.* B157 (1979) 285.

387. E. V. Shuryak, *Phys. Lett.* 136B (1984) 269.
388. A. I. Mil'stein and Yu. F. Pinelis, *Phys. Lett.* 137B (1984) 235.

Polarization operator in the instanton field
389. N. Andrei and D. J. Gross, *Phys. Rev.* D18 (1978) 468.
390. L. Ballieu, J. Ellis, M. K. Gaillard and W. J. Zakrzewsky, *Phys. Lett.* 77B (1978) 290.

Approximate methods for propagators and functional determinants
391. D. I. Diakonov and V. Yu. Petrov, *Zh.E.T.F.(Sov. Phys.-JETP)* 86 (1984) 25.
392. D. I. Diakonov, V. Yu. Petrov and A. Yung, *Phys. Lett.* 130B (1983) 385.
393. S. Chernyshev and M. A. Nowak and I. Zahed [hep-ph/9510326].
394. M. Kacir, M. Prakash and I. Zahed [hep-ph/9602314].

Instanton-induced effects for correlators
395. B. V. Geshkenbein and B. L. Ioffe, *Nucl. Phys.* B166 (1980) 340.
396. V. A. Novikov, M. A. Shifman, A. I. Vainstein and V. I. Zakharov, *Nucl. Phys.* B191 (1981) 301.
397. A. E. Dorokhov and N. I. Kochelev, *Z. Phys.* C 46 (1990) 281.
398. H. Forkel and M. K. Banerjee, *Phys. Rev. Lett.* 71 (1993) 484.

Point-to-point correlators on the lattice
399. M. C. Chu, J. M. Grandy, S. Huang and J. W. Negele, *Phys. Rev.* D48 (1993) 3340; D49 (1994) 6039.
400. C. Michael and P. S. Spencer, *Phys. Rev.* D50 (1994) 7570.
401. D. B. Leinweber, *Phys. Rev.* D51 (1995) 6369, 6383.
402. P. Hasenfratz and I. Montvay, *Phys. Rev. Lett.* 50 (1983) 309.
403. B. Velikson and D. Weingarten. *Nucl. Phys.* B249 (1985) 433.
404. M.-C. Chu, M. Lissia and J. W. Negele, *Nucl. Phys.* B360 (1991) 31.
405. M. Hecht and T. DeGrand, *Phys. Rev.* D46 (1992) 2115.
406. R. Gupta, D. Daniel and J. Grandy. *Phys. Rev.* D48 (1993) 3330.
407. C. Bernard, T. A. DeGrand, C. DeTar, S. Gottlieb, A. Krasnitz, M. C. Ogilvie, R. L. Sugar, and D. Toussaint. *Phys. Rev. Lett.* 68 (1992) 2125.
408. S. Schramm and M.-C. Chu. *Phys. Rev.* D48 (1993) 2279.

The formfactors
409. J. Gasser, H. Leutwyler and M. E. Sainio, *Phys. Lett.* B253 (1991) 260.
410. J. F. Donoghue and C. R. Nappi, *Phys. Lett.* B168 (1986) 105.
411. J. Volmer *et al.*, [The Jefferson Laboratory F_π Collaboration], *Phys. Rev. Lett.* 86 (2001) 1713.
412. J. Gronberg *et al.*, *Phys. Rev.* D57 (1998) 33.
413. H. Forkel and M. Nielsen *Phys. Lett.* B345 (1995) 55.
414. A. Blotz and E. Shuryak, *Phys. Rev.* D55 (1997) 4055–4065 [hep-ph/9606355].
415. P. Faccioli, A. Schwenk and E. V. Shuryak, Instanton contribution to the pion electromagnetic formfactor at Q**2 > 1-GeV**2, PRD in press [hep-ph/0202027].
416. P. Faccioli, A. Schwenk and E. V. Shuryak, *Phys. Lett.* B549 (2002) 93 [hep-ph/0205307].
417. T. Draper, R. M. Woloshyn, W. Wilcox and K. F. Liu, *Nucl. Phys.* B318 (1989) 319.

High energy hadronic collisions — The pomeron phenomenology
418. A. Donnachie and P. Landshoff, *Nucl. Phys.* B303 (1988) 634.
419. S. Donnachie, G. Dosch, O. Nachtmann and P. Landshoff, "Pomeron Physics And QCD".
420. F. Low, *Phys. Rev.* D12 (1975) 163; S. Nussinov, *Phys. Lett.* 34 (1975) 1286.

High energy processes in pQCD

421. V. N. Gribov, L. N. Lipatov, *Yad. Fiz.* 15 (1972) 781–807; *Sov. J. Nucl. Phys.* 15 (1972) 438–450, G. Altarelli, G. Parisi, *Nucl. Phys.* B126 (1977) 298.

422. E. A. Kuraev, L. N. Lipatov and V. S. Fadin, *Sov. Phys. JETP* 45 (1977) 199 [*Zh. Eksp. Teor. Fiz.* 72 (1977) 377]. I. I. Balitsky and L. N. Lipatov, *Sov. J. Nucl. Phys.* 28 (1978) 822 [*Yad. Fiz.* 28 (1978) 1597].

423. V. S. Fadin and L. N. Lipatov, *Phys. Lett.* B429 (1998) 127 [hep-ph/9802290].

424. I. I. Balitsky, in Ref. 1.

The saturation at small x

425. L. V. Gribov, E. M. Levin and M. G. Ryskin, *Phys. Rep.* 100 (1983) 1; J. P. Blaizot and A. H. Mueller, *Nucl. Phys.* B289 (1987) 847.

426. L. D. McLerran, R. Venugopalan, *Phys. Rev.* D49 (1994) 2233–2241 [hep-ph/9309289].

427. A. Krasnitz, Y. Nara and R. Venugopalan, *Phys. Rev. Lett.* 87 (2001) 192302 [hep-ph/0108092].

428. D. Kharzeev, E. Levin and L. McLerran, *Phys. Lett.* B561 (2003) 93–101 [hep-ph/0210332].

429. A. Donnachie and P. V. Landshoff, *Phys. Lett.* B533 (2002) 277–284 [hep-ph/0111427].

430. K. Golec-Biernat and M. Wusthoff, *Eur. Phys. J.* C20 (2001) 313 [hep-ph/0102093].

Semiclassical theory of high energy collisions

431. E. V. Shuryak, *Nucl. Phys.* A717 (2003) 291 [hep-ph/0205031].

432. O. Nachtmann, *Ann. Phys.* 209 (1991) 436 [hep-ph/9609365].

433. E. Meggiolaro, *Nucl. Phys.* Supp. B64 (1998) 191.

434. A. Muller, *Nucl. Phys.* B335 (1990) 115.

435. D. Kharzeev and E. Levin, *Nucl. Phys.* B578 (2000) 351 [hep-ph/9912216].

436. E. Shuryak and I. Zahed, *Phys. Rev.* D62 (2000) 085014; M. Nowak, E. Shuryak and I. Zahed, *Phys. Rev.* D64 (2001) 034008.

437. P. Faccini, Talk at Strange Quark Matter, March 2003.

438. D. Kharzeev, Y. Kovchegov and E. Levin, *Nucl. Phys.* A690 (2001) 621.

439. I. I. Balitsky and V. M. Braun, *Phys. Lett.* B314 (1993) 237; *Phys. Lett.* B346 (1995) 143.

440. F. Schrempp and A. Utermann, [hep-ph/0207052].

441. A. Ringwald, *Nucl. Phys.* B330 (1990) 1; O. Espinosa, *Nucl. Phys.* B343 (1990) 310; V. I. Zakharov, *Nucl. Phys.* B353 (1991) 683; M. Maggiore and M. Shifman *Phys. Rev.* D46 (1992) 3550–3564.

442. D. Diakonov and V. Petrov, *Phys. Rev.* D50 (1994) 266.

443. M. P. Mattis, *Phys. Rep.* 214 (1992) 159.

444. R. Janik, E. Shuryak and I. Zahed, [hep-ph/0206005].

445. E. Shuryak and I. Zahed, *Phys. Rev.* D67 014006 (2003) [hep-ph/0206022].

446. A. Giovannini and R. Ugoccioni [hep-ph/0209040].

447. E. V. Shuryak, *Phys. Lett.* B486 (2000) 378 [hep-ph/0001189].

448. E. Shuryak and I. Zahed, Semiclassical double-Pomeron production of glueballs and eta', PRD in press [hep-ph/0302231].

449. UA8 Collaboration (A. Brandt *et al.*), *Eur. Phys. J.* C25 (2002) 361 [hep-ex/0205037].

450. F. Close and A. Kirk, *Phys. Lett.* B397 (1997) 333; F. Close, *Phys. Lett.* B419 (1998) 387.

451. A. Kirk, New effects observed in central production by the WA102 experiment [hep-ph/9810221].
452. J. Ellis and D. Kharzeev, The glueball filter in central production and broken scale invariance [hep-ph/9811222].
453. N. Kochelev, Unusual properties of the central production of glueballs and instantons [hep-ph/9902203].
454. V. A. Rubakov and D. T. Son, *Nucl. Phys.* B424 (1994) 55 [hep-ph/9401257].
455. M. Luscher, *Phys. Lett.* B70 (1977) 321; B. Schechter, *Phys. Rev.* D16 (1977) 3015.
456. G. W. Carter, D. M. Ostrovsky and E. V. Shuryak, *Phys. Rev.* D65 (2002) 074034.

Reviews on QCD at finite T
457. E. V. Shuryak, *Phys. Rep.* 61 (1980) 71.
458. D. J. Gross, R. D. Pisarski and L. G. Yaffe, *Rev. Mod. Phys.* 53 (1981) 43.
459. J. Kapusta, Finite temperature field theory, Cambridge monographs on mathematical physics, 1989.

Field theory methods in statistical mechanics
460. T. Matsubara, *Progr. Theor. Phys.* 14 (1955) 351.
461. V. M. Galitsky and A. B. Migdal, *Zh.E.T.F* 34 (1958) 139.
462. E. S. Fradkin, *Zh.E.T.F* 34 (1958) 262.
463. A. A. Abrikosov, L. P. Gorkov and I. E. Dzyaloshinsky, *Zh.E.T.F.* 36 (1959) 900, Methods of quantum field theories in statistical physics, 1963, Selected Russian publications in the mathematical sciences.
464. J. W. Negele and H. Orland, Quantum many-particle systems (Frontiers in physics book series).

Few relevant works on the QED plasma
465. M. Gell-Mann and K. Bruekner, *Phys. Rev.* 106 (1957) 364.
466. I. A. Akhieser and C. V. Peletminsky, *Zh.E.T.F.* 38 (1960) 1829 (English translation: *Soviet Physics JETP* 11 (1960) 1316).

Perturbative theory of the quark-gluon plasma
467. J. C. Collins and M. J. Perry, *Phys. Rev. Lett.* 34 (1975) 1336.
468. G. Baim and S. A. Chin, *Nucl. Phys.* A262 (1976) 527.
469. M. B. Kislinger and P. D. Morley, *Phys. Rev.* D13 (1976) 2765, 2771, *Phys. Lett.* B67 (1977) 371.
470. B. A. Freedman and L. D. McLerran, *Phys. Rev* D16 (1977) 1130, 1147, 1169.
471. E. V. Shuryak, *Zh.E.T.F* 74 (1978) 408; *Sov. Phys. JETP* 47 (1978) 212.
472. J. Kapusta, *Nucl. Phys.* B148 (1979) 461.
473. C. G. Kallman and T. Toimela, *Phys. Lett.* B122 (l983) 409.
474. O. K. Kalashnikov and V. Klimov, *Sov. J. Nucl. Phys.* 33 (1981) 443.
475. K. Kajantie and J. Kapusta, *Phys. Lett.* B110 (1982) 299; J Kapusta, D. B. Reiss and S. Rudaz, *Nucl. Phys.* B263 (l986) 207.
476. S Nadkarni, *Phys. Rev.* D33 (1986) 3738.
477. E. Braaten and R. D. Pisarski, *Nucl. Phys.* B337 (1990) 569 J. Frenkel and J. C. Taylor, *Nucl. Phys.* B334 (1990) 199.
478. J.-P. Blaizot and E. Iancu, *Nucl. Phys.* B390 (1993) 589; *Phys. Rev. Lett.* 70 (1993) 3376; *Nucl. Phys.* B417 (1994) 608.
479. P. Arnold and C. Zhai, *Phys. Rev.* D50 (1994) 7603, *ibid.* 51 (1995) 1906; C. Zhai and B. Kastening, *ibid.* 52 (1995) 7232.
480. E. Braaten and A. Nieto, *Phys. Rev.* D53 (1996) 3421.

481. A. Peshier, B. Kämpfer, O. P. Pavlenko and G. Soff, *Phys. Rev.* D54 (1996) 2399; A. Peshier [hep-ph/9809379].
482. J. O. Andersen, E. Braaten and M. Strickland, *Phys. Rev. Lett.* 83 (1999) 2139–2142; *Phys. Rev.* D63 (2001) 105008.
483. J. P. Blaizot, E. Iancu and A. Rebhan, *Phys. Rev. Lett.* 83 (1999) 2906–2909; *Phys. Lett.* B470 (1999) 181–188; *Phys. Rev.* D63 (2001) 65003.

Infrared problems, magnetic screening, dimensional reduction
484. A. M. Polyakov, *Phys. Lett.* B82 (1979) 2410.
485. A. D. Linde, *Phys. Lett.* B96 (1980) 289.
486. E. Braaten, *Phys. Rev. Lett.* 74 (1995) 2164 [hep-ph/9409434]; E. Braaten and A. Nieto, *Phys. Rev. Lett.* 76 (1996) 1417 [hep-ph/9508406]; *Phys. Rev.* D53 (1996) 3421 [hep-ph/9510408].
487. K. Kajantie, M. Laine, K. Rummukainen and Y. Schroder, *Phys. Rev. Lett.* 86 (2001) 10 [hep-ph/0007109].

The low-T pion gas
488. J. Gasser and H. Leutwyler, *Phys. Lett.* B188 (1987) 477; P. Gerber and H. Leutwyler, *Nucl. Phys.* B321 (1989) 387.

The resonance gas
489. E. Beth and G. E. Uhlenbeck, *Physica* 3 (1936) 729.
490. S. Z. Belenky and L. D. Landau, *Nuovo Cim. Suppl 3*, series X (1956) 15.
491. R Hagedorn, *Nuovo Cimento* 35 (1965) 216.
492. E. V. Shuryak, *Yadernaya Fizika* 16 (1972) 395.
493. E. V. Shuryak, *Phys. Lett.* B42 (1972) 357. The Statistical Theory Of Multiple Production Of Hadrons, preprint IYF-74-108, Novosibirsk.
494. R. Venugopalan and M. Prakash, *Nucl. Phys.* A546 (1992) 718.
495. E. V. Shuryak, *Nucl. Phys.* A533 (1991) 761.
496. D. Lissauer and E. V. Shuryak, *Phys. Lett.* B253 (1991) 15.
497. E. V. Shuryak and V. Thorsson, *Nucl. Phys.* A536 (1992) 739.
498. V. L. Eletsky, M. Belkacem, P. J. Ellis and J. I. Kapusta, *Phys. Rev.* C64 (2001) 035202 [nucl-th/0104029].
499. R. Rapp and C. Gale, *Phys. Rev.* C60 (1999) 024903 [hep-ph/9902268].

The pion liquid
500. E. V. Shuryak, *Phys. Rev.* D42 (1990) 1764.
501. R. D. Pisarski and M. Tytgat, *Phys. Rev.* D54 (1996) 2989-2993 [hep-ph/9604404].
502. R. Rapp and J. Wambach, *Phys. Rev.* C53 (1996) 3057.
503. D. T. Son and M. A. Stephanov, *Phys. Rev.* D66 (2002) 076011 [hep-ph/0204226].

Viscosity of QGP
504. G. Baym, H. Monien, C. J. Pethick and D. G. Ravenhall, *Phys. Rev. Lett.* 64 (1990) 1867–1870.
505. P. Arnold, G. D. Moore and L. G. Yaffe, *JHEP* 0011 (2000) 001.

Deconfinement
506. A. M. Polyakov, *Phys. Lett.* B82 (1979) 2413.
507. L Susskind, *Phys. Rev.* D20 (1979) 2610.
508. B. Svetitski and L. G. Yaffe, *Nucl. Phys.* B210 (1982) 423; B Svetitski, *Phys. Rep.* 132 (1986) 1.
509. A. Dumitru and R. D. Pisarski, *Phys. Rev.* D66 (2002) 096003 [hep-ph/0204223].

510. S. Fortunato, F. Karsch, P. Petreczky and H. Satz, *Phys. Lett.* B502 (2001) 321 [hep-lat/0011084].
511. V. Koch and G. E. Brown, *Nucl. Phys.* A560 (1993) 345.
512. D. E. Miller, Gluon condensates at finite temperature [hep-ph/0008031].
513. L. D. Mclerran and B. Svetitsky, *Phys. Lett.* B98 (1981) 195; *Phys. Rev.* D24 (1981) 450.
514. R. V. Gavai and H. Satz, *Phys. Lett.* B145 (1984) 248.
515. T. Celik, J. Engels and H. Satz, *Phys. Lett.* B125 (1983) 411; B129 (1983) 323.
516. J. Kogut, H. Matsuoka, M. Stone, H. Wyld, S. Shenker, J. Shigemitsu and D Sinclair, *Phys. Rev. Lett.* 51 (1983) 869.
517. A. Gocksch and M. Okawa, *Phys. Rev. Lett.* 52 (1984) 1751.
518. M. Gross and J. F. Wheater, *Phys. Rev. Lett.* 54 (1985) 389.
519. T. A. DeGrand and C. E. DeTar, *Nucl. Phys.* B225/FS9/ (1983) 590.
520. P. Hasenfratz, F. Karsch and O. Stamatascu, *Phys. Lett.* B133 (1983) 221.

Chiral symmetry restoration
521. E. V. Shuryak, *Phys. Lett.* B107 (1981) 103.
522. R. D. Pisarski, *Phys. Lett.* B110 (1982) 155.
523. R. D. Pisarski and F. Wilczek, *Phys. Rev.* D29 (1984) 338.
524. F. Wilczek, *Int. J. Mod. Phys.* A7 (1992) 3911.

A restoration of $U(1)_A$?
525. E. V. Shuryak, "Which chiral symmetry is restored in hot QCD?" *Comments Nucl. Part. Phys.* 21 (1994) 235 [hep-ph/9310253].
526. T. D. Cohen, *Phys. Rev.* D54 (1996) 1867 [hep-ph/9601216].

Lattice phase transitions in quenched approximation
527. J. Kogut, M. Stone, H. Wyld, J. Shigemitsu and D. Sinclair, *Phys. Rev. Lett.* 48 (1982) 1140; 50 (1983) 393.
528. J. Engels, H. Satz and F. Karsch (Preprint WI-TP 82/8, Bielefeld).
529. J. Engels and F. Karsch, *Phys. Lett.* B125 (1983) 481.
530. J. B. Kogut *et al.*, *Nucl. Phys.* B225/FS9/ (1983) 93.
531. J Bartolomew *et al.*, *Nucl. Phys.* B230/FS10/ (1984) 222.

Theories with the nontrivial quark representations
532. J. Kogut, J. Shigemitsu and D Sinclair, *Phys. Lett.* B138 (1984) 283.
533. J. B. Kogut *et al.*, *Phys. Lett.* B145 (1984) 239.

Phase transitions with dynamical quarks
534. R. V. Gavai, M. Lev and B. Petersson, *Phys. Lett.* B149 (1984) 492; R. V. Gavai, *Nucl. Phys.* B269 (1986) 530.
535. P. H. Damgaard *et al.*, *Phys. Rev. Lett.* 53 (1984) 2211.
536. J Polonyi *et al.*, *Phys. Rev. Lett.* 53 (1984) 644; J. B. Kogut *et al.*, *Phys. Rev. Lett.* 54 (1985) 1475.
537. H. W. Hamber, *Phys. Rev.* D31 (1985) 586.
538. T. Celik, J. Engels and H. Satz, *Nucl. Phys.* B256 (1985) 670.
539. L. M. Barbour and C. J. Burden, *Phys. Lett.* B161 (1985) 357.

The phase transitions at nonzero baryon density
540. J. Kogut *et al.*, *Nucl. Phys.* B225 (1983) 93.
541. J. Engels and H. Satz, *Phys. Lett.* B159 (1985) 151.
542. F. Karsch, AIP Conf. Proc. 602 (2001) 323–332 [hep-lat/0109017].

Small but non-zero chemical potential

543. Z. Fodor and S. D. Katz, *JHEP* 0203 (2002) 014 [hep-lat/0106002].
544. P. de Forcrand and O. Philipsen, The QCD phase diagram for small densities from imaginary chemical potential [hep-lat/0205016].
545. Z. Fodor, S. D. Katz and K. K. Szabó [hep-lat/0208078].

Instantons in matter

546. D. J. Gross, R. D. Pisarski and L. G. Yaffe, *Rev. Mod. Phys.* 53 (1981) 43.
547. B. Harrington and H. Shepard, *Phys. Rev.* D17 (1978) 2122.
548. R. D. Pisarski and L. G. Yaffe, *Phys. Lett.* B97 (1980) 110.
549. C. A. de Corvalho, *Nucl. Phys.* B183 (1981) 182.
550. V. Baluni, *Phys. Lett.* B106 (1981) 491.
551. A. A. Abrikosov (jr), *Phys. Lett.* B90 (1980) 415; *Nucl. Phys.* B182 (1981) 441.
552. E. V. Shuryak, *Nucl. Phys.* B203 (1982) 140.
553. E. V. Shuryak and J. J. Verbaarschot, *Nucl. Phys.* B364 (1991) 255.
554. V. L. Eletsky, *Phys. Lett.* B299 (1993) 111.
555. D. I. Diakonov and A. D. Mirlin, *Phys. Lett.* 203B (1988) 299.
556. E. V. Shuryak and M. Velkovsky, *Phys. Rev.* D50 (1994) 3323 [hep-ph/9403381].
557. E.-M. Ilgenfritz and E. V. Shuryak, *Phys. Lett.* B325 (1994) 263.
558. M. Velkovsky and E. V. Shuryak, *Phys. Rev.* D56 (1997) 2766 [hep-ph/9603234].
559. T. Schafer and E. V. Shuryak, *Phys. Rev.* D53 (1996) 6522 [hep-ph/9509337].
560. M. C. Chu and S. Schramm, *Phys. Rev.* D51 (1995) 4580 [nucl-th/9412016].
561. C. Gattringer, R. Hoffmann and S. Schaefer, *Phys. Lett.* B535 (2002) 358 [hep-lat/0203013].
562. C. Gattringer, M. Gockeler, P. E. Rakow, S. Schaefer and A. Schafer, *Nucl. Phys.* B617 (2001) 101 [hep-lat/0107016].
563. C. DeTar and J. B. Kogut, *Phys. Rev. Lett.* 59 (1987) 399.
564. V. Koch, E. V. Shuryak, G. E. Brown and A. D. Jackson, *Phys. Rev.* D46 (1992) 3169 [Erratum-ibid. D 47 (1993) 2157] [hep-ph/9204236].

Thermodynamics on the lattice

565. H. Satz, *Ann. Rev. Nucl. Part. Sci.* 35 (1985) 245 General theory.
566. A. Hazenfratz and P. Hazenfratz, *Nucl. Phys.* B193 (1981) 210.
567. J. Engels, F. Karsch, I. Montvay and H. Satz, *Phys. Lett.* B101 (1981) 89; B102 (1981) 332.

Thermodynamics with quarks and nonzero chemical potential

568. T. Celik, J. Engels and H. Satz, *Nucl. Phys.* B256 (1985) 670.
569. P. Hasenfratz, F. Karsch and I. O. Stamatescu, *Phys. Lett.* B133 (1983) 308.
570. J. Kogut *et al.*, *Nucl. Phys.* B225 (1983) 93.
571. R. V. Gavai and A. Ostendorf, *Phys. Lett.* B12 (1983) 137.
572. J. Engels and H. Satz, *Phys. Lett.* B159 (1985) 151.

Lattice evaluation of the magnetic screening length

573. A. Billoire, G. Lasarides and Q. Shafi, *Phys. Lett.* B103 (1981) 450.
574. T. A. de Grand and D. Toussaint, *Phys. Rev.* D25 (1982) 526.

QCD phase transition in Early Universe

575. K. Olive, *Nucl. Phys.* B190 (1981) 483.
576. E. Suhonen, *Phys. Lett.* B119 (1982) 81.
577. E. Witten, *Phys. Rev.* D30 (1984) 272.
578. A. de Rujula and S. L. Glashow, *Nature* 312 (1984) 734.
579. C. J. Hogan, *Phys. Lett.* B133 (1983) 172; *Phys. Rev. Lett.* 51 (1983) 1488.

Relativistic hydrodynamics

580. L. D. Landau and E. M. Lifshitz, Fluid mechanics, Moscow 1953 (Pergamon Press, London, 1959) (Butterworth and Heinemann, 1987).
581. C. Eckart, *Phys. Rev.* 58 (1940) 919. Pioneer works on thermo/hydrodynamical models of hadronic collisions.
582. E. Fermi, *Phys. Rev.* 81 (1951) 683.
583. L. D. Landau, *Izv. Akad. Nauk SSSR, ser. fiz.* 17 (1953) 51. Reprinted in Collected works by L. D. Landau.
584. I. Ya. Pomeranchuck, *Doklady Akad. Nauk (Sov. Physics-doklady)* 78 (1951) 884.
585. F. Cooper and G. Frye, *Phys. Rev.* D10 (1974) 186.
586. C. B. Chin, E. C. G. Sudarshan and K. H. Wang, *Phys. Rev.* 12 (1975) 902.
587. M. I. Gorenshtein, V. A. Zhdanov and Yu. M. Sinjukov, *ZhETF* 74 (1978) 833.
588. J. Bjorken, *Phys. Rev.* D27 (1983) 140.
589. P. Danielewicz, M. Gyulassy, *Phys. Rev.* D31 (1985) 53–62.
590. E. V. Shuryak, *Phys. Lett.* B34 (1971) 509.
591. F. Cooper, G. Frye and E. Shonberg, *Phys. Rev. Lett.* 32 (1974) 862.
592. E. V. Shuryak and O. V. Zhirov, *Yad Fis* 24 (1986) 195.
593. G. Baym, B. L. Friman, J.-P. Blaizot, M. Soyeur and W. Czyz, *Nucl. Phys.* A407 (1983) 541.
594. P. Csizmadia, T. Csorgo and B. Lukas, *Phys. Lett.* B443 (1998) 21–25 [nucl-th/9805006].
595. S. V. Akkelin, T. Csorgo, B. Lukacs, Y. M. Sinyukov and M. Weiner, *Phys. Lett.* B505 (2001) 64 [hep-ph/0012127].
596. D. H. Rischke, *Nucl. Phys.* A610 (1996) 88C [nucl-th/9608024].
597. C. Greiner and D. H. Rischke, *Phys. Rev.* C54 (1996) 1360.

Hydrodynamical model and high energy pp collisions

598. E. V. Shuryak, *Yadernaya Fizika* 16 (1972) 395; 20 (1974) 549.
599. E. L. Daibog, Yu. P. Nikitin and I. L. Rozental, *Yadernaya Fizika* 16 (1972) 1314.
600. P. Curruthers and M. Duong-van, *Phys. Lett.* B41 (1972) 597; *Phys. Rev.* D8 (1973) 859.
601. P. Curruthers, *Ann. NY Acad. Sci.* 229 (1974) 91.
602. G. A. Milekhin, *Zh.E.T.F (Sov Phys.-JETP)* 35 (1958) 1185, Trudy FIAN (Proceedings of Lebedev Physical Institute) 16 (1961) 51.

The role of the QCD phase transition in flow

603. E. V. Shuryak and O. V. Zhirov, *Phys. Lett.* B89 (1979) 253.
604. L. van Hove, *Z. Phys.* C21 (1983) 93; *Z. Phys.* C27 (1985) 135.
605. M. Gyulassy, K. Kajantie, H. Kurki-Suonio and L. McLerran (Preprint LBL-16277, Berkeley, 1983).
606. E. V. Shuryak and O. V. Zhirov, *Phys. Lett.* B171 (1986) 99–102.
607. C. M. Hung and E. V. Shuryak, *Phys. Rev. Lett.* 75 (1995) 4003 [hep-ph/9412360]. *Phys. Rev.* C57 (1998) 1891.
608. D. H. Rischke and M. Gyulassy, *Nucl. Phys.* A597 (1996) 701 [nucl-th/9509040].

Freezeout and resonances

609. Y. M. Sinyukov, S. V. Akkelin and Y. Hama, *Phys. Rev. Lett.* 89 (2002) 052301 [nucl-th/0201015].
610. G. Chanfray, M. Ericson and P. A. Guichon, *Phys. Rev.* C63 (2001) 055202 [nucl-th/0012013].
611. E. V. Shuryak and G. E. Brown, "Matter-induced modification of resonances at RHIC freezeout," *Nucl. Phys.* A. [hep-ph/0211119].

612. P. Fachini (nucl-ex/0211001), Talk at Quark Matter 2002, Nantes, July 2002, to be published in proceedings in *Nucl. Phys.* A.

613. B. L. Ioffe, I. A. Shushpanov and K. N. Zyablyuk, Formation of antideuterons in heavy ion collisions [hep-ph/0302052].

614. H. Bebie, P. Gerber, J. L. Goity and H. Leutwyler, *Nucl. Phys.* B378 (1992) 95–130.

615. D. Teaney [nucl-th/0204023]; F. Kolb, Ralf Rapp [hep-ph/0210222].

616. M. Dey, V. L. Eletsky and B. L. Ioffe, *Phys. Lett.* B252 (1990) 620.

617. M. Dey, V. L. Eletsky and B. L. Ioffe, *Phys. Lett.* B252 (1990) 620.

618. C. Markert, G. Torrieri and J. Rafelski, Strange hadron resonances: Freeze-out probes in heavy-ion collisions [hep-ph/0206260].

619. J. Kapusta, D. Kharzeev and L. D. McLerran, *Phys. Rev.* D53 (1996) 5028 [hep-ph/9507343].

620. E. Shuryak, *Phys. Lett.* B207 (1988) 345.

621. P. Gerber, H. Leutwyler and J. L. Goity, *Phys. Lett.* B246 (1990) 513–519.

622. P. Braun-Munziger, J. Stachel, J. P. Wessels and N. Xu, *Phys. Lett.* B344 (1995) 43.

623. F. Becattini, M. Gadzdzicki and J. Sollfrank, *Eur. Phys. J.* C5 (1998) 143.

624. J. Sollfrank, *J. Phys.* G23 (1997) 1903.

625. J. Cleymans and K. Redlich, *Phys. Rev.* C59 (1999) 354; J. Cleymans, H. Oeschler and K. Redlich, *Phys. Rev.* C59 (1999) 1663.

Flow in heavy ion collisions

626. P. J. Siemens and J. O. Rasmussen, *Phys. Rev. Lett.* 42 (1979) 880.

627. U. Heinz, *Nucl. Phys.* A685 (2001) 414 and references therein.

628. U. Ornik, M. Pluemer, B.R. Schlei, D. Strottman and R. M. Weiner, *Phys. Rev.* C54 (1996) 1381.

629. P. F. Kolb, J. Sollfrank and U. Heinz, *Phys. Rev.* C62 (2000) 054909.

630. P. Huovinen, P. F. Kolb, U. Heinz and H. Heiselberg, *Phys. Lett.* B503 (2001) 58.

631. S. Bass and A. Dumitru, *Phys. Rev.* C61 (2000) 064909.

632. M. Belacem *et al.*, *Phys. Rev.* C58 (1998) 1727.

633. H. van Hecke, H. Sorge and N. Xu, *Phys. Rev. Lett.* 81 (1998) 5764.

634. M. Gyulassy and T. Matsui, *Phys. Rev.* D29 (1984) 419.

635. R. J. LeVeque, *Numerical Methods for Conservation Laws* (Birkhäuser-Verlag, 1990).

636. V. Schneider *et al.*, *J. Comput. Phys.* 105 (1993) 92.

637. D. H. Rischke, Y. Pusun, J. A. Maruhn, *Nucl. Phys.* A595 (1995) 346; *Nucl. Phys.* A595 (1995) 383.

638. P. F. Kolb, U. Heinz, P. Huovinen, K.J. Eskola and K. Tuominen [hep-ph/0103234].

639. H. Sorge, *Phys. Rev.* C52 (1995) 3291.

640. P. F. Kolb, J. Sollfrank and U. Heinz, *Phys. Lett.* B459 (1999) 667.

641. S. Soff, S. A. Bass and A. Dumitru, *Phys. Rev. Lett.* 86 (2001) 3981.

642. J. Sollfrank *et al.*, *Phys. Rev.* C55 (1997) 392.

643. D. Teaney, Ph.D. Thesis, State University of New York at Stony Brook (2001).

644. E. Schnederman, J. Sollfrank and U. Heinz, *Phys. Rev.* C48 (1993) 2462.

645. T. Hirano and K. Tsuda, Collective flow and HBT radii from a full 3D hydrodynamic model with early chemical freeze out [nucl-th/0208068].

646. P. Huovinen, P. F. Kolb and U. Heinz, in Quark Matter 2001 [nucl-th/0104020].

Flow and global observables in heavy ion experiments

647. R. Stock, in Quark Matter 1999, *Nucl. Phys.* A661 (1999) 419c.

648. C. Adler *et al.* [STAR Collaboration], *Phys. Rev.* C66 (2002) 034904 [nucl-ex/0206001].

649. Shunji Nishimura for the WA98 Collaboration, *Nucl. Phys.* A661 (1999) 464c.
650. NA49 Collaboration, H. Appelshäuser *et al.*, *Phys. Rev. Lett.* 80 (1998) 4136.
651. STAR Collaboration, C. Adler *et al.*, *Phys. Rev. Lett.* 87 (2001) 082301.
652. M. van Leeuwen for the NA49, Recent results on spectra and yields from NA49, QM02, Nantes, V. Friese for the NA49, SQM03, [nucl-exp/0305017].
653. NA49 Collaboration, T. Alber *et al.*, *Phys. Rev. Lett.* 75 (1995) 3814.
654. PHOBOS Collaboration, B. B. Back *et al.*, *Phys. Rev. Lett.* 85 (2000) 3100.
655. P. Jacobs and G. Cooper, STAR SN402(1999).
656. STAR Collaboration, C. Adler *et al.*, *Phys. Rev. Lett.* 86 (2001) 4778.
657. PHENIX Collaboration, K. Adcox *et al.*, *Phys. Rev. Lett.* 86 (2001) 3500.
658. PHOBOS Collaboration, B. B. Back *et al.*, submitted to *Phys. Rev. Lett.* [nucl-ex/0105011].
659. STAR Collaboration, C. Adler *et al.*, *Phys. Rev. Lett.* 87 (2001) 182301 [nucl-ex/0107003].
660. NA49 Collaboration, H. Appelshäuser *et al.*, *Phys. Rev. Lett.* 82 (1999) 2471.
661. E. Andersen *et al.*, WA97 collaboration, *Phys. Lett.* B433 (1998) 209; R. Lietava for the WA97 collaboration in Proc. Strangeness 98, *J. Phys. G: Nucl. Part. Phys.* 25 (1999).
662. G. Roland for the NA49 Collaboration, *Nucl. Phys.* A638 (1999) 91c.
663. M. Caldéron, for the STAR Collaboration, talk at Quark Matter 2001.
664. J. Velkovska, for the PHENIX Collaboration, at Quark Matter 2001 [nucl-ex/0105012].

Elliptic flow
665. J.-Y. Ollitrault, *Phys. Rev.* D46 (1992) 229.
666. D. Molnar and M. Gyulassy [nucl-th/0104073].
667. H. Sorge, *Phys. Rev. Lett.* 78 (1997) 2309.
668. H. Sorge, *Phys. Lett.* B402 (1997) 251.
669. H. Sorge, *Phys. Rev. Lett.* 82 (1999) 2048.
670. H. Sorge, *Phys. Rev. Lett.* 82 (1999) 2048.
671. M. Bleicher and H. Stocker [hep-ph/0006147].
672. P. F. Kolb, P. Huovinen, U. Heinz and H. Heiselberg, *Phys. Lett.* B500 (2001) 232.
673. D. Teaney, J. Lauret, and E.V. Shuryak, *Phys. Rev. Lett.* 86 (2001) 4783.
674. S. A. Voloshin and A. M. Poskanzer, *Phys. Lett.* B474 (2000) 27.
675. A. M. Poskanzer and S. A. Voloshin for the NA49 Collaboration, *Nucl. Phys.* A661 (1999) 341c.
676. STAR Collaboration, K.H. Ackermann *et al.*, *Phys. Rev. Lett.* 86 (2001) 402.
677. J.P. Blaizot and J.-Y. Ollitrault, *Nucl. Phys.* A458 (1986) 745.

The limits of hydro
678. D. Teaney, The effects of viscosity on spectra, elliptic flow, and HBT radii [nucl-th/0301099].

Interferometric "microscope"
679. R. Hanbury-Brown and R. Q. Twiss, *Phyl. Mag.* 45 (1954) 663.
680. E. M. Purcell, *Nature* 178 (1956) 1449.
681. E.V. Shuryak, *Zh.E.T.P. (Sov.Physics-JETP)* 67 (1974) 60.
682. G. Goldhaber, S. Goldhaber, W. Lee and A. Pais, *Phys. Rev.* 120 (1960) 300.
683. G. I. Kopylov and M. I. Podgoretsky, *Sov. J. Nucl. Phys.* 15 (1972) 219.
684. E. V. Shuryak, *Phys. Lett.* B44 (1973) 387.
685. G. I. Kopylov and M. I. Podgoretsky, *Sov. J. Nucl. Phys.* 18 (1974) 336.
686. G. Cocconi, *Phys. Lett.* B49 (1974) 459.

687. G. Gyulassy, S. K. Kaufmann and L. W. Wilson, *Phys Rev.* C20(1979)2267.

688. A. N. Makhlin and Y. M. Sinyukov, *Yad. Fiz.* 46 (1987) 637; *Z. Phys.* C39 (1988) 69.

689. B. Tomasik and U. A. Wiedemann, Central and non-central HBT from AGS to RHIC [hep-ph/0210250].

690. U. W. Heinz and B. V. Jacak, *Ann. Rev. Nucl. Part. Sci.* 49 (1999) 529 [nucl-th/9902020].

691. R. Weiner, Bose-Einstein Correlations in particle and nuclear physics (J. Wiley and Sons, 1997).

692. T. Hirano *et al.*, *Phys. Rev.* C65 (2002) 061902 [nucl-th/0110009].

The balance functions

693. S. A. Bass, P. Danielewicz and S. Pratt, *Phys. Rev. Lett.* 85 (2000) 2689 [nucl-th/0005044].

694. STAR Collaboration (J. Adams *et al.*) Submitted to *Phys. Rev. Lett.* 90 (2003) [nucl-ex/0301014].

Correlations of non-identical particles

695. R. Lednický, V. Lyuboshitz, B. Erazmus and D. Nouais, *Phys. Lett.* B373 (1996) 30.

696. R. L. Ray, for the STAR collaboration, Correlations, fluctuations, and flow measurements [nucl-ex/0211030].

Event-by-event fluctuations

697. B. Blaettel, G. Baym, L. L. Frankfurt, H. Heiselberg and M. Strikman, *Phys. Rev.* D47 (1993) 2761.

698. H. Kowalski and D. Teaney, An impact parameter dipole saturation model [hep-ph/0304189].

699. L. Stodolsky, *Phys. Rev. Lett.* 75 (1995) 1044.

700. E. V. Shuryak, *Phys. Lett.* B423 (1998) 9 [hep-ph/9704456].

701. H. Appelshauser *et al.* [NA49 Collaboration], *Phys. Lett.* B459 (1999) 679 [hep-ex/9904014].

702. M. Stephanov, K. Rajagopal and E. Shuryak, *Phys. Rev. Lett.* 81 (1998) 4816 [hep-ph/9806219]. *Phys. Rev.* D60 (1999) 114028 [hep-ph/9903292].

703. S. Mrowczynski, *Phys. Lett.* B314 (1993) 118; *Phys. Rev.* C49 (1994) 2191; *Phys. Lett.* B393 (1997) 26.

704. W. Trautmann, Multifragmentation in relativistic heavy ion reactions [nucl-ex/9611002].

705. M. L. Gilkes *et al.*, *Phys. Rev. Lett.* 73 (1994) 1590; J. Pochodzalla *et al.*, *Phys. Rev. Lett.* 75 (1995) 1040.

706. M. Asakawa, U. Heinz and B. Muller, *Phys. Rev. Lett.* 85 (2000) 2072 [hep-ph/0003169].

707. S. Jeon and V. Koch, *Phys. Rev. Lett.* 85 2076 (2000) [hep-ph/0003168].

708. E. V. Shuryak and M. A. Stephanov, *Phys. Rev.* C63 (2001) 064903 [hep-ph/0010100].

Diagnostics of quark-gluon plasma
Dileptons and photons

709. E. V. Shuryak, *Phys. Lett.* B78 (1978) 150; *Yadernaya Fizika* 28 (1978) 796.

710. E. L. Feinberg, *Izv Akad Nauk SSSR, ser fiz* 26 (1962) 622; *Nuovo dm* A34 (1976) 391.

711. E. V. Shuryak, *Phys. Lett.* B78 (1978) 150; *Yadernaya Fizika* 28 (1978) 796.

712. G. Domokos and J. Goldman, *Phys. Rev.* D23 (1981) 203.

713. K. Kajantie and H. Miettinen, *Z. Phys.* C9 (1981) 341.

714. A. Hasegawa, *Progr. Theor. Phys.* 69 (1983) 689.

715. B. Sinha, *Phys. Lett.* B128 (1983) 91; B160 (1985) 287.
716. J. Kapusta, *Phys. Lett.* B136 (1984) 201.
717. L. D. McLerran and T. Toimela, *Phys. Rev.* D32 (1985) 1109.
718. J. Cleymans, R. V. Gavai and E. Suchonen, *Phys. Rep.* C130 (1986) 217.
719. J. Kapusta, P. Lichard and D. Seibert, *Phys. Rev.* D44 (1991) 2774.
720. G. Q. Li and C. Gale, *Phys. Rev.* C58 (1998) 2914; *Phys. Rev. Lett.* 81 (1998) 1572.
721. R. Rapp and J. Wambach, *Adv. Nucl. Phys.* 25 (2000) 1 [hep-ph/9909229].
722. WA80 collaboration: A. Lebedev *et al.*, *Nuc. Phys.* A566 (1994) 355c.
723. Ceres collaboration: G. Agakiechev *et al.*, *Phys. Rev. Lett.* 75 (1995) 1272.
724. Helios collaboration: M. Masera, *Nuc. Phys.* A590 (1995) 93c.
725. R. Rapp, Thermal lepton production in heavy-ion collisions [nucl-th/0204003] see also [hep-ph/0201101].
726. G. Q. Li, G. E. Brown and C. M. Ko, *Nucl. Phys.* A630 (1998) 563 [nucl-th/9706022].
727. J. I. Kapusta and E. V. Shuryak, *Phys. Rev.* D49 (1994) 4694 [hep-ph/9312245].
728. K. Adcox *et al.* [PHENIX Collaboration], *Phys. Rev. Lett.* 88 (2002) 192303 [nucl-ex/0202002].
729. L. Xiong, E. V. Shuryak and G. E. Brown, *Phys. Rev.* D46 (1992) 3798 [hep-ph/9208206].
730. K. Geiger and B. Muller, *Nucl. Phys.* A544 (1992) 467C.
731. S. J. Parke and T. R. Taylor, *Phys. Rev. Lett.* 56 (1986) 2459.
732. Z. Huang, *Phys. Lett.* B361 (1995) 131.
733. P. Aurenche, F. Gelis, R. Kobes and H. Zaraket, *Phys. Rev.* D60 (1999) 076002.
734. M. M. Aggarwal *et al.* (WA98), *Phys. Rev. Lett.* 85 (2000) 3595.
735. S. Damjanovic and K. Filimonov (CERES), poster P084 at QM2001.
736. B. Lenkeit (CERES), *Nucl. Phys.* A661 (1999) 23c; Ph.D. thesis, Heidelberg 1998.
737. E. Scomparin (NA50), *Nucl. Phys.* A610 (1996) 331c; *J. Phys.* G25 (1999) 235c.
738. R. Rapp and E. V. Shuryak, *Phys. Lett.* B473 (2000) 13 [hep-ph/9909348].
739. B. Kämpfer, K. Gallmeister and O. P. Pavlenko [hep-ph/0102192], *Phys. Lett.* B473 (2000) 20.
740. Jan-e Alam, S. Sarkar, T. Hatsuda, T. K. Nayak and B. Sinha, *Phys. Rev.* C63 (2001) 021901. A. K. Chaudhuri, [nucl-th/0012058]; D. K. Srivastava, B. Sinha, *Phys. Rev.* C64 (2001) 034902.
741. P. Huovinen, P. V. Ruuskanen and S. S. Räsänen [nucl-th/0111052].
742. D. Yu. Peressounko and Yu. E. Pokrovsky, *Nucl. Phys.* A669 (2000) 196.
743. P. Aurenche, P. Chiappetta, M. Fontannaz, J. P. Guillet and E. Pilon, *Nucl. Phys.* B399 (1993) 34; P. Aurenche, F. Gelis, R. Kobes and E. Petitgirard, *Z. Phys.* C75 (1997) 315; P. Aurenche, F. Gelis, R. Kobes and H. Zaraket, *Phys. Rev.* D58 (1998) 085003; P. Aurenche, F. Gelis and H. Zaraket, *Phys. Rev.* D61 (2000) 116001; *Phys. Rev.* D62 (2000) 096012; F. Gelis, *Phys. Lett.* B493 (2000) 282.
744. P. Arnold, G. D. Moore and L. G. Yaffe, JHEP 0111 (2001) 057; JHEP 0112 (2001) 009.
745. M. G. Mustafa and M. H. Thoma, *Phys. Rev.* C62 (2000) 014902; Erratum, *Phys. Rev.* C63 (2001) 069902.
746. D. Neuhauser, *Phys. Lett.* B182 (1986) 289.
747. D. Yu. Peressounko, for WA98, poster at QM02.

Strangeness: the early papers

748. P. Koch, B. Muller and J. Rafelski, *Phys. Rep.* 142 (1986) 167.
749. J. Rafelski and B. Muller, *Phys. Rev Lett.* 48 (1982) 1066; P. Koch and J. Rafelski, *Nucl. Phys.* A444 (1985) 678.

750. T. Biro and J. Zimanyi, *Phys. Lett.* B113 (1982) 6; *Nucl. Phys.* A395 (1983) 525.
751. J. Kapusta and A. Mekjian, *Phys. Rev.* D33 (1986) 1304.

Charmonium suppression

752. T. Matsui and H. Satz, *Phys. Lett.* B178 (1986) 416.
753. M. Peskin, *Nucl. Phys.* B156 (1979) 365.
754. D. Kharzeev and H. Satz, *Nucl. Phys.* A590 (1995) 515C.
755. E. V. Shuryak and D. Teaney, *Phys. Lett.* B430 (1998) 37 [nucl-th/9801016].
756. D. Kharzeev and H. Satz, *Z. Phys.* C60 (1993) 389.
757. L. Grandchamp and R. Rapp, *Phys. Lett.* B523 (2001) 60 [hep-ph/0103124].
758. S. Datta *et al.*, [hep-lat/0208012]; T. Umeda *et al.*, [hep-lat-0211003].
759. By NA50 Collaboration (M. C. Abreu *et al.*) Talk given at 31st International Symposium on Multiparticle Dynamics (ISMD 2001), *Datong 2001, Multiparticle dynamics* 127–131 [hep-ph/0111429].
760. H. Sorge, E. V. Shuryak and I. Zahed, *Phys. Rev. Lett.* 79 (1997) 2775 [hep-ph/9705329].

Equilibration and mini-jets

761. E. V. Shuryak, *Phys. Rev. Lett.* 68 (1992) 3270.
762. L. Xiong and E. V. Shuryak, *Nucl. Phys.* A590 (1995) 589C.
763. R. Baier, A. H. Mueller, D. Schiff and D. T. Son, *Phys. Lett.* B502 (2001) 51 [hep-ph/0009237].

The jet quenching

764. D. Hardtke for STAR Collaboration [nucl-ex/0206006.0212004].
765. High P(T) measurements from PHENIX. (S. Mioduszewski for the collaboration). To appear in the proceedings of Quark Matter 2002, Nantes, France [nucl-ex/0210021].
766. D. A. Appel, *Phys. Rev.* D33 (1986) 717; J. P. Blaizot and L. D. McLerran, *Phys. Rev.* D34 (1986) 2739.
767. M. Gyulassy and M. Plumer, *Phys. Lett.* B243 (1990) 432; X. N. Wang, M. Gyulassy and M. Plumer, *Phys. Rev.* D51 (1995) 3436 [hep-ph/9408344]. G. Fai, G. G. Barnafoldi, M. Gyulassy, P. Levai, G. Papp, I. Vitev and Y. Zhang, Jet quenching as a probe of gluon plasma formation [hep-ph/0111211].
768. M. Gyulassy and X. N. Wang, *Comp. Phys. Comm.* 83 (1994) 307; *Phys. Rev.* D44 (1991) 3501.
769. R. Baier, Y. L. Dokshitzer, S. Peigne and D. Schiff, *Phys. Lett.* B345 (1995) 277 [hep-ph/9411409]; R. Baier, Y. L. Dokshitzer, A. H. Mueller and D. Schiff, JHEP 0109 (2001) 033 [hep-ph/0106347].
770. E. Wang and X.-N. Wang [hep-ph/0202105].
771. E. V. Shuryak, *Phys. Rev.* C66 (2002) 027902 [nucl-th/0112042].
772. J. Casalderrey-Solana and E. V. Shuryak, Jet fragmentation due to a quark/diquark pick-up in high energy heavy ion collisions [hep-ph/0305160].
773. I. B. Khriplovich and E. V. Shuryak, *Zh. Exp. Teor. Fiz.* 65 (1973) 2137; *Sov. Phys. JETP*, 38 (1974) 1067.
774. S. K. Wong, "Field And Particle Equations For The Classical Yang–Mills Field And Particles With Isotopic Spin", Nuovo Cim. A65 (1970) 689.
775. R. Y. Tsien, *Am. J. Phys.* 40 (1972) 46.
776. J. Schwinger, *Phys. Rev.* D7 (1973) 1696.
777. E. V. Shuryak and I. Zahed, *Phys. Rev.* D67 (2003) 054025 [hep-ph/0207163].

Nuclear matter
778. G. E. Brown and A. D. Jackson, The nucleon-nucleon interaction. North-Holland, 1974.
779. B. D. Serot and J. D. Walecka, *Adv. Nucl. Phys.* 16 (1986) 1.

Renormalization group near the Fermi surface
780. R. Shankar, *Rev. Mod. Phys.* 66 (1993) 129; J. Polchinski [hep-th/9210046]; G. Benfatto and G. Galavotti, *Renormalization Group*, Princeton University Press, Princeton, 1995.
781. S. K. Bogner, T. T. S. Kuo and A. Schwenk [nucl-th/0305035]; A. Schwenk, G. E. Brown and B. Friman [nucl-th/0110033].
782. S. R. Beane, P. F. Bedaque, W. C. Haxton, D. R. Phillips and M. J. Savage, in Ref. 1 and also [nucl-th/0008064].

Nuclear matter theory
783. H. A. Bethe, *Phys. Rev.* 103 (1956) 1353; K. E. Bruckner and C. A. Levinson, *Phys. Rev.* 97 (1955) 1344.
784. A. B. Migdal, *Rev. Mod. Phys.* 50 (1978) 107.
785. R. B. Wirinda and V. R. Pandharipande, *Rev. Mod. Phys.* 51 (1979) 821.
786. G. E. Brown and M. Rho, *Phys. Rep.* 269 (1996) 334; *Nucl. Phys.* A596 (1996) 503.

Other hadronic phases
787. A. B. Migdal, *Soviet Phys. JETP* 34 (1972) 1184; *Sov. Phys. JETP* 36 (1973) 1052; R. F. Sawyer, *Phys. Rev. Lett.* 29 (1972) 382.
788. T. D. Lee, *Rev. Mod. Phys.* 47 (1975) 267; T. D. Lee and G. C. Wick, *Phys. Rev.* D9 (1974) 2291.
789. E. V. Shuryak, *Pisma Zh. Eksp. Teor. Fiz.* 29 (1979) 115.
790. V. Soni, *Phys. Lett.* B152 (1985) 231.
791. D. B. Kaplan and A. E. Nelson, *Phys. Lett.* B175 (1986) 57.
792. G. E. Brown, K. Kubodera and M. Rho, *Phys. Lett.* B175 (1987) 57.
793. G. Ropke, L. Munchow and H. Schultz, *Nucl. Phys.* A379 (1982) 536; P. J. Siemens, *Nature* 305 (1983) 410; A. L. Goodman *et al.*, *Phys. Rev.* C30 (1984) 891; D. H. Boal, *Phys. Rev.* C30 (1984) 119, 749.
794. A. M. Baldin in Proceedings of the 16th Int. Conf./ on High Energy Physics, Batavia 1972.
795. J. J. Aubert *et al.*, *Phys. Lett.* B123 (1983) 275.

Metastable quark matter
796. S. A. Chin and A. Kerman, *Phys. Rev. Lett.* 43 (1979) 1292.
797. E. V. Shuryak, *Piswa Zh E.T.F (JETP Letters)* 29 (1979) 115.
798. E. Witten, *Phys. Rev.* D30 (1984) 272
799. A. de Ruhula, *Nucl. Phys.* A434 (1985) 605.
800. E. Farhi and R. L. Jaffe, *Phys. Rev.* D30 (1985) 2379.
801. E. V. Shuryak, Lecture in 1985 LINP winter school, Preprint INP 85-130, Novosibirsk 1985.

Compact stars
802. F. Webber and N. Glendenning [hep-ph/9609074].
803. L. D. Landau, *Doklady Akademii Nauk* 17 (1937) 301; *Nature* 141 (1938) 333.
804. D. C. Backer, S. R. Kulkarni, C. Heiles, M. M. Davis and W. M. Goss, *Nature* 300 (1982) 615.
805. S. E. Thorsett, Z. Arzoumanian, M. M. McKinnon and J. H. Taylor, *Astrophys. J.* 405 (1993) L29.

806. F. M. Walter, J. Lattimer, *Astrophys. J.* 576 (2002) L145-L148; [astro-ph/020419].
807. M. Prakash, Prepared for 23rd School of Theoretical Physics (Ustron 99). Poland, 15–22 Sep. 1999. *Acta Phys. Polon.* B30 (1999) 3187–3209.

Quark Matter in stars
808. U. H. Gerlach, *Phys. Rev.* 172 (1968) 172.
809. N. Itoh, *Prog. Theor. Phys.* 44 (1970) 291.
810. G. Chapline and M. Nauenburg, *Nature* 259 (1976) 377.
811. G. Baym and S. A. Chin, *Phys. Lett.* B62 (1976) 241.
812. E. V. Shuryak, *Sou. Phys. JETP* 47 (2)(1978) 212.
813. W. B. Fechner and P. C. Joss, *Nature* 274 (1978) 347.
814. P. D. Morley and M. B. Kislinger, *Phys. Rep.* 51 (1979) 63.
815. N. K. Glendenning, *Phys. Rev.* D46 (1992) 1274.
816. N. K. Glendenning and C. Kettner [astro-ph/9807155].
817. I. N. Mishustin *et al.*, [hep-ph/0210422].

Color superconductivity
818. S. C. Frautschi, Asymptotic freedom and color superconductivity in dense quark matter, in: Proceedings of the Workshop on Hadronic Matter at Extreme Energy Density, N. Cabibbo, Editor, Erice, Italy (1978).
819. B. C. Barrois, *Nucl. Phys.* B129 (1977) 390.
820. F. Barrois, Nonperturbative effects in dense quark matter, Ph.D. thesis, Caltech, UMI 79-04847-mc (microfiche).
821. D. Bailin and A. Love, *Phys. Rep.* 107 (1984) 325.
822. N. Evans, S. D. Hsu and M. Schwetz, *Nucl. Phys.* B551 (1999) 275 [hep-ph/9808444].
823. N. Evans, S. D. Hsu and M. Schwetz, *Phys. Lett.* B449 (1999) 281 [hep-ph/9810514].
824. T. Schäfer and F. Wilczek, *Phys. Lett.* B450 (1999) 325 [hep-ph/9810509].
825. W. E. Brown, J. T. Liu and H. Ren, *Phys. Rev.* D62 (2000) 054016 [hep-ph/9912409].
826. T. Schäfer, *Phys. Rev.* D62 (2000) 094007 [hep-ph/0006034].
827. M. Alford, K. Rajagopal and F. Wilczek [hep-ph/9711395].
828. R. Rapp, T. Schäfer, E. V. Shuryak and M. Velkovsky, *Phys. Rev. Lett.* 81 (1998) 53 [hep-ph/9711396].
829. D. T. Son, *Phys. Rev.* D59 (1999) 094019 [hep-ph/9812287].
830. T. Schäfer and F. Wilczek, *Phys. Rev.* D60 (1999) 114033 [hep-ph/9906512].
831. R. D. Pisarski and D. H. Rischke, *Phys. Rev.* D61 (2000) 074017 [nucl-th/9910056].
832. D. K. Hong, V. A. Miransky, I. A. Shovkovy and L. C. Wijewardhana, *Phys. Rev.* D61 (2000) 056001 [hep-ph/9906478].
833. W. E. Brown, J. T. Liu and H. Ren, *Phys. Rev.* D61 (2000) 114012 [hep-ph/9908248].
834. T. Schäfer, *Nucl. Phys.* B575 (2000) 269 [hep-ph/9909574].
835. N. Evans, J. Hormuzdiar, S. D. Hsu and M. Schwetz, *Nucl. Phys.* B581 (2000) 391 [hep-ph/9910313].
836. M. Alford, K. Rajagopal and F. Wilczek, *Nucl. Phys.* B537 (1999) 443 [hep-ph/9804403].
837. R. Rapp, T. Schäfer, E. V. Shuryak and M. Velkovsky, *Ann. Phys.* 280 (2000) 35 [hep-ph/9904353].
838. T. Schäfer and F. Wilczek, *Phys. Rev. Lett.* 82 (1999) 3956 [hep-ph/9811473].
839. M. Alford, J. Berges and K. Rajagopal, *Nucl. Phys.* B558 (1999) 219 [hep-ph/9903502].
840. T. Schäfer and F. Wilczek, *Phys. Rev.* D60 (1999) 074014 [hep-ph/9903503].

841. P. Bedaque [hep-ph/9910247].

842. M. Alford, J. Bowers and K. Rajagopal [hep-ph/0008208].

843. T. Schäfer, *Phys. Rev. Lett.* 85 (2000) 5531–5534 [nucl-th/0007021].

844. T. Schafer [nucl-th/0208070].

845. R. Casalbuoni and D. Gatto, *Phys. Lett.* B464 (1999) 111 [hep-ph/9908227].

846. J. B. Kogut, M. A. Stephanov, D. Toublan, J. J. Verbaarschot and A. Zhitnitsky, *Nucl. Phys.* B582 (2000) 477 [hep-ph/0001171].

847. D. T. Son and M. Stephanov, *Phys. Rev.* D61 (2000) 074012 [hep-ph/9910491], erratum: [hep-ph/0004095].

848. M. Rho, A. Wirzba, and I. Zahed, *Phys. Lett.* B473 (2000) 126 [hep-ph/9910550].

849. P. F. Bedaque and T. Schafer, *Nucl. Phys.* A697 (2002) 802 [hep-ph/0105150].

850. D. K. Hong, T. Lee, and D. Min, *Phys. Lett.* B477 (2000) 137 [hep-ph/9912531].

851. C. Manuel and M. H. Tytgat, *Phys. Lett.* B479 (2000) 190 [hep-ph/0001095].

852. D. T. Son, M. A. Stephanov and A. R. Zhitnitsky, *Phys. Lett.* B510 (2001) 167 [hep-ph/0103099].

853. S. R. Beane, P. F. Bedaque and M. J. Savage, *Phys. Lett.* B483 (2000) 131 [hep-ph/0002209].

854. D. V. Deryagin, D. Yu. Grigoriev, and V. A. Rubakov, *Int. J. Mod. Phys.* A7 (1992) 659.

855. E. Shuster and D. T. Son, *Nucl. Phys.* B573 (2000) 434 [hep-ph/9905448].

856. R. Rapp, E. Shuryak and I. Zahed, *Phys. Rev.* D63 (2001) 034008 [hep-ph/0008207].

857. A.W. Overhauser, *Phys. Rev.* 128 (1962) 1437; *Adv. Phys.* 27 (1978) 343.

858. G. Grüner, *Rev. Mod. Phys.* 66 (1994) 1.

859. A. I. Larkin and Yu. N. Ovchinnikov, *Zh. Eksp. Teor. Fiz.* 47 (1964) 1136; translation: *Sov. Phys. JETP* 20 (1965) 762; P. Fulde and R. A. Ferrell, *Phys. Rev.* 135 (1964) A550.

860. W. V. Liu and F. Wilczek, *Phys. Rev. Lett.* 90 (2003) 047002 [cond-mat/0208052].

861. T. Schafer, *Phys. Rev. Lett.* 85 (2000) 5531 [nucl-th/0007021].

862. D. B. Kaplan and S. Reddy, *Phys. Rev.* D65 (2002) 054042 [hep-ph/0107265].

Cosmological variation, alternative universes etc

863. J. D. Bjorken, Cosmology and the standard model [hep-th/0210202] and references therein.

864. L. Smolin, "The Fate of black hole singularities and the parameters of the standard models of particle physics and cosmology", [arXiv:gr-qc/9404011].

865. J. K. Webb V.V. Flambaum, C.W. Churchill, M. J. Drinkwater and J. D. Barrow, *Phys. Rev. Lett.* 82 (1999) 884–887; J. K. Webb, M. T. Murphy, V. V. Flambaum, V. A. Dzuba, J. D. Barrow, C. W. Churchill, J. X. Prochaska and A. M. Wolfe, *Phys. Rev. Lett.* 87 (2001) 091301-1-4; M. T. Murphy, J. K. Webb, V. V. Flambaum, V. A. Dzuba, C. W. Churchill, J. X. Prochaska, J. D. Barrow and A. M. Wolfe, *Mon. Not. R. Astron. Soc.* 327 (2001) 1208 [astro-ph/0012419]; M. T. Murphy, J. K. Webb, V. V. Flambaum, C. W. Churchill and J. X. Prochaska, *Mon. Not. R. Astron. Soc.* 327 (2001) 1223 [astro-ph/0012420].; M. T. Murphy, J. K. Webb, V. V. Flambaum, C. W. Churchill, J. X. Prochaska and A. M. Wolfe, *Mon. Not. R. Astron. Soc.* 327 (2001) 1237 [astro-ph/0012421].

866. V. V. Flambaum and E. V. Shuryak, *Phys. Rev.* D65 (2002) 103503 [hep-ph/0201303].

867. S. R. Beane and M. J. Savage [nucl-th/0208021].

868. V. V. Flambaum and E. V. Shuryak, *Phys. Rev.* D67 (2003) 083507 [arXiv:hep-ph/0212403].

Increasing the number of quark flavors

869. A. A. Belavin and A. A. Migdal, Scale Invariance And Bootstrap In The Nonabelian Gauge Theories, Print-74-0894 (LANDAU INST).
870. T. Banks and A. Zaks, *Nucl. Phys.* B82 (1982) 196.
871. J. A. Gracey, *Phys. Lett.* B373 (1996) 178–184 [hep-ph/9602214].
872. E. V. Shuryak, *Phys. Lett.* B196 (1987) 373.
873. Y. Iwasaki, K. Kanaya, S. Sakai and T. Yoshie, *Nucl. Phys. (Proc. Suppl.)*, B34 (1994) 314.
874. T. Appelquist, J. Terning and L. C. Wijewardhana, *Phys. Rev. Lett.* 77 (1996) 1214 [hep-ph/9602385].
875. T. Appelquist and S. B. Selipsky, *Phys. Lett.* B400 (1997) 364 [hep-ph/9702404].
876. Y. Iwasaki, K.Kanaya and S. Kaya. Phase structure of qcd for general number of flavors, 1996. Talk at Lattice-96, St Louis.

Supersymmetric theories

877. V. A. Novikov, M. A. Shifman, A. I. Vainshtein and V. I. Zakharov, *Nucl. Phys.* B229 (1983) 381.
878. A. I. Vainshtein, V. I. Zakharov, V. A. Novikov and M. A. Shifman, *Sov. J. Nucl. Phys.* 43 (1986) 294.
879. I. Affleck, M. Dine and N. Seiberg, *Nucl. Phys.* B241 (1984) 493.
880. I. I. Kogan and M. Shifman, *Phys. Rev. Lett.* 75 (1995) 2085.
881. N. Seiberg, *Phys. Rev.* D49 (1994) 6857.
882. A. V. Yung, *Nucl. Phys.* B297 (1988) 47.
883. K. Intriligator and N. Seiberg, 1995. Proc. TASI'95 [hep-th/9509066].
884. K. A. Intriligator and N. Seiberg, *Nucl. Phys.* B431 (1994) 551 [arXiv:hep-th/9408155].
885. N. Seiberg and E. Witten, *Nucl. Phys.* B426 (1994) 19.
886. D. Finnel and P. Pouliot, *Nucl. Phys.* B453 (1995) 225.
887. K. Ito and N. Sasakura, *Phys. Lett.* B382 (1996) 95.
888. N. Dorey, V. V. Khoze and M. P. Mattis, *Phys. Rev.* D54 (1996) 2921.
889. H. Aoyama, T. Harano amd M. Sato and S. Wada, 1996. [hep-th/9607076].
890. G. Bonelli and M. Matone, *Phys. Rev. Lett.* 76 (1996) 4107.
891. L. Randall, R. Rattazzi and E. V. Shuryak, *Phys. Rev.* D59 (1999) 035005 [arXiv:hep-ph/9803258].

$N = 4$ and ADS/CFT correpondence

892. J. M. Maldacena, The Large N Limit of Superconformal Field Theories and Supergravity [hep-th/9711200].
893. S.-J. Rey and J.-T. Yee, Macroscopic Strings as Heavy Quarks in Large N Gauge Theories and Anti-de Sitter Supergravity [hep-th/9803001]; J. Maldacena, Wilson Loops in Large N Field Theories [hep-th/9803002].
894. G. T. Horowitz and A. Strominger, *Nucl. Phys.* B360 (1991) 197.
895. S. S. Gubser, I. R. Klebanov and A. A. Tseytlin, *Nucl. Phys.* B534 (1998) 202–222 [hep-th/9805156].
896. S.-J. Rey, S. Theisen and J.-T. Yee [hep-th/9803135]
897. G. Policastro, D. T. Son and A. O. Starinets, *Phys. Rev. Lett.* 87 (2001) 081601 [hep-th/0104066].
898. I. R. Klebanov and A. A. Tseytlin, *Nucl. Phys.* B578 (2000) 123–138 [hep-th/0002159].
899. G. W. Semenoff and K. Zarembo, *Nucl. Phys. Proc. Suppl.* 108 (2002) 106 [hep-th/0202156].

Index

adiabatic paths, 379
asymptotic freedom, 13

balance function, 468
baryon-number violation, by tunneling in the electroweak theory, 310
BFKL Pomeron, 305
Brown-Rho scaling, 522

caloron solution, the instanton at finite T, 381
Casher-Banks relation, 87
Chern–Simons number, 129
Cheshire Cat Principle, 70
chiral Lagrangian, 90, 92
chiral Lagrangian of the order p^4, 92
chiral Lagrangian; the Wess-Zumino term, 92
chiral limit, 33
Chiral Quark-Soliton Model, 70
chiral rotations, 34
chiral symmetry, 83
chiral symmetry Spontaneously broken, 83
Chirality, 34
cocktail model, 390
cocktail model of the instanton ensemble in the mixed phase, with molecules substituting random "atomic" liquid, 390
color quantum number, 12
Color-Flavor-Locked (CFL) phase, 546
covariant, 3
covariant derivative, 3, 4

Debye screening mass and radius, 352
derivatives, 3
diluteness parameter, 153

Euclidean action in quantum mechanics, 19
Euclidean gamma matrices, 31

fermion doublers on the lattice, 205
fermionic doublers, 204
fermions on the lattice: Kogut-Susskind, 205
fermions on the lattice: Wilson, 205
Feynman path integral, 17
fundamental domain, 9

gap equation of the NJL model, 94
gauge invariance
    in non-Abelian gauge theories, 3
    in QED, 3
gauge transformation of the potential, 3, 4
gauge tree for complete gauge fixing, 7
Gell–Mann–Oakes-Renner relation, 89
Gell–Mann–Okubo formula, 89
general thermodynamic relations, 340
Ginsparg–Wilson relation, 206
gluon effective mass, 352
Goldstone theorem, 84
Green function at finite T, 341
Green function in quantum mechanics, 17

heavy quark symmetry, 42
hot glue scenario, 492

instanton in the double well problem, 108

instanton-induced ('t Hooft) effective
    interaction of light quarks, 150
interior gap superconductivity, 551

kaon condensation in the CFL phase, 550
kinetic freezeout, definition, 429

Landau–Pomeranchuck–Migdal effect, 505
lattice link variable, 4, 5
lattice version of chiral symmetry, 206
Link between an $n$dimensional statistical
    system and Euclidean quantum (field)
    theory in $n-1$ dimensions, 20
little bag model, 70
LOFF phase of superconductors, 551

Maldacena conjecture, 571
Matsubara frequencies, 342
Matsubara temperature, 20
Matsubara time, 341
mesoscopic limit of lattice gauge theory,
    198

Nambu-Jona-Lasinio Lagrangian, 93

optical depth factor, 432
overlap matrix in the zero-modes
    subspace, 166

partition function for gauge fields, 8
Pauli-Gursey symmetry, 49
Polyakov line, 370, 371
pomeron, 302
propagator in the multi-instanton
    background, 166

Quark–Gluon Plasma, 338
quasiparticle gas, 360
quenched approximation, 210, 211

Regge poles, 302
region of homogeneity, 464
Resonance gas, 337

saturation scale, 305
screening masses, 394, 395
self-dual fields, 132
semiclassical Pomeron, 308, 310
static quark, 381
static quark effective potential at finite T,
    374
statistical sum, 339
stochastic vacuum model, 79
strange quark matter, 525
strong CP problem, 137
Synchrotron-like QCD radiation, 507

temperature profile function, 479
topological charge, 128
topological current K_μ, 128
truncated eigenmode approach (TEA),
    223

vacuum dominance hypothesis, 79
viscosity: sound attenuation length, 414

Weingarten inequality, 233
Weizsacker–Williams approximation, 313
Wilson loop; the area law, 96
Wilson fermions, 205
winding number, 125
winding or Pontryagin number, 126

Zweig rule, 55

The QCD Vacuum, Hadrons and Superdense Matter

Second Edition

World Scientific Lecture Notes in Physics

ISSN: 1793-1436

*Published titles**

Vol. 65: Universal Fluctuations: The Phenomenology of Hadronic Matter
R Botet and M Ploszajczak

Vol. 66: Microcanonical Thermodynamics: Phase Transitions in "Small" Systems
D H E Gross

Vol. 67: Quantum Scaling in Many-Body Systems
M A Continentino

Vol. 69: Deparametrization and Path Integral Quantization of Cosmological Models
C Simeone

Vol. 70: Noise Sustained Patterns: Fluctuations and Nonlinearities
Markus Loecher

Vol. 71: The QCD Vacuum, Hadrons and Superdense Matter (2nd ed.)
Edward V Shuryak

Vol. 72: Massive Neutrinos in Physics and Astrophysics (3rd ed.)
R Mohapatra and P B Pal

Vol. 73: The Elementary Process of Bremsstrahlung
W Nakel and E Haug

Vol. 74: Lattice Gauge Theories: An Introduction (3rd ed.)
H J Rothe

Vol. 75: Field Theory: A Path Integral Approach (2nd ed.)
A Das

Vol. 76: Effective Field Approach to Phase Transitions and Some Applications
to Ferroelectrics (2nd ed.)
J A Gonzalo

Vol. 77: Principles of Phase Structures in Particle Physics
H Meyer-Ortmanns and T Reisz

Vol. 78: Foundations of Quantum Chromodynamics: An Introduction to Perturbation
Methods in Gauge Theories (3rd ed.)
T Muta

Vol. 79: Geometry and Phase Transitions in Colloids and Polymers
W Kung

Vol. 80: Introduction to Supersymmetry (2nd ed.)
H J W Müller-Kirsten and A Wiedemann

Vol. 81: Classical and Quantum Dynamics of Constrained Hamiltonian Systems
H J Rothe and K D Rothe

Vol. 82: Lattice Gauge Theories: An Introduction (4th ed.)
H J Rothe

*For the complete list of published titles, please visit
http://www.worldscientific.com/series/wslnp